Birds of New Guinea

Second Edition

Princeton Field Guides

Rooted in field experience and scientific study, Princeton's guides to animals and plants are the authority for professional scientists and amateur naturalists alike. **Princeton Field Guides** present this information in a compact format carefully designed for easy use in the field. The guides illustrate every species in color and provide detailed information on identification, distribution, and biology.

Albatrosses, Petrels, and Shearwaters of the World, by Derek Onley and Paul Scofield
Birds of Australia, Eighth Edition, by Ken Simpson and Nicolas Day
Birds of Borneo: Brunei, Sabah, Sarawak, and Kalimantan, by Susan Myers
Birds of Chile, by Alvaro Jaramillo
Birds of the Dominican Republic and Haiti, by Steven Latta, Christopher Rimmer, Allan Keith, James Wiley, Herbert Raffaele, Kent McFarland, and Eladio Fernandez
Birds of East Africa: Kenya, Tanzania, Uganda, Rwanda, and Burundi, by Terry Stevenson and John Fanshawe
Birds of Europe, Second Edition, by Lars Svensson, Dan Zetterström, and Killian Mullarney
Birds of India, Pakistan, Nepal, Bangladesh, Bhutan, Sri Lanka, and the Maldives, by Richard Grimmett, Carol Inskipp, and Tim Inskipp
Birds of Kenya and Northern Tanzania: Field Guide Edition, by Dale A. Zimmerman, Donald A. Turner, and David J. Pearson
Birds of Melanesia: Bismarcks, Solomons, Vanuatu, and New Caledonia, by Guy Dutson
Birds of the Middle East, by R. F. Porter, S. Christensen, and P. Schiermacker-Hansen
Birds of Nepal, by Richard Grimmett, Carol Inskipp, and Tim Inskipp
Birds of New Guinea, by Thane K. Pratt and Bruce M. Beehler
Birds of Northern India, by Richard Grimmett and Tim Inskipp
Birds of Peru, by Thomas S. Schulenberg, Douglas F. Stotz, Daniel F. Lane, John P. O'Neill, and Theodore A. Parker III
Birds of the Seychelles, by Adrian Skerrett and Ian Bullock
Birds of Southeast Asia, by Craig Robson
Birds of Southern Africa, 4th edition, by Ian Sinclair, Phil Hockey, Warwick Tarboton, and Peter Ryan
Birds of Thailand, by Craig Robson
Birds of the West Indies, by Herbert Raffaele, James Wiley, Orlando Garrido, Allan Keith, and Janis Raffaele
Birds of Western Africa, by Nik Borrow and Ron Demey
Carnivores of the World, by Luke Hunter
Caterpillars of Eastern North America: A Guide to Identification and Natural History, by David L. Wagner
Common Mosses of the Northeast and Appalachians, by Karl B. McKnight, Joseph Rohrer, Kirsten McKnight Ward, and Warren Perdrizet
Coral Reef Fishes, by Ewald Lieske and Robert Meyers
Dragonflies and Damselflies of the East, by Dennis Paulson
Dragonflies and Damselflies of the West, by Dennis Paulson
Mammals of Europe, by David W. Macdonald and Priscilla Barrett
Mammals of North America, Second Edition, by Roland W. Kays and Don E. Wilson
Minerals of the World, by Ole Johnsen
Nests, Eggs, and Nestlings of North American Birds, Second Edition, by Paul J. Baicich and Colin J. O. Harrison
Palms of Southern Asia, by Andrew Henderson
Parrots of the World, by Joseph M. Forshaw
The Princeton Field Guide to Dinosaurs, by Gregory S. Paul
Raptors of the World, by James Ferguson-Lees and David A. Christie
Seeds of Amazonian Plants, by Fernando Cornejo and John Janovec
Sharks of the World, by Leonard Compagno, Marc Dando, and Sarah Fowler
Stars and Planets: The Most Complete Guide to the Stars, Planets, Galaxies, and the Solar System, Fully Revised and Expanded Edition, by Ian Ridpath and Wil Tirion
Trees of Panama and Costa Rica, by Richard Condit, Rolando Pérez, and Nefertaris Daguerre
Whales, Dolphins, and Other Marine Mammals of the World, by Hadoram Shirihai and Brett Jarrett

Birds of New Guinea

Second Edition

Thane K. Pratt and Bruce M. Beehler

With editorial contributions by
K. David Bishop, Brian J. Coates, Jared M. Diamond, and Mary LeCroy

Illustrated by
John C. Anderton and Szabolcs Kókay
and
James Coe and Dale Zimmerman

PRINCETON UNIVERSITY PRESS
PRINCETON AND OXFORD

Copyright © 2015 by Princeton University Press
Published by Princeton University Press, 41 William Street, Princeton, New Jersey 08540
In the United Kingdom: Princeton University Press, 6 Oxford Street, Woodstock, Oxfordshire OX20 1TW

press.princeton.edu

Jacket illustration by Szabolcs Kókay

All Rights Reserved

Library of Congress Cataloging-in-Publication Data
Pratt, Thane K.
    Birds of New Guinea / Thane K. Pratt and Bruce M. Beehler ; with editorial contributions by K. David Bishop, Brian J. Coates, Jared M. Diamond, and Mary LeCroy ; illustrated by John C. Anderton and Szabolcs Kókay and James Coe and Dale Zimmerman. — Second edition
    pages cm. — (Princeton field guides)
Previous edition by Bruce M. Beehler, Thane K. Pratt, and Dale A. Zimmerman.
Includes bibliographical references and index.
ISBN 978-0-691-09562-2 (hardback : alk. paper) — ISBN 978-0-691-09563-9 (pbk. : alk. paper) 1. Birds—New Guinea. 2. Birds—New Guinea—Identification.. I. Beehler, Bruce McP. II. Title.
QL691.N48B44 2014
598.0995--dc23   2014009192

British Library Cataloging-in-Publication Data is available

Recommended citation:
Pratt, Thane K., and Bruce M. Beehler. 2015. *Birds of New Guinea, Second Edition.* Princeton University Press, Princeton, NJ, USA

This book has been composed in Minion and Myriad Pro
Printed on acid-free paper. ∞

Printed in China

10 9 8 7 6 5 4

Dedicated to

William J. and Judy Hancock

Will and Judy, formerly teachers at Hawai`i Preparatory Academy, all but adopted Thane from high school onward. Will was instrumental in securing a series of seed grants from the Reid Foundation that allowed us to take the book the quantum leap forward to an entirely new second edition. Their kindness and generosity over the years will always be remembered with much affection. *Mahalo nui loa!*

Russell and Robert Faucett

Russ and Rob joined the project towards the end, and it was their substantial grants through the Faucett Catalyst Fund that enabled us to complete the book by funding artwork, all the maps, and travel to PNG for artist John Anderton so that he would see firsthand the kaleidoscope world of New Guinea birds. Russ and Rob took considerable interest in the project, and they provided encouragement every step of the way. To them we are deeply grateful.

Linda Pratt and Carol Beehler

For their love and support.

Contents

Preface	9
Acknowledgments	11
Abbreviations	13
1. Introduction	14
2. How to Use This Book	17
3. New Guinea Natural History	20
4. In the Field in Search of Birds	33
Selected References	36
Web Sources	39
Plates	40
Species Accounts	262
Index	517

Preface

Twenty-eight years have passed since publication of the first edition of *Birds of New Guinea*. In that time, big changes have taken place in this important bird region. The human population has more than doubled, economic development—particularly mining and logging—has accelerated, and the loss of habitat has increased proportionately. On the brighter side, vast tracts of wilderness remain, several international conservation organizations have joined forces with local governments to protect the forest, birding as a form of ecotourism has blossomed, and a new generation of field researchers has taken to the bush. Exciting new information on bird distribution and biology has poured in. And as if to heighten appreciation for New Guinea's avifauna, modern molecular systematic research has revolutionized the classification of birds and pinpointed New Guinea and Australia as the motherland of the world's most prominent bird lineage, the songbirds (oscine passerines). In light of these changes, a new edition of *Birds of New Guinea* is timely.

A book must suit the purposes of the people using it. In many parts of the world—Australia, Europe, North America—field guides to birds are written mainly for birders (bird-watchers). These books focus solely on how to identify birds to species, how to tell the age and sex of a bird, and how and where to find it. Readers wishing to learn more about birds can turn to other books for information on bird natural history, ecology, reproduction, evolution, geographic variation, classification, and conservation. Unfortunately, much of this information on New Guinea birds can be found only in technical literature that is not readily available.

It is our wish that readers in the field have more basic information at their fingertips. In the new edition of *Birds of New Guinea*, we have expanded the content of the species accounts to include more biological details—behavior, diet, nesting—than are typically covered in a field guide. As for the identification purpose of the book, we have also expanded the species accounts in this second edition to explain how to determine the sex and age of a bird, and how geographic variation is partitioned into subspecies (races). To accomplish all this, we have adopted a "handbook-style" format for the book. Rather than fitting all the information opposite the painted plates—the format typical of most recent field guides—our book presents detailed species accounts in the body of the book separate from the illustrations in the front, with abridged species accounts and maps facing the plates. We hope this additional information will be of use to birders, tour guides, biologists, and conservationists who enjoy New Guinea birds and strive to learn more about them.

Changes have been made in authorship and artwork in the second edition. This time the book has two authors, rather than three. We want to take this opportunity to express our deep appreciation to Dale A. Zimmerman, coauthor and principal artist of the first edition. In many respects, Dale was the "senior partner" who freely offered guidance to the two younger authors embarking on their first major book. We also wish to acknowledge Brian Finch and the late Harry Bell, whose text contributions in the first edition have been brought in one form or another to the new edition. Jared Diamond's descriptions of vocalizations were similarly carried over to the new edition. The text for the new edition has been drafted solely by us (Pratt and Beehler). Early drafts were reviewed and improved by an expert panel that included David Bishop, Brian Coates, Jared Diamond, and Mary LeCroy. These colleagues each contributed knowledge gained over a lifetime of study of the birdlife of New Guinea.

Expanding the scope of the book demanded new artwork, and for that we turned to John Anderton, known for his work in *Birds of South Asia*. Together, authors and artist drew up a plate plan that more than doubled the number of figures and resulted in replacing nearly all original artwork. John painted the majority of the plates. John's observations on bird plumages at the American Museum of Natural History have clarified numerous uncertainties and even made some unexpected discoveries. Toward the end of the project we recruited Szabolcs Kókay to illustrate the shorebirds, fruit-doves, bowerbird bowers, and birds of paradise. James Coe granted permission to use the original tern-and-gull plate in combination with new figures by Szabolcs.

Dale Zimmerman graciously sent us the plate of flying parrots to republish. A new feature of this guide, the range maps, were drafted by Shane Kelly. Lisa Louise Adams drafted the voice diagrams. To these colleagues we extend our most heartfelt thanks.

Field guides are harder to produce than one would think: the amount of detailed information that must be located, organized, condensed, and checked is mind-boggling; the volume of artwork is staggering. Reflecting on this, we wish to acknowledge our editor, Robert Kirk, for patiently guiding us through the long process of publishing our book. He provided funding, answered innumerable questions, freely gave advice, mentored, and was an all-round good guy. Also with Princeton Univ. Press, Dimitri Karetnikov, Mark Bellis, Lucy Treadwell, and Robert Still (WildGuides) did a marvellous job preparing the manuscript for publication. We hope the book has turned out to their satisfaction.

Acknowledgments

In the preface above we acknowledge our immediate collaborators, those who directly took part in the production of the book. In this section we thank a number of friends, correspondents, colleagues, and sponsors who were helpful to our efforts in researching, painting, and writing this book.

We wish to make special mention of the considerable financial support and encouragement given by Will Hancock and the Will J. Reid Foundation (USA) and by Russ and Rob Faucett and the Faucett Catalyst Fund (USA). The contributions of both parties were crucial, and it is with deepest gratitude that we have dedicated the book to them.

Three particularly important grants in support of the book were received from the Herman Slade Foundation—Australia and Pacific Science Foundation (Australia), the Porgera Joint Venture (Papua New Guinea), and Princeton University Press.

For assistance in underwriting the cost of producing the artwork, we owe our gratitude to: Frederick Atwood, Cary Beehler, Ken Berlin, Bishop Museum, Howard Brokaw, George and Rita Fenwick, Will and Judy Hancock, Honolulu Zoo, Bruce Irish, Peter Kaestner, Mary King, Warren King, Allan Keith, James Kushlan, Thomas Lovejoy, Ernst Mayr, John Mitchell, Leo Model Foundation, Frank Parlier, Roger Pasquier, John H. R. Plews, Steve Plewes, Brenda Pratt, C. Dudley Pratt Jr., Bob Pringle, Steve and Melinda Pruett-Jones, Scott K. Robinson, Fred and Dottie Rudolph, Scion Natural Science Association (Jon Fink, his sisters and parents), Thayer Simmons, Nigel Simpson, Carol Sisler, John Swift, Dr. and Mrs. Wargel, Mark Weinberger, John Weske, David Wilcove, and the Zoological Society of San Diego. We also wish to thank John Fitzpatrick, director of the Cornell Lab of Ornithology, for his role in obtaining funding for the project. The American Museum of Natural History awarded Collections Study Grants to TKP and BMB.

For its key role in the development of this project, we have agreed to donate part of the royalties to the Bird Division of the Smithsonian's National Museum of Natural History.

For critical assistance in the examination and loan of study skins for use by the artists, we thank the American Museum of Natural History, in particular Joel Cracraft, Merle Okada, Paul Sweet, Tom Trombone, Peter Capainolo, Matthew Shanley (photographer), and most especially that inestimable host and informant to researchers of New Guinea birds, Mary LeCroy (many New Guinea bird taxa bear the LeCroy eponym, bestowed by taxonomists in honor of Mary's contribution to NG ornithology); Smithsonian National Museum of Natural History, particularly Gary Graves, Storrs Olson, Carla Dove, Helen James, and Chris Milensky; Bishop Museum of Honolulu, in particular Allen Allison, Carla Kishinami, Molly Hagemann, Kathleen Imada, Terry Lopez, and Lydia Garetano; and Naturhistorisches Museum Wien, Anita Gamauf.

The staff of the Macaulay Library, Cornell Lab of Ornithology—Greg Budney, Matthew Medler, and Jay McGowan—kindly processed our voice recordings of New Guinea birds. The library's online collection of recordings and videos proved to be an awesome resource, as did the online recordings library of xeno-canto.

For discussion and advice on the decisions applying molecular studies to systematics and taxonomy we are in debt to Les Christidis, Joel Cracraft, Jack Dumbacher, Rob Fleischer, Leo Joseph, Janette Norman, Dick Schodde, and others.

Of the many individuals who generously provided information on birds of New Guinea and Australia, including voice recordings, we wish to mention Mike Andersen, Brett Benz, Bill Cooper, Lisa Doucette, Jack Dumbacher, Guy Dutson, Brian Finch, Ben Freeman, Cliff and Dawn Frith, Francesco Germi, Phil Gregory, David Holland, Andy Mack, Jen Mandeville, Pam Rasmussen, Frank Rheindt, Robert Simons, Mike Sorenson, Neil Stronach, Katie Faust Stryjewski, Jim Thomas, Bas van Balen, and Iain Woxvold. Until recently, New Guinea seabirds were a big mystery, known almost exclusively from a smattering of nearshore observations; we are especially grateful to Captain

Neil Cheshire for sharing the published results of his extensive and quantitative surveys of seabirds in Papua New Guinea waters and for reviewing drafts of our seabird accounts. Cliff Frith and Ed Scholes III read and commented on early drafts of the birds of paradise accounts. Tim Laman and Ed Scholes were especially generous in sharing information and photographs of birds of paradise.

Michelle Brown, Deidre Fogg, Peter Mumford, and Linda Pratt assisted in the initial compiling of some species accounts.

The U.S. Geological Survey, Pacific Island Ecosystems Research Center, kindly allowed Thane to work part-time on the field guide during his employment there and afterward in emeritus status. For this he is truly grateful and wishes to particularly thank directors Bill Steiner, David Helweg, Loyal Mehrhoff, and Gordon Tribble. Colleagues Rick Camp and Marcos Gorresen offered advice on databases and mapping.

We thank our many friends, colleagues, and acquaintances in the New Guinea Region, who provided hospitality, expertise, guidance, and more. Thane's recent fieldwork in PNG was supported in part by Conservation International's Papua New Guinea office (special thanks to Dave Mitchell), the National Geographic Society, and Trans Niugini Tours. Allen Allison and Fred Kraus of Bishop Museum facilitated fieldwork in Milne Bay Province. He would like to thank the staffs of the PNG National Research Institute (particularly Jim Robins) and Department of Environment and Conservation for obtaining permission to conduct research in the country and for advice, logistical and otherwise. The PNG National Museum and Art Gallery (especially Bulisa Iova) hosted Thane's fieldwork in the country. He also thanks Mike Moore and Lucas Morgan for their assistance in the field. Thane learned from professional birding guides Samuel Kepuknai, Jimmy Warum, Joseph Tano, and Chris Nick novel insights into birds and birding in New Guinea. Bruce would like to recognize the field collaboration over four decades of the following friends and colleagues: Allen Allison, Lisa Dabek, Jack Dumbacher, Christopher Filardi, Ben Freeman, Ninga Kawa, Nev Kemp, Tim Laman, Michael Lucas, Andrew Mack, Jennifer Mandeville, Kurt Merg, Ed Scholes, John B. Sengo, Timmy Sowang, Deb Wright, and a host of others. For both authors, our field experiences in the forests of New Guinea remain with us as some of the most rewarding and remarkable times we have known.

Finally, Thane would like to thank his wife, Linda, for living with this project for so many years. Linda helped with the manuscript, joined travel to PNG and New York, postponed her retirement (one of the reasons), and did a lot of listening, not of the birding variety. Bruce would like to recognize the contribution made by his wife, Carol, who accompanied him in the field on three arduous field trips and assisted and supported in myriad ways, all deeply appreciated then and now recalled with affection.

PLATE CREDITS

John Anderton: Plates 1–9, 12–26, 35–37, 40–93, and 101–110.
James Coe: Plates 10 and 11.
Szabolcs Kókay: Cover, spine, and Figure 2; Plates 10, 11, 27–34, 38–39, 40 (species no. 8), 66 (species no. 4), 67A, 84 (species no. 6), and 94–100.
Dale Zimmerman: Plate 49.

Abbreviations

AU	Australia
cm	centimeters
E	East
I	Island
Is	Islands
N	North
m	meters
mm	millimeters
Mt	mountain
N Amer	North America
NC	New Caledonia
NG	New Guinea Island
NG Region	New Guinea Region
NZ	New Zealand
p	page
PNG	Papua New Guinea
Pen	Peninsula
Pl	plate
R	River
S	South
S Amer	South America
sec	seconds
W	West

1. Introduction

1.1. SCOPE OF THE BOOK

New Guinea is the center of bird diversity in *Australasia* (Australia and New Guinea combined, plus nearby islands). Here lives one of the world's four great tropical avifaunas, separate in its history and evolution from those of Asia, Africa, and the Americas. The region is famous for being home to a rich and distinctive humid forest avifauna characterized by cassowaries, megapodes, pigeons, parrots, cuckoos, kingfishers, owlet-nightjars, and especially the oscine passerines or songbirds. The latter include hundreds of small insectivores belonging to numerous families centered on the region, and most renowned of all, the birds of paradise and bowerbirds. The uniqueness of the ancient passerine lineages that evolved in the region are only now coming to light with the detailed molecular systematic studies that have recently elevated seven New Guinean endemic songbird lineages to full familial status—the satinbirds, typical berrypeckers, painted berrypeckers, the berryhunters, ploughbills, ifrits, and melampittas. These relatively obscure montane forest denizens are fascinating, but overshadowed by the more prodigious songbird lineages that also apparently evolved in New Guinea only to expand out to the forestlands of Australia, Oceania, and Asia.

This volume treats all species of birds known to occur within the New Guinea Region as defined by Mayr (1941), comprising the huge equatorial island of New Guinea and its numerous closely associated satellite islands and island groups (inside cover). This second edition of *Birds of New Guinea* now includes detailed accounts of 779 bird species—70 more than the first edition.

The name New Guinea can cause some confusion. *New Guinea* is a geographic rather than political term that refers to the main island in the region, herein also abbreviated as *NG* or referred to as the *mainland*. The island is not *Papua New Guinea* (here PNG), which is a country that includes both the eastern half of the island of New Guinea and numerous other islands to the north and east, most of them outside the region covered by this book. The western half of the island of New Guinea comprises the *Indonesian provinces of West Papua (Papua Barat) and Papua*, collectively once called West Irian or Irian Jaya. (The name Papua was formerly, but separately, also adopted for a portion of Papua New Guinea prior to the independence of that country). To keep things simple, we'll avoid the name Papua and the adjective Papuan when we mean New Guinea or things New Guinean, although this word is conserved for many bird names.

Aside from the main landmass of New Guinea, the *New Guinea Region* includes numerous islands on the continental shelf or verges thereof: the Raja Ampat Islands, here called the *Northwestern Islands*; islands of Geelvink (Cenderawasih) Bay, here called the *Bay Islands*; the Aru Islands to the southwest; the small fringing islands along the North Coast of PNG; and lastly the islands of Milne Bay Province, here called the *Southeastern Islands*. Politically, the New Guinea Region is made up of two countries, *Indonesia* in the west, and *Papua New Guinea* in the east. Thus, it does not include any of the islands in Torres Strait, which belong to Australia. The area covered extends from the equator to 12°S latitude, and from 129 to 155°E longitude—a region 3100 km long by 850 km wide, and including what is the largest expanse of continuous tropical humid forest in the Asia-Pacific region.

Marine waters we consider to be within the region are all surrounding seas: the Seram, Halmahera, Bismarck, Solomon, Coral, and Arafura Seas, and the open Pacific Ocean to the north, out to approximately 200 km. The term *Oceania* refers to the main body of the tropical Pacific Ocean that lies off the continental shelf, a vast realm of open ocean and small islands, most of it outside the New Guinea Region.

1.2. THE AVIFAUNA

The 779 bird species of the New Guinea avifauna can be classified into four discrete groups: 621 breeding land and freshwater species (some augmented by migrant Australian populations),

20 tropical seabirds (resident or visiting), 60 migrants from eastern Asia, 33 migrants from Australia and New Zealand, 36 vagrants, 5 non-native resident species, and 4 hypothetical species (of uncertain status). By far the richest segment is that comprising land and freshwater birds; these are the rainforest, montane, and alpine species, plus species of more specialized habitats such as mangrove and savannah, that provide the New Guinea Region with such an exotic and varied bird fauna. Of these, 365 species (or 59%) are endemic to the New Guinea Region, meaning they are found nowhere else. It is this primarily endemic element that receives a special emphasis in our book. The seabirds, waders, and other wide-ranging species are also of course included, but the reader should note they are treated more thoroughly in other, specialty guides (examples listed in *Selected References* section, p. 36).

1.3. THE NEW GUINEA REGION IN CONTEXT

The New Guinea Region is geographically complex. It is a region of two different worlds, of vast lowland plains and high mountains—in a sense these are the Amazon and the Andes of the Pacific. The mainland is dominated by a huge, central cordillera called the *Central Ranges* extending 1900 km from northwest to southeast across the length of the island, not including the Bird's Head and Neck. To that are added 11 much smaller, *outlying ranges* that are home to endemic species or subspecies: the Tamrau, Arfak, Fakfak, Kumawa, Wandammen, Van Rees, Foja, Cyclops, North Coastal, Adelbert, and Huon Ranges. The mountain environments and the barriers created by mountains frame the geography of New Guinea. Spreading out from this mountainous framework, and partitioned by it, are the great alluvial basins that support the lowland rainforests. These are aligned into four principal lowland subregions, listed here by size: the Southern Lowlands, the Sepik-Ramu, the Northwestern Lowlands, and the southern lowlands of the Bird's Head and Bird's Neck. Like the mountain subregions, these lowland subregions support their own endemic birds. The reader will find many species' ranges defined by a listing of the mountain groups or lowland basins inhabited. New Guinea also supports extensive coastal mangroves, inland swamp forests, impenetrable karst terrain, and seasonally flooded savannahs. Beyond mainland New Guinea, there lies the island realm, each island or group of islands inhabited by a microcosm of New Guinea birdlife, offering yet more endemic forms.

A number of biogeographic patterns recur when the geography just described is compared with avian distributions, and from these we are able to discern distinct *New Guinea bird regions*. We will refer to these bird regions frequently in the species accounts and facing page text, and to familiarize the reader with the layout of the bird regions, we explain them later in the Introduction (p. 25). Fifteen in all, these bird regions vary in size and homogeneity, but all are defined by some avian endemism at the species or race level.

1.4. SYSTEMATIC AND TAXONOMIC TREATMENT

The list of bird species in this field guide closely follows a new species list we are developing (Beehler and Pratt, in prep.). This regional taxonomic and geographic checklist treats the nomenclature and distribution of all species and subspecies of birds known from the New Guinea Region, and mainly follows the International Ornithological Committee Checklist (Gill and Donsker 2013: IOC World Bird List 3.3); however, in some cases we have sought regional guidance from Christidis and Boles (2008) and Dickinson and Remsen (2013). Ordinal, familial, generic, and species sequences follow the IOC list. Our list is further updated by global molecular systematic studies.

In New Guinea, species limits have traditionally followed the landmark *List of New Guinea Birds* (Mayr 1941), prepared at a time when concepts and methods of defining of species differed from those of today. Recently, many of Mayr's broadly defined polytypic species have been dissected into species-pairs or species-groups, influenced by the arrival of the phylogenetic species concept and new data on morphology, genetics, behavior, and distributions. We follow the biological species concept in this book but acknowledge that the bird species of the world are now more narrowly defined than at the time of the first edition.

This edition of the field guide follows the English names selected in Beehler and Pratt (in prep.). English names were chosen on the basis of (1) widespread prior use, (2) applicability, and (3) brevity. We worked closely with the IOC nomenclature committee to establish congruence in names, but in a few cases names or orthography do not match between the two lists. While striving to maintain nomenclatural stability, we nevertheless had to make updates to reflect changes in the world of ornithology and systematics, as well as political and cultural sensitivities. For species primarily Australian in distribution, we generally accepted the names suggested by Christidis and Boles (2008), and because the fauna treated here is closely linked to that of Australia, proper names of birds reflect Australian spelling. Only as a last resort did we devise any different names for endemic species.

1.5. SUBSPECIFIC VARIATION AND DECIDING WHAT IS A SPECIES

Subspecies reflect regional population variation within species, and birds often display such variation across the diverse landscapes of New Guinea. In developing the taxonomic checklist, we made an effort to reduce the swarm of thinly designated and clinal (changing gradually with distance) subspecies to a minimum; however, this is still a problematic exercise because knowledge of many subspecies is rudimentary.

When is a well-marked subspecies actually better regarded as a "full" species? Quite a few of the more distinct forms treated as subspecies in our first edition are here elevated to full species (e.g., following Frith and Frith (2004), we have split the Flame Bowerbird *Sericulus ardens* from the Masked Bowerbird *S. aureus*). The jury is still out in other examples, mainly because of inadequate fieldwork or genetic (molecular) data. Further field and laboratory studies of well-marked regional populations will help clarify the taxonomic status of certain poorly known groups. To aid field identification, and to promote the study of little-known distinct local forms, we have illustrated nearly all the distinct races found in the region. In addition to marked geographic variation in plumage, some species also exhibit significant variation in voice. When one visits a new region, many songs at first sound unfamiliar; gradually, one develops an ear for the local "dialects" as variations of familiar songs. We encourage fieldworkers to take special note of subspecies and learn as much as possible about their vocalizations and habits. With better study some of these may be discovered to be "cryptic species" that merit recognition.

1.6. SOURCES

Ernst Mayr's *List of New Guinea Birds* (1941) remains a foundational taxonomic and distributional work, incorporating results of the hundreds of preceding publications on New Guinea birds. Austin L. Rand and E. Thomas Gilliard produced the *Handbook of New Guinea Birds* (1967) that served as the best one-volume account of the topic for several decades. Jared M. Diamond's *Avifauna of the Eastern Highlands of New Guinea* (1972) was the first in-depth biogeographic analysis of New Guinea's bird fauna. More recently, Brian J. Coates produced the superbly researched, two-volume, large-format handbook on the *Birds of Papua New Guinea* (1985 and 1990), followed by a very useful small-format photographic field guide to New Guinea and the Bismarck Archipelago (Coates and Peckover 2001). More recently, monographs devoted to specific families of birds and handbooks covering world or regional avifaunas have become leading sources of information. *Handbook of Australian and New Zealand Birds* (Marchant, Higgins, et al. 1990–2006) was our principal source of information for Australian birds, and *Handbook of Birds of the World* (del Hoyo et al. 1992–2011) was the main source for Asian species, when family-specific publications were not available. These volumes and other published reference works can be found in the *Selected References* at the end of this introductory section.

2. How to Use This Book

In contrast to this book's first edition, the facing-page treatments that accompany each painted color plate are more comprehensive and include range maps. This format allows the user to quickly identify the bird in question without, in most cases, having to refer to the species accounts within the main body of the book. Thus, the facing-plate accounts are abridged yet self-contained accounts of each species. For an explanation of the maps, please see Figure 1.

Figure 1. Key to the maps

Range
- Resident
- Migratory from Australia
- Resident and migratory from Australia
- Migratory from Asia
- Resident and migratory from Asia

"1, 2, etc." Species # on the plate
········· Group of occupied islands or areas
→ Occupied small island or range fragment
? Contradictory information or hypothetical record

The expanded species accounts in the body of the book are the main reference and provide more detailed information, including other widely used names, more detailed description consisting of identification aids (including length in centimeters, field marks, and special behavioral or distributional clues), subspecies, similar species, voice, habitat, habits (including nesting data), range, and taxonomy (when relevant).

Measurements. The length measurement is for comparative purposes. It is not intended for in-the-hand species identification, as different populations and sexes often show substantial size variation, and our measurements are from study skins, not living birds. Measurements were derived primarily from Rand and Gilliard (1967), the latest and most comprehensive source of measurements for New Guinea birds. In some cases, these were replaced by measurements from HANZAB for birds originating in Australia, or from other updated sources, or from measurements we made ourselves. For a few species, additional measurements are given to aid comparisons. *Wing length* is the standard measurement of the folded wing, as measured from the carpal bend of the wing to the tip of the wing, with the wing flattened against the ruler (more properly termed *wing arc*). For seabirds and raptors (hawks, eagles, and kin), *wingspan* (distance between the tips of the fully extended wings) measurements were taken from Harrison (1983) and Fergusson-Lees and Christie (2001), respectively, but be aware that min-max values may show too wide a range of values for some species whose geographic distributions extend beyond NG.

Status. Each species account opens with a brief statement on the species' status in the New Guinea Region—residency, migratory pattern, abundance, and main habitats. Generally, five categories of abundance are used, listed here from most to least numerous: *abundant, common, uncommon, rare,* and *vagrant*. An abundant species is seen daily in good numbers in the right habitat, whereas a vagrant species is known from only 1–5 records in the whole NG Region and its distribution does not normally include the region. Most migrants are either *Palearctic migrants* (from northern Eurasia) or *AU migrants* (Australian). *Hypothetical* species reported from the region are of uncertain status and lack adequate documentation. The types of bird habitats are explained on page 21.

Description. The descriptions present points for identification and plumage differences useful for ageing and sexing birds. Key field marks are italicized. For names of the parts of a bird and its plumage, please see Figure 2. Once a year, birds gradually shed all their feathers and replace them with new ones, a process termed *molt*. During molt, the plumage coloration may change, and related species often follow similar sequential patterns. Descriptive information on plumage development in New Guinea birds has been derived mainly by study of museum specimens and from field data, when available.

The plumage categories are as follows:
All Plumages—adult and juvenile indistinguishable.
Adult—sexes indistinguishable as adults; may be indicated as breeding or nonbreeding, if either apply.
Morphs (phases)—in cases of more than one type of normal adult plumage (most species have only a single morph), e.g., Variable Goshawk.
Male and/or Female—adult birds distinguishable by sex.
Imm—(immature) a distinct plumage after the juvenile plumage and before the adult plumage.
Juv—(juvenile) the first plumage following the downy plumage of a hatched chick.
Natal—downy chick.

Figure 2. Parts of a bird

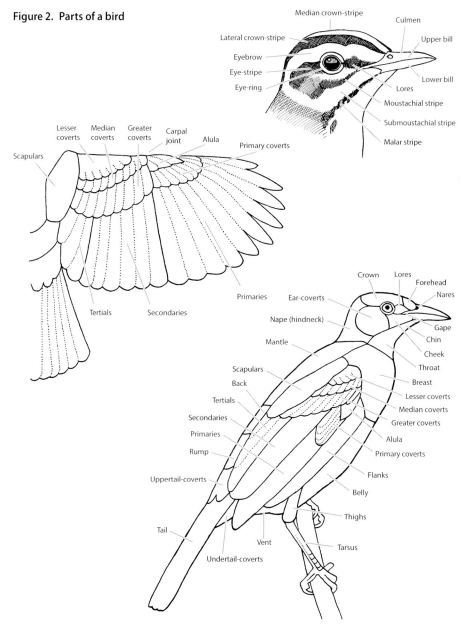

Subspecies. Only subspecies that are obvious or easily discernible are listed. Thinly separated or clinal (varying) subspecies are pooled and identified as a *group*. If the differences among all the subspecies are obscure, they are described as *minor*. If a species also lives outside the New Guinea Region, its subspecies in the region is named, and the relationship to other extraregional subspecies is identified as endemic or shared. If the species does not have subspecies (is monotypic), then the subspecies status is omitted. The format for the Subspecies Section is thus:

> **Subspp** (the number of subspecies, in number of groups): name of a subspecies or group (its geographical range) and key identifying features.

Similar species. This section provides the means to distinguish the focal species from similar-appearing birds.

Voice. Includes songs and calls, when available. These are carried across from the previous edition and modified and/or supplemented by other sources and our unpublished observations (especially those of contributor Jared Diamond). We have extensively compared these descriptions against recordings available at the websites for both xeno-canto Australasia and Cornell Laboratory of Ornithology's Macaulay Library. Collaborators have sent us additional recordings.

Habits. This section details habitat and behavior, including social system, foraging, diet, and nesting. In cases where the bird is better known from somewhere else, usually Australia, then the information is referenced as "AU data" or a similar equivalent for another region.

Range. Describes the known geographic and elevational distribution of the species, including geographic distribution outside the New Guinea Region. We first state whether the species is *endemic*, that is, whether found only in the NG Region. Next is given a list of islands where the species occurs in the region. The species' distribution on the island of NG follows. Elevational distributions are in meters, with the primary elevations of occurrence presented, often followed by marginal occurrences in parentheses. Finally, if the species exists outside the region, that distribution is given.

Taxonomy. Recent or potential changes in classification are given here.

Extralimital spp. These are species living just beyond the New Guinea Region that might be expected to turn up in the region. This category especially pertains to look-alike species (e.g., Black-shouldered Kite and Little Eagle).

3. New Guinea Natural History

3.1. ENVIRONMENT AND BIRD ECOLOGY

New Guinea is the world's second largest island, exceeded in size only by ice-covered Greenland. The world's largest tropical island and world's highest island, New Guinea is topographically diverse and geologically complex. With its equatorial location and oceanic influence, New Guinea has a generally humid and warm climate, but there are some remarkable local exceptions. In this section we address three aspects of the New Guinea environment: geology, climate, and vegetation.

Geology and Geomorphology. First it should be noted that New Guinea is one with Australia, for both rest on the Australian tectonic plate. The island of New Guinea looks distinct on the world map but is separated from mainland Australia by a shallow sea less than 20 m deep that has periodically come and gone over the last million years, as sea levels rose and fell with climate cycles. Australia and New Guinea were last connected less than 8000 years ago—a primary reason their avifaunas are so similar.

New Guinea, in its present form, is geologically young. It is the product of the ongoing impact of two moving portions of the earth's crust, the Australian and Pacific tectonic plates. Collision of these plates has produced the landmass at the northern edge of Australia which is modern New Guinea; moreover, New Guinea has incorporated a series of island arcs over time. These separate components include a series of isolated mountain ranges along the north coast, the Bird's Head, the Huon Peninsula, and parts of the SE Peninsula. These all came together in the Tertiary geologic period, and New Guinea as we know it today is less than 5 million years old. Besides the mountain uplift, we find active volcanism along the northern coast of eastern New Guinea.

Many aspects of New Guinea's topography indicate geological youth: ungraded rivers, V-shaped valleys, waterfalls, cliffs, and frequent land slippage. This unstable topography makes road building difficult and frustrates regional development. The island is highly mountainous, with 66% of the land area more than 300 m above sea level, and 14% higher than 1500 m. The Central Ranges, which form a cordillera that runs unbroken from Milne Bay to the isthmus of the Bird's Head (the Bird's Neck), have only a few passes lower than 1500 m. Furthermore, in several regions the cordillera broadens into a series of parallel ranges separated by high, flat, intermontane valleys. To the west and north of the Central Ranges are 11 outlying ranges of varying height and extent. All support distinct endemic subspecies of birds, and many support endemic species as well.

Although mountains dominate geographically, lowland regions are extensive. The three great river systems of the Fly and Digul, Sepik and Ramu, and Mamberamo (Tariku and Taritatu) meander through broad humid floodplains that are a vast mosaic of forest and swamplands.

To the north and east of New Guinea, the waters are deep, with trenches off the south of New Britain plunging to a depth of 6400 m. These northern deeps are tectonic subduction zones, where one geologic plate is dipping beneath the other. The associated upwellings of cool seawater are the richest waters for pelagic seabirds. The seas to the south and west are shallow, often with depths of less than 100 m and less productive for seabirds. In geologically recent times, when sea levels fell, land bridges connected the main island to six of the large satellite islands (Aru, Misool, Salawati, Batanta, Waigeo, and Yapen) and with Australia. The last connection explains the close faunal similarity of the open-country birds of the Trans-Fly with those in Australia.

Today, small icecaps—remnants of massive Pleistocene glaciers—top Mount Idenburg (Ngga Pilimsit) and Mount Carstensz (Puncak Jaya). The ice caps of Mount Juliana (Puncak Mandala) and Wilhelmina (Puncak Trikora) disappeared historically, and all the remaining mountain glaciers of western New Guinea will be gone in this current century as the climate warms. During the glacial maximum, about 15,000 years ago, the snow line was as low as 3600 m (Löffler 1977). Today a hiker to the higher mountain regions of New Guinea can see glacial lakes, headwalls, and moraines—evidence of an icy past in this tropical region. This alpine terrain shelters a few bird species held over from the ice ages, such as the Snow Mountain Quail, Orange-cheeked Honeyeater, and Snow Mountain Robin.

Climate. New Guinea's equatorial and oceanic position promotes a warm and humid climate; however, mountains capture much of the rainfall and cause considerable variation from site to site. Annual rainfall varies from moderate to very high. For instance, Port Moresby receives about 1200 mm per annum as compared with the foothills of the Purari/Kikori drainage that receives in excess of 10,000 mm per annum. In general, the highest rainfall occurs in association with mountain scarps, especially where coastal mountains catch the precipitation carried in off the warm seas by prevailing winds. Low annual rainfall typifies 4 lowland areas: the Trans-Fly, the Port Moresby region, the north coast of the cape of the peninsular southeast, and the upper Markham-Ramu Valley. Three of these lie in rain shadows created by mountains.

Temperatures range from hot along the coast to frequent nocturnal freezing on the higher mountains (above 3500 m). Temperature decreases at a rate of 0.6 °C per 100 m elevation, and snowfall is regular only above 3800 m. Thus, monthly mean temperatures range from 27 °C in Lae and Port Moresby, to 21 °C in Wau and 19.5 °C at Telefomin.

Fog and cloud cover affect the local climate. In many locales, the mornings are clear and bright, with clouds building by midday, and rain showers often falling in the early afternoon or near dusk. The interior of lowland rainforest reaches 100% humidity on most nights, even during the dry season. In montane regions, clouds often settle on the ridges, and a distinct "cloud line" forms on the mountainside, which varies from locale to locale. This cloud line is usually at or above 2200 m in the Central Ranges, but occurs at lower elevations (often as low as 1200 m) in the smaller, isolated ranges. Forest above the cloud line features heavy growth of bryophytes on trunks and limbs of the trees and is thus called "cloud forest" or sometimes "moss forest."

In most regions, rain falls regularly throughout the year, though with moderately predictable seasons when the likelihood of drought or high rainfall increases. From year to year, however, significant variation exists. One year may have a protracted dry season; another may produce record rainfall, in part because New Guinea is strongly affected by the El Niño/ENSO system that influences much of the weather of the tropical western Pacific. In many regions most precipitation falls between November and March (the Northwest Monsoon), but in some locales (such as Wewak, Lae, Wabo, Tari) the greatest rainfall occurs from May to September (Southeast Trade Winds predominate) with a lesser peak during the end of the calendar year (Northwest Monsoon). Thus the wetter areas get rain from both dominant wind systems, whereas the dryer areas get rain from only the Northwest Monsoon. Temperature shows minor seasonal variation, but there are greater extremes in the dry months, when cloud cover is reduced. The most severe highland frosts occur during the southern winter (July–September).

Vegetation and Habitat. With the generally humid and benign climate, it is not surprising that the native vegetation for most of the region is rainforest.

Lowland Rainforest. These grand forests of the lowland plains and foothills are structurally and floristically complex. A typical one-hectare plot supports from 70 to 220 species of trees with stem diameter greater than 10 cm. This species list would include dozens of families of trees, with mahoganies, nutmegs, dipterocarps, laurels, myrtles, figs, and many other less well-known groups. Canopy height reaches 40 m or higher, and vertical structure is complex, with a subcanopy of saplings, rattans, pandanus, palms, creepers, lianas, and various herbs (especially gingers and ferns).

Hill Forest and Mid-mountain Forest. As one ascends to higher elevations, one finds gradual changes in forest structure and species composition. In mountainous country, the variables of slope, drainage, and natural succession brought about by landslips produce mosaics of vegetation types. Above 750 m, one finds the oak *Castanopsis acuminatissima* forming nearly pure stands on some ridge tops. The towering *Araucaria* "pines" are prominent emergents (70+ m tall), forming patchy stands in some lower montane areas. These montane forests are species-rich, although perhaps they lack the diversity of the lowland and lower hill forests. Oaks, elaeocarps, myrtles, and laurels are particularly common, and figs, while palms and lianas are fewer.

Cloud Forest. Above the cloud line, the forest is often heavily mossed, with the high diurnal humidity promoting luxuriant epiphytic growth. The flora is impoverished in comparison with that at lower elevations, and the canopy of this forest is often broken, with light gaps filled by thick tangles of nearly impenetrable scrambling bamboo *Nastus productus*. One of the conspicuous trees is the southern beech (*Nothofagus*); in some areas it forms dominant stands, especially on ridges. Above 2700 m, the forest is reduced in stature and species-poor, and shows a heavy covering of moss. Conifers (Podocarpaceae) and myrtles increase in importance, and the forest structure is simplified, with a thin canopy and a prominent understory. In protected areas one can find grand podocarp forests with canopies exceeding 40 m. These feature species of *Dacrydium, Phyllocladus, Dacrycarpus,* and *Podocarpus*, all festooned with orange-colored moss and abundant epiphytic orchids, rhododendrons, and ferns.

Timberline and the Subalpine. At elevations varying from 3700 to 4200 m, one encounters the timberline, where alpine shrubbery and grassland begin. Viewed from a distance, the *Deschampsia* grasslands appear inviting; however, this tussock grass habitat is usually swampy or waterlogged, and it is treacherous underfoot. Tree ferns and shrubs such as *Coprosma, Gaultheria, Vaccinium,* and *Rhododendron* are common at forest edge or in protected sites. In many sites, this high-elevation shrubland has been heavily damaged by wildfires set by traditional hunters during El Niño droughts.

This discussion has briefly treated New Guinea's dominant humid forest vegetation along an elevational gradient. Additional specialized habitats of restricted distribution include the following:

Mangrove Forest. Mangroves commonly fringe coastal areas. They form extensive stands from the Gulf of Papua along the south coast to the Bird's Neck, Bintuni Bay on the Bird's Head, and near the mouths of rivers along the north coast.

Seasonal Monsoon Forest. This uncommon forest, typified by the prominent presence of seasonally deciduous trees, occurs at the edges of the drier zones, especially northwest of Port Moresby and in the Trans-Fly.

Savannah and Open Woodland. Dominated by several species of *Eucalyptus* and *Melaleuca* with understory of grass, this vegetation predominates wherever there is a long and severe dry season and regular fires. Main savannah areas are found near Port Moresby and the southern Trans-Fly.

Swamp Forests. These vary in structural and floristic composition and occupy seasonally or permanently inundated tracts. Palm swamps, often dominated by *Metroxylon* (Sago), *Pandanus*, or *Nipa*, are common in the long-inundated zones. Lightly inundated areas often support tall and diverse forest, with particularly large, buttressed canopy tree species.

Swamps and Marshes. These wetlands—some seasonal, others permanent—support mainly grasses, rushes, and reeds and are home to most of the region's waterbirds.

Human-related alteration of the environment has produced significant changes (see Shearman et al. 2009). Most prominent are the grasslands that have developed in long-settled interior valleys of the highlands and, less frequently, in lowland areas. Demands for housing materials, commercial timber, and firewood all act to create large areas of open habitat near permanent settlements. With increasing development, the tracts of grassland surrounding settlements continue to expand, displacing the local forest. The highland valleys of the Ilaga, Baliem, Tari, Wahgi, and Bulolo are now largely treeless, except for the occasional planted stand of *Casuarina, Eucalyptus,* or *Pinus*. Lowland areas in the Sepik, Strickland, Markham, and upper Ramu valleys are dominated by old, human-created grasslands. In both the highland and lowland grasslands, the bird communities are distinctive but much impoverished.

3.2. BIRD GEOGRAPHY

Fundamental to understanding the geography and evolution of New Guinea birds is the notable "break" separating the avifauna of Southeast Asia from that in New Guinea and Australia. This major discontinuity includes both Wallace's Line (further west—separating Borneo from Sulawesi and Bali from Lombok) and Weber's Line (further east—separating Timor from New Guinea and passing west of Halmahera) and relates to the geologically old, deep-water channels separating the islands on the continental shelf of Asia from New Guinea, Australia, and associated islands to the east.

The great land-bridge islands of Sumatra, Java, and Borneo share with mainland Asia the presence of pheasants, barbets, woodpeckers, trogons, and bulbuls. Apart from a few locally introduced species, all these are absent from New Guinea, presumably because they failed to colonize across the imposing and relatively ancient deep-water barrier. New Guinea is instead home to cassowaries, birds of paradise, bowerbirds, honeyeaters, and numerous other bird families, which in turn are all lacking in the Asian fauna. During the Pleistocene glacial periods, when sea levels were lowered by as much as 100 m, the large land-bridge islands of Sumatra, Java, and Borneo formed a vast extension of the Asian mainland. This is often referred to as the *Sundaic Region* or *Sundaland*. In turn, New Guinea and Australia then formed a single landmass, the Australo-Papuan or *Sahul Region* or *Sahul Continent*, a fact reflected in bird distributions today. This separation between the Australasian and Asian faunas dates back tens of millions of years, enough time to allow for the evolution of dozens of bird families and hundreds of bird species in Australia and New Guinea.

Lying between the Sundaic and Sahul Regions are several archipelagoes surrounded by deep water: Sulawesi, the Moluccas, and Lesser Sunda Islands. This region, called *Wallacea* after the famous naturalist Alfred Russel Wallace, contains a mixture of biotas from the Sundaic and Sahul Regions. It has also served as a gateway for Asian bird species entering the New Guinea Region from the west.

The dry habitats of southern New Guinea are floristically similar to those in Australian savannahs to the south, and, not surprisingly, the local bird communities inhabiting these woodlands likewise are similar. Such birds as the Bar-shouldered Dove, Blue-winged Kookaburra, Blue-faced Honeyeater, and Magpielark—widespread in Australia—occur in New Guinea only in a few areas of appropriate habitat. The bird communities of the small "islands" of savannah habitat are impoverished in comparison with those in Australia and may be considered a relict fauna that was richer when the Torres Strait was dry.

Many New Guinea birds are largely restricted to lowland forest. Fruit-pigeons, the large parrots, kingfishers, pittas, meliphagas, monarchs, and manucodes are among those which are poorly represented in or absent from the highlands above 1500 m. In general, lowland forest birds show larger geographic ranges than montane species, reflecting greater habitat contiguity in the lowlands. In many cases, however, regional populations of lowland birds have differentiated, the local isolates achieving species status. Such is the case for the brushturkeys, the crowned pigeons, the streaked lories, the large fig-parrots, the pygmy parrots, and the *Paradisaea* birds of paradise. Typically, populations have differentiated into western, southern, and northern forms centered on the broadest expanses of lowland forest. Thus we find the Black Lory and Western Crowned Pigeon on the Bird's Head, the Yellow-streaked Lory and Southern Crowned Pigeon throughout most of the southern watershed, and the Brown Lory and Victoria Crowned Pigeon in the northern lowlands. The three principal geographical regions into which many lowland birds segregate are separated by barriers defined by low rainfall and/or mountains. It appears probable that speciation occurred when dissected populations were confined to regional refuges of humid forest, segregated from other areas by broad belts of inhospitable habitat during periods when rainfall was reduced.

Differentiation among highland birds is a more complex subject. The Central Ranges of New Guinea extend nearly the length of the island without major breaks. More than 200 breeding bird species are confined to regions above 500 m elevation. Speciation has occurred in many montane taxa, with the result that in typical groups, regional isolates have segregated largely on an east-west basis. For example, three species of astrapia birds of paradise occupy geographically separate ranges along the central cordillera—the Splendid Astrapia in the western half, the Ribbon-tailed in a central stretch, and Stephanie's in the eastern component. The outlying ranges—"mountain islands"—have fewer montane bird species than the Central Ranges, but most support some local specialties. Thus, the mountains of the Bird's Head, which have the most endemic forms of any mountain island, have their own species of forest-rail, two endemic honeyeaters, three birds of paradise, and a bowerbird, among many others.

Certain patterns of distribution remain enigmatic. Some primarily montane species, such as the Black-eared Catbird and White-faced Robin, also occur in lowland forest near the mouth of the Fly River. These lowland isolates appear to be relicts of populations whose ranges formerly extended, unbroken, from Australia to New Guinea. Other species have peculiarly "spotty" or "patchy" distributions within what appears to be continuous habitat. Such species as the Broad-billed Fairywren, Yellow-breasted Satinbird, Painted Quail-thrush, Rusty Whistler, Greater Melampitta, White-rumped Robin, and Banded Yellow Robin all have broad geographic ranges, but often with extremely local distributions—present in one area of forest but absent in another. This seems especially true for hill forest species.

Island Biogeography. Although New Guinea itself is an island, it serves for our purposes as the "mainland," which in turn is surrounded by numerous satellite islands, each of which supports a bird fauna that originated largely from the mainland. A field survey of any coastal island will show that the island bird list is much smaller than that for a comparable forest on the adjacent mainland. But the island list is not a random subset of the mainland avifauna. It is usually a special representation of birds, some of which are wide-ranging species, whereas others are relicts of a former land-bridge connection to the mainland, and a few may be old and sedentary insular endemics.

An important distinction exists between "oceanic" islands, separated from New Guinea by deep water, without past land connections, and "land-bridge" islands, separated from the mainland by water less than 100 m deep and hence connected to New Guinea during periods of low sea level. Examples of oceanic islands are Kofiau, Biak, and the D'Entrecasteaux Islands, whereas land-bridge islands in the region include Aru, Misool, Salawati, Batanta, Waigeo, and Yapen. Hundreds of New Guinea bird species, although perfectly capable of sustained flight, do not disperse over salt water, and hence are absent from all oceanic islands, even large islands and those near to the mainland. Many of these nondispersers occur on one or more of the land-bridge islands, relicts of populations that colonized over the mainland-island "bridges" that appeared at times of low sea level. For instance, the Frilled Monarch is on all 6 of the large land-bridge islands and Australia but inhabits no oceanic island. Insular populations of New Guinea species of scrubwrens, jewel-babblers, pitohuis, and robins occur only on land-bridge islands. The birds of paradise for the most part are confined to land-bridge islands.

The island realm also supports insular "tramp" species, wide-ranging colonists that readily cross open water and are restricted to species-poor, small islands. These tramps are often ubiquitous and common in small-island habitats but are never (or rarely) found on large islands or the mainland. Examples are numerous pigeon species, White-chinned Myzomela, Islet Monarch, and Island Whistler.

In contrast to tramp and land-bridge species, the old insular endemics, having evolved on islands in isolation, are strongly differentiated from their nearest mainland relatives. The larger islands harbor the most endemic species. For instance, Biak Island is home to species absent from nearby Yapen and the mainland of New Guinea, including Biak Scops-Owl, Biak Coucal, Biak Monarch, and many others.

3.3. BIRD REGIONS OF NEW GUINEA

For ease of reference we dissect the area into 15 bird regions of varying size (Figure 3). All 15 bird regions are defined by the presence of endemic bird populations at the species or subspecies level. (Be aware that the avifaunas of some bird regions are incompletely documented, so expect new range records in the future.) The bird regions will be major geographic reference points in discussions of species ranges in the species accounts. We review them, one by one, from west to east.

Northwestern Islands (NW Is). Today's Raja Ampat Islands—"four kings" in Indonesian—comprise Misool, Salawati, and the faunally similar Batanta and Waigeo, plus the smaller Gag, Gebe, and Kofiau Islands, all west of the Bird's Head. The first four are land-bridge islands. Interesting endemics include the Waigeo Brushturkey, Kofiau Monarch and Kofiau Paradise-Kingfisher, and Red and Wilson's Birds of Paradise. Affinities of some birds are with the Moluccas. A birding and diving mecca that is a must-see for adventurous naturalists. Can be accessed by boat from Sorong.

Bird's Head. Also known as the Vogelkop or Doberai Peninsula, this is the mountainous, westernmost peninsula of New Guinea, which includes two compact but high mountain ranges (Tamrau and Arfak) that support many montane endemics (Arfak Astrapia, Western Parotia, and Vogelkop Bowerbird, among others). The southern and western parts of the peninsula are swampy lowland flood-plains that share endemics with the Bird's Neck: Western Crowned Pigeon, Black Lory, and a number of subspecies. The Arfaks are relatively heavily developed and populated, whereas the Tamraus are entirely unpopulated. Both are infested with exotic Rusa Deer. Another mecca for visiting birders. Access is via Sorong and Manokwari.

Bird's Neck. The rugged connecting isthmus linking the Bird's Head with the main body of New Guinea. The Bird's Neck includes remarkable, rugged, and beautiful fjord lands in the southeast (Arguni, Etna, and Triton Bays) and three small isolated mountain ranges, each on its own peninsula (Wondiwoi Mts on the Wandammen Peninsula; Fakfak Mts on the Onin Peninsula; and Kumawa [or Kumafa] Mts on the Bomberai Peninsula). Although the Bird's Neck has no known endemic species, it shares with the Bird's Head many of its specialities. The avifaunas of all three mountain ranges remain inadequately surveyed and are home to undescribed forms such as an uncrested *Amblyornis* bowerbird that builds an unusual maypole bower, a smoky honeyeater, and a paradigalla bird of paradise.

Bay Islands. Situated between the Bird's Head and the northern cape of the main body of western New Guinea, Geelvink (Cenderawasih) Bay features the oceanic Biak and Numfor Islands, remarkable for their many distinctive endemic species: Numfor Paradise-Kingfisher, Biak Paradise-

Figure 3. New Guinea Bird Regions

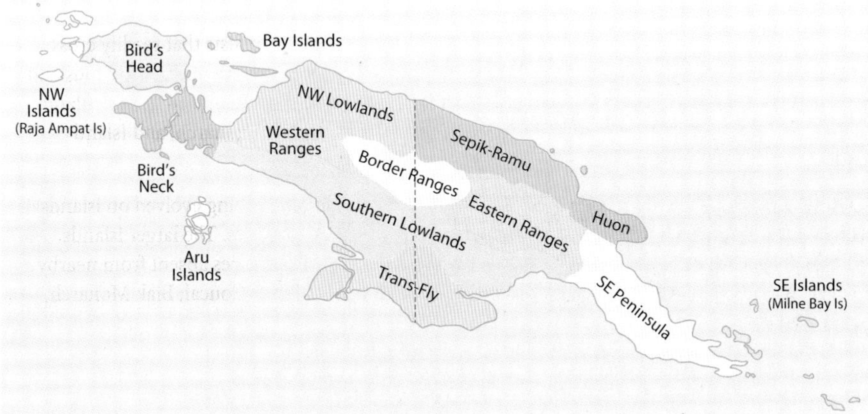

Kingfisher, and Biak Monarch, to mention a few. Without question, Biak supports the most localized endemism of any island in the New Guinea Region. The bird region also includes Yapen Island (Japen or Jobi I), a land-bridge island with no endemic species but many endemic subspecies.

Northwestern Lowlands. Bounded on the west by Geelvink Bay, on the east by the Cyclops Mts in the north and the Idenburg/Sepik divide in the south, the NW Lowlands include large expanses of lowland forest and swamp, and dozens of large meandering rivers—principally the Mamberamo, Rouffaer (Tariku), and Idenburg (Taritatu). The interior lowlands of the Mamberamo Basin constitute a vast wilderness with no roads and sparse populations. This region also encompasses the Van Rees (1400 m), Foja (2150 m), and Cyclops (2100 m) Mountains—the north coastal ranges of Indonesian New Guinea. The lowland avifauna has affinities with that of Sepik-Ramu. Endemics are Salvadori's Fig-Parrot and Brass's Friarbird, and the Pale-billed Sicklebill is a near-endemic. Golden-fronted Bowerbird, Wattled Smoky Honeyeater, and Bronze Parotia are restricted to the Foja Mts.

Sepik-Ramu. This near mirror image of the Northwestern Lowlands centers on the lowland basin of the Sepik and Ramu Rivers and is bordered to the west by hills near the Indonesian border and to the south by the Central Ranges. Includes the North Coastal Ranges (1900 m) and Adelbert Mts (1600 m). The sole endemic species is the Fire-maned Bowerbird of the Adelbert Mts. The region also includes numerous bird species that reach the easternmost edge of their ranges here. Unfortunately the rivers and lakes are now heavily populated with introduced fish species that have devoured the aquatic plants upon which many waterbirds depend.

Southern Lowlands. Bounded by the steep southern scarp of the Central Ranges, south coast, the Trans-Fly, and the Purari Delta, this region is a vast lowland flood-plain extensively dissected by fast-flowing, silt-laden rivers that rush out of New Guinea's highest ranges, then become wide, muddy, and meandering as they approach the coast. Timika, the lowland base of operations for the Grasberg Mine (see Western Ranges, below), lies in the Indonesian part of the region. On the PNG side, the towns of Tabubil and Kiunga support another giant mining operation and are important birding destinations. Avian specialties include Southern Crowned Pigeon, Starry Owlet-nightjar, Little Paradise-Kingfisher, Flame Bowerbird, White-bellied Pitohui, Greater Bird of Paradise, and Yellow-eyed Starling.

Trans-Fly. This is the dry savannah country south of the S Lowlands and includes the swampy island of Dolak (Frederik Hendrik, Kimaan, Kolepom, or Yos Sudarso). It offers the most ecologically distinct regional avifauna in New Guinea, a small piece of Australia isolated in southern New Guinea. Expanses of savannah, grasslands, reeds, and monsoon forest support numbers of Australian specialties as well as true endemic species (e.g., Fly River Grassbird and Grey-crowned and Black Mannikins). The Trans-Fly is a famous haven for wintering Australian dry country birds and waterbirds. It also includes patches of remnant rainforest that are home to isolated lowland populations of certain montane forms. Introduced Rusa Deer have created a parklike expanse in parts of the area. Bensbach is a popular birding and fishing destination.

Aru Islands. Faunally very close to the S Lowlands and Trans-Fly, the Aru Islands are hilly karst islands with a lowland biota. No endemic species are found here, but specialties include Chestnut Rail, Wallace's Fruit-Dove, and Elegant Imperial Pigeon.

Western Ranges. Western quarter of the Central Ranges was historically named the Snow Mts. It includes the ranges west of the Baliem Valley, from west to east: the Tiyo (Charles Louis), Kobowre (Weyland), and Sudirman (Snow and Nassau Mts) Ranges. Here rise the very highest summits in the tropical Pacific, with many peaks of more than 4000 m and two summits capped by permanent snow. Large intermontane valleys lie among the ranges, including the famous Baliem Valley.

These mountains contain the world's largest gold mine, the Grasberg Mine (Freeport-McMoRan). Endemic species include the Snow Mountain Quail, Orange-cheeked Honeyeater, Snow Mountain Robin, and Black-breasted Mannikin.

Border Ranges. The short middle segment of the Central Ranges is bounded on the west by the Baliem River gorge and on the east by headwaters of the Strickland River, thus including, from west to east, the Jayawijaya (Oranje), Star, Victor Emanuel, and Hindenburg Mts. Although they hold no endemic species, these mountains include the easternmost populations of some western montane species (Short-bearded Melidectes, Splendid Astrapia) and also lack a few species belonging to the Western Ranges.

Eastern Ranges. This is the central segment of the 4 montane bird regions defining the main cordillera. It is a complex of ranges constituted by the Schrader, Bismarck, Kubor, and Kratke Ranges, and includes Papua New Guinea's highest peak, Mt Wilhelm (4450 m), and a number of Pleistocene volcanoes (Mt Bosavi, Mt Giluwe, Mt Hagen, Crater Mt, Mt Karimui, and others). The two avian endemics are the Long-bearded Melidectes and Ribbon-tailed Astrapia. The region is known for its diversity of birds of paradise and is highly populous, accessible, and oft-traveled. This is a great place to visit to see tribal sing-sings and montane birds.

Huon. Four closely connected mountain groups (west to east: Finisterre, Saruwaged, Rawlinson, Cromwell Ranges) form this small, distinct, and isolated high mountain system separated from the Eastern Ranges by the lowland grassland barrier of the Markham and Ramu Rivers. Endemic species include Huon Bowerbird, Spangled Honeyeater, Huon Melidectes, Wahnes's Parotia, Huon Astrapia, and Emperor Bird of Paradise.

Southeastern Peninsula (SE Pen). The slender and mountainous peninsular region whose northwest boundary is roughly the head of the Huon Gulf (Lae) and the Watut/Tauri drainages, which form a break in the cordillera. A mountainous province, with no large lowland expanses, the SE Pen has peaks of more than 4000 m, two patches of dry savannah country (Port Moresby and southeast of Popondetta), and specialties such as Streaked Bowerbird, Elegant Meliphaga, and Eastern Parotia.

Southeastern Islands (SE Is). These include three island groups: D'Entrecasteaux (main islands of Goodenough, Fergusson, Normanby), Trobriands (main islands: Kiriwina, Kaileuna, Vakuta), and Louisiades (Tagula, Rossel, Misima, and many smaller isles), plus scattered, even more remote islands, the largest being Woodlark. Some islands are high volcanic remnants, others are low coral platforms. Numerous endemic species include Curl-crested Manucode and Goldie's Bird of Paradise. Another important destination because of its diving as well as birding potential. Access is via Alotau.

3.4. ECOLOGY

In this section we touch on some key features of the ecology of New Guinea birds.

Elevation (Altitude). New Guinea's forest bird communities fundamentally vary with elevation. To begin with, the lowland and hill forest communities are richest in species, those of the mountains becoming gradually less species-rich with increasing elevation. Thus along a sampled transect in eastern New Guinea, the forest community at 1500 m had about 100 species, one at 2500 m supported 75 species, and one at 3500 m had but 25 species. Explanations for gradual impoverishment increasing with elevation include several factors that are probably interrelated: reduction of available land area, decline in air temperature, increase in daily incidence of cloud cover and fog, and increase in frequency of physical disturbances caused by land slippage.

Accompanying these physical changes are unfavorable biotic factors: decline in productivity, decrease in floral species diversity, plus lower canopy height, and otherwise reduced structural diversity of the vegetation.

Second, each species has its own elevational range, and for some species this range can be quite narrow. As one proceeds from the lowlands up into the mountains, individual species drop out and new ones appear, so that by the time the highest peaks are reached, an entirely different species community is present. Thus, the species list changes with elevation, and the elevation where one is searching for birds very much determines the selection of species that may be present. Know your elevation—carry a map, altimeter, or GPS device.

Vegetation. Each bird species prefers forest or nonforest habitat, rarely both. Since most of New Guinea is forested, it is not surprising that most bird species are found in forest, and it is these communities that are species-rich. By contrast, nonforest habitats are species-poor. In addition, grasslands, mangroves, and savannah habitats support communities that are relatively small in New Guinea compared with those in regions where such habitats are widespread, such as Australia.

Diet. New Guinea is unusual in supporting large numbers of fruit and nectar feeders. This includes numbers of obligate frugivores, relatively uncommon in other parts of the world. In comparison with a comparable lowland forest community in Peru, the New Guinea fauna has twice as many fruit feeders and nearly twice as many nectar feeders, as measured by proportion of the fauna. Of special interest are the fruit-pigeons, amazingly diverse and abundant, and the brush-tongued lories and lorikeets, remarkable for their nearly exclusive diet of pollen and nectar. It is not uncommon to find noisy assemblages of bird species in a fruiting or flowering tree.

New Guinea's assemblage of ground-feeding forest birds is likewise considerable. With 3 cassowaries, 11 species of mound-building megapodes, 5 large ground-pigeons, plus an array of forest rails and other smaller species, it appears that a considerable radiation in the terrestrial avifauna has occurred. The ground-feeding passerines in particular include some of the most interesting components of the avifauna, birds such as logrunners, jewel-babblers, mouse-warblers, bellbirds, melampittas, and ground-robins.

In spite of its overall richness, gaps do exist in the ecological composition of the New Guinea avifauna. Vultures, a conspicuous element in other tropical faunas, are absent from New Guinea. The "woodpecker" niche is only partly filled by a few bark-gleaning species of treecreeper, sittellas, and various birds of paradise. The remarkable ant-following guild of the Neotropics is absent in New Guinea (as is the army ant phenomenon).

Nonetheless, mixed interspecific flocks are important in New Guinea, not only because of their ecological significance but also because they are crucial to observing many songbird (passerine) species. Interspecific flocking is most conspicuous in the lowlands but also occurs in montane forest. One may encounter practically no birds for an hour and then be overwhelmed by dozens of vocal individuals of many species. The forest becomes quiet again after the flock passes. There are two main types of flocks, which usually move separately but occasionally join.

The lowland "brown and black" flocks are composed of medium to large omnivorous species whose plumage is predominately brown or rusty, or black. These are led by groups of Papuan Babblers and pitohui species. A key "sentinel" species is the Spangled Drongo, which follows the group, sallies rather than gleans, and steals prey from other species. Followers include the Black Cicadabird, Tawny Straightbill, a number of birds of paradise, and various less-regular flock associates.

The "warbler/flycatcher" flocks are composed of small insect-eaters. The leaders are usually gerygones or Island Leaf-Warblers, with regular followers including monarchs, boatbills, fantails, and whistlers, and occasional followers such as white-eyes, shrikethrushes, and bronze cuckoos, among others.

Nesting. As in other tropical forest regions, nesting for New Guinea birds appears to be a risky business, although few studies have been conducted to support this belief. A large proportion of nesting attempts fail owing to depredation. This generalization probably varies depending on the species, its nest type, and placement. Birds are thought to evolve different kinds of nests and nest placement to escape the attentions of predators that would rob their eggs and young. Nesting habits of New Guinea birds are indeed diverse. The scrubfowl along the coast builds a mound of sand and decaying vegetation and lays its eggs inside, whereas other ground-nesting birds, such as the Pheasant Pigeon, may hide their nest at the base of a tree trunk. Many non-passerine birds nest in tree cavities, including parrots, owls, owlet-nightjars, and kingfishers, but few songbirds do. The nest of the Buff-faced Pygmy Parrot and some kingfishers is usually a cavity excavated within an active termite nest. The frogmouths, treeswifts, cuckooshrikes, peltopses, and flycatchers build such a small and cryptic nest atop a canopy branch that the nest hardly shows beneath the incubating bird. Songbirds excel at *building* nests, and this allows them to hide the egg or place it out of reach. For instance, the Papuan Babbler constructs a pendent, domed nest, some 2 m long, suspended from the spiny lawyer-vine *Calamus*—an effective defense against marauding snakes and rats. In the species accounts for the field guide, we have described the nests that are known. See if you can discover one not yet reported.

The median clutch size for New Guinea forest birds is two. Some groups, such as pigeons and birds of paradise, commonly lay only one egg per nest, but few regularly produce three or more. Although active nests can be found in any month, there are apparently optimal nesting seasons for certain groups. Most songbirds breed just prior to the rainy season, whereas many pigeons nest during the rains. These seasons vary regionally. In many areas, where the rains come at the end of the year, most nesting occurs between August and December, during the austral spring and early summer.

Molt. Little is known about patterns of molt (the shedding and replacement of feathers) in New Guinea birds. Perhaps most adult birds molt after successfully nesting, during the austral summer and autumn. For example in the Wau region, insectivorous songbirds revealed a molting peak between November and April, with much lower percentages of individuals in molt from May to October. In this area, the passerine (songbird) nesting season centers on the last months of the year, thus timing of molt seems to follow nesting.

Migration within New Guinea. Many New Guinea birds are sedentary, rarely leaving their established patch of forest or savannah; however, some species offer conspicuous exceptions and travel locally within New Guinea. Not surprisingly, such birds are those feeding on patchy and seasonal resources such as fruits and flowers. The Papuan Mountain-Pigeon is commonly observed in flocks, flying swiftly at high altitudes, but little is known about the pattern and seasonality of its travels. Lories and lorikeets form conspicuous flocks that wander in search of flowers. The jewel-like Red-collared Myzomela generally lives at high elevations on Mt Kaindi, above the Wau Valley, but during certain seasons, it descends the mountain to forage in town gardens at 1100 m. These small-scale movements regionally and with elevation merit further investigation.

Intercontinental Migration. New Guinea serves as winter quarters for long-distance migrants from both the Northern and Southern Hemispheres. Birds from the North, the *Palearctic migrants*, move south into the New Guinea Region in August–October, spend their winter, and leave in March–May. The largest assortment of Palearctic migrants consists of waders and seabirds, and these either settle in the region to overwinter or pass through on their way further south to Australia (waders) or temperate seas (seabirds). Most of the waders winter along coastal locations. Only a few species of northern land birds reach New Guinea, and then usually only as far as the western portion of the region. These move southward and eastward from mainland Asia and the Indonesian islands to arrive in the New Guinea Region. Among them are the Oriental Cuckoo, Gray's Grasshopper-Warbler, Grey-streaked Flycatcher, and Grey and Eastern Yellow Wagtails.

The greatest volume of land-bird movement occurs between Australia and New Guinea, here termed *AU migrants*. A number of Australian breeding land bird species spend the nonbreeding season during the austral winter (March–October) in New Guinea, but how many birds migrate and when they arrive depend on both the timing of the winter season and unpredictable drought conditions in Australia. Some, like the Rainbow Bee-eater and Sacred Kingfisher, are among New Guinea's best-known garden birds. Most of these migrants occupy nonforest or forest-edge habitats. Such zones may offer the benefit of reduced competition from resident New Guinea species. Some migrants do winter in the forest proper: Buff-breasted Paradise-Kingfisher, Red-bellied Pitta, Black-faced Monarch, and Rufous Fantail. The nature of their interactions with resident relatives is not known.

A remarkable interchange of waterbirds occurs between the open but well-watered country of the Trans-Fly and Australia. Tens of thousands of waterfowl, ibises, egrets, pelicans, and others spend the austral winter in this southern part of New Guinea.

There are also great migrations of southern seabirds. Most abundant are the *trans-equatorial migrants*, for example the Short-tailed Shearwater and Wilson's Storm-Petrel, that breed in the southern oceans during austral summer and afterwards escape winter by passing through the NG Region in autumn to spend their winter season feeding in the waters of the north Pacific and Indian Oceans during the northern summer. (A few northern seabirds do just the reverse; for example, jaegers breed in the Arctic and winter in southern oceans.)

Historical Changes. Although the tropics have been considered a zone of great stability, change is very much a part of the New Guinea Region's history. As recently as 10,000 years ago, most of the higher mountaintops were encrusted with snow, and forest zonation must have been considerably different from that existing today. As the snowcaps continue to recede, it is clear that climatic and other environmental changes are occurring. Bird populations inevitably will respond to these fluctuations, but it is difficult to monitor changes in a region with few long-term observers.

The most rapid shifts we see are in relation to man's transformation of local habitats. Coinciding with continued clearing of forest in mid-mountain valleys are major changes in local bird faunas. The new clearings are prime targets for opportunistic avian colonists, either from the lowlands or from other open habitats. The Grey Shrikethrush, known primarily along the southern coastal lowlands, recently has invaded the Eastern Ranges to elevations of 1500 m. Documenting similar changes in the forest fauna is much more difficult, and more study of this phenomenon would be useful.

3.5. CONSERVATION AND FUTURE

Compared with many tropical regions, New Guinea is blessed in that most of its native vegetation remains intact. This happy condition, however, is changing rapidly.

Natural Resource Development. Not surprisingly, since publication of the first edition of this field guide, many changes have come to the region. Industrial logging has expanded a great deal, especially in Papua New Guinea, focusing on the lowland and foothill forests. Because most logging in New Guinea is selective, the forests affected are left standing to regenerate over time.

Oil palm cultivation is another story altogether. These large-scale monoculture plantations are mainly in the flat lowlands. Establishing oil palm plantations requires clear-felling expanses of some of New Guinea's very richest lowland tropical forests and spells doom for local populations of forest-dependent wildlife that cannot live in oil palm plantations. There is no business more destructive of New Guinea's forests and birdlife than oil palm.

Large mining operations with inadequate tailing management damage watersheds and river systems in the highlands regions wherever gold, nickel, and copper are found in rich deposits. Oil and natural gas projects are also expanding in the region. These lead to new road networks and creation of new population centers and further habitat loss.

All this resource development comes at an environmental cost. It is important that the leaders of Papua New Guinea and Indonesian New Guinea balance economic advancement with smart environmental planning at the scale of major ecosystems. Administrators and decision makers need to think big and think long-term—making environmental planning an integral part of regional or national development. Economic development can lead to gains on the environmental front so long as smart planning and trade-offs are considered. If a mine is developed, the negative environmental impacts need to be properly *offset* by counter-balancing environmental actions—for instance, setting aside forests or other natural environments as permanent conservation areas. Economic development should provide the engine for good environmental management.

Population Growth. The human population on the island of New Guinea is growing rapidly. In most rural areas, population growth leads to permanent deforestation of lands near any population center as demand for new gardens and fuel-wood rise. Most readers probably believe that industrial logging is the main cause of deforestation in New Guinea, but this is not so—it is agricultural expansion in association with population growth. Family planning is needed to ensure rural communities do not outgrow their resource base. As we have seen in dry regions of Africa, such resource loss leads to a downward cycle of poverty. Communities with rich natural resources are well off because of the many noncash benefits provided by nature—clean drinking water, rich agricultural soils, abundant wood and fiber for building and fuel, and rich assemblages of species for game and spiritual well-being.

Climate Change. Climate change is now recognized as a significant phenomenon in New Guinea. It is most obvious to subsistence farmers, who report changes in seasons and conditions. Coconut palms in the 1950s were found only in the hot lowlands, but now are being planted in villages at 1200 m and higher. Rainfall seasons are now reported to be much less predictable, with heavy rains in the dry season and periodic droughts in the rainy season. Research on future climatic conditions for New Guinea predicts three major changes: (1) increased annual rainfall in the wetter areas, (2) annual temperatures above those now known for the region in the interior basins, and (3) greater seasonal extremes in certain areas. It is not clear how these changes are impacting the birdlife aside from the obvious upslope advance of many species. This century will undoubtedly be a time of great change, and although some bird species will benefit, others will lose; some may disappear altogether. Habitat for alpine and montane bird species in particular will shrink.

Threatened Bird Species. Although New Guinea hosts a fair share of rare species, most of these are naturally rare and not endangered with extinction by human-caused threats at present. Thus far no species in the region has become extinct historically. The only species listed as *critically endangered* on the Red List (2014) for the New Guinea Region is actually a vagrant seabird. Four species on the Red List are categorized as *endangered*, and an additional 22 species are listed as *vulnerable*. Many others are listed as near threatened. The majority of threatened species have small ranges, and many are endemic to the satellite islands.

Hunting of birds of paradise for plumes is now in decline, both because shotgun ownership has been much restricted and because traditional demand for plumes for ceremonies seems to be declining as New Guinean communities westernize. Populations of common species remain common, even near towns and villages. Today, it is considerably easier to observe a displaying male Black Sicklebill or Blue Bird of Paradise in the Tari valley than it was two decades ago.

We suspect the species that will be most at risk over the next several decades will be harmed by the impacts of forest loss and global climate change, and these will probably be poorly known forest-dwelling species, not necessarily the glamorous species that formerly worried us. For instance, the

Table 1. New Guinea's Threatened Birds.
This is the complete IUCN Red List of threatened bird species for the New Guinea Region (http://www.iucnredlist.org, accessed 20 February 2014). Not shown are species listed as Near-threatened or Least Concern.

CRITICALLY ENDANGERED	
Christmas Island Frigatebird—vagrant	
ENDANGERED	
Waigeo Brushturkey—endemic	Biak Scops-Owl—endemic
Abbott's Booby—vagrant	Biak Monarch—endemic
VULNERABLE	
Southern Cassowary—resident	New Guinea Vulturine Parrot—endemic
Northern Cassowary—endemic	Black-winged Lory—endemic
Moluccan Scrubfowl—vagrant	Kofiau Paradise-Kingfisher—endemic
Biak Scrubfowl—endemic	Giant Wattled Honeyeater—endemic
Salvadori's Teal—endemic	Long-bearded Honeyeater—endemic
Heinroth's Shearwater—migrant	Kofiau Monarch—endemic
Eastern Curlew—migrant	Wahnes's Parotia—endemic
Great Knot—migrant	Black Sicklebill—endemic
New Guinea Harpy-Eagle—endemic	Blue Bird of Paradise—endemic
Western Crowned Pigeon—endemic	Fly River Grassbird—endemic
Southern Crowned Pigeon—endemic	Grey-banded Mannikin—endemic

Fire-maned Bowerbird is vulnerable because its range is so circumscribed: it occurs in only a few remaining patches of montane forest in the Adelbert Range northwest of Madang. This bowerbird probably has a smaller population than any other New Guinea mainland species. Most of its primary forest habitat has been lost to subsistence agriculture. Climate change will drive this bowerbird upslope into an ever-smaller refuge from the rising temperature.

Protected Areas. Conservation of natural habitats is one way to address climate change threats. It is thus important for both Indonesia and Papua New Guinea to develop an extensive network of natural protected areas (marine, aquatic, terrestrial) that can act as storehouses of New Guinea's biodiversity as well as providers of ecosystem services and resilience. Indonesia has created a rather larger network of "paper parks"—conservation areas with little management or protection. Papua New Guinea has a mix of conservation areas of uncertain provenance, and those are provided minimal oversight. Both countries should be able to do much better. Indonesia and Papua New Guinea are signatories of various international treaties on climate change and biodiversity. This obligates them to take action to address climate issues and protect their wildlife and forests. These two nations need to be held responsible for the future of the rich and wonderful natural wealth in their care. It is our obligation to help these governments to do the right thing—to preserve the earth's natural legacy for the good of the earth but also to provide tangible and lasting benefits to the nations that achieve this.

4. In the Field in Search of Birds

Whether a resident or a visitor, you will find bird-watching in New Guinea a challenge, albeit an exciting one. Thus it pays to plan and to be prepared. Birding in New Guinea can produce amazing rewards but also offers up no shortage of hazards and frustrations. One recommendation for the international birder is that it is probably wise to save a planned birding trip to New Guinea until after successful birding ventures to other key destinations such as Australia, South Africa, India, or Costa Rica. New Guinea is best considered a specialty destination, to be savored after gaining a seasoned grasp of birding in the tropical world.

The big pluses offered by New Guinea include (1) essentially a new bird fauna, particularly for songbirds; for example, the wonderful birds of paradise, bowerbirds, fruit-doves, pigeons, parrots, and kingfishers; (2) its stupendous human cultural and ethnic diversity; and (3) its lovely natural environments. The drawbacks are (1) the long distance from the world's population centers, (2) the substantial expense of traveling to New Guinea, (3) the major expense of traveling within New Guinea, and (4) the logistically difficult conditions.

The resident novice or newly arriving visitor to New Guinea will soon discover that although birds of the open country are conspicuous and easy to study, the forest-dwelling species typically are often less common, more secretive, and difficult to observe. This phenomenon is common to tropical humid forest environments around the world. Patience and repeated effort provide the only antidote to the frustrations created by those rare species that defy locating on one's first few field trips.

Equipment. Here is a short list of tools needed for field study: (1) Waterproof binoculars of about 8-power are best for maximum light-gathering in the shade of the forest. (2) A tripod-mounted spotting scope (telescope) is valuable in open habitats and along roadsides, especially for birds in the canopy. (3) A lightweight digital sound recorder, with a speaker suitable for playback and a directional microphone, greatly aids in learning vocalizations and luring shy vocal species into view. (4) A field notebook and pencil or pen with waterproof ink are advisable for notes on field marks, habits, calls, and unusual observations. (5) Loose-fitting, lightweight clothing and a hat with some sort of visor to shield the eyes from sun (or rain). It is often very cool at dawn but hot at midday; wear layers that can be easily removed. Loose long pants and lightweight, good quality walking boots are often the most comfortable combination. Conditions in the mountains are much cooler, so prepare to keep warm. (6) Insect repellent is required in some localities. Guard against bush mites (chiggers or scrub-itch) by rubbing repellent onto legs and socks and around the waist, where these burrowing pests readily attack. Leeches are abundant locally in rainforest but can be deterred in the same way as for bush mites. Mosquitoes are troublesome in some sites; keeping mosquitoes off reduces the risk of contracting malaria. (7) A lightweight poncho or rain jacket. It is sometimes useful to carry a larger umbrella (especially if you have a local guide who is helping with gear).

Where and when. In nearly all New Guinea habitats, bird activity is greatest at dawn and shortly thereafter. It pays to be afield early. By 11 a.m. it is usually quiet, especially in the lowlands, unless rain brings cooler conditions.

Good sites for encountering birds are ecotones—forest edges, riversides, and verges of marshes or swamps. Luckily, these are the places most accessible to mobile birders. For forest birding, optimal conditions exist along roadways and paths through the forest or at any break in the canopy. If one encounters a mixed feeding flock, stay with it as long as possible—that is where the birds are!

Whereas birds are universally more visible, more numerous, and easiest to see at habitat edges, it is necessary to enter the forest (preferably using a trail) to see those species that inhabit only the dark forest interior (pittas, paradise-kingfishers, forest-rails, and so on). We believe that the most enriching experience in New Guinea bird study can result from spending time in undisturbed forest searching for endemics seen by few.

It rains a lot in New Guinea, but one should not completely despair when the weather turns wet: (1) More often than not, it will break sometime during the day. (2) Even if it doesn't, wait for the rain to slacken, then from your bush camp scan for bird activity at the forest edge or in the treetops—birds must find food, and they will seek the margins of vegetation where they may be able to dry out a bit.

Identifying birds. In the forest, one hears 10 birds for each one actually seen. To truly master bird identification in New Guinea, one must learn the vocalizations. It is a difficult task. Although our color plates provide a ready assessment of bird appearances, text descriptions of calls and songs are less definitive. In addition, bird song is often complex and variable; one species may have many different vocalizations. When in the field, train the ear to distinguish between known calls and those not yet identified. When possible, pursue and identify the latter, or make recordings to send to others who might help with identification. Knowledgeable local guides and field assistants can be enormously helpful in identifying or locating birds by voice, a skill they have learned since childhood.

Geography is all-important—knowledge of locality and elevation can quickly eliminate expectations of finding some species and raise hopes for others. Are you looking in mountain forests of Eastern Ranges or lowland rainforest of the Sepik-Ramu? The two locations have very different bird faunas. Habitat choice can be critical in finding desired species. The rare Archbold's Bowerbird inhabits only uplifted plateau-like, high-elevation basins, where repeated frost kills have created a peculiar open-canopied forest formation with abundant pandanus, scrambling bamboo, and emergent podocarp trees. A search for this bowerbird in any other type of montane forest will probably prove fruitless. Additional species of New Guinea birds possess equally specific habitat requirements. Similar-appearing species often segregate within a habitat. Does the bird live in the forest canopy or in low thickets? Habitat factors can be valuable clues in identification.

Visiting from Overseas. The visitor desiring to see birds of paradise or other specialties of the New Guinea bird fauna must plan the trip in detail (or join a professionally led birding tour). In all respects, Papua New Guinea is more readily accessible than Indonesian New Guinea. The latter requires special papers (a "surat jalan") to travel to many locations, and the Indonesian Government does not encourage foreigners to visit Papua or Papua Barat as tourists. It is best to visit western New Guinea by joining an organized bird-tour group (which can be easily found online). By contrast, in Papua New Guinea one can purchase a visa upon arrival at the airport in Port Moresby, although here too it is better to have arranged a tour or at minimum an in-country host.

Within the region, travel is primarily by air, secondarily by automobile, foot, or boat. Air travel is expensive but well organized. There are hundreds of airstrips scattered across the island. Air travel is decent and predictable, and most larger towns have hotels or guest houses. Many centers have rental cars, and four-wheel vehicles may be available, but expensive. Prices in PNG are more expensive than those in Australia; those in Indonesian New Guinea are less expensive (although Indonesian New Guinea is considerably more expensive than western Indonesia).

Health. New Guinea is not the disease-filled island most first-time visitors imagine it to be. Nonetheless, some precautions are advised. Take appropriate antimalarial medication when traveling anywhere in the region. Also ensure that all cuts and insect bites are kept washed, disinfected, and covered. It really helps to wear long-sleeved shirts and long pants, despite the discomfort in hot weather. Otherwise tropical sores can develop with surprising speed, creating great discomfort and threat of systemic infection. Water in most restaurants is safe, as it is boiled and filtered. Bottled water is widely available. Stomach complaints, while not unknown, are much less a problem than in most other tropical countries.

Field Research. Anyone desiring to conduct scientific research should note that specific research permits are required from the relevant authority. Expect it to take a minimum of 6 to 12 months to obtain clearance, for it is not a simple process. In PNG contact the PNG National Research Institute (visit their website). In Indonesia, one must apply for a research permit from the government's research office, RISTEK; they can be located on the Internet. It is best to discuss plans with someone who has recently done research in either country, as conditions and requirements may have changed. Also plan on engaging with an in-country partner institution for the research project—this is usually required. Realize that delays in clearance are commonplace.

Etiquette for Visitors. It is often difficult to quickly learn the customs and mores of a new culture. We stress one point above all others: No one appreciates having perfect strangers tramping about in one's backyard; landowners in New Guinea are no exception. In the New Guinea setting, however, the definition of "backyard" takes on a new meaning. What to any outsider's eye is nothing but trackless rainforest is, in fact, someone's family property. All land is owned, even if the nearest village is 10 km away, so it is always polite to ask permission rather than trespassing. Explain your objectives; tell the owner you wish to look at birds, not take anything away, and in most instances, you will be welcomed.

It is usually best to hire a responsible local guide—not just to find birds and show you around, but also to keep you from getting lost or coming to harm, and to satisfy your host's wishes and concerns. The expense is small, and the potential gain is high. New Guinea is famous for its excellent village naturalists, and help from them will ensure you see great birds you will not find on your own. Use this field guide as a means of communication—jointly examining the plates and asking the guide to tell stories about particular birds. Instead of pointing at a species on a page and asking, "Is this species here?" it is better to open the book to a plate and ask the informant to look it over and indicate which species he knows and what he knows about each. In that way you can informally test the informant's knowledge without giving away the answer.

New Guinean Culture and Birds. Birds form a significant part of the cultural life of the people of New Guinea. The larger species are important sources of protein in the diet. Among birds, the cassowaries, pigeons, megapodes, and even small birds are valued as food and hunted with bow and arrow, spear, snare, or net. Rich folklore and local custom surround those birds that are regularly hunted (see Majnep and Bulmer 1977).

New Guinea's famous ritual dances ("sing-sings") are often dependent on the special use of a variety of bird feathers for embellishment of customary dress. Plumes of birds of paradise (Lesser, Raggiana, Blue, Superb, and King of Saxony BoPs, and Stephanie's Astrapia); tail feathers or whole skins of lorikeets; parrot, cockatoo, and eagle wing and tail feathers; and cassowary body plumes all figure prominently. These feathers are treasured, hoarded, traded, and often used year after year and passed on from one generation to next. This is an expression of a deep cultural attachment to New Guinea birdlife.

Hazards to Avoid. A number of potential pitfalls await the traveler to New Guinea. Artifact collectors should be warned that export of any animal material (boar tusk, feather, tanned skin), whether attached or unattached to local artifacts, requires a permit. This is probably most important to visitors to Papua New Guinea. International convention forbids export of plumes of birds of paradise, cassowaries, crowned pigeons, and a number of other species. Permits for nonprotected material are simple to obtain. No artifacts manufactured prior to 1960 may be exported from PNG.

In highland areas, some people, especially if richly ornamented, will expect to receive a gratuity for posing for tourist photographs. Local landowners may charge fees for photography or for bird-watching on traditionally owned lands.

In some parts of Papua New Guinea, mainly near cities and in the highlands, personal security remains a problem, and travelers should avoid walking about alone at night or leaving their valuables unprotected. Always inquire locally about security and be aware of local custom. Taking extra precautions, based on local knowledge, will help make one's visit to New Guinea a success.

Selected References

Barker, F. K., G. F. Barrowclough, and J. G. Groth. 2001. A phylogenetic hypothesis for passerine birds: taxonomic and biogeographic implications of an analysis of nuclear DNA sequence data. *Proceedings of the Royal Society of London* B 269:295–308.

Barker, F. K., A. Cibois, P. Schikler, J. Feinstein, and J. Cracraft. 2004. Phylogeny and diversification of the largest avian radiation. *Proceedings of the National Academy of Sciences* 101:11040–11045.

Beehler, B. M., and B. W. Finch. 1985. *Species-Checklist of the Birds of New Guinea*. Australasian Ornithological Monographs, No. 1. Royal Australian Ornithologists, Moonee Ponds, Victoria, AU.

Beehler, B. M., and T. K. Pratt. In prep. *Birds of New Guinea: Distribution, Taxonomy, and Systematics*. Princeton University Press, Princeton, NJ, USA.

Beehler, B. M., T. K. Pratt, and D. A. Zimmerman. 1986. *Birds of New Guinea*. Princeton University Press, Princeton, NJ, USA.

Chandler, R. 2009. *Shorebirds of North America, Europe, and Asia: A Photographic Guide*. Princeton University Press, Princeton, NJ, USA.

Cheke, R. A., C. F. Mann, and R. Allen. 2001. *Sunbirds: A Guide to Sunbirds, Flowerpeckers, Spiderhunters, and Sugarbirds of the World*. Christopher Helm, London, UK.

Cheshire, N. 2010. Procellariiformes observed around Papua New Guinea including the Bismarck Archipelago from 1985 to 2007. *South Australian Ornithologist* 36:9–24.

Cleere, N. 2010. *Nightjars, Potoos, Frogmouths, Oilbirds, and Owlet-nightjars of the World*. Princeton University Press, Princeton, NJ, USA.

Coates, B. J. 1985 and 1990. *Birds of Papua New Guinea*. Dove Publications, Alderley, Queensland, AU.

Coates, B. J., and K. D. Bishop. 1997. *Field Guide to the Birds of Wallacea: Sulawesi, The Moluccas and Lesser Sunda Islands, Indonesia*. Dove Publications, Alderley, Queensland, AU

Coates, B. J., and W. S. Peckover. 2001. *Birds of New Guinea and the Bismarck Archipelago*. Dove Publications, Alderley, Queensland, AU.

Cristidis, L., and W. E. Boles. 2008. *Systematics and Taxonomy of Australian Birds*. CSIRO Publishing, Collingwood, Victoria, AU.

Diamond, J. M. 1972. *Avifauna of the Eastern Highlands of New Guinea*. Nuttall Club Publication #12, Cambridge, MA, USA.

Dickinson, E. C., and J. V. Remsen Jr. (Eds.). 2013. *The Howard and Moore Complete Checklist of the Birds of the World*, 4th ed. Vol. 1, Non-passerines. Aves Press, Eastbourne, UK.

Edwards, S. V., and W. E. Boles. 2002. Out of Gondwana: The origin of passerine birds. *Trends in Ecology and Evolution* 17:347–348.

Erritzoe, J., K. Kampp, K. Winker, and C. B. Frith. 2007. *The Ornithologist's Dictionary*. Lynx Edicions, Barcelona, Spain.

Ferguson-Lees, J., and D. A. Christie. 2001. *Raptors of the World*. Christopher Helm, London, UK.

Forshaw, J. M. 2010. *Parrots of the World*. Princeton University Press, Princeton, NJ, USA.

Frith, C. B. 1979. *Ornithological Literature of the Papuan Subregion, 1915 to 1976: An Annotated Bibliography*. Bulletin of the American Museum of Natural History, 164: 377–465.

Frith, C. B., and B. M. Beehler. 1998. *The Birds of Paradise, Paradisaeidae*. Oxford University Press, Oxford, UK.

Frith, C. B., and D. W. Frith. 2004. *The Bowerbirds, Ptilonorhynchidae*. Oxford University Press, Oxford, UK.

Geering, A., L. Agnew, and S. Harding. 2007. *Shorebirds of Australia*. CSIRO Publishing, Collingwood, Victoria, AU.

Gibbs, D., E. Barnes, and J. Cox. 2001. *Pigeons and Doves: A Guide to the Pigeons and Doves of the World*. Pica Press, London, UK.

Gill, F., and D. Donsker (Eds.). 2013. IOC World Bird List (v 3.3). Available at http://www.worldbirdnames.org (accessed 1 May 2013).

Gressitt, J. L. (Ed.). 1982. *Ecology and Biogeography of New Guinea*. Two volumes. Junk, The Hague, Netherlands.

Harrison, P. 1983. *Seabirds: An Identification Guide*. Houghton Mifflin, Boston, MA, USA.

Hayman, P., J. Marchant, and T. Prater. 1986. *Shorebirds: An Identification Guide to Waders of the World*. Houghton Mifflin, Boston, MA, USA.

Hollands, D., and C. Minton. 2012. *Waders: the Shorebirds of Australia*. Bloomings Books, Melbourne, AU.

Holyoak, D. T. 2001. *Nightjars and Their Allies: The Caprimulgiformes*. Oxford University Press, Oxford, UK.

del Hoyo, J., et al. (Eds.). 1992–2011. *Handbook of the Birds of the World*. 16 volumes. Lynx Edicions, Barcelona, Spain.

Jones, D. N., R. W. R. J. Dekker, and C. S. Roselaar. 1995. *The Megapodes*. Oxford University Press, Oxford, UK.

Jønsson, K. A., P.-H. Fabre, R. E. Ricklefs, and J. Fjeldsa. 2011. Major global radiation of corvoid birds originated in the proto-Papuan archipelago. *Proceedings of the National Academy of Sciences* 108:2328–2333.

Juniper, T., and M. Parr. 1998. *Parrots: A Guide to Parrots of the World*. Yale University Press, New Haven, CT, USA.

Laman, T., and E. Scholes. 2012. *Birds of Paradise: Revealing the World's Most Extraordinary Birds*. National Geographic and Cornell Lab of Ornithology, Washington, DC, USA.

Löffler, E. 1977. *Geomorphology of Papua New Guinea*. The Commonwealth Scientific and Industrial Organization, Canberra, AU.

Majnep, I. S., and R. Bulmer. 1977. *Birds of My Kalam Country*. Auckland University Press, Auckland, NZ.

Marchant, S., P. J. Higgins, et al. (Eds.). 1990–2006. *Handbook of Australian, New Zealand and Antarctic Birds*. 7 volumes. Oxford University Press, Melbourne, AU.

Morcombe, M. 2004. *Field Guide to Australian Birds, Complete Compact Edition*. Steve Parish Publishing, Archerfield, Queensland, AU.

Marshall, A. J., and B. M. Beehler (Eds.). 2007. 2 parts. *The Ecology of Papua*. Periplus Editions, Singapore.

Mayr, E. 1941. *List of New Guinea Birds: A Systematic and Faunal List of the Birds of New Guinea and Adjacent Islands*. American Museum of Natural History, New York, NY, USA.

Mayr, E., and J. Diamond. 2001. *The Birds of Northern Melanesia: Speciation, Ecology, and Biogeography*. Oxford University Press, Oxford, UK.

Message, S., and D. Taylor. 2005. *Shorebirds of North America, Europe, and Asia: A Guide to Field Identification*. Princeton University Press, Princeton, NJ, USA.

Muruk: Journal of the Papua New Guinea Bird Society. 1986–2009. Published periodically and distributed from various locations in Papua New Guinea and eventually Kuranda, Queensland, AU.

Newsletter of the New Guinea [later *Papua New Guinea*] *Bird Society*. 1965–1997. Published periodically and distributed from Port Moresby.

Onley, D., and P. Schofield. 2007. *Albatrosses, Petrels, and Shearwaters of the World*. Princeton University Press, Princeton, NJ, USA.

Payne, R. B. 2005. *The Cuckoos*. Oxford University Press, Oxford, UK.

Pizzey, G., and F. Knight. 2013. *The Field Guide to the Birds of Australia,* 9th ed. HarperCollins, Sydney, New South Wales, AU.

Rand, A. L., and E. T. Gilliard. 1967. *Handbook of New Guinea Birds*. Weidenfeld and Nicolson, UK.

Restall, R. 1996. *Munias and Mannikins*. Pica Press, London, UK.

Schodde, R., and I. J. Mason. 1999. *The Directory of Australian Birds: Passerines*. CSIRO Publishing, Collingwood, Victoria, AU.

Shearman, P. L., J. Ash, B. Mackey, J. E. Bryan, and B. Lokes. 2009. Forest conversion and degradation in Papua New Guinea 1972–2002. *Biotropica* 41:379–390.

Simpson, K., and N. Day. 2010. *Birds of Australia,* 8th ed. Princeton University Press, Princeton, NJ, USA.

Slater, Peter, Pat Slater, and R. Slater. 2009. *The Slater Field Guide to Australian Birds, 2*nd *ed*. Reed New Holland, Sydney, AU.

Strange, M. 2001. *A Photographic Guide to the Birds of Indonesia,* 1st ed. Princeton University Press, Princeton, NJ, USA.

Taylor, B. 1998. *Rails: A Guide to the Rails, Crakes, Gallinules and Coots of the World*. Pica Press, London, UK.

White, C. M. N., and M. D. Bruce. 1986. *The Birds of Wallacea (Sulawesi, the Moluccas, and Lesser Sunda Islands, Indonesia)*. B.O.U. Checklist No. 7. British Ornithologists' Union, London, UK.

Web Sources

Australian Bird Image Database
(http://www.aviceda.org/abid).

Bird Checklists for 610 Melanesian Islands
(http://birdsofmelanesia.net/).
A unique and useful reference; sources and records should, however, be confirmed independently in all cases.

Cornell Lab of Ornithology, Macaulay Library
(http://macaulaylibrary.org/browse/taxa/aves)
Voice recordings and videos.

GeoNames
(http://www.geonames.org)
For geographic names (towns, mountains, rivers, etc.).

Kukila—The Journal of Indonesian Ornithology
(http://kukila.org/index.php/KKL).

IOC World Bird List
(http://www.worldbirdnames.org)
A frequently updated list of birds of the world.

IUCN Red List of Threatened Species
(http://www.iucnredlist.org)
International Union for Conservation of Nature.

Muruk
(http://eng.sicklebillsafaris.com/index.php/muruk)
A journal for ornithology in New Guinea.

NGA GEOnet Names Server (NGS)
(http://earth-info.nga.mil/gns/html/index.html)
Another site for geographic names.

Oriental Bird Images: A Database of the Oriental Bird Club
(http://orientalbirdimages.org).

xeno-canto
(http://www.xeno-canto.org/world-area.php?area=australasia)
Recordings and sonograms of bird voices.

PLATE 1 CASSOWARIES

There are 3 species of cassowaries, but rarely more than 1 at a given location
—note how they sort out geographically.

1 Dwarf Cassowary *Casuarius bennetti*   p. 262
1.0–1.1 m. Montane forests. Smallest cassowary. Lacks pendulous
wattles. Wedge-shaped head. Black casque triangular and flattened
in back. Adult black; diagnostic broad black mask; rest of head
and neck variably blue and red. Juv brown. Imm intermediate.
Subspp facial and neck skin varies in the placement of red patches:
Races from Central Ranges with red-and-blue cheek;
Bird's Head (*westermanni*), white cheek patch.

2 Northern Cassowary *Casuarius unappendiculatus*   p. 263
1.2–1.5 m. Northern lowland forest. Brown or bluish casque
flattened in back. Single, fingerlike wattle at middle of throat.
Adult black; throat color variable, but usually with red (or yellow)
encircling the base of the neck; often with dull yellow eye.
Juv brown.

3 Southern Cassowary *Casuarius casuarius*   p. 263
1.3–1.7 m. Southern lowland forest. Dark brown casque high, thin,
bladelike, curved back. Double wattles; these vary in placement
on neck. Adult black, with red wattles and patch at back of neck.
Juv brown.

PLATE 2 BRUSHTURKEYS

1   Wattled Brushturkey *Aepypodius arfakianus*   p. 264
38–46 cm. Montane forest. Common locally where not hunted.
Adult with maroon-brown rump and dark legs. Male with
conspicuous comb and bare skin of face and neck bluish white.
Female and Juv comb reduced, skin of face and neck greenish grey.
Chick with deep bill and dark legs. **Subsp shown**: *arfakianus*
(NG and Bay Is: Yapen) comb red (blue in race *misoliensis* from
Misool I in NW Is).

2   Waigeo Brushturkey *Aepypodius bruijnii*   p. 265
41–46 cm. Waigeo I only, in upland forest. Adult differs from
Wattled BT by pink head and neck and chestnut-brown
underparts. Illustration shows displaying male with wattles
extended. Female has a much-reduced comb and other wattles
essentially absent; face and neck with more black, bristlelike hairs.

3   Red-billed Brushturkey *Talegalla cuvieri*   p. 265
53–58 cm. Bird's Head and Neck. In lowland, hill forests.
Adult with orange legs, orange-red bill, bright yellow eye,
and dull yellow face. Chick similar to Red-legged BT chick.

4   Red-legged Brushturkey *Talegalla jobiensis*   p. 265
53–61 cm. Northern watershed mostly. Lowlands to mid-
mountain forest. Adult with red legs and drab red facial skin.
Chick with reddish legs, upperparts faintly barred.

5   Yellow-legged Brushturkey *Talegalla fuscirostris*   p. 265
50–53 cm. Southern watershed. Lowland forests, avoids hills.
Adult has bright yellow legs, and the dark face is bluish grey;
black eye and bill. Chick dark with yellow legs; upperparts
uniform dusky with little or no barring.

PLATE 3 SCRUBFOWL

1 Moluccan Scrubfowl *Eulipoa wallacei*   p. 266
30 cm. Vagrant on Misool (NW Is); hill forest.
Adult has banded brown saddle, short olive legs, and yellow bill.
Chick heavily banded above.

2 Dusky Scrubfowl *Megapodius freycinet*   p. 266
34–41 cm. NW Is (all) in lowland and hill forest. Adult sooty black
with black legs. Chick dark brown with dark legs.

3 Melanesian Scrubfowl *Megapodius eremita*   p. 266
35–39 cm. Karkar and Bagabag Is forest, where it hybridizes with
New Guinea SF. Adult dark, with bare forehead and short crest.
Skin of face and neck red; legs greenish yellow, olive, or blackish.
Chick similar to Dusky SF. A "pure" Melanesian SF is shown.

4 Biak Scrubfowl *Megapodius geelvinkianus*   p. 266
30 cm. Bay Is in forest and scrub. Adult small and blackish with
bright red face and variably reddish legs. Chick, no information—
presumably dark with reddish legs.

5 New Guinea Scrubfowl *Megapodius decollatus*   p. 267
33–35 cm. The widespread northern scrubfowl. Common in forest
from lowlands to mountains. Adult differs from Orange-footed
SF by its dark legs that are either yellowish, greenish, brownish,
or black; the plumage is also darker and crest shorter. Chick with
brown legs.

6 Orange-footed Scrubfowl *Megapodius reinwardt*   p. 267
37 cm. The widespread southern scrubfowl. All types of forest
mainly in lowlands, but also hills in SE Pen. Adult and chick have
orange legs. **Subsp shown**: *reinwardt* (mainland NG and Aru Is)
rufous brown wings and back.

PLATE 4 QUAIL AND BUTTONQUAIL

1 King Quail *Excalfactoria chinensis* p. 268
11–12 cm. A tiny, secretive quail found locally in weedy places at the edge of grasslands or gardens, especially in mid-mountains. Male with black-and-white bib, blue-grey body. Female with buffy face cut by a dark "moustache line." Juv similar but spotted or streaked below, rather than barred.

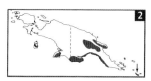

2 Red-backed Buttonquail *Turnix maculosus* p. 308
12–15 cm. Grasslands from lowlands to mid-mountains.
Adult has buff wing-coverts with dark spots; unpatterned throat, breast, and belly. White iris; yellowish base of the bill. Female with orangish face and breast. Male greyer, less buffy and orange than Female; no rufous collar. Juv heavily speckled black and white above and down breast; iris dark.

3 Brown Quail *Coturnix ypsilophora* p. 268
15–19 cm. Larger and duller than buttonquail and King Quail. Finely and evenly textured markings. Females of all races similar; darker overall, more coarsely marked than Male.
Juv more dark-spotted below. **Subspp groups shown**:
LOWLAND RACES most sexually dichromatic—Male all grey: *dogwa* (S Lowlands); *plumbea* (Huon and SE Pen); *saturatior* (NW Lowlands and Sepik-Ramu).
MID-MONTANE RACES with Male buff and has finer markings than Female: *mafulu*, *lamonti* (E Ranges, SE Pen).
MONTANE RACE Male essentially identical to Female: *monticola* (alpine SE Pen).

4 Snow Mountain Quail *Anurophasis monorthonyx* p. 267
25–28 cm. A large, partridge-sized quail of alpine grasslands and scrub; no other quail shares its habitat. Note barred, brown back. Male underparts cinnamon, barred. Female underparts pale, more heavily barred. Juv less heavily marked than Adult.

PLATE 5 DUCKS

1 Hardhead *Aythya australis* p. 272
46–49 cm. Sporadically disperses from AU to NG, where some breed. Dark brown with contrasting white undertail. Male's eye white, Female's and Juv's brown. In flight, a white blaze across flight-feathers of upperwing; white underwing and belly.

2 Eurasian Wigeon *Anas penelope* p. 271
45–51 cm. Palearctic vagrant to lowland wetlands during austral summer. Male has rich chestnut head with buff cap. Female and Juv have an unpatterned head and peaked forehead; breast cinnamon. In flight, note white forewing.

3 Garganey *Anas querquedula* p. 272
37–41 cm. Palearctic migrant locally common on lowland wetlands during austral summer. Male's head maroon brown with prominent white eyebrow; underparts white. Female and Juv have a dark eye-stripe. In flight, blue-grey shoulders.

4 Pacific Black Duck *Anas superciliosa* p. 271
48–56 cm. A widespread resident on all sorts of freshwater wetlands; also a likely AU visitor. Adult is a large, blackish duck with boldly striped pale face. Juv with streaky underparts. In flight shows green speculum and white underwing-coverts.

5 Grey Teal *Anas gracilis* p. 272
36–43 cm. A widespread resident; also a visitor from AU. Adult is a small, plain duck with domed head, pale throat, and red eye. Juv paler, with streaky breast. In flight, upperwing has white wedge, underwing with white triangle.

6 Wandering Whistling Duck *Dendrocygna arcuata* p. 270
40–45 cm. Locally common resident and visitor from AU. Adult is orange-brown with a black cap and white flanks. Juv duller, less boldly marked. Note chestnut shoulders.

7 Plumed Whistling Duck *Dendrocygna eytoni* p. 270
40–45 cm. A rare AU visitor to wetlands in savannah country. Adult pale, with diagnostic, elongate flank plumes and lacking a dark crest. Juv paler, with indistinct breast barring. In flight, the wings are same pale color as back.

8 Spotted Whistling Duck *Dendrocygna guttata* p. 269
43–50 cm. Common resident on freshwater marshes in lowlands. Adult has white-spotted flanks. Juv duller, with white flank spots and streaks. In flight shows unpatterned wings and is plain overall, but look for flank spots.

48

PLATE 6 GREBES, DUCKS, AND RAILS

1 Australasian Grebe *Tachybaptus novaehollandiae* p. 279
25 cm. More common and widespread than Tricoloured G. Resident. AU visitors likely. On lakes and ponds. Yellow eye. White wing blaze on primaries as well as secondaries. Breeding adult has black throat. Nonbreeding adult plain grey-brown, darker above; note pale belly. Juv similar to Nonbreeding adult, but head striped black and white.

2 Tricoloured Grebe *Tachybaptus tricolor* p. 278
25 cm. Uncommon resident. Dark red eye. White on secondaries only. Breeding adult has extensive red-brown on throat and foreneck; black restricted to chin. Nonbreeding plumage and Juv darker than Australasian G; orangish neck and a dark belly.

3 Cotton Pygmy Goose *Nettapus coromandelianus* p. 271
33–36 cm. Northern watershed resident; in freshwater with floating vegetation. No scalloping on the flank. Male with white head and neck; breeding male with distinct breast band; in flight, white band across both primaries and secondaries. Female has thick white eyebrow; in flight, wing mostly dark. Juv like Female but back paler.

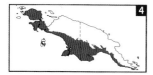

4 Green Pygmy Goose *Nettapus pulchellus* p. 271
30–36 cm. Southern watershed resident; common in lowland freshwater with water lilies. Scalloped flank. In flight, a conspicuous white panel in the secondaries. Breeding male with green head and white cheek patch (obscure in Nonbreeding male). Female has dark cap and white cheek. Juv like Female but back and wings less iridescent.

5 Raja Shelduck *Tadorna radjah* p. 270
48 cm. Locally common resident in lowland wetlands. Large. Adult with all-white head and underparts. Juv crown and nape brown. In flight, note white shoulder patch and green speculum. **Subsp shown**: *radjah* (NG except Trans-Fly) with black back.

6 Salvadori's Teal *Salvadorina waigiuensis* p. 270
43 cm. Uncommon endemic on rocky mountain streams and alpine lakes. Brown head with orange-yellow bill; body profusely barred. Male's eye red, Female's brown.

7 Purple Swamphen *Porphyrio porphyrio* p. 307
38–43 cm. Lowlands; locally in montane valleys. A large, purplish-blue swamp dweller. Black with blue underparts and red frontal shield and bill. Juv duller, bill dusky red.

8 Dusky Moorhen *Gallinula tenebrosa* p. 307
25–32 cm. Lowland wetlands only. Less common than swamphen. Smaller and more compact; all black rather than purplish blue. Juv duller and throat whitish, bill and legs dusky.

9 Eurasian Coot *Fulica atra* p. 307
36–41 cm. Resident in mountain lakes and swamps; vagrant or migrant to lowland wetlands. Usually in small flocks swimming in open freshwater. Prominent white bill and forehead. Juv greyer, lacks frontal shield.

PLATE 7 PETRELS AND SHEARWATERS

1 Black-bellied Storm-Petrel *Fregetta tropica* — p. 278
20 cm. Rare migrant to Coral Sea in austral winter and spring. Dark upperparts and hood that extends to upper breast; white rump, breast, and belly; black midline or mottling on breast; white underwing-coverts with smudgy margins.

2 Matsudaira's Storm-Petrel *Oceanodroma matsudairae* — p. 277
24–25 cm. Uncommon passage migrant to N coast (and probably W coast) during austral winter. Largest NG storm-petrel. All sooty brown with pale wing-coverts and forked tail; in good light, note diagnostic pale shaft streaks on primaries.

3 Wilson's Storm-Petrel *Oceanites oceanicus* — p. 277
15–19 cm. Common pelagic migrant during austral winter along NG continental shelf in E Seram Sea; reported as rare elsewhere along S coast. Sooty brown with large white patch around base of square tail. Feet project beyond tail tip.

4 Tahiti Petrel *Pseudobulweria rostrata* — p. 275
38–40 cm; wingspan 84 cm. Common visitor. Dark hood and upperparts, white breast and belly. Underwing dark with pale central stripe.

5 Beck's Petrel *Pseudobulweria becki* — p. 275
29 cm. Likely to occur in NG waters. Similar to Tahiti P, but smaller, and bill less robust.

6 Heinroth's Shearwater *Puffinus heinrothi* — p. 276
27 cm. Rare visitor. Small—size of Black Noddy. Bill long and thin. Tail short. Dark grey with pale underwing-coverts. Some have a pale patch on breast and belly.

7 Bulwer's Petrel *Bulweria bulwerii* — p. 276
26–27 cm. Uncommon visitor. Small, dark petrel with long, tapered tail. Pale upperwing bar.

8 Short-tailed Shearwater *Puffinus tenuirostris* — p. 275
41–43 cm. Locally common passage migrant in austral autumn. From dark morph Wedge-tailed S by its paler underwings, shorter rounded tail, and dark bill and legs. Toes project beyond tail.

9 Wedge-tailed Shearwater *Puffinus pacificus* — p. 275
41–46 cm. The common NG shearwater. Both a breeding resident and migrant in austral winter. Medium-large, slender, and with broad wings; long, narrow, wedge-shaped tail; slim grey bill; and pinkish legs. Two color morphs: dark morph (more common) all dark brown; pale morph with white underparts.

10 Tropical Shearwater *Puffinus bailloni* — p. 276
31 cm. Likely to occur on the northeastern fringe of the NG Region. A small shearwater showing black upperparts contrasting with clear white underparts.

Hutton's Shearwater *P. huttoni*, rare in Coral Sea during austral winter, similar but slightly larger and with a dark, almost complete breast band.

11 Herald Petrel *Pterodroma heraldica* — p. 274
34–39 cm. Rare visitor. Much variation. All morphs have a distinctive underwing pattern with white panel on the outer wing that contrasts with the mostly dark coverts.

12 Streaked Shearwater *Calonectris leucomelas* — p. 275
48 cm. Abundant Palearctic migrant in austral summer. A large, slender-bodied, brownish shearwater with white underparts. Pale face and bill. Head with white streaking.

13 Flesh-footed Shearwater *Puffinus carneipes* — p. 276
41–45 cm. Rare austral passage migrant. From dark morph Wedge-tailed S by its heavier build, pale bill with a dark tip, and the shorter, less wedge-shaped tail.

PLATE 8 BOOBIES AND TROPICBIRDS

1 Red-tailed Tropicbird *Phaethon rubricauda* — p. 279
46 cm w/o streamers. Pelagic vagrant. Adult looks all white and short-tailed at distance. At closer range note red bill, red tail-streamers, and thin black streaks on the outer primaries. Juv lacks tail-streamers and has black, V-marked feathers on mantle and wing-coverts; bill is grey with black tip; note fine black shaft streaks on outer primaries.

2 White-tailed Tropicbird *Phaethon lepturus* — p. 279
38–40 cm w/o streamers. Uncommon pelagic visitor. Adult more slender than Red-tailed T and wing pattern shows striking black patches. Look for white tail-streamers, yellowish bill. Juv resembles juv Red-tailed T but shows black patch on primaries.

3 Red-footed Booby *Sula sula* — p. 289
66–77 cm. Common visitor near coast or out at sea. Adult in white, brown, and intermediate morphs, all with diagnostic white tail. Juv all pale brown; dark bill.

4 Brown Booby *Sula leucogaster* — p. 289
64–74 cm. Common visitor and rare breeding resident. Frequents coasts, river mouths, harbors, islands, and open ocean. Adult chocolate brown with clean white breast and belly. Facial skin blue in Male, yellow in Female. Juv paler with mottled underparts.

5 Masked Booby *Sula dactylatra* — p. 289
81–92 cm. Rare visitor. Adult larger than Red-footed B and has a dark mask and black tail. Juv differs from Brown B adult by white collar and upper breast and dark bill. Imm has brown head, but less brown in plumage than Juv.

PLATE 9 FRIGATEBIRDS AND JAEGERS

1 Great Frigatebird *Fregata minor* p. 288
86–100 cm. Uncommon visitor. All plumages lack the white underwing patch ("armpits") of Lesser FB. (In some Great FB, the white breast patch creeps toward the underwing.) Male all black. Female blackish with grey chin and throat, white breast. Juv head white or tawny, separated from white breast by a dark band. Imm intermediate.

2 Lesser Frigatebird *Fregata ariel* p. 288
71–81 cm. Common visitor. All plumages have diagnostic white "armpits"—a flank patch that extends as a streak onto the dark underwing-coverts. Male blackish with white flank patch. Female has all-black head, with black chin and throat. Juv can be difficult to distinguish from Great FB, but look for greater amount of white extending onto underwing. Imm intermediate.

3 Pomarine Jaeger *Stercorarius pomarinus* p. 331
65–78 cm. Fairly common overwintering Palearctic migrant. Larger, heavier, and somewhat shorter-winged than other jaegers. Pale or dark morphs. Nonbreeding adult has short streamers, wider and more blunt than other jaegers, and barred rump, flanks, and undertail. Breeding adult with tail-streamers broad, twisted, and round-tipped. Juv without streamers; pale or dark morphs, much variation but all show barring in underwing. Imm intermediate.

4 Arctic Jaeger *Stercorarius parasiticus* p. 331
46–67 cm. Rare passage Palearctic migrant. Smaller and more falcon-shaped than Pomarine J, with longer, narrower wings and swift flight. Pale, intermediate, and dark morphs. Nonbreeding plumage has much shorter streamers that may not be visible, and heavy light-and-dark scalloping on rump and tail-coverts. Breeding adult has tail-streamers narrow, pointed, and of moderate length. Juv has pointed tail-streamers that barely project and underwing-coverts heavily barred. Imm intermediate.

5 South Polar Skua *Stercorarius maccormicki* p. 331
53 cm. Rare austral passage migrant. Large, robust, and broad-winged. Lacks projecting central tail feathers. Only pale morph expected. Adult pale morph with diagnostic pale grey-buff head and underparts contrasting with dark underwing and upperparts. Juv more greyish, and bill with grey base.

6 Long-tailed Jaeger *Stercorarius longicaudus* p. 332
50–58 cm. Rare passage Palearctic migrant. Smallest jaeger, slender with long, narrow wings. Only pale morph reported. In all plumages, upperwing with only 2 white shafts in outer primaries (many primary shafts white in other jaegers). Adult underwing has all-dark primaries, like rest of underwing (pale patch in juv Long-tailed and adult/juv other jaegers). Breeding adult has very long tail-streamers. Juv similar to juv Arctic J, but paler, especially on the head.

PLATE 10 TERNS AND GULLS

1 Sooty Tern *Onychoprion fuscatus* — p. 328
43–45 cm. Common visitor over the open ocean far from shore, yet occasionally wanders inland. Adult wholly black above except for white forehead triangle. Juv distinct—black with white breast and belly and white spots on back and wing-coverts.

2 Bridled Tern *Onychoprion anaethetus* — p. 328
35–38 cm. Common resident and local breeder, usually near coast and islands. Adult's grey-brown back and longer eyebrow distinguish it from Sooty T. Juv has shorter tail-streamers and the mantle, back, and wing-coverts with white and buff markings.

3 Grey-backed Tern *Onychoprion lunatus* — p. 328
35–38 cm. Rare over ocean far from shore. Resident of tropical Pacific. Adult paler grey than Bridled T, with less white in outer tail feathers. Juv slightly paler than juv Bridled.

4 White Tern *Gygis alba* — p. 326
28–33 cm. Rare oceanic visitor off N coast. Adult all white with moderately forked tail. Large dark eye and dark, upturned bill. Juv has brown scalloping on upperparts and wings.

5 Black Noddy *Anous minutus* — p. 326
35–39 cm. Locally abundant resident over offshore waters. Adult smaller than Brown N and more compact. Overall blackish with all-dark wings. Gleaming white cap extends onto nape. Juv white cap extends only to crown.

6 Brown Noddy *Anous stolidus* — p. 326
40–45 cm. Locally common resident over offshore waters. Adult larger than Black N, with lankier shape; upperwing-coverts distinctly paler than rest of wing; underwing pale greyish. Juv has thin white forehead mark.

7 Caspian Tern *Hydroprogne caspia* — p. 327
48–59 cm. A scarce visitor mainly to coastal Trans-Fly. A gull-sized, robust tern with thick, red bill. Nonbreeding plumage with white frosting on black crown. Breeding plumage with heavy black cap. Juv similar to Nonbreeding adult, but with black markings on the mantle, back, and wings; darker primaries.

8 Black-headed Gull *Chroicocephalus ridibundus* — p. 327
38–43 cm. Palearctic vagrant to harbors and sewage ponds. Nonbreeding plumage with white head marked with black smudges around eye and on ear-coverts. Juv similar but with black markings on upperparts. Breeding plumage with blackish-brown head.

9 Silver Gull *Chroicocephalus novaehollandiae* — p. 327
38–43 cm. Occasional AU visitor to S coast. Adult has all-white head and bright red bill and legs. Wing tips black with white "mirror." Juv similarly white-headed, but with dark bill and legs and black markings on upperparts.

PLATE 11 TERNS

1 Little Tern *Sternula albifrons* p. 328
20–28 cm. Inshore coastal waters, lagoons, and swamps. Common Palearctic migrant during austral summer. Smallest NG tern. Dark leading primaries. Nonbreeding plumage with white cap and black bill. Breeding plumage with white forehead, short white eyebrow, yellow bill with black tip. Juv with dark scalloping on the mantle and back.

2 Black-naped Tern *Sterna sumatrana* p. 329
30–32 cm. Outer reefs and atolls near breeding islands, where locally common. Adult pure white but for a crisp, crescent-shaped black band crossing the nape and a thin black leading edge to the outermost primary. Forked tail lacks streamers. Juv has sharp, black scalloping on mantle, back, and wings, and grey outer primaries.

3 White-winged Tern *Chlidonias leucopterus* p. 330
22–25 cm. Over shallow fresh and salt water. Uncommon Palearctic migrant during austral summer. Nonbreeding plumage similar to nonbreeding Whiskered T except has white rump and paler tail contrasting with the slightly darker back; also darker head markings. Breeding plumage unique. Juv has mostly black mantle and back.

4 Whiskered Tern *Chlidonias hybrida* p. 330
25–26 cm. The common tern over shallow freshwater. An AU migrant during austral winter. Small with shallow-forked tail. Nonbreeding plumage looks evenly whitish except for strong black eye-stripe. Breeding plumage unique. Juv mantle and back marked with black and buff; darker wings.

5 Crested Tern *Thalasseus bergii* p. 327
43–48 cm. A widespread and common sea tern that breeds locally. Very large, with greenish-yellow bill and black, shaggy crest. Nonbreeding plumage with extensive white forehead merging with black cap. Breeding plumage with a black cap sharply demarcated from white forehead. Juv with sooty black cap; black markings on mantle, back, and wings; and dark primaries.

6 Lesser Crested Tern *Thalasseus bengalensis* p. 328
38–43 cm. Widespread but uncommon coastal visitor. Similar to Crested T, but smaller, more compact, slightly darker above, and with a smaller, orange bill. In breeding plumage, black cap extends to bill. Juv dorsal markings not as bold.

7 Gull-billed Tern *Gelochelidon nilotica* p. 327
35–43 cm. Tidal flats, estuaries, coastal wetlands, and rivers. Uncommon austral winter migrant; also a Palearctic vagrant. All white with diagnostic stout black bill. Nonbreeding plumage has black smudge across face. Breeding plumage has black cap. Juv mantle and back have black markings. **Subsp shown** is *macrotarsa* (AU).

8 Common Tern *Sterna hirundo* p. 329
32–38 cm. Mainly coastal, where it is the predominant tern. Common to locally abundant Palearctic migrant during austral summer, with nonbreeders remaining. Adult has dark outer primaries; outer tail feather with thin, black edge; black bill and legs. Nonbreeding plumage with white forehead and well-forked tail but no streamers. Breeding plumage with black cap and long tail-streamers. Juv has darker carpal and forewing and diagnostic darker trailing edge to secondaries.

9 Roseate Tern *Sterna dougallii* p. 329
35–43 cm. Found offshore over coral reefs and clear, blue water near nesting islands. A local and uncommon breeder. Long, slender bill; tail always lacks darker edge of Common T. Adult has long, all-white tail-streamers. Nonbreeding plumage with white forecrown, black bill. Breeding plumage with all-black cap, red bill. Juv has black scalloping on upperparts; forked tail lacks streamers; black legs and feet.

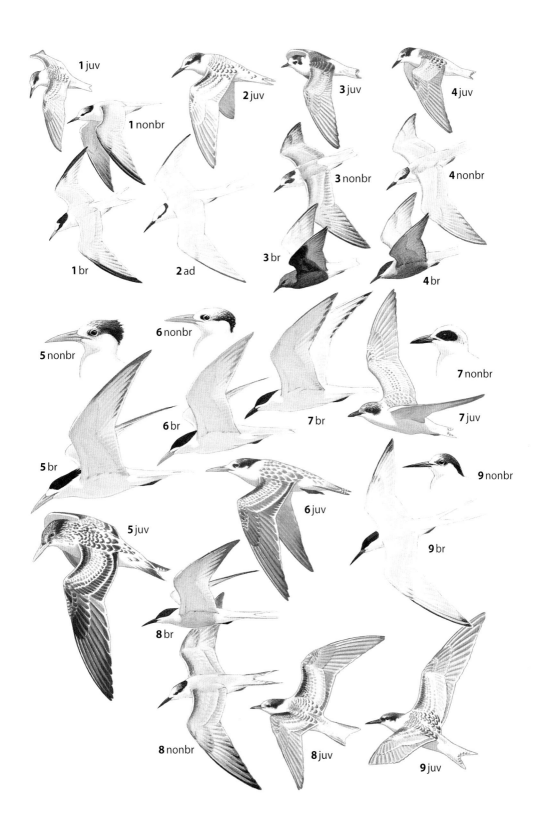

PLATE 12 CORMORANTS, DARTER, PELICAN, AND MAGPIE GOOSE

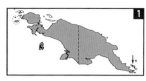

1 Little Black Cormorant *Phalacrocorax sulcirostris* p. 290
61 cm. Common resident, and likely AU visitor, in wetlands.
Adult is a small, all-black cormorant with a black face; green eye.
Juv is brownish black; brown eye.

2 Little Pied Cormorant *Microcarbo melanoleucos* p. 290
58 cm. Common resident, also visits from AU. Adult is a small,
black-and-white cormorant with short yellowish bill; white
around the dark eye. Juv has a black cap extending down to eye
and a black flank mark.

3 Great Cormorant *Phalacrocorax carbo* p. 290
81 cm. Rare visitor from AU. Adult is large, with pale bill,
yellow facial skin, white gular patch, and when breeding, a white
patch on flank; green eye. Juv brown with white chin; brown eye.

4 Australasian Darter *Anhinga novaehollandiae* p. 291
90 cm. Locally common resident on large bodies of freshwater in
the lowlands; also an AU visitor. Diagnostic dagger-shaped bill,
snaky neck, and white facial stripe. Male black. Female has whitish
throat and underparts. Juv paler than Female.

5 Magpie Goose *Anseranas semipalmata* p. 269
70–90 cm. Common breeding resident and AU migrant to swampy
floodplains of the Trans-Fly. Adult unmistakable—a very large
black-and-white gooselike bird. Note high-peaked crown and
hook-tipped bill. Male larger on average, with a higher crown.
Juv mantle black; white flanks mottled with dark grey.

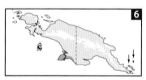

6 Australian Pelican *Pelecanus conspicillatus* p. 287
150–160 cm. Common migrant from AU to southern Trans-Fly;
perhaps occasionally breeds. Adult unique—a very large, white-
and-black bird with a huge pinkish bill and pouch. Male larger
on average. Juv has brown feathers where Adult has black.

PLATE 13 IBISES, SPOONBILLS, STORK, CRANE, AND BUSTARD

1   Glossy Ibis *Plegadis falcinellus* p. 281
55–65 cm. Common migrant; locally present year-round and may breed. Inhabits lowland swamps. Appears blackish at a distance. Adult has metallic sheen. Breeding plumage chestnut brown. Nonbreeding plumage similar but heavily streaked white on the head and neck. Juv nearly without iridescence; face and throat mottled with white.

2   Yellow-billed Spoonbill *Platalea flavipes* p. 281
76–92 cm. Vagrant to Trans-Fly. Similar to Royal S, but with dull yellow bill and legs. Adult has pale eye and legs; wings entirely white. Breeding plumage with strawlike plumes on lower foreneck and dark filamentous scapular plumes. Juv has dark eye and dusky legs; dark wing tips.

3   Royal Spoonbill *Platalea regia* p. 281
74–81 cm. Common migrant and local breeding resident. Frequents lowland swamps and estuaries. Black, spatula-shaped bill and black legs. Adult has red forehead patch and yellow eyespot. Breeding plumage with thick crest. Juv similar to crestless Nonbreeding adult but lacks forehead patch and eyespot; dark wing tips.

4   Australian White Ibis *Threskiornis moluccus* p. 280
65–75 cm. Common migrant and locally breeding resident. Mainly in lowland swamps. Adult has bare, black head and shaggy black tertials covering the tail. Breeding plumage with filamentous white plumes on the foreneck. Juv has head and neck with scattered white feathers; black tertials not shaggy.

5   Straw-necked Ibis *Threskiornis spinicollis* p. 281
60–70 cm. Common migrant to Trans-Fly savannahs (and suspected breeder); vagrant elsewhere. Glossy black wings and back. Adult foreneck with yellowish, strawlike plumes; legs reddish. Green breast band incomplete in Male, complete in Female. Juv and Imm lack straw-neck and are less glossy; legs blackish.

6   Black-necked Stork *Ephippiorhynchus asiaticus* p. 280
110–137 cm. Locally common breeder along margins of swamps, lagoons, rivers, and coastal mudflats. A tall, long-necked wading bird with pied plumage and huge black bill. Adult has black head, neck, and wings. Male's eye dark, Female's pale. Juv brownish.

7   Brolga *Grus rubicunda* p. 308
114–124 cm. Common resident in Trans-Fly marshes and dry grassland. Adult has distinctive profile showing jowls and feather "bustle" over the tail; bare head red and grey. Juv head dull-colored.

8   Australian Bustard *Ardeotis australis* p. 302
91–119 cm. Locally common resident in Trans-Fly savannah. Note black cap and grey neck. Male huge and has elongated feathers on neck. Female much smaller with greyer neck. Juv similar to Adult but smaller and with pale dorsal spotting.

PLATE 14 BITTERNS AND HERONS

1 Australian Little Bittern *Ixobrychus dubius* p. 282
27–36 cm. Fairly common during austral winter, mainly in the South. Likely both a resident and AU migrant. In reeds. Tiny. Buff wing patch contrasts with dark back that is black in Male and dark brown in Female. In flight, Female has grey-brown primaries (black in adult Yellow B). Juv in flight shows rufous brown tips to the primaries and secondaries (lacking in juv Yellow B).

2 Yellow Bittern *Ixobrychus sinensis* p. 283
30–40 cm. Locally abundant Palearctic migrant during austral summer to lowland wetlands, mostly in the North. From Australian Little B by the buff-brown back and longer and narrower bill. Male, unmarked underparts. Female, neck striped cinnamon and white. Adults have black primaries contrasting with pale buff wing-coverts and back. Juv similar to juv Australian Little, but paler, with a longer, narrower bill.

3 Cinnamon Bittern *Ixobrychus cinnamomeus* p. 283
40–41 cm. Local resident of westernmost NG. Overall cinnamon without contrasting pale wing. Female has streaked neck. Juv similar but duller.

4 Black Bittern *Ixobrychus flavicollis* p. 283
54–66 cm. Common resident along the forested margins of streams, wooded swamps, and mangroves. Dark with diagnostic buff blaze along the side of the neck. Male has blue-black upperparts. Female has chocolate brown upperparts with faint scaling. Rare cinnamon morph also. Juv paler and more rufous than Female, with heavily buff-scaled upperparts.

5 Striated Heron *Butorides striata* p. 284
38 cm. Common resident along the coast in mangroves and into the interior along rivers. A small, squat, big-headed heron with black crown and grey back. Juv shows adult colors beneath white dorsal spotting and pale streaking on the neck. **Subspp shown**: *papuensis* (NW Is, Bird's Head, Bay Is, Aru Is) dark, with brown underparts; *idenburgi* (NW Lowlands and maybe Sepik-Ramu) pale, with grey underparts. Not shown is *macrorhyncha* (S NG), which is pale with brown underparts.

6 Nankeen Night-Heron *Nycticorax caledonicus* p. 283
55–65 cm. Common resident and visitor from AU, in freshwater and coastal habitats with trees nearby. Crepuscular and nocturnal. Adult is a robust heron with rufous plumage and black cap. Juv drab brown with copious whitish streaking and spotting.

7 Forest Bittern *Zonerodius heliosylus* p. 282
65–71 cm. A rare heron of forest streams in lowlands and foothills. Shy and seldom seen. Bulky and heavily barred all over. Darker, more heavily barred birds with green facial skin may be Adults. Paler, finely barred birds with yellow facial skin possibly Juv.

PLATE 15 HERONS AND EGRETS

1 Little Egret *Egretta garzetta* p. 286
55–65 cm. AU visitor to freshwater in lowlands. Small and very active. Slender black bill and yellow facial skin. Breeding plumage with nape streamers and plumes on back and breast. Nonbreeding plumage has plumes only on back. Juv lacks plumes.

2 Intermediate Egret *Ardea intermedia* p. 285
56–70 cm. AU visitor and local breeding resident in freshwater wetlands. Shaped like an E Great Egret; size of a Cattle Egret. Differs from E Great by smaller size, shorter bill compared with head, shorter neck (head and neck as long as body), neck less often kinked, and gape not extending past eye. Differs from nonbreeding Cattle E by more slender shape with a longer neck and thinner head. Breeding plumage has filamentous plumes on back and lower neck. Nonbreeding plumage and Juv without plumes; the bill is yellow, never black-tipped like many E Great.

3 Cattle Egret *Ardea ibis* p. 284
70 cm. A common visitor; breeding expected. Associated with grassland and cattle, but also uses ponds. A small, stocky egret with neck shorter than body, and a thick head. Breeding plumage with orange-buff, lacking in Nonbreeding and Juv.

4 Eastern Great Egret *Ardea modesta* p. 285
83–103 cm. The most common NG egret, inhabiting freshwater and mangrove waterways. Both a resident and AU visitor. Long, kinked neck stretches longer than body length. Gape extends well past the eye. Breeding plumage has filamentous plumes on back; bill is mostly or all black. Nonbreeding plumage and Juv lack plumes; bill mostly yellow.

5 Eastern Reef-Egret *Egretta sacra* p. 286
60–65 cm. The common heron of seashore, reef flats, and offshore islands. Resident. Distinctive elongate shape with short legs. Two color morphs—dark grey and white. Breeding plumage has lanceolate plumes on nape, back, and foreneck. Juv and Imm grey morph are brown-grey with a dark bill.

6 White-necked Heron *Ardea pacifica* p. 284
76–106 cm. AU vagrant to freshwater wetlands. A large, dark-winged heron with white head and neck. Breeding plumage has maroon lanceolate plumes on scapulars. Juv has spots scattered over the neck (not just in 2 rows) and there are more of them.

7 Great-billed Heron *Ardea sumatrana* p. 285
100–115 cm. A scarce resident in mangroves, lagoons, and rivers with tree cover. Adult is very large and uniformly grey-brown; impressive bill is grey-black. Breeding plumage has lanceolate silvery plumes on nape and back. Juv is rusty brown; bill brown.

8 White-faced Heron *Egretta novaehollandiae* p. 286
66–68 cm. An AU visitor and uncommon resident in the South, strays elsewhere. Adult is all grey with a white face. Breeding plumage has a ragged crest and longer lanceolate plumes on back and breast. Juv and Imm, grey face and white throat.

9 Pied Heron *Egretta picata* p. 286
43–55 cm. Locally common AU visitor to freshwater wetlands in lowlands. Small. Adult slaty black with white neck. Breeding plumage with longer nape streamers; lanceolate plumes on back. Juv, brown mottling on crown and neck. Imm, all-white head.

PLATE 16 OPEN COUNTRY KITES AND HAWKS

1   Black-winged Kite *Elanus caeruleus*   Pl. 24, p. 293
30–33 cm. Mainly over mid-mountain grasslands and farmland. Rare and local. Adult unique: white head and underparts; prominent black shoulder; dark-shadowed red eye. Female larger on average. Juv tinged brown and eyes black.

2   Pacific Baza *Aviceda subcristata*   Pl. 22, p. 292
36–41 cm. Forest edge and in tall second growth in lowlands and hills. Uncommon. Only crested NG raptor. Adult's grey head contrasts with whitish underparts boldly barred black. Note brown scapulars. Female larger on average and slightly browner. Juv darker with pale-scalloped upperparts and paler underparts.

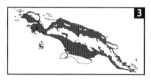

3   Bat Hawk *Macheiramphus alcinus*   Pl. 24, p. 293
43–51 cm. Very rare. Forest edge in lowlands and montane valleys. Active at dawn and dusk. Adult is large, dark, and falcon-like. White throat marked with a black stripe. Note large yellow eye, wide mouth, and small bill. Female larger. Juv thought to have more white in underparts.

4   Brahminy Kite *Haliastur indus*   Pl. 18, p. 294
41–43 cm. A ubiquitous raptor of seacoasts, rivers, and forest edge. Adult unmistakable with its simple white-and-chestnut pattern. Female larger on average. Juv indistinctive but note dark eye-patch and when perched folded wings often extend beyond tail.

5   Whistling Kite *Haliastur sphenurus*   Pl. 18, p. 294
53–58 cm. A common scavenging kite of open country, roadsides, towns, and harbors. Adult has pale buff head, underparts, and tail. Tail long with rounded tip. Female larger on average. Juv back and wing-coverts with feathers broadly tipped whitish.

6   Black Kite *Milvus migrans*   Pl. 18, p. 294
48–53 cm. Ecologically similar to the more widespread Whistling Kite. Only NG raptor with a forked tail. Adult dark brown. Female larger on average. Juv overall paler, with pale edges to feathers yielding a more streaked or scalloped appearance.

7   Pygmy Eagle *Hieraaetus weiskei*   Pl. 17 and 18, p. 299
38–48 cm. Seen soaring over forest from hills to mid-mountains. Uncommon and local. Robust and buteo-like. From Brahminy Kite juv by shape—perched eagle has shorter wings, so tail is same length or slightly projects; legs powerful, feet large. Only NG hawk with feathered tarsus and thick feathering on thighs. Female larger. Pale and dark morphs. Juv upperparts with pale feather tips.

8   Grey-faced Buzzard *Butastur indicus*   Pl. 18, p. 298
38–48 cm. Rare Asian migrant to NW Is during austral summer. A chunky, sluggish hawk. Adult has white throat with black chin-stripe and rufous breast. Sexes same size. Juv also has chin-stripe, but underparts white with dark streaking, heaviest on breast; fresh plumage shows white feather edges on wing-coverts.

PLATE 17 EAGLES AND OSPREY

1 Eastern Osprey *Pandion cristatus*   Pl. 18, p. 291
46–53 cm. Seacoasts and fringing islands, occasionally lowland rivers and lakes. A fish hawk with a thick black bar through the eye and white underparts marked with an obscure breast band. Female larger; breast band thicker. Juv has pale tips to dark back and wing-coverts.

2 White-bellied Sea-Eagle *Haliaeetus leucogaster*   Pl. 18, p. 294
69–79 cm. Seacoasts and rivers, where uncommon. Adult is the only grey-and-white eagle. Short, wedge-shaped tail. Female larger. Juv and Imm same shape, but mottled brown.

3 Gurney's Eagle *Aquila gurneyi*   p. 299
74–86 cm. Seen soaring over forest in lowlands and adjacent hills. Uncommon. Adult huge and blackish brown with long wings and a round-tipped tail. Female larger. Juv paler and tawny, especially below.

4 Wedge-tailed Eagle *Aquila audax*   Pl. 18, p. 299
76–100 cm. Savannah country of the Trans-Fly, where uncommon. From Gurney's E by its pointed, wedge-shaped tail and tawny hackles. Adult blackish. Female larger. Juv paler and tawny on nape, shoulders, and rump.

5 Pygmy Eagle *Hieraaetus weiskei*   Pl. 16 and 18, p. 299
38–48 cm. For comparison. Note much smaller size than other eagles. Adult pale morph illustrated.

PLATE 18 KITES AND EAGLES IN FLIGHT

1 Brahminy Kite *Haliastur indus* — Pl. 16, p. 294
Wingspan 110–125 cm. Adult unique—orange-brown with white head and breast. Juv not distinctive, but note pale window in the primaries, dark wing tips, and rather short tail.

2 Whistling Kite *Haliastur sphenurus* — Pl. 16, p. 294
Wingspan 120–146 cm. Slender body, long wings, and long tail with rounded tip. Pale buff head, underparts, and tail. Pale "window" of inner primaries framed by blackish flight-feathers on either side. Glides with distal half of wing bowed downward.

3 Black Kite *Milvus migrans* — Pl. 16, p. 294
Wingspan 120–153 cm. Forked tail diagnostic (triangular when fanned). Nondescript dark brown plumage.

4 Grey-faced Buzzard *Butastur indicus* — Pl. 16, p. 298
Wingspan 101–110 cm. Long, narrow wings and tail. Unique tail pattern with 3 bands on the central feathers only, the outer feathers plain. White throat with black chin-stripe. Adult with rufous breast. Juv lacks rufous breast and instead underparts are dark-streaked.

5 Pygmy Eagle *Hieraaetus weiskei* — Pl. 16 and 17, p. 299
Wingspan 101–136 cm. Buteo-like. Medium length, rounded wings held flat when soaring; fanned tail. Dark-streaked breast. Brahminy Kite juv has bigger wings and unbarred wings and tail.

6 Eastern Osprey *Pandion cristatus* — Pl. 17, p. 291
Wingspan 127–174 cm. Square tail and long wings with a slight bend at the wrist and a diagnostic black smudge at the hand of the wing. White below with blurry breast band.

7 Wedge-tailed Eagle *Aquila audax* — Pl. 17, p. 299
Wingspan 182–232 cm. Adult huge and blackish, with pointy, wedge-shaped tail and wings held in a V when soaring. Juv tawny above; underside of wings and tail narrowly barred.

8 Gurney's Eagle *Aquila gurneyi* — Pl. 17, p. 299
Wingspan 165–185 cm. Adult huge and dark, with rounded tail and long wings held flat when soaring. Juv paler and tawny below; underside of wings and tail narrowly barred.

9 White-bellied Sea-Eagle *Haliaeetus leucogaster* — Pl. 17, p. 294
Wingspan 178–218 cm. Adult unique: white below with black flight-feathers. Barrel-chested profile with small head and brief, wedge-shaped, white tail; broad wings held in a shallow V. Juv and Imm have same distinctive shape; they show banded underwing-coverts and white patch at base of primaries.

PLATE 19 HARRIERS

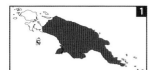

1 Swamp Harrier *Circus approximans spilothorax*   p. 295
Resident race: 48–53 cm; wingspan 118–145 cm. The most frequently seen harrier. Uncommon resident over grassland and marshes. Male pied morph uniquely patterned black, grey, and white. Black morph Male rare; note greyish tail. All other plumages are inadequately known, and separation of Female and Juv/Imm birds from migratory AU race is as yet uncertain. Female larger than Male, has pale rump and tail with clear dark barring (tail rarely without bands, like Male's) and often pale face; both pale and black morphs known. Juv and Imm are blackish brown (some with rusty wash) with large, pale nape patch, and underwing with limited silvery base to primaries.
In flight: A large harrier with slim body on long wings held in a shallow V; look for whitish rump patch.

2 Swamp Harrier *Circus approximans approximans*   p. 295
Migratory race from Australia: 46–58 cm; wingspan 118–145 cm. Rare during austral winter to grassland and marshes.
Male separable from female of resident race by tail either all grey or with faint, broken bands. Female from female resident race by obscurely barred tail (not clearly dark barred). Juv and Imm similar to resident race, but often rusty brown below, with less pale streaking on nape.

3 Black Kite *Milvus migrans*   Pl. 16 and 18, p. 294
48–53 cm; wingspan 120–153 cm. For comparison.
When perched on the ground, such as an airstrip, the kite has much shorter legs than a harrier.

PLATE 20 GOSHAWKS AND SPARROWHAWK

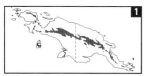

1 Black-mantled Goshawk *Accipiter melanochlamys*   Pl. 22, p. 297
30–43 cm. Montane cloud forests. Uncommon. Adult uniquely black and orange-brown. Female larger. Juv dark above with brown scalloping; white below with sharp, black streaking and no barring; smaller than Chestnut-shouldered G.

2 Grey-headed Goshawk *Accipiter poliocephalus*   Pl. 22, p. 297
30–38 cm. Lowland and hill forest. Common but secretive. Identified by large, dark eye and orange eye-ring. Cere, gape, legs, and feet also bright orange. Adult common morph grey and white; also rare dark morph (illustration conjectural as no specimens or photos exist). Female larger. Juv similar but brownish grey above and finely dark-streaked below.

3 Chinese Sparrowhawk *Accipiter soloensis*   Pl. 22, p. 295
25–30 cm. Rare Asian migrant in austral summer to open habitats and swamps in the West. Tiny. Diagnostic white undertail-coverts without markings. Male pale grey with whitish underparts washed rufous across breast. Female larger; pale rufous breast obscurely barred. Juv streaked below.

4 Collared Sparrowhawk *Accipiter cirrocephalus*   Pl. 22, p. 297
28–36 cm. Forest and lightly wooded habitats from lowlands to mid-mountains. Uncommon. Adult and Juv plumages similar to Brown Goshawk and difficult to separate. Collared best identified by smaller size, longer thin legs, lighter build, square tail, and lack of heavy brow-ridge. Female larger.

5 Brown Goshawk *Accipiter fasciatus*   Pl. 22, p. 296
30–43 cm. Open country from lowlands to mid-mountains. Common. Similar to Variable G's common morph, except finely and evenly barred below and with rufous collar (sometimes obscure), greenish-yellow cere, and pale yellow legs. See also Collared Sparrowhawk. Female larger than Male. Juv has streaked breast and barred lower breast and flanks; look for greenish-yellow cere. **Subsp shown** is *polycryptus* (most of NG range) with belly barred rufous brown. Not shown is *dogwa* (Trans-Fly) with much paler barred belly.

6 Variable Goshawk *Accipiter hiogaster*   Pl. 22, p. 296
33–48 cm. The common goshawk of forest edge from lowlands to mid-mountains. Variable, but all plumages have bright orange-yellow cere, legs, and feet. Common morph plain rufous brown below—not barred or at most faintly barred; lacks distinct rufous collar. Grey morph is the only all-grey NG hawk. White morph is the only all-white NG hawk. Female larger. Juv common morph white or buff below with variable streaking and barring (barring sometimes absent); look for orange-yellow cere and legs. **Subsp shown** is *leucosomus* (NG mainland and some islands) usually unbarred below and lacking a collar. Not shown are 3 island races for Biak and SE Is.

PLATE 21 FOREST HAWKS AND EAGLES

1 Chestnut-shouldered Goshawk *Erythrotriorchis buergersi* Pl. 22, p. 298
43–53 cm. Hill and lower montane forest. Rare. Tail with many dark bands (9–10 bands). Adult common morph boldly patterned; black above with shoulder feathers rufous-edged; white below with sharp black streaking. Adult black morph all black with pale eye. Female larger. Juv common morph is bright rufous brown with bold streaking below and dark centers to feathers on back and wing-coverts.

2 Meyer's Goshawk *Accipiter meyerianus* Pl. 22, p. 297
48–56 cm. Hill and montane forest. Rare. A large, powerful goshawk. Tail with few thick, dark bands (4–5 visible). Adult common morph black above and white below with fine markings; dark red eye. Adult black morph all black; note dark eye. Female larger. Juv common morph underparts tawny buff with fine dark streaking and saddle brownish black with buff scaling; eye may be pale.

3 New Guinea Harpy-Eagle *Harpyopsis novaeguineae* Pl. 22, p. 298
76–89 cm. Forest interior from lowlands to timberline. Uncommon. Does not soar. An enormous, short-winged, long-legged, and long-tailed raptor with white underparts. Female much larger. Juv has prominent pale tips on the wing-coverts; more and narrower bars on wing and tail; tail lacks distinct, dark subterminal bar of Adult.
Do not confuse with the more common Long-tailed Buzzard.

4 Long-tailed Buzzard *Henicopernis longicauda* Pl. 22, p. 293
51–61 cm. This is the hawk most often seen soaring over forest of hills and mountains. Wings and tail boldly patterned with thick black bars of even width; Adult with 3 black tail bars, Juv with 4 bars. Head and breast finely streaked. Pale bill and face, and yellow eye. Female larger on average.

5 Doria's Hawk *Megatriorchis doriae* Pl. 22, p. 298
51–69 cm. Forest interior from lowlands to mid-mountains. Rare and inconspicuous. Does not soar. Adult has a very long tail with many narrow black tail bars. Note thick black eye-stripe. Female much larger. Juv lacks prominent eye-stripe (head may be very pale); black barring narrower and more profuse throughout.

PLATE 22 ACCIPITERS AND FOREST HAWKS IN FLIGHT

1   Pacific Baza   *Aviceda subcristata*   Pl. 16, p. 292
Wingspan 80–105 cm. Broad, rounded wings prominently banded; black-barred breast
(faint in Juv); square tail with thick, black terminal band separated by a gap from a few thin bars.

2   Brown Goshawk   *Accipiter fasciatus*   Pl. 20, p. 296
Wingspan 60–98 cm. Rounded tail. Adult finely barred below; greenish cere and pale yellow legs.
Juv has streaked breast and barred lower breast. Longer-winged than Variable G.

3   Collared Sparrowhawk   *Accipiter cirrocephalus*   Pl. 20, p. 297
Wingspan 53–77 cm. Square-tipped or notched tail. Smaller and more lightly built than
Brown Goshawk, which is otherwise essentially identical in adult and juv plumages.

4   Chinese Sparrowhawk   *Accipiter soloensis*   Pl. 20, p. 295
Wingspan 52–62 cm. Tiny. Diagnostic white underwings with dark tips.
Juv streaked and scaled below.

5   Black-mantled Goshawk   *Accipiter melanochlamys*   Pl. 20, p. 297
Wingspan 65–80 cm. Black and orange-brown. Juv underwing much paler than
Chestnut-shouldered G.

6   Variable Goshawk   *Accipiter hiogaster*   Pl. 20, p. 296
Wingspan 55–80 cm. Rarely soars. Rather short-winged. Three morphs—common, white,
and grey—are all smooth-textured without barring. Juv similar to juv Brown G.

7   Grey-headed Goshawk   *Accipiter poliocephalus*   Pl. 20, p. 297
Wingspan 56–65 cm. Grey and white with black eye and orange eye-ring, cere, and legs.
Juv with barred flight-feathers and finely streaked breast. Usually seen crossing a forest opening.

8   Chestnut-shouldered Goshawk   *Erythrotriorchis buergersi*   Pl. 21, p. 298
Wingspan 85–109 cm. Wings and tail distinctly barred; tail with many dark bars.
Adult common morph thickly streaked below. Juv bright rufous brown with bold streaking.

9   Meyer's Goshawk   *Accipiter meyerianus*   Pl. 21, p. 297
Wingspan 86–105 cm. Wings and tail obscurely barred; tail with few bars.
Juv underparts buff brown, finely streaked.

10   Long-tailed Buzzard   *Henicopernis longicauda*   Pl. 21, p. 293
Wingspan 105–140 cm. Usually seen soaring. Distinctive silhouette with small head,
long rounded tail, and large, "fingered" wings thrown forward. Wings and tail boldly patterned—
Adult with thick black bars; Juv with thinner bars.

11   Doria's Hawk   *Megatriorchis doriae*   Pl. 21, p. 298
Wingspan 88–106 cm. Never soars. Broad wings and very long tail.
Adult tail with many narrow black bars. Juv barring narrower and more profuse.

12   New Guinea Harpy-Eagle   *Harpyopsis novaeguineae*   Pl. 21, p. 298
Wingspan 121–157 cm. Never soars. Huge. Broad wings, long tail.
Diagnostic white underparts without streaking or barring.
Juv with more bars in wing and tail.

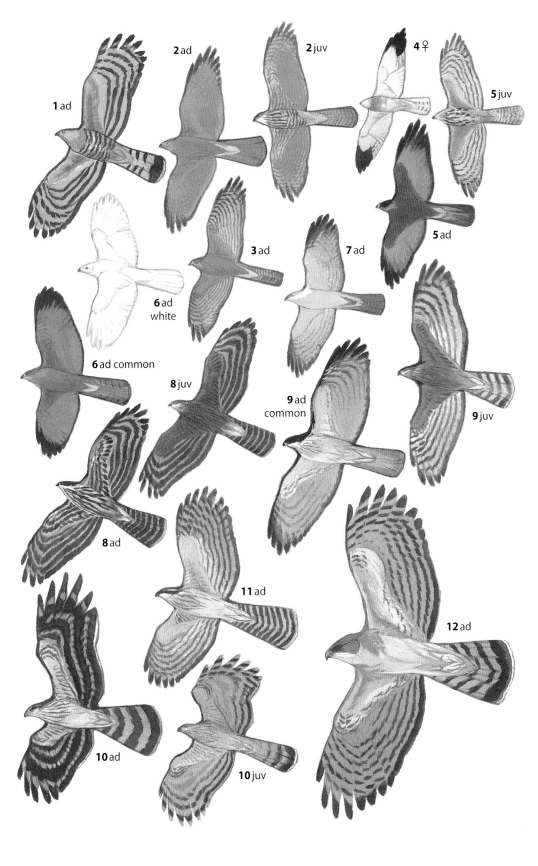

PLATE 23 FALCONS

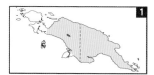

1   Nankeen Kestrel *Falco cenchroides*   Pl. 24, p. 300
30–36 cm. The widespread NG kestrel, uncommon in open habitats. Small, slender, with rufous upperparts and black subterminal tail-band. Juv more heavily marked. **Subspp shown**: *baru* (alpine resident in W Ranges) darker and orange-buff below, sexes similar, but Female larger and with more dorsal spotting; *cenchroides* (AU migrant, low to mid-elevations) whitish below, Female and Male differ.

2   Spotted Kestrel *Falco moluccensis*   Pl. 24, p. 300
28–33 cm. NW Is only, where resident and uncommon in open country. Similar to Nankeen K, but darker reddish brown and heavily spotted and streaked. Male with grey tail; lacks grey cap of Nankeen K. Female and Juv similar to Male, but tail barred.

3   Oriental Hobby *Falco severus*   Pl. 24, p. 301
25–30 cm. Rare throughout NG forests from the lowlands to mid-mountains. Note stubby profile. Adult black and orange-brown; all-dark head framed by buff half-collar and throat. Female larger. Juv underparts heavily dark-spotted.

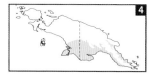

4   Australian Hobby *Falco longipennis*   Pl. 24, p. 301
30–38 cm. Rare Australian migrant overwintering in open habitats. Adult similar to Oriental H but larger and with noticeably longer wings and tail. Usually paler, with barring on the tail. Note white lores, lacking in Oriental H. Whitish half-collar wraps around the black cheek patch. Female larger. Juv has rufous lores.

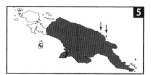

5   Brown Falcon *Falco berigora*   Pl. 24, p. 301
41–53 cm. A rather large, sluggish falcon of open country, mainly in the mountains. Uncommon. Black bar below eye. Variation in plumage color poorly understood for NG birds, but likely to indicate age. Juv plumage blackish; underparts heavily blotched.

6   Peregrine Falcon *Falco peregrinus*   Pl. 24, p. 302
34–51 cm. A rare falcon with resident and migratory races that can turn up anywhere. A long-winged, streamlined, swift flier. Adult with barred underparts and prominent black hood. Female larger. Juv darker, browner, and heavily streaked below. **Subsp shown**: *ernesti* (resident NG Region) head, back, and wings blackish. Not shown are migratory races from Eurasia and AU that are greyer above.

PLATE 24 FALCONS AND KITES IN FLIGHT

1 Bat Hawk *Macheiramphus alcinus* — Pl. 16, p. 293
Wingspan 95–120 cm. Large, black, and falcon-like. Actively hunts bats and swiftlets at dawn and dusk. White throat marked with a unique black medial stripe. Large yellow eye. Juv thought to have more white in underparts.

2 Black-winged Kite *Elanus caeruleus* — Pl. 16, p. 293
Wingspan 77–92 cm. Hovers. White below with pointed, dark-tipped wings; grey above with black shoulders. Underside of the flight-feathers varies from slaty grey to whitish grey. Juv similar, but head, breast, and back tinged brown.

3 Nankeen Kestrel *Falco cenchroides* — Pl. 23, p. 300
Wingspan 66–78 cm. Hovers. Rufous above, pale below, and with black subterminal tail-band. Underparts whitish in migratory AU race *cenchroides*, orange-buff in resident alpine race *baru*.

4 Spotted Kestrel *Falco moluccensis* — Pl. 23, p. 300
Wingspan 59–71 cm. Hovers. Similar to Nankeen K, but underparts darker and brownish; breast and flanks heavily spotted and streaked. NW Is only.

5 Australian Hobby *Falco longipennis* — Pl. 23, p. 301
Wingspan 66–83 cm. Similar to Oriental H but larger and with noticeably longer wings and tail. Usually paler, with bold barring on the wing feathers and tail.

6 Oriental Hobby *Falco severus* — Pl. 23, p. 301
Wingspan 61–71 cm. Adult dark—black above, deep rufous brown below. Underwing not barred, tail obscurely barred. Juv underparts heavily spotted.

7 Brown Falcon *Falco berigora* — Pl. 23, p. 301
Wingspan 88–115 cm. Wings shorter and broader than other falcons, and flight involves slow flapping, soaring (wings held in a V), and hovering. Dark underwing-coverts contrast with pale flight-feathers in all plumages. Adult color varies from rufous brown above and white with dark shaft streaking below, to all dark. Juv blackish.

8 Peregrine Falcon *Falco peregrinus* — Pl. 23, p. 302
Wingspan 79–114 cm. Adult with barred underparts and prominent black hood. Juv underparts darker, browner, and heavily streaked. Resident race (*ernesti*) blackish above; migratory races grey above.

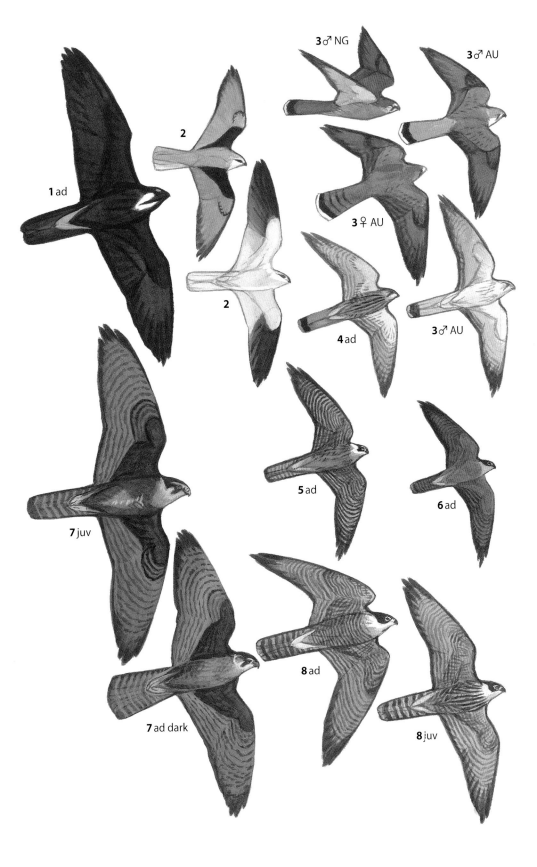

PLATE 25 FOREST-RAILS AND OTHER RUFOUS RAILS

1 Chestnut Forest-Rail *Rallicula rubra*   p. 303
18–23 cm. Cloud forest. Smaller and more rufous than other forest-rails, and generally at higher elevation. Male entirely rufous brown. Female's black back and wings are marked with white spots (buff spots in Forbes's FR); flanks are pale-spotted rather than dark-barred. Both sexes lack barring on tail. Juv uniform sooty brown.

2 Mayr's Forest-Rail *Rallicula mayri*   p. 304
20–23 cm. Foja, Cyclops, and N Coastal Mts, where it is the only forest-rail. Darker than other forest-rails. Male entirely dark reddish brown. Female much like female Forbes's FR.

3 Forbes's Forest-Rail *Rallicula forbesi*   p. 304
20–25 cm. Mid-mountain and cloud forest of Central Ranges. Generally occupies lower elevations where it shares its range with Chestnut FR. Differs by larger size and usually having the tail barred black. The back and wings are not quite black. Female's back and wings are spotted with buff rather than white; belly and lower flanks barred brown and black, not pale spotted. Juv uniform sooty brown.

4 White-striped Forest-Rail *Rallicula leucospila*   p. 304
20–23 cm. Montane forest of Bird's Head and Neck, where it co-occurs with smaller Chestnut FR. Male has white-striped back and wings. Female differs from other female forest-rails by the white-spotted pattern extending up the back of the neck.

5 Red-necked Crake *Rallina tricolor*   p. 304
25–30 cm. Lowland and hill forest, especially near streams. Local. Most active at dusk and dawn. Quite different from the forest-rails. Adult brownish black with rufous head and breast, green bill. Juv dark with a buff face.

6 Bare-eyed Rail *Gymnocrex plumbeiventris*   p. 305
30–33 cm. An uncommon, medium-sized, and long-billed rail of lowland and hill forest. Adult shows unique combination of chestnut foreparts, olive wings, and black tail, rump, and belly. When running, the vertical, black, finlike tail sticks up. Juv undescribed.

7 Chestnut Rail *Eulabeornis castaneoventris*   p. 305
51–58 cm. Found only in mangroves on Aru Is. Secretive and shy, but ventures onto mudflats at low tide. A very large, orange-brown, grey-headed rail. Juv likely similar to Adult.

PLATE 26 CRAKES AND RAILS

1 Spotless Crake *Porzana tabuensis* p. 306
17–20 cm. Widespread and common through highlands, local in lowlands. Swampy grassland and marsh-edged lakes and rivers. A small, black, red-legged crake. Juv with much white on chin and throat.

2 Baillon's Crake *Porzana pusilla* p. 306
15–18 cm. Lowlands to mountain valleys, in marshy wetlands. A miniature crake that keeps hidden in marsh grasses and is little known. Adult with streaky upperparts and plain grey underparts; note short bill. Juv has buffier underparts.

3 Lewin's Rail *Lewinia pectoralis* p. 304
18–20 cm. Mid-mountain grasslands and swamps. Probably common, but shy and seldom seen. Note long bill. Adult with rich brown cap and nape, and grey breast. Juv duller, head darker and lacks chestnut.

4 White-browed Crake *Porzana cinerea* p. 306
15–18 cm. Lowlands to mid-mountain valleys, in marshy habitats. Common. Emerges at dusk or in cloudy weather to forage atop floating vegetation. Adult has white-striped facial pattern and silvery grey underparts. Juv, obscure facial pattern and buffy underparts.

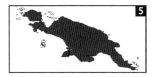

5 Buff-banded Rail *Gallirallus philippensis* p. 305
25–28 cm. A widespread, common rail of tall grasses, scrub, and gardens, seen along roads and other verges. Adult prettily marked with black-and-white barring and spotting and brown streaking on mantle and back. Whitish eyebrow cuts across rusty cap; hallmark buff band on chest. Juv duller with bill brown, not red.

6 Barred Rail *Gallirallus torquatus* p. 305
33–36 cm. Far West. Coastal lowlands, in swampy areas with grass and trees. Adult has black face and throat, contrasting white "moustache," and heavy barring below. Juv darker with white throat and underparts washed buff.

7 Rufous-tailed Bush-hen *Amaurornis moluccana* p. 306
22–30 cm. Lowlands to mid-mountains, in tall, wet, grassy and shrubby habitats, such as garden fallow and forest edge. Common. A plain-colored, stocky rail. Juv paler with more white on throat.

8 New Guinea Flightless Rail *Megacrex inepta* p. 307
36–38 cm. Uncommon in lowland swamp forests, wet thickets, and mangroves. Attracted to sites of sago harvesting. Robust with massive head and long, powerful grey legs and feet. Long, stout, yellowish bill. Dark smudging around eyes. Juv similar, though with less brown. **Subsp shown:** *inepta* (S Lowlands) rusty brownish sides of neck and body. Not shown is *pallida* (NW Lowlands to Sepik-Ramu), paler with buff sides of neck and body.

PLATE 27 LARGE ODDBALL WADERS

1 Bush Stone-curlew *Burhinus grallarius*   p. 309
54–59 cm. Trans-Fly savannah woodlands. Cryptic, streaked plumage. Large, yellow eye. Juv has a wider pale patch on folded wing, incorporating the median wing-coverts that are blackish in Adult. Nocturnal. At night, fluttering wings show white patches.

2 Beach Stone-curlew *Esacus magnirostris*   p. 309
51–56 cm. Remote, undisturbed coastlines with sandy beaches. Large, robust, boldly patterned. Note black-and-white face, yellow eye, and massive, upturned bill. Adult without streaking. Juv more mottled. Crepuscular and mainly nocturnal. Black-and-white wings prominent in flight.

3 Australian Pied Oystercatcher *Haematopus longirostris*   p. 310
42–50 cm. Coastal, especially on exposed mudflats and rubble. Resident on Aru Is; rare AU visitor to southern mainland coast. Adult wings and back black, eye red. Juv wings and back brownish black with notched pale margins, eye brown, and bill duller.

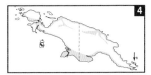

4 Black-winged Stilt *Himantopus himantopus*   p. 310
33–37 cm. Shallow freshwater lagoons, occasionally salt water. Locally common visitor from AU during austral winter; occasionally stays to breed. Adult has white head and neck with black nape-stripe that broadens to a half-collar. Juv with variable dingy grey cap and nape-stripe; back brownish black; legs pale pink.

5 Masked Lapwing *Vanellus miles*   p. 311
30–37 cm. Open habitats in the lowlands. Both a common resident and AU visitor. Noisy. Yellow facial wattle and bill. Adult has large wattle, mantle and wing-coverts clean grey-brown without markings, and legs pink. Juv has smaller wattle, mantle and wing-coverts with dark scalloping, and legs pinkish grey. **Subspp shown**: *miles* (resident in NG and perhaps some visitors from tropical N AU); *novaehollandiae* (AU vagrant) darker, black nape and sides of breast, wattle smaller.

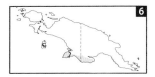

6 Red-kneed Dotterel *Erythrogonys cinctus*   p. 312
17–19 cm. Lowland wetlands. Rare, sporadic AU visitor. A small waterside plover with striking color pattern. Compact and large-headed. Adult with black cap and breast band. Juv with dark grey-brown cap, no breast band. In flight, black wings with thick, white trailing edge and white underwing-coverts.

PLATE 28 PLOVERS

1 Grey Plover *Pluvialis squatarola*   p. 312
28–30 cm. Coastal flats and beaches. An uncommon, overwintering Palearctic plover. Larger than similar Pacific Golden P, more robust, and grey without any buff. In flight shows a broad white wing-stripe, white rump, and black "armpits."

2 Pacific Golden Plover *Pluvialis fulva*   p. 312
23–25 cm. Fields of short grass (e.g., airstrips) and coasts. A ubiquitous overwintering Palearctic migrant. Nonbreeding adult and Juv have a buff-colored pale face, large dark eye, and mantle and wings spangled gold and black. Breeding female less black in underparts than Male. In flight shows only an obscure pale wing-stripe.

3 Oriental Plover *Charadrius veredus*   p. 313
23–25 cm. Fields of dry, open, sparse, short grass. Uncommon Palearctic migrant in passage to its AU wintering grounds. Slender, with long, yellowish legs. Nonbreeding similar to Pacific Golden P, but mantle and wings uniform, without spangling. Breeding male unique. Breeding female duller, with darker neck and breast band. Juv mantle and wings with wider buff scalloping. In flight lacks wing-stripe.

4 Greater Sand-Plover *Charadrius leschenaultii*   p. 313
20–23 cm. Coastal mudflats and beaches. Common Palearctic migrant. Very similar to Lesser SP, differing subtly in larger size, more slender shape, longer legs, bigger head, and longer, heavier, more pointed bill. Breeding adult differs from Lesser by the paler orange-rufous nape and throat-band, and the throat band is narrower, not extending onto breast. Juv like Nonbreeding adult, but feathers of upperparts pale-fringed. In flight, the feet project beyond tail tip (they don't in Lesser SP) and tail distinctly dark-tipped (not so in Lesser).

5 Lesser Sand-Plover *Charadrius mongolus*   p. 313
19–20 cm. Habitat and status same as for Greater Sand-Plover. Smaller and more compact, with short, stout, rather blunt-tipped bill. Color differences in breeding plumage—see account of Greater Sand-Plover for comparison.

6 Little Ringed Plover *Charadrius dubius*   p. 312
14–17 cm. A small plover with conspicuous yellow eye-ring, white collar, and black breast band. **Subspp shown**:
dubius, uncommon resident along gravelly rivers; Breeding and Nonbreeding plumages more similar, with thick yellow eye-ring (locally red when breeding). Voice: *chit*.
curonicus, rare Palearctic visitor to muddy shores of wetlands. Breeding plumage has narrower eye-ring, narrower black frontal band, darker upperparts, and smaller pink/yellow spot at base of lower bill that does not include gape and edge of upper bill. Nonbreeding facial pattern and breast band much reduced, blurred, and brownish; feathers of wing-coverts and back pale-edged. Voice: *pee-u*. Juv similar, but breast band interrupted and mantle and wings with fine, buff scalloping.

PLATE 29 WOODCOCK, SNIPE, DOWITCHER, AND GODWITS

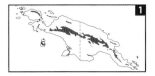

1 New Guinea Woodcock *Scolopax rosenbergii* p. 315
30 cm. High-elevation cloud forest and at timberline. Secretive, yet gives vocal display-flights at dusk and dawn. Snipelike but plumper and color pattern differs: all plumages dusky with thick barring across the crown and nape (instead of stripes) and no sharp, pale dorsal streaking. Wings broad and dark underneath.

2 Swinhoe's Snipe *Gallinago megala* p. 316
27–29 cm. The common NG snipe. A Palearctic winter visitor to moist grassland and wetlands from sea level to alpine bogs. Snipe species are difficult to tell apart. All have striped head, creamy scapular striping, and long straight bill. Swinhoe's tail has typically 20–22 [18–26] feathers, and the outer 5–6 are very narrow. In flight, toes project beyond tail (do not project in Latham's). Wing 130–150 mm, tail 50–63 mm.

3 Latham's Snipe *Gallinago hardwickii* p. 315
29–33 cm. Rare passage and overwintering Palearctic migrant to same habitats as Swinhoe's S. Mainly overwinters in AU. Because of its longer wings and tail, Latham's body shape is somewhat more elongated in the rear. There are 16–18 tail feathers; outer tail feathers not so narrow as in Swinhoe's S. Wing 146–158 mm, tail 58–72 mm.

4 Pin-tailed Snipe *Gallinago stenura* p. 316
25–27 cm. Hypothetical. Smaller and more squat. Tail barely projects beyond folded wing tips (projects well in Swinhoe's and Latham's). Tail feathers 24–28; 7 pairs are thin.

5 Asian Dowitcher *Limnodromus semipalmatus* p. 316
33–36 cm. Rare Palearctic migrant to coastal mudflats and sandflats. Smaller and more compact than a godwit, with straight, more robust, all-dark bill and longer face. Breeding plumage variably rusty, with Female duller and frosty; scapulars and wing-coverts not pale-notched as in godwits. Juv buffier than Nonbreeding adult; cap darker; scapulars and wing-coverts darker with sharp, creamy edging. In flight, similar color as Bar-tailed Godwit but shape differs.

6 Black-tailed Godwit *Limosa limosa* p. 317
40–44 cm. Mudflats and beaches of coasts and wetlands. Common Palearctic passage and overwintering migrant. Bill only slightly upturned, pink at base. Nonbreeding adult, unstreaked upperparts. Breeding female much paler than Male, both have barred underparts. Juv buffier than Nonbreeding adult; feathers of wings and back with dark markings. In flight, white bar on black tail and flashy black-and-white wings.

7 Bar-tailed Godwit *Limosa lapponica* p. 317
37–39 cm. Same habitat as Black-tailed G. Uncommon Palearctic migrant. Larger and stockier than Black-tailed, bill obviously upturned. Nonbreeding adult has streaked upperparts. Breeding adult rusty below, lacks barring; Female much paler. Juv buffier than Nonbreeding; dark scapulars with pale-notched edges. In flight, finely barred white tail; unmarked wings; only toe tips project beyond tail.

PLATE 30 CURLEWS AND SHANKS

1 Eastern Curlew *Numenius madagascariensis* p. 318
60–66 cm. Estuaries, beaches, mangroves, and salt marshes. Uncommon Palearctic migrant, mainly in passage, though some overwinter. Largest curlew in the NG Region. Heavyset, with amazingly long bill. Nonbreeding adult buff-grey. Breeding adult buffier with rufous highlights on the upperparts. Female larger. Juv bill shorter; scapulars and tertials with sharp, pale notching. In flight, lacks white rump.

2 Little Curlew *Numenius minutus* p. 317
28–31 cm. Expanses of short grass. Locally abundant, passage Palearctic migrant on arrival (Sep–Dec); rare on return. Widespread, but concentrates in Trans-Fly. Only a bit larger than a golden plover. Smaller, daintier than Whimbrel, and bill shorter, slim, and slightly decurved. All plumages resemble Whimbrel but buffier, with incomplete dark loral stripe. Female larger on average. In flight lacks white rump.

3 Whimbrel *Numenius phaeopus* p. 317
40–45 cm. Estuaries, mudflats, beaches, reef flats, less often fields. Common Palearctic migrant, both on passage and overwintering. Nonbreeders present year-round. The midsize curlew. Large with long, downcurved bill. Head prominently striped. Female larger on average. Juv, bill shorter. In flight, white rump forms wedge up the back.

4 Common Greenshank *Tringa nebularia* p. 318
30–35 cm. Coastal and inland mudflats and marshes. Common Palearctic migrant, both on passage and overwintering. A large sandpiper, lanky and large-headed with distinctive upturned, greenish-grey bill. Nonbreeding adult similar to Marsh S. Breeding plumage heavily streaked on the head, neck, and flanks. Juv similar to Nonbreeding, but scapulars and wing-coverts darker with broad buff or white edges. In flight, similar to Marsh S but only feet project beyond tail, not shanks.

5 Marsh Sandpiper *Tringa stagnatilis* p. 318
22–26 cm. Primarily in shallow marshes and freshwater lagoons. Common Palearctic migrant, both on passage and overwintering. A smaller, delicate version of Common Greenshank, with a straight, needlelike bill. Nonbreeding adult plain grey above, white below. Breeding adult with black dorsal markings and speckled head, neck, and breast. Juv like Nonbreeding, but upperparts quite dark with broad, buff edges. In flight, white wedge up the back, feet project beyond tail.

6 Common Redshank *Tringa totanus* p. 318
27–29 cm. Coastal flats and nearby wetlands. Vagrant or rare Palearctic migrant.
The only brownish, red- or orange-legged wader. Nonbreeding adult brownish grey above, dingy white below. Breeding adult, upperparts light-and-dark notched, underparts finely streaked, barred. Juv like breeding, more brightly spotted above, paler with no barring below. In flight, white trailing edge to wing and white rump.

PLATE 31 SMALL TRINGAS AND RELATIVES

Tattlers show uniform grey upperparts (diagnostic in flight) and short, yellowish legs. The 2 tattlers are difficult to tell apart.

1 Wandering Tattler *Tringa incana* p. 319
27 cm. Primarily on offshore islands and rocky reefs along mainland coast. Uncommon Palearctic migrant mainly to eastern NG Region. Identified by flight call: a whimbrel-like *li li li . . .* of 4–10 notes. From nearly identical Grey-tailed T: slightly larger size; darker overall; obscure white eyebrow; darker grey tail; flanks with a thicker, obvious grey margin in Nonbreeding and Juv plumages; thicker barring on underparts in breeding plumage; and longer wing-tip projection beyond the tail in resting posture; but all of these variable. Nasal groove more than half the length of bill.

2 Grey-tailed Tattler *Tringa brevipes* p. 319
25 cm. Sheltered coasts in a variety of habitats. Common Palearctic migrant in passage and overwintering. Flight call is a plover-like, *too eee*. From nearly identical Wandering T by: typically paler coloration, bolder white eyebrow contrasting with dark lores, paler grey tail, whiter flanks in Nonbreeding and Juv plumages, finer barring and more white on belly in breeding plumage, and wing tip in resting posture as long as or only slightly projecting beyond the tail tip. Juv tattlers have pale-dotted fringes to scapulars and wing-coverts. In the hand, see text account.

3 Terek Sandpiper *Xenus cinereus* p. 320
22–24 cm. Coastal shores and freshwater lagoons. Common Palearctic migrant in passage and overwintering. Hyperactive. The only sandpiper with short, orange-yellow legs and long, upturned bill. Nonbreeding adult plain. Breeding adult with black bar on scapulars; streaky head and neck. Juv wing-coverts with narrow buff tips (white in Adult). In flight, white trailing edge to secondaries, feet do not project beyond tail.

4 Wood Sandpiper *Tringa glareola* p. 319
19–23 cm. Muddy, freshwater habitats and inundated grassland. Common Palearctic migrant in passage and overwintering. Dark back speckled white, yellowish legs. Nonbreeding adult speckled white on wing-coverts and scapulars; smooth grey neck. Breeding adult more mottled above, streaked and barred below. Juv like Nonbreeding but speckled buff on darker wing-coverts and scapulars, streakier neck. In flight, white rump patch square, does not extend up back as a wedge.

5 Common Sandpiper *Actitis hypoleucos* p. 320
19–21 cm. Only sandpiper commonly found along streams and rivers in the interior; also coasts. Palearctic migrant in passage and overwintering. Teeters rear end up and down. Diagnostic white crescent before folded wing. Nonbreeding adult, black vermiculation on shoulders. Breeding adult, faint dark streaking on head and neck, black markings on scapulars and wing-coverts. Juv similar to Nonbreeding, but more buff fringing on upperparts. In flight, note distinctive, pulsing wing flutter and white wing flashes, with wings held low.

PLATE 32 KNOTS, TURNSTONE, AND STINTS

1 Great Knot *Calidris tenuirostris* — p. 320
26–28 cm. Coastal mudflats. A locally abundant Palearctic migrant in passage and overwintering. Medium-sized, robust. Very similar to Red K, but has longer bill and legs, and smaller head. Always has spotting across breast and flanks. Juv similar to Nonbreeding adult but saddle blackish (not greyish), head and neck more streaky. In flight, white rump.

2 Red Knot *Calidris canutus* — p. 321
23–25 cm. Same habitat as Great K. In passage to coasts of the Trans-Fly with some birds overwintering; elsewhere rare or absent. Differs in more compact shape and proportionately shorter bill. Juv similar to Nonbreeding adult, but grey saddle feathers distinctly fringed with black and white. In flight, finely barred rump.

3 Ruddy Turnstone *Arenaria interpres* — p. 320
22–24 cm. Variety of coastal habitats; prefers rocky shores, reefs, open fields, and small islands. Common Palearctic migrant in passage and overwintering. Distinctive stocky shorebird with black breast band and orange legs. Nonbreeding adult and Juv like much darker, duller versions of breeding plumage. In flight: pied wings and tail.

4 Sanderling *Calidris alba* — p. 321
20–21 cm. Beaches. Rare Palearctic migrant in passage and overwintering. A small, sturdy sandpiper similar to but larger than Red-necked Stint. Bill stout, straight, and black; legs black. Nonbreeding adult has unique black shoulder patch. Juv similar to Nonbreeding adult, but saddle checkered black and white and crown dark and streaky. In flight, shows the most prominent wing-stripe of any small sandpiper.

5 Red-necked Stint *Calidris ruficollis* — p. 321
13–16 cm. Common Palearctic migrant, mainly in passage. Sheltered coasts with mudflats, sometimes inland. The most often seen NG stint (small sandpiper); other stints are rarities. Short, tapered bill; short, black legs. Breeding adult shows diagnostic rufous neck. Juv like Nonbreeding adult, but has dark, scaly saddle with rufous tinge.

6 Little Stint *Calidris minuta* — p. 322
12–14 cm. Hypothetical vagrant; probably overlooked. Difficult to separate from Red-necked S. Smaller, more compact, and bill longer. Call diagnostic: a short *tip* or *tit*, vs lower-pitched, longer *kreet* or disyllabic *chirit* of Red-necked. Breeding plumage separable: throat white, not rufous; orange-rufous areas of face and neck always finely streaked or spotted. Juv has brownish wing-coverts (vs greyish in juv Red-necked) and a few white lines down the mantle and back (these obscure or absent in juv Red-necked).

PLATE 33 STINTS AND SANDPIPERS

1 Pectoral Sandpiper *Calidris melanotos*   p. 322
19–24 cm. Grassy verges of wetlands. Rare Palearctic migrant, mainly in passage in Oct–Dec. Similar to Sharp-tailed S, but averages larger, less rufous, longer bill, diagnostic sharp demarcation between the streaked bib and clear white lower breast and belly. Male larger. Nonbreeding adult has soft, grey-brown tone. Breeding adult more brownish, and Male bib darker than Female's. Juv similar to Breeding adult, more rufous-capped, closer to Sharp-tailed.

2 Sharp-tailed Sandpiper *Calidris acuminata*   p. 323
17–22 cm. Mostly along the muddy margins of freshwater wetlands, also tidal mudflats and short grass. Common passage and overwintering Palearctic migrant. Dark-streaked, rusty cap offset by whitish eyebrow. Dark, shadowy eye-stripe. Buffy wash on breast with only obscure streaking. Male is larger. Nonbreeding adult, plain, brownish-grey tone. Breeding adult, bright rufous tone on back and throat, dark arrow marks on throat and breast. Juv similar to Breeding adult but lacks arrow marks on throat and breast. In flight, very little white in wing, white at either side of dark rump.

3 Curlew Sandpiper *Calidris ferruginea*   p. 323
18–23 cm. Margins of freshwater and brackish wetlands, less often on coastal mudflats. Uncommon Palearctic migrant in passage. Long, slender decurved bill and long legs. Juv like Nonbreeding adult but feathers of upperparts with more scaled appearance and creamy buff wash on breast. In flight, all-white rump.

4 Baird's Sandpiper *Calidris bairdii*   p. 322
14–18 cm. A vagrant N American stint. Favors dry edges of mudflats and short grassland. Unusually long-winged—folded wings project well beyond tail. Long, fine-tipped bill. Legs dark. Buffy breast band in all plumages. Nonbreeding plumage shows grey scapulars and wing-coverts with long, dark shaft streaks that are obscure in Juv.

5 Broad-billed Sandpiper *Limicola falcinellus*   p. 323
16–18 cm. Mudflats along coast; freshwater and brackish lagoons. Uncommon Palearctic migrant in passage with some birds overwintering. Long, stout bill, drooped at the tip. Short-legged. Snipelike head pattern with unique white-striped crown as well as white eyebrow. Juv more colorful than Breeding adult, neater, and streaky above; dark streaks and dots (rather than arrow marks) on brown-tinged breast. In flight, wing with dark leading edge; rump dark; feet do not project past tail.

6 Long-toed Stint *Calidris subminuta*   p. 322
13–16 cm. Freshwater or brackish wetlands. Rare Palearctic migrant, mainly in passage, Sep–Dec. Much smaller than the similar-patterned but more common Sharp-tailed S. Full-bodied with longish neck, long legs, fine-tipped bill. Pale yellow-olive legs and streaked breast. Cap is always dark, often rufous. Nonbreeding adult brownish grey. Breeding adult with rufous wash; Juv similar but plumage less worn and brighter.

PLATE 34 RUFF, PHALAROPE, JACANA, AND PRATINCOLES

1 Australian Pratincole *Stiltia isabella* p. 325
21–24 cm. Open fields of short grass. Common to abundant AU migrant that spends the austral winter in NG. Sandy-colored and ternlike. Note dark chestnut flanks. Adult buffy; bill red with back tip. Juv obscurely mottled; much paler in worn plumage; all-dark bill; flank patch duller, smaller, more smudgy than Adult's.

2 Oriental Pratincole *Glareola maldivarum* p. 325
23–25 cm. Expanses of sparse, short grass. Scarce Palearctic passage migrant and overwintering visitor. A ternlike wader with pale throat patch. Adult Nonbreeding plumage has throat patch with thin, streaked border; pale lores and eye-stripe; dark bill. Breeding plumage has buffy tinge, a throat patch with a black border, black lores, red bill. Juv mottled, lacks black throat border.

3 Ruff *Philomachus pugnax* p. 324
Male 26–32 cm; Female 20–26 cm. Muddy margins of freshwater swamps and lagoons, less often on brackish or salt water. Rare Palearctic passage and overwintering migrant. Shape distinctive: a tall, robust, medium-sized wader with small head. Male much larger than Female. Nonbreeding adult pale grey-brown above, white below obscurely mottled grey on breast. Breeding male not seen in NG Region; unmistakable with thick neck ruff and puffy crests; color highly variable. Juv similar to breeding female but with buffy face, foreneck, breast; legs brownish.

4 Comb-crested Jacana *Irediparra gallinacea* p. 314
20–23 cm. Lowland lagoons and swamps with floating vegetation. Common and conspicuous. Amazingly long toes; walks on water lilies and other floating plants. Adult has conspicuous red comb, white face, and black breast band. Female larger. Juv paler with all-white breast, lacking dark band; comb barely evident; cap and nape rufous.

5 Red-necked Phalarope *Phalaropus lobatus* p. 324
18–19 cm. Open ocean far from land. Common Palearctic migrant on seas NW, N, and NE of NG, rarely on freshwater. No other shorebird found swimming at sea. Look for flocks of small pied birds bobbing on sea or flying low over water. Bill black, straight, needle-thin. Note the thick black eye-mask of Nonbreeding adult. Female larger on average; more colorful in breeding plumage. Juv similar to Nonbreeding, but has buffy fringes on saddle feathers. In flight, white wing-stripe.

PLATE 35 LONG-TAILED DOVES

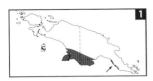

1 Bar-shouldered Dove *Geopelia humeralis* p. 336
26–28 cm. Savannah edge. Common. Adult with rufous hindneck and unbarred grey throat. Juv with white spectacles, dark eye, no rufous neck patch. In flight, bright rufous primaries unique.

2 Peaceful Dove *Geopelia placida* p. 336
18–20 cm. Towns and savannahs in drier lowlands. Common. Barred throat. Adult with blue eye skin, pale eye, and dark scalloping on upperparts. Juv with white "spectacles" around dark eye; pale scalloping on back and wings.

3 Spotted Dove *Streptopelia chinensis* p. 333
28–30 cm. Recently introduced to Indonesian towns of Biak and Sentani; to be looked for elsewhere. Towns and roadsides in disturbed habitats. A medium-sized, grey-brown dove with moderately long tail. Black outer tail feathers with white tips. Adult has rosy breast and spotted collar. Juv lacks these features.

4 Black-billed Cuckoo-Dove *Macropygia nigrirostris* p. 334
29–30 cm. Forest and edge, especially in mountains. Smaller than Brown CD. Short, stout, black bill. Tail finely barred with black. Male rich rusty brown. Female heavily barred with black. Juv like Female but tail irregularly barred.

5 Mackinlay's Cuckoo-Dove *Macropygia mackinlayi* p. 335
27–30 cm. Karkar I only. Tail not barred, and neck feathers have spotted appearance. Adults similar. Juv with distinct barring.

6 Brown Cuckoo-Dove *Macropygia amboinensis* p. 334
34–37 cm. Forest and open country, less common montane forest. Geographically variable, but look for unbarred tail. Male has rosy grey head and neck with greenish iridescence. Female brown with barred head and breast. Juv has mantle with black subterminal bars and rusty fringes to feathers. **Subspp**: Male nearly unbarred below in East to heavily barred below in West and Louisiades Is (SE Is). Shown are Male and Female *cinereiceps* (central NG) and Female *maforensis* (Bay Is: Numfor).

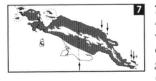

7 Great Cuckoo-Dove *Reinwardtoena reinwardti* p. 334
48–53 cm. Foothills to mid-mountain forest. Largest cuckoo-dove, with distinctive elongate shape. Pale grey head and underparts; dark brown upperparts. Juv sooty brown. **Subspp shown**:
griseotincta (NG range except Biak);
brevis (Bay Is: Biak) smaller, with white head and underparts and black wing patch.

PLATE 36 GROUND-DOVES AND EMERALD DOVES

1 Bronze Ground-Dove *Alopecoenas beccarii* p. 336
18–20 cm. Montane forest. Uncommon. Adult a tiny, compact ground-dove, dark with grey head. Male has glossy maroon shoulder patch. Female overall duller, no shoulder patch. Juv has brown edges to feathers. **Subspp shown**:
beccarii (NG mountains) dark olive green;
johannae (Karkar I, also Bismarck Is) dark brown with prominent white eye-ring.

2 Cinnamon Ground-Dove *Gallicolumba rufigula* p. 335
22–24 cm. Lowland to mid-montane forest. Smallest of the lowland ground-doves; common and widespread. Adult clay brown with bright white underparts stained yellow. Large, beady eye against pale face. Four grey wing bars. Juv with brown underparts.

3 Common Emerald Dove *Chalcophaps indica* p. 338
23–25 cm. Western islands. Adult resembles Pacific ED but has a white forehead patch (smaller in Female), grey crown, and reduced shoulder patch. **Subspp shown**:
indica (NW Is);
minima (Bay Is) underparts darker purplish brown.

4 Pacific Emerald Dove *Chalcophaps longirostris* p. 338
23–25 cm. Lowland and hill forest, especially edge and second growth. Overall grey-maroon. Both back and wing-coverts green or bronze-green. White or grey shoulder patch. Two grey bars on rump. Male has grey-maroon head and underparts; sooty rump and tail. Female has grey-brown head and underparts; dingier shoulder patch; brownish rump and tail. Juv has reduced green; breast obscurely barred; bill black.

5 Stephan's Emerald Dove *Chalcophaps stephani* p. 339
24–26 cm. Lowland and hill forest, especially forest interior. Overall rich, dark reddish brown. Back brown with green restricted to the wing. Rump bands buff (not grey). Male's forehead white; plumage with a purplish cast. Female's forehead grey; plumage more brownish. Juv has dark bill and brown, not green, secondaries.

6 White-bibbed Ground-Dove *Alopecoenas jobiensis* p. 335
24–25 cm. Hill and lower montane forest. Nomadic and unpredictable. Adult has distinctive pattern. Male has bright white throat and breast. Female variable, some like Male, others duller with white areas dingy. Juv grey-brown with rufous-edged feathers.

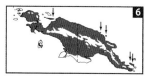

7 New Guinea Bronzewing *Henicophaps albifrons* p. 335
33–36 cm. Lowland to lower montane forest. Adult is large and dusky. Unusually long bill, conspicuous white forehead, and bronze wing. Forehead white in Male, dingy white to buff in Female. Juv lacks iridescence.

8 Thick-billed Ground-Pigeon *Trugon terrestris* p. 336
32–36 cm. Lowland and hill forest. Adult has distinctive face: white-tipped bill, white cheek, and short crest. Bright buff flanks. Juv bill darker; wing feathers edged rufous.

PLATE 37 CROWNED PIGEONS AND PHEASANT PIGEON

1 Pheasant Pigeon *Otidiphaps nobilis*   p. 337
44–48 cm. Hill to lower montane forest. Uncommon.
Fowl-like shape and tail. Adult iridescent blackish with
red-brown back and wings. Juv duller. **Subspp shown**:
nobilis (NW Is, Bird's Head to W Ranges) crested, opal nape,
underparts dark purple;
cervicalis (Border Ranges east to SE Pen) no crest, white nape,
underparts green;
insularis (D'Entrecasteaux Is: Fergusson) no crest or nape patch,
underparts purple;
aruensis (Aru Is) no crest, neck patch white and large,
underparts green and purple.

2 Western Crowned Pigeon *Goura cristata*   p. 337
61–71 cm. Lowland forest, especially near rivers and streams.
Adult from other crowned pigeons by its grey breast (lacking
maroon) and maroon mantle (not grey). Juv duller, wing-coverts
with grey or buffy edging.

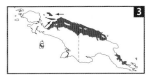

3 Victoria Crowned Pigeon *Goura victoria*   p. 338
58–74 cm. From other crowned pigeons by white-tipped crest
and grey wing patch (not white).

4 Southern Crowned Pigeon *Goura scheepmakeri*   p. 337
71–79 cm. Combination of all-grey crest, maroon breast, white
wing patch, and by the chin being grey (not black). **Subspp shown**:
scheepmakeri (southern SE Pen) grey shoulder and foreneck;
sclaterii (S Lowlands) maroon shoulder and foreneck.

PLATE 38 FRUIT-DOVES 1

1 Wompoo Fruit-Dove *Ptilinopus magnificus* p. 339
29–33 cm. Lowland and hill forest. Common. A large, long-tailed fruit-dove with purple breast, pale grey head, and band of pale spots on the shoulder.

2 Wallace's Fruit-Dove *Ptilinopus wallacii* p. 341
27–29 cm. Aru Is mainly. Large, with red cap, grey breast and back, and yellow-orange belly.

3 Pink-spotted Fruit-Dove *Ptilinopus perlatus* p. 340
25–27 cm. Lowland and hill forest. Common. Mustard-brown neck band and diagnostic pink spotting on shoulder. **Subspp shown**: *perlatus* (most of range) head yellowish green; *plumbeicollis* (NE: Sepik-Ramu to N coast of SE Pen) head grey.

4 Ornate Fruit-Dove *Ptilinopus ornatus* p. 340
23–26 cm. Montane cloud forests when breeding; afterward wanders in groups to lowlands. Common. Similar to Pink-spotted FD but has purple bar on the shoulder, throat band darker, and a broad, yellowish tail-band. **Subspp shown**: *ornatus* (Bird's Head) wine-red head; *gestroi* (most of range) mustard-yellow head.

5 Orange-fronted Fruit-Dove *Ptilinopus aurantiifrons* p. 340
23 cm. Disturbed lowland forests, gardens, and towns, less common in expanses of primary forest and foothills. Unique orange-yellow forecrown. Diagnostic white throat (looks grey), grey breast, and green belly.

6 Superb Fruit-Dove *Ptilinopus superbus* p. 341
20–23 cm. The most widespread and prevalent fruit-dove in forests from the lowlands up to mid-mountains. Diagnostic white belly and undertail. Conspicuous grey tail-band and white belly in flight. Female has grey-green throat and breast and dark blue hindcrown.

7 Rose-crowned Fruit-Dove *Ptilinopus regina* p. 341
20–23 cm. Vagrant AU species. Pale green with yellow-tipped tail. Purple-and-orange belly.

8 Beautiful Fruit-Dove *Ptilinopus pulchellus* p. 341
18–20 cm. Wet hill forest and lowlands. Common. Red crown; throat and breast white and pale grey offset from orange belly by a wine-red band. **Subspp shown**: *pulchellus* (Bird's Head, S Lowlands, and SE Pen); *decorus* (NW Lowlands and Sepik-Ramu) grey breast feathers heavily tipped white.

9 Coroneted Fruit-Dove *Ptilinopus coronulatus* p. 341
18–19 cm. Lowland forest, less common in hills. Prefers drier forest than Beautiful FD. Common. Dark green, with lilac crown and yellow undertail. **Subspp shown**: *coronulatus* (S NG and Aru Is) crown purple, belly lacks orange; *geminus* (N NG, including Bird's Head) crown paler, belly with orange.

PLATE 39 FRUIT-DOVES 2

1 Yellow-bibbed Fruit-Dove *Ptilinopus solomonensis* p. 343
18 cm. Bay Is. Small size, bright yellow undertail-coverts, and unique, bare, pale greenish skin around the eye. Male has yellow breast band (not white) and lacks purple cap, but note purple spot in front of eye.

2 Dwarf Fruit-Dove *Ptilinopus nainus* p. 344
13–15 cm. Foothills and nearby lowland forest. Rare to common. Surprisingly small, all green with yellow undertail and unique yellow edging to wing-coverts and secondaries. Male has small purple breast patch and grey spot on side of neck, both lacking in Female.

3 Claret-breasted Fruit-Dove *Ptilinopus viridis* p. 343
18–20 cm. Hill forest of N NG; also in lowlands on islands. Uncommon. Small, with grey face. Undertail-coverts striped green and white. Male, and usually also Female, shows a grey wing patch and a unique maroon patch on throat. **Subspp shown**:
pectoralis (NW IS, Bird's Head and Neck);
geelvinkiana (Bay Is: Biak, Mios Num, and Numfor);
salvadorii (N coast, including Foja, Cyclops, and N Coastal Mts; Yapen I);
vicinus (SE Is: D'Entrecasteaux and Trobriand Is).

4 Orange-bellied Fruit-Dove *Ptilinopus iozonus* p. 343
20–22 cm. Common and widespread; prefers open lowland and hilly habitats. Stocky, stubby-tailed, overall green with orange belly and white undertail. Bright eye. **Subspp groups shown**:
iozonus (species' range except the following) shoulder patch grey;
humeralis (NW Is, Bird's Head and western S Lowlands) shoulder patch purple.

5 Knob-billed Fruit-Dove *Ptilinopus insolitus* p. 344
22–24 cm. Extralimital species (Bismarck Is), unique for its large, red knob atop the bill. Otherwise similar to Orange-bellied FD, but larger. One sighting near Madang.

6 Moluccan Fruit-Dove *Ptilinopus prasinorrhous* p. 342
22 cm. Small western islands. From Yellow-bibbed and White-bibbed FDs by green undertail. Female green with an all green head.

7 White-bibbed Fruit-Dove *Ptilinopus rivoli* p. 342
20–24 cm. SE Is and Bay Is. Bright yellow undertail and lower belly. Male lacks purple belly patch. From Yellow-bibbed FD by larger size and yellow-green loral stripe.

8 Mountain Fruit-Dove *Ptilinopus bellus* p. 342
23–26 cm. Mainland NG, in montane forests. A stolid, dark green fruit-dove—look for green belly and undertail and bright yellow-green loral stripe. Female from Moluccan and White-bibbed by bluish cast to crown and face.

PLATE 40 ISLAND PIGEONS

1 Pied Imperial Pigeon *Ducula bicolor*   p. 347
41–43 cm. Restricted to the far West, where it replaces the
Torresian IP. Lacks black markings on undertail-coverts and
thighs, and the bill is blue-grey with a dark tip.

2 Torresian Imperial Pigeon *Ducula spilorrhoa*   p. 347
40–43 cm. The widespread white pigeon on small islands,
coastlines, mangroves, and savannahs. From Pied IP by the
black markings on undertail-coverts and by the yellowish bill.
Juv has obscure black markings on the undertail. Trans-Fly
birds tinged grey.

3 Elegant Imperial Pigeon *Ducula concinna*   p. 345
43–45 cm. Small islands in far West. Large size. All-grey
underparts, except brown undertail-coverts. Slightly swollen cere
gives forehead a flatter profile. Striking yellow iris, but no white
eye-ring. Blackish underwing in flight—looks broad-winged.

4 Spectacled Imperial Pigeon *Ducula perspicillata*   p. 344
44 cm. Kofiau I only. White eye-ring, dark grey head, green
extending partway up back of neck (not sharply demarked at
the shoulder), and steep forehead profile. Undertail-coverts grey,
not brownish. Greyish underwing in flight.

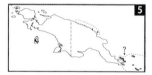

5 Floury Imperial Pigeon *Ducula pistrinaria*   p. 346
41–44 cm. Prevalent and common on small and midsize islands
on PNG side. No black bill knob. Note white ring of feathers
around the eye and at the base of the bill. Subtly pink-tinted
throat contrasting with the grey breast.

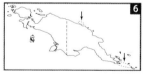

6 Pacific Imperial Pigeon *Ducula pacifica*   p. 345
34–37 cm. Highly localized—restricted to a few islands in the East.
A small imperial pigeon. Black bill knob, especially prominent in
Male. No white eye-ring. Underparts entirely soft rose-grey.
In flight, underwing uniformly dark.

7 Spice Imperial Pigeon *Ducula myristicivora*   p. 345
41–43 cm. NW Is. Mostly on small islands. Whitish head and
neck, black bill knob, white eye-ring (variable), and dark red iris.
Head actually has a pale rosy tint, hard to see at a distance.
Grey throat contrasts with rose-grey breast.

8 Geelvink Imperial Pigeon *Ducula geelvinkiana*   p. 345
38 cm. Bay Is. Distinctive yellow iris. All-grey head lacks bill knob
and white eye-ring. Body plumage resembles that of Spice IP.

9 Nicobar Pigeon *Caloenas nicobarica*   p. 338
32–33 cm. A large, dark, fowl-like pigeon of remote, forested
offshore islets. Local. Travels to larger islands to feed. A terrestrial
bird, unlike the arboreal imperial pigeons. Male cere knob larger,
hackles longer, upperparts greener than the more bronze-tinted
Female. Adult has clean white tail. Juv dingier, lacks hackles, and
tail black.

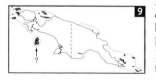

118

PLATE 41 MAINLAND PIGEONS

1 Purple-tailed Imperial Pigeon *Ducula rufigaster* p. 345
33–39 cm. Lowland and hill forest interior. Common but inconspicuous. Usually solitary. Smallest and most brightly colored of the mainland imperial pigeons. Conspicuous terminal tail-band. **Subsp shown:** *rufigaster* (NW Is, Bird's Head, S Lowlands, SE Pen) back and rump purplish red. Not shown is *uropygialis* (NW Lowlands to Huon) these parts more green, underparts paler.

2 Rufescent Imperial Pigeon *Ducula chalconota* p. 346
37–39 cm. Montane forest interior. Uncommon and inconspicuous. Solitary. Note green rump and grey "cowl" over the head and sides of neck; lacks a white eye-ring. **Subsp shown:** *smaragdina* (Central Ranges and Huon) upperparts all green. Not shown is *chalconota* (Bird's Head and Neck) green upperparts tinged purple-brown.

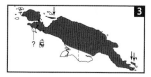

3 Papuan Mountain-Pigeon *Gymnophaps albertisii* p. 348
33–36 cm. Seen in flocks flying above the forest. Nests in the mountains, commutes downslope to forage in montane, hill, and lowland forest. Loud whistling of the birds' wings is diagnostic. Simple color scheme—dark with a pale breast and a band at the tip of the tail. Breast patch white in Male, grey in Female. Juv lacks red eye patch.

4 White-throated Pigeon *Columba vitiensis* p. 333
36–37 cm. Widespread but sparsely distributed through lowland and montane forest. A large, blackish pigeon with a clear white throat patch. Note maroon-grey breast and glossy green scalloping on upperparts. Juv duller than Adult, nearly without iridescence.

5 Zoe's Imperial Pigeon *Ducula zoeae* p. 347
38–41 cm. Lowland and hill forest. Common. Chunky and short-tailed. Easily identified by plumage pattern.

6 Pinon's Imperial Pigeon *Ducula pinon* p. 346
44–48 cm. Common and conspicuous in lowland and hill forest. Largest imperial pigeon in NG Region and powerfully built. Note narrow white band across the dark tail. **Subspp shown:** *pinon* (all of range except the following) wings deep grey; *jobiensis* (Yapen I; NW Lowlands to Huon) wings boldly scalloped; *salvadorii* (SE Is) like *pinon* but lacking the white eye-ring.

7 Collared Imperial Pigeon *Ducula mullerii* p. 347
38–41 cm. A seasonally common pigeon of mangroves and riverine lowland forest. Similar to Pinon's IP but smaller, slimmer, and appears darker. Tail-band broader. Diagnostic black collar contrasting with white throat and neck. **Subsp shown:** *mullerii* (S Lowlands and southern SE Pen) larger and darker. Not shown is *aurantia* (NW Lowlands and Sepik-Ramu) smaller and paler.

PLATE 42 COCKATOOS AND LARGE PARROTS

1 Eclectus Parrot *Eclectus roratus* p. 360
38 cm. The most often seen large parrot of the lowlands and foothills; common in forest and openings. Male and Female unmistakable if seen well. Juv resembles Adult but with brown-and-orange beak and dark iris. In flight, either sex may look all dark against the sky; note long, rounded, steadily beating wings; short, squared tail; and rather slow flight; Male's underwing red, Female's blue.

2 Palm Cockatoo *Probosciger aterrimus* p. 348
51–63 cm. Lowland and hill forest. A huge, blackish parrot with high, erectile crest and red cheeks. Massive, hooked beak. Adult all black. Juv has beak tipped white; belly with faint pale barring. In flight note the long, rounded wings and tail; undertucked beak and flattened crest; slow, straight, and level flight with ~4 flaps, then a glide.

3 Sulphur-crested Cockatoo *Cacatua galerita* p. 349
38–51 cm. Lowland and hill forest and open country.
A large and vocal white cockatoo with high yellow crest and black beak. Male iris dark brown; Female's same or red-brown. In flight note heavy body; wings rounded, underwing pale yellow; wingbeats rapid and shallow, followed by a glide.

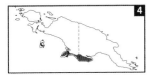

4 Little Corella *Cacatua sanguinea* p. 349
36–38 cm. Trans-Fly woodland, savannah, and agricultural areas. Now uncommon. A medium-sized, white cockatoo with short crest and pale beak. Large, bluish eye-patch. Adult has eye-patch larger and possibly paler than that of Juv. Fast, direct, pigeon-like flight with quick, deep wingbeats; wings narrower than Sulphur-crested C.

5 New Guinea Vulturine Parrot *Psittrichas fulgidus* p. 349
46 cm. Primary forest of lower montane zone, hills, and nearby lowlands. Uncommon and local. Hunted for its red feathers; widely extirpated. A noisy, black, crow-sized parrot with vulture-like head; red on belly, rump, and wings. Male has small red streak behind eye, lacking in Female. In flight look for red in wings and underparts, not always visible; note the profile of narrow head and broad wings; rapid, shallow wingbeats followed by a short glide.

PLATE 43 GREEN LORIKEETS

1 Pygmy Lorikeet *Charmosyna wilhelminae*   Pl. 49, p. 352
11–13 cm. Local in hill and montane forest. Smallest lorikeet. Look for tiny size and short, pointed tail. Male has red rump and underwing. Juv beak and iris brown.

2 Striated Lorikeet *Charmosyna multistriata*   Pl. 49, p. 352
19 cm. Uncommon or rare in hill and lower montane forest. Adult dark green; yellow streaking on whole of face and underparts; unusual blue-and-orange beak.

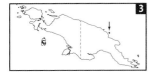

3 Red-chinned Lorikeet *Charmosyna rubrigularis*   Pl. 49, p. 352
18–20 cm. The only small lorikeet on Karkar I. Red chin and yellow tail tip.

4 Red-fronted Lorikeet *Charmosyna rubronotata*   Pl. 49, p. 353
15–18 cm. Lowland and hill forest of Bird's Head and N NG; infrequently found. Differs from Red-flanked L by tail black above and red rump patch. Male has red crown and red breast patch near bend of wing. Female, traces of yellow streaking behind eye.

5 Red-flanked Lorikeet *Charmosyna placentis*   Pl. 49, p. 353
15–18 cm. The prevalent green lorikeet of lowland and hill forest and open habitats. Tail with red-orange tip and yellow underside. Compare with previous species. **Subspp group shown**: *placentis* (species' range exclusive of next) blue rump patch. Not shown is group *subplacens* (Sepik-Ramu to SE Pen and SE Is) with all-green rump.

6 Goldie's Lorikeet *Psitteuteles goldiei*   Pl. 49, p. 356
19 cm. Mountain forests and some highland towns; occasionally descends to foothills. Gregarious, nomadic, unpredictable. Two-toned: dark green above, pale green below. Unique dark-streaked underparts. Juv, red forehead only.

7 Plum-faced Lorikeet *Oreopsittacus arfaki*   Pl. 49, p. 352
15–18 cm. Cloud forest, especially at higher elevation. Tiny, slender, dark green. Look for red-tipped tail. Male has red cap. Juv with dusky scalloping. **Subsp shown**: *grandis* (E Ranges, Huon, SE Pen) all green below. Not shown is subspp group *arfaki* (Bird's Head, W Ranges) orange-red patches on breast and flank.

8 Yellow-billed Lorikeet *Neopsittacus musschenbroekii*   Pl. 49, p. 354
19–23 cm. Mountain forests and around highland settlements. From Orange-billed L: Flanks yellowish green contrasting with emerald green wings, undertail orange-yellow, beak larger and yellow. Juv has darker beak, less red on breast.

9 Orange-billed Lorikeet *Neopsittacus pullicauda*   Pl. 49, p. 355
18–20 cm. Cloud forest and openings. From Yellow-billed L: Flanks and wings both emerald green; beak smaller, more orange; underside of tail dull green with red at the base. Juv, as for Yellow-billed. **Subspp groups shown**: *pullicauda* (E Ranges to Huon and SE Pen) red breast; *alpinus* (Western and Border Ranges) orange breast, darker red belly.

PLATE 44 RED LORIES AND LORIKEETS

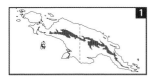

1 Stella's Lorikeet *Charmosyna stellae*   Pl. 49, p. 354
36–41 cm. Widespread in montane cloud forest (but absent from Bird's Head). Unmistakable. Sleek, with long tail-streamers. Red and black morphs. Male's lower back is red; Female's bright yellow (green in black morph). Juv, dusky beak (not red), feathers with dark-scalloped edges, much shorter tail.
Subspp groups shown:
stellae (Central Ranges and Adelbert Mts) breast all red;
wahnesi (Huon) yellow band across breast.

2 Papuan Lorikeet *Charmosyna papou*   p. 354
36–41 cm. Bird's Head counterpart of Stella's L. Red morph only, sexes alike, blue and black patch on crown (rather than nape), red nape patch, black crescent on hindneck, and two yellow patches on each side. Juv as for Stella's L.

3 Josephine's Lorikeet *Charmosyna josefinae*   Pl. 49, p. 353
23–25 cm. Hill and lower mid-mountain forest. Uncommon. Intermediate in size and looks between Stella's/Papuan Ls and Fairy L. Tail all red with yellow tip; tail length less than or equal to head-and-body length. Lower back red in Male, yellow or greenish yellow in Female. Juv with dusky beak (not red), dark-scalloped feathers.

4 Fairy Lorikeet *Charmosyna pulchella*   Pl. 49, p. 353
16–19 cm. Hill and montane forest, locally common.
The little red lorikeet. Yellow streaks on breast; tail with yellow tip. Male rump green, Female's with yellow patches. Juv, green replaces much of red; beak dark, not orange. **Subspp shown**:
pulchella (most of NG, excluding range of next subsp) underparts red;
rothschildi (N slope of W Ranges, plus Foja and Cyclops Mts) green patch on breast.

5 Black-capped Lory *Lorius lory*   Pl. 49, p. 355
25–30 cm. The widespread large red lory with black cap; common in forest from lowlands to mid-mountains.
Most populations possess a black hind-collar and a blackish "vest." Skin around the nostril (cere) dark. Beak orange. Juv has dark beak; back patch greenish rather than purple. **Subspp groups shown**:
lory (NW Is, Bird's Head and Neck) has purple-black vest, red underwing-coverts;
erythrothorax (S Lowlands, SE Pen) breast mostly red, hindneck collar blue-black;
somu (foothills of upper Fly R east to Purari R) black collar absent or much reduced;
cyanauchen (Biak I) dark underwing-coverts, blue nape and black crown merge.
See plate 49 for race *jobiensis* (NW Lowlands to Sepik-Ramu) dark underwing-coverts (red in previous races).

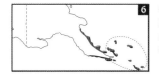

6 Purple-bellied Lory *Lorius hypoinochrous*   Pl. 49, p. 355
25–28 cm. Locally distributed along coast of SE Pen; ubiquitous on SE Is. From Black-capped L by lack of a black hind-collar and presence of white cere. Adult beak orange, Juv beak dark.

PLATE 45 DARK LORIES AND LORIKEETS

1 Violet-necked Lory *Eos squamata* Pl. 49, p. 357
28 cm. NW Is, especially small ones. Common and primarily coastal in a wide range of habitats, including villages and coconut plantations. Much variation in the smudgy purple collar. Adult beak orange. Juv beak dark; dark-scalloped appearance.

2 Black-winged Lory *Eos cyanogenia* Pl. 49, p. 357
30 cm. Bay Is. Locally common in all forest types and open habitats. Black saddle, blue ear patch. Adult beak orange. Juv beak dark; red plumage with dark scaling.

3 Dusky Lory *Pseudeos fuscata* Pl. 49, p. 356
28 cm. Lowlands to mid-mountains in forest and near settlements. Nomadic and unpredictable; when present can be common to abundant. Highly gregarious. Adult medium-sized, dusky brown with orange bands and conspicuous bright orange beak and facial skin. Distinctive white rump patch. Individuals vary in hue: red, orange, or yellow. Juv, darker beak, facial skin, and iris.

4 Rainbow Lorikeet *Trichoglossus haematodus* Pl. 49, p. 357
25–30 cm. The most often seen NG parrot. Ubiquitous in towns and settled areas, also forest, from coast to mid-mountains. Green and red with pointed tail, dark blue head, and yellow hind-collar. Beak orange in Adult, dark in Juv. **Subspp groups shown**: *haematodus* (excluding range of other subspp) maroon head, blue face, green belly;
rosenbergii (Biak I) wide yellow-orange collar, heavily barred breast, blackish belly;
nigrogularis (Trans-Fly, Aru Is) all-blue head, breast barring obscure, blackish belly.

5 Yellow-streaked Lory *Chalcopsitta scintillata* Pl. 49, p. 357
30 cm. Lowland forest and semi-open habitats. Locally common. A large, dusky green lory with red forehead (reduced in Juv); yellow streaking often difficult to see. This genus has distinctive slow flight with shallow wingbeats. **Subspp shown**:
scintillata (Bird's Neck to lower Fly R) red underwing-coverts;
rubrifrons (Aru Is) broad orange streaking on breast.
See plate 49 for race *chloroptera* (upper Fly R to southern SE Pen) with green underwing-coverts.

6 Brown Lory *Chalcopsitta duivenbodei* Pl. 49, p. 357
30 cm. Lowland forest and semi-open habitats. Uncommon. A large, brown lory with distinctive yellow and violet markings; black beak framed by yellow forehead and yellow crescent around the mouth. Rump, undertail-coverts, and primaries violet.

7 Black Lory *Chalcopsitta atra* Pl. 49, p. 356
30 cm. Lowland forest and semi-open habitats. Uncommon. A large, black lory with long, rounded tail. Note black beak.
Subspp shown:
atra (W Bird's Head; NW Is of Batanta, Salawati) all black, red only in base of tail;
insignis (E Bird's Head, Bird's Neck) red in face, flanks, thighs. Not shown is *bernsteini* (Misool Is); forehead usually tinged red, thighs red.

PLATE 46 MEDIUM-SIZED PARROTS

1 Red-winged Parrot *Aprosmictus erythropterus* p. 360
30 cm. Trans-Fly savannah and open woodlands. Locally common. Brilliant yellow-green with red wing patch. Male has large, scarlet wing patch; black back. Female has red only on greater wing-coverts. Juv similar to Female but iris dark brown, beak yellow.

2 Red-cheeked Parrot *Geoffroyus geoffroyi* p. 361
20–25 cm. The common, medium-sized, green parrot throughout the lowlands. Veering rapid flight overhead with musical vocalizations. Male has purple head with red cheeks. Female, dull brown head. Juv, green crown, olive cheeks. Sky blue underwing-coverts noticeable in flight. **Subspp groups shown**: Southern and SE subspp with green rump (*aruensis* shown); Northern subspp with reddish-brown rump (*pucherani* shown).

3 Blue-collared Parrot *Geoffroyus simplex* p. 361
23–24 cm. The only parrot, other than lorikeets, making flights high over mountain forest. Vocal, flocking, nomadic. Distinctive chiming flight call. Entirely green, with pale iris. Male has bluish collar. Juv with bluish cast to head, no collar, dark iris. Blue underwing visible in flight.

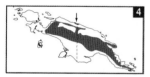

4 Papuan King-Parrot *Alisterus chloropterus* p. 360
38 cm. The widespread king-parrot, locally common within forest interior from foothills to mid-mountains. Green with long, blackish tail and red underparts. Male has red head, yellow-green shoulder stripe. Female, see subspp. Juv similar to Female but more green on breast, pink tips to tail feathers, darker beak and iris.
Subspp: Female very different from Male in first 2 races.
chloropterus (SE Pen, Huon) Male with blue nape, Female green with red breast;
callopterus (S slopes of W, Border, and E Ranges) not illustrated; Male with red nape, Female similar to previous;
moskowskii (N slopes of W and Border Ranges) Male similar to previous, Female resembles Male except that the mantle and back are green.

5 Moluccan King-Parrot *Alisterus amboinensis* p. 360
38 cm. The far-western counterpart of the Papuan KP. Adult differs from Papuan in the entirely dark green wing and blue mantle. Juv indistinguishable from Papuan.

6 Great-billed Parrot *Tanygnathus megalorynchos* p. 362
38 cm. A large parrot of the NW Is. All green with tapered body, long wings, and pointed tail. Big head and beak. Adult has orange-red beak and shoulders marked gold and black. Juv has smaller, yellowish beak and the black-and-gold pattern nearly lacking. In flight, wings turquoise above, yellow below. Compare with Eclectus Parrot, also in NW Is (plate 42).

PLATE 47 FIG-PARROTS AND TIGER-PARROTS

1 Salvadori's Fig-Parrot *Psittaculirostris salvadorii* p. 358
18–19 cm. NW Lowlands. Lowland and hill forest, edge, and villages. Local and uncommon. Turquoise highlights and yellow-streaked cheeks. Note turquoise-green crown and blue mark behind the eye. Male has orange breast band, Female's is turquoise-blue.

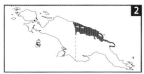

2 Edwards's Fig-Parrot *Psittaculirostris edwardsii* p. 358
18–19 cm. Sepik-Ramu and Huon. Lowland and hill forest, edge, and villages. Locally common to abundant. Unique head pattern: red cheek bounded by blackish eye-line and dark collar. Conspicuous yellow ear patch. Male has red breast, Female's is green.

3 Large Fig-Parrot *Psittaculirostris desmarestii* Pl. 49, p. 358
18–19 cm. Bird's Head, S Lowlands, SE Pen. Forest and edge of lowlands, hills, and mid-mountain basins. Uncommon. Unique fiery orange crown; cheeks green, yellow, or orange, by race. Sexes alike in most races. **Subspp groups shown**: *desmarestii* (Bird's Head and Neck) blue spot below eye, cheeks green or yellow; *blythii* (Misool I) head all orange (*godmani* of western S Lowlands, yellow cheeks); *cervicalis* (eastern S Lowlands to SE Pen) face golden orange, blue-and-purple collar.

4 Madarasz's Tiger-Parrot *Psittacella madaraszi* p. 351
14–15 cm. Mountain forests. Uncommon. Madarasz's generally lives at elevations below Modest TP. Male shows much yellow spotting on the back of the head; brown of head merges into green on breast at bend of wing. Female and Juv have blue-green forehead and crown; little or no barring on the breast.
Subsp group shown: *madaraszi* (Central Ranges) Female with red in nape (lacking in *huonensis* of Huon).

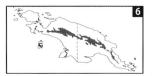

5 Modest Tiger-Parrot *Psittacella modesta* p. 351
14–15 cm. High-mountain cloud forests. Locally common. Male has brown of head merging into green on breast well below bend in wing, usually has yellow collar on hindneck. Female and Juv have barred, orange breast. **Subspp groups shown**:
modesta (Bird's Head) Male head all brown;
collaris (Central Ranges) Male with yellow nape-collar.

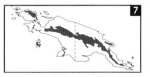

6 Painted Tiger-Parrot *Psittacella picta* p. 351
17–19 cm. Cloud forest and at timberline. Local and uncommon. Identified by the blue patches on cheeks and/or breast. Rump reddish or yellowish. Male has blue or bluish breast patch and yellow neck crescent. Female and Juv, breast barred. **Subspp shown**: *lorentzi* (W Ranges) crown dull brown, Male's breast blue-green, rump yellow; *excelsa* (E Ranges) resembles next subsp but head dull brown; *picta* (SE Pen) head red-brown, rump red, Male with blue breast.

7 Brehm's Tiger-Parrot *Psittacella brehmii* p. 351
21–24 cm. The most often seen tiger-parrot; common in cloud forest. Note large size; all-brown head sharply defined from the green underparts; red in undertail only. Male has yellow neck crescent. Female and Juv, breast barred.

PLATE 48 PYGMY PARROTS AND SMALL FIG-PARROTS

1 Orange-fronted Hanging Parrot *Loriculus aurantiifrons* p. 359
9–10 cm. Lowland and hill forest. Uncommon and frequently overlooked. Detect by voice: high-pitched, rapid *tseo tseo tseo* (2–4 notes). Bright green, with red rump and gorget. Male has yellow cap, white iris. Female face bluish, dark iris. Juv like Female, but no red gorget. Fast-flying and bullet-shaped; bright blue underwing.

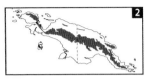

2 Red-breasted Pygmy Parrot *Micropsitta bruijnii* p. 363
9 cm. Montane forest above 1000 m. Uncommon and local. Male unique. Female and Juv with blue cap and pale face; pale eyebrow is narrow, not as broad as Buff-faced PP.

3 Buff-faced Pygmy Parrot *Micropsitta pusio* p. 362
8–9 cm. Lowland and hill forest. Common. Adult face buff or brown, with blue cap; dark beak. Juv has greenish blue cap and bicolored beak. **Subspp groups shown**:
pusio (southern SE Pen and SE Is) face buff-brown;
beccarii (NW Lowlands to northern SE Pen) face brown.

4 Yellow-capped Pygmy Parrot *Micropsitta keiensis* p. 362
9 cm. Lowland and hill forest. Common. Adult with dark brown face and dingy yellowish cap; beak dark. Juv with pale yellowish beak, dark-tipped. **Subspp shown**:
keiensis (Aru Is, S Lowlands) entirely green underparts;
chloroxantha (NW Is, Bird's Head and Neck) Male has red streak on breast and belly.

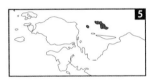

5 Geelvink Pygmy Parrot *Micropsitta geelvinkiana* p. 362
9 cm. Numfor and Biak Is. Blue-and-brown head. Male has yellow patch down breast and belly. Female's face usually paler and more mottled. Juv has pale face and beak. **Subspp shown**:
geelvinkiana (Numfor I);
misoriensis (Biak Is).

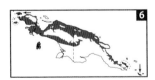

6 Double-eyed Fig-Parrot *Cyclopsitta diophthalma* Pl. 49, p. 359
14–15 cm. Mainly foothills and lower mountains. Locally common in forest and edge. Yellow flank-stripe diagnostic. Male cheeks red, lacking in Female and Juv. **Subspp shown**:
diophthalma (N NG, Bird's Head to SE Pen) Female crown red;
aruensis (S Lowlands east to Fly R; also Aru Is)
Female with blue-and-grey face;
virago (D'Entrecasteaux Is) Male lacks eye-spot,
Female like next race, cheek blue;
inseparabilis (Tagula I) sexes alike, showing only a red-and-blue forehead spot.

7 Orange-breasted Fig-Parrot *Cyclopsitta gulielmitertii* p. 359
11–13 cm. Lowland and foothill forest and edge. Black-and-white face unique; breast often orange. Male has white or yellowish face, without orange; usually with large orange breast patch.
Female and Juv have orange ear-coverts, breast generally green.
Subspp groups shown:
gulielmitertii (Bird's Head, NW Lowlands to Sepik R);
amabilis (Ramu R to northern SE Pen);
melanogenia group (S Lowlands to southern SE Pen; Aru Is)
illustrated by subspp *fuscifrons* (S Lowlands) and *suavissima* (southern SE Pen).

Bird's Head, NW   NE   S Lowlands   SE

PLATE 49 FLYING LORIES, LORIKEETS, AND FIG-PARROTS

1 Dusky Lory *Pseudeos fuscata* — Pl. 45, p. 356
Dusky with orange bill, yellow or orange banding on breast, rounded tail.

2 Rainbow Lorikeet *Trichoglossus haematodus nigrogularis* — Pl. 45, p. 357
Green with blue head, red breast, and yellow wing-stripe.

3 Yellow-streaked Lory *Chalcopsitta scintillata* — Pl. 45, p. 357
Blackish with red forehead and flame-colored undertail; underwing-coverts green in race *chloroptera* (SE) or red in *scintillata* (SW).

4 Brown Lory *Chalcopsitta duivenbodei* — Pl. 45, p. 357
Very dark but for yellow flash in wing, from underwing-coverts; yellow undertail.

5 Black Lory *Chalcopsitta atra* — Pl. 45, p. 356
W race *atra* all black; E race *insignis* red in face and underwing-coverts.

6 Fairy Lorikeet *Charmosyna pulchella pulchella* — Pl. 44, p. 353
Little red lorikeet with moderately long, pointed tail that is yellowish underneath.

7 Josephine's Lorikeet *Charmosyna josefinae* — Pl. 44, p. 353
Larger than Fairy L and tail longer, but much shorter than Papuan or Stella's Ls.

8 Violet-necked Lory *Eos squamata* — Pl. 45, p. 357
NW Is. Mostly red; no blue eye patch or black saddle.

9 Black-winged Lory *Eos cyanogenia* — Pl. 45, p. 357
Bay Is. Red with blue ear patch and black saddle.

10 Yellow-billed Lorikeet *Neopsittacus musschenbroekii* — Pl. 43, p. 354
Bright red underwing, yellow bill, orange-yellow undertail.

11 Stella's Lorikeet *Charmosyna stellae stellae* — Pl. 44, p. 354
Red or black with long tail-streamers.

12 Orange-billed Lorikeet *Neopsittacus pullicauda pullicauda* — Pl. 43, p. 355
Bright red underwing, orange bill, yellowish green undertail with red at base.

13 Black-capped Lory *Lorius lory* — Pl. 44, p. 355
Stocky. Short wings flash yellow. Underwing-coverts black in race *jobiensis* (N), red in *erythrothorax* (S).

14 Purple-bellied Lory *Lorius hypoinochrous* — Pl. 44, p. 355
Large red lory of SE coasts and islands. Note white cere and lack of black collar.

15 Pygmy Lorikeet *Charmosyna wilhelminae* — Pl. 43, p. 352
Smaller than other lorikeets; entire underwing red in Male, green in Female.

16 Red-chinned Lorikeet *Charmosyna rubrigularis* — Pl. 43, p. 352
Green underwing; yellow stripe across wing feathers; yellow undertail.

17 Striated Lorikeet *Charmosyna multistriata* — Pl. 43, p. 352
All green, without flashes of color. Streaking and unique bluish beak may be visible.

18 Red-fronted Lorikeet *Charmosyna rubronotata* — Pl. 43, p. 353
Similar to Red-flanked L but lacks red in tail; yellow wing-stripe obscure.

19 Red-flanked Lorikeet *Charmosyna placentis* — Pl. 43, p. 353
Tail with red-orange tip and yellow underside. Underwing-coverts red (Male) or green (Female); yellow streak across primaries.

20 Goldie's Lorikeet *Psitteuteles goldiei* — Pl. 43, p. 356
Yellow stripe across wing. Underparts pale green with dark streaks. Purplish head.

21 Plum-faced Lorikeet *Oreopsittacus arfaki grandis* — Pl. 43, p. 352
Only lorikeet with all-red undertail and tail tip. Underwing-coverts also red.

22 Large Fig-Parrot *Psittaculirostris desmarestii* — Pl. 47, p. 358
Yellow wing-stripe, blue underwing-coverts.

23 Double-eyed Fig-Parrot *Cyclopsitta diophthalma* — Pl. 48, p. 359
Yellow flank-stripe and wing-stripe, blue-green underwing-coverts.

PLATE 50 LARGE CUCKOOS AND HORNBILL

1 Biak Coucal *Centropus chalybeus* p. 363
44–46 cm. The only coucal on Biak I. Inhabits forest. Adult has a yellow iris. Juv dark-eyed and washed rufous (no barring).

2 Lesser Black Coucal *Centropus bernsteini* p. 364
41–51 cm. Grassland, scrub, and forest edge. Common. Geographic range partly overlaps that of Pheasant C, with which it might be confused. Lesser Black C smaller with shorter tail and has a dark brown iris. Adult all black (no barring). Juv black with faint barring of rufous or buff; head pale spotted, not streaked as in Pheasant C.

3 Pheasant Coucal *Centropus phasianinus* p. 364
43–60 cm. Most often seen coucal of grassland and scrub. Adult blackish but always with some brown barring in the wing, even if faintly on leading edge of primaries only; iris red. Juv with much pale streaking and barring; iris brown.
Subspp groups shown:
thierfelderi (S Lowlands and Trans-Fly) rufous barring on wing and tail;
nigricans (N NG and SE Pen) obscure barring on wing (can be hard to see).

4 Greater Black Coucal *Centropus menbeki* p. 363
56–69 cm. The forest-dwelling coucal. Keeps to thick cover from the ground up high into the canopy. Large with an oversized tail. Bill ivory with dark grey base. Adult with red iris. Juv black with pale tan iris and rufous-barred tail. **Subsp shown**: *menbeki* (species' range excluding Aru Is) blue-green gloss. Not shown is *aruensis* (Aru Is) purple gloss.

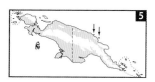

5 Channel-billed Cuckoo *Scythrops novaehollandiae* p. 366
51–58 cm. Uncommon austral winter migrant to open habitats in lowlands. An unmistakable giant cuckoo with loud grating call. Shape unique—massive bill, long pointed wings, and long tail with black-and-white tip. Adult grey with red eye and orbital ring; Female has smaller bill with more cream on tip (>30%). Juv head and underparts buff, wings and tail spotted buff.

6 Blyth's Hornbill *Rhyticeros plicatus* p. 394
76–91 cm. Lowland and hill forest. Common where not heavily hunted. A huge, black, white-tailed bird. Male has tawny head; Female's head black. Juv like Male, but casque with just 1 wrinkle.

PLATE 51 LARGE PARASITIC CUCKOOS

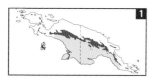

1 Fan-tailed Cuckoo *Cacomantis flabelliformis*   p. 370
26–28 cm. Migratory AU race (*flabelliformis*) rare in open habitats: Adult paler, especially on breast; tail with obvious white notching on margins of dorsal surface. Resident NG race (*excitus*) uncommon in montane forest: larger than Chestnut-breasted, duller brown below and with hint of extremely fine barring on breast and belly—visible only at close range. Female of both races with paler underparts, faintly barred on flanks. Juv variable, but obviously barred below.

2 Chestnut-breasted Cuckoo *Cacomantis castaneiventris*   p. 370
20–23 cm. Common resident of forest interior on hilly terrain, from lowlands to mid-mountains. Smaller than Fan-tailed C. Adult with blue-grey upperparts and chestnut underparts. Juv underparts white, buff, or brown without barring.

3 Brush Cuckoo *Cacomantis variolosus*   Pl. 52, p. 370
19–23 cm. Resident and AU migrant. Gardens and forest edge of lowlands and mid-mountains, less frequent in forest interior. Male dull brownish grey with grey throat and inconspicuous eye-ring. Female polymorphic: *plain morph*, like Male; *intermediate morph*, plain above and faintly barred below (also plate 52); and *barred morph* underparts heavily barred and upperparts with feathers pale-notched and barred. Juv upperparts buff-mottled and barred; underparts darkly mottled and barred.

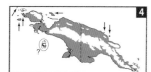

4 Oriental Cuckoo *Cuculus optatus*   p. 371
29–34 cm. Regular Palearctic migrant in austral summer to open habitats. Male has grey upperparts and white breast barred with black, often with pale iris. Most Females resemble Male but have faint rufous wash on neck and breast; rare *hepatic morph* with upperparts rufous brown and heavily barred throughout. Juv with pale scalloping on upperparts. Himalayan Cuckoo (*Cuculus saturatus*), distinguishable only by song and slightly smaller size, also recorded from NW NG Region.

5 Pallid Cuckoo *Heteroscenes pallidus*   p. 369
31–32 cm. Rare austral migrant to Trans-Fly savannah. Note dark line behind the eye and down the neck and the pale nape spot in all plumages. Iris dark. Some do not show white notching. Female variably browner and more heavily marked; others like Male.

6 White-crowned Cuckoo *Caliechthrus leucolophus*   p. 369
30–36 cm. Resident of hills and lower mountain forests. Uncommon and difficult to see, but vocal. Adult identified by white crown-stripe. Note white-tipped undertail-coverts. Juv all black.

7 Long-tailed Cuckoo *Urodynamis taitensis*   p. 366
36–39 cm. Vagrant in austral winter to small islands off N NG and the SE Is. Slender profile with long tail. Note bicolored bill, conspicuous pale eyebrow, and streaked underparts. Adult upperparts entirely barred, iris yellow. Juv upperparts spotted, iris dark, but young birds in NG Region should be in Adult or intermediate plumage.

8 Eastern Koel *Eudynamys orientalis*   p. 365
38–42 cm. Lowland and lower montane forest. Slender shape and size distinctive. Red iris and pale bill in Adult. Female with white malar streak, boldly spotted upperparts, and many-barred tail; some have black head and darker upperparts. Juv with black eye-patch and white malar streak; wing-coverts barred and spotted; dark iris. **Subspp shown:** *rufiventer* (NG resident); *subcyanocephalus* (AU migrant) Female much paler, some with black head.

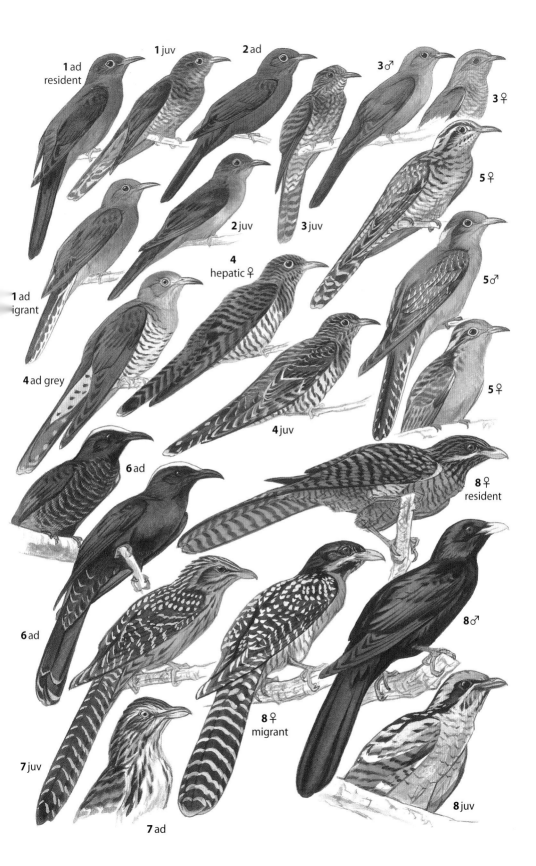

PLATE 52 SMALL PARASITIC CUCKOOS

1 Little Bronze Cuckoo *Chalcites minutillus* p. 369
14–17 cm. The common bronze cuckoo of lowland forests. Adult bronze-brown above, densely barred below, with variable rufous cast. Tail may show rufous. Red eye in Male; dark brown in Female. Juv unbarred below or with obscure flank barring. **Subspp shown**: *poecilurus* (NG resident) forehead brown or green, breast often has a rufous tinge; *barnardi/minutillus* (AU migrant) forehead white, underparts always white.

2 Rufous-throated Bronze Cuckoo *Chalcites ruficollis* p. 367
15–17 cm. The bronze cuckoo of cloud forests. Uncommon. Adult with rufous face and throat; upperparts with rufous highlights. Juv greyish green above, grey fading to white below and faintly barred.

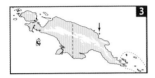

3 Shining Bronze Cuckoo *Chalcites lucidus* p. 368
15–17 cm. Migrant in austral winter to savannah, second growth, and forest edge. Lowlands to intermontane valleys. Uncommon. Adult glittering green with white-speckled face and thick bars in the underparts. Juv with barred flanks. **Subspp shown**: "*plagosus*" (AU race, prevalent throughout NG) both sexes with brown cap; "*lucidus*" (NZ race, rare in SE Is) wider bill; green cap in Male, brown in Female.

4 Horsfield's Bronze Cuckoo *Chalcites basalis* p. 367
15–20 cm. Rare austral winter migrant to open country of Trans-Fly and Aru Is. A rather dull-colored bronze cuckoo. Adult has diagnostic combination of pale eyebrow, sooty ear patch, and scaled feathers of upperparts. Note that barring on the underparts does not meet in the middle of the breast. Juv with same field marks, but dorsal scaling seldom visible and underparts lack barring.

5 White-eared Bronze Cuckoo *Chalcites meyerii* p. 368
14–15 cm. The forest bronze cuckoo of hills and mid-mountains; uncommon. Adult has white underparts with thick dark barring, white ear, and bright rufous primaries. Female has unique chestnut forehead. Eye red in Male, dark in Female. Juv lacks barring below, but note rufous patch in primaries (usually lacking in juv Little BC).

6 Black-eared Cuckoo *Chalcites osculans* p. 367
19–22 cm. Vagrant in austral winter to open country in Aru Is. Adult pale grey with broad white eyebrow contrasting with black patch through the eye. Underparts buff or white. No barring anywhere. Pale rump in flight. Juv is dull version of Adult.

7 Brush Cuckoo *Cacomantis variolosus* Pl. 51, p. 370
19–23 cm. For comparison. Int. morph Adult female—plain, but faintly barred below and with white tips to tail feathers.

8 Dwarf Koel *Microdynamis parva* p. 365
20–22 cm. Lowland and lower montane forest. Canopy-dwelling fruit eater. Short, thick, hooked bill and distinctive white streak below eye. Male with blue-black crown and black malar streak. Female duller. Juv with obscure barring. **Subsp shown**: *parva* (species range except next) back brown, underparts buff. Not shown is *grisescens* (Sepik-Ramu to northern SE Pen) upperparts greyer, underparts rufous brown.

9 Long-billed Cuckoo *Rhamphomantis megarhynchus* p. 366
18–19 cm. A rare cuckoo of lowland forests. Suggests a Tawny-breasted Honeyeater, but tail has unique, faint barring on the underside. Male with black cowl and striking red eye. Female with grey head, dark eye. Juv with greyish-white patch framing dark eye and dark ear coverts.

PLATE 53 BARN-OWLS AND FROGMOUTHS

1 Australian Masked Owl *Tyto novaehollandiae* p. 372
38–43 cm. Savannah owl of the Trans-Fly. Largest NG barn-owl, powerfully built. Similar to Australian Barn-Owl, but much larger, back and wings marked with dark grey, underparts heavily speckled black, and facial mask outlined with a thick dark border. Individuals vary in degree of darkness. Female larger.

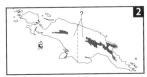

2 Eastern Grass-Owl *Tyto longimembris* p. 373
30–36 cm. Mid-mountain grasslands. Patchily distributed. Long-faced and small-eyed compared with other barn-owls. Dark grey-brown dorsally; face and underparts buff or white. Female larger on average.

3 Sooty Owl *Tyto tenebricosa* p. 372
33–38 cm. The forest barn-owl of NG. Forest from the lowlands to well above timberline in subalpine grasslands. The only sooty-colored NG owl, with some individuals darker than others. Female is larger. Voice a drawn-out, decending whistle, likened to a falling bomb or teakettle. Other *Tyto* owls screech.

4 Australian Barn-Owl *Tyto delicatula* p. 372
30–33 cm. Frequents disturbed habitats—roadsides, gardens, plantations, and forest edge. Lowlands to mid-mountains. A small and rather slender barn-owl. Very pale, with white breast (faintly spotted at most). White facial disc with only a trace of a black border. Female larger on average and tends to be more buffy.

5 Papuan Frogmouth *Podargus papuensis* p. 375
46–53 cm. Lowland and hill forest and openings, also open habitats such as towns and parks. Common. Identified by voice and large size, otherwise resembles Marbled F. Variable. Song 9–15 notes, low, resonant, monotonous: *oom oom oom. . . .* Male grey or brown, underparts marbled with black-bordered white blotches; scapulars patterned with white spots and black markings. Female usually greyish brown, underparts lacking marbling or with only a few pale blotches on belly; scapulars buffy, smooth textured and relatively unpatterned. Juv may exhibit bits of white down.

6 Marbled Frogmouth *Podargus ocellatus* p. 375
33–38 cm. Same habitats as preceding species, also common. From Papuan F by voice and size (Marbled two-thirds the size of Papuan). Song 4–6 notes, slow and hummed with insistent quality: *hoooa hoooa hoooa hoooa*. Much plumage variation. Male usually overall grey or brown, with much coarse white spotting on throat and breast. Female usually overall rufous brown and plainer (finer patterned). Juv retains bits of down. **Subspp shown**:
ocellatus (all of species' NG range, except SE Is);
meeki (Tagula I) no rufous-brown birds known;
intermedius (D'Entrecasteaux Is, Trobriand Is) largest race.

PLATE 54 OWLS

1 Papuan Boobook *Ninox theomacha* p. 374
25–28 cm. The common New Guinea owl. In forest and edge from sea level to lower cloud forest. Small, chocolate-colored owl, located by *hyu-hyu* call. **Subspp groups shown**:
theomacha (NG, NW Is) without markings;
goldii (SE Is) coarsely spotted with white below.

2 Biak Scops-Owl *Otus beccarii* p. 373
25 cm. Biak forests, where it is the only owl. Only scops-owl in NG Region. Note small size, ear tufts, yellow eyes. Tawny and grey morphs. Located by croaking voice.

3 Southern Boobook *Ninox novaeseelandiae* p. 374
25–28 cm. Trans-Fly savannah. A reddish-brown owl with white spots on wings and messy white-and-brown mottling on breast. Smaller and proportionately chunkier than Barking Owl; note darker mask. Voice says its name, *boo-book*.

4 Rufous Owl *Ninox rufa* p. 373
41–51 cm. Rare in lowland, hill, and mid-mountain rainforest. Large, dark owl barred throughout. Adults vary in color from rufous to brown. Downy Juv has black mask that offsets whitish head and breast. Generally quiet; voice deep, slow *woo-hoo* or *mumph-mumph*.

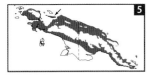

5 Papuan Hawk-Owl *Uroglaux dimorpha* p. 374
30–33 cm. Poorly known, but probably widespread in lowland and hill rainforest. Small-headed, long-tailed, rather pale owl. Adult with unique combination of streaked breast and profusely barred upperparts. Downy Juv with white face and underparts. Voice a series of slowly repeated *hoo-hoo* notes.

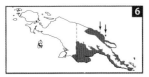

6 Barking Owl *Ninox connivens* p. 374
33–36 cm. Lowland forest edge and open habitats. Dark above with pronounced, dark streaking below. Voice, doglike barking: *wuf-wuf*.

PLATE 55 NIGHTJARS AND LARGE OWLET-NIGHTJARS

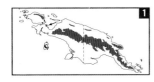

1 Feline Owlet-nightjar *Aegotheles insignis* p. 379
25–30 cm. Montane forests. Uncommon. Large, with striking white or buff stripes on face and underparts. Adult rufous or brown; body plumage always with some dark, stippled barring. Juv resembles rufous Adult but body plumage lacks any barring. Voice 3 hoarse whistles, *whor . . . whor, whor.*

2 Starry Owlet-nightjar *Aegotheles tatei* p. 379
25–30 cm. Low hilly country, where rare in tall forest along rivers. Slender. Bright rufous. Tail with white dorsal bars. Adult may show frosty barring on the breast and belly and starry spots on back and scapulars, lacking in Juv. Voice similar to Feline, but higher pitched and often of 4 notes.

3 Archbold's Nightjar *Eurostopodus archboldi* p. 377
26–30 cm. The only high-mountain nightjar. Rare in cloud forest, alpine heaths, moors. Adult blackish, appears lichen-encrusted, with silvery-grey blotches and golden-buff spots. Prominent grey eyebrow. No white patches in primaries or tail. Juv ashy grey with black spots and diamonds on upperparts and black eye-patch. Voice unknown.

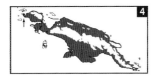

4 Papuan Nightjar *Eurostopodus papuensis* p. 377
25–28 cm. Uncommon lowland forest nightjar in gaps, edge, along rivers. Adult dark, marked with grey tertials and buff scapular patch that contrasts with dark wing. Wings and tail lack white markings. Breast with horizontal rows of buff spots. Juv milk chocolate, with black markings much reduced. Voice a rapid *coo coo coo coo.*

5 Spotted Nightjar *Eurostopodus argus* p. 376
27–35 cm. Vagrant to Aru Is and possibly Trans-Fly.
Austral winter migrant from AU to open habitats. Adult pale with buffy band of sharp streaking across scapulars. White patch on the primaries, but not on tail.

6 White-throated Nightjar *Eurostopodus mystacalis* p. 376
30–35 cm. Uncommon austral winter migrant from AU to open habitats in lowlands to mid-mountain valleys. Adult large, dark, and greyish. Thick, grey band along scapulars. Long wings nearly reach tail tip. Wing, tail lack white markings.

7 Large-tailed Nightjar *Caprimulgus macrurus* p. 377
25–28 cm. The common open-country, roadside nightjar from lowlands to mid-mountain valleys. Shoulder with diagnostic rows of buff spotting. Unique white-based rictal bristles, white gape streak. Folded wings fall short of tip of long tail. Male has large, white wing and tail patches. Female patches smaller, buff. Voice a prolonged, rhythmic knocking, *tok tok tok . . . .*

8 Grey Nightjar *Caprimulgus jotaka* p. 377
25–28 cm. Vagrant, Bird's Head. Palearctic winter migrant in open habitats. Adult plumage evenly textured, no bold markings. Folded wing nearly reaches tail tip.

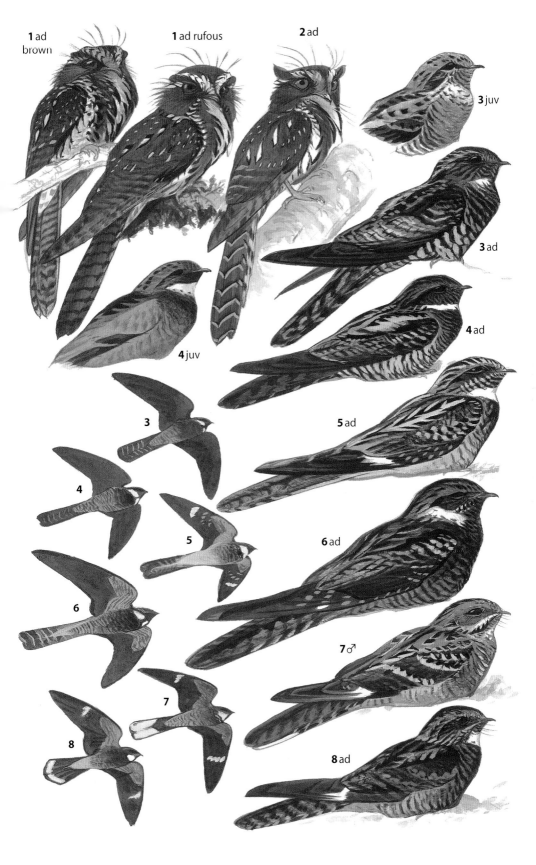

PLATE 56 SMALL OWLET-NIGHTJARS

1 Wallace's Owlet-nightjar *Aegotheles wallacii* p. 379
20–23 cm. Hill and lower montane forest, plus flat lowland forest near foothills. Rare and sparsely distributed. All plumages blackish above without distinct pattern. Buffy chin, throat contrast with dark neck. Note white, streaky spots on scapulars and wing-coverts, lacking in Barred ONJ. Voice a series of rapid, squeaky whistles, *per pew-pew*, or *peer pew*, or variation.

2 Mountain Owlet-nightjar *Aegotheles albertisi* p. 380
18–20 cm. Mountain forests. Common. Lacks white facial and breast streaks of the larger Feline ONJ. Plumages highly variable. Breast with irregular dark spots and streaks. Compared with Archbold's ONJ, Mountain is usually more finely patterned, often less color-saturated, with barring on sides of neck, and tail bars usually broken. Voice a steady series of explosive squeaks, ~1/sec, *ee! . . . ee! . . . ee! . . . .*

3 Archbold's Owlet-nightjar *Aegotheles archboldi* p. 380
18–20 cm. Counterpart of Mountain ONJ in high mountains of W and Border Ranges. Compared with Mountain, more richly colored and coarsely mottled; breast more heavily marked and barring obscure; tail bars usually solid.

4 Allied Owlet-nightjar *Aegotheles affinis* p. 381
23 cm. Lower montane forests of Bird's Head. Like Barred ONJ but with brownish cast, and bird thickly feathered, producing plump shape with large head. Head coarsely patterned. Most birds show a distinct pale collar. Voice undescribed.

5 Barred Owlet-nightjar *Aegotheles bennettii* p. 380
20–23 cm. The common owlet-nightjar of lowland and hill forest and edge. Slender. Dark charcoal grey, finely barred. (A few mainland birds with brownish cast; D'Entrecasteaux Is birds always with buffy cast.) Lacks brown on cheeks, unless plumage brownish. Voice an unevenly paced series of muffled yapping barks, suggests a small dog, *ap . . . ap . . . ap . . . ap. . . .*
Subspp groups shown:
bennettii (mainland NG) usually charcoal grey and lacks streaking down breast;
terborghi (E Ranges: Karimui) similar to *bennettii* but much larger;
plumifer (SE Is: D'Entrecasteaux Is) buffy; breast with shaft streaking and barring (upper left corner of plate).

6 Australian Owlet-nightjar *Aegotheles cristatus* p. 381
23–24 cm. The pale owlet-nightjar of southern savannahs. Medium grey above (paler than Barred ONJ), with mostly white breast and belly (these usually more darkly barred in Barred ONJ). Some birds with brown wash. Always some buff on cheeks. Birds of Port Moresby area are same size as Barred; Trans-Fly birds are larger. Voice a repeated shriek.

PLATE 57 SWIFTS AND TREESWIFT

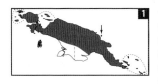

1 Glossy Swiftlet *Collocalia esculenta* p. 382
9 cm. Common and widespread, from the foothills to above timberline; avoids extensive flat country and savannah. Smallest NG swift. The only swiftlet with glossy blue upperparts and whitish belly. Underwing all dark (see Papuan Spinetailed Swift).

2 Moustached Treeswift *Hemiprocne mystacea* p. 381
28–30 cm. Lowlands and foothills, at forest edge. Common. A very large swift with long, thin wings and long, deeply forked tail. Adult grey with white eyebrow and moustache. Male has chestnut on ear-coverts, lacking in Female. Juv mottled with brown.

3 Mountain Swiftlet *Aerodramus hirundinaceus* p. 382
11–13 cm. The grey-brown swiftlet most often seen in the highlands, but not separable from Uniform S in the field. In the hand, Mountain S larger, tarsus usually feathered.

4 Uniform Swiftlet *Aerodramus vanikorensis* p. 383
13 cm. The grey-brown swiftlet most often seen at low elevations, indistinguishable from Mountain S in the field, but of smaller size and tarsus usually not feathered.

5 Three-toed Swiftlet *Aerodramus papuensis* p. 383
14 cm. Uncommon and local in foothills and mountains. Large size, closest to Bare-legged S, but head smaller, body slimmer like Mountain S. The only swiftlet with 3 toes.

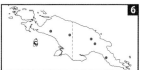

6 Bare-legged Swiftlet *Aerodramus nuditarsus* p. 383
14 cm. A rare montane swiftlet. Largest NG swiftlet, with large-headed profile. Grey-brown above, evenly lead-grey below. Tarsus bare. White fringe on eyebrow is unique.

7 Papuan Spinetailed Swift *Mearnsia novaeguineae* p. 383
11 cm. Uncommon in lowlands. Unique stubby shape and short tail. Color resembles Glossy S, but note pale stripe in the underwing of S race. **Subspp shown**:
novaeguineae (S Lowlands and SE Pen) throat grey;
buergersi (NW Lowlands and Sepik-Ramu) throat dark, like back.

8 Fork-tailed Swift *Apus pacificus* p. 384
18 cm. A Palearctic migrant. Common over Trans-Fly savannah, rare elsewhere. Long, thin, curved wings and long tail. White rump and long, forked tail diagnostic. Chin and throat white; rest of underparts scaled white.

9 White-throated Needletail *Hirundapus caudacutus* p. 384
19 cm. An uncommon Palearctic migrant, mainly in lowlands. Largest NG swift. Adult shows unique combination of white throat and undertail, otherwise dark below. Juv has smaller white patches on forehead and scapulars; upperparts less glossy.

PLATE 58 SWALLOWS

1 Sand Martin *Riparia riparia* p. 495
12 cm. Vagrant from Asia during austral summer. Adult a small brown swallow with clean white underparts divided by a breast band. Juv upperparts pale-scalloped and throat buff-tinged.

2 Fairy Martin *Petrochelidon ariel* p. 496
12 cm. A rare AU migrant to S NG during austral winter. Small swallow with white rump and rufous cap. Tail notched. Adult crown bright rufous, back glossy blue-black. Juv crown paler with some black streaking, back dull brownish black.

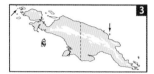

3 Tree Martin *Petrochelidon nigricans* p. 496
12 cm. A common and widespread AU migrant during austral winter. Small, black-and-white swallow with white rump patch. Adult forehead rufous, back glossy black. Juv forehead paler buff, back dull brownish black.

4 Pacific Swallow *Hirundo tahitica* p. 495
13 cm. The only resident NG swallow, common in cities, towns, and open areas. Adult upperparts glossy black; throat rufous brown blending into grey breast and belly. Transition is gradual between dark throat and pale breast, unlike sharp contrast in Barn S. Tail short and notched, lacking streamers, same in both sexes. Juv not glossy. **Subsp shown**: *frontalis* (most of NG) darker underparts, tail spots very small. Not shown is *albescens* (Trans-Fly to SE Pen) paler underparts, tail spots larger.

5 Welcome Swallow *Hirundo neoxena* p. 496
13–17 cm. Australian vagrant to NG south coast. Adult similar to Pacific S but with long tail wires (longest in Male) and broader white subterminal tail-band. Juv similar to Pacific S.

6 Barn Swallow *Hirundo rustica* p. 495
15–18 cm. A rare to locally common Asian migrant. Adult with long tail wires (longest in Male), clean white underparts, dark reddish face, and obvious black throat band. Juv overall duller with paler face and shorter tail.

7 Red-rumped Swallow *Cecropis daurica* p. 496
16–17 cm. An uncommon and irregular Asian migrant during austral summer. A fork-tailed swallow with orange rump and pale underparts with dark streaking. Adult with back glossy blue-black, rump orange, long tail wires (longest in Male). Juv back brownish black without gloss, rump buff, outer tail feathers pointed and protruding, but no wires.

PLATE 59 PARADISE-KINGFISHERS

1 Little Paradise-Kingfisher *Tanysiptera hydrocharis* p. 386
31 cm. Riverine forests of S NG and Aru Is. Uncommon and local. Adult nearly identical to Common PK, but smaller, outer tail feathers all blue, and crown and wing-coverts dark blue (not pale blue). Juv distinguished by size and dark blue crown, if present.

2 Common Paradise-Kingfisher *Tanysiptera galatea* p. 385
33–43 cm. The familiar paradise-kingfisher of mainland NG lowland forest. Usually common, inconspicuous, yet vocal. Adult has white breast and contrasting pale blue crown and wing-coverts; note mostly white outer tail feathers on the long, flag-tipped tail. Juv mostly brown with some crown feathers edged blue; bill all or partly black.

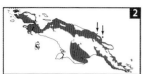

3 Rossel Paradise-Kingfisher *Tanysiptera rosseliana* p. 386
35–36 cm. Rossel I. Common in lowland forest. Adult similar to Common PK, but much less contrast between crown, back, and wings, and base of tail all white. Shorter tail.

4 Kofiau Paradise-Kingfisher *Tanysiptera ellioti* p. 386
33 cm. Kofiau I. Common in lowland forest. Adult similar to Common PK, but tail shorter, thicker, and all white.

5 Biak Paradise-Kingfisher *Tanysiptera riedelii* p. 386
35–36 cm. Biak I. Common in lowland forest and second growth. Adult with pale blue head.

6 Numfor Paradise-Kingfisher *Tanysiptera carolinae* p. 386
36–38 cm. Numfor I. Common in forest and gardens. Adult dark blue. Juv duller; underparts whitish or rufous marked with black; bill starts out dark.

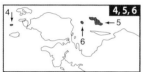

7 Buff-breasted Paradise-Kingfisher *Tanysiptera sylvia* p. 387
37 cm. A rare, local resident in monsoonal lowland and hill forests of SE Pen. Also an uncommon austral winter migrant to much of lowland NG, in monsoonal forest and rainforest. Note whitish patch on dark back. Adult with buffy orange breast. Female has blue margins on central tail feathers and variable dusky streaking on white back and rump. Juv duller with dark-marked breast and buff-scalloped wing-coverts.

8 Red-breasted Paradise-Kingfisher *Tanysiptera nympha* p. 387
32 cm. Patchy range. Locally common mainly in hill forest, also lowlands and mangroves. Adult has red-orange breast and blue crown. Juv similar to juv Brown-headed PK but shows blue in crown.

9 Brown-headed Paradise-Kingfisher *Tanysiptera danae* p. 387
28–30 cm. Locally common in forest mainly in foothills, occasionally nearby lowlands. Adult has red-orange breast and all-brown head. Juv has red-orange wash on breast and pink rump and lacks blue on head.

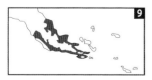

PLATE 60 DOLLARBIRD, BEE-EATERS, AND LARGE KINGFISHERS

1   Oriental Dollarbird *Eurystomus orientalis*   p. 384
25–28 cm. Open habitats in lowlands and hills. Resident race uncommon. Migratory AU race common during austral winter. Chunky with large head and broad bill. In most situations appears all dark. Wings have pale "silver dollar" marks. Juv blackish; bill blackish. **Subspp shown** (distinguished with care): *waigiouensis* (resident) brighter greenish blue; *pacificus* (migratory, breeds AU) paler grey-blue.

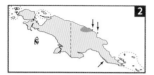

2   Rainbow Bee-eater *Merops ornatus*   p. 394
25 cm. Widespread, common AU migrant during March–October. Very local resident breeder. Occupies open habitats. Adult has black throat bar, much rufous in wing, and squared, black tail with thin tail wires. Juv lacks throat bar and tail wires.

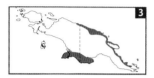

3   Blue-tailed Bee-eater *Merops philippinus*   p. 393
27–29 cm. Local resident in lowland grassland and other open habitats. Common. Adult compared with Rainbow BE: wings all green; tail longer, more tapered and blue; brown throat blending into breast without a black collar. Juv duller than Adult and lacks tail wires.

4   Rufous-bellied Kookaburra *Dacelo gaudichaud*   p. 389
28 cm. The common, lowland forest kookaburra. Male has a blue tail, Female brown. Juv bill blackish and underparts with dark scalloping.

5   Shovel-billed Kookaburra *Clytoceyx rex*   p. 388
32 cm. Rare and local in interior of forest from lowlands to mid-mountains. Often flushed from the ground where it has been digging. Bill broad and rounded at the tip. Male tail blue, Female's brown. Juv similar to Female, but upperparts with rusty-tip feathers, underparts with dark scalloping.

6   Blue-winged Kookaburra *Dacelo leachii*   p. 388
38–40 cm. Savannah and edge of monsoon forest. Largest NG kingfisher. Male's tail blue; Female's barred brown and blue. Juv buffier with heavier barring in underparts.

7   Spangled Kookaburra *Dacelo tyro*   p. 389
33 cm. Monsoonal forest and adjacent savannah in Trans-Fly and Aru Is. Common. Smaller than Blue-winged K. Adult head and mantle darker and spangled; iris dark; tail all blue, lacking white in outer tail feathers. Female, blue parts greenish blue. Juv head and back darker than Adult; underparts darker-scaled. **Subsp shown**: *archboldi* (Trans-Fly) whitish spangling and underparts. Not shown is *tyro* (Aru Is) with buff spangling and underparts.

8   Hook-billed Kingfisher *Melidora macrorrhina*   p. 388
25 cm. Interior of lowland and hill forest and patchy regrowth. Common. An inveterate caller at dawn and dusk, but difficult to observe. Unique striped facial pattern. Male's crown scalloped blue, Female's darker. Juv breast darker-scalloped.

PLATE 61 MEDIUM KINGFISHERS

1 Blue-black Kingfisher *Todiramphus nigrocyaneus* p. 389
23 cm. Rare in lowland forest, often near streams.
Deep, glowing blue with diagnostic black mask and back. Male belly dark blue or rufous. Female underparts white with a blue breast band. Juv upperparts like Adult but duller; underparts dingy white with rufous breast band, faintly scaled.
Subspp shown (only Males differ):
stictolaemus (central S Lowlands to SE Pen) underparts blue-black;
quadricolor (Yapen I, NW Lowlands to Sepik-Ramu) rufous belly;
nigrocyaneus (NW Is, Bird's Head to western S Lowlands) white throat and breast band.

2 Forest Kingfisher *Todiramphus macleayii* p. 390
20–21 cm. Resident and austral winter migrant in open habitats in lowlands and hills. Uncommon. Crown deep blue. White forehead spots and underparts. Unique white wing patch. Female's nape blue, not white. Juv upperpart buff-edged; underparts scalloped.
Subspp shown:
macleayii (resident) bright blue back;
incinctus (AU migrants) greenish-blue back.

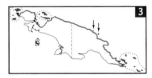

3 Beach Kingfisher *Todiramphus saurophagus* p. 390
28 cm. Seacoasts, especially on small, offshore islands. Local. Inhabits strand, mangroves, coconut groves. Only white-headed kingfisher in NG Region. Massive bill. Female mantle sometimes more greenish. Juv head and underparts buffy; upperparts buff-scaled.

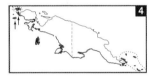

4 Collared Kingfisher *Todiramphus chloris* p. 390
20–25 cm. Mangroves and other coastal habitats. Uncommon. Adult heavy-billed with white collar and a blue-green crown and back. Underparts white. Female often duller. Juv with faint buff scaling in upperparts; obscure dark scaling on underparts and collar. **Subspp shown**:
chloris (NW Is, Bird's Head and Neck, also Wallacea) coastal; bright bluish green;
sordidus (NG south coast and Aru Is, also AU) mangroves; dusky olive green;
colonus (SE Is) small islands; darker, smaller than *sordidus* and Sacred K.

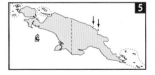

5 Sacred Kingfisher *Todiramphus sanctus* p. 391
21 cm. Common austral winter migrant to open habitats at low to mid-elevations, where it is the most common kingfisher. Adult underparts and collar buffy white with variable degree of dark edging. Cap and back greenish. Juv has buff scaling in upperparts.

PLATE 62 RIVER AND YELLOW-BILLED KINGFISHERS

1 Azure Kingfisher *Ceyx azureus* p. 393
16 cm. Along any body of freshwater or tidal creek with adequate cover. Common. A deep blue kingfisher with evenly dark blue crown and upperparts and black bill with pale tip. Breast variable—rufous to cream. Juv has blackish forehead and sides of breast.

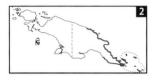

2 Common Kingfisher *Alcedo atthis* p. 392
14–15 cm. Local and habitat-specific in the NG Region. Along coasts, mangroves, rivers. Smaller and brighter than Azure K. Sky blue streak up the back, obvious in flight. Female has red in lower bill. Juv duller, with some dusky-edged feathers on breast.

3 Papuan Dwarf Kingfisher *Ceyx solitarius* p. 392
12 cm. Not associated with water. Lowland and hill forest. Common. Adult crown and wing-coverts with glittering, sequined feathering; back and rump paler blue. Juv duller, may have scaling on breast.

4 Little Kingfisher *Ceyx pusillus* p. 393
11 cm. Mangroves and small, lowland creeks and pools. Common. Smallest NG kingfisher. Adult is the only small kingfisher with white underparts. Juv's greenish-blue crown appears barred; hint of buff in underparts and black edging.

5 Yellow-billed Kingfisher *Syma torotoro* p. 391
18–20 cm. Lowland and hill forest. Common. Yellow bill lacks black markings. Male crown unblemished yellow-tan; nape has 2 black patches. Female crown has black patch; nape patches often joined. Juv bill entirely or partly black; white nape patch.
Subspp shown:
torotoro (widespread—all range except next);
ochracea (SE Is: D'Entrecasteaux Is) much larger, underparts more ochre.

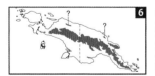

6 Mountain Kingfisher *Syma megarhyncha* p. 391
21–23 cm. Montane counterpart to Yellow-billed K. Uncommon. Nearly identical, but larger, with proportionately longer bill. Black marking atop bill (except on Huon). Juv ear-coverts blackish.
Subspp groups shown:
megarhyncha (Central Ranges) bill with black tip;
sellamontis (Huon) bill all yellow.

PLATE 63 LARGE, GROUND-DWELLING SONGBIRDS

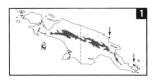

1 Island Thrush *Turdus poliocephalus* p. 506
23–25 cm. Common in alpine shrubland and subalpine woods. Adult unmistakable. Juv dark above, spotted with rufous; below rufous spotted with black. **Subspp groups shown**:
papuensis (NG) all brownish black;
canescens (SE Is: D'Entrecasteaux) grey head.

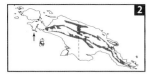

2 Russet-tailed Thrush *Zoothera heinei* p. 506
21–23 cm. Uncommon, rarely observed; forest floor in hills and mid-mountains. Robust with elongate shape. Large, dark eye; pink legs. Heavily scaled; buff wing bars. In flight note black-and-white patch under the wing and white outer tail feathers.

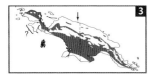

3 Papuan Scrub-Robin *Drymodes beccarii* p. 490
20 cm. A lanky, thrushlike terrestrial robin that inhabits forest interior mainly in hills and lower mountains, also locally in S Lowlands. Adult with black vertical bar through eye, white throat, white wing bars, long tail, and long pink legs. Juv has crown feathers edged black, throat and breast flecked brown.
Subspp groups shown:
beccarii (Bird's Head to NW Lowlands and Sepik-Ramu) upperparts blackish brown;
brevirostris (S Lowlands, Aru Is, SE Pen) upperparts cinnamon.

4 Lesser Melampitta *Melampitta lugubris* p. 472
18 cm. An uncommon, reclusive ground bird of cloud forests. Distinctive double-click call. An entirely velvet black, pitta-shaped bird. Male's eye red, Female's dark brown. Juv dull, sooty black with brownish cast in underparts.

5 Red-bellied Pitta *Erythropitta erythrogaster* p. 395
15–17 cm. Mainly hill forest. Common, secretive, but vocal resident. Adult has brown head, red belly. Juv brown and mottled, with pale or white throat patch. **Subspp shown**:
macklotii (most of NG Region) greenish back and wings;
finschii (SE Is: D'Entrecasteaux Is) blue back and wings.

6 Hooded Pitta *Pitta sordida* p. 395
15–17 cm. Lowland and hill forest. Common but secretive resident. Sings often. Adult has black hood, green body, shining turquoise shoulder patch. Glittering blue rump patch in Male, reduced or absent in Female. Juv mostly sooty. **Subspp groups shown**:
novaeguineae (NG Region, except next) blue-green flanks, glittering throat-band;
rosenbergii (Bay Is) deep blue flanks, lacks glittering throat-band.

7 Noisy Pitta *Pitta versicolor* p. 396
17–18 cm. Resident in monsoon forest and mangroves of Trans-Fly. Some birds may be migrants from AU. Adult has yellow-tan breast and flanks with black blotch in center. Juv duller and lacking dark chin and throat and dark belly patch.

PLATE 64 WHIPBIRDS, JEWEL-BABBLERS, AND LOGRUNNER

1 Painted Quail-thrush *Cinclosoma ajax* p. 438
23 cm. Very local in foothill and gallery forests.
A grey-brown bird that shows a white facial stripe or throat.
Look for white tail tips when flushed. Male boldly patterned.
Female has white eyebrow and throat. Juv like Female, but
iris paler.

2 Papuan Logrunner *Orthonyx novaeguineae* p. 430
18 cm. Rarely encountered. A scurrying, quail-shaped ground
dweller of wet mid-mountain and cloud forest. Note prominent
wing bars and spiny tail. Throat white in Male, orange in Female.
Juv mottled throughout. **Subsp shown**: *victorianus* (Central
Ranges) darker, and black breast border present in Female.
Not shown is very similar *novaeguineae*, lacking breast border
(Bird's Head).

3 Papuan Whipbird *Androphobus viridis* p. 436
16–17 cm. Rare and seldom seen in cloud forest thickets.
Much smaller than Australian whipbirds. Male has white
moustachial stripe, black throat. Female all moss green (do not
confuse with female Fantailed Berrypecker or juv Rufous-naped
Bellbird). Juv darker green with blackish face and underparts.

4 Spotted Jewel-babbler *Ptilorrhoa leucosticta* p. 436
20 cm. Cloud forest. Uncommon. Adult has white cheek patch
on black face and rows of white spots on shoulder.
Juv washed olive; wing spots dull. **Subspp groups shown**:
sibilans (Cyclops Mts) green-backed with green breast band;
centralis (Bird's Neck to Border Ranges) brown-backed.
Not shown are green-backed *leucosticta* (Bird's Head) and
loriae (E Ranges, Huon, SE Pen).

5 Chestnut-backed Jewel-babbler *Ptilorrhoa castanonota* p. 437
23 cm. Mid-mountain forests. Common. White throat, chestnut
back, and blue underparts. Male has blue rump and eyebrow.
Female rump usually chestnut like back (blue in race *uropygialis*).
Juv dark brownish grey with white throat. **Subspp groups shown**:
castanonota (Bird's Head, Batanta I) underparts blue, Female with
blue eyebrow; *pulchra* (E Ranges, SE Pen, Huon) Female eyebrow
white, tan, or pale blue; Not shown are *saturata* (W Ranges) and
uropygialis (N slope of W Ranges and N Coastal Mts).

6 Dimorphic Jewel-babbler *Ptilorrhoa geislerorum* p. 437
22–23 cm. Common in lowland and hill forest; hills only
where it meets Blue JB. Male similar to male Blue JB, but crown
has a brownish tinge. Female brown with white throat.

7 Blue Jewel-babbler *Ptilorrhoa caerulescens* p. 437
22 cm. Lowland forest. Common. All dull blue with white throat.
Sexes generally alike, but see subspp for exceptions. Juv browner.
Subspp groups shown: *caerulescens* (Bird's Head to Sepik-Ramu)
with blue undertail; *nigricrissus* (S Lowlands, SE Pen) blackish
undertail, Female has white eyebrow.

PLATE 65 MOUSE-WARBLERS AND GROUND-ROBINS

1 Rusty Mouse-Warbler *Crateroscelis murina* p. 423
12 cm. The mouse-warbler of lowland and hill forest; common. White throat; pale rusty breast. Male, crown usually darker than back. Female and Juv, crown same as back (dark Juv illustrated).
Subspp groups shown:
murina (species' range except for next) cinnamon breast;
monacha (Aru Is) similar but underparts mostly white.
Not shown is *pallida* (S Lowlands) very pale cinnamon flanks and belly.

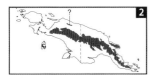

2 Bicoloured Mouse-Warbler *Crateroscelis nigrorufa* p. 423
12 cm. Mid-mountain forest interior. Rare. Narrow elevational range lies above Rusty MW and overlaps lower range of Mountain MW. Adult is uniquely black and rufous. Juv's extent of rufous varies from only throat to all of underparts.

3 Mountain Mouse-Warbler *Crateroscelis robusta* p. 423
12 cm. The mouse-warbler of high mountain forest; common. Subspp in 3 groups by plumage; the geography is quite specific for each, but widely scattered:
WHITE-THROATED DICHROMATIC GROUP:
White throat and belly divided by a dark breast band in Male, band lacking or faint in Female. Iris of Male white, grey, yellow, orange, red-brown, or brown; of Females mostly red-brown or brown. Subspp *robusta* (eastern E Ranges, Huon, SE Pen east to Owen Stanley Mts); *pratti* (far eastern SE Pen) much smaller, darker, stronger pattern; *diamondi* (Foja Mts) large, also dark with strong pattern.
Birds from western and central E Ranges (e.g., Mt Hagen) less patterned, approaching buff-breasted group.
PALE-WASHED MONOMORPHIC GROUP:
All plumages with underparts uniformly pale
(*peninsularis*, Bird's Head; *deficiens*, Cyclops Mts).
BUFF-BREASTED MONOMORPHIC GROUP:
All plumages with underparts uniformly rusty brown.
Subspp *sanfordi*, W Ranges and Border Ranges;
bastille, N Coastal Mts.

4 Lesser Ground-Robin *Amalocichla incerta* p. 487
14–15 cm. Cloud forest; widespread, common, and vocal. Similar to Greater GR but smaller and more dainty. Adult, clean brownish breast and conspicuous white forehead spot, throat. Juv, dark scalloped buffy plumage.

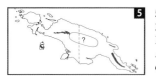

5 Greater Ground-Robin *Amalocichla sclateriana* p. 487
20 cm. Subalpine forest floor; usually rare and local. Robust and long-billed. Note pitta-like shape and thrush size. Adult has whitish throat, grey breast and flanks, with vague, dark scalloping on throat and breast. Juv rufous with dark mottling and scaling. **Subsp shown**: *sclateriana* (SE Pen). Not shown: *occidentalis* (W Ranges) browner.

PLATE 66 MAYPOLE BOWERBIRDS AND SATINBIRDS

NOTE: To locate a bowerbird, locate its bower first. These are pictured on plate 67a.

1 Vogelkop Bowerbird *Amblyornis inornata* p. 398
25 cm. Bird's Head and Neck, where it is the only maypole bowerbird. Common in montane forest. Male crestless like Female. Indistinguishable from female Macgregor's BB.

2 Golden-fronted Bowerbird *Amblyornis flavifrons* p. 398
24 cm. Foja Mts, where it is the only maypole bowerbird. Common in montane forest. Male has unique golden orange crest, and the forehead is golden too, unlike Macgregor's.

3 Streaked Bowerbird *Amblyornis subalaris* p. 397
22 cm. Lower mts of the far SE Pen. Locally common. Shares its range with Macgregor's BB, which lives mostly at higher elevation. Identified by obscure streaking on throat and breast, smaller size and Male's shorter crest.

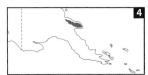

4 Huon Bowerbird *Amblyornis germana* p. 397
25 cm. Mountains of the Huon Peninsula. Plumages identical to those of Macgregor's BB, but Male's crest shorter.

5 Macgregor's Bowerbird *Amblyornis macgregoriae* p. 397
26 cm. The common and widespread gardener bowerbird, distributed throughout the Central Ranges. Male's crest orange. All plumages show an unmarked throat and breast. Vocal, but shy and elusive.

6 Archbold's Bowerbird *Archboldia papuensis* p. 398
35–37 cm. A rare and elusive fruit eater of high mountain forest of the Central Ranges. Patchily distributed. Note Male's black plumage, longish forked tail, and flat, golden crown. Female, Juv, and Imm more greyish than Male; tail is shorter and squared.

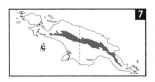

7 Loria's Satinbird *Cnemophilus loriae* p. 430
20–23 cm. A common fruit eater of the cloud forest canopy. Male is compact and all black with bluish iridescence on forehead and a pale yellow gape. Female and Imm are olive green with brownish wings. Juv is dull grey. *Amblyornis* bowerbirds are much more brown. Look for Male at his song perch on an exposed branch atop the canopy.

8 Crested Satinbird *Cnemophilus macgregorii* p. 431
25 cm. A quiet, plump, and retiring frugivore of high mountain cloud forest and subalpine thickets. Uncommon. Seen at fruiting trees. Male unique. Female and Imm brownish olive; note diagnostic forehead groove and grey iris. Juv dull grey.
Subspp shown:
sanguineus (E Ranges) Male deep orange;
macgregorii (SE Pen) Male yellow-orange.

9 Black Pitohui *Melanorectes nigrescens* Pl. 83, p. 452
22–23 cm. For comparison. Female deceptively similar. Note hooked bill. (Upper left corner of plate.)

PLATE 67 AVENUE BOWERBIRDS AND SATINBIRDS

NOTE: To locate a bowerbird, locate its bower first. These are pictured on plate 67a.

1 Flame Bowerbird *Sericulus ardens* p. 399
25–26 cm. Lowland and foothill forest of S Lowlands.
Attracted to patches of regrowth and rubber plantations.
Uncommon. Male showy and unmistakable, lacks black mask.
Female, Juv, and Imm with yellow breast, but note yellow chin and
line down throat, which lack mottling and scalloping; dark iris.
Imm Male similar but has white iris.

2 Masked Bowerbird *Sericulus aureus* p. 398
24–25 cm. Lower montane forests of the North and West.
Uncommon. Male unmistakable with orange plumage and
black face. Female, Juv, and Imm differ from Flame BB
(note that ranges abut) by the dark-scalloped chin and throat
(plate shows 2 variations); Imm Male has white iris.

3 Fire-maned Bowerbird *Sericulus bakeri* p. 399
26–27 cm. Lower montane forests of the Adelbert Range.
Uncommon. Look for it at large, fruit-bearing fig trees.
Male unmistakable. Female, Juv, and Imm dull grey-brown,
pale below with dark scalloping; Imm Male has white iris.

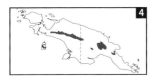

4 Yellow-breasted Satinbird *Loboparadisea sericea* p. 431
16–18 cm. Rare and patchily distributed. A seldom-seen fruit
eater of mid-mountain forest. Male is rich brown above, silky
yellow below with yellow rump; a unique, bulbous, blue-green
wattle sits atop bill. Female duller than Male and lacks wattles.
Juv underparts buffy and streaked.

5 Fawn-breasted Bowerbird *Chlamydera cerviniventris* p. 400
28–30 cm. Mainly lowlands in transitional thickets between
grassland and forest, also edges of mangroves and towns.
In mid-mountains of SE Pen and Huon. Note orange-buff
breast and belly. Crown pale-streaked. Habits similar to
Yellow-breasted BB.

6 Yellow-breasted Bowerbird *Chlamydera lauterbachi* p. 399
25–28 cm. Margins of grassland, scrub, and forest edge
of mid-mountain valleys and Sepik-Ramu lowlands.
Locally common, but shy and wary. Most often seen
when crossing openings with distinctive, head-up,
undulating flight. All plumages have a yellowish breast
and belly and an unstreaked crown. **Subspp shown**:
uniformis (western NG and parts of Sepik) dull greenish crown;
lauterbachi (Sepik-Ramu and Huon) orange-washed crown.

PLATE 67A BOWERBIRD BOWERS

Male bowerbirds build their bowers to attact females for mating. The bower is a stage for his display; the female actually builds her nest elsewhere. Note that bowers vary in size among species; they are not illustrated to scale here.

1 Vogelkop Bowerbird p. 398 and plate 66
Bird's Head and Neck. On a ridge crest or flank in cloud forest.
Giant, tented bower.

2 Streaked Bowerbird p. 397 and plate 66
SE Pen. On ridge flank, not crest, in mid-mountain forest.
Hutlike rather than the maypole of co-occurring Macgregor's BB.

3 Macgregor's Bowerbird p. 397 and plate 66
Central Ranges and Adelbert Mts. On a ridge crest in cloud forest.
Maypole bower of small twigs built up around a sapling pole surrounded
by a circular mossy base of ~1 m diameter with an elevated rim.

4 Golden-fronted Bowerbird p. 398 and plate 66
Foja Mts. Set atop ridge spine.
Bower similar to Macgregor's but for presence of blue fruit and lack of an
elevated perimeter rim.

5 Huon Bowerbird p. 397 and plate 66
Huon Mts. Set below crest of ridge.
Bower lacks perimeter ridge and has a basal mossy column hung with small,
colored decorations.

6 Archbold's Bowerbird p. 398 and plate 66
Central Ranges, in high-elevation cloud forest.
Bower a broad stage of matted ferns shrouded by orchid stems on overhanging branches.

7 Flame Bowerbird p. 399 and plate 67
S Lowlands. Near a forest opening.

  Masked Bowerbird p. 398 and plate 67
  Lower mountain forests of N and W.

  Fire-maned Bowerbird p. 399 and plate 67
  Lower mountain forests of the Adelbert Range.

Bowers of all three are small structures with 2 parallel walls of sticks.

8 Fawn-breasted Bowerbird p. 400 and plate 67
Widespread.
Two-walled avenue bower placed in thicket and decorated with green berries.

9 Yellow-breasted Bowerbird p. 399 and plate 67
Widespread.
Four-walled avenue bower placed in thicket and decorated with berries and pebbles.

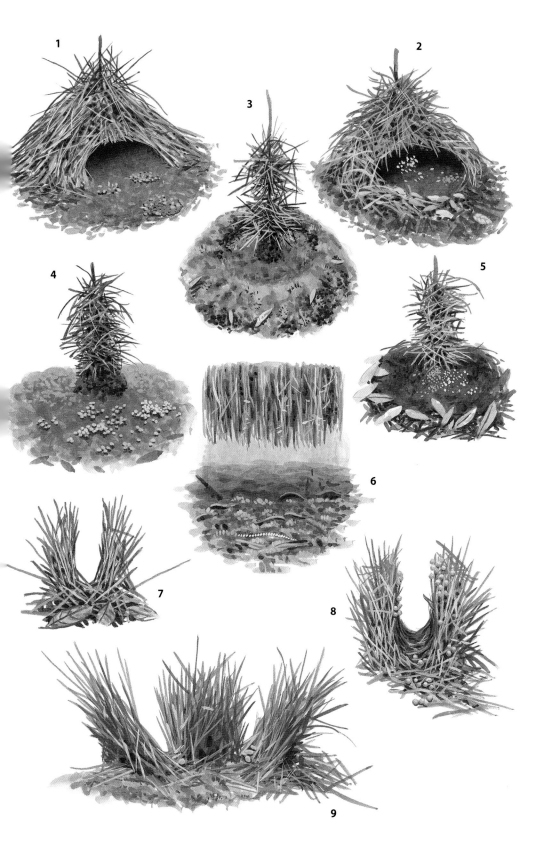

PLATE 68 FAIRYWRENS

1 White-shouldered Fairywren *Malurus alboscapulatus*   Pl. 105, p. 402
10–13 cm. Grasslands and low shrubby thickets. Common and widespread. Male jet black with white shoulder patch. Female color varies geographically. All differ from Male in white partial eye-ring or brow, and black Females have brownish wings. Juv similar to Imm, but duller. Imm more similar to Female, but all-black subspp have some white on chin and abdomen. **Subspp groups shown**:
alboscapulatus (NG, except for next)
Female black, locally with white below;
lorentzi (Trans-Fly, S Lowlands, southern W Ranges)
Female brown.

2 Emperor Fairywren *Malurus cyanocephalus*   p. 402
13–16 cm. Thickets in abandoned gardens and at lowland forest edge. Common. Male intense dark blue. Female boldly patterned with blue-black head. Juv similar to Female, but duller; head dusky black, not blue.

3 Wallace's Fairywren *Sipodotus wallacii*   p. 401
11–12 cm. Hill forest. Uncommon. Holds its body horizontal, tail not cocked. A tiny, uniquely patterned warbler-like bird with all-white underparts and wing bars. Throat white in Male, buff-white in Female. Juv duller; crown dark grey and spotted buff; wing bars buff.

4 Broad-billed Fairywren *Chenorhamphus grayi*   p. 401
12–14 cm. Fond of small patches of regrowth appearing after tree falls and landslides within primary forest of hills and nearby lowlands. Rare. Note soft blue color. Differs from Campbell's FW in Male's mottled, blue crown, and blue mantle in both sexes. Belly blue in Male, white in Female. Juv, brown feathering in the crown and underparts.

5 Campbell's Fairywren *Chenorhamphus campbelli*   p. 402
11–12 cm. Replaces Broad-billed FW in S Lowlands. Distinguished from that species by smaller size, crown black in both sexes, and mantle brown. Sexes alike.

6 Orange-crowned Fairywren *Clytomyias insignis*   p. 401
14–16 cm. Cloud forest thickets. Rare. Unique orange-rufous cap. Juv duller and darker, ochre-brown, yet with brownish-orange cap.

PLATE 69 HONEYEATERS—MYZOMELAS

1 Elfin Myzomela *Myzomela adolphinae* p. 405
9 cm. Mid-mountain forest canopy and highland towns.
Common. Smallest myzomela. Bill short, decurved.
Male red-headed. Female and Juv have pink chin, but not forehead.

2 Red-headed Myzomela *Myzomela erythrocephala* p. 405
11–12 cm. Mangroves of S NG only. Locally common.
Larger than Elfin M, but plumage similar. **Subspp shown**:
infuscata (S NG and Aru Is), darker than nominate race (SE Pen).

3 Red Myzomela *Myzomela cruentata* p. 404
10–11 cm. Hill and mid-mountain forest canopy, less frequent
in lowlands. Male uniquely all red. Female, look for red in tail.
Juv male similar, but may have reddish wash above.

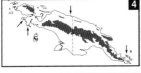

4 Red-collared Myzomela *Myzomela rosenbergii* p. 403
11–12 cm. Cloud forest. Male unique. Female mottled brown,
with black face and red breast and rump. Juv mottled brown
all over, lacking red and black. **Subspp shown**:
rosenbergii (NG mainland) Female brownish with black face;
longirostris (SE Is: Goodenough) longer bill; Female paler, greener,
red collar.

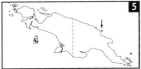

5 Sclater's Myzomela *Myzomela sclateri* p. 405
11–12 cm. Only on Karkar I, off Sepik-Ramu coast, where it is
the only myzomela. Mostly in mountains. Male red-throated.
Female and Juv have streaked throat, red chin-spot.

6 Papuan Black Myzomela *Myzomela nigrita* p. 404
12–13 cm. Hill and lower montane forest canopy; also Trans-Fly
woodland. Male black with white underwing-coverts (compare
with Black Sunbird, Plate 79). Female with reddish forehead and
throat, but not tail. In SE, many Females are all black, duller than
Male. Juv resembles grey-brown Female or is darker.
Subspp groups shown: *nigrita* (NG and Aru Is) Male all black;
forbesi (SE Is: D'Entrecasteaux) Male has red cap.

7 Ruby-throated Myzomela *Myzomela eques* p. 404
14–15 cm. Lowland and hill forest and edge.
Uncommon. Large and dark with red throat-stripe.
Male larger, more red on throat. Juv duller.
Subspp shown: *primitiva* (N NG).

8 White-chinned Myzomela *Myzomela albigula* p. 403
13–14 cm. The drab, grey-brown myzomela of select SE Is.
Common in all wooded habitats. Male larger, exhibits more
distinct, red throat-streak (hard to see). **Subspp shown**:
pallidior (small islands only) white throat present in both sexes;
albigula (Rossel I) darker and with white throat patch missing
in Male.

9 Dusky Myzomela *Myzomela obscura* p. 404
13–14 cm. The common, drab, coastal myzomela.
Male larger, more greyish. **Subspp shown**:
fumata (NG and Aru Is) grey-brown;
rubrobrunnea (Biak I) washed with red.

PLATE 70 HONEYEATERS—STRAIGHTBILLS AND STREAKED HONEYEATERS

1 Little Shrikethrush *Colluricincla m. megarhyncha* Pl. 85, p. 450
For comparison. More likely seen than Tawny Straightbill.
Note robust bill and dark leg color.

2 Tawny Straightbill *Timeliopsis griseigula* p. 414
18 cm. Rare in lowland forest. Met in mixed-species flocks of
other rusty brown birds. Climbs and searches dead leaves.
Note short, straight, pointed bill; bright iris; and pinkish grey legs.
Adult iris reddish. Male larger. Juv iris grey or dark brown.

3 Olive Straightbill *Timeliopsis fulvigula* p. 414
13–14 cm. Mid-mountain forest and lower cloud forest.
Uncommon. Scolding call. Warbler-like, with green wing edgings
and a sharp, straight bill. Adult eye orange; Juv's grey or brown.

4 Green-backed Honeyeater *Glycichaera fallax* p. 405
10–12 cm. Lowland and hill forest canopy and midstory, often
seen at forest edge. Uncommon. Warbler-like, with straight bill.
Petite and nondescript but for whitish iris and narrow pale
eye-ring. Male larger. Juv iris brown. **Subspp shown**:
pallida (NW Is: Batanta, Salawati, and Waigeo) paler and greyish;
fallax (rest of range) greener and underparts pale yellow.

5 Grey-streaked Honeyeater *Ptiloprora perstriata* p. 407
19–20 cm. Cloud forest and subalpine. A big, dark, boldly streaked
Ptiloprora with no rufous on back. Adult iris green. Male larger.
Juv with greenish wash below, iris brown.

6 Mayr's Streaked Honeyeater *Ptiloprora mayri* p. 407
19–20 cm. Restricted to several N coastal ranges: Bewani,
Cyclops, and Foja Mts. Inhabits cloud forest. Larger and
duller than Rufous-backed H.

7 Rufous-backed Honeyeater *Ptiloprora guisei* p. 407
16–18 cm. Cloud forest. Abundant. Rufous mantle diagnostic.
Adult has clear, grey streaking below; iris bright green.
Male larger. Juv duller, greenish tint on breast, dull iris.

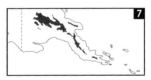

8 Rufous-sided Honeyeater *Ptiloprora erythropleura* p. 406
16–17 cm. Western *Ptiloprora* of mid-mountain and cloud forest.
Uncommon. Russet flanks, all-grey back, and reddish iris.
Some birds with green iris. Male larger.

9 Yellow-streaked Honeyeater *Ptiloprora meekiana* p. 406
16–17 cm. Patchily distributed and rare in cloud forest.
A medium-sized, yellowish-streaked honeyeater. Male larger.
Iris pale grey.

10 Leaden Honeyeater *Ptiloprora plumbea* p. 406
14–15 cm. Patchily distributed and rare. Mid-mountain forest.
A small grey bird with a thin, slightly curved bill, and a silver-grey
iris. Male larger.

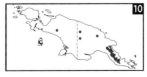

PLATE 71 HONEYEATERS—FRIARBIRDS AND OTHER LARGE, BROWN HONEYEATERS

1 Plain Honeyeater *Pycnopygius ixoides* p. 407
18 cm. Forest and edge from lowlands to mid-mountains. Locally common. Uniquely lacks any conclusive field characters. Note small head, small bill, and upright posture. Male larger. Compare with Tawny H. **Subspp groups shown**: *finschi* (SE Pen, east of Kumusi R in N and Port Moresby in S) variably rufous brown; *ixoides* (most of range, except that of previous race) brown.

2 Marbled Honeyeater *Pycnopygius cinereus* p. 408
22–23 cm. Mid-mountains, in oak forests and second growth. Uncommon. Short bill, unique white crescent below the eye, and white cheek-stripe. Adult has greenish edges to wing and tail feathers; in Juv these tinged brown. Male larger. **Subspp shown**: *cinereus* (Bird's Head to Weyland Mts of W Ranges) faint scaling below; *marmoratus* (most of range, exclusive of previous) pronounced scaling on breast.

3 Streak-headed Honeyeater *Pycnopygius stictocephalus* p. 408
20–22 cm. Lowland forest edge and gardens. Sings loudly from high, open perch. Uncommon. Smaller than a friarbird; much smaller bill. Adult shows diagnostic white cheek-stripe and speckled crown. Male larger. Juv shows only indistinct white cheek-stripe.

4 Long-billed Honeyeater *Melilestes megarhynchus* p. 413
22–23 cm. Forest interior from lowlands to mid-mountains. Common but furtive. Note long, curved bill. Adult has plain, dark underparts and orange eye. Male larger. Juv has paler, streaky breast (variable) and pale eye-ring.

5 Brass's Friarbird *Philemon brassi* p. 410
21–22 cm. NW Lowlands. Found locally along rivers and associated lagoons, in dense, second-growth trees and cane grass. Much paler below than Meyer's F. Whitish chin, black face, dark grey neck. Imm has white scaling on back and wings.

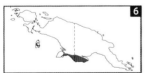

6 Little Friarbird *Philemon citreogularis* p. 410
24–25 cm. Savannah and monsoon forests of Trans-Fly. Common. Smallest of the 3 savannah friarbirds. Lacking bill knob. White upper throat, blue-grey skin on the face. Male larger. Imm with throat and sides of breast yellow-tinged.

7 Helmeted Friarbird *Philemon buceroides* p. 411
33–35 cm. A conspicuous bird of towns, gardens, savannah, and forest edge from the lowlands into the mid-mountains. Pugnacious habits and raucous calls. Adult plain grey-brown. Male larger. Imm upperparts pale-scalloped. **Subspp groups shown**: *jobiensis* (NW Lowlands and Yapen I to Huon) bill-knob small or missing; *novaeguineae* (remainder of range, mainly southern) prominent bill-knob.

8 Noisy Friarbird *Philemon corniculatus* p. 411
30 cm. Trans-Fly savannah. Common. Unique bald, black head and neck ruff. White tail tip. Male larger. Juv head partly feathered. Imm has white-scaled wings and back.

9 Meyer's Friarbird *Philemon meyeri* p. 410
21–22 cm. Lowland and hill forests and openings. Uncommon. Plain, dark, sooty brown. Note the long, pointed, slightly decurved bill. Juv upperparts scaled whitish, underparts tinged yellow. Imm like Adult but for bronzy yellow wash across upper breast.

PLATE 72 HONEYEATERS—OF GARDENS AND OPEN COUNTRY

1 Tawny-breasted Honeyeater *Xanthotis flaviventer*   p. 409
18–21 cm. Forests and second growth from lowlands to mid-mountains. Common. Distinctive shape—a medium-sized, robust honeyeater with medium-long bill. Bare eye and a small, flamelike ear-stripe. Male larger. **Subspp groups shown**: GREEN GROUP, *fusciventris* (NW Is) and race *flaviventer* (Bird's Head) intermediate; TAWNY GROUP, *visi* (Bird's Neck, S Lowlands, SE Pen) tawny brown, grey throat; DUSKY GROUP, *meyeri* (NW Lowlands to Sepik-Ramu) dark grey brown, dark eye-skin; SPOTTED GROUP, *spilogaster* (SE Is: D'Entrecasteaux) spotted below, spectacled.

2 Yellow-tinted Honeyeater *Ptilotula flavescens*   p. 419
12–13 cm. Only in SE. Eucalypt savannahs and nearby town gardens, where locally common. A small, pale, yellowish honeyeater with black-and-gold ear-stripe. Dark eye set against yellow face. Adult bill all black, Juv bill pale with dark tip. Male larger.

3 Rufous-banded Honeyeater *Conopophila albogularis*   p. 414
13 cm. Open habitats near coast and rivers, including towns, roadsides. Locally common. Small. Adult has pale rufous breast band, white throat, grey hood. Male larger. Juv, note whitish throat, may have an eye-ring.

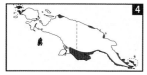

4 Brown-backed Honeyeater *Ramsayornis modestus*   p. 415
11 cm. Another coastal species, common near water in open habitats; in some towns. No other has white throat and dark barring or pinkish bill and legs. Juv streaky below.

5 Silver-eared Honeyeater *Lichmera alboauricularis*   p. 409
15–16 cm. Open coastal and riverine habitats. Gregarious and noisy. Like a small, grey-brown meliphaga with silvery ear-spot and dark-spotted pale breast. Male larger. Juv, obscure ventral spotting and ear-spot.

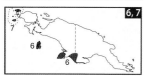

6 Brown Honeyeater *Lichmera indistincta*   p. 408
13–14 cm. Savannah and other open, wooded, lowland habitats near water. Locally common. Loud, cheery song. Small yellow spot behind eye and yellow-green edges of wings and tail. Gape line yellow, becoming black in breeding male. Male larger, with greyish crown. Juv head tinged yellowish, eye-spot absent. See Dusky Myzomela (plate 69).

7 Olive Honeyeater *Lichmera argentauris*   p. 408
14–15 cm. NW Is. Small, meliphaga-like. White-and-yellow ear-spot. Male larger.

8 White-throated Honeyeater *Melithreptus albogularis*   p. 412
12–13 cm. Savannah, in eucalyptus and paperbark trees. Common, noisy, gregarious. Unmistakable. Male larger.

9 Blue-faced Honeyeater *Entomyzon cyanotis*   p. 411
30 cm. Trans-Fly savannah. Common. Unmistakable. Wings flashing white while flying. Adult has bright blue face patch, Imm yellowish-green patch. Male larger.

PLATE 73 HONEYEATERS—VARIOUS GREEN, SMOKY, AND GIANT HONEYEATERS

1 Spotted Honeyeater *Xanthotis polygrammus* p. 409
16–17 cm. Canopy dweller in hill and lower montane forests; also monsoon forest and well-wooded savannah in Trans-Fly. Uncommon. Profusely dark-spotted below. Big, dark eye; colorful, bare eye-ring; yellow ear-streak (most races). Male larger. Juv duller.

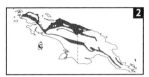

2 Obscure Honeyeater *Caligavis obscura* p. 415
18–19 cm. In hill forest and adjacent lowlands. Uncommon and unobtrusive, though vocal. Resembles a meliphaga, but the face is more colorful. Diagnostic bicolored ear-spot and yellow-washed throat. Male larger. Juv, brownish rump.

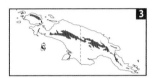

3 Black-throated Honeyeater *Caligavis subfrenata* p. 415
20–21 cm. Cloud forests and subalpine shrubbery. Uncommon, except in flowering trees and shrubs. Vocal and active. Looks like a miniature melidectes; however, the bicolored (yellow/white) ear-spot and orange legs are diagnostic. Male larger.

4 Orange-cheeked Honeyeater *Oreornis chrysogenys* p. 416
24–26 cm. Timberline forest and shrubbery on highest peaks of W Ranges. Locally common. Distinctive. A big, olive-green honeyeater with orange cheek patch and yellowish legs. Male larger.

5 Varied Honeyeater *Gavicalis versicolor* p. 418
19–22 cm. Strictly coastal. Strand forests, mangroves, small islands, and many seaside towns. Common but shy. A familiar, melodious voice. Adult has streaky underparts, black mask, and yellow-and-white ear-patch. Male larger. Juv paler, duller. **Subspp shown**: *sonoroides* (N coast of NG, NW Is, SE Is) green above, strong yellow wash below; *versicolor* (S coast) grey and green above, whitish below.

6 Spangled Honeyeater *Melipotes ater* p. 413
31–32 cm. Huon mountains. Common. Size of bird of paradise. Long tail. White spots on breast.

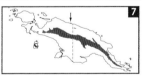

7 Common Smoky Honeyeater *Melipotes fumigatus* p. 413
21–22 cm. The widespread smoky honeyeater. Common in montane forest. Yellow facial skin can flush red. From other SHs by the grey chin and grey-edged breast feathers. Male larger. **Subspp shown**: *fumigatus* (all range except that of next race) typical for the species, size varies. Not shown is *kumawa* (Kumawa Mts) wings white-spotted, breast white-scalloped.

8 Wattled Smoky Honeyeater *Melipotes carolae* p. 413
21–22 cm. Foja Mts. Common. Pendent wattle below the eye and dark chin.

9 Western Smoky Honeyeater *Melipotes gymnops* p. 412
21–22 cm. Bird's Head. Montane forest, edge. Abundant. Black chin and white streaking below.

10 Giant Wattled Honeyeater *Macgregoria pulchra* p. 412
38–40 cm. Highest mountains, though patchily distributed. Inhabits alpine forest and shrubland at timberline. Unmistakable. Noisy wings aid detection. Adult iris red-brown. Male larger, with bigger wattle. Juv duller; iris brown. Subsp *pulchra* (SE Pen) shown.

PLATE 74 HONEYEATERS—MELIDECTES AND SOOTY HONEYEATERS

1 Ornate Melidectes *Melidectes torquatus* p. 417
22–23 cm. Mid-mountain forests, disturbed areas with trees, and highland towns. Common. A medium-sized, colorful melidectes with banded throat and breast. Yellow-orange eye-skin and buffy breast. Male larger. Juv duller. **Subspp groups shown:** *emilii* (SE Pen) small white throat patch, throat wattles large; *torquatus* (rest of range) large white throat patch, throat wattles small or absent.

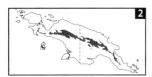

2 Sooty Honeyeater *Melionyx fuscus* p. 416
23–24 cm. High mountain cloud forest and shrubland. Common in the East, rare in the West. A slender, sooty black honeyeater with a white, tear-shaped eye wattle and thin black bill. Forages with tail cocked. Adult black. Male larger. Juv browner.

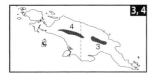

3 Long-bearded Honeyeater *Melionyx princeps* p. 416
26–27 cm. Cloud forest at timberline and shrubbery in alpine grassland. Common away from people. Unusually long, slim bill and conspicuous, wispy "beard." Beard white in Adult, yellowish in Juv. Male larger.

4 Short-bearded Honeyeater *Melionyx nouhuysi* p. 416
27–28 cm. The western counterpart of the Long-bearded H. Larger. Shorter bill and beard.

5 Vogelkop Melidectes *Melidectes leucostephes* p. 417
26 cm. The large, blackish melidectes of Bird's Head and Neck. Mid-mountain forest and edge. White-spotted breast and exaggerated white facial markings.

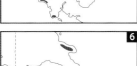

6 Huon Melidectes *Melidectes foersteri* p. 417
28–29 cm. The large melidectes of Huon mountains. Largest of all melidectes. Blue eye-skin, white brow and ear-stripe, and unique red gape wattles.

7 Cinnamon-browed Melidectes *Melidectes ochromelas* p. 417
24–25 cm. Occupies mid-mountain forest and edge. Locally common. A medium-size, dark melidectes with pale greenish-yellow orbital skin. Lacks a white throat-stripe, so the ruby-red throat wattles stand out alone against the dark throat. Male larger.

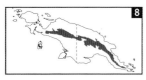

8 Belford's Melidectes *Melidectes belfordi* p. 418
27–28 cm. The cloud forest melidectes of the Central Ranges, living at mostly higher elevations than Yellow-browed M, but locally hybridizing with it. Common, noisy, and conspicuous. "Pure" Belford's distinguished by the blackish bill and legs, black forehead, white brow and ear-streak, blue eye-skin, and absence of throat wattles. Male larger. Juv similar to Adult but blacker, with reduced white facial markings.

9 Yellow-browed Melidectes *Melidectes rufocrissalis* p. 418
26–27 cm. The large, mid-mountain melidectes of the E and Border Ranges. Locally common. Originally a forest bird, but now most often seen in casuarina groves and other man-made habitats of intermontane valleys, below the forested ridges and peaks occupied by Belford's. Hybridizes with Belford's M, so identification often not possible. "Pure" Yellow-browed has pearly grey bill and legs; white forehead; yellow brow and ear-streak; extensive, pale, greenish-yellow facial skin; and presence of a large, pale gape wattle and a pair of red throat wattles.

PLATE 75 HONEYEATERS—THE MIMIC MELIPHAGAS

1 Elegant Meliphaga *Meliphaga cinereifrons* p. 420
14–16 cm. Lowland and hill forests, particularly second growth, edge, and other open wooded habitats and mangroves. Small. Dark sides of face; often blackish hind edge of yellow ear-spot. Ear-spot large and round, often with a projection reaching upward behind the eye. Underwing-coverts ochraceous. **Subsp shown is** *cinereifrons* (S watershed) shown.

2 Graceful Meliphaga *Meliphaga gracilis* p. 421
14–16 cm. Lowland forests of all types; predominantly in second growth, edge, mangroves, swamp forest, and savannah. Resembles Mimic M, but smaller with more slender bill. Underwing pale yellowish to white without ochraceous wash.

3 Mimic Meliphaga *Meliphaga analoga* p. 421
16–19 cm. The standard meliphaga—ordinary in plumage and shape, present nearly everywhere in lowland and hill forests of all sorts, easy to observe. Medium to large size; long slender bill. Greenish upperparts; unmottled underparts. Ear-spot yellow, large, and elongate. Sides of head not so dark. Shown are former subspp *flavida* (NW Lowlands, Sepik-Ramu) and *analoga* (rest of range).

4 Puff-backed Meliphaga *Meliphaga aruensis* p. 419
16–18 cm. Lowland and hill forest and edge; common in midstory and understory. Large, chunky, stout-billed. Diagnostic thick rump-tuft shows white fringe on the side, but usually hidden under the wing. Variation shown.

5 Mountain Meliphaga *Meliphaga orientalis* p. 421
14–16 cm. Mid-mountain forests, where usually the only meliphaga. Small, with a slender bill. Variable—best identified by habitat and size. **Subspp shown:** *facialis* (species range exclusive of next 2 races) less dusky face, reduced mottling; *orientalis* (eastern SE Pen) dark sides of face, coarse-textured below; *citreola* (N slopes of W Ranges to N Coastal Mts) face green, underparts yellower.

6 White-eared Meliphaga *Meliphaga montana* p. 420
16–17 cm. Hill forests of the North. Common in forest interior. Adult has unique combination of brown plumage and white ear-spot. Note mottling of underparts. Juv also brown, but ear-spot pale yellow. Extremes from Bird's Head and Bantana I shown.

7 Scrub Meliphaga *Meliphaga albonotata* p. 422
16–17 cm. A scrub and garden specialist. Numerous in mountains, less so in lowlands. Adult is the only green meliphaga with white ear-spot (rictal streak obscure); underparts lack mottling. Juv has pale yellow ear-spot.

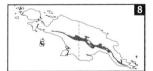

8 Mottled Meliphaga *Meliphaga mimikae* p. 420
16–18 cm. Southern, yellow-eared counterpart of White-eared M. Look for mottled breast, more heavily marked than any other meliphaga. Narrow, pale eye-ring.

9 Yellow-gaped Meliphaga *Meliphaga flavirictus* p. 420
15–16 cm. Least common of the meliphagas. Lowland and hill forest canopy. Prominent bright yellow rictal streak and a small, pale yellow ear-spot (colors reversed in the very similar Mimic M); face not dusky; diagnostic yellowish-olive legs in S race. A rather pale species with smooth-textured underparts. **Subspp shown**: *crockettorum* (W and N NG) darker and brighter yellow-green, legs greyish; *flavirictus* (Trans-Fly to SE Pen) paler, legs yellowish olive, ear-spot palest yellow.

10 Tagula Meliphaga *Meliphaga vicina* p. 422
17–18 cm. Tagula I (SE Is) endemic. Common in all forests and disturbed habitats. Large size; long, heavy bill. Closest to Mimic M, except has ochraceous underwing-coverts.

PLATE 76 SCRUBWRENS

1 Buff-faced Scrubwren *Sericornis perspicillatus* p. 425
10 cm. Montane forest. Common. Buff face set with a dark, beady eye and capped by a contrasting greyish crown.

2 Vogelkop Scrubwren *Sericornis rufescens* p. 425
9–10 cm. Endemic to the Bird's Head mts. Pronounced buff eye-ring.

3 New Guinea Thornbill *Acanthiza murina* p. 426
9–10 cm. Subalpine forest. Common. Diagnostic whitish iris and grizzled cheek. Dingy white underparts and short tail with a dark subterminal bar contrasting with pale tip.

4 Large Scrubwren *Sericornis nouhuysi* p. 425
12–14 cm. The large, common scrubwren of montane forest. Olive- or rufous-brown with a contrast between the upperparts and underparts, especially evident at the eye-line. Lacks dark subterminal tail-band. Compare with Papuan S. **Subspp shown**: *nouhuysi* (W Ranges) more rufous brown; *oorti* (Huon, SE Pen) more olive above and yellowish below; *cantans* (Bird's Head) similar in wing markings to Tropical S.

5 Tropical Scrubwren *Sericornis beccarii* p. 424
10–13 cm. Lowland and hill forest. Locally common. Usually more olive than Large S, and shows white tips to wing-coverts and usually a broken white eye-ring. The subspp show a confusing geographic pattern and appear to be the result of hybridization with Large S. **Subspp groups shown**:
"Perplexing group 1" tends toward Large S (shown is *virgatus* of N Coastal Range);
"Perplexing group 2" tends toward typical Tropical (shown are *idenburgi* of N slope of W Ranges and *wondiwoi* of Wandammen Mts);
"Typical Tropical" olive and brightly patterned (shown is *randi* of S Lowlands).

6 Papuan Scrubwren *Sericornis papuensis* p. 426
10–11 cm. Cloud forest, mainly at higher elevations than Buff-faced S. Common. From Large S: smaller, with buffy orange eye-ring and forehead band. Unique pale spot at the base of the primaries. Dark subterminal tail-band. Juv olive green. **Subspp shown**: *papuensis* (SE Pen) olive brown (*meeki* of W Ranges is similar); *buergersi* (E Ranges) brown.

7 Grey-green Scrubwren *Sericornis arfakianus* p. 426
9–10 cm. Lower montane forest. Elevational range narrow, fits between Pale-billed S below and Buff-faced S above. Common. A small, grey-green scrubwren. Juv slightly more brownish. Shown is variation in the degree of streaking below.

8 Pale-billed Scrubwren *Sericornis spilodera* p. 424
10–12 cm. Hill forest and locally in lowlands. Common. Pale bill diagnostic. Usually greenish with dark-spotted pale throat (spots lacking in Juv). Clinal variation. **Subspp shown**: *spilodera* (N NG from Bird's Head to Sepik-Ramu) dark and strongly marked; *guttatus* (SE Pen) green crown; *aruensis* (Aru Is) very pale, without spotting. Not shown: *ferrugineus* (NW Is, rufous crown) and *granti* (S Lowlands, green crown).

PLATE 77 GERYGONES

1 Grey Thornbill *Acanthiza cinerea* p. 426
9 cm. Cloud forest. Uncommon. All plumages grey above, white below. Iris brown.

2 Large-billed Gerygone *Gerygone magnirostris* p. 428
10 cm. Lowland forests along streams, rivers, and swamps in mainland NG; in forest away from water on islands. Common. Rather plain. Note broken eye-ring and sometimes a loral spot; red iris. Juv throat yellowish; brown iris. **Subspp groups shown**:
magnirostris (all NG Region except next) underparts variable whitish to pale buffy;
hypoxantha (NW Is: Biak) underparts washed pale yellow.

3 Brown-breasted Gerygone *Gerygone ruficollis* p. 428
9–10 cm. Mountains. Common in highland villages and second growth with tall trees; less common in forest. Adult brown above, whitish below; white band in outer tail feathers; silver-edged tertials; red iris. Juv with yellowish underparts, brown iris.

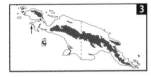

4 Yellow-bellied Gerygone *Gerygone chrysogaster* p. 427
10 cm. Lowland forest interior. Common. Adult with two-toned underparts: white breast, yellow flanks. Contrasting brown tail. Juv ear-coverts greyer. **Subspp groups shown**:
chrysogaster (most of range except next)
upperparts olive, breast white, bill dark;
notata (Bird's Head; NW Is of Salawati and Misool)
upperparts greener, bill pale;
neglecta (NW Is: Waigeo and Batanta)
breast and belly with yellow wash, bill pale.

5 Mangrove Gerygone *Gerygone levigaster* p. 428
9–10 cm. Mangrove specialist of S NG; uncommon.
Adult with white eyebrow and cheeks. Juv has a yellowish eyebrow and throat; iris brown.

6 Green-backed Gerygone *Gerygone chloronota* Pl. 105, p. 427
8–9 cm. Lowland and hill forest canopy and edge. Locally common. Often heard, rarely seen. Tiny, short-tailed, stubby. Adult with distinctive 3-colored pattern: white underparts, grey head, and green back. No eye-ring. Juv duller, has broken eye-ring.

7 Fairy Gerygone *Gerygone palpebrosa* p. 428
10 cm. Common in forest of foothills and lower mountains, less so lowlands. Male, unique. Female green above, yellow below with white throat and loral spot. Juv similar but underparts all yellowish. **Subspp shown**:
wahnesi (N NG, from NW Lowlands to SE Pen),
Male has black crown, nape.
palpebrosa (S NG, from NW Is to SE Pen),
Male has green crown, nape.

8 White-throated Gerygone *Gerygone olivacea* p. 427
9–10 cm. Savannahs of Port Moresby region.
Adult brown above, yellow below with white throat and loral spot.
Juv entirely pale yellow below; note pale eyebrow.

9 Island Leaf-Warbler *Phylloscopus poliocephalus* Pl. 105, p. 497
9–10 cm. For comparison. Strong eyebrow and crown-stripes.

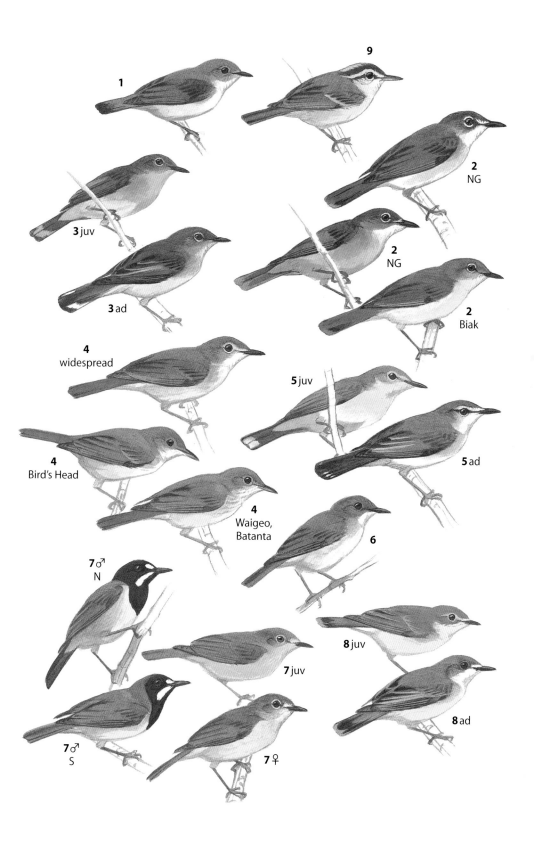

PLATE 78 BERRYPECKERS AND PAINTED BERRYPECKERS

1 Spotted Berrypecker *Rhamphocharis crassirostris* p. 433
11–12 cm. Rare in mid-mountain forest. Bill long and straight.
Female spotted. **Subspp groups shown**:
crassirostris (Bird's Head to Border Ranges)
much shorter bill; dark ventral spotting;
piperata (E Ranges to SE Pen and Huon)
longer bill; white ventral spotting.

2 Black Berrypecker *Melanocharis nigra* p. 432
11–12 cm. The common berrypecker of lowland and hill forest.
Note white pectoral tuft (yellow in N Huon). Female and Juv green
with white pectoral tuft. **Subspp groups shown**:
nigra (NW Islands, Bird's Head and Neck)
Male black above, grey below;
chloroptera (S Lowlands and Aru Is)
Male similar, but has green wing;
unicolor (NW Lowlands to SE Pen)
Male all black.

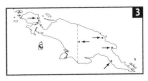

3 Obscure Berrypecker *Melanocharis arfakiana* p. 432
11–12 cm. Rare in hill forest. Both sexes similar to female
Black BP. Identify by prominent orange gape and tan-horn
bill color. Pectoral tufts lemon yellow. Head grey.

4 Mid-mountain Berrypecker *Melanocharis longicauda* p. 433
13 cm. The common berrypecker of mid-mountain forest.
Tail longer than that of Black Berrypecker. Note yellow pectoral
tuft and yellow wash on belly. Male black above. Female and Juv
green above.

5 Streaked Berrypecker *Melanocharis striativentris* p. 433
13–14 cm. Usually rare in mid-mountain forest. All plumages
show streaky breast, broad buff eye-ring, and orange gape.
Female and Juv with longer tail than Male.

6 Fan-tailed Berrypecker *Melanocharis versteri* p. 433
14–15 cm. Common in cloud forest. Note long, graduated tail.
Male metallic blue-black above, grey below. Female and Juv
green above with white pectoral tuft and yellow gape.

7 Tit Berrypecker *Oreocharis arfaki* p. 435
12–14 cm. Montane forests. Male unique. Female and Juv have
a diagnostic combination of green upperparts, grey throat, and
dark-scalloped yellowish flanks. Note yellow wing spots.

8 Eastern Crested Berrypecker *Paramythia montium* p. 435
20–22 cm. High cloud forest and subalpine shrubland.
Adult with distinctive black mask, drooping crest, and long
white eyebrow with greyish cast. Blue and yellow underparts.
Juv duller with obscured pattern and blue restricted to side of neck.
Nominate subspecies shown.

9 Western Crested Berrypecker *Paramythia olivacea* p. 435
20–22 cm. Differs by upcurved crest; short, all-white eyebrow;
and olive (vs bright green of Eastern CB) back.

PLATE 79 SUNBIRDS, LONGBILLS, AND FLOWERPECKERS

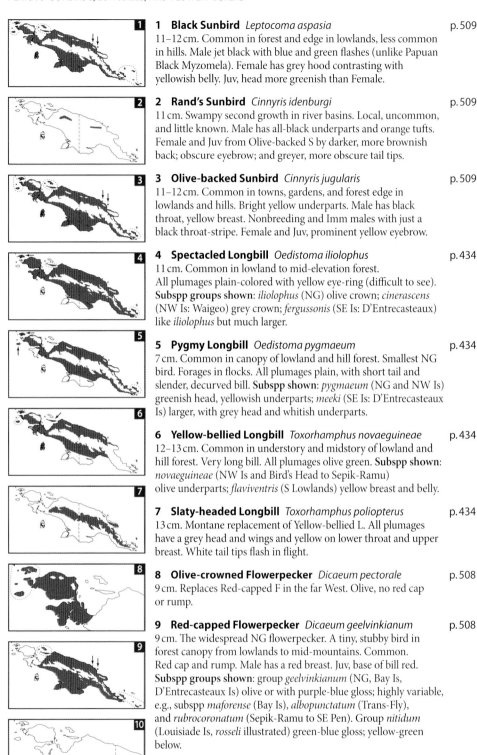

1 Black Sunbird *Leptocoma aspasia* p. 509
11–12 cm. Common in forest and edge in lowlands, less common in hills. Male jet black with blue and green flashes (unlike Papuan Black Myzomela). Female has grey hood contrasting with yellowish belly. Juv, head more greenish than Female.

2 Rand's Sunbird *Cinnyris idenburgi* p. 509
11 cm. Swampy second growth in river basins. Local, uncommon, and little known. Male has all-black underparts and orange tufts. Female and Juv from Olive-backed S by darker, more brownish back; obscure eyebrow; and greyer, more obscure tail tips.

3 Olive-backed Sunbird *Cinnyris jugularis* p. 509
11–12 cm. Common in towns, gardens, and forest edge in lowlands and hills. Bright yellow underparts. Male has black throat, yellow breast. Nonbreeding and Imm males with just a black throat-stripe. Female and Juv, prominent yellow eyebrow.

4 Spectacled Longbill *Oedistoma iliolophus* p. 434
11 cm. Common in lowland to mid-elevation forest. All plumages plain-colored with yellow eye-ring (difficult to see). **Subspp groups shown**: *iliolophus* (NG) olive crown; *cinerascens* (NW Is: Waigeo) grey crown; *fergussonis* (SE Is: D'Entrecasteaux) like *iliolophus* but much larger.

5 Pygmy Longbill *Oedistoma pygmaeum* p. 434
7 cm. Common in canopy of lowland and hill forest. Smallest NG bird. Forages in flocks. All plumages plain, with short tail and slender, decurved bill. **Subspp shown**: *pygmaeum* (NG and NW Is) greenish head, yellowish underparts; *meeki* (SE Is: D'Entrecasteaux Is) larger, with grey head and whitish underparts.

6 Yellow-bellied Longbill *Toxorhamphus novaeguineae* p. 434
12–13 cm. Common in understory and midstory of lowland and hill forest. Very long bill. All plumages olive green. **Subspp shown**: *novaeguineae* (NW Is and Bird's Head to Sepik-Ramu) olive underparts; *flaviventris* (S Lowlands) yellow breast and belly.

7 Slaty-headed Longbill *Toxorhamphus poliopterus* p. 434
13 cm. Montane replacement of Yellow-bellied L. All plumages have a grey head and wings and yellow on lower throat and upper breast. White tail tips flash in flight.

8 Olive-crowned Flowerpecker *Dicaeum pectorale* p. 508
9 cm. Replaces Red-capped F in the far West. Olive, no red cap or rump.

9 Red-capped Flowerpecker *Dicaeum geelvinkianum* p. 508
9 cm. The widespread NG flowerpecker. A tiny, stubby bird in forest canopy from lowlands to mid-mountains. Common. Red cap and rump. Male has a red breast. Juv, base of bill red. **Subspp groups shown**: group *geelvinkianum* (NG, Bay Is, D'Entrecasteaux Is) olive or with purple-blue gloss; highly variable, e.g., subspp *maforense* (Bay Is), *albopunctatum* (Trans-Fly), and *rubrocoronatum* (Sepik-Ramu to SE Pen). Group *nitidum* (Louisiade Is, *rosseli* illustrated) green-blue gloss; yellow-green below.

10 Mistletoebird *Dicaeum hirundinaceum* p. 509
9 cm. Only flowerpecker on Aru Is. Adults have a red vent.

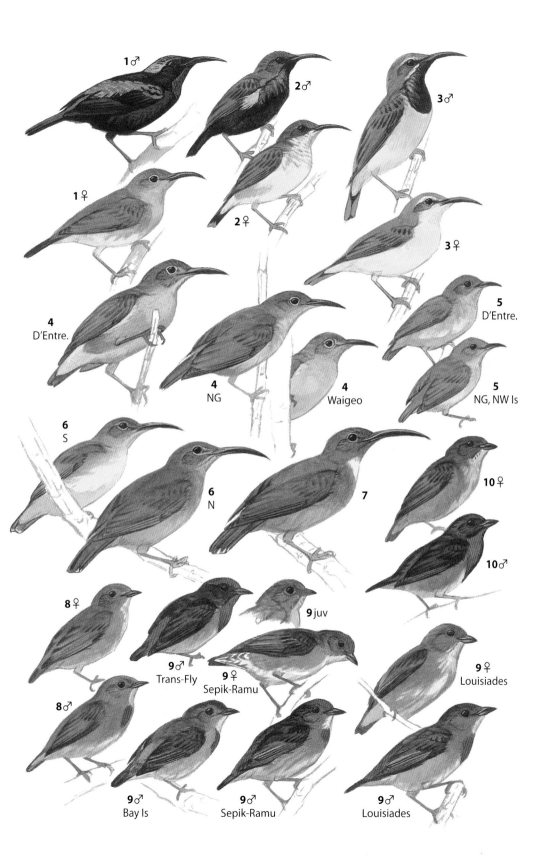

PLATE 80 BUTCHERBIRDS AND ALLIES

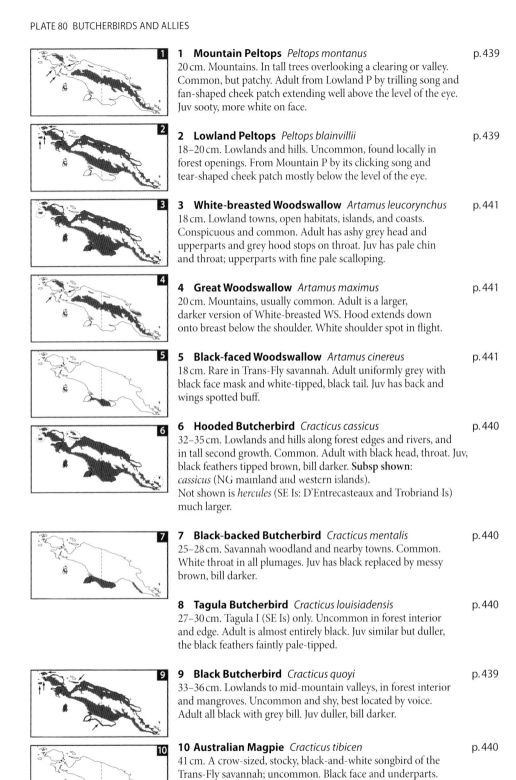

1 Mountain Peltops *Peltops montanus* p. 439
20 cm. Mountains. In tall trees overlooking a clearing or valley. Common, but patchy. Adult from Lowland P by trilling song and fan-shaped cheek patch extending well above the level of the eye. Juv sooty, more white on face.

2 Lowland Peltops *Peltops blainvillii* p. 439
18–20 cm. Lowlands and hills. Uncommon, found locally in forest openings. From Mountain P by its clicking song and tear-shaped cheek patch mostly below the level of the eye.

3 White-breasted Woodswallow *Artamus leucorynchus* p. 441
18 cm. Lowland towns, open habitats, islands, and coasts. Conspicuous and common. Adult has ashy grey head and upperparts and grey hood stops on throat. Juv has pale chin and throat; upperparts with fine pale scalloping.

4 Great Woodswallow *Artamus maximus* p. 441
20 cm. Mountains, usually common. Adult is a larger, darker version of White-breasted WS. Hood extends down onto breast below the shoulder. White shoulder spot in flight.

5 Black-faced Woodswallow *Artamus cinereus* p. 441
18 cm. Rare in Trans-Fly savannah. Adult uniformly grey with black face mask and white-tipped, black tail. Juv has back and wings spotted buff.

6 Hooded Butcherbird *Cracticus cassicus* p. 440
32–35 cm. Lowlands and hills along forest edges and rivers, and in tall second growth. Common. Adult with black head, throat. Juv, black feathers tipped brown, bill darker. **Subsp shown**: *cassicus* (NG mainland and western islands).
Not shown is *hercules* (SE Is: D'Entrecasteaux and Trobriand Is) much larger.

7 Black-backed Butcherbird *Cracticus mentalis* p. 440
25–28 cm. Savannah woodland and nearby towns. Common. White throat in all plumages. Juv has black replaced by messy brown, bill darker.

8 Tagula Butcherbird *Cracticus louisiadensis* p. 440
27–30 cm. Tagula I (SE Is) only. Uncommon in forest interior and edge. Adult is almost entirely black. Juv similar but duller, the black feathers faintly pale-tipped.

9 Black Butcherbird *Cracticus quoyi* p. 439
33–36 cm. Lowlands to mid-mountain valleys, in forest interior and mangroves. Uncommon and shy, best located by voice. Adult all black with grey bill. Juv duller, bill darker.

10 Australian Magpie *Cracticus tibicen* p. 440
41 cm. A crow-sized, stocky, black-and-white songbird of the Trans-Fly savannah; uncommon. Black face and underparts. Male with white back. Female and Imm black above with white nape and rump. Juv like Female, but greyish.

PLATE 81 CUCKOOSHRIKES AND TRILLERS

1 Black-browed Triller *Lalage atrovirens* p. 444
18–19 cm. Forests of the North, in lowlands and hills. Common.
No white eyebrow. Vent white. Male glossy black, plain white
below. Female and Juv duller above and barred below.

2 Varied Triller *Lalage leucomela* p. 445
18 cm. Forests of the South and Southeast, common in lowlands
and hills. White eyebrow. Vent buff or cinnamon. Rump barred.
Male boldly patterned, usually barred below. Female duller;
underparts all barred. Juv pale-mottled above. **Subsp shown** is
polygrammica (NG except Trans-Fly; also Aru Is).
See text for other races.

3 Biak Triller *Lalage leucoptera* p. 445
18–19 cm. Biak I. Forests and edge.
Large white wing patch, and Female not barred below.

4 White-winged Triller *Lalage tricolor* p. 444
16–18 cm. Savannah and open habitats. Patchy and scarce.
Breeding male has large white patch across wing, grey rump.
Nonbreeding male like Female, but primaries and tail black,
rump grey. Female drab; pale eyebrow, pale edges to wing-coverts
and secondaries, and greyish rump. Juv drab, obscurely mottled
and scalloped.

5 Golden Cuckooshrike *Campochaera sloetii* p. 444
20 cm. Foothills and nearby lowlands. Locally common.
Orange and black. Male, black face. Female, grey face.
Juv yellow face and greenish central tail. **Subsp shown**:
flaviceps (S NG: S Lowlands to SE Pen) crown olive.
Not shown is *sloetii* (N NG: Bird's Head to Sepik) crown grey.

6 White-bellied Cuckooshrike *Coracina papuensis* p. 443
25–26 cm. Midsize cuckooshrike of open habitats and towns
from lowlands to highland valleys. Common and conspicuous.
Adult has thick, black lores with crisp border. Juv has lores less
distinct, breast obscurely scalloped. **Subspp groups shown**:
papuensis (N NG: Bird's Head to Huon) evenly grey;
hypoleuca (S NG: S Lowlands to SE Pen, and Aru Is)
throat and breast white.

7 Stout-billed Cuckooshrike *Coracina caeruleogrisea* p. 442
33–37 cm. Large cuckooshrike in lowland to mid-mountain forests.
Long, stout bill; long tail. Male black lores. Female all-grey face.
Juv partly white or just white-tipped tail.

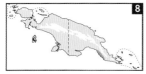

8 Black-faced Cuckooshrike *Coracina novaehollandiae* p. 443
32–35 cm. Large, elongate cuckooshrike of open country and
towns. Adult's black throat grades into grey breast. Juv, small
black mask. **Subsp shown**: *melanops* (breeds Port Moresby area,
also austral winter migrant from AU).

9 Hooded Cuckooshrike *Coracina longicauda* p. 442
33 cm. Large, long-tailed cuckooshrike of cloud forests.
Uncommon. In family parties. Male, black hood. Female and Imm,
black face. Juv like Female but with messy, white feather edges.

202

PLATE 82 CUCKOOSHRIKES AND CICADABIRDS

1 Barred Cuckooshrike *Coracina lineata* p. 442
23 cm. Hill forest and edge; uncommon. Fig specialist.
Note pale eye. Male all grey in common race. Female of all races
barred below. Juv like Female but with obscure barring on chin
and throat and with white-tipped tail feathers. **Subspp shown**:
axillaris (NG range, exclusive of next two) Male all grey, not barred;
maforensis (Numfor I) Male darkly barred below. Not shown is
lineata (rare in Trans-Fly, also AU) larger, both sexes barred.

2 Boyer's Cuckooshrike *Coracina boyeri* p. 443
22 cm. Lowland forest canopy. Common. Small size, short bill.
Pale crown. Male, black lores and chin. Female, pale grey lores.
Juv like Female but marked. **Subspp shown**: *boyeri* (N NG: NW Is to
SE Pen) underwing rufous, Female with pale lores; *subalaris* (S NG:
S Lowlands to SE Pen) underwing buff, Female with plain grey face.

3 Common Cicadabird *Edolisoma tenuirostre* p. 446
24–27 cm. Lowlands and hill. Forest edge, second growth,
savannah; also forest on islands. Uncommon. Compare with
the following 2 spp. Common CB larger, more elongate; wing
with strongly contrasting silver-grey (not medium grey) edges
on wing-coverts and secondaries. Female and Imm drab brown,
barred below; black eye-stripe. Juv, brown above; whitish below,
streaked brown. **Subspp groups shown**: *tenuirostre* (AU migrant)
widespread in winter (resident *aruense* of S NG smaller);
rostratum (Rossel I); *meyerii* (Biak I); *numforanum* (Numfor I).

4 Papuan Cicadabird *Edolisoma incertum* p. 445
21–22 cm. Hill and lower montane forest and edge. Common.
Male more compact than Common CB, often with darker throat,
and wing with poorly contrasting dark grey edges on coverts and
secondaries. Female like Male but paler and black only in lores.
Juv like Female, but obscurely pale-marked.

5 Grey-headed Cicadabird *Edolisoma schisticeps* p. 446
21–22 cm. Hills and nearby lowlands. Locally common in forest.
Male wing unpatterned grey. Female and Juv rufous, lacking pale
eyebrow of female Black CB. **Subspp shown**: *schisticeps* (NW Is,
Bird's Head and Neck) Female with grey crown; *vittatum* (SE Is:
D'Entrecasteaux Is) Female underparts barred; *reichenowi* (NW
Lowlands and Sepik-Ramu) Female with all-brown head.
Not shown is *poliopsa* (S Lowlands and SE Pen) Female with
all-grey head.

6 Black-bellied Cicadabird *Edolisoma montanum* p. 445
24 cm. Montane forest. Common. Male grey above, black below.
Female and Imm, all-black tail and flight-feathers.
Juv, white-edged secondaries and tail tip.

7 Black Cicadabird *Edolisoma melas* p. 447
23 cm. Lowland and hill forest. Common. Forages below the
canopy in mixed-species flocks. Male black. Female and Imm
brown or rufous with pale eyebrow.

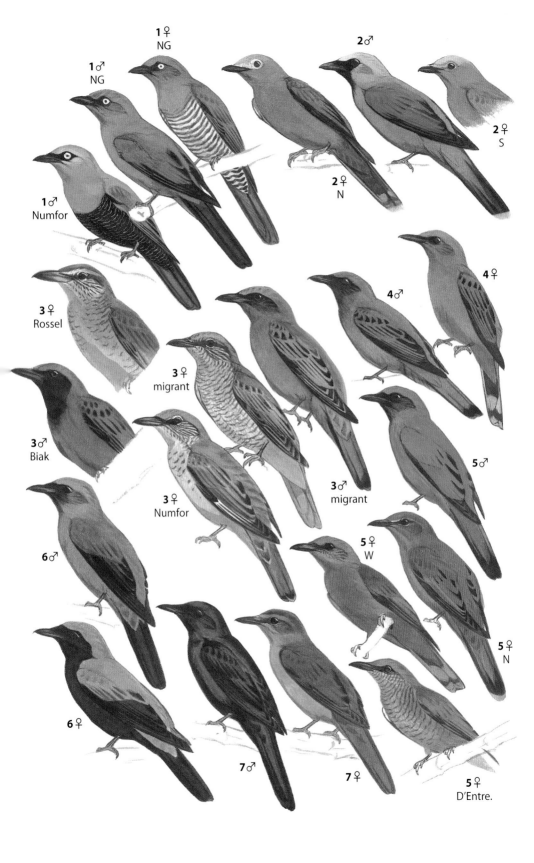

PLATE 83 BIRDS THAT CLIMB AND MISCELLANEOUS MONTANE WHISTLERS

1 Black Sittella *Daphoenositta miranda* p. 448
12 cm. Rare in cloud forest of highest mountains. In flocks.
Adult black; pink face, tail tip. Male iris, legs dark; yellow in
Female. Juv blackish with rusty spotting and face.

2 Varied Sittella *Daphoenositta chrysoptera* p. 447
11–12 cm. Mid-montane and cloud forest. Uncommon.
Twittering flocks forage on branches in treetops. Small, compact,
streaked; white rump. Male, see Subspp. Female's head white or
grey, plain or streaked. Juv, white-speckled. **Subspp groups shown**:
papuensis (Bird's Head) top of Male's head black;
alba (northern W Ranges) Male's head white;
toxopeusi (southern W Ranges and E Ranges) Male's head streaked.
Not shown is *albifrons* (SE Pen); similar to *toxopeusi*, but top of
head darker.

3 Papuan Treecreeper *Cormobates placens* p. 400
14–15 cm. Montane forest. Locally common. Creeps up tree
trunks. Female has rufous moustachial stripe, more scaly
markings on the belly. Juv breast and flanks indistinctly mottled.
In flight, pale buff wing-stripe. **Subspp groups shown**:
placens (Bird's Head to western E Ranges) greyish olive,
Female and Juv shown;
meridionalis (SE Pen) brownish olive, Male shown.

4 Blue-capped Ifrit *Ifrita kowaldi* p. 466
16–17 cm. Cloud forest. Common. Chunky bird with
creeping habits. Blue-and-black crown and pale eye-stripe.
Male's eye-stripe white, Female's buff. Juv forecrown brown.

5 Wattled Ploughbill *Eulacestoma nigropectus* p. 449
13–14 cm. Cloud forest, especially in climbing bamboo.
Uncommon. Unique bill. Male has black breast and wings,
pink gape wattles. Female all olive green, lacks wattles. Juv grey-
green with chestnut. Imm like Female but with chestnut wings.

6 Rufous-naped Bellbird *Aleadryas rufinucha* p. 448
16–18 cm. Mid-mountain and cloud forest understory. Common.
Adult unique. Juv shows much green above, chestnut below.
Imm like Adult but duller. **Subsp group shown**: *gamblei* (all range
excluding next) white forehead, rufous nape patch large. Not
shown is *rufinucha* (Bird's Head), grey forehead, small nape patch.

7 Sooty Shrikethrush *Colluricincla tenebrosa* p. 451
18–19 cm. Rare and local in mid-mountain forest.
Very poorly known. Shy, elusive. A robust, dark bird, entirely
brownish black. Shape and size similar to Little S, but tail longer.

8 Black Pitohui *Melanorectes nigrescens* Pl. 66, p. 452
22–23 cm. Mid-mountain and cloud forest. Uncommon.
Note large, strongly-hooked, black bill. Male black. Female brown;
compare with bowerbirds in plate 66. **Subspp shown**:
nigrescens (Bird's Head) Female with greyish cap;
schistaceus (rest of range) Female uniform rufous- or grey-brown.

PLATE 84 PITOHUIS AND GREATER MELAMPITTA

1 Raja Ampat Pitohui *Pitohui cerviniventris* p. 457
20–21 cm. Waigeo and Batanta Is (NW Is) only. A small, insular pitohui. All plumages unique, with pale brownish-grey head and upperparts; underparts tan-brown.

2 Northern Variable Pitohui *Pitohui kirhocephalus* p. 456
20–25 cm. Lowland and hill forest, most common in edge and second growth. Adult and Juv alike. From Southern Variable P either by grey head, wings, and tail or by entirely rusty brown plumage; however, 1 subsp is black and chestnut. Eye usually dark; bill grey-brown. **Subspp groups shown**: *kirhocephalus* (Bird's Head and Neck) grey head, wings, tail (illustrated are subspp *kirhocephalus* of NE Bird's Head and *decipiens* of Bird's Neck: Onin Pen); *jobiensis* (NW Lowlands and Yapen I) all rufous brown. Not shown are *dohertyi* (Bird's Neck: Wandammen Pen) black head, wings, tail and *brunneicaudus* (Sepik-Ramu), like *kirhocephalus* but underparts paler.

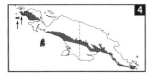

3 Piping Bellbird *Ornorectes cristatus* p. 449
24–25 cm. Foothills and nearby lowlands. Locally common in forest interior where it lives on the ground and in understory. Extremely shy. Amazingly protracted, monotonous song is key to discovery. Adult is a bushy-crested, brown bird. Some have blackish face. Juv, rufous edges to wing-coverts.

4 Southern Variable Pitohui *Pitohui uropygialis* p. 457
22–25 cm. Southern counterpart of Northern Variable P. Range divided by Northern Variable at Bird's Neck. Usually separable from that species either by black head, wings, and tail, or by these parts dusky brown, and black bill. (Black-and-chestnut race of Northern does not co-occur.) **Subspp groups shown**: *aruensis* (western S Lowlands, Aru Is) black with chestnut back, flanks, belly; *brunneiceps* (eastern S Lowlands) dusky brown head, wings, and tail, brown body; *uropygialis* (NW Is, W Bird's Head) and *meridionalis* (southern SE Pen) resemble Hooded P but for black rump.

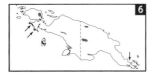

5 Hooded Pitohui *Pitohui dichrous* p. 457
22–23 cm. Common in mid-mountain forests, second growth, gardens; understory to canopy. Adult has striking pattern of orange-brown and black. Note brown uppertail-coverts, to separate from black-and-brown variable pitohuis (they have black coverts, except race *dohertyi* of Northern Variable). Juv, brownish edges to feathers of throat.

6 Greater Melampitta *Megalampitta gigantea* p. 472
29 cm. Associated primarily with hilly, limestone terrain and sinkholes, in which the bird shelters. Local, rare. Shy and difficult to see; best located by its loud, distinctive song. The only large, dark, terrestrial songbird. Adult is all black. Juv and Imm resemble a pitohui with black head and brown body, but note stockier shape, long legs, and shorter tail.

PLATE 85 SHRIKETHRUSHES, PITOHUIS, AND AUSTRAL BABBLERS

1 Little Shrikethrush *Colluricincla megarhyncha* Pl. 70, p. 450
17–19 cm. Widespread, common in forests and scrub of all sorts, from lowlands to mid-mountains. Actually a complex of species, some identical in appearance, see text. A stocky, medium-sized songbird modestly colored in shades of brown or olive-green. Juv has rufous edges to wing-coverts and secondaries.
Subspp (9 groups; 5 illustrated, for more see text):
tappenbecki (Sepik-Ramu, N Huon) like *despecta*, grading with *obscura* in West;
megarhyncha (Bird's Head to S Lowlands) rufous (W), like *despecta* (E);
despecta (SE Pen) above olive-brown, below pale rufous, throat mottled darker;
obscura (NW Lowlands) above olive or brown-grey, below greyish, bill black;
fortis (D'Entrecasteaux and Trobriand Is) greenish olive with grey head, breast.

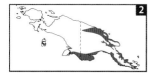

2 Grey Shrikethrush *Colluricincla harmonica* p. 451
23 cm. Savannahs, towns, and open lowland country, following human settlement into the highlands. Known for its powerful and musical song. Grey with white face and belly; note contrasting dark eye. Male has dark bill, immaculate underparts. Female bill greyish, underparts obscurely streaked. Juv underparts more heavily streaked, rufous wash on wings.

3 White-bellied Pitohui *Pseudorectes incertus* p. 451
22–23 cm. Riverine floodplain forest in high-rainfall lowland areas. In shy, noisy groups joined by other species. Adult is pale tan; dark beady eye prominent against the pale face. Pale bill. Juv, no information.

4 Rusty Pitohui *Pseudorectes ferrugineus* p. 451
25–28 cm. Lowland and hill forest. Common. In small groups, often with mixed-species flocks. Adult is large, rusty brown with white eye. Juv eye dark. **Subspp groups shown:**
leucorhynchus (NW Is: Batanta, Waigeo) yellowish bill, uniform dark rusty;
clarus (eastern S Lowlands, Huon, SE Pen) black bill, pale ochre breast;
ferrugineus (W and N NG) black bill, plumage uniform rusty.

5 Grey-crowned Babbler *Pomatostomus temporalis* p. 429
24 cm. Locally common in Trans-Fly woodland. In noisy flocks. A dark bird with whitish head and sooty eye-stripe. Cocks black tail with white tip. Adult has pale iris. Juv has dark iris; wing-coverts edged rufous.

6 Papuan Babbler *Garritornis isidorei* p. 429
25 cm. Common in forest tangles in lowlands and hills, though may be locally absent. Always found in noisy flocks and often in mixed-species flocks. A rufous brown bird with curved, yellow-orange bill. Yellow-eyed birds are Adult. Some or all dark-eyed birds are Juv.

PLATE 86 GOLDEN WHISTLERS AND GOLDENFACE

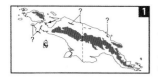

1 Goldenface *Pachycare flavogriseum* p. 422
13 cm. A yellow-faced warbler with a whistler-like song. Uncommon, reclusive in the canopy of hill and lower montane forest. Female, sooty ear patch. **Subspp shown**: *subaurantium* (W, Border, and western E Ranges) orange-yellow; *lecroyae* (N Coastal Mts) blaze orange.
Not illustrated are yellow *flavogriseum* (Bird's Head and Neck) and *subpallidum* (eastern E Ranges, Huon, SE Pen).

2 Louisiade Whistler *Pachycephala collaris* p. 453
15–16 cm. Coastal forest on SE Is. Green-edged wing feathers and greenish tail in both sexes.

3 Mangrove Golden Whistler *Pachycephala melanura* p. 454
16–17 cm. Mangroves on mainland NG, forest on small SE Is. Wing margins silvery (not green). Male tail black (not green). Female has grey-tan breast band; tail partly black.

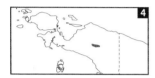

4 Baliem Whistler *Pachycephala balim* p. 453
15–16 cm. Montane forests of Baliem Valley. Green-edged wing. Male's tail black, Female's green. From co-occurring Sclater's W by Male's yellow collar; Female may not be separable. Juv rufous. Imm like Female, but retains brownish juv wing feathers.

5 Golden-backed Whistler *Pachycephala aurea* p. 454
15–16 cm. Second growth and tall cane grass bordering rivers and lakes in foothills and nearby lowlands. Spotty distribution. Adult has black-and-gold pattern. White throat patch obscure in some birds. Imm, olive wash.

6 Sclater's Whistler *Pachycephala soror* p. 453
15 cm. Mid-mountain forest. Common. Male lacks yellow hind-collar. Female lacks extensive grey on the head, has broad, olive-brown breast band. Juv mostly chestnut brown. Imm similar to Female but with some juv feathering. **Subsp group shown**: *soror* (all of range except Fakfak Mts) Male head black with thick black breast band.

7 Regent Whistler *Pachycephala schlegelii* p. 452
15–16 cm. Cloud forest and subalpine shrubs. Common. Male has broad black breast band, orange-brown breast, yellow nape patch. Female head grey, chin speckled pale grey and white, grey border above olive-green breast band. Juv with much rufous brown.

8 Lorentz's Whistler *Pachycephala lorentzi* p. 454
15–16 cm. Common in W cloud forest and subalpine scrub. Adult like female Regent W but lacks throat speckling and the olive-green breast band. Juv mostly rusty.

9 Vogelkop Whistler *Pachycephala meyeri* p. 454
14–15 cm. Bird's Head, plus sightings in Foja Mts. Forest and thickets. Small size, small bill. Adult with pale, buff-grey ear-coverts; same color patch on sides of breast that merges with pale yellowish abdomen. Juv, no information.

PLATE 87 PLAIN WHISTLERS

1 Black-headed Whistler *Pachycephala monacha* p. 456
15–16 cm. Widely scattered in hills and mid-mountain valleys, rarely in lowlands. Found in tall trees in open habitats. Hybridizes with White-bellied W. Male black with white lower breast and belly. Female dark grey and white. Juv rufous brown above, dingy white below. **Subsp shown**: *lugubris* (NG) upperparts sooty black.

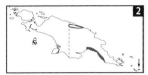

2 White-bellied Whistler *Pachycephala leucogastra* p. 456
14–15 cm. Scarce, local in open-wooded habitats in E NG; common in forest on Rossel I (SE Is). Male black-hooded with white throat, breast. Female grey with grey breast band. Juv has streaked breast and white eye-ring, washed buff in race *meeki*. Imm male resembles Female but often shows indistinct dark breast band. **Subspp shown**:
leucogastra (NG mainland)
paler; Male, side of breast white; Female, whitish breast;
meeki (SE Is: Rossel)
darker; Male, side of breast grey; Female, breast washed buff.

3 Grey Whistler *Pachycephala simplex* p. 455
14–15 cm. The only whistler in lowland forest and regrowth, where common. Extends up into lower mountains. Small, dainty. Adult plain; diagnostic pale eyebrow and pale throat. Juv largely rufous. **Subspp groups shown**:
simplex (SE Pen and D'Entrecasteaux Is)
brown with pale brownish breast;
griseiceps (rest of NG range)
olive with pale yellowish belly and flanks, greyish cap.

4 Brown-backed Whistler *Pachycephala modesta* p. 455
14 cm. Eastern cloud forest and subalpine scrub. Common. Small size. Adult has grey crown, white throat. Breast off-white (East) to pale grey (West). Juv rufous brown.

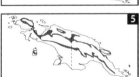

5 Rusty Whistler *Pachycephala hyperythra* p. 455
15 cm. Hill forest midstory. Locally common. Adult brown with tan or orange-tan underparts. Grey cap; white or pale grey throat. Juv rufous. **Subspp groups shown**:
hyperythra (Bird's Head to western S Lowlands)
rufous brown above, indistinct throat;
sepikiana (NW Lowlands to Sepik-Ramu)
olive brown above, orange-tan below;
salvadorii (eastern S Lowlands, Huon, SE Pen)
grey-brown above, tan below.

6 Island Whistler *Pachycephala phaionota* p. 455
16–17 cm. On tiny islets in Indonesia, in all wooded habitats. Adult brown with greyish head, white throat, and dull white underparts. Juv, top and sides of head olive brown.

7 Mottled Berryhunter *Rhagologus leucostigma* p. 448
15–16 cm. Montane forest interior. Elusive, but locally common. Locate by loud, slow-paced, whistled song. Female and Juv more ornate: dark olive with rusty face; streaked and spotted; Juv iris dark. Male drab; differs by subspecies. **Subspp shown**:
obscurus (Border Ranges to SE Pen)
Male greyish olive with faint white throat spotting;
leucostigma (Bird's Head to W Ranges) sexes alike but Male duller.

PLATE 88 MYIAGRA FLYCATCHERS

1 Shining Flycatcher *Myiagra alecto* p. 461
16–18 cm. Lowland thickets near water in swamp forest, mangroves, second growth. Common. Forages near the ground. Male shiny blue-black; note angular head. Female bright orange-rufous with black cap and white underparts (may be tinged buff). Juv, dull black cap. Imm male similar to Female but often with black sides of neck.

2 Biak Black Flycatcher *Myiagra atra* p. 460
13–14 cm. Biak and Numfor only. In all types of forest, though most common in hilly country. Small.
Male entirely blue-black. Female and Juv slate grey above, paler below.

3 Restless Flycatcher *Myiagra inquieta* p. 461
16–18 cm. Savannah, edge of gallery forest, and scrub bordering rivers, entering sedge beds at riverside. Common. Hovers near ground while hunting, uttering distinctive call. Adult glossy blue-black above, white below, including throat. Juv duller and shows white wing bars and streak across scapulars.

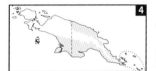

4 Satin Flycatcher *Myiagra cyanoleuca* p. 461
15 cm. Rare overwintering AU migrant to open country and forest edge, from lowlands to mid-mountain valleys. Male's black-white division on breast is concave, running under the wing at a sharp angle. Female from Leaden F female by darker upperparts with solidly dark wings and grey-brown undertail (not grey). Imm Male resembles Female but often with dark smudging.

5 Broad-billed Flycatcher *Myiagra ruficollis* p. 460
15–16 cm. Rare mangrove specialist; also locally in thickets along river oxbows; usually under cover and only a few meters above the ground or water. Adult resembles female Leaden F; distinguished by broad bill, tips of tail feathers not overlapping completely, and glossy crown. Female differs from male Broad-billed or female Leaden by paler throat and lores. Juv from juv Leaden by wide bill, graduated tail, and association with parents.

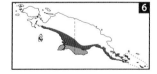

6 Leaden Flycatcher *Myiagra rubecula* p. 460
13–16 cm. Locally common in open lowland habitats—savannah, mangroves, forest edge, settlements. Year-round resident; also an AU migrant during austral winter. Male from male Satin F by dull, slaty blue-grey upperparts; grey-white division cuts across breast in a bulging line and meets the wing at a right angle. Female and Imm, from female Satin F by paler upperparts and pale margins to the secondaries; from Broad-billed F by the narrower bill and tail feathers of even length. Juv grey-brown above with pale eyebrow, white streak across the scapulars; whitish below. NG and AU races indistinguishable in the field.

PLATE 89 MONARCHS OF THE FOREST INTERIOR

1 Spot-winged Monarch *Symposiarchus guttula* p. 464
14–15 cm. Lowland and hill forest interior. Common. Small. Flashy white tail spots. Adult grey and white with black face; white wing spots. Juv/Imm as shown.

2 Kofiau Monarch *Symposiachrus julianae* p. 464
15 cm. Kofiau I (NW Is) only, where Spot-winged M is absent. Common in forest. Adult, blackish upperparts, no wing spots (or obscure).

3 Biak Monarch *Symposiachrus brehmii* p. 464
15–16 cm. Biak (Bay Is) only. Rare in lowland forest. Adult has unique bold pattern of pale yellow and black; note distinctive, pale, tear-shaped patch on the ear-coverts. Juv resembles juv Spectacled M but with mostly whitish outer tail feathers.

4 Hooded Monarch *Symposiachrus manadensis* p. 463
15–16 cm. Forest of foothills and adjacent lowlands. Uncommon and local. Adult's black breast-bib framed by white sides. Juv and Imm as shown.

5 Spectacled Monarch *Symposiachrus trivirgatus* p. 464
15 cm. Resident on SE islands in lowland and hill forest. Also rare AU migrant to Trans-Fly monsoon forest. Adult from Spot-winged M by cinnamon breast, spotless wing, and shape of face mask. Note white in tail. Juv and Imm from juv Spot-winged by grey chin instead of throat. **Subspp shown**: *melanopterus* (resident SE Is) shoulders black, breast pale rufous; *gouldii* (AU migrant to Trans-Fly) shoulders grey (same as back), breast rich rufous.

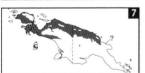

6 Fantailed Monarch *Symposiachrus axillaris* Pl. 92, p. 463
15–16 cm. Mid-mountain forest. Uncommon. Fantail-like. Small white tuft at the bend of wing. Male satiny black. Female duller. Juv has browner flight-feathers; lacks white tufts.

7 Rufous Monarch *Symposiachrus rubiensis* p. 463
18 cm. Lowland and foothill forest interior. Uncommon and patchily distributed. A large monarch, entirely rich rufous brown. Face black in Male, rufous in Female and Juv.

8 Frilled Monarch *Arses telescopthalmus* p. 462
15–16 cm. Lowland and hill forest interior. Blue, fleshy eye-skin. Male with snowy white ruff and white patch on back. Female and Juv from Ochre-collared by rufous collar same color as back and wing. **Subspp groups shown**: *telescopthalmus* (excluding range of next subsp) Female belly white; *henkei* (SE Pen) Female belly buff in the North, white or buff in South.

9 Ochre-collared Monarch *Arses insularis* p. 462
15–16 cm. Northern counterpart of Frilled M. Male has ochre throat and ruff. Female and Juv have rufous collar contrasting with grey-brown saddle and white belly.

PLATE 90 MONARCHS OF THE FOREST CANOPY

1 Golden Monarch *Carterornis chrysomela* p. 465
13–14 cm. The common treetop monarch in lowland and hill forest. Sexes differ, but both show the unique, white tear-spot below eye. Male shows striking black-and-yellow pattern; note yellow wing patch. Female yellowish olive, look for white tear-spot. Juv similar to Female, but bill black with yellowish base. **Subspp groups shown**: *chrysomela* (all of range except for Biak) Female mostly all green; *kordensis* (Biak I) Female resembles Male, but lacks black face, Male deep orange.

2 Black-breasted Boatbill *Machaerirhynchus nigripectus* p. 438
13–14 cm. The highland boatbill; common in forest and edge. Black patch in the middle of the breast and yellow (not white) throat. Female duller black, medium yellow below, breast patch smaller and more ragged, black eye-stripe extends behind eye. Juv female duller and may have scalloped breast.

3 Yellow-breasted Boatbill *Machaerirhynchus flaviventer* p. 438
11–13 cm. The boatbill of lowland and hill forest. Uncommon and inconspicuous. Small, yellow-breasted bird with white throat. Female dusky olive above, paler yellow below, with faint dusky markings on throat and flank; black eye-stripe passes through eye. Juv resembles the Adult of its sex but shows reduced wing bars; Male's underparts with faint dusky markings, Female's often heavily dark-scalloped. **Subspp groups shown**:
albifrons (NW Is, Bird's Head to Sepik-Ramu) forehead and eyebrow white;
xanthogenys (S Lowlands, Huon, SE Pen) these parts yellow.

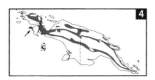

4 Black-winged Monarch *Monarcha frater* p. 466
18 cm. The common monarch of treetops and edge of hill and mid-mountain forests. A resident species, though AU race overwinters in NG somewhere. From Black-faced M by striking contrast between pale grey body and black wings, tail, and face. Adult has well-defined black mask; blue bill. Imm, black mask small and poorly defined, wings brownish grey (not black), bill dark.
Subspp groups:
frater (Bird's Head, northern W Ranges) and *canescens* (AU) eye and mask separate;
kunupi (Weyland Mts) and *periophthalmicus* (remainder of NG) mask joins with eye.

5 Islet Monarch *Monarcha cinerascens* p. 465
17–19 cm. Small-island specialist. Forest and edge, coastal strand. Common. A large, sluggish monarch. Adult dull grey with cinnamon breast; note beady eye. Juv has buffy wash from throat to breast, brownish edges to wings and dark bill.

6 Black-faced Monarch *Monarcha melanopsis* p. 465
17–19 cm. Breeds AU, migrates to eastern NG and SE Is during austral winter; some imm birds remain over summer. Common in all forest types and edge, mainly in lowlands. Diagnostic pale grey eye-patch in all plumages. Adult has black forehead and throat. Imm lacks black mask, is blue-grey like Adult, but retains brown-grey juv wings; note pale loral spot; some may be inseparable from same-age Island M.

PLATE 91 GREY FANTAILS

1 Northern Fantail *Rhipidura rufiventris* p. 471
17–18 cm. The common fantail of forest and edge of lowlands, hills, and mid-mountains. Adult grey or black, lacks white wing markings. Juv duller, wing-coverts tipped buff. **Subspp shown**: *kordensis* (Biak) black with white belly and strong white eyebrow; *gularis* (widespread NG form) grey with buffy belly; *nigromentalis* (SE Is: Misima, Tagula)
black chin, white tail tips larger.
Not shown is *vidua* (NW Is: Gag, Gebe, Kofiau) breast band pale-spotted, belly white.

2 Mangrove Fantail *Rhipidura phasiana* p. 471
13–14 cm. In mangroves only; locally common. Adult is a small, pale grey fantail with white eyebrow and white wing bars. Juv is paler, with upperparts suffused with buff.

3 Chestnut-bellied Fantail *Rhipidura hyperythra* p. 470
14–15 cm. Hilly lowlands to mid-mountain forest. Common. Adult slaty grey above, rich rufous below. Juv paler, duller below; wing-covert spots rufous.

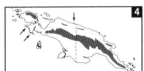

4 Friendly Fantail *Rhipidura albolimbata* p. 470
14–15 cm. Cloud forest interior, edge, and openings. Common. Adult sooty-colored with white facial markings, wing bars. Juv wing bars and underparts buffy.

5 Willie Wagtail *Rhipidura leucophrys* p. 468
17–20 cm. Towns and gardens; originally savannah, beaches, lagoons, and riverside clearings. Wags entire body, calling persistently. Adult black except for white belly and eyebrow. Juv duller; bigger white eyebrow; buff tips to wing-coverts.

6 Black Thicket-Fantail *Rhipidura maculipectus* p. 468
18–19 cm. Lowland forest edge, swamp forest, mangroves, and second growth. Local and uncommon, secretive. Closely resembles White-bellied TF, except overall more blackish, and the belly is black; also, the central tail feathers lack white tips. Female duller, with reduced wing-spotting. Juv sooty black with much reduced white markings; tail pattern same.

7 White-bellied Thicket-Fantail *Rhipidura leucothorax* p. 468
18 cm. Forages near the ground in thickets in gardens, forest edge, and disturbed forest in lowlands and hills. Locally common and often heard, but keeps to cover. Adult's black face marked with a prominent white eyebrow and white submoustachial patch; belly white; *all* tail feathers white-tipped. Juv greyer; all-dark bill; black breast band spotted white.

8 Sooty Thicket-Fantail *Rhipidura threnothorax* p. 469
17–18 cm. Thickets within forest interior in the lowlands. Common and vocal, but secretive. Tail all black, lacking white tips. All-white throat. Male darker; Female browner. Juv paler, with breast and belly grey marked by pale shaft streaks.

PLATE 92 BROWN FANTAILS

1 Rufous-backed Fantail *Rhipidura rufidorsa*   p. 469
13–14 cm. Lowland and hill forest interior. Common.
The plainest and smallest lowland fantail. Adult, grey head
and dingy white underparts. Juv similar, tail shorter.

2 Arafura Fantail *Rhipidura dryas*   p. 470
14–15 cm. This species and Rufous F are very similar but
have different ranges. They share a rufous patch at the base of
the tail and a black throat "necklace." Arafura F differs in the
small rufous patch at base of tail and the tail tip is always white.
Subspp shown:
squamata (forests of NW Is, Aru Is; also Banda Is)
black forehead, white eyebrow;
streptophora (mangroves S Lowlands and possibly Gulf of Papua)
extensive rufous forehead, difficult to separate from Rufous F.

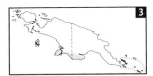

3 Rufous Fantail *Rhipidura rufifrons*   p. 469, 470
14–15 cm. Common resident on small SE Is; uncommon
AU migrant to Trans-Fly; both inhabit forest. Adult, broad
rufous tail patch; narrow pale tail tip. Juv duller, browner,
with rufous tips to wing-coverts, necklace reduced.
Subspp groups shown:
louisiadensis (SE Is, endemic) white tail tip;
rufifrons (migrates to Trans-Fly from AU) grey tail tip.

4 Dimorphic Fantail *Rhipidura brachyrhyncha*   p. 469
15–16 cm. Cloud forest and subalpine shrubland. Common.
Small, pale brown fantail. Rufous rump and lower back.
DARK MORPH: Blackish tail with broad rufous tip and
dull ochraceous underparts;
PALE MORPH: All ashy-grey tail and pale grey-brown underparts.

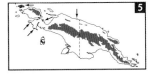

5 Black Fantail *Rhipidura atra*   p. 470
16–17 cm. Montane forest. Common in undergrowth.
Male velvety black with small white eyebrow.
Female and Juv rufous with black central tail feathers.

6 Fantailed Monarch *Symposiachrus axillaris*   Pl. 89, p. 463
15–16 cm. For comparison with co-occurring Black F.
Note bluish bill and white flank patch.

PLATE 93 MANUCODES AND CROWS

Manucodes are birds of paradise in which the males and females look the same. They live in the lowlands and lower mountains and feed mainly on fruit, especially figs. Their exaggeratedly floppy and undulating flight is diagnostic.

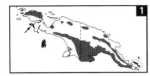

1 Trumpet Manucode *Phonygammus keraudrenii*   p. 474
28–33 cm. Lower montane forest and S Lowlands. Smaller than other manucodes. Adult with diagnostic shaggy nape crest; iris red. Male larger. Juv and Imm dull blackish with short hackles; iris grey-brown. **Subspp shown**:
jamesii (S NG: S Lowlands to SE Pen, Aru Is) lowlands, greenish blue, *ooo-uh* call;
purpureoviolaceus (E Ranges and SE Pen) montane, large, purplish, deep voice;
hunsteini (SE Is: D'Entrecasteaux)
very large, less glossy, with slightly twisted tail.

2 Curl-crested Manucode *Manucodia comrii*   p. 474
43 cm. Common, crow-sized frugivore found in all wooded habitats within its restricted island range. Adult distinctive; iris red-brown. Male larger. Juv and Imm blacker, duller, lack curling feathers; iris brown. Shares habitat with smaller Trumpet M.

3 Jobi Manucode *Manucodia jobiensis*   p. 475
30–36 cm. Lowlands. Nearly identical to Crinkle-collared M but lacks "bump" above eye and has slightly smaller bill, shorter tail, and subtle color differences (see account).

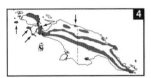

4 Crinkle-collared Manucode *Manucodia chalybatus*   p. 475
33–37 cm. Hill and lower montane forest; also scrub in highland valleys. Common but shy. Adult has a diagnostic "bump" above each eye; crinkled feathers of breast and mantle show as alternating bands of black and iridescence; iris red. Male larger. Juv and Imm blacker, duller, with feather-crinkling absent; iris brown.

5 Glossy Manucode *Manucodia ater*   p. 476
Male 38–42 cm, Female 33–37 cm. Lowland forest edge and open woods. Largest mainland manucode, with longest tail. Sleek, thin-necked. Breast, mantle glossy, lacking crinkled feathers. **Subspp shown**:
ater (widespread) green, blue, or purple gloss;
song a whistled tone, *eeeEEEEee*;
alter (SE Is: Tagula) larger with more massive bill, purple gloss, and deep song *mmm*.

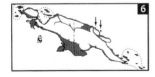

6 Torresian Crow *Corvus orru*   p. 472
46 cm. The common black crow of coasts, settlements, and open lowland country. Adult shiny black. Juv, Imm duller, iris matures from grey, to brown, to hazel, to white.

7 Brown-headed Crow *Corvus fuscicapillus*   p. 471
46 cm. Uncommon and very patchy in lowland forests of the West. Robust, with large, high-arched bill and short, square tail. Brownish. Juv plumage paler; bill all yellow.

8 Grey Crow *Corvus tristis*   p. 472
51–56 cm. Lowland and hill forest. Common, in noisy family groups. A lanky, long-tailed crow with bare, pink face. Adult sooty-brown; bill slaty grey. Juv paler greyish; bill pink.

PLATE 94 BIRDS OF PARADISE—PAROTIAS

Parotias are moderately large birds of paradise that dwell in mid-mountain forest, where they are locally common. The 6 flagged, wiry plumes projecting from behind the eyes gave rise to the name "six-wired birds of paradise." Both males and females are best recognized by their chunky shape and yellow or blue-and-yellow eyes. Parotias feed mainly on fruit but also forage by hopping along branches and picking invertebrates from the bark and epiphytes. The male displays at a "court" on the ground that he has cleared of leaves. The court is situated in a tree-fall gap. Male calls of Carola's and Bronze Ps are higher pitched and ringing; calls of the other species are explosive, harsh, cockatoo-like screams.

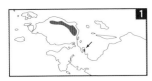

1 Western Parotia *Parotia sefilata* p. 476
30–33 cm. Bird's Head and Neck. Male from other parotias by the combination of medium-length tail and white forehead. Female, Juv, or Imm distinguished by the paler underparts and darker upperparts that show less contrast with the black head. Compare with Superb BoP female race *superba*, plate 97.

2 Lawes's Parotia *Parotia lawesii* p. 477
25–27 cm. E Ranges and SE Pen. Common. Male has a long, gleaming white patch above bill and very short tail. Female, Juv, and Imm are black-headed, blue-eyed, and deep brown on the body. Compare with Eastern P and female Superb BoP races *latipennis* and *minor*, plate 97.

3 Eastern Parotia *Parotia helenae* p. 477
25–27 cm. Northern SE Pen. Both sexes nearly identical to Lawes's P but differ in forehead profile, which is steep and more concave in Eastern vs sloped and more tapered in Lawes's. Eastern thus has a rounder head-shape. Male has small, bronze-colored patch over the bill. Female, Juv, and Imm have reduced feathering over the bill, leaving half or more of bill tip exposed (less exposed in Lawes's), culmen (dorsal ridge of the bill) sharply keeled (angular or slightly rounded in Lawes's).

4 Wahnes's Parotia *Parotia wahnesi* p. 477
Male 43 cm, Female 36 cm. Huon Pen (locally common) and Adelbert Mts (nearly extinct). Tail longest of all parotias. Male with golden frontal crest. Female, Juv, and Imm are russet dorsally in contrast to the black head; note pale eye-stripe and moustachial streak.

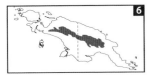

5 Bronze Parotia *Parotia berlepschi* p. 478
Male 26 cm. Foja Mts sister-species of Carola's P. Differs by the pale grey-blue iris of both sexes and the Male's blackish face and throat and overall bronzed cast.

6 Carola's Parotia *Parotia carolae* p. 478
25–26 cm. Note yellow eye. Male shows diagnostic white flank plumes and buffy face and throat. Female, Juv, and Imm recognized by their cream-colored iris and the pale stripes above and below eye. Compare with female Superb BoP race *feminina*, plate 97.

PLATE 95 BIRDS OF PARADISE—ASTRAPIAS

Astrapias are long-tailed birds of paradise with short bills. The 5 species replace each other regonally across the high mountains of New Guinea, with overlap only between Ribbon-tailed and Stephanie's. They feed on fruit (especially *Schefflera*) and methodically forage on branches for insects and other small animal life. Males' wings hiss loudly in flight—a good means of detection because otherwise the birds are mostly silent. As far as is known, all species display in treetops on ridges and at edges of clearings. Males display while hanging beneath a branch or in flight.

Astrapias form 1 of 2 genera of long-tailed birds of paradise, the other being the montane sicklebills. Female-plumaged astrapias and sicklebills are quite similar and often feed together. Sicklebills have long, curved bills and more pointed tails, and the females show a brown cap.

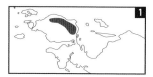

1   Arfak Astrapia *Astrapia nigra*   p. 482
Male 76 cm, Female 50 cm. Arfak mts. Uncommon. Male blackish with green breast and extremely long, broad, round-tipped tail. Unique paired nape crest and bronze neck-stripe. Female, Juv, and Imm with the least amount of pale barring below of any astrapia.

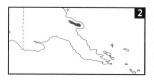

2   Huon Astrapia *Astrapia rothschildi*   p. 482
Male 69 cm, Female 47 cm. Huon mts. Common. Similar to Arfak A, but Male lacks bronze neck-stripe and has a purplish nape shield. Female, Juv, and Imm show more pronounced narrow whitish bars on breast and belly. See Spangled Honeyeater (plate 73).

3   Ribbon-tailed Astrapia *Astrapia mayeri*   p. 482
Male 125 cm, Female 53 cm. Western sector of the E Ranges. Common. Male has a pompom atop bill, and a stupendous pair of narrow, white, ribbonlike central tail feathers. Female, Juv, and Imm show a distinctive greenish sheen on head with a small tuft over the base of the bill, and the central tail feathers are pointed and much longer than the rest. These feathers may show some white on the shaft and as blotches near the base, rarely being extensive like the Male's. Range overlaps that of Stephanie's A, with which Ribbon-tailed hybridizes. Ribbon-tailed above ~2700 m, Stephanie's below.

4   Stephanie's Astrapia *Astrapia stephaniae*   p. 483
Male 84 cm, Female 53 cm. Eastern cordillera, overlapping in the West with Ribbon-tailed. Common. Male is black with iridescent green head, bronze breast band, and spectacular, long, all-black, central tail feathers. Female, Juv, and Imm have a black head and buff-brown underparts with fine blackish barring. Compared with female Ribbon-tailed A, head lacks obvious green sheen and tuft over base of bill, and the tail is evenly graduated with blunt feather tips and does not show any white.

5   Splendid Astrapia *Astrapia splendidissima*   p. 482
Male 39 cm, Female 37 cm. Western cordillera. Common. Male predominantly greenish and highly iridescent with unique paddle-shaped, black-and-white tail. Female, Juv, and Imm much duller but also show white in the base of the tail, which is evenly graduated (central feathers much longer in Ribbon-tailed A).

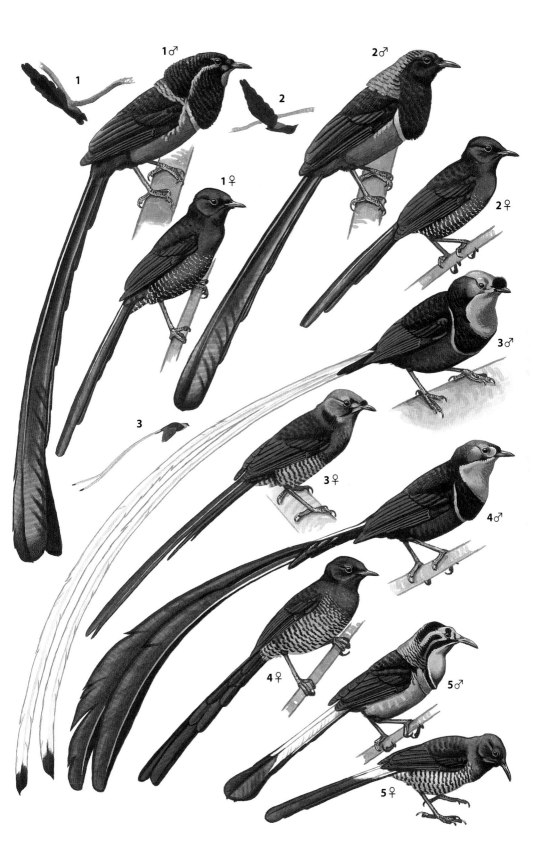

PLATE 96 BIRDS OF PARADISE—SICKLEBILLS

The Black and Brown Sicklebills are large birds of paradise with a long, downcurved bill and a very long, pointed tail. They live in the high mountains, where the males maintain large territories. The males are black with flank-plumes and iridescent highlights. They display on a snag or branch, stretching themselves horizontally. Sicklebills often associate with astrapias, which are superficially similar but have a short, straight bill and blunt-tipped tail, and the male's wings hiss in flight.

The Black-billed and Pale-billed Sicklebills live at lower elevations, and males are also territorial. These sicklebills have a medium-length tail and are similar in size and shape to other co-occurring birds of paradise, from which they can be distinguished by the distinctive sickle-shaped bill and rounded, pale buff or cinnamon tail.

1 Black Sicklebill *Epimachus fastosus*   p. 481
Male 110 cm, Female 55 cm. Inhabits the transition from mid-mountain to cloud forest. Rare and local. Seen foraging on trunks and branches in the canopy or midstory or at fruit trees. Male Black S differs from Brown S in darker breast and red (not pale blue) iris. His song is best means of discovery: a far-carrying *KWINK! KWINK!* Sings in early morning from a prominent branch. Female, Juv, Imm separated from Brown S by rusty wing edges contrasting with the rest of the upperparts and the dark brown iris. **Subsp shown**: *atratus* (Bird's Neck and Central Ranges).

2 Brown Sicklebill *Epimachus meyeri*   p. 480
Male 96 cm, Female 52 cm. Mossy cloud forest, above the elevations of other sicklebills. Common. Male differs from Black S by the brown underparts, pale blue iris, and jackhammer song in 2 bursts: *TAT-AT-AT TAT-AT-AT*. Female, Juv, Imm from Black S by the olive-brown wing edge (same as rest of wing) and pale blue iris.

3 Pale-billed Sicklebill *Drepanornis bruijnii*   p. 480
34–35 cm. Lowland forest. Easily located by song, but difficult to observe in the canopy. Joins mixed flocks. Long, curved, ivory-colored bill; large, bare eye-patch; and "Mohawk haircut." Chestnut tail. Male is dusky-bodied. See account for unique song of varied hollow whistles. Female, Juv, Imm barred below. Slightly larger, darker, and more compact than Black-billed S of the mountains.

4 Black-billed Sicklebill *Drepanornis albertisi*   p. 480
33–35 cm. Mid-mountain forests. Patchily distributed. Uncommon and easily overlooked. In flight, diagnostic buff tail. Black bill distinguishes it from lowland-dwelling Pale-billed S. Male has a diagnostic pattern from below: smooth grey breast contrasts with white belly and vent. Song is a series of loud, piping notes. Female, Juv, Imm have abundant fine barring below.

PLATE 97 BIRDS OF PARADISE—MISCELLANEOUS HIGHLAND SPECIES

Other highland birds of paradise include the parotias, astrapias, and sicklebills.

1 King of Saxony Bird of Paradise *Pteridophora alberti*   p. 476
20–22 cm. A small, uncommon canopy dweller of the interior of cloud forest. Male black with a pale ochre breast and belly, and two remarkable, stiff head-plumes that can be moved around like the antennae of a long-horned beetle. Broad, orange-buff wing band in flight. The territorial Male sings in early morning and late afternoon from a high, open perch, usually a dead branch, but displays to Female on a vine in understory. Female, Juv, Imm are obscure-looking—the small bill, grey plumage, and fine, scalloped barring below suggest something other than a bird of paradise, perhaps a whistler or bowerbird.

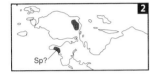

2 Long-tailed Paradigalla *Paradigalla carunculata*   p. 481
35–37 cm. The paradigalla on the Bird's Head. From Short-tailed P by the longer, graduated tail and blue-and-red gape wattles. Male with iridescent crown; Female duller. Juv unknown, but believed to be similar to that of Short-tailed P.

3 Short-tailed Paradigalla *Paradigalla brevicauda*   p. 481
22–23 cm. Middle and upper stories of mossy cloud forest. Distinctive compact shape with narrow bill and short tail. Yellow forehead wattles; blue gape wattles. Male crown exhibits greenish, scaly iridescence. Female duller, longer-tailed, and smaller. Juv very dull, lacking iridescence. Polygynous, but display unknown. Songs/calls include a melodious *hoo-ee?* or rising *zheee*. Also a high-pitched, mournful, 4-note whistling.

4 Superb Bird of Paradise *Lophorina superba*   p. 479
25–26 cm. Mid-mountain forest, disturbed areas, and even casuarina and oak copses in highland valleys. Common. Small size. Male unique for blue-green breast shield that is shaped like a pair of wings and the long, thick, black nape-plume (actually a folded cape). Sings from the canopy—a harsh series of 5–8 *shree* notes—but displays on a log on or near the ground. Female, Juv, Imm regionally variable, and color pattern resembles the parotia species in their area. **Subspp groups shown**:
superba (Bird's Head and Neck) Male lacks the black spots on breast shield and song differs, Female has very dark head and upperparts;
feminina (W and Border Ranges) Female's crown wreathed and speckled;
latipennis (E Ranges to Wau and Herzog Mts) Female variable and intermediate;
Not shown is *minor* (SE Pen) Female has black head, brown back, and usually lacks an eyebrow.

PLATE 98 BIRDS OF PARADISE—MISCELLANEOUS LOWLAND AND HILL SPECIES

Other lowland and hill BoP include the manucodes, Pale-billed Sicklebill, and paradisaeas.

1 Magnificent Riflebird *Ptiloris magnificus* p. 479
Male 34 cm, Female 28 cm. The 2 riflebird species are nearly identical and are separated mainly by range and voice. Both are vocal inhabitants of the interior of lowland and hill forest. Males are terribly shy. The female-plumaged riflebirds are often seen in mixed flocks of other rufous species. Both sexes recognized by distinctive shape: robust with longish neck, tapered head, and long bill. Males' wings rustle in flight. He displays on a branch or thick horizontal vine. Female, Juv, Imm: note whitish eyebrow and moustachial stripe and dark legs. Male Magnificent's song is 2–3 colossal, upsweeping, musical whistles.

2 Growling Riflebird *Ptiloris intercedens* p. 479
Male 34 cm, Female 28 cm. Song is a 2-noted growl. This species' feathered culmen and Male's shorter flank plumes are difficult to discern in the field—see species account.

3 Twelve-wired Bird of Paradise *Seleucidis melanoleucus* p. 478
33–35 cm. Swampy lowland forest and regrowth. Best seen along rivers. The wary Male may be tracked down by his powerful song and hissing wings, or in early morning at his display site atop a dead spire. Song similar to paradisaeas, short series of *harnh, haw*, or *koi* notes. The more approachable Female is often seen in mixed flocks. Female, Juv, Imm diagnostic with reddish-brown back and tail, black cap, and finely barred underparts. Unique red iris and pink legs. Long chisel-bill is prominent.

4 Magnificent Bird of Paradise *Diphyllodes magnificus* p. 484
19 cm. Hill and lower mid-mountain forest. Common but seldom seen except at fruiting trees or display court—a cleared space on the ground around a thin sapling. Song a series of ~8 loud, downslurred *chur* notes rapidly delivered. The only small mainland bird of paradise with bright blue-grey bill and legs. Male colors muted in the forest gloom; look for tail wires and contrasting dark ventral and light dorsal pattern. Female, Juv, Imm identified by the blue-grey eye-stripe, bill, and legs.

5 Wilson's Bird of Paradise *Diphyllodes respublica* p. 484
16 cm. Hill forest on Waigeo and Batanta Is. Unique bald, cobalt-blue crown with black lines. Male's red back easy to spot. Habits and voice similar to Magnificent BoP.

6 King Bird of Paradise *Cicinnurus regius* p. 483
Male 16 cm, Female 19 cm. Lowland and foothill forest, edge, and second growth. Common but difficult to observe. The smallest mainland BoP. Male unique. He sings and displays from a subcanopy tangle of shaded vines in the forest interior. Song is a slow, swelling series of nasal, paradisaea-like notes: *rahn rahn rahn rahn . . .* or a rapid *ki kyer kyer kyer kyer kyer . . . .* Female, Juv, Imm lack facial markings but have a diagnostic head shape with long, tapered profile, and pale yellow bill.

PLATE 99 BIRDS OF PARADISE—PARADISAEAS WITH SMALL RANGES

The "plumed birds of paradise" are the common and characteristic birds of paradise in lowland and hill forest, edge, and regrowth, including near human settlement. Their songs are heard persistently through the forest: a loud *WAU WAU WAU*... and similar notes. The plumed birds can be divided into 2 groups: the 4 species with small ranges (this page) and the 3 "typical" plumed birds (next page). Males of all but the Blue BoP display in communal leks high in the treetops; the male Blue BoP displays alone in the midstory.

1 Goldie's Bird of Paradise *Paradisaea decora*   p. 485
Male 33 cm, Female 29 cm. Hill forests of Fergusson and Normanby Is, locally in the lowlands. Male has reddish flank plumes, yellow crown and back, and unique grey breast. Female, Juv, Imm show a dull yellow crown, chocolate throat, and finely barred breast, unusual for a *Paradisaea*.

2 Red Bird of Paradise *Paradisaea rubra*   p. 485
Male 33 cm, Female 30 cm. Waigeo, Batanta, and some nearby islands. Male has curved red flank plumes and a pair of prominent, long, curling, black tail-streamers. Female, Juv, Imm are chocolate-faced and with broad breast band of dull yellow that connects to yellow nape.

3 Emperor Bird of Paradise *Paradisaea guilielmi*   p. 484
31–33 cm. Mid-mountain forests of the Huon Pen, mainly above the elevational range of the Raggiana and Lesser BoPs (hybridizes with both). Male has dark green crown and extensive breast-bib; the mostly white flank plumes are short and sparse. Female, Juv, Imm differ from Raggiana in that the dark face includes the entire crown; the yellow nape color extends far onto the back; and the dark iris (not yellow) at all ages.

4 Blue Bird of Paradise *Paradisornis rudolphi*   p. 484
30 cm. Mid-mountain oak forests and regrowth. Uncommon and best located at Male's morning calling perch or by observation at a favored fruiting tree. A dark BoP recognized by the broken white eye-ring, prominent white bill, and blue wings. Male, blue breast and blue-and-amber flank plumes; long black tail-streamers. Song is a slow, plaintive series, more nasal and higher pitched than those of a typical paradisaea: *wahr wahr wahr*.... Male sings from a high, open perch and displays near the ground in a concealed site. Female, Juv, Imm, brown breast, variably barred.

PLATE 100 BIRDS OF PARADISE—PARADISAEAS WITH LARGE RANGES

These are the iconic birds of paradise, and one—the Raggiana Bird of Paradise—is the national bird of Papua New Guinea. Each of these 3 closely related species is confined to an exclusive geographic area of mainland New Guinea. Males display in leks high in the treetops.

1 Greater Bird of Paradise *Paradisaea apoda* p. 486
Male 43 cm, Female 35 cm. SW NG and Aru Is. Male has yellow-and-white flank plumes; from male Lesser BoP by larger size, brown back sharply contrasting with yellow nape, and prominent blackish breast cushion. Female, Juv, Imm are entirely brown; note bright yellow eye and pale blue bill to separate it from the other large, all-brown songbirds with which is may associate. Hybridizes with Raggiana BoP at Fly R (Kiunga).

2 Lesser Bird of Paradise *Paradisaea minor* p. 485
Male 32 cm. The plumed BoP of the North and West.
Male has yellow-and-white flank plumes; from male Greater BoP by smaller size, yellow mantle and wing-coverts, and absence of blackish breast-cushion. Female, Juv, Imm have unique snowy white breast and belly. Hybridizes with Raggiana BoP in upper Ramu R and Huon Pen, and with Emperor BoP.

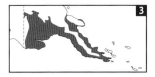

3 Raggiana Bird of Paradise *Paradisaea raggiana* p. 486
33–34 cm. E NG. Male has reddish or orange flank plumes. Female, Juv, Imm exhibit diagnostic yellowish crown and nape framing the dark face, dark brown breast, and maroon-brown belly.
Subspp shown:
salvadorii (S NG east to SE Pen; E Ranges)
brown mantle, brick red flank plumes;
augustaevictoriae (Huon and northwestern SE Pen)
yellow dorsally, orange plumes.
Not shown are *raggiana* (far SE NG) yellow mantle, reddish flank plumes and *intermedia* (N coast of SE NG) like previous but much yellow dorsally.

PLATE 101 LOWLAND ROBINS, TORRENTLARK, AND MAGPIELARK

1 Mangrove Robin *Peneothello pulverulenta* p. 492
14–15 cm. Mangroves along NG coasts and some rivers. Common but patchily distributed. Adult is grey with white throat, white in tail, and all-dark wings. Juv brown above, whitish below, streaked and flecked throughout. **Subsp shown:** *pulverulenta* (NG). Not shown is *leucura* (Aru Is and AU), head and breast greyer.

2 Jacky Winter *Microeca fascinans* p. 489
14 cm. Port Moresby savannah. Scarce and local. Adult dingy brown above, whitish below, with diagnostic white outer tail feathers. Juv similar but spotted.

3 Black-chinned Robin *Heteromyias brachyurus* p. 491
14–15 cm. Northern lowlands and hills. Uncommon in forest interior. Black-and-white Adult has thick white eyebrow and black chin. Note pale legs, compared with dark legs of the more common Black-sided R. Juv, like Adult but heavily smudged with brown. **Subsp shown:** *dumasi* (NW Lowlands and Sepik-Ramu) back black. Not shown is *brachyurus* (Bird's Head and Neck, far W Ranges) back dark grey.

4 Black-sided Robin *Poecilodryas hypoleuca* p. 492
13–15 cm. The common robin of lowland and hill forest. Adult is black and white with distinctive facial pattern; note black mark on side of the breast. Juv brown with white belly and undertail-coverts.

5 Torrent Flycatcher *Monachella muelleriana* p. 490
14–15 cm. Forages over rocky streams and rivers from foothills up to the transition to cloud forest. In both primary forest and disturbed habitat. Locally common. Adult is whitish with bold black cap, wings, and tail. Juv shows same pattern but lacks the white supraloral spot and is faintly pale-spotted and brownish-mottled.

6 White-rumped Robin *Peneothello bimaculata* Pl. 104, p. 492
13–14 cm. Hill forest understory. Uncommon and local. Adult jet black with white on rump. Juv sooty brown with reduced amounts of white. **Subspp shown:** *bimaculata* (species range except next) belly white; *vicaria* (Adelbert Mts to N slopes of SE Pen) Male's belly black, Female's partly white.

7 Torrentlark *Grallina bruijnii* p. 462
20 cm. Along fast-flowing creeks and rivers in the mountains. Prefers primary forest and steep terrain. Scarce but easily located noisy pairs and family groups. Bobs body and wags its tail. Male with black face and underparts. Female with white brow and breast. Juv like Female but forehead black and eyebrow white.

8 Magpielark *Grallina cyanoleuca* p. 463
26 cm. S Trans-Fly. Common in grassland with scattered trees and along river courses. Struts about wagging its tail. Male with white eyebrow and black throat. Female with white forehead and throat. Juv with white eyebrow and throat and dark eye.

PLATE 102 HIGHLAND ROBINS

1 Snow Mountain Robin *Petroica archboldi* p. 488
14 cm. Restricted to a few peaks in the W Ranges. Inhabits rocky alpine scree well above the tree line. Conspicuous and noisy. Adult with red breast patch, lacking in Juv.

2 Subalpine Robin *Petroica bivitatta* p. 488
11–12 cm. Subalpine forest, shrubland. Uncommon, local. Black with white breast and forehead spot. Male black. Female sooty black. **Subsp shown:** *bivittata* (E Ranges, SE Pen) outer tail mostly white. Not shown is *caudata* (W Ranges), little white in tail.

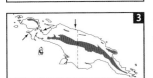

3 Blue-grey Robin *Peneothello cyanus* p. 493
14–15 cm. The common understory robin at the transition between mid-mountain and cloud forest. Adult bluish grey. Juv grey with brown shaft streaking and spotting.

4 Smoky Robin *Peneothello cryptoleuca* p. 493
14–15 cm. Mid-mountain and cloud forest in the West. Common. Adult smoky grey. **Subsp shown:** *cryptoleuca* (Bird's Head, Foya Mts) underparts entirely grey. Not shown are *albidior* (W Ranges) similar and *maxima* (Kumawa Mts) underparts white.

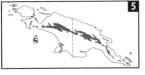

5 White-winged Robin *Peneothello sigillata* p. 492
14–15 cm. Cloud forest and timberline. Common. Adult velvety black with a white wing patch. Juv plumage similar but overlaid with brown markings. **Subspp groups shown:** *sigillata* (species range except next); *quadrimaculata* (W Ranges) white patch on side of breast.

6 Ashy Robin *Heteromyias albispecularis* p. 491
15–16 cm. Bird's Head representative of Black-capped R, but overall grey and less patterned.

7 Black-capped Robin *Heteromyias armiti* p. 491
15–18 cm. Forest floor and understory of lower cloud forest. Uncommon. A medium-large robin with long, pale legs. Black cap, white eyebrow, white throat patch. Male bill tipped white. **Subsp shown:** *armiti* (SE Pen, Huon) greenish-olive back. Not shown is *rothschildi* (W Ranges to E Ranges) olive brown back.

8 White-eyed Robin *Pachycephalopsis poliosoma* p. 488
15–16 cm. The common and vocal grey robin of mid-montane forest understory, but difficult to observe. Adult with white iris and throat. Juv tinged brown. **Subspp shown:** *poliosoma* (SE Pen) large white throat patch and undertail grey, not buffy; *hypopolia* (Huon and Adelbert Mts) a small, dark race, with buff in the undertail.

9 Black-throated Robin *Plesiodryas albonotata* p. 491
18–19 cm. Mid-mountain and cloud forest midstory. Uncommon. High, tonelike song. Adult with black face marked by a white streak. Juv pale cinnamon. **Subspp shown:** *albonotata* (Bird's Head) and *correcta* (SE Pen, Huon) underparts mostly white; *griseiventris* (W Ranges to E Ranges) underparts mostly grey.

PLATE 103 YELLOW AND GREEN ROBINS AND FLYCATCHERS

1 Olive Flycatcher *Kempiella flavovirescens* p. 489
13–14 cm. Lowland and hill forest. Common in midstory.
Adult is yellowish olive and plain but for its yellow lower bill
and legs. Has the most obvious eye-ring of any flycatcher.
Juv brown above with pale spots; white eye-ring, yellowish legs.

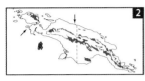

2 Yellow-legged Flycatcher *Kempiella griseoceps* p. 489
12–13 cm. Mid-mountain forest and Trans-Fly.
Frequents subcanopy and midstory. Local, uncommon, and
inconspicuous. Adult a small robin with unique combination
white throat, pale yellow breast, and yellow legs.
Juv brown with pale spotting.

3 Papuan Flycatcher *Devioeca papuana* p. 489
12–13 cm. Cloud forest. Common. Adult has rich yellow
underparts and bright orange legs. Juv brown and green above,
heavily pale-spotted; yellow legs.

4 Lemon-bellied Flycatcher *Microeca flavigaster* p. 490
12–13 cm. Savannah and large clearings in the forest, from
lowlands to mountain valleys. Patchily distributed but locally
common. Adult differs from other yellow-breasted flycatchers
by its all-black legs and bill (may be pale at base of lower bill).
Juv brown and pale-spotted above; white with brown markings
below. Subspp poorly understood but varying in brightness of yellow.

5 Garnet Robin *Eugerygone rubra* p. 488
10–11 cm. A petite, fantail-like robin, local and rare in montane
cloud forest. Perches horizontally; nervously jerks about and
flashes wings, showing white blaze across flight-feathers and white
in tail. Male has maroon-red upperparts, green in Female. Juv
reddish brown, paler below.

6 White-faced Robin *Tregellasia leucops* p. 493
11–13 cm. Mid-mountain forest. A common bird of the understory
and midstory. There is also a population in S Lowlands. Adult has
white face. Juv is extensively smudged with brown; only a hint of
adult facial pattern. **Subspp groups shown**:
leucops (Bird's Head to S slope of W Ranges) no white eye-ring,
black forehead line;
melanogenys (mts of NW Lowlands to Huon) white eye-ring,
bill mostly black.
Not shown are *wahgiensis* (E Ranges, S Lowlands) white eye-ring,
bill mostly orange and *albifacies* (SE Pen) white eye-ring, black
forehead stripe, bill mostly black.

7 Banded Yellow Robin *Gennaeodryas placens* p. 493
14–15 cm. Hill forest. This understory robin rarely encountered
because of patchy distribution and restriction to a narrow
elevational band. Adult has unusual color pattern with brilliant
yellow half-collar. Juv similar but tinged brown.

8 Green-backed Robin *Pachycephalopsis hattamensis* p. 487
15 cm. Common in understory of mid-mountain forest of W NG.
Adult a white-eyed robin with whistler-like plumage pattern.

PLATE 104 OPEN COUNTRY SONGBIRDS

1 Long-tailed Shrike *Lanius schach* p. 458
25 cm. Montane grasslands and gardens. Common. Adult unique.
Juv has black mask.

2 Brown Shrike *Lanius cristatus* p. 458
18 cm. Vagrant Palearctic migrant to open habitats in far West.
Black mask, white eyebrow.

3 Blue Rock-thrush *Monticola solitarius* p. 507
20–23 cm. Vagrant Palearctic migrant to rocky places and shores.

4 Siberian Rubythroat *Luscinia calliope* p. 507
14–16 cm. Vagrant Palearctic migrant to open country and thickets.

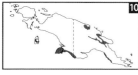

5 Grey-streaked Flycatcher *Muscicapa griseisticta* p. 508
13–14 cm. Overwintering Palearctic migrant in lowlands to
mid-mountains. On high, open perches at forest edge, gardens.
Dark streaked below, white spot in front of eye.

6 Pied Bushchat *Saxicola caprata* p. 507
14–15 cm. Grasslands, gardens, edge of towns. Montane, locally
in lowlands. Common. Male black with white rump and shoulder.
Female dull grey; note white rump and undertail. Juv similar with
buff spots above and buff wing bars.

7 White-rumped Robin *Peneothello bimaculata* Pl. 101, p. 492
14–15 cm. For comparison.
Completely different habitat: hill forest interior.

8 Eurasian Tree Sparrow *Passer montanus* p. 510
14 cm. Not native; spreading. Established in many ports and towns,
mainly on Indonesian side. All plumages show black spot on
ear-coverts, chestnut crown.

9 House Sparrow *Passer domesticus* p. 510
15 cm. Not native; spreading. Common in a few ports, towns, and cities.
Male facial pattern unique. Female and Juv drab with pale eye-stripe.

10 Horsfield's Bushlark *Mirafra javanica* p. 494
13 cm. Short grass, roadsides, airfields from lowlands to mid-
mountains. Hesitant, mothlike flight unique. Conical, stubby bill;
much streaking; rufous patches on wings; white outer tail feathers.
Plumage darker with wear.

11 Australasian Pipit *Anthus novaeseelandiae* p. 516
15 cm. Montane grasslands and dry lowlands of NE.
Slender profile, thin bill, pinkish legs, white outer tail feathers.
Streaked breast. Juv, pale edges to feathers of upperparts.

12 Alpine Pipit *Anthus gutturalis* p. 516
18 cm. Alpine grasslands. A heavyset, dark pipit.
Adult, dark unstreaked breast, white "spectacles" and throat,
black "eye shadow." Juv darker than Australasian P.

13 Eastern Yellow Wagtail *Motacilla t. tschutschensis* p. 515
16–17 cm. Uncommon Palearctic migrant, mainly in lowlands.
Short grass and bare areas. Black legs; tail not longer than body.
Lacks yellow rump and white wing-stripe in flight. Nonbreeding adult
brown-olive above, pale yellow below. Breeding adult has grey cap.
Imm like Nonbreeding adult but more variable; often white below.

14 Green-headed Yellow Wagtail p. 515
Motacilla tschutschensis taivana
16–17 cm. Hypothetical. Yellowish eyebrow. Breeding adult's crown
green, same as back.

15 Grey Wagtail *Motacilla cinerea* p. 516
18–19 cm. Palearctic migrant, mainly in the mountains. Near
water: creek beds, wet gravel roads. Pale legs, tail longer than body.
Yellow rump. Flashes white wing-stripe and tail feathers in flight.
Nonbreeding adult and Imm have yellow breast, white throat.
Breeding adult is bright yellow below, and Male has black throat.

PLATE 105 "SYLVIID" WARBLERS

1 Numfor Leaf-Warbler *Phylloscopus maforensis* p. 497
9–10 cm. The only leaf-warbler on Numfor I. Dull grey; head pattern obscure.

2 Biak Leaf-Warbler *Phylloscopus misoriensis* p. 497
9–10 cm. The only leaf-warbler on Biak I. Note orange legs and bill. Relatively long-legged.

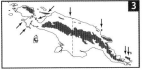

3 Island Leaf-Warbler *Phylloscopus poliocephalus* Pl. 77, p. 497
9–10 cm. Montane forests. Common. Note yellowish wash on underparts, dark eye-line and crown, pale eyebrow.
Subspp groups shown: *poliocephalus* (Bird's Head and Neck) grey crown lacks median stripe; *giulianettii* (mainland minus range of previous) grey crown with pale mid-stripe; *hamlini* (D'Entrecasteaux Is) blackish crown with pale mid-stripe.

4 Gray's Grasshopper-Warbler *Locustella fasciolata* p. 501
17–18 cm. Rare Palearctic migrant. Skulks in dense underbrush and tall grass. From reed-warblers by stocky build and pale, pinkish legs. Adult greyish below, Juv yellowish.

5 Oriental Reed-Warbler *Acrocephalus orientalis* p. 501
18 cm. Rare Palearctic migrant in cane, reeds, sedges bordering wetlands. Distinguish with care from Australian RW by song and dusky streaking on sides of throat.

6 Australian Reed-Warbler *Acrocephalus australis* p. 501
16–17 cm. Resident, patchily distributed. A retiring songster of tall reeds and other grasses, often near water. Plain brown with faint pale eyebrow and dark legs.

7 Golden-headed Cisticola *Cisticola exilis* p. 503
10–11 cm. Widespread, common. Tiny, long-legged warbler in grassland. Buffy nape and flanks. Breeding male: head buff, tail short. Female and Nonbreeding: crown streaked.

8 Zitting Cisticola *Cisticola juncidis* p. 503
10–11 cm. Trans-Fly. Grass hummocks. From Golden-headed C: zitting or clicking song; more streaked above; less buffy; and conspicuous pale tail tip. Breeding male: crown not streaked, tail short; note eyebrow. Female and Nonbreeding: crown streaked.

9 Little Grassbird *Megalurus gramineus* p. 503
14–15 cm. Local and little known. Hides in marsh vegetation. Small, with dark, streaked crown.

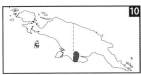

10 Fly River Grassbird *Megalurus albolimbatus* p. 502
14–15 cm. Trans-Fly marshes. Local, but common where found. Clean white underparts; bright, unstreaked rufous cap; prominent white eyebrow; shortish tail.

11 Tawny Grassbird *Megalurus timoriensis* p. 502
18–21 cm. Trans-Fly grasslands. From Papuan GB: tawnier, cap streaked. Juv from adult Tawny by reduced eyebrow, diffuse streaking; from juv Papuan by lack of yellowish wash.

12 Papuan Grassbird *Megalurus macrurus* p. 502
20–23 cm. Widespread except Trans-Fly. Grassland. Longer tail tapers to finer point. Adult's cap unstreaked, breast white. Juv cap and breast streaked. **Subspp groups shown**:
macrurus (widespread, exclusive of next race) not streaked below; *alpinus* (Central Ranges above 2500 m) streaked flanks.

13 Green-backed Gerygone *Gerygone chloronota* Pl. 77, p. 427
8–9 cm. For comparison. Gerygones differ from leaf-warblers by absence of crown striping.

14 White-shouldered Fairywren Pl. 68, p. 402
Malurus alboscapulatus
12 cm. For comparison. Female in S NG similar to Fly River Grassbird.

PLATE 106 WHITE-EYES

1 Green-fronted White-eye *Zosterops minor* p. 499
11 cm. Northern counterpart of Black-fronted WE. Common. Differs by eye-ring absent or nearly so, forehead always green, and upperparts brighter yellow-green.

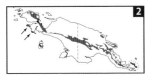

2 Black-fronted White-eye *Zosterops atrifrons* p. 498
11 cm. Hill and lower montane forests. Common. From New Guinea WE: dark forehead and yellow throat sharply delineated from white breast, except on Bird's Head where Black-fronted may have green forehead and New Guinea WE also has sharply defined yellow throat. Here, look for Black-fronted's brighter white flanks, breast, and belly, whereas those of New Guinea WE are dingy.
Subspp groups shown:
chrysolaemus (mostly in South) eye-ring medium width, forehead black or dark green;
delicatulus (southern SE Pen, D'Entrecasteaux Is) broad eye-ring, black mask.

3 New Guinea White-eye *Zosterops novaeguineae* p. 500
11 cm. To separate from Black-fronted WE, see above account.
Subspp (4 groups):
wuroi (Trans-Fly, Aru Is) small, with yellow throat sharply defined, eye-ring wide;
crissalis (mts from E Ranges to SE Pen) yellow throat grading into white breast.
Not shown are: *novaeguineae* (mts of Bird's Head/Neck) see Black-fronted account, above; and *magnirostris* (coast at Ramu R) breast with prominent yellow streak.

4 Tagula White-eye *Zosterops meeki* p. 499
11 cm. The only white-eye on Tagula I (SE Is). Sister species to Black-fronted WE, with similar song. Uncommon in hill forests. All-white throat and breast.

5 Lemon-bellied White-eye *Zosterops chloris* p. 498
12 cm. On a few islets and small islands in the NW Is and Aru Is. Only white-eye where it occurs; abundant. Bright yellow underparts.

6 Biak White-eye *Zosterops mysorensis* p. 499
11 cm. The only white-eye on Biak. Uncommon. Lacks white eye-ring. Note yellow vent.

7 Capped White-eye *Zosterops fuscicapilla* p. 499
11 cm. The uniformly olive white-eye of W mts. Locally common in forest, regrowth, and gardens. Olive-green with black cap.

8 Oya Tabu White-eye *Zosterops crookshanki* p. 499
11 cm. The montane, cloud forest white-eye in the D'Entrecasteaux Is. Locally common. Differs from Capped WE by the wider eye-ring and silver-grey iris. Crown and face dark green on Goodenough I (shown), blackish on Fergusson I.

9 Louisiade White-eye *Zosterops griseotinctus* p. 500
11–13 cm. Small SE Is. Abundant. Large, olive green with large bill.
Subspp groups shown:
griseotinctus (small SE Is) greyish legs, plumage variably olive to yellow-green;
pallidipes (Rossel I) orangish legs, plumage yellow-green.

PLATE 107 DRONGOS AND STARLINGS

1 Pygmy Drongo *Chaetorhynchus papuensis* p. 467
20 cm. Interior of hill and mid-mountain forests, where it is the only black flycatcher-like bird with upright posture and unforked tail. Common. Juv duller; white gape.

2 Spangled Drongo *Dicrurus bracteatus* p. 466
25–28 cm. Ubiquitous in forest and edge from lowlands to mid-mountains. Sits upright; note unique "fish tail." Adult glossy with red iris. Juv duller; dark iris. **Subspp shown**: *bracteatus* (AU migrant, winters mainly in Trans-Fly) larger, tail shorter and less forked, bill longer, Juv with white tips to undertail-coverts;
carbonarius (widespread resident).

3 Common Myna *Acridotheres tristis* p. 506
23–25 cm. Hypothetical introduction. Reported from Alotau (SE Pen). A bird of cities and towns. Seen perched on wires or walking on lawns and roadsides.

4 Sooty-headed Bulbul *Pycnonotus aurigaster* p. 494
15–20 cm. Introduced to Biak I and Jayapura. A conspicuous, garrulous, easily recognized bird of towns and countryside. Pale rump and white tail tip. Juv face and cap brownish.

5 Singing Starling *Aplonis cantoroides* p. 504
19–22 cm. Lowland towns, deforested areas, seacoast, forest edge. Common. Short, square tail. Adult black; iris orange-red. Juv brown above, whitish and streaked below; iris dark brown. Imm similar, darker above, more heavily marked, iris red.

6 Moluccan Starling *Aplonis mysolensis* p. 505
20 cm. NW Is only, mainly small islands. Local and uncommon in forest and open habitats. Wedge-shaped tail, dark eye. Adult all black. Juv brown above, white below with dark streaking.

7 Metallic Starling *Aplonis metallica* p. 504
22–24 cm. Lowland and hill forest and edge. Common. Pointed tail. Adult with green and purple gloss, hackles, and red iris. Juv dark brown above, white below with streaking; iris brown. Imm upperparts like Adult, underparts like Juv; iris red. **Subsp shown**: *metallica* (widespread, range exclusive of next race). Not shown is *inornata* (Bay Is of Biak and Numfor) smaller and less iridescent.

8 Yellow-eyed Starling *Aplonis mystacea* p. 504
20 cm. Uncommon relative of Metallic S, with which it associates. White iris is the best field mark. Other differences harder to see include the scant crest and stubbier profile. Plumages same as for Metallic S.

9 Long-tailed Starling *Aplonis magna* p. 505
28–41 cm. Bay Is only. Abundant. Very long, lax tail. Dark iris. Female smaller. Juv duller and tail shorter. **Subspp shown**:
magna (Biak I) larger, with longer tail;
brevicauda (Numfor I) smaller, with shorter tail.

PLATE 108 CATBIRDS, ORIOLES, AND MYNAS

1 Black-eared Catbird *Ailuroedus melanotis*   p. 396
28–29 cm. Catbirds are green bowerbirds that do not build bowers; they are vocal but very shy. The species differ in habitat. This one found in 2 habitats: lower mountains and stands of rainforest/monsoon forest in the Trans-Fly. Black ear-patch and blackish chin. Tail with white tip, lacking in White-eared CB. Breast scalloped rather than spotted. Much regional variation.
Shown are 2 of the 8 subspp:
melanotus (Trans-Fly and Aru Is);
arfakianus (Bird's Head and Neck).

2 White-eared Catbird *Ailuroedus buccoides*   p. 396
24–25 cm. The catbird of lowland and foothill forest interior. Black-spotted breast, white throat and ear-patch, prominent uniform cap, and lacks white tail tip. **Subspp groups shown**:
geislerorum (NW Lowlands to northern SE Pen) pale brown crown;
buccoides (widespread, exclusive of previous subsp) dark crown.

3 Brown Oriole *Oriolus szalayi*   p. 459
25–28 cm. Lowlands into mid-mountains. Common in forest canopy and regrowth. Adult has blackish face patch and blood-red iris and bill. Juv has white eyebrow, black face, dark brown iris, and blackish bill; don't mistake it for a friarbird.

4 Olive-backed Oriole *Oriolus sagittatus*   p. 459
25–28 cm. Trans-Fly savannah, where locally common. Upperparts grey-olive and underparts white with sharp, dark streaking. Adult, red iris and bill. Juv greyer; dark iris and bill, pale eyebrow, and buff edges to wing feathers.

5 Green Oriole *Oriolus flavocinctus*   p. 459
25–28 cm. Trans-Fly, where common in all wooded habitats. Adult yellow-green; dark wings and tail tipped white; iris and bill red. Rare grey morph. Juv duller olive with heavier streaking; pale eyebrow; dark iris and bill.

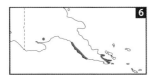

6 Australasian Figbird *Sphecotheres vieilloti*   p. 458
27 cm. Open wooded habitats and trees in towns. Locally common. Male unique. Female and Juv whitish below and heavily streaked; dark eye and grey facial skin.

7 Golden Myna *Mino anais*   p. 505
24 cm. Lowland forest, where uncommon. Adult black with orange neck. Juv head black, underparts mottled black and yellow. Rump yellow (not white). **Subspp groups shown**:
orientalis (main trunk of NG) yellow crown, black nape, lacks yellow eye patch;
anais (NW Is, W Bird's Head) head black, pale patch behind eye.

8 Yellow-faced Myna *Mino dumontii*   p. 505
25–26 cm. Lowland and hill forest canopy, particularly at edge of clearings or along river courses. Common, vocal, and conspicuous. Adult has bright yellow-orange facial skin and white rump, conspicuous in flight. Juv facial skin paler.

PLATE 109 FINCHES 1

1 Crimson Finch *Neochmia phaeton* — p. 511
13 cm. Lowlands, where locally common in tall grass bordering rivers, marshes, and lakes. A slender, red-and-grey finch with long, pointed, red tail. Adult has red face and beak; breast red in Male, grey in Female. Juv has brown head and breast, dark beak.

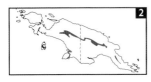

2 Mountain Firetail *Oreostruthus fuliginosus* — p. 511
13 cm. Subalpine forest edge and glades, where inconspicuous but locally common. Seen on the ground or in low shrubbery. Adult has red flanks and beak. Throat of Male is same dark brown as rest of head; orange-brown in Female, with upper beak dark. Juv brown except for red rump; beak black.

3 White-spotted Mannikin *Lonchura leucosticta* — p. 513
10 cm. Uncommon in grassy habitats in lowlands, also bamboo and forest edge. Adult is warm brown with white chin and cheeks and fine white speckling on head and foreparts. Juv, spots on wing-coverts; from juv Streak-headed M by pale chin.

4 Grey-headed Mannikin *Lonchura caniceps* — p. 513
10 cm. Grasslands of lowlands and mid-mountains. Common. Adult grey head contrasts with dark beak and beady eye. Juv plain tan with dark bill. **Subspp groups shown**: *caniceps* (lowlands) darker, belly and flanks blackish, rump orange; *scratchleyana* (mountains) paler, belly and flanks buff, rump yellow.

5 Grey-banded Mannikin *Lonchura vana* — p. 513
10 cm. Mid-mountain grasslands of Bird's Head. Adult has white face; unique chestnut breast bordered above by grey band. Juv plain tan; note orangish rump and belly.

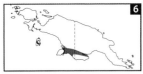

6 Grey-crowned Mannikin *Lonchura nevermanni* — p. 513
11 cm. Trans-Fly marshes, riverine grasses, and nearby savannah. Local, but common where found. Adult has white cap, black throat, rich brown underparts. Head often darker in Female. Juv tan with pale bill.

7 Papuan Parrotfinch *Erythrura papuana* — p. 512
13 cm. Same habitats as the nearly identical Blue-faced PF, with which it may co-occur, but patchily distributed and very rare. Difficult to separate in the field. Differs in larger size and more massive beak with somewhat more swollen shape.

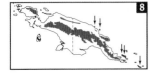

8 Blue-faced Parrotfinch *Erythrura trichroa* — p. 511
12 cm. Hills and mountains, in forest and edge; common but difficult to observe. Listen for its call—an insect-like, short, descending, high-pitched trill, *t-t-t-t*. From Papuan PF by smaller size and smaller, narrower, straight-edged, conical beak. Adult has blue face; Female paler green below. Juv face mostly green, tail brownish red.

PLATE 110 FINCHES 2

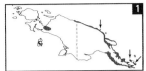

1 Chestnut-breasted Mannikin *Lonchura castaneothorax* p. 514
10 cm. Grassy areas, mostly in lowlands, also mid-mountains. Locally common, nomadic. Adult has tan breast, black breast band, white belly; pale nape contrasts with black face. Juv has buff breast band and blue-grey bill. **Subspp groups shown**:
sharpii (NW Lowlands to Sepik-Ramu) grey crown;
ramsayi (SE Pen) head nearly all dark.
Not shown is *boschmai* (western W Ranges) brown scalloping on flank.

2 Hooded Mannikin *Lonchura spectabilis* p. 514
10 cm. Grassy areas, lowlands to mid-mountains. Locally common, nomadic. Adult has simple pattern: black hood, brown saddle, and breast and belly of a contrasting color (varies by locale). Juv has dark ear-patch, very pale underparts, no breast band. Much geographic variation, but poorly understood.

3 Western Alpine Mannikin *Lonchura montana* p. 515
11–12 cm. Alpine grassland and moors. Adult has black crown and face, buff breast, barred flanks. Juv larger than juv Black-breasted M; lacks the obscure breast streaking.

4 Black-faced Mannikin *Lonchura molucca* p. 512
10–11 cm. Only finch on the islands of Gag and Kofiau (NW Is). Adult has black throat and bib; unique white, scaly breast and rump. Juv tan, paler below, with cream rump.

5 Eastern Alpine Mannikin *Lonchura monticola* p. 515
11–12 cm. Alpine grasslands and moors. Locally common. Adult has white breast, black lower breast band and flank-stripe. Juv, large; blackish face; faintly streaked breast band.

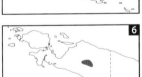

6 Black-breasted Mannikin *Lonchura teerinki* p. 514
11 cm. Mid-mountain grasslands of W Ranges. Common. Adult has black head, breast, and flank streak. Juv smaller than juv Western Alpine M and has streaked breast.

7 Grand Mannikin *Lonchura grandis* p. 513
12 cm. Open grasslands and marshes of lowlands and lower mountains. Uncommon, local. Large. Big-headed. Note massive, bluish-white beak. Adult has black head and underparts, chestnut wings and flanks. Juv head dark, breast with smudgy streaks.

8 Black Mannikin *Lonchura stygia* p. 514
11 cm. Wetland grasses of W Trans-Fly. Local common. Adult black with yellow tail and rump. Juv creamy white below with smudgy streaking on breast; note streaky head.

9 Streak-headed Mannikin *Lonchura tristissima* p. 512
10 cm. The mannikin of forest edge and second growth in lowlands to mid-mountains. Uncommon. Adult blackish with yellow rump and black tail. Juv uniform dull brown. **Subsp group shown**:
tristissima (N watershed) no spotting in underparts. Not shown is *bigilalei* (southern SE Pen) whitish spots along flanks.

CASSOWARIES: Casuariidae

This primarily New Guinean family of ostrichlike birds (ratites) includes the 3 species of fearsome-looking cassowaries. Cassowaries are most closely related to the Emu of Australia, and their range extends south to the wet tropics of Australia. Species also have been introduced to the islands of Seram and New Britain. Cassowaries are forest dwellers, preferring old growth forest that offers a wide variety of fruits, the staple of their diet. These huge flightless birds stand from 1.0 to 1.7 m tall and weigh as much as 60 kg. The species are much alike, with shiny black, coarse, shaggy plumage, and bare face and neck that are pigmented blue, black, red, orange, and in some, yellow or white. The head is adorned with a helmetlike casque of hard, spongy tissue encased in a horny sheath. The casque is small in young birds but grows over a period of years to be grotesquely large in the adults of the two lowland species. Its purpose may be a social ornament of age and status. The legs and feet are massive, and the inner toe is dangerously armed with a long, sharp, daggerlike nail. Cassowaries fight and have attacked people by leaping and kicking their victim, inflicting serious wounds.

The adult female is larger than the male and supposedly exhibits brighter coloration of the bare skin, although in most cases the sex of a cassowary can't be determined with confidence. Skin color and casque shape also vary individually and geographically. Young chicks are pale grey-brown, striped with darker brown, presumably as disruptive coloration. They attain adult size within a year, during which they acquire the plain, dull brown juvenile plumage and later the glossy black adult plumage. Cassowaries are solitary and territorial. Males alone care for the eggs and young. The green eggs are laid on the forest floor with little if any attempt to make a nest.

Wild cassowaries are rarely encountered and then typically in the close space of a jungle track or streamside, the startled bird bolting into the forest, leaving the observer with a fleeting image and a racing heartbeat. In places not regularly visited by people, cassowaries can be amazingly fearless; everywhere else they are very wary. More frequently found are their impressive dinosaurian footprints and loose droppings of partially digested fruit.

Cassowaries are more likely to be seen in captivity, and in villages they are raised for their meat and feathers. In PNG their pidgin name is *muruk*. Captured as small chicks, they are not bred. Tame birds are traded long distances and so can be found in villages and towns outside their natural range. In some situations, these domestic birds are allowed to roam freely. Although wild cassowaries usually avoid people, captive birds can be ill-tempered and pugnacious. Whether a village bird or a wild one, a cassowary is best kept at a distance.

Despite pressure on cassowary populations from hunting of adults and the capture and trade of their chicks, deforestation is the greatest threat to the future of all cassowary populations.

Dwarf Cassowary *Casuarius bennetti* Pl 1
(**Little** or **Mountain Cassowary**, *C. bennettii*)
1.0–1.1 m. *Montane forest*; lowlands where other cassowary species are absent. Uncommon and shy but ubiquitous. Smallest cassowary, *lacking pendulous neck wattles* (some have a small nub on the throat). *Casque is low, triangular,* flattened in back, yielding a *wedge-shaped facial profile.* **Adult**: *Black-faced* with black of casque and facial skin extending down around eyes. Neck skin mostly blue, varying in placement of red patches. **Juv**: May have indication of a black mask. **Imm**: intermediate. **Subspp (no.?)**: Facial and neck skin varies in the placement of red patches; *westermanni* (Arfak Mts, Bird's Head), white cheek patch. **Similar spp:** The two larger, wattled cassowary species, both with high casques, occupy lowland habitats avoided by the Dwarf C; the skin of both is blue from the casque down to eyes and ear opening, although some Southern C may have black between the bill and eye. Their adult-sized, brown-plumaged young have a low, triangular casque similar to Dwarf's, but they sport an obvious, small, pendent neck wattle. **Voice:** Quiet glugging noise sometimes given by foraging bird. Fleeing bird gives a loud, staccato chugging. Other calls are low, booming grunts and a loud *mwaaaaa!* Bowed threat display with erected rump feathers includes a low rumbling growl and rapid bill-clapping. **Habits:** Traverses steep slopes and thick forest.

Nest a slight depression sparingly lined with sticks and leaves on the forest floor, sometimes between tree buttresses or by a fallen log. Eggs (2–5) green. **Range:** Endemic. NG and Yapen (Bay Is), 0–3300 m. NG mts (records from all but Fakfak, Kumawa, Wandammen, and Cyclops Mts), locally in lowlands (e.g., northern SE Pen). A possible sighting on Batanta I. Also introduced to New Britain I. **Taxonomy:** Intraspecific taxonomy needs revision. Westermann's C (*westermanni*) of N Bird's Head is sometimes considered a separate species.

Southern Cassowary *Casuarius casuarius* Pl 1
(Double-wattled, Two-wattled, or Australian Cassowary)

1.3–1.7 m. Southern lowland forests and savannah, but overlaps with Northern C on the Bird's Head and Neck. Common in remote forests, but rare or missing near population centers and from hilly coasts. Tallest cassowary. The *brown casque is narrow and high, with dorsal profile curved backward*; sometimes lopsided. *Two neck wattles* arise from folds of skin, one on each side of the throat. The wattles may take various forms: short wattle on each side of the neck, or a long double or a single split wattle at the front of the neck; rarely just 1 wattle, or as many as 3. **Adult:** Face and neck blue with red restricted to wattles in front and patch low at the back of the neck (rarely yellow); skin blue between the casque and eye (often black between bill and eye). **Juv:** Brown. Note the start of *2 neck wattles*. **Imm:** Intermediate. **Subspp (no.?):** Usually the wattle is forked, at front of neck, but birds may have wattle on each side of neck. **Similar spp:** Dwarf C avoids lowlands. Be aware that young Southern C in either brown or black plumage has a short, narrow, triangular casque, similar to Dwarf's, but will always have 2 wattles or a split wattle. Co-occurs with Northern C only in S Bird's Head lowlands, where Southerns can be identified by the 2 wattles vs the Northern's single wattle. **Voice:** Deep rumbling and booming; hisses on approach. **Habits:** Inhabits lowland rainforest, gallery forest, swamp forest, and savannah. Best seen at forest edge, riverbanks, or crossing small savannahs. Swims, crossing large rivers. Nest is a scrape lined with grass and leaves (AU data). Eggs (3–8) green. **Range:** NG and Aru Is, 0–800 m (rarely higher). NG: absent from NW Lowlands and Sepik-Ramu; presence inadequately documented and uncertain in western NW Lowlands, Huon, and N coast of SE Pen. Also AU and Seram I (where presumably introduced).

Northern Cassowary *Casuarius unappendiculatus* Pl 1
(Single- or One-wattled Cassowary)

1.2–1.5 m. Northern lowland forests. Uncommon. Similar to the Southern C in size, with the *high, narrow casque flattened at an angle from the back* and *single finger- or club-shaped neck wattle* arising from the middle of the throat (sometimes 2 wattles). Casque brownish or *bluish*. Casque profile variable: curved at the front like Southern C, or more triangular and steeply peaked, or so flattened from the back as to appear to be shoved forward; the apex is usually flattest and broadest. May have unique large blue wattle at the gape or a dull yellow iris (rather than brown). **Adult:** Head and neck usually differ from those of Southern C: mostly blue, with red (or yellow) usually wrapping completely around the base of the neck, or more rarely restricted to the front and back of the neck; reddish-brown nape patch (most birds?); facial skin blue, sometimes black between bill and eye. **Juv:** Note the start of the single neck wattle. **Imm:** Intermediate. **Subspp (no.?):** Poorly understood but variation involves casque shape, skin color of head and neck, and iris color. **Similar spp:** Dwarf C avoids the lowland habitat of Northern C and differs by black face and absence of a wattle (or at most a nub). Co-occurs with twin-wattled Southern C only in the Bird's Head lowlands. **Voice:** Similar to other species. **Habits:** Lowland rainforest and swamp forest. Wild birds are very shy; least known of the cassowaries. No information on nest. Egg (clutch?) green. **Range:** Endemic. NG, NW Is (Salawati; the cassowary on Batanta is presumed to be this species), and Bay Is (Yapen), 0–700 m. Northern lowlands of NG, from Bird's Head east to Ramu and Gogol R drainages. Details of overlap with Southern C on the Bird's Head and Neck need to be clarified.

MEGAPODES: Megapodiidae

Megapodes are an ancient family of gallinaceous (chicken-related) birds confined mainly to the Australasian region, where they are believed to have originated. New Guinea is the center of their distribution, with 11 of the 22 species. The name "megapode" refers to the birds' oversized feet, which they deploy scraping leaves and other debris on the forest floor to uncover food and to excavate or build their nest, whether an incubator mound or a burrow in volcanically heated soil.

The family is composed of 2 groups, brushturkeys and scrubfowl. Brushturkeys are black and possess a large folded tail, and they build incubation mounds of leaves and sticks. The *Aepypodius* brushturkeys have a fleshy crest and wattle and a maroon rump, whereas *Talegalla* brushturkeys are larger, more slender, longer tailed, and conspicuous for their noisy, far-carrying calls. Scrubfowl are brown or black and the bird appears tailless; the mound is built of leaves and soil, or the eggs are laid in sand or soil and incubated by volcanic or solar heat.

Megapodes inhabit forests. One species, the Wattled Brushturkey, resides entirely in the mountains. Brushturkeys are poor fliers and have not ventured beyond the continental shelf. By contrast, scrubfowl, even as hatchlings, readily cross open water and have colonized all but the most remote islands. All megapodes share a broad diet of invertebrates, small vertebrates, seeds, and fruit.

Megapodes are best known for their unusual nesting behavior. Unlike all other birds, they do not incubate their eggs with body heat. Instead they find a source of environmental heat, such as sunbaked sand or warm volcanic soil, and there bury and tend their eggs, laying one every few days. Most megapode species take the process a step further and create their own incubation environment by building a large mound of compost that generates warmth as the heaped leaves and sticks decay. Megapodes lay very large eggs. When an egg hatches, the chick digs itself out and immediately leaves the nest site and its parents to set out on a life on its own. The remarkably advanced hatchling can run, fly, feed, and fend for itself, although its furtive life remains largely a mystery. Meanwhile, the mother, free from incubation duty and chick rearing, can lay many more eggs; however, there is a trade-off: the nests are conspicuous targets for predators, especially varanid lizards and people. Megapode eggs are prized as food, and some populations have been depleted and even extirpated.

The state of knowledge regarding megapodes, particularly their distribution, systematics, and conservation status, has improved greatly since the first edition of this book. In the accounts that follow, we have relied heavily on information from Jones et al. (1995). Note that because young chicks are found disassociated from their parents, we have chosen to illustrate them as well. Hatchlings molt into an adultlike, but usually separable, juvenile plumage before acquiring immature plumage that is practically indistinguishable from the adult.

Finding megapodes is usually not difficult. *Talegalla* brushturkeys and scrubfowl are noisy birds, and their honking or cackling calls are frequently heard both day and night. *Aepypodius* brushturkeys are usually quiet. All are most often encountered near their nesting mounds but are usually very shy.

Wattled Brushturkey *Aepypodius arfakianus* Pl 2

38–46 cm. *Montane forests* of NG. Locally common where not hunted. Shy. Flushes with thrashing of heavy wingbeats. **Adult:** The only mainland brushturkey with a *maroon-brown rump and dark (olive) legs. Bare face and neck bluish white (Male) or greenish grey (Female).* **Male:** Facial colors brightest when breeding; conspicuous red comb and throat sac; pale blue nape shield; note swollen throat. **Female:** Facial colors duller than Male's; comb reduced with more dark feathering on the crown; throat not swollen; wattles smaller; nape shield absent or smaller; size smaller. **Chick:** Deep bill and dark legs; uniformly dark brown (except ochre face), only faintly barred. **Juv:** Similar to Female (separable?) but facial skin duller, more olive; iris, bill, and legs darker. **Subspp (2):** *arfakianus* (NG and Yapen) with comb red; comb blue in *misoliensis* (NW Is: Misool). **Similar spp:** Sympatric *Talegalla* brushturkeys have a black rump and bright-colored legs; these lowland species drop out at ~1500 m. Pheasant Pigeon shows chestnut wings. Waigeo BT does not co-occur. **Voice:** Rather quiet. Harsh, explosive crowing, usually near mound. Six notes, all downslurred: *kyew, kyew-kyew-kyew-kyew, kyew*, with slight hesitations after first and fifth notes. For captive Misool birds, 7 notes: *hja hja hjahjahjahja hja*, lasting 3 sec. **Habits:** Solitary or in pairs. Omnivorous, feeding on fruits, seeds, and invertebrates. Male territorial, builds his own mound. Female mates with more than one Male, laying eggs in their mounds. Mound smaller (1.5–2.0 m high and 3.0–3.5 m diameter) and often more steep-sided than those of other megapodes and composed of dead leaves and sticks. Egg white. **Range:** Endemic. NG, NW Is (Misool), and Bay Is (Yapen). All NG ranges, 1000–2700 m (rarely down to 750 m). On Misool local in limestone hills, 300–840 m.

Waigeo Brushturkey *Aepypodius bruijnii* Pl 2
(Bruijn's Brush-turkey)
41–46 cm. *Only brushturkey on Waigeo I*, where it inhabits upland forest. **Adult:** Differs from Wattled BT by its *pinkish face and neck, papillose red cap and throat wattle*, and *chestnut-brown underparts*. **Male:** Bears a characteristic *bright pink nape shield* behind the comb and pendulous pink throat sac, both contracted when not displaying. **Female:** Comb much reduced; nape shield and throat sac small or absent; face and neck instead with more black, bristlelike hairs. **Chick:** Undescribed. **Juv:** Little known; possibly not separable from Female. **Similar spp:** Dusky Scrubfowl also black, but smaller; appears tailless; has a short, pointed crest and black (not maroon) rump; nests in lowlands. **Voice:** Normally silent. Two call types: a repeated cluck and a series of raucous cackles. **Habits:** Behavior and mound similar to Wattled BT. Egg not described. **Range:** Endemic. Waigeo (NW Is), 600–970 m (summit peaks).

Red-billed Brushturkey *Talegalla cuvieri* Pl 2
53–58 cm. Lowland and hill forests of mainly Bird's Head and NW Is, where common.
Adult: *Orange-red bill, bright yellow eye, and yellow face*. Note *pale orange legs*. **Chick:** Note orangish legs and large dark eye (enhanced by black eye-ring); inseparable from Red-legged BT, but Yellow-legged BT has yellow legs and is darker and less marked with brown. **Juv:** Like Adult, but more dark feathers on head and overall plumage brownish black with rufous nape. **Similar spp:** Yellow-legged BT has yellow legs, dark face, and black bill. Red-legged BT has red face and legs, and dark bill. Mountain-dwelling Wattled BT has maroon-brown rump, dark legs, and blue-white face with wattles. **Voice:** 2 to 6 very loud notes, *kok* or *wok*, penetrating and with the quality of a donkey's bray; harsher than calls of Red-legged, the first notes a series of ascending upslurs, the last note delayed and on a lower pitch. **Habits:** Where co-occurs with the Yellow-legged, Red-billed lives at higher elevation. Social system and breeding behavior unknown. Mound low and broadly sloped, made of leaves and sticks. Egg pinkish buff. **Range:** Endemic. NG and NW Is (Misool, Salawati), 0–1500 m. NG: principally Bird's Head and Neck, enters S Lowlands east to Utakwa R (eastern extent unclear in both N and S).

Yellow-legged Brushturkey *Talegalla fuscirostris* Pl 2
(Black-billed Brush-turkey)
50–53 cm. Southern watershed, where common in lowland rainforest, generally avoiding hills.
Adult: All black with *bright yellow legs*. Dark face bluish grey; black eye and bill. **Chick:** Legs yellow; otherwise similar to chicks of Red-billed and Red-legged BTs, but upperparts more uniform dusky, with little or no barring. **Juv:** Like Adult, but more dark feathers on head and overall plumage brownish black, often with rufous nape. **Similar spp:** See Red-billed, which overlaps in the W. Overlaps with Red-legged broadly but locally in the southern foothills of PNG. **Voice:** Usually 4 or 5 very loud honking or braying notes, typical of *Talegalla*, but with the initial 3 (or 4) notes higher and upslurred, the final note slightly delayed and lower-pitched, upslurred or downslurred: *wah wah wah—wuh*. Call oft repeated. **Habits:** Similar to Red-billed BT. **Range:** Endemic. NG and Aru Is, 0–800 m. NG: Bird's Neck, S Lowlands eastward to about Port Moresby.

Red-legged Brushturkey *Talegalla jobiensis* Pl 2
(Brown-collared or **Collared Brush-turkey)**
53–61 cm. Northern watershed (very local in south); forests from lowlands to mid-mountains. Common and noisy rainforest inhabitant; in mountains favors ravines. **Adult:** *Diagnostic red legs, brown collar, dull red facial skin, and flat bushy crest of blackish feathers*. **Chick:** Inseparable from Red-billed BT; Yellow-legged BT chick with yellow legs and darker plumage; Wattled BT with heavier bill and dark legs. **Juv:** Like Adult, but head more feathered, and overall plumage brownish black.

Similar spp: Other BTs lack collar, have orange or yellow legs, and olive or grey face. Where it overlaps with smaller Yellow-legged BT, Red-legged found in mountains only. **Voice:** 2–3 very loud notes, with a honking nasal quality, last note slightly higher pitched: *wah wah weh?* or last note lowest and delayed: *waa waa—woh*. Calls day or night. **Habits:** Similar to Red-billed BT. **Range:** Endemic. NG and Yapen (Bay Is), 0–1800 m. NG: NW Lowlands east to SE Pen; also locally in hills of S Lowland and southern SE Pen.

Moluccan Scrubfowl *Eulipoa wallacei*   Pl 3
(**Wallace's Scrubfowl, Moluccan Megapode**, *Megapodius wallacei*)
30 cm. Misool (NW Is), where rare, possibly vagrant. **Adult:** A small, secretive scrubfowl with *banded brown wings and back, short olive legs, and yellow bill*. **Chick:** Upperparts heavily banded; co-occurring Dusky SF chick with upperparts darker, only faintly barred brown. **Juv:** Similar to Adult but face more heavily feathered. **In flight:** Contrasting white bar in underwing (greater under-primary coverts) may show in either Adult or Chick. **Similar spp:** Co-occurring Dusky SF larger and sooty black with longer, black legs. **Voice:** Various series of sharp, nasal notes. Duetting not reported. **Habits:** Found in foothill forest on Misool; mainly montane forest elsewhere in its range. Away from nesting grounds encountered singly or in small groups during the daytime. Female visits coastal nesting grounds at night; thought to fly to nest site, crossing ocean straits between islands. Nests in its own sand burrow colonially along coastal beach. Nesting not reported on Misool. Egg buff to brown, inseparable from Dusky SF. **Range:** 1 record from Misool (NW Is), at the far eastern edge of species' range (Moluccas), 0–600? m.

Dusky Scrubfowl *Megapodius freycinet*   Pl 3
(**Common Scrubfowl**)
34–41 cm. NW Is. Common in lowland and hilly rainforest of large and small islands. Hybridizes with Orange-footed SF locally along western coast of Bird's Head. **Adult:** *Sooty black with black legs*; bare face dull red. **Chick:** Dark brown; note dark legs; Orange-footed SF paler and browner, has orange legs. See very rare Moluccan SF. **Juv:** Small version of Adult. **Similar species:** Orange-footed SF browner, with orange legs. Biak SF similar but does not co-occur, has red legs. Moluccan SF, rare on Misool I, is browner, with banded wings and olive legs. **Voice:** Duet, first part laughing *kejowowowowowowow* . . . second part, 2 notes: *keya keyauw*. . . . **Habits:** In pairs. Nest mounds are built of soil or sand, with leaves. The large mound is a broad, gently sloped cone often with a summit crater, suggesting a miniature volcano. Egg brown to buff. **Range:** NW Is (all) and NG, where very local along west coast of the Bird's Head, 0–500 m. Also N Moluccas. **Taxonomy:** Formerly included with the following 4 species as Common Scrubfowl (*M. freycinet*).

Biak Scrubfowl *Megapodius geelvinkianus*   Pl 3
30 cm. *Only megapode on Bay Is*; in forest and scrub. A small blackish megapode with bright red face and variably red to reddish-black legs. **Adult:** *Dark like Dusky SF* (does not co-occur), *but small*, somewhat more brownish (not black), and with brighter red face and *reddish legs*. Possible introgression on smaller islands with larger New Guinea SF, which has olive-brown upperparts and dark olive legs. May co-occur with New Guinea SF on Yapen I and with larger, rufous brown Orange-footed SF near Manokwari. **Chick:** No information. (Presumed to have red legs.) **Juv:** No information. **Voice:** Duet, one bird calling *keryak-keyer*, the second bird simultaneously softly clucking, *urrr kuh-kuh-kuh* . . . ; very vocal. **Habits:** Habits and nest undescribed, but probably like Dusky SF. **Range:** Endemic. Bay Is (Biak, Mios Num, Numfor, and nearby small islands; doubtfully Yapen). **Taxonomy:** In some treatments lumped with Dusky SF, and formerly in all-inclusive Common SF.

Melanesian Scrubfowl *Megapodius eremita*   Pl 3
35–39 cm. *Karkar and Bagabag Is* off Sepik-Ramu, where it hybridizes with New Guinea SF. Inhabits forest. **Adult:** *A very dark scrubfowl with obviously bare forehead and insignificant, short, broad crest*. Bare skin of face and neck red; legs vary from greenish yellow, to olive, to blackish. New Guinea SF slightly larger and with paler olive-brown upperparts, feathered forehead, and more obvious pointed crest. **Chicks:** Similar to Dusky SF. **Juv:** Like Adult, but lower wings and back, flank, and belly sometimes browner. **Voice:** On New Britain, duet begins with a short, laughing *keyououourrr*, the partner joining in shortly after with 2 notes, *keyou keyourr*. Various series of *ko* or *kio* notes. Also a *kee-yah* call. **Habits:** Similar to Dusky SF. Nest mounds found on Karkar; elsewhere also nests in burrows in volcanically warmed soil or in sunbaked beach sand. **Range:** Karkar and Bagabag Is

off Sepik-Ramu, 0–700 m. Also Admiralty, Bismarck, and Solomon Is. **Taxonomy:** Formerly included in Common SF.

New Guinea Scrubfowl *Megapodius decollatus* Pl 3
(*Megapodius affinis*)
33–35 cm. Widespread *northern* scrubfowl. Common in lowland and montane forests; also locally on south slope of Central Ranges where it replaces very similar Orange-footed SF in hills and mountains. Hybridizes with Melanesian SF on Karkar and Bagabag Is. **Adult:** From Orange-footed SF by its *dark legs that are yellowish, greenish, brownish, or black*; wings and back darker and less rufous; crest shorter and face somewhat more sparsely feathered; shape different, with tail shorter and legs longer. Hybridizes with Melanesian SF (refer to those species). **Chick:** From Orange-footed by dark brown legs (not bright orange). **Subspp:** None, but montane birds from SE Pen (Wau) darker above. **Voice:** Duet initiated with *kok, nyacal, nailleue* ascending to a high screech; the partner joins in with a stuttering *nu-nu-nu-nu-nu*. **Habits:** Little known, apparently similar to better-known Orange-footed. Nests in mounds identical to Orange-footed; egg similar. Sometimes lays eggs in mounds of brushturkeys. **Range:** Endemic. NG, Yapen (Bay Is), and small nearshore islands off Sepik-Ramu, including Manam and Karkar Is, 0–2950 m. NG: NW Lowlands to SE Pen; S Lowland hills at upper Otakwa basin, Moroka district, and elsewhere. **Taxonomy:** Formerly included in Common SF.

Orange-footed Scrubfowl *Megapodius reinwardt* Pl 3
37 cm. The widespread *southern* scrubfowl. Lowlands only, except hills in SE. Common in all wooded habitats. A *large* megapode identified at all ages by its *bright orange legs*. Hybridizes with New Guinea SF on W Bird's Head. **Adult:** *Crest long and pointed*; throat well feathered. Wings and back usually more rufous than other scrubfowl. **Chick:** Orange legs; plumage variable, generally rufous brown with obscure barring. **Subspp (2):** *macgillivrayi* (SE Is) compared with *reinwardt* (NG); smaller, darker, and more olive, forehead partly bare, but legs orange. Birds in montane valleys of SE Pen similar to *macgillivaryi*. **Voice:** Includes *eukeu-keu-keu-keuw* or *kokauw-krauw . . . kokauw-krauw* from Bird's Head and *crou-o-o-o-ou* from SE Pen. **Habits:** Seems to be monogamous and territorial. Builds a volcano-like nesting mound of sand, dirt, and leafy debris reaching 2 m in height. Nesting most active during monsoonal wet season. Egg brown to buff. **Range:** NG, Batanta (NW Is), Aru, and SE Is (all), 0–2600 m, but usually lowlands. NG: Bird's Head and Neck, S Lowlands, and SE Pen, including N coast. Also Lesser Sunda Is and AU. **Taxonomy:** Formerly included in Common SF.

PHEASANTS, PARTRIDGES, AND QUAIL: Phasianidae

New Guinea's 3 species of quail belong to the large, primarily Asian and African family of pheasants and partridges (182 spp). Quail are small, compact ground dwellers identified by their short, stubby bill and seemingly tailless appearance. They are colored to blend in with their grassy habitat, although in some the male's plumage stands out. Downy chicks are usually brownish with buff face; they pass through a juvenile plumage that resembles the adult female. Unlike most of their relatives, the *Coturnix* quail are capable of flying long distances and known to disperse to new habitat when the need arises. Perhaps this adventurous behavior brought them to the Australasian region originally; however, long distance movements are probably rare, as the daily life of quail generally involves no flight at all. These cryptic little birds live as monogamous pairs or small coveys in grassy and weedy vegetation, where they remain hidden most of the time. They venture out onto paths or other clearings to feed, offering the best chance to see them. If flushed they erupt in swift, whirring flight, traveling only a short distance (and often calling, unlike the similar-looking buttonquail) before dropping down in safer cover.

Snow Mountain Quail *Anurophasis monorthonyx* Pl 4
(*Coturnix monorthonyx*)
25–28 cm. *A large, partridge-sized quail of alpine grasslands and scrub in the W Ranges*. Uncommon and shy. Upperparts heavily cross-barred rufous brown and black. **Male:** Underparts chestnut-brown with dark barring. **Female:** Similar, but underparts white or buff and more heavily barred. **Juv:** Similar to Adult, but barring less distinct. **Similar spp:** Does not co-occur with other quail. Brown Q much

smaller, streaked rather than barred. **Voice:** A loud squeal, *queeU* or *queeah*, often repeated 4 or 5 times when alarmed. **Habits:** Frequents grassland and the edges of heavy alpine scrub with trees and brush. When flushed, noisily shoots off at a sharp angle, calling, then alighting 50–100 m away. Encountered singly or in parties of 2–3. Feeds on flowers, fruit, seeds, and foliage, also insects. Nest concealed beneath a spreading tussock grass; nest a shallow depression on a pad of grass. Eggs (3) pale brown with dark chocolate spots. **Range:** Endemic. NG: grassland plateaus on the highest peaks of the W Ranges, including Mts Jaya (Carstensz), Trikora (Wilhelmina), and Mandala, 3200–3800 m. **Taxonomy:** Relationships of this quail to *Coturnix* spp are unstudied.

Brown Quail *Coturnix ypsilophora* Pl 4
(*Synoicus* or *Coturnix australis*)

15–19 cm. *Dullest colored* of the quail-like birds. Locally common throughout lowland and montane grasslands of eastern and central NG. Generally dull grey or brown with rather *finely and evenly textured markings*. More sexually dichromatic in lowlands. Iris of both sexes red or brown in the lowlands, orange or yellow in the mountains. **Male:** *In the lowlands, all grey* with few dark markings. In mountains, more similar to Female. **Female:** Brown. All races similar; darker overall, more coarsely marked than Male. **Juv:** Similar to Female but more dark-spotted than barred below. **Subspp (6, in 3 groups):** *plumbeus* (NW and S Lowlands, eastward), 3 lowland races with sooty-colored Male; *mafulu* (E Ranges, SE Pen), 2 mid-mountain races, Male warm buff above with lightly barred underparts grey-brown or chestnut; *monticola* (alpine SE Pen) similar to Female. **Similar spp:** Female King Q is much smaller, with a pale buff face bisected by a dark line below eye. Red-backed Buttonquail has a distinct pale area on the forewing, more pointed wings in flight (sharper angle at elbow), and unmarked underparts. Much larger Snow Mountain Q does not co-occur. Scrubfowl chicks inhabit forest and are much darker, with longer legs and neck and plumage more uniform. **Voice:** Song, an interrogatory, whistled *wi weeeeeeii*, the second part of the call rising in pitch. In alpine moors of E Ranges: a 3-note squeaky whistle, *cuh cuh queeh* or 2 notes, *cuk whee*. When flushed, a harsh nasal but musical *chak chak chak*. **Habits:** Readily colonizes new grassy clearings; expected anywhere in suitable habitat. Feeds quietly in dense grass, often in coveys. Feeds on seeds, herbs, and insects. Runs rapidly when approached. Rarely flies far when flushed. Nests on ground in grass or weeds; a shallow depression lined with grass, surrounding material pulled in to form hood. Eggs (4–6) pale, usually freckled. **Range:** All NG except W mts and Bird's Head, 0–3700 m. Distribution of alpine populations in need of study, known as far west as Tari and Porgera areas. Also from Lesser Sundas to AU. **Taxonomy:** Relationship between the NG lowland vs montane races and those elsewhere is yet to be resolved and could indicate more than one species in NG.

King Quail *Excalfactoria chinensis* Pl 4
(**Asian Blue**, **Blue-breasted**, or **Chinese Quail**, *Coturnix chinensis*)

11–12 cm. *A tiny quail*; widespread but local and rarely observed. Mainly in extensive mid-mountain clearings where it inhabits weedy places at the edges of grasslands or gardens. **Male:** Unmistakable with *black throat, white bib*, blue-grey breast and flanks contrasting with deep chestnut belly and undertail. **Female:** Note *clear buffy face and chin*, with a dark "moustache line" from bill to ear-coverts; buffy underparts with dark crescent-shaped barring on lower throat and flanks. **Juv:** Similar to adult Female but spotted or streaked below, rather than barred. **Subspp (2):** Minor. **Similar spp:** Brown Q larger and lacks buffy face. Red-backed Buttonquail has clear, unmarked underparts. **Voice:** A high-pitched descending whistle of 3 notes; quite vocal at dusk. When flushed: *peep peep peep* (AU data). **Habits:** Favors cleared areas grown up with dense weeds or crops rather than pure grassland. Remains concealed in typical quail-fashion, flushing at close range. Feeds on small seeds and herbs. Singly, in pairs, but sometimes in larger groups. Nests in dense cover; nest a depression lined with grass, sometimes with hood (AU data). Eggs (4–7) pale brown, darkly spotted. **Range:** NG (plus unconfirmed sightings from NW Is of Batanta and Salawati), 0–2300 m. Also SE Asia, Philippines, Indonesia, Bismarck Is, and AU.

MAGPIE GOOSE: Anseranatidae

This archaic family is only distantly related to the true geese, and its sole surviving member, the Magpie Goose, is confined to Australasia. A large and ungainly looking waterbird seen in flocks of its own kind.

Magpie Goose *Anseranas semipalmata* Pl 12
70–90 cm. Common breeding resident and AU migrant to swampy floodplains of the Trans-Fly.
Adult: *Unmistakable—a very large black-and-white gooselike bird with a long, black neck.* The head exhibits *a high-peaked crown*; note also the *long, hook-tipped bill*; the pink face is bare of feathers.
Male: Larger on average with a higher crown. **Juv:** Mantle black, not white; white flanks mottled with dark grey; crown not swollen. **In flight:** Vaguely crane- or storklike, with broad wings, fingered primaries, long neck, and pink forehead; legs much shorter, though. **Voice:** A resonant honking, given in flight and by feeding birds. Male voice louder and deeper pitched than Female.
Habits: Highly social and noisy, in small or large flocks. Numbers build during the austral summer wet season. Inhabits lakes, swamps, and river margins. Grazes and dabbles in shallows. Digs with its bill for tubers. Flies in lines between roosting and foraging sites. Breeds as water level drops, nesting in a marsh as a colony. A cooperative breeder with the primary pair typically joined by another female or rarely by more auxiliary birds. Nest a trampled and woven platform of grasses with central cup. Eggs (1–16) white, laid by 1–2 females (AU data). Colonies disperse after breeding.
Range: NG: Trans-Fly and adjacent S Lowlands, 0–100 m. Resident, but also migrates from AU across Torres Strait.

DUCKS, GEESE, AND SWANS: Anatidae

Of the 172 species of waterfowl worldwide, 15 occur in the New Guinea Region. There are 9 resident species, but only one, Salvadori's Teal, is endemic; the rest are shared with nearby islands and Australia. Six species are nonbreeding visitors to the region—3 from Australia and 3 from northern Asia. Perhaps all 11 species found in both New Guinea and Australia travel between the two landmasses to some extent, driven mainly by winter drought and summer flooding in northern Australia.

This is a family of small to large waterbirds, variously colored and patterned. Wing patterns are often useful for identifying look-alike species. Most are long-necked, with short legs and webbed feet for swimming. Nearly all are sociable, gathering into feeding congregations, especially after breeding. Flight is fast, with rapid wingbeats. Waterfowl are highly mobile and often disperse widely. New Guinea species prefer freshwater swamps, highland lakes, and rivers, yet most are prevalent in the lowlands, particularly in the South. Their diet is primarily of waterweeds, algae, and aquatic invertebrates. All species are monogamous and nest near water. Nests of the various species are placed in a tree hollow or on the ground under cover. Sewage ponds or wetlands that are drying up are good places to find waterfowl.

Spotted Whistling Duck *Dendrocygna guttata* Pl 5
(Spotted Tree Duck)
43–50 cm. Common resident on freshwater marshes in lowlands. **Adult:** A *grey-brown* whistling duck with *white-spotted flanks* and *dark crown and wings*. Note short nuchal crest. **Juv:** Duller, with white flank spots and streaks. **In flight:** *Unpatterned wings and plain overall, but look for flank spots*. **Similar spp:** Other whistling ducks have bold, white flank streaks, rather than spots. **Voice:** A wheezy, nasal *zzeou*, different from the twittering of the other two whistling ducks. **Habits:** Usually in flocks. Inhabits most lowland marshy habitats including mangroves. Usually feeds at night or early morning, arriving at feeding site in twilight. Dabbles in water or grazes on banks; also dives. Takes small seeds and snails, probably other plant material and aquatic invertebrates. Roosts in trees by day. Flies in loose groups. Reputedly nests in tree cavity. **Range:** NG and Aru Is, plus sightings from NW Is (Batanta, Kofiau), 0–200 m. Also E Wallacea, Philippine Is, Bismarck Is, and AU.

Plumed Whistling Duck *Dendrocygna eytoni* Pl 5
(Grass Whistle-Duck, Plumed Tree Duck)
40–45 cm. Apparently a rare AU visitor to wetlands in savannah country, although status in NG Region unclear. Found among other whistling ducks. **Adult:** A *pale grey-and-buff* tree duck with *diagnostic cream-colored flank plumes that are elongated and protruding*; lacks dark crest or cap. **Juv:** Paler. Breast barring indistinct; plumes shorter. **In flight:** *Unpatterned wings same pale color as back*; narrow white rump patch. **Similar spp:** Wandering WD has much shorter plumes and is more boldly colored with black cap and chestnut shoulders. **Voice:** Calls include a shrill whistle, *whew*, and wheezy twittering in flight. **Habits:** Gregarious and noisy; in NG Region usually in small numbers. Rests on the banks of muddy waterholes. Herbivorous, grazing on short grass on land and dabbling in water. **Range:** NG: Trans-Fly and Port Moresby region of SE Pen, 0–200 m. Also AU.

Wandering Whistling Duck *Dendrocygna arcuata* Pl 5
(Water Whistle-Duck, Whistling Tree Duck)
40–45 cm. Locally common resident and visitor from AU. **Adult:** *A brightly colored whistling duck with distinctive orange-brown plumage, white flanks, and black cap*. **Juv:** Duller, less boldly marked. **In flight:** *Chestnut shoulders and large white rump patch.* **Subspp (1):** *australis* (also AU). **Similar spp:** Other whistling ducks plainer-looking, without black cap, chestnut shoulder, and conspicuous white rump. **Voice:** A cheerful whistle and twittering, descending series of 4–6 notes. **Habits:** Gregarious. Inhabits freshwater lagoons, swamps, and rivers. Flocks move about in reaction to local water conditions and may gather into large feeding aggregations. Active day or night. Feeds on aquatic vegetation and seeds taken by dabbling in shallow water; also tips and dives, but does not graze. Rests on swamp edges and rarely settles in trees. Nests on the ground sheltered by shrubs or grass; nest a mound of trampled and woven grass with depression in the center, no down (AU data). Eggs (6–15) creamy white. **Range:** NG, 0–500 m, rarely to E Ranges. Also Greater Sundas and Philippines to New Britain I and AU.

Black Swan *Cygnus atratus*
110–140 cm. An Australian icon, vagrant to Trans-Fly wetlands. **Adult:** *An unmistakable, enormous black waterfowl with extraordinarily long, curved neck and red bill.* **Juv:** Dusky grey. **In flight:** White primaries. **Voice:** Normally silent. Call a musical trumpeting. **Habits:** Drawn to large expanses of freshwater with vegetation to feed on. **Range:** NG: 2 records from Trans-Fly. Also AU, introduced to NZ.

Raja Shelduck *Tadorna radjah* Pl 6
(White-headed Shelduck, Burdekin Shelduck)
48 cm. Locally common resident in lowland wetlands. **Adult:** *Only large duck with all-white head, neck, and underparts.* Note narrow chestnut breast band. **Juv:** Similar but with brown crown and nape; back dusky, not chestnut. **In flight:** White shoulder patch and green speculum. **Subspp (2):** *radjah* (also Moluccas) with blacker dorsal plumage than browner *rufitergum* (Trans-Fly and AU), though some birds intermediate. **Similar spp:** Cotton Pygmy Goose much smaller with black on crown and no white shoulder. **Voice:** Vocal. Male's call is a thin, rattling whistle given alternately with Female's lower, harsh rattling note. **Habits:** In pairs or small groups on freshwater lagoons, swamps, muddy pools, and less commonly on mangrove creeks and estuaries. Eats mollusks and other aquatic invertebrates taken by dabbling while swimming or wading. May roost in trees. Flight is slow, with measured wingbeat. Nests in a tree hollow (AU data). Eggs (clutch?) white. **Range:** NG, where mainly from Trans-Fly to Port Moresby area; island records include NW Is (Batanta, Gag, Gebe, Kofiau, Salawati, Waigeo), Bay Is (Yapen), Aru Is, and SE Is (D'Entrecasteaux Is: Goodenough and Fergusson), 0–200 m. Also Moluccas and AU.

Salvadori's Teal *Salvadorina waigiuensis* Pl 6
(Salvadori's Duck, *Anas waigiuensis*)
43 cm. Uncommon resident in two different habitats—rocky montane streams and rivers, and alpine lakes. NG's unique duck. **Adult:** *Chocolate brown head with orange-yellow bill* and body profusely barred black and white. Cocks its tail when alarmed. **Male's** eye red, **Female's** brown. **In flight:** Green speculum bordered with white bars. **Similar spp:** Pacific Black D larger and all dark with white facial stripes. Northern Pintail male (a vagrant) has white neck and bluish bill. **Voice:** Usually silent, even when flushed. Call is a froglike clicking series of 7–8 notes for 3–4 sec, *kek kek . . . kek*. **Habits:** Solitary or in pairs in alpine lakes and fast-flowing rocky streams down to the fall line at the

foot of the ranges. Feeds on aquatic insects, tadpoles, and plants by dabbling and diving. Flight is fast, often low over the course of the river. Nests on a stream or lake bank above drift line; nest a grass-lined depression. Eggs (~3) pale brown. **Range:** Endemic. NG mts, probably including Bird's Head and Neck, 100–4100 m. **Taxonomy:** Once placed in the genus *Anas*, Salvadori's Teal is now regarded as having no close relatives and is instead an "old endemic."

Australian Wood-Duck *Chenonetta jubata*
(Maned Duck)
44–56 cm. AU vagrant. A large duck with gooselike posture. **Male:** *Head dark brown, maned; grey body with black tail.* **Female:** Similar, but has paler head with white lines above and below eye and heavy black-and-white spotting on breast. **Juv:** Like Female, but flanks white-spotted. **In flight:** White secondaries. **Habits:** Frequents wetlands. Grazes on plants away from water. **Range:** NG: 1 record from Port Moresby (SE Pen). Also AU.

Cotton Pygmy Goose *Nettapus coromandelianus*   Pl 6
(White Pygmy-Goose)
33–36 cm. Resident of *northern watershed* where once locally common in quiet freshwater with floating vegetation. Current status uncertain owing to habitat destruction by introduced herbivorous fish species. *A tiny green-and-white waterfowl.* Distinguished from the more widespread Green PG by *lack of dark scalloping on the flank.* **Male:** Breeding plumage with white head and neck. Nonbreeding plumage similar but black breast collar indistinct. In flight, white band across both primaries and secondaries. **Female:** Thick white eyebrow; short dark eye-stripe does not meet cap. In flight, thin white trailing edge on secondaries. **Juv:** Like Female, but back paler and breast faintly mottled rather than barred. **Subspp (1):** *coromandelianus* (Asia to NG). **Similar spp:** Green PG. **Voice:** Male cackles, Female quacks. **Habits:** Similar to Green PG. Feeds on aquatic vegetation, particularly *Hydrilla* and pondweed (AU data). **Range:** NG: Sepik-Ramu and Lake Kandep in E Ranges; rarely SE Pen (Port Moresby), 0–2250 m. Also Asia to AU.

Green Pygmy Goose *Nettapus pulchellus*   Pl 6
30–36 cm. Resident of *southern watershed* where common in lowland freshwater with water lilies. *A tiny waterfowl with dark green upperparts and white cheek.* Separated from Cotton PG by *barred flank* and in flight by a *conspicuous white panel in the secondaries.* **Male:** Breeding plumage with *dark green head and white cheek* patch. Nonbreeding plumage similar but cheek patch obscure. **Female:** *Dark cap, white cheek.* The short, narrow eyebrow is visible only at close range; the long, dark eye-stripe connects with the black cap on the nape. **Juv:** Like Female, but back and wings duller, less iridescent. **Similar spp:** See Cotton PG. **Voice:** Male contact call *pee whit.* Female contact call *pee yew.* Alarm call *whit* or *whit whit.* Also a musical trill. **Habits:** Usually in flocks. Prefers still, shallow wetlands with water lilies and other floating and emergent vegetation. Dabbles and dives, feeding on live plants, particularly water lily seeds and flowers (AU data). Flight swift and agile. Nests in tree hollow; presumably lined with down (AU data). Eggs (clutch?) white. **Range:** NG: Bird's Head and Neck, S Lowlands, Trans-Fly, southern SE Pen, and rarely Sepik-Ramu, 0–500 m. Also Wallacea and AU.

Eurasian Wigeon *Anas penelope*   Pl 5
45–51 cm. Palearctic vagrant to lowland wetlands during austral summer. **Male:** *Rich chestnut head with buff cap, pinkish breast, and grey body.* **Female and Juv:** A dark brown duck with *unpatterned head and peaked forehead; breast cinnamon, with a sharp demarcation from white underparts.* **In flight:** White forewing, dark flight-feathers, and green speculum. **Voice:** Male gives a whistled *wheeoo.* Female makes a low growl. **Habits:** Frequents open bodies of freshwater with grassy margins. Feeds by dabbling and tipping; grazes on land. **Range:** NG: records from Bird's Head and SE Pen. Breeds in Palearctic, winters southward.

Pacific Black Duck *Anas superciliosa*   Pl 5
(Black or Grey Duck, Australian Black Duck)
48–56 cm. A widespread but usually thinly distributed resident on all sorts of freshwater wetlands. Also a likely visitor from AU. **Adult:** *A large, blackish duck with boldly patterned pale face marked by a black crown and heavy black eye-stripe.* **Juv:** Similar but with streaky underparts. **In flight:** Green speculum with black border. White underwing-coverts. **Subspp (2):** Minor. **Similar spp:** Other dabbling ducks are paler overall and lack the bold striped face. Garganey female has similar striped

facial pattern but is paler overall and much smaller. Hardhead is also dark, but has smooth-textured (not scaly) plumage, dark face, and more sloped head profile. **Voice:** Male gives a *raaaehb* call, Female a loud quacking (AU data). **Habits:** Singly, in pairs, or in small parties; mixes with other waterfowl. On any freshwater site from sea level to alpine lakes, though prefers the lowlands and quiet water. Dabbles, tips, and grazes for plant material, especially seeds, and aquatic invertebrates. Usually rests on shore; forages in shallows or on open water far from cover. Flight fast and powerful. Nests in tree hollows or on the ground under cover; down-lined nest may have no other material or is fashioned from loose vegetation (AU data). Eggs (7–14) white. **Range:** NG and SE Is (Misima and Woodlark), but to be expected on other satellite islands, 0–3700 m. Also Greater Sundas, Wallacea, Oceanea, AU, and NZ.

Grey Teal *Anas gracilis* Pl 5
(*A. gibberifrons*)
36–43 cm. Occurrence similar to Pacific Black D: a widespread and adaptable resident. Also visits from AU. **Adult:** *A small, plain brownish duck with a domed head profile, a distinctly darker crown and nape, and pale throat.* Red eye. **Juv:** Paler and with a somewhat streaky breast. **In flight:** Speculum black (with some green), but note white wedge in front; white triangle in underwing. **Similar spp:** Other dabbling ducks have marked facial patterns. Female Eurasian Wigeon's body feathers lack strong pale scalloping. **Voice:** Noisy. Male gives a *gedee-oo* call and a whistle followed by a grunt; Female gives a *cug-cug-cug* call and a harsh *wreee-ak* alarm call (AU data). **Habits:** Singly, in pairs, or in small parties, usually gregarious. On any area of freshwater, from sea level to alpine lakes, but prefers large areas of open water; also brackish water in mangroves, rarely tidal flats. Feeds by dabbling and tipping, mainly for seeds. Swift flight. Nesting similar to Pacific Black D. **Range:** NG and Aru Is, plus sighting from NW Is (Salawati), 0–3200 m. Also Bismarck Is, Solomon Is, AU, and NZ. **Taxonomy:** Separated from Sunda Teal (*A. gibberifrons*). **Extralimital spp: Chestnut Teal** (*Anas castanea*) of AU, one hypothetical record of a bird collected in the Trans-Fly; whereabouts of specimen unknown.

Northern Pintail *Anas acuta*
50–65 cm. Vagrant. Eurasian migrant during austral summer. Note *blue-grey bill*. **Male:** *Dark chocolate head strongly contrasts with white neck and breast*; upperparts grey. **Female and Juv:** A large, all-brown duck with *long neck*. **In flight:** *Speculum bronzy and white-bordered*. **Subspp (1):** *acuta* (Northern Hemisphere). **Similar spp:** Grey Teal much smaller, short-necked, and with black-and-white speculum. **Habits:** A typical dabbling duck. **Range:** NG: a few records, but could turn up anywhere. Breeds in Arctic and boreal Eurasia and N Amer; winters to the south. **Extralimital spp: Northern Shoveler** (*Anas clypeata*), another long-distance northern migrant, has a wide, extra-long bill that is orange-brown in the drab female; male has green head and chestnut-and-white underparts.

Garganey *Anas querquedula* Pl 5
37–41 cm. The only Palearctic duck that migrates to the NG Region in fair numbers. Locally common on lowland wetlands during austral summer. **Male:** *Head maroon-brown with prominent white eyebrow; underparts white.* **Female and Juv:** A small, brown duck with dark eye-stripe. **In flight:** Both sexes show *blue-grey shoulders* and white-bordered, green speculum. **Similar spp:** Other small ducks lack prominent white eyebrow. Pacific Black D has similar facial pattern but is much larger and darker overall, without white below. **Voice:** Male gives a mechanical-sounding rattle. **Habits:** Singly or in small flocks, frequenting a wide variety of wetlands. Feeding typical of dabbling ducks, taking aquatic plants and invertebrates. **Range:** NG. Breeds in N Eurasia, winters in Africa and S Asia to AU.

Hardhead *Aythya australis* Pl 5
(Australian White-eyed Duck)
46–49 cm. Sporadically disperses from AU to NG, and some remain to breed. Overall a *dark brown duck with contrasting white undertail* and belly patch. Distinctive head profile shows *long, sloped forehead and bill*. **Male's** eye white, **Female's and Juv's** brown. **In flight:** *White blaze across flight-feathers of upperwing*; white underwing and belly. **Similar spp:** Pacific Black D also dark and with white underwing, but has pale face, scaly plumage, and lacks white belly. **Voice:** Generally silent. Male gives a wheezy whistle, Female a loud, harsh rattle. **Habits:** Usually in pairs or small groups in deep open freshwater, where it dives for aquatic plants and shellfish. Nests on the ground in thick cover near water; nest a dish-shaped platform of plant material (AU data). Eggs (6–18?) white. **Range:** NG and possibly NW Is, but expected on other satellite islands, 0–3200 m. Also Java, Wallacea, Bismarck Is, Solomon Is, and AU.

PETRELS AND SHEARWATERS: Procellariidae

Petrels and shearwaters are medium-sized to large seabirds that are long-winged and have tubular nostrils. With their hook-tipped bills, they capture fish, squid, and smaller marine life. These are birds of the open ocean that have mastered the winds with a distinctive type of arcing flight called dynamic soaring. The family consists of 90 species ranging throughout the world's seas.

The actual number of species that frequent New Guinea waters is a matter of debate, reflecting the confusing diversity of similar species that might be present and the paucity of fieldwork conducted in the open oceans in the region. One species is known to breed locally, and two others may do so; however, most species in the region are migrants or vagrants, of which one species nests along the rim of the north Pacific and the rest breed on islands in the southern oceans. We have written accounts for species recorded as specimens or from published field identifications, particularly the extensive and quantitative survey by Cheshire (2010). Be aware that petrels and shearwaters wander, and even species from far beyond the western Pacific may turn up as vagrants. Readers especially interested in petrels, shearwaters, and storm-petrels should study field guides for Australia and guides that specialize in seabirds, such as Harrison (1983) and Onley and Scofield (2007). Unfortunately, some species cannot be separated with confidence at sea, and identification must often be based on beach-washed specimens, of which there have been few for New Guinea.

To see petrels and shearwaters well you must be on a boat in the open ocean. Here petrels and shearwaters fly low over the water and sometimes rest on the sea. They are seldom numerous, apart from when they gather to feed in association with tuna or when they migrate in seasonal flocks, traveling in a single direction.

Scant information exists on seabird colonies in the New Guinea Region, particularly for petrels and shearwaters. These burrow nesters seem especially vulnerable to predators such as small mammals, snakes, and people. Exploration of remote, small islands might yield new colonies of procellariids. Another potential nesting habitat is on the mountaintops of the medium-sized islands, for example the D'Entrecasteaux Is, where fields of matted ferns may offer refuge for nesting petrels and small shearwaters, as they do on other Pacific islands. Petrels and shearwaters are monogamous, with both parents sharing nesting duties. They typically dig a nesting burrow and lay a single egg.

Southern Giant Petrel *Macronectes giganteus*
86–100 cm; wingspan 185–205 cm. Vagrant. *Albatross-sized, stocky and long-winged.*
Adult: Polymorphic. Most birds are mottled greyish brown and have a whitish head. Rare white morph mainly white with asymmetrical black marks. *Stout bill appears all pale at a distance;* bill yellowish horn with pale green tip. **Juv:** All black; bill color same as Adult. **Habits:** Flight is slow, typically with much flapping. Follows ships in open seas; sits on water. Feeds on carrion and squid. **Range:** One band recovery from Gulf of Papua. Circumpolar; breeds in Antarctica and sub-antarctic islands, disperses over southern oceans, normally far south of NG. **Extralimital spp: Northern Giant Petrel** (*M. halli*) similar, but has a dark crown and red-brown tip to pale bill; pale underparts contrast with dark upperparts. **Black-footed Albatross** (*Phoebastria nigripes*) of tropical N Pacific is of similar size but arcs and glides on long wings; bird all sooty black; note black bill edged with white feathers.

Slender-billed Prion *Pachyptila belcheri*
25-26 cm; wingspan 56 cm. Vagrant. **All plumages:** *Blue-grey with obscure M pattern across upperwings and a narrow dark tip to tail* usually confined to central feathers; underparts white, including underwing. Note broad white eyebrow. **Habits:** Not gregarious; flies close to water, scooping up plankton from the surface. Flight fast, erratic, frequently dropping feet into the water. **Range:** One specimen from Gulf of Papua. Breeds on sub-antarctic and temperate islands of Southern Hemisphere and winters mainly in subtropical water south of 30°S latitude. Occasionally irrupts into the tropics.
Taxonomy: The NG specimen was originally identified as *P. turtur*, but now has been re-identified as *P. belcheri*. Molecular studies suggest *P. belcheri* may be a subspecies of the Broad-billed Prion *P. vittata*. Field identification of prions poses a major challenge.

Providence Petrel *Pterodroma solandri*
40 cm; wingspan 94 cm. Vagrant. **All plumages:** *One of 2 all-dark gadfly petrels reported from NG Region (the other is Kermadec P). Long, wedge-shaped tail and white panel in the underwing at base of primaries.* Head and wings darker than body; white feathering around base of the bill. **Similar spp:** Wedge-tailed Shearwater has longer, narrower bill and lacks white panels in underwing. Kermadec P has shorter, blunter tail and shows essentially no contrast between the back vs wings and head; upperwing with pale primary shafts. Herald P dark morph not recorded from NG waters; it has shorter tail, and the underwing shows more white that extends onto the secondaries and coverts toward the body. **Range:** 1 record from N Solomon Sea. Breeds on Lord Howe I between AU and NZ, winters north-central Pacific, passing east of NG Region.

Kermadec Petrel *Pterodroma neglecta*
38 cm; wingspan 92 cm. Vagrant. **All plumages:** *Plumage variable but note short, blunt tail and short, thick bill.* Dark morph similar to Providence P but *upperwing shows pale shafts on primaries.* Pale morph has whitish head and underparts contrasting with dark upperparts and wings. **Subspp:** No information. **Similar spp:** Providence P has all-dark upperwing and longer, wedge-shaped tail. **Habits:** Solitary; does not follow ships. **Range:** A few records from Bismarck Sea. Breeds in subtropical S Pacific, disperses to central tropical Pacific.

Herald Petrel *Pterodroma heraldica* Pl 7
(*P. arminjoniana*)
34–39 cm; wingspan 88–102 cm. Rare visitor. **All plumages:** Much individual variation. All morphs have a *distinctive underwing pattern with white panel on the outer wing* that contrasts with the *mostly dark coverts.* Pale morph *recorded from NG seas:* upperparts sooty to ash brown with *variable amounts of white in face,* underparts white with a variable smudgy, dark collar. Intermediate morph with dark head and upper breast, white lower breast and belly, similar to Tahiti P, but note underwing pattern. *Dark morph all dark and difficult to distinguish from related dark petrels.* **Habits:** Little known, although reported feeding with Wedge-tailed Shearwaters. **Range:** A few sight records from Coral Sea. Breeds on islands in tropical western and central Pacific with a few pairs on Raine I SE of Cape York, AU.

Collared Petrel *Pterodroma brevipes*
(Gould's Petrel, *P. leucoptera*)
30 cm; wingspan 71 cm. Vagrant. **All plumages:** *A small, compact petrel* with plumage morphs varying from pale to dark. *Unique combination of dark hood, a dark M across the upperwings contrasting with the grey back, and dark tip to the tail.* Underparts range from white to dark grey. Face with *white forehead and throat. Distinctive white underwing-coverts marked by a conspicuous black leading edge turning diagonal across the wing from the wrist.* Upperparts and sides of head and neck dark, becoming a *collar around the throat in all but the palest birds.* **Similar spp:** Gould's and Pycroft's Ps. **Habits:** Flight usually a slow flap-and-glide progression low to the water, like that of a small shearwater; also swiftly banks and arcs more like a petrel. **Range:** One sighting near Budibudi (Laughlin) Is, east of Woodlark I (SE Is). Breeds from New Hebrides to Fiji Is and possibly in the Samoan, Cook, and Solomon Is; disperses eastward in equatorial Pacific. **Taxonomy:** Split from extralimital Gould's Petrel (*P. leucoptera*), which breeds off E AU and on NC and winters in tropical central and E Pacific and is similar to Collared P but always white below.

Pycroft's Petrel *Pterodroma pycrofti*
26 cm; wingspan 66 cm. Vagrant. **All plumages:** Small petrel with *grey upperparts* (darker on crown); grey on back of neck extending to sides of neck and upper breast in *a neat, well-defined patch;* dark open *M across wings and back;* also, *dark eye patch and tail tip.* Underparts white. *Underwing white with thin, dark, diagonal line.* **Similar spp:** Collared P pale morph with similar underwing pattern but has dark hood and is larger. **Range:** A band recovery from SE Is. Breeds in NZ; wintering distribution unknown. **Taxonomy:** Split from Stejneger's Petrel (*P. longirostris*) which breeds in SE Pacific and winters in subtropical N Pacific. **Extralimital spp: Cook's Petrel** (*P. cookii*), a trans-equatorial migrant, also breeds in NZ but winters in NE Pacific; nearly identical, but shape differs subtly with tail shorter and bill heavier; crown and nape paler grey, extending down sides of neck as a paler and more diffuse patch.

Tahiti Petrel *Pseudobulweria rostrata* Pl 7
(*Pterodroma rostrata*)
38–40 cm; wingspan 84 cm. Common visitor, but status unclear owing to possible confusion with Beck's P. **All plumages:** *Dark hood sharply delineated from white breast.* All sooty brown, with white breast and belly. *Both surfaces of wings dark,* although underwing has a narrow pale stripe along center. **Subspp:** No information; either *rostrata* (Society and Marquesas Is) or *trouesssarti* (New Caledonia) with larger bill. **Habits:** Seen singly, avoids ships. Flies in high-banked arcs and glides, interspersed with deep wingbeats, the "wrists" bent. Drawn to schools of fish and squid in open seas, and thought to seize prey at the surface. **Range:** Seen north and south of NG. Breeds in Society and Marquesas Is and New Caledonia; disperses to tropical and subtropical Pacific. **Extralimital spp: Beck's Petrel** (*Pseudobulweria becki*) Pl 7, sometimes classified as a race of Tahiti P, is thought to breed in the Bismarck and Solomon Is and likely occurs in NG waters; separation from Tahiti P problematic at sea, but Beck's is distinguished by its smaller size (29 cm) and less robust bill. **Phoenix Petrel** (*Pterodroma alba*) of tropical central Pacific similar to Tahiti P but has white throat.

Streaked Shearwater *Calonectris leucomelas* Pl 7
48 cm; wingspan 122 cm. Abundant Palearctic migrant during austral summer.
All plumages: A large, slender-bodied, brownish shearwater with white underparts. Underwing white with dark flight-feathers. *Pale face (forehead, lores, and eye-ring) surrounding dark eye; pale bill.* Head finely streaked. **Similar spp:** Wedge-tailed S pale morph and other shearwaters have a dark face. **Habits:** Congregates in flocks. Flight consists of long glides interspersed with irregular wingbeats, wing slightly bent. Usually flies low over water, but towers into wind above horizon. Feeds on fish and squid snatched at the surface or captured after a shallow dive and pursuit under water. **Range:** All NG seas, but especially off N coast. Breeds on islands off Japan, Korea, and China; winters in equatorial W Pacific, mainly seas N of NG, Arafura Sea, and South China Sea.

Wedge-tailed Shearwater *Puffinus pacificus* Pl 7
(*Ardenna pacifica*)
41–46 cm; wingspan 97–104 cm. The only resident shearwater in the region, with numbers augmented by migrants from AU during austral winter. Locally common, but otherwise rare or absent. **All plumages:** Medium-large, slender, and with *broad wings; long, narrow, wedge-shaped tail; slim grey bill; and pinkish legs.* Two color morphs: dark morph *(more common)* all dark brown and *pale morph* with white underparts. **Similar spp:** Flesh-footed S similar to dark morph but has pale bill with dark tip and the tail is not as long and tapered. Short-tailed and Sooty Ss, compared with dark morph Wedge-tailed, have narrower, more rapidly beating wings that are paler on the underside, short rounded tails, and dark feet. Streaked S similar to pale morph Wedge-tailed but with pale face. **Voice:** At night from the nest gives a ghostly crooning *kaa whooo aahhh* or chuckling *kuku oo kuk*. **Habits:** Banding returns from the Philippines suggest some AU birds may be passing through north NG seas as trans-equatorial migrants. Numbers increase off the north coast during austral winter, when Streaked S is mostly absent. When migrating, flies fast in high arcs and glides, interspersed with wingbeats. Feeds mainly on fish taken on the wing by dipping, but also makes shallow dives and pursues prey underwater. The most successful tropical shearwater because of its ability to reproduce, albeit with losses, on rat-infested islands. Nests in sandy soil back from the beach wherever it is protected from predators, including humans. Visits nest after dark. Digs a nest burrow. Egg (1) white. **Range:** Possibly all coastal waters in NG Region; nests on small islands off coast, although distribution incompletely known. Range includes tropical and subtropical Pacific and Indian Oceans.

Short-tailed Shearwater *Puffinus tenuirostris* Pl 7
(*Ardenna tenuirostris*)
41–43 cm, wingspan 97–100 cm. Probably a locally common austral passage migrant in April–June, but status unclear because of confusion with Sooty S. **All plumages:** *From dark morph Wedge-tailed S by its paler underwings, shorter rounded tail, and dark bill and legs.* Note that the tips of the toes project beyond the tail. **Habits:** After breeding on islands near SE AU, huge flocks make an annual migration proceeding on a clockwise course around the rim of the Pacific. Flight involves high banking arcs and glides on *stiff wings*, interspersed with *rapid wingbeats*; sustained *flight is fast and directional.* Feeds on fish, crustaceans, and squid mainly by diving and pursuit under water. **Range:** A passage migrant in NE seas of NG Region, but needs better documentation. Breeds off AU; trans-equatorial migrant

in Pacific. **Extralimital spp: Sooty Shearwater** (*P. griseus*) with main population in NZ and small AU population has migratory path further to the east, although there are questionable sight records from Bismarck and Solomon Seas. Has paler underwing than Short-tailed S, seen as a silvery flash, but the two species are difficult to distinguished at sea.

Flesh-footed Shearwater *Puffinus carneipes* Pl 7
(*Ardenna carneipes*)
41–45 cm; wingspan 99–106 cm. Rare austral passage migrant. **All plumages:** *From dark morph Wedge-tailed S by its heavier build, straighter wings in flight, and the shorter, less wedge-shaped tail.* Note darker, sooty-brown plumage and *pale bill with dark tip.* **Habits:** Heavy flight of short glides interspersed with powerful wingbeats. Feeds on fish and squid taken by shallow dives and pursuit under water. **Range:** A passage migrant in NG Region, but very few records. Breeds off AU and NZ; trans-equatorial migrant in Pacific and Indian Oceans.

Hutton's Shearwater *Puffinus huttoni*
36–38 cm, wingspan 72–78 cm. Probably a rare passage migrant or winter resident to Coral Sea and Torres Strait during austral winter. **All plumages:** *A small, black-and-white shearwater* similar to Tropical S but slightly larger and with a *dark, almost complete breast band.* **Range:** Collected from SW Torres Strait; tentative sight record from Cape Suckling (SE Pen, NW of Port Moresby). Breeds in NZ and disperses to AU coast including Torres Strait.

Tropical Shearwater *Puffinus bailloni* Pl 7
(**Audubon's Shearwater**, *P. lherminieri*)
31 cm, wingspan 69 cm. Rare on the NE fringe of the NG Region, where possibly resident. To be looked for in the SE Is. **All plumages:** *A small, distinctive shearwater with short-winged* profile showing *black upperparts* contrasting with *clear white underparts;* underwing white with dark flight-feathers. **Subspp (1):** *dichrous* (Pacific) expected. **Habits:** Solitary or in small groups. Typically found over deep water off breeding grounds. Fluttering wingbeats followed by smooth gliding flight. Feeds by pursuit diving, plunging, and surface-seizing. Breeds on atolls or rocky islets, nesting in soil burrows or rock cavities. Egg (1) white. **Voice:** At night when at nest, a rasping *shoo-kree.* **Range:** Sightings off Bismarck and Solomon Is and Pockington Reef (Solomon S), far from mainland NG. Also tropical Pacific and Indian Oceans.

Heinroth's Shearwater *Puffinus heinrothi* Pl 7
27 cm. Rare visitor, status poorly known. **All plumages:** *Smallest shearwater in the region—size of Black Noddy. Tail short; bill long and thin. Overall sooty grey with pale underwing-coverts.* Individuals vary, some having a whitish oval patch on breast and belly. The bluish-grey eye differs from dark eye of other shearwaters. **Similar spp:** Short-tailed and Sooty S distinctly larger with short bills; never have pale patch on belly as some Heinroth's do. **Habits:** Little-known. Flies low to water with fluttering wingbeats and glides, seldom arcing. Typically seen in mixed seabird flocks over feeding tuna, sometimes near shore. **Range:** Recorded off Madang in Bismarck Sea. Known principally from around the Bismarck and Solomon Is, with circumstantial evidence of breeding on Bougainville I. **Extralimital spp: Christmas Shearwater** (*P. nativitatis*) of tropical Pacific is larger and darker; all dark with dark feet and mostly dark underwing; shorter, more rounded tail; and fast, stiff wingbeats.

Bulwer's Petrel *Bulweria bulwerii* Pl 7
26–27 cm; wingspan 65–70 cm. Uncommon, mainly in Indonesian waters; probably a visitor rather than a resident. **All plumages:** *Small blackish petrel (larger than storm-petrel) with thin, tapered tail.* Tail shows wedge shape when fanned. When gliding, holds thin, bladelike wings in a crescent. *Prominent pale, diagonal bar across upperwing-coverts.* **Similar species:** Noddy terns have white cap that in juveniles is scarcely noticeable, but they have a pointed, not hooked, beak. **Habits:** Solitary. Flies low to the water, with graceful, gliding flight on twisting path, marked by frequent arcing and banking. Feeds mainly at night, seizing fish and squid from the surface. **Range:** Sight records from NW Is and Gulf of Papua; probably more widespread. Worldwide; breeds on oceanic islands, mainly in the northern oceans, disperses widely.

NORTHERN STORM-PETRELS: Hydrobatidae

All storm-petrels were previously classified in a single family, but they have been recently split into two families, the northern and southern storm-petrels. The two families differ structurally in that northern storm-petrels have a longer forearm resulting in a longer, narrower wing that is held with a noticeable bend at the elbow, whereas the southern storm-petrels have a shorter, broader, straighter wing and are given to more fluttering flight, and their legs are longer and used for pattering on the surface of the water. Although both families occupy all oceans, northern storm-petrels breed mainly in northern seas, whereas the southern storm-petrels nest mostly in southern ones. There are no breeding species of either type of storm-petrel in the NG Region, only visitors. As with visiting petrels and shearwaters, knowledge of the diversity of species in NG waters is scanty and awaits future study. The one species of northern storm-petrel reported migrates through the region on its way to and from its nesting islands in the subtropical western Pacific.

Storm-petrels are the smallest and most inconspicuous of seabirds. They resemble miniature petrels, but have long, dangling legs and large webbed feet for pattering on the sea surface. With butterfly-like flight, they flutter close to the water, picking up small morsels of food. Sometimes they follow ships, permitting close examination. They nest in burrows on islands where they are safe from predators and lay a single white egg.

Matsudaira's Storm-Petrel *Oceanodroma matsudairae* Pl 7
(*Halocyptena* or *Hydrobates matsudairae*)
24–25 cm; wingspan 56 cm. Uncommon passage migrant to N coast (and probably W coast) during austral winter. **All plumages:** Largest NG storm-petrel. *All sooty brown with pale wing-coverts and forked tail; in good light note diagnostic pale shaft streaks on primaries.* **Habits:** Slow flap-and-glide flight. **Range:** Sight records from the Bismarck Sea and equatorial seas to the north; expected elsewhere along the N and W coasts. Breeds on islands south of Japan and migrates to tropical W Pacific and Indian Oceans.

SOUTHERN STORM-PETRELS: Oceanitidae

See account for northern storm-petrels for information on both families. Of the 8 species of southern storm-petrels, 3 species visit New Guinea and breed on islands in the southern temperate or Antarctic oceans.

Wilson's Storm-Petrel *Oceanites oceanicus* Pl 7
15–19 cm; wingspan 38–42 cm. Common pelagic migrant found during austral winter along NG continental shelf in E Seram Sea; reported as rare elsewhere along S coast. **All plumages:** *Sooty brown with large white patch around base of square tail.* Feet project beyond tail. **Subspp:** No information. **Habits:** Singly or in flocks. Concentrates at edge of continental shelf, at upwellings, and at mixing water where currents meet. Strong, swallow-like traveling flight of rapid wingbeats and glides. Flight slow when foraging; wings held in V over back and legs dangling; patters and dips for crustaceans and fish. Follows ships. **Range:** Poorly known. Reported as eastward-moving flocks in seas between Bomberai Pen and Aru Is in Aug; also several sight records from Arafura and Coral Seas; possible in other NG waters. Breeds on sub-antarctic islands, dispersing as a trans-equatorial migrant to the Pacific, Indian, and Atlantic Oceans. **Extralimital spp:** The very similar **Leach's Storm-Petrel** (*Oceanodroma leucorhoa*), a species of northern storm-petrel, has been recorded as a vagrant from north of Bismarck Is. It usually has a divided white rump patch; the wings are long and often held with an obvious bend at the wrist; tail forked (hard to see when folded) and the feet do not project beyond it; usually with faster, more direct flight than Wilson's SP.

White-faced Storm-Petrel *Pelagodroma marina*
20 cm; wingspan 42 cm. Vagrant. **All plumages:** *Unique white face with dark eye-stripe and ear-coverts*; brownish-grey upperparts, pale grey rump, and white underparts. Feet extend well beyond tail. **Subspp:** No information. **Range:** One sight record from the Solomon Sea. Breeds on small islands in temperate S Pacific and Indian Oceans and in Atlantic off NW Africa; winters in tropical Indian, E Pacific, and Atlantic Oceans.

Black-bellied Storm-Petrel *Fregetta tropica* Pl 7
20 cm; wingspan 46 cm. Uncommon migrant mainly to Coral Sea during austral winter and spring. **All plumages:** *Stocky with short wings. Dark upperparts and hood; white breast, belly, and rump; usually with a central black line or mottling on the breast.* Compared with White-bellied SP (which lacks dark midline on breast), the black hood extends further down onto breast, and the pale underwing is duskier with blurred margins. Feet usually project beyond the tail. **Subspp:** No information. **Range:** Sightings from Coral Sea and specimens from the Solomon Sea. Breeds on islands in sub-antarctic and antarctic waters and disperses to warmer southern oceans. **Taxonomy:** White-bellied Storm-Petrel (*F. grallaria*) now usually regarded as a species separate from Black-bellied. Two records from Coral Sea, both questionable. Breeds mostly in subtropics of southern oceans and winters in tropics. Similar in appearance, but black hood stops at the throat, lacks black line down belly, has crisp margins to the white underwing, and the feet usually do not project beyond the tail.

GREBES: Podicipedidae

The grebes are distributed nearly worldwide, with a center of diversity and perhaps place of origin in South America. Of the world's 23 species, only 2 inhabit New Guinea. Grebes resemble small, swimming rails or ducks, yet they have no close relatives other than the flamingos. Well adapted for aquatic life, they rarely come ashore, preferring to float on the water and dive for food. When swimming, they are propelled by a uniquely structured foot, which rather than being webbed has lobed toes that come together on the outward stroke. The dumpy, seemingly tailless body supports a small head on a thin neck. Flight appears weak, but grebes are capable of traveling great distances. Grebes are monogamous and build a floating nest of pond vegetation.

Tricoloured Grebe *Tachybaptus tricolor* Pl 6
(**Little Grebe**, *Tachybaptus* or *Podiceps ruficollis*)
25 cm. An uncommon resident on lakes and ponds, mainly in lowlands. *Co-occurs with the very similar Australasian G.* Both species are small, buoyant, nearly tailless waterbirds that dive for food, and both have a yellow spot at base of bill in all plumages. *The main distinctions of the Tricoloured are its dark red iris and reduced white blaze on the extended wing, visible in flight*, field marks difficult to see at a distance. **Adult:** *Breeding plumage sooty black with red-brown throat and foreneck* (Australasian G has black throat). Nonbreeding plumage plain grey-brown, with *orangish neck and darker belly* (Australasian G paler with a pale belly). **Juv:** Similar to Nonbreeding adult but head striped black and white. **In flight:** White blaze *restricted to secondaries only*—does not extend onto primaries as with Australasian G. **Subspp (1):** *tricolor* (N NG, also Wallacea); *collaris* (Huon and NE NG, also Bismarck and Solomon Is) darker, with less white in wing, intergrades with previous race.
Voice: A whinnying trill. Alarm call one or more *whit* notes. **Habits:** Solitary or in pairs, but often in gatherings spread out across open freshwater. Readily seeks out new habitat and can be expected anywhere. Feeds on quiet bodies of water of all sizes and on slow-moving streams. Dives for small fish and aquatic invertebrates. Nest built on water, anchored to vegetation; nest a platform of vegetation. Eggs (~4) pale blue. **Range:** NG and NW Is (Kofiau, Misool, Salawati), 0–1500 m. NG: Bird's Head to Sepik-Ramu and Huon, and W and E Ranges. Also Wallacea, Bismarck Is, and Solomon Is.
Taxonomy: Split from Little Grebe (*T. ruficollis*).

Australasian Grebe *Tachybaptus novaehollandiae* Pl 6
(*Podiceps novaehollandiae*)
25 cm. Resident, but likely some birds are visitors from AU. More common and widespread than Tricoloured G, but similar in habitat and habits. Differs by having a *bright yellow eye and white blaze extending the length of the wing, visible in flight*. **Adult:** *Breeding plumage has black throat* (Tricoloured has red-brown throat). Nonbreeding plumage paler than Tricoloured G and with pale (not dark) belly. **Juv:** Similar to juv Tricoloured G. **In flight:** *White blaze extends across both secondaries and primaries*. **Subspp (2):** Minor. **Voice and Habits:** Similar to Tricoloured G. **Range:** NG (throughout) and Woodlark (SE Is), 0–3200 m. Also AU and NZ, N to Java, Wallacea, and Solomon Is.

TROPICBIRDS: Phaethontidae

Tropicbirds are an ancient family of seabirds with no close relatives; their fossil record goes back 55 million years. Three species grace the tropical oceans of the world, and two are rare visitors to New Guinea waters. Easily recognized and admired for their gleaming white plumage, cool "sunglasses," and long tail-streamers, tropicbirds are iconic inhabitants of the tourism literature, although they are much harder to find at sea. Truly oceanic, they never approach land except when breeding or storm-blown. Tropicbirds nest on small islands but wander widely. They soar high above the sea scouting for fish and make spectacular plunges from 10–50 m to catch their prey. In flight, tropicbirds resemble large terns, but their flight combines bouts of rapid flapping with periods of graceful soaring and gliding. Tropicbirds are usually solitary, although pairs or small gatherings may be seen at nesting sites. Both species nest on cliff faces or on level ground where predators are absent, and they typically visit these sites at midday, conspicuously flying around and calling.

Red-tailed Tropicbird *Phaethon rubricauda* Pl 8
46 cm (not including the tail-streamers that project ~35 cm); wingspan 104 cm. Pelagic vagrant. **Adult:** *All white with red bill* and *black streaks in the outer primaries*. Tail-streamers red, but so thin as to be invisible at a distance ("No-tailed Tropicbird"). Bird has pinkish cast when breeding. **Juv:** Bill grey with dark tip; black, V-marked feathers on mantle and wing-coverts; *note fine black shaft streaks on outer primaries*, like Adult; no tail-streamers. **Similar spp:** White-tailed T. **Voice:** An exuberant barking around nest cliffs. **Habits:** Similar to White-tailed T. **Range:** Mostly off SE NG coast. Tropical and subtropical Pacific and Indian Oceans.

White-tailed Tropicbird *Phaethon lepturus* Pl 8
38–40 cm (not including the tail-streamers that project ~45 cm); wingspan 89–96 cm. Uncommon pelagic visitor; may breed on small islands. **Adult:** Mainly white, separated from the larger, bulkier Red-tailed T by its *wing pattern, showing black patch on primaries and stripe across the secondary coverts and tertials; white tail; yellowish bill*. **Juv:** From juv Red-tailed by the black patch on primaries. **Subspp (1):** *dorotheae* (tropical Pacific). **Voice:** Call around nest site a sharp clicking, like slowly picking the teeth of a comb. **Habits:** Feeds on fish, especially flying fish, and squid. Nests on cliffs and on the ground, and sometimes in tree hollows. **Range:** Anywhere off NG coasts. All tropical oceans.

STORKS: Ciconiidae

Storks are very large wading birds related to ibises and herons. This family of 19 species has a nearly worldwide distribution centered on Africa and Asia. Only one species reaches New Guinea.

Black-necked Stork *Ephippiorhynchus asiaticus* Pl 13
(**Jabiru**, *Xenorhynchus asiaticus*)
110–137 cm. Locally common breeding resident along margins of swamps, lagoons, rivers, and coastal mudflats. *An unmistakable, tall, long-necked wading bird with pied plumage and huge black bill.* **Adult:** Black head, neck, and wings. Legs pink. **Male's** eye dark, **Female's** pale yellowish white. **In flight:** Distinctive black band running the length of the white wing. Long neck and legs held extended. **Juv:** Head dingy brown, and plumage pattern muted; legs blackish. **Subspp (1):** *australis* (also AU). **Similar spp:** Magpie Goose has similar plumage pattern but is smaller and has a much shorter bill and shorter legs. **Voice:** Normally silent. **Habits:** Solitary or in a pair or small family group; will join other waders. Walks sedately with long strides. Wades in shallow water when feeding. Captures prey by probing, sweeping, or stabbing with the bill. Feeds on aquatic vertebrates (mainly fish), crustacea, and insects. Slow flapping flight broken by short glides; often soars. Monogamous. Nests high in tree; nest a stick structure lined with grass, etc. Eggs (2–4) white. **Range:** NG: Trans-Fly, 0–100 m. Also AU and S Asia.

IBISES AND SPOONBILLS: Threskiornithidae

Some 35 species of tropical and temperate wading birds constitute this worldwide family, with 5 species recorded from New Guinea. Ibises and spoonbills feed on insects, crustaceans, small terrestrial vertebrates, and fish caught by two very different techniques. Ibises have a long, slim, downcurved bill used for probing soil or mud, whereas spoonbills have a peculiar spatula-shaped bill that they sweep through the water snapping up prey. Both ibises and spoonbills inhabit freshwater wetlands. Highly social, they are usually encountered in flocks and nest colonially. All NG species migrate between Australia and New Guinea, where they occur primarily in the swamps and lagoons of the Trans-Fly and southern coast. Only the Australian White Ibis and Royal Spoonbill are known to breed in the New Guinea Region; other species eventually may be found to do so.

Australian White Ibis *Threskiornis moluccus* Pl 13
(**Sacred Ibis**, *Threskiornis aethiopicus*)
65–75 cm. Local breeding resident and common migrant; inhabits mainly lowland swamps, less frequently grassland, and seldom coasts. *A large white bird with decurved bill, black head and neck, and black wing tips.* **Adult:** Bare head and neck black, often with red nape patch; shaggy black tertials cover tail. Breeding plumage with filamentous white plumes on the foreneck; bare skin of the underwing bright red, not pink. **Male:** Averages larger. **Juv:** Head and neck dark with scattered white feathers; tertials black but not shaggy. **Subspp (2):** *moluccus* (rare resident NW Is, Bird's Head to Sepik-Ramu; also Moluccas) shafts of secondaries white; *strictipennis* (common and mainly migratory, Trans-Fly to Port Moresby, also AU) secondary shafts black and body size larger. **Similar spp:** Other ibises much darker. Royal Spoonbill also white but has a straight, spatula-tipped bill and white rear end. **Voice:** Call when taking flight a loud *urk*. **Habits:** Present year-round at Bensbach with peak numbers in the wet season; a dry season visitor elsewhere. Gregarious—sometimes alone, but usually in flocks, often large ones. Roosts communally. Mingles with other waterbirds. Walks and wades slowly, probing for food on dry land or in the water. Feeds on all sorts of small creatures, aquatic and terrestrial. Socially monogamous. Nests colonially in trees, often with other species; nest a stick platform lined with leaves (AU data). Eggs (~3) white. **Range:** NG and NW Is (Salawati, Waigeo), 0–300 m. Most of mainland NG, although absent from northern and eastern SE Pen. Also Moluccas to AU and Solomon Is. **Taxonomy:** Split from African Sacred I (*T. aethiopicus*).

Straw-necked Ibis *Threskiornis spinicollis*   Pl 13
60–70 cm. A common migrant from AU to Trans-Fly savannahs (suspected of breeding in Trans-Fly); a vagrant elsewhere. *Glossy black wings and back contrast with white foreneck, underparts, and tail.* **Adult:** Foreneck with yellowish, strawlike plumes—these obscure a dark, glossy green breast band that is incomplete in the **Male**, complete in the **Female** (which also has a shorter bill and smaller body size on average). Legs reddish. **Juv and Imm:** Lack straw-neck and are less glossy; legs blackish grey. Adult plumage attained gradually. **Similar spp:** Australian White I has white back and neck. **Voice:** Croaks when taking flight. **Habits:** Present year-round at Bensbach where numbers build during the wet season. Said to be mainly a dry season visitor elsewhere. Habits similar to Australian White I but frequents dry grassland and is less associated with water. Follows grass fires in search of prey. Not known to breed in NG Region. Nests in bushes or reed beds surrounded by water; nest a platform of reeds (AU data). Eggs (2–5) white. **Range:** NG: Trans-Fly (vagrant elsewhere), 0–300 m. Also AU.

Glossy Ibis *Plegadis falcinellus*   Pl 13
55–65 cm. Seasonally common, although present year-round locally. Known as an AU migrant, mainly to Trans-Fly in austral winter and Port Moresby area somewhat later, and suspected of breeding. Inhabits swampy freshwater habitats and avoids dry ground. An *all-dark ibis* that appears blackish at a distance, with a curlew-like profile. **Adult:** With metallic sheen of bronze, green, and purple. Breeding plumage chestnut brown. Nonbreeding plumage dark brown, heavily streaked white on the head and neck. White line bordering bill evident year-round. **Male:** Larger. **Juv:** Nearly without iridescence; face and throat mottled with white. **Similar spp:** Curlews are paler brown and smaller. **Voice:** The seldom-given flight call is a guttural chattering. **Habits:** Similar to other ibises. Gregarious. Flying bird holds neck extended but sagging slightly below body; flies with stiff wingbeats and glides, like a cormorant. Flocks fly in sinuous Vs or lines. Feeds mostly on invertebrates and insects. Nests colonially, often with other species, in low bushes or small trees; nest a platform of sticks and leaves (AU data). Eggs (~3) greenish blue. **Range:** NG: Trans-Fly to Port Moresby area (SE Pen) and vagrant elsewhere, 0–300 m. Also nearly worldwide.

Royal Spoonbill *Platalea regia*   Pl 13
74–81 cm. Common migrant from AU during austral winter and locally breeding resident. Frequents swamps and estuaries, mainly in Trans-Fly and southern SE Pen, also S Lowlands to Bird's Head. *A large, all-white wading bird with a long, black, spatula-shaped bill and black legs.* **Adult:** Note red forehead patch and yellow spot above the eye. *Breeding plumage sports a thick, bushy, swept-back crest.* **Male:** Larger on average. **Juv:** Similar to the crestless Nonbreeding adult but lacks forehead patch and eye-spot; black-margined wing tips. **Similar spp:** Australian White I has a curved, narrow black bill and black tertials covering the tail. Yellow-billed S has pale, dull yellow bill and legs. **Voice:** Quiet when not breeding. Soft grunts and groans at nest. **Habits:** Gregarious, traveling in chevron-forming flocks. Same flap-and-glide flight pattern as for ibises. Roosts socially. Captures fish and shrimp while wading in the water and immersing its open bill vertically; sweeps bill slowly from side to side, snapping up prey. Nests in small colonies with other wading birds in trees, shrubs, or reeds/rushes; nest a bowl of sticks lined with leaves (AU data). Eggs (~3) white with dark blotches. **Range:** NG and Aru Is, 0–300 m. NG: S coast from Bird's Head to SE Pen, recorded breeding in Trans-Fly; vagrant elsewhere. Also AU, Sunda Is (Java) through Wallacea, Solomon Is, and NZ.

Yellow-billed Spoonbill *Platalea flavipes*   Pl 13
76–92 cm. Vagrant to Trans-Fly. Similar to Royal S, but dingy white and with pale, *dull yellow bill and legs.* **Adult:** Very pale eye and legs; wings entirely white. Breeding plumage with strawlike plumes on lower foreneck and dark filamentous scapular plumes. **Male:** Larger and with more pronounced nuptial plumes. **Juv:** Dark eye and dusky leg; dark tips to primaries. **Habits:** Similar to Royal S. **Range:** NG: Sight records from Trans-Fly (Bensbach). Also AU.

HERONS AND BITTERNS: Ardeidae

The herons and bitterns occur worldwide with a total of 72 species. Many Australian species and certain Asian ones fly long distances to take advantage of seasonal wetlands or to escape drought or winter, and these migrations bring them to New Guinea. Of the 16 New Guinea species, 9 are entirely or mostly migrants from Australia, whereas 2 are known to be Asian visitors, with the possibility of additional vagrant species. Surprisingly little is known about the location of heron nesting colonies in New Guinea. For only 9 heron or bittern species is there definite or even circumstantial evidence of breeding in New Guinea, and it seems likely that more of the listed species actually do nest here. The sole endemic species is the Forest Bittern. Herons and bitterns in New Guinea are largely confined to the lowlands where they prefer open aquatic environments. Two resident species keep to the coast, and several of the migrants stray into the highlands.

Herons and bitterns are small to very large, long-legged, broad-winged wading birds that forage by spearing aquatic prey with their long, pointed bills. The neck is extraordinarily long, even in bitterns and other herons that normally assume a hunched, compact posture. In sustained flight, the neck is tucked into an S shape, not outstretched. These birds typically molt directly from juvenile into adult plumage, although some pass through an immature plumage first. Most adult herons have nuptial plumes when nesting, and also in many the facial skin changes color. All species are monogamous as far as is known. They build simple stick nests, and some species gather in colonies to breed. Herons feed in the open and are easily observed on wetlands or fields. Bitterns, however, skulk inside vegetation, although their loud advertisement calls may give away their presence; upon detection they "freeze" with neck extended.

Forest Bittern *Zonerodius heliosylus* Pl 14
(**New Guinea Tiger Heron**, **NG Zebra Bittern**, or **NG Zebra Heron**)
65–71 cm. A rare and local resident of rainforest streams in lowlands and foothills. Shy and seldom seen. *A bulky heron heavily barred all over.* Note dark cap; white belly and rump. Variation in plumage color poorly understood, but darker, more heavily barred birds of both sexes with green facial skin may be Adults, and paler, finer-barred birds with yellow facial skin are possibly Juv. **Similar spp:** Nankeen Night-Heron juv is mostly streaked, not barred except on the wings. Black Bittern juv has scaly upperparts and pale neck patch. **Voice:** Normally silent.
Habits: An enigmatic bird in need of study. Solitary and very shy. Forages at edge of forest creeks and pools. Its varied diet is known to include fish, reptiles, crustaceans, and insects. Flushes to perch in overhanging vegetation. Nest is built in forest midstory and is reported to be a bulky stick structure. No egg information. **Range:** Endemic. NG, Salawati (NW Is), and Aru Is, 0–1400 m.
Taxonomy: A New Guinea specialty, the Forest Bittern is provisionally classified with the similar-looking tiger-herons of South America and Africa, but this arrangement has not been tested by molecular genetic research.

Australian Little Bittern *Ixobrychus dubius* Pl 14
(**Black-backed** or **Little Bittern,** *I. minutus*)
27–36 cm. Fairly common during austral winter, mainly in the South. Circumstantial evidence for both local breeding and migration from AU. Occupies reeds and other emergent vegetation in swamps and marshes. The *region's smallest heron* is usually flushed from reeds or seen posing motionless, with neck outstretched. **Adult:** *A pale buff wing patch contrasts with dark back. Dark central throat stripe.* Male: Back all black. Female: *Back dark brown* with some pale scalloping; look for *blackish central throat stripe* and *grey-brown primaries in flight* to distinguish from Yellow B female (primaries black). Juv: Streaked buff and brown; look for *rufous brown tips to the primaries and secondaries* (these narrowly white-tipped or all dark in juv Yellow B). **Similar spp:** See Yellow B. **Voice:** Usually silent, but vocal when breeding. Song a far-carrying *rok rok rok* . . . 2 notes/sec in a series of up to 10 notes, uttered throughout the morning and renewed at dusk. Alarm call when flushed, a soft, low *koh!*
Habits: Solitary or in pairs. Dates of observation suggest it is a wintering migrant from AU, but there is some evidence of local breeding. Feeds mainly at dawn and dusk, perhaps also at night; food taken by waiting or stalking, then jabbing; preys on small aquatic crustaceans, insects, and fish. Monogamous. Nest built in reeds is a small platform of reeds (AU data). Eggs (4–6) white. **Range:** NG: mainly Lake

Daviumbu (Trans-Fly) and Port Moresby area (SE Pen), 0–500 m. Also AU. **Taxonomy:** Formerly combined with Little B (*I. minutus*) of Eurasia and Africa.

Yellow Bittern *Ixobrychus sinensis* Pl 14

30–40 cm. Locally abundant Palearctic migrant mostly in the north during austral summer; some birds possibly reside and breed. **Adult:** Distinguish with care from Australian Little B. *Uniform buff-brown back (not dark)*, lacks dark central throat stripe, longer and narrower bill, and larger size on average. **Male:** *Overall yellowish brown*; underparts unmarked. **Female:** Similar, but *neck and underparts broadly striped cinnamon and white*. **In flight:** *Both Adults have black primaries and secondaries that strongly contrast with the pale buff wing-coverts.* Australian Little B female has a brown back and the primaries and secondaries are grey-brown. **Juv:** Like Adult, but streaked. From juv Australian Little by its paler coloration with paler streaking, *all-black primaries and secondaries lacking of rufous brown tips*, and bill shape. **Voice:** Alarm call is a croak, *ar!* **Habits:** Similar to Australian Little B but perhaps less attached to reed beds and in other parts of its range extends into mangroves, rice fields, and marginal wetlands, scrub, and rank grass. Likely to appear anywhere in suitable habitat, including small islands. **Range:** NG, Waigeo (NW Is), Biak (Bay Is), and expected from other satellite islands, 0–500 m. NG: NW Lowlands, Sepik-Ramu, and Port Moresby area (SE Pen). Breeds in Asia, Philippine Is, Micronesia, and Solomon Is; winters from Great Sunda Is through to Solomon Is.

Cinnamon Bittern *Ixobrychus cinnamomeus* Pl 14

40–41 cm. Appears to be a local resident in westernmost NG. **Adult:** From Yellow B by its *overall cinnamon coloration without contrasting pale wing*. From cinnamon morph Black B by much smaller size and presence of only a small whitish blaze along neck. **Female:** Has streaked neck and may show obscure spotting on mantle, back, and wings. **Juv:** Similar to Female but duller. **Habits:** As for Yellow B. **Range:** NG: Bird's Head (Sorong). Breeds SE Asia, from India to China, south to Wallacea. **Extralimital spp: Von Schrenck's Bittern** (*I. eurhythmus*) of E Asia winters in Wallacea. Male is grey-brown above and has buff wing-coverts and underparts; female's mantle, back, and wings show unique, fine, white-and-black spotting.

Black Bittern *Ixobrychus flavicollis* Pl 14
(*Dupetor flavicollis*)

54–66 cm. A fairly common resident, inhabiting the forested margins of streams, wooded swamps, and mangroves. *A small, blackish heron with diagnostic buff blaze along the side of the neck.* At close range, note white streaking down the front. **Male:** Blue-black upperparts. **Female:** Dark chocolate brown upperparts with faint scaling. Some birds show brighter cinnamon upperparts; sex of these birds unknown, but possibly Female. **Juv:** Paler and more rufous than Female with heavily buff-scaled upperparts. **Subspp (1):** *australis* (also AU, Moluccas, Bismarck Is). **Similar spp:** Nankeen Night-Heron juv is streaked and spotted with white; bill and legs yellow. Forest Bittern is barred all over. Striated Heron smaller, not as dark, and with yellow legs (not olive). **Voice:** Song is a booming reminiscent of a pigeon or frogmouth, but much louder. Sings at dawn and dusk, a cooing *croo rrorh* or single *rrorh* repeated regularly. Gives a croaking call when flushed. **Habits:** Solitary or rarely in pairs. Flushes from wooded stream banks and flies into a tree and "freezes" with bill pointed skyward. Feeds mostly at dawn or dusk or at night. Waits patiently at the water's edge, then lunges when prey draws near. Feeds mainly on fish and crustaceans, also frogs, lizards, and insects. Nests on a branch in a leafy tree overhanging water; nest is a small pad of sticks and reeds. Eggs (2) white. **Range:** NG, NW Is (Gebe, Kofiau, Misool, and Salawati), Aru Is, SE Is (Misima, Normanby, Tagula, Woodlark), and expected on other satellite islands, 0–2100 m. Also SE Asia to AU and Solomon Is. Unknown whether birds move between AU and NG Region.

Nankeen Night-Heron *Nycticorax caledonicus* Pl 14
(*Rufous Night-Heron*)

55–65 cm. A common resident and visitor from AU, in a wide range of freshwater and coastal habitats with trees nearby. **Adult:** *A distinctive, medium-sized, robust heron with hunched posture and overall rufous plumage* topped by a black cap and white crest plume. **Juv:** Drab brown with copious whitish streaking and spotting. **Subspp (1):** *australasiae* (also Java to AU). **Similar spp:** Forest Bittern and Black B lack the thick whitish streaking and spotting of juv night-heron. **Voice:** Call a loud, harsh, short croak, *kwok* or *kyok*, heard mainly at night. **Habits:** Gregarious, though often feeds alone.

Nocturnal, roosting in trees or cane grass during day, flying out to feed at dusk or on overcast days. Will roost in single-species flock or with other heron species. Its catholic diet includes all sorts of small animal life captured in or near the water. Nests in the crown of a tree; nest a stick platform (AU data). Eggs (3) blue-green. **Range:** NG, NW Is (Batanta, Kofiau, Misool, Salawati, Waigeo), Bay Is (Biak, Yapen), and Aru Is, 0–1600 m. Also Greater Sunda Is and Philippine Is through to the Solomon Is and AU.

Striated Heron *Butorides striata* Pl 14
(**Mangrove Heron**, *Ardea* or *Ardeola striata*)
38 cm. A common resident sparsely distributed throughout the coast in mangroves and into the interior along riverbanks. **Adult:** *A small, squat, big-headed heron with black crown and grey back.* Coloration varies geographically from grey to muddy brown, see Subspp. Color morphs within populations are known from AU but not as yet from NG; these are brown and grey morphs. **Juv:** From other juv herons by size, shape, and color pattern, which shows Adult coloration beneath white dorsal spotting and heavy pale streaking on the neck. **Subspp (3):** *papuensis* (NW Is, Bird's Head, Bird's Neck, Bay Is, Aru Is, also Moluccas) dark, with brown underparts; *idenburgi* (NW Lowlands and maybe Sepik-Ramu) pale, with grey underparts; *macrorhyncha* (S Lowlands, Trans-Fly, southern SE Pen; also N AU) pale, with brown underparts. **Similar spp:** Black Bittern is larger and much darker with olive legs (not yellow). Nankeen Night-Heron juv is similar to Juv Striated, but much larger. **Voice:** Alarm call when flushed is an explosive squawk. **Habits:** Solitary. Forages by waiting at water's edge for prey. Feeds on fish, crustaceans, and insects taken from the water. Nests in shrub or tree, usually over water or tidal flat; nest a stick platform. Eggs (2–3) pale blue-green. **Range:** NG, NW Is (Kofiau, Misool, Salawati, Waigeo), Bay Is (Biak, Numfor, Yapen), and Aru Is, 0–200 m. Also, AU, Asia, Africa, and S Amer.

Cattle Egret *Ardea ibis* Pl 15
(**Eastern Cattle Egret;** *Ardeola/Egretta coromanda* or *Bubulcus coromandus*)
70 cm. A newcomer to the NG Region. A locally common visitor; breeding expected. Associated with grassland and cattle, but also uses other agricultural lands and ponds. **Adult:** *A small, rather stocky egret with the neck shorter than the body and a large jowl. Breeding plumage distinctive, with orange-buff head, neck, and back* and orange-red bill and legs. Nonbreeding plumage similar to other egrets—white with short yellow bill and greenish-yellow to blackish legs. **Juv:** Same as Nonbreeding adult. **Subspp (1):** *coromandus* (Asia and Australasia). **Similar spp:** See account for Intermediate E, which is confusingly similar and sometimes follows cattle. White morph of Eastern Reef E of marine habitats is shaped differently, with a longer neck and body, longer bill, and shorter greenish-yellow legs. **Voice:** Silent away from breeding colonies. At colonies, calls include a variety of croaks, squawks, and chatter. **Habits:** Gregarious. Usually forages, travels, roosts, and breeds in flocks, potentially in big numbers. Flocks commute long distances daily, ranging widely and unpredictably. This egret characteristically forages alongside grazing cattle and other livestock, capturing insects and small vertebrates disturbed by the herbivores. But beware that they also wade into ponds to fish in a more typical egret-like fashion and are then more likely to be confused with other egret species. In AU, nests colonially in mangroves and other trees over water; the nest is a shallow platform of twigs. Eggs (~3) pale green. **Range:** NG, NW Is (Batanta, Waigeo), and SE Is (Misima). NG: Bird's Head, Trans-Fly, and SE Pen, expected elsewhere and may eventually move into highland valleys, 0–500 m. NG birds originally from Asia via Wallacea or AU or both. **Taxonomy:** Sometimes treated as a separate species, Eastern Cattle Egret (*A. coromanda*). **Extralimital spp:** Javan Pond-Heron (*Ardeola speciosa*) of SE Asia to Wallacea is similar to Cattle E but has buff head and neck and blackish back, rest of body white.

White-necked Heron *Ardea pacifica* Pl 15
(**Pacific Heron**)
76–106 cm. An AU vagrant visiting freshwater wetlands. **Adult:** *A large, blackish-winged heron with white head and neck.* Breeding plumage has maroon lanceolate plumes on scapulars. Nonbreeding plumage has a greyish cast to the head and neck, more prominent neck spots, and may lack scapular plumes. **Juv:** Like Nonbreeding plumage but the spots are scattered over the neck (not just in 2 rows) and there are more of them. **Similar spp:** Pied Heron juv also has white head and neck, but the bird is only half the size and has yellow (not dark) bill and legs. **Voice:** Silent, except for guttural alarm call. **Habits:** Solitary, foraging in wet grass or freshwater shallows. Feeds mainly on crustaceans, mollusks,

and other invertebrates, less often vertebrates such as fish. **Range:** NG: Trans-Fly, E Ranges, Huon, and SE Pen, but could appear anywhere. Also AU. **Extralimital spp:** Two large, Asian herons are possible vagrants. **Grey Heron** (*A. cinerea*) is grey with a pale head and thick, black eyebrow. **Purple Heron** (*A. purpurea*) is brownish grey with black cap and rusty neck with black striping.

Great-billed Heron *Ardea sumatrana* Pl 15
100–115 cm. A scarce coastal resident in mangroves, estuaries, and lagoons, and follows rivers inland; shy and stays close to or in tree cover. **Adult:** *Very large and long-necked; uniformly dusky grey-brown* with whitish throat. Breeding plumage has long, lanceolate, silvery grey plumes on nape and back. Facial skin blue-green when breeding, yellow when not. Bill grey-black. **Juv:** Rusty brown with rufous feather tips and brown bill. **Similar spp:** No other heron is so large and plainly colored. **Voice:** Normally silent, but utters a series of loud, intimidating roars at dusk during breeding season. **Habits:** Solitary and elusive; avoids people. Sedentary and presumably territorial. Crepuscular when not breeding, resting by day in mangroves or riverside trees. Stands in the water to capture fish. Nests alone in mangroves over tidal mud; nest a stick platform (AU data). Eggs (2) bluish. **Range:** NG, NW Is (Batanta, Kofiau, Misool, Salawati, Waigeo), Bay Is (Biak), Aru Is, and SE Is (D'Entrecasteaux Is: Normanby), 0–200 m. Also SE Asia to AU.

Eastern Great Egret *Ardea modesta* Pl 15
(**Great Egret**, *Casmerodius albus* or *Egretta alba*)
83–103 cm. The most common egret in the NG Region, inhabiting freshwater wetlands, estuaries, and mangrove waterways. Both a resident and an AU visitor, mainly in lowlands. **Adult:** *A large, white heron with a much-tapered head and long, kinked neck stretching to longer than body length.* At close range note that *gape extends well past the eye*. Breeding plumage with filamentous plumes on back, green facial skin, and bill mostly or all black. Nonbreeding plumage lacks dorsal plumes; facial skin is yellow; and the bill is mostly yellow. **Juv:** Same as Nonbreeding adult. **Similar spp:** Other egrets are smaller and have proportionately smaller bill and head and shorter necks without such an exaggerated kink. Intermediate E close in size and confusingly similar—see that account. **Voice:** Generally silent. When startled, growls *ar-ar-ar-ar*. Croaks at nest. **Habits:** Singly or in small groups, though hundreds may congregate at a good feeding site. Also dispersive, individuals commonly appearing in any open area of the lowlands and on satellite islands, occasionally in highlands. Hunts in shallows of freshwater and wet grassland, less commonly in marine habitats. Stands in water and stabs suddenly at prey; otherwise movements slow and deliberate. Takes fish and other small animals. Nests colonially in trees standing in water; nest a stick platform (AU data). Eggs (3–4) pale blue-green. **Range:** NG, NW Is (Batanta, Gag, Misool, Salawati), Bay Is (Biak, Yapen), Aru Is, and SE Is (Misima, Normanby); probably all satellite islands; 0–500 m, rarely to 2200 m. E Asia to AU and NZ. **Taxonomy:** Split from Western Great Egret (*A. alba*).

Intermediate Egret *Ardea intermedia* Pl 15
(**Plumed** or **Lesser Egret**, *Egretta intermedia*)
56–70 cm. Common; an AU visitor and local breeding resident; freshwater wetlands mostly in the lowlands. *This species is the source of identification woes regarding egrets. It is shaped like an E Great Egret but is the size of a Cattle Egret.* **Differs from E Great Egret:** Smaller size, bill shorter compared with head, head profile more rounded (less tapered), neck shorter (Intermediate E head and neck about as long as body; E Great head and neck 1.3× body), neck less often kinked, and gape not extending past eye. **Differs from nonbreeding Cattle Egret:** Overall more slender with a longer neck (Cattle E head and neck less than length of body) and thinner head and neck. **Adult:** Breeding plumage with filamentous plumes on back and lower neck, face green, bill red, legs dark with red blush. Nonbreeding plumage without plumes, face yellow, bill yellow (never black-tipped like many E Great) legs black with some olive. **Juv:** Same as Nonbreeding. **Subspp (1):** *intermedia* (Asia to AU). **Similar spp:** Eastern Reef E and Little E, both quite different. **Voice:** Silent but for a few soft calls at the nest. **Habits:** Habits and habitat similar to E Great. Occasionally forages on land like Cattle Egret. Nesting details similar to E Great. **Range:** All NG (breeds in Trans-Fly), NW Is (Batanta, Gag, Kofiau, Salawati), Bay Is (Numfor), Aru Is, and SE Is (Misima), but probably visits all satellite islands; 0–500 m, rarely to 2200 m. Africa and E Asia to AU.

Pied Heron *Egretta picata* Pl 15
(**Pied Egret**, *Ardea* or *Notophoyx picata*)
43–55 cm. Locally common AU visitor to freshwater wetlands in lowlands; not known to breed in NG Region. **Adult:** *Smallest of the NG egrets*—a dainty, lovely bird. *Slaty black with white throat and neck.* In Breeding plumage, the nape streamers are longer, reaching the blackish basal third of the hindneck; lanceolate plumes on back. **Juv:** Variable brown mottling on crown and neck. **Imm:** *All-white head; brown-and-white underparts.* **Similar spp:** White-necked H also white-headed, but twice as large and with dark (not yellow) bill and legs. **Voice:** Alarm call a muffled *orrk*. Cooing at nest. **Habits:** Usually in flocks of varying size, feeding in shallows of freshwater or in fields. Also picks over burned savannah and follows livestock. Deploys a full repertoire of heron hunting tactics; feeds mainly on insects, but also frogs and a few fish. **Range:** NG (S Lowlands, Trans-Fly, southern SE Pen, and Sepik-Ramu) and Aru Is, 0–500 m. Also Wallacea, AU.

White-faced Heron *Egretta novaehollandiae* Pl 15
(*Ardea* or *Notophoyx novaehollandiae*)
66–68 cm. An AU visitor and uncommon resident in the South; strays into the highlands; vagrant elsewhere. **Adult:** *A medium-sized, all-grey heron with a white face and upper throat.* Breeding plumage with ragged crest and longer lanceolate plumes on back and breast. **Juv and Imm:** Face grey, throat white. **In flight:** Dark flight-feathers contrast with paler wing-coverts. **Similar spp:** Eastern Reef E grey morph lacks white face; the bill is longer and buff-orange (not black); the legs are shorter; and the underwing is uniformly dark. **Voice:** Social call is a repeated *graaw*; alarm call a brief croak, *ooooaaark*; numerous other calls when nesting. **Habits:** Solitary by nature, occasionally as a pair or in a scattered group. Feeds in the shallows of freshwater bodies, ditches, and fields; less often on saltwater and tidal inlets. Forages in the same opportunistic manner as Pied H, and its diet includes a mix of small animal life, terrestrial and aquatic. No information on nesting in NG, but breeding suspected. Typically a solitary nester; nests in a tree near or far from water; nest a flimsy platform of twigs (AU data). Eggs (3–5) pale green. **Range:** NG and Aru Is, and vagrant to SE Is (Rossel, Woodlark), 0–1500 m. NG: mainly S Lowlands, Trans-Fly, and southern SE Pen; uncommon in Central Ranges; scattered records elsewhere. Also Wallacea to AU and NZ.

Little Egret *Egretta garzetta* Pl 15
(*Ardea garzetta*)
55–65 cm. Locally common AU visitor to freshwater wetlands in lowlands, especially in the South; strays into highlands; not known to breed in NG Region. **Adult:** *A small and very active white egret with slender black bill contrasting with yellow facial skin.* Breeding adult has 2 long nape streamers and filamentous plumes on back and breast. Nonbreeding plumage lacks nape streamers and breast plumes and shows plumes on back only. **Juv:** Similar to Nonbreeding adult but lacks plumes altogether. **Subspp (1):** *nigripes* (SE Asia to AU). **Similar spp:** Other white egrets are larger and usually have yellow or grey bill, not black. **Voice:** Alarm call a croaking *kark*. Gargling and harsh calls at nest. **Habits:** Usually solitary; mixes with other herons and ibises. Inhabits fresh- and saltwater habitats feeding energetically in the shallows. Forages impatiently, stepping about and dashing after prey; flutters wings at times. Pursues mainly fish, but also frogs and insects. Nests colonially with other herons, building its nest in the lower half of a tree; nest a platform of sticks (AU data). Eggs (3–4) pale blue-green. **Range:** NG, NW Is (Batanta, Kofiau, Misool, Waigeo, Salawati), and Aru Is, and expected for other satellite islands, 0–500 m, strays higher. Also Eurasia, Africa, and Australasia.

Eastern Reef-Egret *Egretta sacra* Pl 15
(**Pacific Reef Egret**, *Ardea sacra*)
60–65 cm. The common resident heron of seashore, reef flats, and islands. *Distinctive shape, with a long bill and neck and short legs.* **Adult:** *Two color morphs—one plain dark grey, the other white.* (Rare pied morph shows patches of grey and white.) On NG mainland, most birds are grey. The white morph differs from other white egrets by its long, thick, buff-orange bill with some dark shading and greenish-yellow legs and feet. Breeding bird has lanceolate plumes on nape (forming a short, ragged crest), back, and lower foreneck. **In flight:** Only the feet project past the tail; in other egrets the legs show as well. The grey morph has a uniformly dark wing. **Juv and Imm:** Grey morph more brownish than Adult, and the bill is dark; white morph possibly same as Adult. **Subspp (1):** *sacra* (Asia to AU). **Similar spp:** White-faced Heron similar to grey

morph but has white face, dark bill, and longer legs. Cattle E and Intermediate E similar to white morph but both have shorter bills and longer legs. **Voice:** Usually silent. Gives harsh squawks in alarm and in territorial chases. **Habits:** Solitary and territorial, rarely seen even in pairs. Defends its own patch of coast by chasing away other reef-egrets. Will mix with other species of egrets when foraging. Frequents all marine coastal habitats but prefers to forage in the open, not in vegetation. On islands it may venture inland to forage on streams and ponds. Forages in shallows using a variety of techniques to capture fish and crustaceans. Roosts and nests alone or in small colonies. The well-hidden nest is built in a tree or on a rock outcrop; nest a platform of sticks and other plant material (AU data). Eggs (2–3) bluish white. **Range:** Entire NG Region, 0–100 m. Also E Asia to AU, NZ, and Oceania.

PELICANS: Pelecanidae

The pelicans have a worldwide distribution in temperate and tropical areas. Of the 8 species, 1 breeds in Australia and migrates to New Guinea and perhaps breeds there. Pelicans are large aquatic birds with a huge bill featuring an expandable pouch used for catching fish and collecting rainwater to drink. They are good swimmers, with short legs set well back on the body and large webbed feet. They are awkward on land, but in the air they are masterful fliers, soaring on thermals, often in spectacular formations. Pelicans are monogamous and nest on sandy islands in large bodies of water, laying a clutch of 2 white eggs.

Australian Pelican *Pelecanus conspicillatus* Pl 12

150–160 cm. Common migrant from AU during dry season to southern Trans-Fly; numbers vary from year to year; perhaps occasionally breeds. Found on open freshwater and on mudflats along the coast. **Adult:** *A very large, unmistakable, white-and-black bird with a huge pinkish bill and pouch.* **Male** is larger on average. **Juv:** Similar, but brown where Adult is black. **Voice:** Call a 3-note grunt. **Habits:** Subject to vast movements when drought in AU forces birds to leave. Occasional single birds stray to the interior of NG, and they are later remembered in village lore long after they are gone. Gregarious, often in very large flocks; commonly seen fishing in a group or soaring in formation. Pelicans fish by sitting on the water, slowly paddling forward, and then lunging to scoop up prey in the pouch. They feed alone or communally, comrades abreast driving the fish in front of them toward the shallows. Nests colonially on sand islands or spits; nest on the ground is a platform built of plant material and debris (AU data). Eggs (1–2) white. **Range:** NG, Aru Is, and SE Is (Goodenough, Trobriands), 0–2000 m; expected for entire region. NG: mainly Trans-Fly and SE Pen, but wanders widely. Breeds AU; disperses to Wallacea, NG Region, Solomon Is.

FRIGATEBIRDS: Fregatidae

There are 5 species of frigatebirds worldwide, with 2 regularly in New Guinea and a third species vagrant. Frigatebirds are very large, mainly black seabirds with extremely long, slender wings and a long, forked tail. They have a long, powerful, hooked-tip bill and short legs. Supremely adapted for life in the sky, frigatebirds are most often seen soaring high on motionless wings. Nevertheless, they are capable of swift action and can chase down boobies and other, smaller seabirds to steal their food. More often, they fish for themselves, deftly plucking their prey from the sea surface or taking their favorite food, flying fish, on the wing. They never dive or sit on the water. Frigatebirds are commonly seen over coastal New Guinea and in nearshore waters. Normally silent, they are solitary but may gather to fish, travel, or roost on small islands. Females tend to be larger than males on average, and both sexes pass through a series of molts as they mature, a process that takes a few years. They nest on small, tropical oceanic islands, including islands around northern Australia and the Coral Sea, but not in the NG Region.

Christmas Island Frigatebird *Fregata andrewsi*
89–100 cm; wingspan 206–230 cm. Vagrant. Both sexes have *extensive white in belly* where other frigatebirds have black. **Male:** Black except for well-defined white belly patch. **Female:** Similar to smaller Lesser FB but with white extending onto belly and a *"clean cut" black shoulder spur extending down onto the breast and axillaries*. **Juv:** Differs from juv Lesser FB in having white belly. Similar to juv Great FB but has white extending onto underwing-coverts. **Similar spp:** Great FB imm may have mottled white on the breast or belly, not to be mistaken for the bright, well-defined white belly of Christmas Island FB. **Range:** Two likely sight records from S NG coast: Port Moresby and mouth of Fly R. Breeds on Christmas I (Indian O) and disperses into seas around Greater Sundas and Malay Pen.

Great Frigatebird *Fregata minor*   Pl 9
86–100 cm; wingspan 206–230 cm. Uncommon visitor. All plumages *lack the white underwing patch of Lesser FB*. **Male:** All black. **Female:** Blackish with *pale grey chin and throat* and white breast. (Beware that for some individuals the white breast patch creeps toward the underwing, suggesting markings of a Lesser FB.) **Juv:** Head white or tawny, separated from white breast by a dark band. **Imm:** Intermediate between Juv and Adult plumage. **Similar spp:** Lesser FB. **Range:** All NG coasts and seas, perhaps more frequent in the South. Tropical Pacific, Indian, and S Atlantic Oceans.

Lesser Frigatebird *Fregata ariel*   Pl 9
71–81 cm; wingspan 175–193 cm. Common visitor. All plumages have *diagnostic white "armpits"—a flank patch that extends as a streak onto the dark underwing-coverts*. **Male:** Blackish with *white flank patch*. **Female:** *All-black head without grey chin and throat*. **Juv:** Can be difficult to distinguish from Great FB, but look for the greater amount of white extending onto underwing. **Imm:** Intermediate between Juv and Adult plumage. **Similar spp:** Great FB. **Range:** All NG coasts and seas. Tropical Pacific, Indian, and S Atlantic Oceans.

BOOBIES: Sulidae

Ten species of boobies and gannets inhabit the oceans of the world, with 4 species in New Guinea waters. They are large seabirds with long, heavy bodies tapered at both ends and stiff tapered wings. The thick-based bill is long and pointed. Flight is an alternating series of wingbeats and short glides. Boobies feed by diving from considerable height, 15 m or more, necks outstretched and wings thrown back at the last moment, the bird entering the water like a dart. Boobies feed on fish and squid, which they capture mostly underwater, and occasionally taking flying fish on the wing. When not foraging or traveling, they will rest on the water. Boobies frequent both open seas and the coast, usually staying outside the reefs. They nest on small islands remote from humans.

Abbott's Booby *Papasula abbotti*
(*Sula abbotti*)

71 cm. Vagrant or possibly rare visitor to western seas in the region, where found at the edge of the continental shelf. **All plumages:** *Unique pattern of dark upperwing and tail*; white body and underwing. Large bill, long head and neck, thin body, and long wings all combine to give a *lankier shape than other boobies*. **Female** bill pink; **Male** and **Juv** bill pinkish grey. **Habits:** Poorly known at sea, but apparently similar to other boobies. **Range:** Sightings from open seas between Bird's Neck and Aru Is made in Feb–Mar. Sole remaining population of a few thousand birds nests on Christmas I in the Indian Ocean, and disperses as far as the Banda Sea.

Masked Booby *Sula dactylatra* Pl 8
(Blue-faced Booby)

81–92 cm; wingspan 152 cm. A rare visitor. Open seas, occasionally seen fishing fairly close to the coast. **Adult:** A *large white booby* with both black-and-white wings and *black tail* (Red-footed B has white tail). Note *dark facial mask*. **Juv:** Differs from Brown B adult by white collar and upper breast and dark bill. **Imm:** Has brown head, but otherwise less brown in plumage than Juv. **Subspp (1):** *personata* (W and central Pacific). **Habits:** Preys mostly on fish, especially flying fish. **Range:** Probably regionwide; breeds in Torres Strait. Tropical seas worldwide.

Red-footed Booby *Sula sula* Pl 8

66–77 cm; wingspan 91–101 cm. A common visitor near coast or out at sea. **Adult:** Three color morphs, all with a diagnostic *white tail*. *White morph* white with black on wings; lacks dark facial mask of Masked B. *Brown morph* pale brown with white belly, rump, and tail. *Intermediate morph* white with brown back and wings. **Juv:** Uniform pale brown, note dark bill. **Subspp (1):** *rubripes* (W and central Pacific and Indian Oceans). **Similar spp:** Masked B juv and Brown B juv, compared with Juv Red-footed, have distinct, dark brown heads with white (Masked) or pale brown (Brown) underparts. **Habits:** Takes fish, especially flying fish, and squid. **Range:** Probably regionwide; breeds in Torres Strait. Tropical seas of the world.

Brown Booby *Sula leucogaster* Pl 8

64–74 cm; wingspan 132–150 cm. Common visitor and rare breeding resident. Frequents coasts, river mouths, harbors, islands, and open ocean. **Adult:** *Dark chocolate brown with clean white breast and belly*. Facial skin blue in **Male**, yellow in **Female**. **Juv:** Similar to Adult, but underparts mottled, pale brown. **Subspp (1):** *plotus* (central Pacific to Indian Ocean). **Similar spp:** Masked B juv has white collar. See Red-footed B juv. **Habits:** Singly or in small groups. Feeds on a variety of fish, especially flying fish, and squid. Drawn to flocks of seabirds diving for baitfish driven to the surface by tuna schools. Also feeds alone at the coast, making shallow, oblique dives to catch prey. Nests on the ground; nest a small depression lined with grass and other vegetation. Eggs (2) white. **Range:** Breeds on at least one of the SE Is and in Torres Strait; also a regionwide visitor. Tropical seas of the world.

CORMORANTS: Phalacrocoracidae

There are 41 species of cormorant worldwide. The 4 New Guinea species dwell mainly in freshwater wetlands and are shared with Australia. Populations between the two continents are presumably linked by the seasonal movement of birds leaving Australia when habitat there dries up. Cormorants are heavy-bodied, blackish or black-and-white birds with a long neck and tail, hooked bill, and pouched throat. They swim well and capture prey underwater. They spend much time perched, sunning and preening their wings. Cormorants fly with the neck extended, flocks usually in V-formation. The sexes cannot be separated by plumage, and the male is only slightly larger on average than the female. Cormorants breed in monogamous pairs and nest colonially, usually in dead trees.

Little Pied Cormorant *Microcarbo melanoleucos* Pl 12
(*Phalacrocorax melanoleucos*)
58 cm. The most widespread cormorant in NG—locally common in lowland wetlands, uncommon in the mountains, and even ranges into the alpine zone. Little evidence of breeding, and many birds may be visitors from AU. **Adult:** A small *black-and-white* cormorant with *short, yellowish bill. All-white face and contrasting black crown.* Eye dark brown. **Juv:** Black cap extends down to eye; black flank mark. Eye grey-brown. **Subspp (1):** *melanoleucos* (Wallacea to Solomon Is and AU). **Similar spp:** Australian Pied Cormorant, a vagrant, much larger; has a longer, grey bill; and on the flank a black bar extends down the leg. **Voice:** Usually silent. Call a cooing *keh keh keh keh* (AU data). **Habits:** Singly or in small flocks, often with Little Black C. In swamps, along rivers, and on seacoasts; perches in trees along the shore. Swims underwater to capture crustaceans and occasionally fish. Nests in a shrub or tree near water; nest a stick platform lined with leaves (AU data). Eggs (~4) pale blue. **Range:** NG, NW Is (Misool, Waigeo, Salawati), Bay Is (Numfor), Aru Is, and SE Is (Fergusson), 0–3200 m. Also, Wallacea to Solomon Is, AU and NZ.

Little Black Cormorant *Phalacrocorax sulcirostris* Pl 12
61 cm. A common breeding resident in lowland wetlands; occasionally found in the mountains. Some birds may be AU visitors. **Adult:** *A small, all-black cormorant with a black face.* Eye green. **Juv:** Brownish black. Eye brown. **Similar spp:** Great Cormorant is larger, with a pale chin and cheek. **Voice:** Usually silent. Guttural calls at nest. **Habits:** Gregarious, especially common in open waterways. Seen perched in dead trees overlooking water, or swimming with head slightly uptilted. Catches fish while swimming under water. Flocks will "herd" fish. Nests in trees over water; nest is a platform of sticks (AU data). Eggs (~4) bluish or greenish white. **Range:** NG, NW Is (Kofiau, Misool), Aru Is, and SE Is (Fergusson), 0–500 m, rarely higher. Also Greater Sundas to AU.

Australian Pied Cormorant *Phalacrocorax varius*
75 cm. Vagrant from AU. **Adult:** *From Little Pied C by its larger size; longer, grey bill; and black bar from the flank down the leg.* Eye green. **Juv:** With dirty scalloping on white underparts. Eye brown. **Subspp (1):** *hypoleucos* (also AU). **Habits:** Prefers large expanses of open fresh or salt water, where it catches fish. **Range:** NG: 1 record from Port Moresby (SE Pen). Breeds AU and NZ.

Great Cormorant *Phalacrocorax carbo* Pl 12
81 cm. A rare, sporadic visitor from AU. **Adult:** *A large black cormorant with pale bill, yellow facial skin, white gular patch,* and when breeding, a white patch on flank. Eye green. **Juv:** Similar, but plumage brown with white chin. Eye brown. **Subspp (1):** *novaehollandiae* (also AU, NZ).
Similar spp: Little Black C much smaller, lacks any facial pattern, and bill blackish. **Habits:** Singly or in small groups in marine and lowland freshwater habitats. Takes fish. **Range:** Probably throughout NG, plus a sighting from Batanta (NW Is). Also Eurasia, Africa, AU, NZ, and N Amer.

DARTERS: Anhingidae

The 4 species of darters inhabit the warmer parts of the world, 1 species to a continent. The New Guinea species is confined to Australasia. Related to cormorants, darters are diving birds with a distinctive, slender shape: slim head, sharp pointed bill, and long neck and tail. They inhabit lowland waterways and swamps, often in company with cormorants. They perch at the water's edge with wings outstretched to dry. When fishing, they swim under water and spear their prey with the dagger-shaped bill. Darters are monogamous and nest colonially in trees.

Australasian Darter *Anhinga novaehollandiae* Pl 12
(Darter, *A. melanogaster*)
90 cm. Locally common resident of lowland swamps, lagoons, and slow streams and rivers. Also an AU visitor. Wandering individuals can turn up anywhere. *An elongated cormorant-like bird with diagnostic dagger-shaped bill and snaky neck.* Both sexes show a *white facial stripe.* **Male:** Blackish. **Female:** Upperparts somewhat paler; throat and underparts white or pale buff. **Juv:** Paler than Female and facial stripe obscure. **Similar spp:** Cormorants are not as slender and have a shorter neck and hook-tipped bill. **Voice:** Noisy. Both sexes produce an easily recognized rattling call that speeds up: *kak, kak, kak . . .* for several secs. Male gives an explosive *KAH!* when breeding. **Habits:** Usually singly or in small groups, perched on trees close to water or swimming gracefully, with kinked neck holding the head uptilted. Dives for fish. Its flight, a few flaps followed by a glide, is distinctly different from that of a cormorant. Commonly seen soaring. Nests in a tree standing in water; nest a stick cup. Eggs (4) pale green. **Range:** NG, NW Is (sighting from Salawati), and SE Is (Fergusson), 0–300 m. Also Wallacea, AU, and Bismarck Is. **Taxonomy:** Has been classified with the Oriental Darter (*A. melanogaster*).

OSPREYS: Pandionidae

This worldwide family is composed of only 2 similar species of fishing hawks. Molecular and fossil studies have confirmed the ospreys to be an old lineage distinct from other raptors.

Eastern Osprey *Pandion cristatus* Pl 17, 18
(Osprey, *Pandion haliaetus*)
46–53 cm; wingspan 127–174 cm. Thinly distributed along coastlines and fringing islands, occasionally inland along lowland rivers and lakes. **Adult:** A fish hawk with a *thick black bar through the eye* and *white below* with blurred breast band. **Female:** Larger; breast band thicker. **Juv:** Pale tips to dark feathers of back and wings. **In flight:** *Square tail. Long wings with a slight bend at the wrist* and *a diagnostic black smudge at the bend of the wing.* **Similar spp:** White-bellied Sea-Eagle much larger; has brief, wedge-shaped tail and uptilted wings. **Voice:** Usually silent. Call is a repeated weak whistle, *seee* or double note squeal. **Habits:** Most often encountered on small islands, at harbor entrances, and over reef lagoons. Solitary or paired. Dives from the air, plunging into the water to take fish and sea snakes. Nests inland, building a robust stick platform in tree crown. Eggs (2–3) cream with dark spots and blotches (AU data). **Range:** All NG Region, mainly coastal, 0–500 m. Also Wallacea east to Solomon Is and AU. **Taxonomy:** Eastern O recently split from Western Osprey, *P. haliaetus*.

HAWKS AND EAGLES: Accipitridae

This is the largest family of diurnal birds of prey, with 256 species worldwide and 25 species in the New Guinea Region. Two species are migratory from Asia, and at least one other migrates from Australia. Most resident species of hawks and eagles in New Guinea and Australia have relatives in Eurasia, suggesting they originally came to our region from that vast neighboring continent. A few of the endemic species seem of ancient origin, having long resided in Australasia. The most striking example, the New Guinea Harpy-Eagle, is related to the Harpy Eagle (*Harpia harpyja*) of South America.

All hawks and eagles are predators or scavengers, taking birds, mammals, arthropods, and carrion. Prey is captured with taloned feet and consumed whole if possible or torn into pieces that can be swallowed. Hawks and eagles occupy all New Guinea but are more diverse and abundant in open habitats, forest edge, or near water. Species that reside mainly in the forest interior are difficult to find and observe, and these are among the least known of New Guinea birds. Most hawks and eagles are solitary, although they actually live in pairs with both members sharing nesting duties. The nest is usually built of sticks in the crown of a tree. After fledging, the young remain with the parents for a period of months and depend on them for food.

In most species of hawks and eagles, male and female have similar plumages, but females are larger, in some species much larger. Because size is often the only way to distinguish the sexes, we have indicated size difference in three classes: female larger on average (size the same or larger), larger (usually larger), and much larger (larger by >20% length of folded wing). Size difference is best observed when a pair of birds is together.

The often-noisy juveniles are typically paler and have well-marked upperparts characterized by pale-tipped feathers. In some species, it takes more than a year for young birds to attain adult plumage. Identifying juveniles to species can be difficult, particularly for goshawks, not just because species can look alike but also because of individual variation—so, identify juveniles with caution. Finally, locating and studying hawks and eagles in open country is not difficult, but any encounter with a forest raptor is an opportunity not to be missed!

Pacific Baza *Aviceda subcristata* Pl 16, 22
(Crested Hawk)
36–41 cm; wingspan 80–105 cm. An uncommon but distinctive hawk of forest edge and tall second growth in lowlands and hills. Smooth grey head contrasts with *bold black barring on whitish underparts*. *Brown scapulars* stand out against otherwise grey upperparts. *Yellow eye* set in evenly grey face. Note *spiky crest*—this is the only crested NG raptor. Buffy "pantaloons." **Female**: Larger on average, slightly browner. **Juv**: Browner yet, with pale-scalloped dark upperparts; underparts paler than Adult. **In flight**: *Broad, rounded wings prominently banded; black-barred breast; square tail with thick black terminal band separated by a gap from a few thin medial bars.* **Subspp** (4): Minor—differ in size and intensity of barring on underparts. **Similar spp**: Long-tailed Buzzard similar in flight because of general shape and barred wings and tail, but the breast is streaked rather than barred, the wing is more rounded and has fewer thicker bars, and on the tail the thick terminal band shows less contrast with the medial tail bars. Flying goshawks show thinner bars in wings and tail. **Voice**: A distinctive, repeated, high-pitched, disyllabic call, *pee-peow . . . pee-peow*, or *peecho . . . peecho*, the first note rising, second descending and more explosive. **Habits**: Solitary or paired. Sluggish. Seen perched or soaring at moderate heights. Swoops on prey in the forest canopy: arthropods, frogs, lizards, snakes, and sometimes fruit. Striking display flight: soars steeply upward, flaps a few times with wings held in a high V and tail flexed down, then glides downward and repeats. Nest built in tree crown; a small flimsy cup of sticks lined with leaves (AU data). Eggs (2–3?) white, sometimes with light spotting. **Range**: NG, NW Is (Batanta, Kofiau, Misool, Salawati, Waigeo), Bay Is (Biak, Yapen), Aru Is, Karkar I, SE Is (D'Entrecasteaux: 3 main islands), 0–1300 m, rarely higher. Also Wallacea east to Solomon Is and AU.

Long-tailed Buzzard *Henicopernis longicauda* Pl 21, 22
(Long-tailed Honey Buzzard)
51–61 cm; wingspan 105–140 cm. This is *the* hawk most often seen soaring over hill and montane forests. **In flight, all plumages:** Distinctive silhouette with *small head, slender body, long rounded tail, and large rounded wings thrown forward with spreading, fingered primaries. Wing and tail boldly patterned with black bars* of rather even width. Head and breast finely but clearly streaked. At close range note pale bill and face, and yellow eye. **Adult:** 3 thick black bars in wing and tail (a fourth is hidden). **Female:** Larger on average. **Juv:** Paler; 4 narrower black bars in wing and tail. Wing-coverts show 2 whole black bars exposed instead of just 1. **Similar spp:** New Guinea Harpy-Eagle—never seen soaring—is much larger and more robust; unstreaked whitish below; very broad, short wings. Pacific Baza, see account. Doria's Hawk doesn't soar and is slimmer, with short wings and longer and many-barred tail. **Voice:** Usually silent, but gives a goshawk-like series during display.
Habits: Singly or in pairs. Sails slowly over the forest, circling; also patrols open corridors beneath the canopy, slowly picking its way and stopping to perch and look about. Conspicuous and usually not shy. Feeds mainly on insects (including wasps and their larvae), also lizards and birds' eggs. Stick nest in tree crown. No egg information. **Range:** Endemic. NG, NW Is (Batanta, Misool, Salawati, Waigeo), Bay Is (Biak, Yapen), Aru Is, SE Is (D'Entrecasteaux: Fergusson, Normanby), 0–3000 m.
Extralimital spp: Crested or **Oriental Honey Buzzard** (*Pernis ptilorhynchus*), an Asian migrant, is vagrant to AU and likely to turn up in NG; it shows 2 main tail-bands and overall bears little resemblance to the Long-tailed B. **Taxonomy:** Affiliated with two endemic Australian genera, *Hamirostra* and *Lophoictinia*.

Bat Hawk *Macheiramphus alcinus* Pl 16, 24
(*Machaerhamphus alcinus*)
43–51 cm; wingspan 95–120 cm. An unusual, *dark, falcon-like hawk* known from few localities. Found at forest edge in lowlands and montane valleys. Generally overlooked, except for its dawn and dusk activity. *White throat marked with a unique black stripe*; dark underparts unevenly marked with small amounts of white. Head shape unlike other raptors' with wide mouth, sloped forehead, and tiny bill—suggestive of dollarbird or treeswift. *Large, piercing yellow eye accentuated by partial white eye-ring.* **Adult:** Nearly all black. **Female:** Larger. **Juv:** Presumably has more white in underparts. **Subspp (1):** *papuanus* (endemic). **Similar spp:** Brown Falcon same size and can be nearly all dark, but has dark eye and "moustache" and lacks Bat Hawk's white throat with black stripe. Flying foxes also crepuscular but have broader wings and slower wingbeats. **Voice:** A rapidly repeated *ki ki ki . . .* , or a slow *keek keek keek. . . .* **Habits:** During the day usually found roosting at the forest edge, perched on a branch in the shade of the canopy. Forages in the twilight with swift, agile flight over forest near caves and through trees in pursuit of bats and swiftlets. Captures prey in talons and transfers it to mouth whence it is swallowed whole. Abroad in daylight when breeding, giving vocal display flights. Singly or in pairs. No nesting information for NG, but elsewhere builds a large stick nest in treetops. Eggs (1–2) white with a few grey markings. **Range:** NG, with scattered records throughout (although none as yet from Trans-Fly), 0–1000 m. Specimens and sightings from SE Pen; additional sightings from Bird's Head, W Ranges (Baliem Valley), NW Lowlands (Tiri R and Vanimo), S Lowlands (Asmat, Middle Fly R, and Lake Kutubu), and Huon. Also Africa and Asia.

Black-winged Kite *Elanus caeruleus* Pl 16, 24
30–33 cm; wingspan 77–92 cm. Rare and local, mainly over mid-mountain grasslands and farmland. Usually seen perched in the open on a bare branch or wire or characteristically hovering with body at a 45° angle to the ground. **Adult:** Elegant—uniquely patterned grey, white, and black. *Prominent black shoulder*. Dark-shadowed *red eye*. **Female:** Larger on average. **Juv:** Shares characteristic black shoulder and pale tail, but head, breast, and back tinged brown, back and wing feathers tipped white, and eyes black. **In flight:** Color of the underside of the flight-feathers varies greatly, from slaty grey to whitish grey. **Subspp (1):** *hypoleucos* (also Sundas, Philippines, and Wallacea). **Voice:** Usually silent, but gives shrill whistles during display flight; also *kree it*.
Habits: Usually solitary. Buoyant flight over grasslands with rowing wingbeats, gliding with wings in a shallow V. Hovers stationary high above the grass before gliding down and pouncing on its prey. Often hunts from exposed perches, sometimes dropping to the ground in the manner of a shrike. Preys on small rodents and lizards. No nesting information for NG; elsewhere a bulky stick nest in treetops. Eggs (3–4) white blotched brown. **Range:** NG E Ranges and SE Pen; a few records from Trans-Fly and

Sepik-Ramu, 0–2300 m. Also Africa, Eurasia. **Extralimital spp: Black-shouldered Kite** (*E. axillaris*) could turn up in South as an AU vagrant. Very similar, but with proportionately narrower wings and longer tail, paler upperparts, and when seen in flight from below, a small black patch at the hand of the underwing.

Black Kite *Milvus migrans*   Pl 16, 18, 19

48–53 cm; wingspan 120–153 cm. A common scavenging kite of open country, roadsides, towns, and harbors, mainly in the North and East. Only NG raptor with a *forked tail*—folded tail obviously forked or triangular when fanned. **Adult:** Dark brown, somewhat streaky and individually variable. **Female:** Larger on average. **Juv:** Overall paler, with pale edges to feathers yielding a more streaked or scalloped effect. **Subspp (1):** *affinis* (Wallacea to AU). **Similar spp:** Whistling K paler and has a rounded, sandy-colored tail. Brahminy K juv has shorter, rounded tail. **Voice:** A feeble, high-pitched, tremulous, nasal whistle. **Habits:** Solitary or in loose parties; may associate with other kites. Drawn to grass fires in search of fleeing or scorched prey. Often abundant near towns. Follows roads, patrolling for carrion. When foraging, flies at low altitude, but also soars higher occasionally. Bulky stick nest in tree crown. Eggs (2–3) dull white with obscure dark markings (AU data). **Range:** NG and SE Is (D'Entrecasteaux: 3 main islands), 0–2200 m. NG: eastern NW Lowlands eastward to SE Pen, also Border and E Ranges. Also Africa, Eurasia, and AU.

Whistling Kite *Haliastur sphenurus*   Pl 16, 18

53–58 cm; wingspan 120–146 cm. Occupies same habitats as preceding species and is ecologically similar and also common but more widespread. **Adult:** *Pale buff head, underparts, and tail.* Tail long, with rounded tip. **Female:** Larger on average. **Juv:** Wings and back with feathers more broadly tipped whitish. **In flight:** *Pale "window" of inner primaries* framed by blackish primaries on either side. Note slender body and unusually long wings; glides with distal half of wing bowed downward. **Similar spp:** Black K darker, has forked tail. Brahminy K juv slightly bulkier, has shorter wing and tail and pale underwing with only tips of the flight-feathers black. **Voice:** A high, shrill whistling—first note is leisurely and descending, followed by a quick rising series, *peeer pipipipipi*. Also a descending trill. **Habits:** Similar to Black Kite but more frequent near water. Singly or in loose flocks; gregarious. Snatches food from water as well as dry land. Nest built in tree; large stick structure. Eggs (1–3) dull white, spotted and blotched (AU data). **Range:** NG and SE Is (Goodenough), 0–700 m. Throughout mainland NG except Bird's Head to western S Lowlands and perhaps the western NW Lowlands. Also AU, New Caledonia.

Brahminy Kite *Haliastur indus*   Pl 16, 18

41–43 cm; wingspan 110–125 cm. A ubiquitous raptor of seacoasts, rivers, and forest edge, where it is a common sight. **Adult:** An unmistakable *white-and-chestnut* hawk. **Female:** Larger on average. **Juv:** Mottled brown; note *streaky front* and *dark eye-patch*; when perched *folded wings usually extend beyond tail tip*. **In flight, all plumages:** *Dark wing tips* and rather short tail. **Subspp (1):** *girrenera* (Moluccas to Bismarck Is and AU). **Similar spp:** See Whistling K and Pygmy Eagle. **Voice:** Call note is a surprisingly weak, nasal downslur like the bleat of a lamb or the call of a bush-hen: *peeyah*. **Habits:** Usually singly or in pairs. Seen perched conspicuously in treetops or soaring. Plucks prey from the water, foliage, ground, or out of the air. Takes insects, fish, amphibians, reptiles, birds, and mammals. Also scavenges. Nests in treetops; bulky stick structure. Eggs (1–2) white with spots, lines, or blotches (AU data). **Range:** Entire NG Region, 0–2200 m. India to Solomon Is and AU.

White-bellied Sea-Eagle *Haliaeetus leucogaster*   Pl 17, 18
(White-breasted Sea Eagle)

69–79 cm; wingspan 178–218 cm. Uncommon along seacoasts and rivers. **Adult:** *The only large grey-and-white raptor.* **Female:** Larger. **Juv and Imm:** Mottled brown. **In flight:** *Short, wedge-shaped tail.* Distinctive, misshapen profile with the small head and brief tail on either end of a massive body carried aloft on *huge wings held in a shallow V.* Adult from below all white with black flight-feathers. Juv and Imm show banded underwing-coverts and *white patch at base of primaries.* **Similar spp:** Gurney's E and Wedge-tailed E (both slightly larger) have much longer tails, uniform underwing-coverts, and although both can have a pale window at the base of the primaries, it is not nearly as conspicuous as in juv and imm sea-eagle. Kites are much smaller and have longer rounded or forked tail. **Voice:** Vocal when nesting. Strident, very loud and far-carrying, gooselike clanking, *ank—ank—ank . . .* , often by 2 birds together, perched or in flight. A variation is a harsh, high-pitched scream: *schuwee—schuwee—schuwee!* **Habits:** Regularly soars, singly or in pairs. Forages over reefs and lagoons and along the seacoast, lowland rivers, and other large bodies of freshwater. Diet includes fish, snakes, and carrion, occasionally stolen from smaller raptors. Perches conspicuously in the open on a branch or rock pinnacle. Nest in tree is an enormous stick structure. Egg (1) white (AU data). **Range:** All NG Region, 0–500 m. Also India to Bismarck Is and AU.

Swamp Harrier *Circus approximans* Pl 19
(Includes **Papuan Harrier**, *Circus spilothorax*)
48–53 cm; wingspan 118–145 cm. Widespread but uncommon resident over grassland and marshes from sea level to above timberline. Also a seemingly rare migrant from AU to lowland grassland and marshes during austral winter, but status uncertain owing to confusion with Female and Juv of the resident race. Be aware that plumage development and variation are inadequately known for the resident race; plumage sequence likely to pass through several plumages, as in AU race of Swamp H. **In flight:** A large harrier with slim body on long wings held in a shallow V; look for *whitish rump patch*. **Male resident ssp:** *Pied morph*, handsomely patterned with *black head, mantle, and wing tips* contrasting with *bright white underparts and grey tail*. *Black morph* rare, all black with grey tail. **Female resident ssp:** Larger. *Pale morph*, brown with streaked underparts, pale rump; from AU race by *tail with clear, dark barring and face often with pale streaking* (tail without bands in 1 specimen). *Black morph* similar to Juv but may have white rump—difficult to know if these are early breeding birds in Imm plumage. **Juv and Imm resident ssp:** *Charcoal with brownish cast*, can be rusty brown on belly (underparts not all rusty brown as in many juv AU race) with *large, pale nape patch*. Older Imm pied morph like adult Male of AU race but with darker wing pattern suggestive of adult pied morph. **Subspp (2):** *spilothorax* (endemic resident); *approximans* (vagrant from AU to NG Trans-Fly and possibly SE Pen, lowlands) Male similar to Female of resident race, but tail either all-grey or with faint, broken bands; Female difficult to separate from resident Female—tail often has more obscure bands, sometimes reduced to spots, and face usually all brown; Juv and Imm often rusty brown below, with less pale streaking on nape than resident race. (Iris color not helpful to identification.) **Similar spp:** Kites also do low gliding, but lack white rump and usually do not hold wings in a V. **Voice:** Usually silent. Call a shrill *whieeuw*. **Habits:** Usually solitary. Quarters low over grassland or marsh. Drops into vegetation to catch prey. Takes rodents, reptiles, birds, and probably frogs. Nest built in tall grass and marsh; nest a mound of sticks, grass stems, and other vegetation. Eggs (3) white. **Range:** NG, except Bird's Head and Neck, 0–3500 m. Also AU, NZ, Oceania. **Taxonomy:** Some authorities treat the resident race as a separate species, the Papuan H. Both the NG and AU populations sometimes included in the Eastern Marsh-Harrier (*C. spilonotus*) or all three together as races of a more inclusive Marsh-Harrier (*C. aeruginosus*). **Extralimital spp: Pied Harrier** (*C. melanoleucos*), a migratory E Asian species, with probably erroneous sightings of males from NG. Similar to Swamp H, but pied male with large white patch at leading edge of the shoulder and black head with sharp edge at white breast (instead of streaky edge). Female with vague pale shoulder patch.

> **Goshawks and Sparrowhawks.** The 8 species of these short-winged hawks, called "accipiters," form 5 sets in their distributions and habits. One is a tiny, migratory Asian species that overwinters in the far West; all the rest are resident species. Three lowland species—Variable Goshawk, Brown Goshawk, and Collared Sparrowhawk—are confusingly similar in appearance. Among them, the Variable G lives at the forest edge, is a still hunter, and rarely soars; Brown G lives in open country and soars; and Collared S is commoner in the forest interior than in the open and soars. The fourth lowland species, Grey-headed Goshawk, is distinctive in appearance, lives both in the forest interior and at the edge, and doesn't soar. Of the 3 montane species, Black-mantled Goshawk lives in the forest interior and seldom soars. The remaining 2 species are big montane forest goshawks that soar; Meyer's Goshawk is usually at higher elevations than Chestnut-shouldered.

Chinese Sparrowhawk *Accipiter soloensis* Pl 20, 22
(**Chinese Goshawk**, **Frog Hawk**)
25–30 cm; wingspan 52–62 cm. A rare Asian migrant in austral summer to open habitats and swamps mainly in NW Is, but also coastal W NG. *Tiny—smallest NG hawk*. Note diagnostic *white undertail-coverts with no markings*. **Male:** Pale grey with whitish underparts washed rufous across the breast. **Female:** Similar but larger and pale rufous breast obscurely barred. **Juv:** Typical juv accipiter, but look

for white undertail-coverts. **Similar spp:** Collared S slightly larger, lankier with longer legs and tail, and has squared tail tip; in adult, note heavy barring on the wings and more rufous in underparts and rufous collar; in juv, undertail-coverts marked, and tail feathers with numerous obscure bars. **Habits:** Little known in NG Region. Mostly solitary, in open habitats and swampland. Elsewhere in its wintering range feeds heavily on cicadas. In its breeding range, drawn to wetlands where it forages mainly for frogs but also other small vertebrates and insects. Typically perches in open and scans for prey. **Range:** NG, NW Is (Batanta, Gag, Waigeo, and probably all), and Bay Is (sightings on Biak), 0–200 m. NG: sightings from Bird's Head, southwest S Lowlands (Timika), and Trans-Fly (Dolak I). Breeds E Asia, wintering from Indochina to Wallacea.

Variable Goshawk *Accipiter hiogaster* Pl 20, 22
(**Varied Goshawk; formerly Grey Goshawk**, *A. novaehollandiae*)
33–48 cm; wingspan 55–80 cm. The common accipiter of forest edge in lowlands and hills; follows disturbed habitats into the mountains. Rarely soars; a still hunter that watches from a perch. At least *3 color morphs* exist, but all have *bright orange-yellow cere, legs, and feet*. **Adult:** *Common morph* grey above, *plain cinnamon brown below with no barring* or at most obscure barring; *lacks a distinct rufous collar* (SE Is races differ). *White morph is the only all-white NG raptor*. *Grey morph* (smooth monochromatic slate-grey) is *the only all-grey NG hawk with pale eye*. **Female:** Larger. **Juv:** *Common morph* difficult to identify; brown above, white or buffy below with variable streaking and barring that is sometimes absent; look for orange-yellow cere, legs. *White morph* presumably all white. Grey morph, no information. **Subspp (4):** *leucosomus* (most of region) typical race large with all 3 morphs of which the common morph is unbarred or faintly barred below, and lacks a distinct rufous collar; *misoriensis* (Bay Is: Biak) tiny sparrowhawk-sized race with common morph only, showing plain underparts and finely barred tail; *pallidimas* (SE Is: D'Entrecasteaux Is: 3 main islands) large race, common morph only, pale with finely barred flanks and belly, has obscure rufous collar; *misulae* (SE Is: Misima, Tagula; presumably this form on Woodlark I, Alcester I, and Trobriands) large, common morph only, not pale but belly barred and with traces of a rufous collar. **Similar spp:** Three hawks are difficult to separate: Variable G, Brown G, and Collared Sparrowhawk. See account for Brown G, whose range overlaps broadly with the typical and usually unbarred and collarless race of Variable G. Collared S more delicately built with proportionately longer legs and square-tipped tail; chestnut collar in adult. Grey-headed G dark morph has dark eye with orange eye-ring. **Voice:** An unhurried series of 8–10 thin, high, weak notes, either upslurred or downslurred. **Habits:** Solitary, occasionally in a pair. Perches in shade trees. Crosses clearings at canopy height. Takes small vertebrates and arthropods. Conspicuous white morph could be passed over as a cockatoo. Nests in a tree crown; bulky stick and twig structure. Egg, no information. **Range:** Throughout NG Region, except a few islands (Mios Num, Karkar, Manam, Rossel), 0–1500 m. Also Wallacea, Admiralty, Bismarck, and Solomon Is. **Taxonomy:** Previously included in Grey Goshawk (*A. novaehollandiae*) of AU.

Brown Goshawk *Accipiter fasciatus* Pl 20, 22
(**Australian** or **Australasian Goshawk**)
30–43 cm; wingspan 60–98 cm. Common in open country in E half of NG from lowlands to mid-mountains. **Adult:** Similar to Variable G's common morph, except *finely and evenly barred below* and with *rufous collar* (sometimes indistinct and hard to see) and *greenish-yellow cere and pale yellow legs*. **Female:** Larger. **Juv:** Streaked breast and barred lower breast and flanks; difficult to separate from juv Variable G, but note difference in color of cere and legs. **Subspp (2):** *polycryptus* (most of NG range) belly barred rufous brown; *dogwa* (Trans-Fly) barred belly much paler. **Similar spp:** Variable Goshawk, see above. Also, Variable G has shorter wings that when folded project about a third of the way down the tail (about halfway down the tail in Brown G). Collared Sparrowhawk nearly identical but smaller, more delicate with proportionately longer and thinner legs, squared tail tip, and lacks a heavy brow line; Collared S occurs in closed forest whereas Brown G does not. **Voice:** A high-pitched, rapid *kikikiki* . . . a series of more than 12 thin notes rising in pitch. **Habits:** Solitary. Occupies savannah, scrub, suburban gardens; hangs around water sources to ambush thirsty prey. Forages for birds, other small vertebrates, and arthropods. Perches for long periods. Soars on slightly uptilted wings. Nests in a tree; stick structure (AU data). Eggs (3) bluish white, some sparingly marked. **Range:** NG: Sepik-Ramu and Trans-Fly eastward, 0–1900 m. Also AU, Wallacea to Fiji.

Black-mantled Goshawk *Accipiter melanochlamys* Pl 20, 22
30–43 cm; wingspan 65–80 cm. The goshawk of cloud forests; uncommon. **Adult:** Unmistakable, handsome *black and chestnut* with *black head* and *bright yellow cere and legs*. **Female:** Larger. **Juv:** Blackish brown above with brown or chestnut scalloping, *clean white below with sharp streaking and no barring*; crown pale and dark-streaked. **Similar spp:** No other goshawk shares its habitat except occasionally Meyer's G, which is larger and black-and-white or all black. Chestnut-shouldered G suggestive of Juv Black-mantled (which has rufous-edged wing-coverts and black-streaked white underparts) but the larger Chestnut-shouldered has a black crown and 9–10 tail bars vs 5–6 in juv Black-mantled. **Voice:** Typical goshawk call. A high, weak series of rapidly delivered notes. The series rises, then falls in pitch. **Habits:** Inhabits forest interior as well as edge. Solitary or as a pair. Seen crossing forest clearings or flying through the canopy; infrequently soars. Hunts for birds and other small vertebrates and arthropods. Nest is built in a tree; no other information. **Range:** Endemic. NG: Bird's Head, Central Ranges, and Huon, 1800–3100 m.

Grey-headed Goshawk *Accipiter poliocephalus* Pl 20, 22
30–38 cm; wingspan 56–65 cm. Common but secretive in lowland and hill forest. In all plumages identified by *large dark eye surrounded by fleshy orange ring; cere, gape, legs, and feet also bright orange*. **Adult:** *Common morph* is an easily recognized grey-and-white hawk. *Dark morph* (rare) is greyish black. **Female:** Larger. **Juv:** Similar but brownish grey above and finely dark-streaked below. Wing and tail feathers barred. **Similar spp:** No other hawk has this plumage pattern and striking orange soft parts. **Voice:** A rapid series of very high, thin upslurs, 2.5–3/sec: *sluweek, sluweek, sluweek. . . .* More rapidly delivered than other NG goshawks. **Habits:** Inhabits forest interior and edge. Usually solitary. Inconspicuous—seen perched at midstory levels in the forest or flying low across a clearing. Does not soar. Takes lizards and arthropods from the forest floor. Nest built in tree crown; a compact mass of sticks, twigs, and leaves. Egg (1) white blotched buff and olive. **Range:** Endemic. NG, NW Is (Batanta, Misool, Salawati, Waigeo), Bay Is (Yapen), Aru Is, and SE Is (Fergusson, Normanby, Misima, Tagula), 0–1500 m.

Collared Sparrowhawk *Accipiter cirrocephalus* Pl 20, 22
(**Australian Sparrow Hawk**, *A. cirrhocephalus*)
28–36 cm; wingspan 53–77 cm. Uncommon in forest and other wooded habitats in lowlands, hills, and mid-mountains. *A small, lightly built hawk with long thin legs, square-tipped or notched tail, a small head, and indistinct brow line yielding a staring expression rather than scowling one.* **Adult, Juv:** Plumages similar to corresponding plumages of Brown Goshawk and also long-winged; best separated by size and square tail. Only a hint of a brown collar in some birds. **Female:** Larger. **Subspp (2):** Minor. **Similar spp:** See Brown Goshawk. Variable G often lacks barring in underparts and shows rounded tail. **Voice:** A 3–4 sec series of forced high notes, many per sec. **Habits:** Co-occurs with similar Brown and Variable Gs at forest edge and savannah, but unlike them, Collared S inhabits forest interior, where it leads a life of stealth and is seldom seen. Ambushes prey, especially birds. Nests in a tree crown; a stick platform (AU data). Eggs (~3) white, unmarked or with speckles or blotches. **Range:** NG, NW Is (Salawati, Waigeo, and possible audio record from Batanta), Bay Is (Yapen), Aru Is, and SE Is (Rossel), 0–2500 m. Also AU.

Meyer's Goshawk *Accipiter meyerianus* Pl 21, 22
48–56 cm; wingspan 86–105 cm. Rare in hill and montane forest. A *large, powerful goshawk with 2 color morphs. Tail with only 4–5 thick dark bands*, although these are sometimes obscure (other goshawks have many more bars, or no bars). **Adult:** *Common morph* has *white underparts with fine dark streaking and barring*; dark, reddish eye. *Black morph* all black; note dark eye. **Female:** Larger. **Juv:** *Common morph*, from other juv goshawks by large size, tawny buff underparts with fine dark streaking, and 4–5 tail bars; iris yellow becoming reddish. *Black morph* Juv no information. **Similar spp:** Variable G grey morph smaller, paler, has pale iris, and does not show barring in tail. Chestnut-shouldered G juv is orange-brown on the back as well as the front, whereas Meyer's Juv has a dark brown back with buff scaling. Pygmy Eagle smaller and differently proportioned with longer, fingered wings and shorter, unbarred tail. **Voice:** A slow series of loud, nasal upslurs, ~3 notes in 4 secs: *whi-i-yu* (faster and less forced than the upslurs of Chestnut-shouldered G). Also described as a slurred screech, *ka-ah*. **Habits:** Poorly known. Solitary. Usually seen soaring low over canopy. Hunts in the forest and along rivers. Prey includes birds. No nesting information.

Range: NG (widely scattered localities, likely throughout; no reports from Bird's Head), Yapen (Bay Is), and Karkar I, 0–2700 m. Also Moluccas to Solomon Is.

Chestnut-shouldered Goshawk *Erythrotriorchis buergersi* Pl 21, 22
(**Buergers's Goshawk**, *Accipiter buergersi*)
43–53 cm; wingspan 85–109 cm. NG's least known hawk. Rare in hill and lower montane forest. *Tail with 9–10 dark bars.* **Adult:** *Common morph* boldly patterned—*black above with shoulder feathers rufous-edged; white below with sharp black streaking. Black morph all black but note pale eye.* **Female:** Larger. **Juv:** *Common morph* bright rufous brown with dark streaking; iris brownish becoming yellow. *Black morph* no information. **In flight:** All plumages show primary and secondary *wing feathers with numerous, distinct, narrow bars,* whereas Meyer's G has indistinct barring on flight-feathers. **Similar spp:** No other goshawk as strongly patterned, but see Meyer's G (especially compare juvs) and Doria's Hawk. **Voice:** Big, forced, nasal, high-pitched upslurs, 1/sec. **Habits:** Little known. Haunts forest interior and seldom soars. Catches birds. No nesting information. **Range:** Endemic. NG, imperfectly known: Central Ranges, Foja Mts, N Coastal Mts, Adelbert Mts, Huon, and SE Pen, 0–1600 m.

Doria's Hawk *Megatriorchis doriae* Pl 21, 22
(**Doria's Goshawk**, *Accipiter doriae*)
51–69 cm; wingspan 88–106 cm. Rare and inconspicuous in forest interior from lowlands to mid-mountains. **Adult:** *Tail very long, with 10–12 narrow black bars.* Wings also finely barred. Profusely barred black above, streaked below. *Note thick, black eye-stripe.* **Female:** Much larger. **Juv:** Lacks prominent eye-stripe (head can be very pale); black barring narrower and more profuse in wings, wing-coverts, and tail. **Similar spp:** Long-tailed Buzzard has many fewer, thicker tail bars; has a pale face; and soars frequently. **Voice:** A short series of high-pitched, forced screams; ~4 notes delivered 1/sec. **Habits:** Does not soar. Keeps below forest canopy; flies with slow wingbeats and perches quietly. Preys on birds ambushed in the forest depths. Nest in tree crown; a bulky stick structure. Eggs, no information. **Range:** Endemic. NG and Batanta (NW Is), 0–1600 m.

Grey-faced Buzzard *Butastur indicus* Pl 16, 18
38–48 cm; wingspan 101–110 cm. NW Is. Rare Asian migrant during austral summer. A chunky, sluggish hawk perched at the forest edge. **Adult:** Distinctive pattern—*white throat with black chin stripe contrasts with rufous breast and brown barred belly.* Unique tail pattern with *3 broad bands on the central feathers only,* the outer feathers plain. Sexes same size. **In flight:** Underwing pattern plain, but look for distinctive chin and breast pattern and bands in the central tail feathers. **Juv:** Pattern similar (has chin stripe), but underparts white with dark streaking heaviest on breast, and fresh plumage shows white feather edges on wing-coverts. **Similar spp:** None, but compare with juv Pacific Baza and Long-tailed Buzzard. **Voice:** A tremulous, 2-note whistle, *whick-wee.* **Habits:** Little known in NG Region. Elsewhere migrates in flocks, the birds flying in lines. In Asia said to prefer hills. Waits for prey from a high perch, then flies down to ground to make its capture. Takes mainly frogs, reptiles, and rodents. **Range:** NW Is (Salawati, Waigeo) and possibly NG mainland, especially Bird's Head, plus 1 sighting from Trans-Fly, 0–200 m. Breeds in NE Asia and Japan, winters in SE Asia to far western NG Region.

New Guinea Harpy-Eagle *Harpyopsis novaeguineae* Pl 21, 22
(**New Guinea** or **Papuan Eagle**, **Kapul Eagle**)
76–89 cm; wingspan 121–157 cm. Uncommon in forest expanses from lowlands to timberline. Does not soar. Seen flying only when crossing a ravine or changing perches in the canopy. Best means of finding one is to scan the treetops along a ridge crest or at the forest edge—look for a huge, pale-breasted raptor perched upright and motionless, watching its domain. **Adult:** *An enormous forest raptor with white underparts.* Shape distinctive with *short broad wings, long rounded tail, and long bare legs.* When erected, full head ruff adds to the distinctive profile. **Female:** Much larger. **In flight:** *Whitish below* without streaking or barring. Slow wingbeats. **Juv:** Prominent pale tips to wing-coverts; wings and tail have more and narrower bars; tail lacks distinct dark subterminal bar of Adult. **Similar spp:** All other eagles and Long-tailed Buzzard usually seen soaring. Long-tailed Buzzard obviously lighter weight, smaller, and has streaked underparts. Gurney's E juv buffy below with longer wings. White-bellied Sea-Eagle also white below, but tail absurdly short in comparison with that of New Guinea Harpy. Doria's Hawk smaller, more abundantly barred. **Voice:** Arresting and unforgettable. Unlike any other raptor—a far-carrying, low,

explosive *UUMH!*, like the sound produced by releasing a taut bowstring. This may be answered by a higher-pitched *OHK! OHK!*, given by Juv or a second Adult. Eagles call throughout the day, and also before dawn. Greater Black Coucal gives a call of similar quality to the *uumh* call, but usually more musical and in a characteristic, slowly descending series. **Habits:** Solitary or in dispersed pairs. Home range believed to be enormous, but no data for Adults as yet. Perches for long periods in the top of a tree along a ridge crest or at the edge of the forest. Flies at or below canopy level from perch to perch to locate prey in trees or on ground. Hunts mainly for larger mammals (especially possums) but also birds and reptiles. Swoops on its prey, or uses claws to extract prey from tree cavity or excavate epiphyte clump to expose prey. A nimble climber and runner, will pursue prey on foot. Nests in the crown of a large tree; nest built of sticks atop an existing clump of epiphytes or moss cushion growing on a horizontal branch or fork. Egg (1?) undescribed. Only 1 young fledges. **Range:** Endemic. NG, 0–3000 m. **Taxonomy:** Molecular studies confirmed a close relationship between the NG Harpy-Eagle and the Harpy E and Crested E of South America, but not with the Philippine E.

Gurney's Eagle *Aquila gurneyi* Pl 17, 18
(*Spizaetus gurneyi*)
74–86 cm; wingspan 165–185 cm. An uncommon but widely distributed forest eagle seen soaring over lowlands and adjacent hills. **Adult:** *Huge and blackish brown; tail not wedge-shaped.* **Female:** Larger. **In flight:** Wings long and broad; tail medium length, slightly rounded. All dark underside; in better light note *pale-tinted, dark-tipped primaries.* **Juv:** Paler and tawny, especially below. Tail and secondaries finely barred. **Imm:** Approaching Adult plumage, but paler head and legs. **Similar spp:** New Guinea Harpy-Eagle, the other forest eagle, mostly white below and never soars. Wedge-tailed E and White-bellied Sea-E have strongly uptilted wings and prominent wedge tails. Little E much smaller, shorter-winged, and shorter-tailed. **Voice:** A medium high-pitched, piping, dowslurred note, occasionally repeated. **Habits:** Not well known. Solitary or paired. Patrols the forest canopy from above, slowly soaring or circling on a still sunny day, or gliding on the wind above ridges. Seems more prevalent near the coast, perhaps because of onshore breezes. Often flies quite low in hilly country, affording great views. Reputedly preys on arboreal mammals such as possums. No nesting information. **Range:** NG (mostly absent from Trans-Fly), NW Is (Batanta, Misool, Salawati, Waigeo), Bay Is (Biak sightings, Yapen), Aru Is, and SE Is (D'Entrecasteaux: 3 main islands), 0–1000 m. Also Moluccas.

Wedge-tailed Eagle *Aquila audax* Pl 17, 18
76–100 cm; wingspan 182–232 cm. The iconic Australian eagle, restricted in NG to savannah country of the Trans-Fly, where it is an uncommon resident. **Adult:** From Gurney's E by its *pointy, wedge-shaped tail,* tawny hackles, and its *wings held in a V* when soaring. **Female:** Larger. **Juv:** Paler and tawny on nape, shoulders, and rump (dark in Adult); in flight, underside of wings and tail narrowly barred. **Subspp (1):** *audax* (also AU). **Similar spp:** Gurney's E. **Voice:** A weak, upslurred disyllable, *peer yu.* **Habits:** Solitary, paired, or gathering at a large carcass; sometimes 2 or more birds cooperate to chase down prey (AU data). Locates food by soaring, low quartering, or from a perch. Takes wallabies, bandicoots, and other small mammals, reptiles, and birds. Also consumes carrion. Nest in a tree; bulky stick structure. Eggs (2) white with dark spots or blotches. **Range:** NG: Trans-Fly, 0–130 m. Also AU.

Pygmy Eagle *Hieraaetus weiskei* Pl 16, 17, 18
(**Little Eagle**, *Hieraaetus morphnoides, Aquila weiskei*)
38–48 cm; wingspan 101–136 cm. Seen soaring over forest from hills to mid-mountains; uncommon and local. *Robust and buteo-like, small for an eagle.* The most *buteo*-like raptor in the region. Same size as Brahminy Kite, juv of which difficult to separate at a distance. *When eagle is perched look for tail extending to the tip of the folded wing or slightly beyond* and *longish, powerful, fully feathered legs and large feet.* This is the only NG hawk with feathered tarsus. Dark iris. **Female:** Larger. **In flight:** Medium length, rounded wings held flat when soaring; fanned tail. **Adult:** Plumage variable. *Pale morph* more common, brown above, buffy white below with dark-streaked breast. *Dark morph* heavily streaked below. **Juv:** Upperparts with pale feather tips; iris may be paler (needs confirmation). **Similar spp:** Brahminy Kite juv has longer (and broader) wings that extend beyond the tail tip; note that tail has essentially no barring (Pygmy E tail barred); broader wings with an S-curve trailing edge; short bare legs and small feet. **Voice:** Usually silent, but utters a whistling *sip sip seeee* during display flight. **Habits:** Solitary. Soars over forest or edge;

quarters low over canopy; perches in treetop in plain view—somewhat inquisitive. Dives on prey in the canopy, often birds. No nesting information. **Range:** NG mts (Bird's Head, Central Ranges, and Foja, Adelbert, and Huon Mts) and possible sighting on Batanta I (NW Is), 0–2000 m. Also sightings from Moluccas. **Extralimital spp: Little Eagle** (*H. morphnoides*) of AU, which could visit NG, differs subtly, including narrower and fainter streaking on the breast, dark crown strongly contrasting with rufous nape (these dark brown and streaked in Pygmy E), short nuchal crest (Pygmy E lacks crest), and for pale morph in flight, underwing with rufous leading edge (Pygmy E underwing-coverts all white).

FALCONS: Falconidae

Falcons, falconets, forest-falcons, and caracaras are a diverse family of raptorial birds (61 spp). Their relationship to the hawks and eagles is still controversial, and the 2 groups may not be at all related. They show a worldwide distribution centered on South America. Only the falcons reach Australasia, with 6 species in the New Guinea Region. Supreme aerial predators, falcons typically chase down their prey on the wing, although the kestrels and Brown Falcon pounce on theirs on the ground. Birds, small mammals, reptiles, and arthropods compose their diet. Most species prefer open country, but the Peregrine Falcon hunts above the forest. Falcons live in pairs and nest on cliff ledges, in tree hollows, or in old nests of other birds.

Spotted Kestrel *Falco moluccensis*   Pl 23, 24
(Moluccan Kestrel)
28–33 cm; wingspan 59–71 cm. In region, known only from Gag and Kofiau in the NW Is, where resident and uncommon in large clearings and other open country. Similar to Nankeen K, but *darker reddish brown rather than pale orange-brown, and heavily spotted and streaked*; sexes more alike. **Male:** Note grey tail with black subterminal band, but lacks grey cap of Nankeen K. **Female:** Similar to Male, but grey tail is barred. Larger on average. **Juv:** Similar to Female, but darker and rufous-washed tail with heavier bars. **Subspp (1):** Sightings indicate *moluccensis* (Moluccas). **Similar spp:** Nankeen K, especially juv. **Voice:** A slow, irregularly repeated series of short, high-pitched notes, *ki ki ki ki*. **Habits:** Similar to Nankeen K. Nest on an old stick nest or in a tree hollow. Eggs (4, data extralimital). **Range:** NW Is (Gag, Kofiau), 0–200 m. Also Java and Wallacea.

Nankeen Kestrel *Falco cenchroides*   Pl 23, 24
(Australian Kestrel)
30–36 cm; wingspan 66–78. The widespread NG kestrel. Two distinct populations. Uncommon resident restricted to a small area of alpine grasslands above the Baliem Valley in the W Ranges. Also scarce migrant from AU visiting during austral winter in open habitat in lowlands in the South and in mid-mountains of Central Ranges. A small, slender falcon with *rich orange-rufous upperparts and black subterminal tail-band*. Differences between the sexes blurred by individual variation—look for the tail color and don't rely solely on size or cap color. **Male:** Grey cap and tail; tail lacks bars; dorsal spotting hardly shows. **Female:** AU race has cap and tail rufous like back and wings; dorsal spotting obvious, so is streaking in the underparts; tail barred; size larger on average. **Juv:** Similar to Female and sometimes not distinguishable. Usually much more heavily marked with black spotting above and streaking below. **Subspp (2):** *cenchroides* (widespread AU visitor) whitish below; *baru* (endemic to alpine W Ranges) darker, with orange-buff breast, sexes alike. **Similar spp:** See Spotted K. Some Brown Falcons may approach kestrel in color pattern but are much larger and lack tail bar. Other falcons darker and lack tail bar. **Voice:** High-pitched staccato chatter *kee-kee-kee . . .* (AU data). **Habits:** Resident bird flies high over alpine grasslands and frequents cliffs; feeds on lizards and insects. Migratory race turns up at airstrips, fields, and savannah country. Usually as single birds; most are Females. Irruptive, numbers varying from year to year. Perches on wire and bare branch from which it launches itself to dive on prey on the ground. Also flies about and hovers before diving. One nest reported on a ledge in a deep sinkhole. **Range:** NG, Aru Is, and probably other satellite islands. Resident above 3200 m in the Lake Habbema-Puncak Trikora highlands of W Ranges. Overwintering migrants mainly in South and the Central Ranges, 0–1500 m. Also breeds in AU.

Oriental Hobby *Falco severus* Pl 23, 24

25–30 cm; wingspan 61–71 cm. Rare throughout NG forests and especially forest edge from the lowlands to mid-mountains. Usually seen flying or perched high on a bare snag. **Adult:** A small, dark, black and orange-brown falcon with a stubby profile. *Slaty black above, rich orange-brown below*, with a *buff or white half-collar* and throat and all-dark head. Essentially without streaking on underparts. Feet bright orange-yellow. **Female:** Larger. **Juv:** Underparts heavily dark-spotted. **Similar spp:** See Australian H. Other falcons larger, sleeker, not as rufous below. **Voice:** A very sharp, high-pitched *ki ki ki . . .* in a series of 5–10 notes. **Habits:** Little known. Perches habitually on high, dead limbs at forest edge or in garden trees, from which it sails out on graceful flights. Hawks for insects, birds, and bats, mainly in early morning and evening. Nest (elsewhere in range) built on old stick nest or in tree hollow; one tree-hollow nest observed in NG. Eggs (3–4) buff with dark speckles and blotches. **Range:** NG, NW Is (Salawati, plus sighting from Batanta), Bay Is (Yapen, plus 1 sighting from Biak), 0–1350 m. Also SE Asia through to Solomon Is.

Australian Hobby *Falco longipennis* Pl 23, 24
(Little Falcon)

30–38 cm; wingspan 66–83 cm. A rare Australian migrant to open habitats in NG lowlands during austral winter. **Adult:** *Similar to Oriental H but larger and with noticeably longer wings and tail.* Usually paler, with *barring on the tail* (tail evenly black without evident barring in Oriental H). Note *white lores* (black, same as crown in Oriental H). Whitish *half-collar extends further up the side of the head*, wrapping around the black cheek patch, rather than being simply a flat bar. Underparts duller brown with more extensive dark markings, some birds with almost no markings. **Female:** Larger. **Juv:** Underparts darkly blotched; upperparts scalloped brown. Look for *rufous lores.* **Subspp (1):** *longipennis* (SE and SW AU); much paler *murchisonianus* of northern and inland AU possible. **Similar spp:** Oriental Hobby, see above. Peregrine Falcon much larger, with pale underparts barred. **Voice:** Similar to Oriental H. **Habits:** An aggressive falcon seen perched conspicuously or pursuing small birds (including swiftlets), bats, or insects. Hunts mainly at dawn or dusk. Most specimens collected in NG were Females. **Range:** NG, mainly in Trans-Fly east to SE Pen, 0–500 m. Also AU and Wallacea.

Brown Falcon *Falco berigora* Pl 23, 24
(Brown Hawk, *Ieracidea berigora*)

41–53 cm; wingspan 88–115 cm. *A rather large sluggish falcon of open country*, mainly in the mountains; generally uncommon. *Black bar below eye.* Plumage color ranges from rufous brown above and white below to all dark (uncommon). Initially thought to be color morphs, this plumage variation in AU is related to age and sex. Plumage variation for the endemic NG race poorly understood. **In flight:** Wings slightly shorter and broader than other falcons. Dark underwing-coverts contrast with paler flight-feathers in all plumages. **Adult:** In NG Region, both sexes show a wide range of plumage colors. **Female** is larger. **Juv:** Plumage sooty black without shaft streaking on breast; underparts heavily blotched. **Subspp (1):** *novaeguineae* (endemic); AU race *berigora* not reported, but a possible vagrant; pale morph differs in showing less bold and even shaft streaking (probably not separable in the field). **Similar spp:** Bat Hawk is larger, with white chin and pale eye, and is absent from grassland. Nankeen Kestrel much paler and smaller, with black terminal tail bar. Peregrine F has barred or streaked underparts. **Voice:** A very rapid series of hoarse notes on 1 pitch. **Habits:** Often perches on an exposed limb overlooking grassland; takes prey from the ground, much like an oversized kestrel. Does not act like a typical falcon; *soars or flaps leisurely more like a harrier*, with glides on uplifted wings and ponderous hovering. Feeds on reptiles, rodents, arthropods. No nesting information from NG. In AU, reuses old stick nests of other birds. Eggs (~3) buff, spotted and blotched. **Range:** NG (W Ranges eastward to Huon and SE Pen), plus Manam I and Karkar I, 0–1800 m, rarely to 2800 m. **Extralimital spp: Grey Falcon** (*Falco hypoleucos*) from AU dry country reported without descriptions from Trans-Fly and Port Moresby savannahs. A pale grey falcon; compare with Brown F. Adult grey above, white below with markings so faint as to be visible only at close range; eye-ring and cere yellow-orange. Juv darker and faintly scaled above, more streaked below; eye-ring, cere grey.

Peregrine Falcon *Falco peregrinus* Pl 23, 24
34–51 cm; wingspan 79–114 cm. A rare falcon with resident and migratory races that can turn up anywhere. A long-winged, streamlined, powerful flier. **Adult:** Profuse barring on underparts and prominent black hood. **Female:** Larger. **Juv:** Darker, browner, and underparts heavily streaked black and buff, rather than barred black and white. **Subspp (2+?):** *ernesti*, resident (NG Region, also Wallacea and Philippines to Solomon Is) Adult *head, back, and wings blackish*; underparts *heavily barred* black against *grey background*; Juv upperparts black; migratory race or races (Eurasia or AU) as rare vagrants, usually with *grey back and wings* and underparts *narrowly barred against white or buff*; 2 possible, *peregrinus* group (N Asia) palest, with white further up cheek forming a narrow black moustachial streak, and *macropus* (AU) black-hooded like *ernesti*. **Similar spp:** No other falcon has the dark barring or heavy streaking on the pale underparts. Hobbies smaller, slimmer, and have tawny underparts and underwing. **Voice:** A fairly high-pitched, slightly hoarse, weak, upslurred screech *ka ah* repeated some 7 times in 5 sec, or *chek chek chek . . .* in a series.
Habits: Most often seen in swift direct flight or soaring high over the forest or gorges in the mountains. Typically preys on flying birds, swooping from great height. Takes pigeons and other species that travel above the treetops. No nesting information from NG Region, but elsewhere occupies cliff ledges or reuses old stick nests of other birds. Eggs (2–4) buff heavily spotted and blotched. **Range:** Probably regionwide. NG, NW Is (sightings from Batanta, Kofiau, Salawati), Bay Is (Yapen), Aru Is, SE Is (mostly sightings: Misima, Normanby, Tagula, Woodlark), 0–3500 m. Also nearly worldwide.

BUSTARDS: Otididae

The 26 species of bustard form a Eurasian and African family with one outlying species in Australasia. Bustards are large terrestrial birds of dry, open country. The males are typically much larger than females, and they attract their mates with dramatic displays enhanced by bizarre ornamentation.

Australian Bustard *Ardeotis australis* Pl 13
(**Plains Turkey**, *Choriotis* or *Eupodotis australis*)
91–119 cm. Uncommon resident in open grassy plains, light woodland, and edges of swamps in the Trans-Fly. Particularly vulnerable to hunting. *A very large, robust, long-legged, and long-necked bird.* Note *black cap and grey neck*; the *head is often tilted upward*. **Male:** Huge by comparison, has elongated feathers on neck that cover a pronounced black breast band. **Female:** Much smaller, with neck appearing greyer, caused by heavier dark lines. **Juv:** Similar to Adult but smaller and pale-spotted dorsally. **In flight:** Wings show much white. **Voice:** Usually silent except in display, when *Male roars*—a surprisingly deep sound to come from a bird. Other calls include croaking and deep hissing.
Habits: Solitary or in small groups. Wary. When disturbed, it often remains motionless before stalking away; may take flight while still at distance. Flight strong with slow beats; low flying. On the ground, movements slow and cautious; walks sedately. Normally feeds in grass on dry ground, but will also wade into deep water or follow grass fires looking for prey. Feeds on insects and small vertebrates, and some plant material. Male performs a spectacular nuptial display: standing tall while expanding his throat sac until the feathers of the neck and breast nearly reach the ground, he droops his wings and struts about calling. Nests in dense grass; eggs laid on bare ground (AU data). Eggs (1–2) buff with dark markings. **Range:** NG: Trans-Fly, 0–100 m. Also AU.

RAILS: Rallidae

This cosmopolitan family includes 151 extant species, with 18 species in New Guinea. As a family, rails are rather easily recognized ground birds adapted for living in thick vegetation—such as marshes and other wetlands, forest thickets, and garden fallow—or on the water. Many are secretive and keep well hidden; however, their fondness for watery habitats necessitates that they must occasionally disperse to new living quarters as wetlands dry up or flood, or as young birds strike out on their own.

Despite their seeming reluctance to fly and their weak progress when they do, rails have a remarkable ability to get around. Thirteen species of New Guinea rails extend beyond our region and 12 are shared with Australia, leaving only 5 species endemic to the region. Rails likely disperse from Australia to New Guinea during drought, but this phenomenon remains unconfirmed.

Over the millennia, rails have dispersed well to isolated oceanic islands. Virtually every tropical island is, or once was, home to endemic rail species. More than a hundred species evolved on islands, and many insular forms evolved flightlessness. Sadly, nearly all of these island endemics are now extinct, unable to withstand threats accompanying human settlement. The islands offshore of New Guinea seem unexpectedly poor in rails, although this observation may be misleading because some secretive species could be overlooked.

The phylogeny and classification of rails has been in dispute for years. Although the issue would be much enlightened by molecular studies, such information is lacking, and thus the classification of rails could change substantially. New Guinea rails of particular interest are the forest-rails (4 *Rallicula* spp). Endemic to the NG mountains, they nevertheless share plumage coloration and unusual ball-shaped nests with genera of "primitive" rails in Africa.

Rails can be characterized as having substantial bills, long legs, and large feet. They feed on vegetable matter (leaves, shoots, seeds) and small animal life (insects, spiders, crustaceans, mollusks, and vertebrates such as fish, frogs, and lizards). Although some species inhabit reed beds along the shores of lakes and rivers, others occupy grasslands or forest in tropical New Guinea. Males are typically somewhat larger than females, but the two sexes typically look alike, except for the forest-rails. Rails live in pairs or small family groups that may cooperate in raising young. They defend territories that are advertised by noisy calling. There is still much to be learned about New Guinea rails, particularly about distribution.

Because of their secretive habits, most rails are often difficult to find and observe; however, their voices give away their presence, and by waiting quietly and patiently the observer may be rewarded by a view of the bird itself—it may even be inquisitive and approach. Realize that some rails are more likely to step out in the open at dawn or dusk, or under cloud cover after heavy rains. Rails, such as forest-rails, are readily lured in by playback of their calls.

Chestnut Forest-Rail *Rallicula rubra*   Pl 25
(*Rallina rubra*)

18–23 cm. A small, rufous rail, shy but common in montane forest. Generally *smaller, more rufous, and at higher elevation* than other forest-rails. *Tail rufous without black barring*. **Male:** *Entirely and uniformly chestnut* (flight-feathers black). **Female:** Similar to other female forest-rails but note the *black* wings and back with *white* spotting and *flanks pale-spotted (rather than dark-barred)*. **Juv:** Uniform sooty brown. **Subspp (3):** Minor. **Similar spp:** White-striped and Forbes's FRs co-occur with Chestnut FR. Mayr's FR does not co-occur. **Voice:** Higher pitched than other forest-rails. A long series (50–150 notes) of harsh screaming; notes vary between songs, transcribed as *kreel, krill, keow,* or *kee*. Starts slow, gradually accelerating to 1 note/sec, then decelerating. Two birds may duet or counter-call, alternating notes in lockstep. **Habits:** Lives in pairs or small parties that rummage about on the forest floor foraging for insects and small vertebrates. Domed nest of moss, leaves, and ferns built in pandanus crown ~2 m high. Egg (1) white with a few pale tan spots. **Range:** Endemic. NG: Bird's Head eastward to E Ranges (Mt Hagen), 1500–3050 m. More prevalent in W Ranges where Forbes's FR is rare and local.

White-striped Forest-Rail *Rallicula leucospila* Pl 25
(**White-striped Chestnut Rail**, *Rallina leucospila*)
20–23 cm. *Bird's Head and Bird's Neck*. Co-occurs only with smaller Chestnut FR. White-striped FR is the Bird's Head counterpart of Forbes's FR. **Male:** Differs from other forest-rails by having *white-striped wings and back*. **Female:** From Chestnut by barred tail and by pale-spotted pattern of the wings and back *extending up the back of the neck*. No nesting information. **Voice:** Similar to Forbes's FR. **Habits:** Similar to other forest-rails. **Range:** Endemic. NG: Bird's Head and Neck (Wandammen Mts), 1450–1600 m.

Forbes's Forest-Rail *Rallicula forbesi* Pl 25
(**Forbes's Chestnut Rail**, *Rallina forbesi*)
20–25 cm. Co-occurs only with Chestnut FR and is more common in the East, where Chestnut is confined to highest elevations. Rare and local in W Ranges and Border Ranges, stronghold of the Chestnut. Differs from Chestnut by being larger, having *dark-barred tail* (barring nearly lacking in some) and occupying lower montane forests wherever the 2 species co-occur (elevational range may overlap). **Male:** *Dark wings and back lacks white stripes or spots*. **Female:** Wings and back *dark but not black* as in female Chestnut and with *spotted buff* (not white); *belly and lower flanks barred brown and black* (not pale-spotted). **Juv:** Uniform sooty brown. **Subspp (4, in 2 groups):** *forbesi* (Central Ranges) wings and back nearly black; *dryas* (Huon and Adelbert Mts) wings and back dark olive. **Voice:** Song pattern similar to Chestnut FR, but notes are a froglike *gua* or quacking. Alarm call is muffled clucking. **Habits:** Similar to Chestnut FR; nesting and egg also similar. **Range:** Endemic. NG: Central Ranges plus Adelbert and Huon Mts, 1150–3000 m.

Mayr's Forest-Rail *Rallicula mayri* Pl 25
(**Mayr's Chestnut Rail**, *Rallina mayri*)
20–23 cm. Confined to montane forests of *Foja, Cyclops, and N Coastal Mts*, where it is the only forest-rail. Darker than other forest-rails. *Uniformly reddish brown, including wings and back (not blackish)*. *Tail barred or not*. **Male:** Wings and back lack spots. **Female:** Wings and back with buff spots. **Subspp (2):** *mayri* (Cyclops Mts) wings and back reddish brown, similar to Chestnut FR; *carmichaeli* (Foja and N Coastal Mts) wings and back browner/darker. **Similar spp:** None in its range. Despite resemblance of Male to Chestnut FR, Mayr's is actually more like Forbes's FR but with wings and back brown rather than black. **Voice:** Series of froglike croaks similar to that of Forbes's FR. **Habits:** Similar to Forbes's FR. No nesting information. **Range:** Endemic. NG: Foja, Cyclops Mts, and N Coastal Mts, 1100–2200 m. Recent searches have not found it in the Cyclops Mts. **Taxonomy:** Perhaps only a race of Forbes's FR.

Red-necked Crake *Rallina tricolor* Pl 25
(**Red-necked Rail**)
25–30 cm. Quite different from the forest-rails. Inhabits dense lowland and hill forest and second growth, especially near streams and rivers. Locally common but shy; heard, rarely seen. Most active at dusk and dawn **Adult:** Brownish black with *rufous head and breast. Green bill*. **Juv:** Brownish black with a buff face. **Similar spp:** The much smaller forest-rails occupy cool montane forest. Bare-eyed Rail lacks the blackish wings and underparts, its bill is much longer, and the legs are red. **Voice:** Song a loud, repeated *nark nak nak nak . . .* accelerating toward the end like a bouncing ball. Also repeated *clock* and *plop* sounds. Sings most often at dawn and dusk. Calls are squeals, clucking, and grunts. **Habits:** Nervously skulks under cover, venturing no further than the forest edge. Occurs singly, in pairs, or in small groups. Seems to shift home range in response to availability of water. Nests on the ground or up to 2 m in vegetation; nest a shallow saucer of dried leaves. Eggs (5) white. **Range:** NG, NW Is (Batanta, Gag, Kofiau, Misool, Salawati, Waigeo), Bay Is (Yapen), Aru Is, Karkar I, and possibly the D'Entrecasteaux Is (SE Is), 0–1200 m. Also AU, Wallacea, and Bismarck Is.

Lewin's Rail *Lewinia pectoralis* Pl 26
(**Slate-breasted Rail**, *Dryolimnas, Gallirallus*, or *Rallus pectoralis*)
18–20 cm. The small, long-billed rail of mid-mountain grasslands and swamps. Probably common, but very shy and seldom seen. **Adult:** *Rich brown cap and nape* contrast with grey cheek, throat, and breast. **Juv:** Duller, head darker and lacks chestnut. **Subspp (2):** *mayri* (NG range minus SE Pen); *alberti* (SE Pen) much smaller. **Similar spp:** Buff-banded R larger and more robust, with white eyebrow

and proportionately shorter bill. **Voice:** A loud song repeated 1 note/sec, 10–20 times, transcribed as *crek*, *jik*, or *tree-eek*, likened to 2 coins being tapped together (AU data). **Habits:** Poorly known, especially in New Guinea. Nests in grass usually near water; a small saucer woven of grass (AU data). Eggs (3–5) cream or buff, dark spotted, streaked, or blotched. **Range:** NG: Bird's Head and Central Ranges, 1000–2600 m. Also Wallacea and AU.

Barred Rail *Gallirallus torquatus* Pl 26
(*Rallus torquatus*)
33–36 cm. A larger, darker relative of the more familiar Buff-banded R. Enters New Guinea in the NW Is and Bird's Head coastal lowlands. Prefers swampy areas with grass and trees but will emerge onto roadsides; also rice fields. **Adult:** *Black face and throat with contrasting white "moustache."* Note heavy barring below. **Juv:** Darker; throat white; underparts washed buff. **Subspp (1):** *limarius* (endemic). **Similar spp:** Buff-banded R smaller, lacks black throat. **Voice:** No information for NG Region, but elsewhere said to give a loud, harsh croaking for several seconds. **Habits:** Little information, none for NG Region. Nests on the ground. Eggs (4). **Range:** Bird's Head on NG, plus Salawati (NW Is), 0–200 m. Also Philippine Is and Wallacea.

Buff-banded Rail *Gallirallus philippensis* Pl 26
(*Banded Landrail*, *Rallus philippensis*)
25–28 cm. A common rail of tall grasses, shrubby regrowth, and gardens; most often encountered along roads and other verges. **Adult:** A medium-sized rail, prettily marked with black-and-white barring and spotting, streaked brown above to blend with the grass. *Whitish eyebrow cuts across rusty cap*; note hallmark *buff band on chest*. **Juv:** Duller with bill brown, not red. **Subspp (2):** Minor. **Similar spp:** Lewin's R, with longer bill, lacks white eyebrow and buff band. Barred R much larger and darker, restricted to far West. **Voice:** Not especially vocal. Most often gives a harsh squeak at dawn and dusk. **Habits:** Usually seen alone. Feeds on seeds and invertebrates. Shy and skittish but may pluck up its courage to venture into the open. Nest a small, round, grassy cup. Eggs (4–6) buffy pink with dark spots. **Range:** Throughout NG mainland, but absent from islands in the region, 0–3600 m. Also Philippine Is, Wallacea, AU, east to NZ and Polynesia.

Chestnut Rail *Eulabeornis castaneoventris* Pl 25
(*Chestnut-bellied Rail*, *Gallirallus castaneoventris*)
51–58 cm. Aru Is only. *A very large, orange-brown, grey-headed rail inhabiting mangroves.* Secretive and shy, but ventures onto mudflats at low tide. **Adult:** Male on average larger than Female. **Juv:** Likely similar to Adult. **Subspp (1):** *sharpei* (endemic). **Similar spp:** New Guinea Flightless R, absent from Aru, has grey, not yellowish legs and white breast. **Voice:** Territorial song a *loud, resonant trumpeting, possibly a duet*; consists of usually alternating notes rapidly repeated many times: the first note resembles alarm screech of Sulphur-crested Cockatoo, the second is a drumming. **Habits:** Virtually unknown in the region. One observation along a beach at the edge of mangroves. In AU, lives in pairs on large territories. Feeding dependent on low tide. Gleans and probes for crabs and other crustaceans. Nest in mangrove tree is a bulky flat-topped pyramid of sticks that is refurbished and reused between years. Eggs (4–5) pinky white, dark-spotted and splotched. **Range:** Aru Is, sea level. Also AU north coast. **Taxonomy:** DNA studies have shown a close relationship with the genus *Gallirallus*.

Bare-eyed Rail *Gymnocrex plumbeiventris* Pl 25
(*Eulabeornis plumbeiventris*)
30–33 cm. An uncommon and seldom seen medium-sized and long-billed rail of lowland and hill forest and wet grassland bordering freshwater. **Adult:** Unique 3-color combination of *chestnut foreparts, olive wings, and black tail, rump, and belly*. In horizontal running posture, the vertical, *black, finlike tail* is prominent. Throat grey or chestnut (unknown significance). Bare *eye skin and legs pinkish red*. **Juv:** Undescribed. **Similar spp:** Red-necked Crake co-occurs but is smaller, darker, has blackish legs. Chestnut R much larger, has grey head and yellow legs. **Voice:** Possible territorial song is loud gulping *wow wow wow wow* heard at start of the wet season. Also piglike grunting calls while foraging. **Habits:** Usually seen singly, but lives in pairs or small parties. Very active by day; normally shy, but inquisitive if observer remains still. Extends from lowlands into mid-mountain valleys. Seems to vacate local habitat during long dry spells. Takes insects. No nesting information. **Range:** NG, NW Is (Misol), Aru Is, and Karkar I, 0–1200 m. Also Wallacea and Bismarck Is.

White-browed Crake *Porzana cinerea* Pl 26
(*Amaurornis cinerea* or *Poliolimnas cinereus*)
15–18 cm. A common small rail of marshy habitats and wetland margins from lowlands to mid-mountain valleys. Hidden by day, it emerges at dusk or in cloudy weather to forage on floating vegetation. **Adult:** *White-striped facial pattern and silvery grey underparts.* **Juv:** Similar, but underparts buff and facial pattern weaker. **Similar spp:** Other small rails lack the distinctive facial pattern and have darker underparts. Comb-crested Jacana, with which it regularly coexists, is larger and patterned black and white. **Voice:** Noisy and loud; sometimes several birds will call together. Most obvious song is a chattering note *cutchee* repeated 10 times or more. A variety of other calls include squeaks, grunts, and a *charr-r* alarm note. **Habits:** A crepuscular feeder, taking small plant and animal life. Walks and runs on floating and submerged vegetation; will swim. Nests in grass or reeds. Nest a cup of woven grass. Eggs (3–6) pale brown with dark spots. **Range:** NG, NW Is (sightings from Batanta and Misool), and SE Is (Fergusson), likely more wide ranging, 0–1700 m. Also SE Asia through Australasia to Polynesia.

Baillon's Crake *Porzana pusilla* Pl 26
(**Marsh** or **Little Crake**)
15–18 cm. A *tiny rail* that keeps hidden in marsh grasses and is little known. **Adult:** *Streaky upperparts and plain grey underparts.* Short bill. **Juv:** Underparts buffier. **Subspp (2).** Minor? **Similar spp:** Lewin's Rail is larger, has longer, pinkish bill, chestnut crown and nape. **Voice:** Probably the only reliable means of detection, but not known for New Guinea population. In AU, the loud territorial "ratchet" song sounds like fingernail rapidly picked along a comb and is described as a purring, whirring, or trilling; alarm call *kek kek*. **Habits:** Feeds on small seeds and animal life. Small, flimsy cup nest, hooded or not, in marsh grasses (AU data). Eggs (4–7) olive brown with/without dark flecks. **Range:** NG, incompletely known: W Ranges (Paniai Lakes), E Ranges, and SE Pen (Port Moresby), 0–2500 m. Also Africa, Eurasia, AU, and NZ.

Spotless Crake *Porzana tabuensis* Pl 26
(**Sooty Rail**)
17–20 cm. A small, black rail of swampy grassland and marsh-edged lakes and rivers. Widespread through highlands, local in lowlands. Common, but secretive. **Adult:** *All sooty black with red eye and leg.* **Juv:** Slightly paler, with white extensive on chin and throat; dark eye, legs pinkish. **Similar spp:** None. **Voice:** Territorial song a loud, trilling purr, rapidly decreasing and descending, lasting 1–3 sec (AU data). Other calls. **Habits:** Poorly known in New Guinea. Keeps hidden in dense, tall grass, rarely venturing out at dawn or dusk. Feeds on seeds, leaves, and small animal life. Usually in pairs. Nests in marsh grass; nest a grass cup (AU data). Eggs (3–4) pale brown with darker flecks. **Range:** NG: Bird's Head and Central Ranges, 0–3150 m. Also Philippine Is and Wallacea through to parts of Oceania, AU, and NZ.

Rufous-tailed Bush-hen *Amaurornis moluccana* Pl 26
(**Bush-hen**, **Pale-vented** or **Common Bush-hen**, *A. olivaceus*, *Gallinula olivacea*)
22–30 cm. A plain-colored, stocky rail with upright posture. Common and widespread in tall, wet, grassy and shrubby habitats. Noisy. **Adult:** *Patterned olive above, grey below, with buff undertail—but the net effect is nondescript.* Green-and-red bill color brightens during breeding. Note olive-yellow legs. **Juv:** Somewhat paler with more white on throat. **Subspp (1):** *moluccana* (also Moluccas to AU). **Similar spp:** None, but be aware of scrubfowl. **Voice:** A memorable voice from jungle regrowth, heard at dawn, dusk, and sometimes night. Unmistakable duet: one bird calls a nasal catlike higher-pitched note, the other a similar lower-pitched note; the 2 birds call back and forth, slightly more than 1 sec between each pair of notes. Contact calls include clucking, a single low piping note, and clicks. **Habits:** Mainly crepuscular. In pairs or small groups in garden fallow, forest edge, overgrown ditches, and road verge. Often near human settlement. Usually keeps to thick vegetation, rarely venturing into the open. Feeds on seeds, leaves, insects, and small animals. Nests in tall grass, bending down leaves and stems to form a cup-shaped platform, sometimes with a dome (AU data). Eggs (4–7) white with dark speckling. **Range:** NG, NW Is (Misool, plus sighting on Batanta), Bay Is (Biak), Karkar I, and SE Is (Normanby, Misima), 0–1500 m. Also Wallacea, Bismarck Is, Solomon Is, and AU. **Taxonomy:** Split from *A. olivacea*, which is now Plain Bush-hen.

New Guinea Flightless Rail *Megacrex inepta* Pl 26
(*Amaurornis inepta, Habroptila inepta*)
36–38 cm. An uncommon inhabitant of lowland swamp forests, wet thickets, bamboo, and mangroves. Attracted to sites of sago harvesting—the best chance to see one (from a hide). **Adult:** *A huge (swamphen size) brown rail with massive head and long, powerful grey legs and feet.* Brown body with greyish wings and back. Note grey face (some individuals showing a brief black mask) marked with a pale streak behind the eye and *long, stout, greenish-yellow bill.* Throat and belly white, though does not always show well. Seems tailless. Flicks wings upward while walking. **Juv:** Similar, though with less brown. **Subspp (2):** *inepta* (S Lowlands) rusty brownish sides of neck and body; *pallida* (NW Lowlands to Sepik-Ramu) paler with buff sides of neck and body. **Similar spp:** Rufous-tailed Bush-hen smaller, more typically proportioned, has olive yellow legs. Chestnut R inhabits Aru Is, where NG Flightless R is absent; its body is all chestnut, and its legs are yellowish. Scrubfowl have somewhat similar body shape and large powerful legs, but the head is completely different, they have a tail, and they don't wing-flick. **Voice:** Harsh but shrill *aaah-aaah*, not unlike the squeal of a pig. **Habits:** Little known. Feeds on insects. Clambers up into low trees to roost. No nesting information. **Range:** Endemic. NG: S and NW Lowlands and Sepik-Ramu, perhaps more widespread, 0–500 m.

Purple Swamphen *Porphyrio porphyrio* Pl 6
38–43 cm. A familiar large, purplish-blue swamp dweller. Widespread and common, mainly in the lowlands, but occupies montane valleys where habitat is favorable. **Adult:** *Appears black with blue underparts and red frontal shield and bill.* White undertail shows from behind. **Juv:** Duller; bill dusky red. **Subspp (1):** *melanotus* (also AU). **Similar spp:** Dusky Moorhen smaller, all black. **Voice:** A variety of loud calls. Mostly a high-pitched explosive honk or crowing. Also, a yip, squawk, and other sharp calls. **Habits:** Known to migrate between NG and AU. Associated with grassy and shrubby wetlands, but makes forays into neighboring fields and gardens. Lives in pairs or small territorial groups; the complex social organization studied elsewhere in its range may apply to NG birds as well. Congregates where food is plentiful. Feeds on vegetable matter, insects, and small animal life. Nests in swamps, trampling down reeds to form a saucer platform (AU data). Eggs (2–6) buff with dark spots and blotches. **Range:** NG, NW Is (sighting from Salawati), Aru Is, and SE Is (Goodenough, Misima, Tagula, Trobriands, Woodlark), 0–1800 m. Also Eurasia east to AU, NZ, and Polynesia. **Taxonomy:** When split into multiple species by some authors, the NG population belongs in *P. melanotus*.

Dusky Moorhen *Gallinula tenebrosa* Pl 6
25–32 cm. Less common and more local than the swamphen. Smaller and more compact. More closely tied to water and restricted to lowlands. **Adult:** *Black* rather than purplish blue. Also has white undertail. **Juv:** Duller and throat whitish; bill and legs dusky. **Subspp (2):** *frontata* (most of region) wings and back suffused olive; *neumanni* (NW Lowlands, Sepik-Ramu) wings and back blackish, also smaller. **Similar spp:** Purple Swamphen. **Voice:** Territorial call a single loud *krik* (often at night). **Habits:** Usually seen swimming or loafing near the water's edge. Forages in vegetation and water for aquatic plant and small animal life. In pairs or small territorial breeding groups. Nest built in reeds and other vegetation over or near water; a platform or saucer of reeds and other vegetation (AU data). Eggs (~8) buff with dark spots and blotches. **Range:** NG, 0–400 m. Also Borneo, Wallacea, AU.

Eurasian Coot *Fulica atra* Pl 6
36–41 cm. Breeds locally in montane lakes; rarely turns up in the lowlands. Usually seen in small flocks swimming in open freshwater. **Adult:** *All black with prominent white bill and forehead.* **Juv:** Greyer, with frosty feather tips on breast, lacks frontal shield. **Subspp (2):** Probably not separable in the field; *lugubris* (resident on montane lakes; also on Java) often very blackish, white tips to secondaries prominent, reduced, or lacking; *australis* (not yet recorded as an AU vagrant, but likely explains rare sightings of coots in NG lowlands) paler, with white tips to secondaries reduced or absent. **Similar spp:** Swamphen and moorhen with red forehead and bill. **Voice:** A single loud *kowk*. Also 2–5 unpleasant, creaking, downslurred brays, sounding like a squeaky hinge. **Habits:** Gregarious. Takes food mainly in the water by swimming and diving, but also grazes on land. Feeds almost entirely on aquatic vegetation (algae, leaves, shoots, seeds), with small animal life a minor component of diet. Breeds in territorial pairs sometimes with helpers; flocks when not breeding. Nests in emergent aquatic vegetation or on shore; nest a mound of grass and sticks with depression at center (AU data). Eggs (~6) whitish with dark speckles, few blotches. **Range:** NG, resident in Bird's Head and Central Ranges, 1500–3500 m; vagrant or migrant in lowlands. Also Eurasia, Africa to AU and NZ.

CRANES: Gruidae

> Cranes are tall wading birds related to rails. There are 15 species, all rather similar, and the family is distributed nearly worldwide except for South America and most islands. Only one species, the Brolga, occurs in New Guinea, and it is typical of the family in its appearance and habits. The Brolga is shared with Australia, where it is a familiar bird and featured on the Queensland State coat of arms.

Brolga *Grus rubicunda* Pl 13
(Australian Crane)
114–124 cm. Common resident in Trans-Fly savannah country, in marshes and dry grassland. Unknown whether birds move between AU and NG. **Adult:** *A large, grey, heron-like bird with distinct profile* showing sagging, dark *jowls* and feather *"bustle"* formed by tertials protruding over the tail; *bare head red and grey*; pale iris. **Male:** Larger on average. **Juv:** Similar to Adult, but head feathered in grey, no red skin color or jowls, iris dark, and legs paler. **Imm:** Gradually assumes Adult features over a period of years. **Similar spp:** Great-billed Heron and juv Black-necked Stork look cranelike but have different profiles and lack red on head. **Voice:** A loud trumpeting *ga ronggg*, the second syllable rising, then falling (not to be confused with Channel-billed Cuckoo). **Habits:** In pairs or small parties; congregates in the hundreds around water during the dry season. Forages by wading in water, probing with bill; also feeds on land. Omnivorous, taking a variety of small animal life and seeds and tubers. The birds dance together by gracefully jumping in the air and flapping. Pairs monogamous over many years. During the rainy season, they disperse to breed in the marshes. Nest a grass platform in shallows or on ground (AU data). Eggs (2) white with dark spots and blotches. **Range:** NG: Trans-Fly, 0–200 m (not counting a vague early record for Sepik). Also AU. **Extralimital spp: Sarus Crane** (*G. antigone*), a potential visitor from AU, is deceptively similar, but adult has bare red skin extending down neck (confined to head in Brolga) and has pinkish-grey legs (not dark grey); juv head feathered in cinnamon-brown (not grey).

BUTTONQUAIL: Turnicidae

> The 17 species of buttonquail form a unique family of small, quail-like birds. Although formerly classified near the rails, buttonquail have recently been placed with shorebirds and their relatives. Buttonquail range from Africa through Asia to Australia, which holds the most species. Only one species inhabits the New Guinea Region, perhaps because the open habitats they favor are limited in extent. Buttonquail differ from true quail in lacking a hind toe; they also show reversed sexual dimorphism, meaning that the female is larger and more brightly colored than her mate. The male incubates the eggs and cares for the young. Chicks are all black, differing from quail chicks, which are not as dark and show a buffy face. Chicks molt into a distinctive juvenile plumage before attaining adult plumage. In New Guinea, the Red-backed Buttonquail stays hidden in its grassy habitat and is seldom seen. Hearing the female's booming song is probably the best means of discovery.

Red-backed Buttonquail *Turnix maculosus* Pl 4
12–15 cm. An uncommon and secretive, *tiny, quail-like bird* of lowland and mid-mountain grasslands. In breeding season, Female calls persistently from cover, indicating the species' presence. *Pale buff wing-coverts with dark spots and generally unpatterned throat, breast, and belly*. Note *white iris* of Adult and *yellowish base of the bill*. Female in South has *rufous collar on hindneck*. **Male:** Greyer, less buffy and orangish than Female, lacks obvious rufous collar. **Female:** Brighter, with chestnut collar (S race only) and orangish throat and breast. **Juv:** Heavily speckled above and down breast; iris dark. **In flight:** Buff shoulder patch contrasts with dark flight-feathers. **Subspp (4):** All poorly known. **Similar spp:** Co-occurring quail (Brown and King Qs) have a barred or grey breast, dark or red iris, all-dark or horn-colored bill, wing-coverts uniform with rest of wing, and lack a chestnut collar. King Quail is the same size as buttonquail. **Voice:** Female gives a deep, carrying *ooom ooom ooom . . .* from hiding, at odd times during the day. In AU, call notes monotonous, delivered 1/sec. Silent when

flushed, unlike quail. **Habits:** Prefers dense grassland and pastures. Hides in thick grass; seen only when flushed or crossing a track. In flight, shows more sharply angled wings than true quail. Feeds on insects and seeds. Nests in grass, a shallow depression lined with grass or leaves (AU data). Eggs (4) pale grey, densely speckled. **Range:** NG and Tagula (SE Is), 0–2400 m. NG: locally distributed in the lowlands of eastern S Lowlands, SE Pen, and Huon, extending into mid-mountain valleys of E Ranges. Probably more widely distributed. Also Philippines and Wallacea eastward to Solomons and northern AU.

STONE-CURLEWS: Burhinidae

Nearly worldwide, the stone-curlews or thick-knees comprise 10 species, with 2 in the New Guinea Region. They are large shorebirds adapted to a nocturnal, terrestrial life. Their large eyes and slow, cautious movement give them an otherworldly appearance. Resting and cryptic by day, stone-curlews become active at dusk, and their eerie, night-time wailings are often the only clue to their presence. Stone-curlews feed on a variety of small animals taken by gleaning and stalking. They are monogamous and territorial. Their ground-nesting habits probably expose them to depredations by dogs, pigs, and people, a vulnerability that may explain why, in New Guinea, both species are found only in places remote from humans.

Bush Stone-curlew *Burhinus grallarius* Pl 27
(**Bush Curlew** or **Thick-knee**, *Burhinus magnirostris*)
54–59 cm. A large, long-legged nocturnal plover-like bird locally resident in the Trans-Fly savannah woodlands. The *cryptic, streaked plumage* blends in with grassy background. Note the *large, staring, yellow eye set between the white eyebrow and black facial streak*. **Adult:** White-and-black wing-coverts give the folded wing a banded look. **Juv** has a wider pale patch on folded wing that incorporates the median wing-coverts, which are blackish in Adult. **In flight:** Fluttering wings of a fleeing bird in torchlight show *white patches*, suggestive of a nightjar. **Similar spp:** Beach SC bulkier, lacks streaking, has a massive bill, inhabits coastlines. **Voice:** An eerie night sound of the Australian bush. The bird says its name—a loud, wailing *cur-leew*. Call usually repeated or may be taken up in chorus by other birds. Wail can be followed by shrieks, growls, e.g., *koi leee* then a feverish bubbling and wailing *keeEEE* reaching a climax and falling away. **Habits:** Inhabits open, dry woodlands and nearby savannah. Usually quiet during the day, resting on ground near cover. Moves with hesitation, a few steps at a time, then pauses. Wingbeats are stiff and shallow; usually flies only a short distance. Picks invertebrates and small vertebrates from the ground. Alone or in monogamous pairs. Nests in open situations, often next to a log, shrub, or stump; nest little more than a scrape on the ground (AU data). Eggs (2) color variable, spotted and blotched. **Range:** NG: Trans-Fly near the Indonesian-PNG border; known from Wasur National Park east to Bensbach R, but may be more widespread, 0–100 m. Also AU.

Beach Stone-curlew *Esacus magnirostris* Pl 27
(**Beach Curlew** or **Thick-knee**, *E.* or *Burhinus neglectus, E. giganteus*)
51–56 cm. *A large, robust, boldly patterned shorebird* of remote, undisturbed coastlines with sandy beaches. Crepuscular and mainly nocturnal. Large head with *black-and-white, harlequin face; yellow eye; and massive, upturned bill.* **Adult:** Overall tan-grey; folded wing grey. Note black shoulder bar with crisp white margins. **Juv:** Plumage more mottled; white-tipped greater wing-coverts form a distinct bar; and the white eyebrow may be broken in two by a black bar projecting above the eye. **In flight:** Wings show a striking pattern of black, white, and grey. **Similar spp:** Bush SC streaked and lives in savannah. **Voice:** Vocal mainly at night. A shrill series of repeated, wailing *weer loo* notes, accelerating and rising in pitch. Higher pitched and less musical than that of Bush SC. Alarm call a high, strident *teu-teu!* **Habits:** Inhabits undisturbed beaches and small islands, moving from one to another. Very shy, readily taking flight. Forages unobtrusively on beaches, mudflats, and reefs; shelters in the shadows of mangroves at high tide. Feeds by gleaning and stalking for crabs and other invertebrates. Solitary or in pairs. Nests above shoreline; nest a shallow depression, sometimes lined with plant material (AU data). Egg (1) pale brown or grey with dark spots and blotches. **Range:** Regionwide, sparingly distributed along the coast. Also Andaman Is and Greater Sunda Is to AU and Solomon Is.

OYSTERCATCHERS: Haematopodidae

The cosmopolitan family of oystercatchers has 12 species, of which one Australian species is marginally present in New Guinea. Oystercatchers are sturdy, black-and-white shorebirds (some are all black) with a bright red, bladelike bill for stabbing and prying open hard-shelled seafood. They occupy most coastal habitats, where they forage mainly in the intertidal. Although moderately gregarious, they separate into territorial, monogamous pairs to breed. They nest on the ground near the shore.

Australian Pied Oystercatcher *Haematopus longirostris* Pl 27
(Pied Oystercatcher)
42–50 cm. Status unclear—reported as resident on Aru Is and occasionally seen on southern mainland coast where possibly only visits from AU. *An unmistakable, large, black-and-white shorebird with long red bill and bright pink legs.* **Adult:** Wings and back black; eye red (iris and eye-ring). **Juv:** Similar, but feathers of wings and back brownish black with notched pale margins, eye brown, and bill duller. **Voice:** Piping call a loud, piercing series of piping then trilling notes. Flight call a loud *peep peep*. **Habits:** Most mainland records in Sep–Nov following breeding season in N AU; but also seen in other months. Gregarious, yet independent of other waders. Diurnal and nocturnal. Forages in intertidal zone, especially on exposed mudflats and rubble. Probes mud and sand for mollusks and marine worms; opens bivalves by stabbing with bladelike bill. Breeds in long-term, monogamous pairs. Nests in a variety of open situations above tidal zone; nest a ground scrape, often lined with small objects (AU data). Eggs (2) color and markings variable. **Range:** NG and Aru Is, plus 1 record from Misima (SE Is). NG: entire S coast from Bird's Head to SE Is, but most frequently reported from coastal Trans-Fly. Also AU and Kai Is (Wallacea). **Extralimital spp: Sooty Oystercatcher** (*H. fuliginosus*) of AU is all black.

STILTS AND AVOCETS: Recurvirostridae

This worldwide shorebird family of 9 species is represented in the NG Region by 2 visiting Australian species. Characteristic features include the remarkably long legs, slender bill, and pied plumage. Stilts have straight bills, whereas those of avocets are turned upward (recurved). Living in shallow wetlands of either fresh or salt water, these birds use their stiltlike legs for wading and occasionally swimming. Stilts and avocets feed on small invertebrates. Gregarious at all times, they form flocks made up of pairs or family groups. They breed as monogamous pairs and nest close to water.

Black-winged Stilt *Himantopus himantopus* Pl 27
(Sometimes split as a separate species: **White-headed Stilt**, *H. leucocephalus*)
33–37 cm. Locally common visitor from AU to S Trans-Fly and Port Moresby area, occasionally breeds; rare elsewhere. Present mainly in austral winter. Inhabits shallow freshwater lagoons, occasionally salt water. *A slim, black-and-white wader with very long, spindly pink legs.*
Adult: White head and neck with black nape-stripe that broadens to a half-collar at the base of the neck, glossy black mantle and wings, and reddish-pink legs. **Juv:** Head and neck all white with variable dingy grey cap and nape-stripe; back brownish black; legs pale pink. **Imm:** Similar to Juv but cap and back blackish, legs reddish pink. **Subspp (1):** *leucocephalus* (Java, Philippine Is, AU, NZ). **Voice:** Adult call is a puppy-like yapping, *kyap kyap kyap*. Juv call is a feeble whistling. **Habits:** Wanders in response to rain and availability of shallow wetlands. Gregarious, gathering in large groups year-round. Diurnal and nocturnal. Wades, but also ventures into water so deep it must swim. Feeds on small invertebrates by pecking or probing mud below waterline. Monogamous. Nests on the ground close to water; nest usually a cup or mound of vegetable matter (AU data). Eggs (3–6) olive or brown with spots and blotches. **Range:** NG, plus a few scattered island records. NG: most records from mainland along S coast of Trans-Fly and SE Pen (Port Moresby), 0–100 m (0–1650 m). Worldwide.

Red-necked Avocet *Recurvirostra novaehollandiae*
40–48 cm. Vagrant from AU to coastal Trans-Fly. *A pied shorebird with chestnut-brown head and long, thin, upturned bill.* Larger than White-headed Stilt and has a different pied pattern, diagnostic recurved bill, and grey legs (not pink). **Adult:** Head evenly dark, rich brown; body plumage immaculate white and jet black. **Juv:** Paler brown head, white scapulars with dingy markings, and black plumage dull and scalloped with buff. **Imm:** Intermediate. **Voice:** Barking call and wheezing whistle. **Habits:** Similar to White-headed Stilt. To be looked for among stilts and other waders. **Range:** NG: 1 record from coastal Trans-Fly. Also AU.

PLOVERS AND LAPWINGS: Charadriidae

Plovers are small to medium-sized waders of shoreline or terrestrial habitats. The 67 species are distributed worldwide. Various odd plover genera in South America and Australasia may indicate that the family originated in those regions (and possibly Antarctica). The 9 species of plovers in New Guinea include 2 resident species, 5 Palearctic visitors, and 2 Australian visitors.

The family has been traditionally divided into 2 groups: lapwings and plovers. Lapwings are larger and more strikingly colored; there are 2 species in New Guinea, the Masked Lapwing and Red-kneed Dotterel. The remaining 7 NG species are true plovers. Most plovers and lapwings are gregarious and encountered in flocks. Their typical foraging behavior is to run forward, abruptly halt to look and listen, then pluck their prey from the ground. They feed mainly on insects, crustaceans, mollusks, and other invertebrates. They are usually monogamous and nest on the ground. Male and female are the same size in nearly all species and usually cannot be distinguished in the field.

The plumage sequence of the Palearctic species is complex. Adults arriving in New Guinea in August–September to stay for the austral summer (northern winter) are molting from breeding plumage into nonbreeding plumage. Before they depart in April–May, they molt back into breeding plumage. Breeding plumage is colorful and species-specific, whereas wintering plumage is plain and looks much the same from one species to the next. Young birds (those that hatched in the northern summer) arrive in the region in juvenile plumage, which typically resembles the adult nonbreeding plumage except that the feathers of the back and wing-coverts are fringed buff, giving them a richer appearance. During the austral summer, juveniles molt first into an adultlike, first nonbreeding plumage and then, just before they migrate, into a partial, first breeding plumage. In this field guide, we provide descriptions of the three main plumages—adult nonbreeding, adult breeding, and juvenile—omitting the hard-to-recognize first nonbreeding and first breeding plumages. Readers can find more detailed information on all these plumages in books specializing in waders/shorebirds (see *Selected References*, p. 36).

Masked Lapwing *Vanellus miles* Pl 27
(**Masked Plover, Spur-winged Plover,** *Lobibyx miles,* includes *V. novaehollandiae*)
30–37 cm. Both a common resident and AU visitor. An unmistakable *sandy-brown-and-white shorebird of open habitats in the lowlands.* Big and noisy. *Yellow facial wattle and bill.* **Adult:** Large wattle, mantle and wings clean grey-brown without markings, and legs pink. **Male** larger on average. **Juv:** Smaller wattle, mantle and wings with dark scalloping, and legs pinkish grey. **In flight:** Black wing and tail feathers and white band at base of tail. **Subspp (2):** *miles* ("Masked," NG and tropical N AU); vagrant hybrids are intermediate between *miles* and *novaehollandiae* ("Spur-winged," most of AU and NZ), a subsp having numerous differences: larger size, darker overall, black nape and sides of breast, shorter bill, and facial wattle smaller and not extending past eye. **Voice:** Alarm call is grating and unpleasant: a loud, metallic note rapidly and continually repeated—*keer kik ki ki ki. . . .* **Habits:** In territorial pairs when nesting. Sociable outside the breeding season, aggregating in flocks in open country—airfields, plowed fields, paddocks, and water edges. Forages singly in typical plover fashion (walk, stop, pluck). Feeds on invertebrates, seeds, and the occasional frog. Noisy and belligerent, readily diving at intruder approaching the nest. The yellow wing-spurs are weapons for fighting other lapwings. Monogamous. Nests in a variety of open situations; nest a shallow depression, unlined or lined with whatever material is handy. Eggs (3) olive with dark spots and blotches. **Range:** NG and Aru Is, 0–500 m. NG: Trans-Fly,

Sepik-Ramu to SE Pen. Also AU and NZ. **Taxonomy:** Some authorities treat the two forms as different species.

Red-kneed Dotterel *Erythrogonys cinctus* Pl 27
(*Charadrius cinctus*)
17–19 cm. Rare, sporadic AU visitor. A *small waterside plover* with unique and striking color pattern. *Compact and large-headed, with diagnostic dark cap and snowy white throat.* Bill reddish with dark tip; upper legs reddish. **Adult:** Strikingly patterned—*black cap and breast band contrast with white underparts and flank-stripe.* **In flight:** *Black wings with thick, white trailing edge and white underwing-coverts; much white in tail.* **Juv:** Cap dark grey-brown, no breast band. **Imm:** Similar to Adult, but breast band may not be as distinct. **Similar spp:** Other small plovers lack the black or dark brown cap. **Voice:** Alarm call always disyllabic, *wit wit.* Flight call a low, rolling trill, quite unlike that of any other wader. **Habits:** Alone, in pairs, or small parties. Aquatic: forages in shallow water or along margins in low vegetation of shallow lagoons and river edges. Gleans and probes mud; eats small invertebrates and seeds. Bobs head strongly. Flight is buoyant with steep downstrokes. Nests near water's edge, often under cover; nest a depression lined with plant material (AU data). Eggs (3) cream, grey, or pale brown, dark-lined, spotted, and blotched. **Range:** NG: A few records from Trans-Fly and Port Moresby area (SE Pen), 0–100 m. Also AU.

Pacific Golden Plover *Pluvialis fulva* Pl 28
(**Lesser Golden Plover**, *P. dominica*)
23–25 cm. A common and ubiquitous overwintering Palearctic migrant on short grass, marine coasts, and freshwater edges. **In flight:** Only an obscure pale wing-stripe. **Adult:** Nonbreeding—*buff-colored plover with pale face and large dark eye under strong, buffy white eyebrow;* black bill framed by pale feathers around the mouth; and *blackish mantle and wings spangled with gold and white.* Breeding—**Male:** *Black underparts bordered by white lateral stripe.* **Female:** Less black in underparts. **Juv:** Similar to Nonbreeding adult, separable with difficulty. **Imm:** Breeding—few black feathers in underparts, similar to Adult in prebreeding molt. **Similar spp:** See Grey P. Oriental P, also on short grass, is more slender and has yellowish legs (not greyish); nonbreeding plumage has uniform upperparts (no spangling). **Voice:** Alarm call is a melodious, slurred, double-noted *tlu-eee.* Also various, similar 3-noted calls. **Habits:** Stops on airfields, playing fields, paddocks, and dry lagoon basins; less commonly on coastal beaches, mudflats, and reefs. Sociable when resting or traveling. Birds spread out when foraging, establish individual feeding territories on winter grounds. Feeds mainly on small invertebrates. **Range:** Regionwide, 0–1700 m. Breeds in Palearctic and Alaska; winters along Pacific and Indian coasts and islands. **Taxonomy:** Lesser Golden Plover was split as Pacific GP and American GP (*P. dominica*).

Grey Plover *Pluvialis squatarola* Pl 28
(**Black-bellied Plover**)
28–30 cm. An uncommon overwintering Palearctic plover of coastal flats and beaches, rare near freshwater. *Larger than Pacific Golden P, more robust, and grey without any buff.* **In flight:** All plumages show unique combination of *a broad white wing-stripe, white rump, and a black patch in the "armpits."* **Adult Nonbreeding, Juv and Imm:** Similar to Pacific Golden P, but grey, not buff. **Adult Breeding:** Upperparts spangled white-and-black; underparts black. **Voice:** Generally silent. Call is a diagnostic, drawn-out, wavering, and plaintive *peeowee.* **Habits:** Coastal, always near water. Rather solitary, feeding warily along the shore. Feeding behavior similar to Pacific Golden P, but more lethargic. **Range:** Regionwide at sea level. Breeds in Palearctic and Nearctic; winters along warm coasts of all continents but Antarctica.

Little Ringed Plover *Charadrius dubius* Pl 28
(**Little Ringed Dotterel**)
14–17 cm. A *small plover of interior wetlands*, represented by two races: one an uncommon resident breeder along gravelly rivers, the other a rare Palearctic visitor, mainly to mud banks along freshwater or tidal wetlands. *Bold white collar and black breast band.* **Adult:** Breeding—black mask with *conspicuous yellow eye-ring* (locally red in resident race) and white forehead. Nonbreeding (migratory race)—facial pattern and breast band much reduced, blurred, and brownish; feathers of wing-coverts and back pale-edged. **Juv:** Similar to Nonbreeding adult, but breast band interrupted in the middle

and wings and back with fine, buff scalloping. **Subspp (2):** *dubius* (resident NG, Bismarcks, and Philippines) Breeding and Nonbreeding plumages are more similar; *curonicus* (a Palearctic migrant) differs by voice, darker upperparts, and distinct Breeding and Nonbreeding plumages, with Breeding plumage that differs from *dubius* by the narrower yellow eye-ring, smaller pink or yellow spot at base of lower bill that does not include gape and edge of upper bill, narrower black frontal band, and the black tail-band extending onto the outer tail feathers. **Similar spp:** Other small plovers lack white collar, black breast band, and yellow orbital ring. **Voice:** Alarm call a harsh *chit chit chit* (resident race) or *pee-u* (migratory race). **Habits:** Resident race found on gravel flats along rivers and nearby open habitats, including gravel roads, car parks, and quarries, also short grass, e.g., airstrips, playing fields, paddocks. Mobile. Singly, in pairs, rarely in small parties. Feeds on small invertebrates. Nest a depression in gravel, lined or not (Indian data). Eggs (3–4) cream to buff, spotted and blotched. **Range:** Regionwide, 0–1400 m. Also breeds in Eurasia, wintering in Africa, SE Asia through to NG and AU. **Extralimital spp: Common Ringed Plover** (*C. hiaticula*), a Palearctic vagrant to AU, similar but chunkier; lacks distinct eye-ring; has brighter, orange legs; and in flight shows a thick, white wing-stripe.

Red-capped Plover *Charadrius ruficapillus*
(**Kentish Plover**, *C. alexandrinus*)

14–16 cm. Hypothetical. Expected visitor to southern beaches. Possibly once a vagrant from AU, but no recent records. *A tiny plover, pale with black legs and fine bill.* **Male:** *Orange-brown cap* bordered by black crescent on neck; black eye-stripe. **Female:** Cap dull orange-brown and lacking crescent; brown eye-stripe. **Juv:** Facial pattern obscure and only a trace of rust on the nape. **Voice:** Contact call, *wit-wit-wit*. Alarm call, *twink*. **Habits:** Gregarious on sandy and shelly beaches. Very active when foraging—runs, legs a blur, pauses. **Range:** Old, doubtful reports from S NG. Breeds AU.

Lesser Sand-Plover *Charadrius mongolus*   Pl 28
(**Mongolian Plover**)

19–20 cm. Common Palearctic migrant on coastal mudflats and beaches. *A small brown-backed plover with a hunched posture and short, stout, rather blunt-tipped bill.* **In flight:** *A distinct white wing bar* in all plumages. **Adult:** Nonbreeding—white underparts with indistinct brown breast band, dark smudge around eye, white forehead and eyebrow, and sandy brown cap. Breeding—bright rufous breast and collar, black eye-patch, white forehead and throat. **Juv:** Like Nonbreeding, but feathers of upperparts pale-fringed. **Subspp (1 group):** *mongolus* (breeds E Asia; winters SW Pacific, NG, AU). **Similar spp:** See Greater SP. **Voice:** A distinct, rattling *trrrrt* often repeated. **Habits:** While primarily a coastal plover, it will move inland to airstrips and playing fields and during migration may turn up on the shores of freshwater wetlands. Found in flocks; sociable, mixing with other species, including Greater SP, although solitary when feeding (for invertebrates). **Range:** Regionwide, coastal. Breeds in Asia, winters along the rims of Indian and W Pacific Oceans.

Greater Sand-Plover *Charadrius leschenaultii*   Pl 28
(**Large Sand-Plover**)

20–23 cm. Common Palearctic migrant *very similar in all respects to Lesser SP*. Differs subtly in size and proportions: *larger and more slender shape, longer legs, bigger head, and longer, heavier, more-pointed bill (bill length usually greater than distance from base of bill to back of eye)*. Legs may be slightly paler or more yellowish. **Adult:** Nonbreeding—upperparts slightly paler than Lesser SP. Breeding—rufous nape and throat-band paler and more orange, and throat band narrower, not extending onto breast. **In flight:** Feet project beyond tail tip (they don't in Lesser SP) and tail distinctly dark-tipped (not so in Lesser). **Subspp (1):** *leschenaultii* (breeds in deserts of E Asia, winters in Australasia). **Voice and Habits:** Similar to Lesser SP. **Range:** Regionwide at sea level. Breeds in central Asia, winters along coasts of Indian and W Pacific Oceans.

Oriental Plover *Charadrius veredus*   Pl 28
(**Oriental Dotterel**, *Eupodia veredus*, *C. asiaticus*)

23–25 cm. A Palearctic migrant, mainly in passage to its AU wintering grounds. Local and uncommon in areas of dry, open, sparse, short grass. Cryptic and easily overlooked. *Like a golden plover, but slender, with diagnostic long, pale yellowish legs.* **In flight:** Long-winged, no white wing-stripe. **Adult Nonbreeding:** *Similar to nonbreeding Pacific Golden P, but upperparts uniform, without spangling.*

Male breeding: *Distinctive whitish face and neck and a pale rufous breast band with black lower border.* **Female Breeding:** More like Nonbreeding plumage, but with darker neck and breast band. **Juv:** Like Nonbreeding, but upperparts with wider buff scalloping. **Similar spp:** Pacific Golden P. Lesser Sand P and Greater Sand P are smaller with darker legs. **Voice:** Flight call a soft, repeated *tink*. Social call a low trill. **Habits:** Favors dry habitats—playing fields, airfields, paddocks, occasionally freshwater swamp margins, rarely beaches. Surprisingly inconspicuous in such open habitat. Gregarious, in small parties, sometimes with other plovers and pratincoles. Active during cooler
parts of the day and at night. Inactive during the heat of the day, when rests in groups, some birds on low, elevated perches such as cow-pats. Forages with deliberate movements, feeds on invertebrates. **Range:** NG and Aru Is, 0–200 m. Breeds in arid NE Asia, winters in AU.

JACANAS: Jacanidae

This pantropical family comprises 8 species, 1 of which inhabits New Guinea and is shared with Australia. Somewhat rail-like in appearance, jacanas specialize by living on floating aquatic vegetation, especially water lilies. Their long toes permit them to walk over this yielding substrate. They feed on the abundant invertebrates and seeds such habitat provides. Their unusual breeding system has the male as the main care provider. He holds a small territory, builds the nest, incubates the eggs, and rears the young, while the larger female moves on to another mate. The male can carry small chicks under his folded wing.

Comb-crested Jacana *Irediparra gallinacea* Pl 34
(Lily-trotter, Lotusbird)
20–23 cm. Common and conspicuous resident in lowland ponds, lagoons, and swamps where there is floating vegetation. A small black-and-white rail-like bird with *amazingly long toes*; *walks on water lilies* and other floating plants. Absent where non-native, plant-eating fish have been introduced. **Adult:** Conspicuous *red comb*, white face, and black breast band. **Female:** Larger. **Juv:** Paler with all-white breast, lacking dark band; comb barely evident; cap and nape rufous. **In flight:** Flutters along on conspicuous *black wings*; note pale face and *dangling legs*. **Subspp:** None, but wings and back darkest in North. **Similar spp:** Rails lack comb and white face. **Voice:** Agonistic calls vary, often shrill piping and trilling. Alarm call when flushed a shrill trumpeting. **Habits:** Highly mobile and dispersive. Sociable, congregating at good foraging areas. Active, quarrelsome, and noisy. Flies readily. Forages by walking with deliberate motions over lily pads and other waterweeds, leaning forward to peer under upturned leaves. Takes small invertebrates and seeds. Polyandrous—Female lays successive clutches in nests of multiple Males, each with its own territory. Male assumes nesting duties. Builds nest on vegetation, usually water lilies; nest a platform or shallow cup of vegetation (AU data). Eggs (2–3) pale brown marked with wavy markings. **Range:** NG, NW Is (Misool, plus sighting from Salawati), Aru Is, and SE Is (D'Entrecasteaux Is: Goodenough, Fergusson), 0–500 m. Also AU, Borneo, Philippines, Wallacea.

SANDPIPERS AND SNIPES: Scolopacidae

Of the 96 species of sandpipers and snipes worldwide, 32 occur in the New Guinea Region, almost exclusively as migratory visitors from Asia. They are a varied group of long-billed and long-legged wading birds. Most species feed by picking or probing in short grass, mudflats, or shallow water, taking small invertebrates and seeds. Breeding habitat includes the margins of deserts, wetlands, and tundra and in winter, the seashore, estuaries, wetlands, and fields of short grass. Most species are monogamous and nest on the ground; soon after hatching, the precocial chicks leave the nest and follow their parents.

There is only 1 resident species, the endemic New Guinea Woodcock, a cloud forest dweller that has departed from the usual shorebird way of life. All other NG shorebirds breed in the colder northern latitudes during the northern summer and migrate thousands of kilometers south to overwinter on coasts and lowlands of New Guinea and Australia. While snow and ice cover the nesting grounds, a seasonal flush of food attracts the shorebirds to the southern hemisphere for the austral summer. New Guinea and Australia are the main wintering grounds for certain species—the Little Curlew, Eastern Curlew, Great Knot, and Sharp-tailed Sandpiper; for many others, a large portion of their population winter in Australasia. How large these NG shorebird populations are and where they reside have been only partly documented, and future surveys of the great river deltas and tidal bays could well discover important new concentrations of shorebirds (Bishop 2006).

Several aspects of shorebird migration are worth mentioning. The long journey south in autumn (Sep–Nov) and north in spring (Apr–May) is hurried, with most species making a brief stopover in New Guinea before moving on. The numbers of birds seen in passage (migration) are usually greater on the journey south than when northward bound. Shorebirds arrive in NG during the monsoonal dry season when the margins of wetlands are exposed and optimal for foraging. However, with the coming of the rains in December, rising water levels flood this habitat and most shorebirds dependent on wetlands move on, presumably south to Australia. Nevertheless, significant populations of coast-loving species spend the winter in New Guinea, and a few nonbreeding individuals stay the whole year.

Shorebirds, including plovers, can be difficult to identify—there are so many species, and their plumages are confusingly similar. Learn the common species first to compare later with the more rare species. Pay particular attention to body size and overall shape, length of bill, legs, and wings (how far the wing tip projects beyond the tail when the bird is resting), as well as examining plumage pattern and leg color. There are 3 main plumages: nonbreeding (plain grey or grey-brown), breeding (often brownish and more colorful), and juvenile. Juveniles arriving in NG typically have fresh plumage and all feathers of equal age and wear, whereas adults have both new and old, worn feathers. Several excellent field guides devoted to shorebirds can aid the identification of New Guinea shorebirds (Hayman et al. 1986, Message and Taylor 2005, Geering et al. 2007, Chandler 2009, Hollands and Minton 2012). These guides will also prepare the reader for recognizing new vagrant species from Asia or N America. (We mention in the "Extralimital spp" sections those vagrant species that turn up in Australia with some frequency.)

New Guinea Woodcock *Scolopax rosenbergii* Pl 29
(**Rufous Woodcock**, **East Indian Woodcock**, *S. saturata*)
30 cm. Rare in high-elevation cloud forest and at timberline. Secretive. Best detected when making its vocal display flights at dusk or dawn, but occasionally flushed in daylight or seen scurrying away. *Snipelike but plumper and color pattern differs.* Note big eyes set far back on head. **All plumages:** *Dark, with thick barring across the crown and nape* (instead of stripes) and *no sharp, whitish streaking on back.* **In flight:** Wings broad, rounded, and dark underneath. **Similar spp:** Snipe, which may visit subalpine bogs but not the forest, are paler and have creamy striped head and back pattern; their wings are longer, narrower, and pale underneath. **Voice:** Song-flights, known as "roding," performed by Male—bird flies a regular course low over forest canopy or grasslands, periodically singing and landing on a tree snag. Song *cree, quo, quo, quo. . . .* **Habits:** Solitary. Crepuscular or nocturnal, spending the day asleep on the forest floor or a mossy tree branch inside the forest. Emerges to feed at damp patches in forest openings at dusk. Probes the ground for worms, insect larvae, etc. One nest situated ~1.6 m above ground on a leaf-and-stick-covered horizontal branch. Eggs said to be (2), well-marked. **Range:** Endemic. NG: Bird's Head, Central Ranges, Foja Mts, and Huon, 2400–3800 m (rarely lower). **Taxonomy:** The New Guinea W and Javan W were formerly combined as the Rufous Woodcock (*S. saturata*).

Latham's Snipe *Gallinago hardwickii* Pl 29
(**Japanese Snipe**)
29–33 cm. Rare passage and overwintering Palearctic migrant to similar habitats as Swinhoe's S. The snipe species are usually impossible to tell apart in the field. All are recognized as snipe by the camouflage plumage of rich brown, black, and rufous, with *striped head, creamy scapular striping, and long straight bill.* Juveniles are said to be separable from Adults with practice. Species cannot be confidently identified by plumage differences, and besides there is much individual variation. Surest means is by examination of a bird in the hand, specifically body size and tail configuration, including the total number of tail feathers, number of narrow outer "pin" tail feathers, and color

pattern. If possible, take photos to aid/confirm identification. **All plumages:** Body shape of Latham's *more attenuated toward the rear* (Swinhoe's shape more truncated); in folded wing, tertials entirely cover primaries, which, if they project, do so for only a few millimeters (obviously project in Swinhoe's); if present, broad straight barring across lower breast diagnostic. **In flight:** Toes normally do not project beyond tail (tips of toes project in Swinhoe's). **In the hand:** *Large size with wing length 146–158 mm. Tail 58–72 mm with 16–18 tail feathers* and *the outer 5–6 not as narrow as Swinhoe's.* Tail shows more white on the tips of the feathers than Swinhoe's. There are numerous, other ill-defined differences; see a specialty guide. **Voice:** Alarm call when flushed, an explosive *krek*. **Habits:** Singly or in small groups. Wary and skulking. Active mainly at dawn and dusk. Thrusts bill into mud to find and extract worms, etc. Upon being flushed, often follows a zigzag path. **Range:** NG, but probably regionwide, 0–3500 m. Breeds Japan and far NE Asia, winters AU, with few records in between suggesting a nonstop migration.

Pin-tailed Snipe *Gallinago stenura* Pl 29

25–27 cm. Hypothetical. Probably a rare passage migrant and visitor to NG Region, where sightings have been reported but not verified. *Smaller and squatter than the other 2 snipe species. Short tail that barely projects out from under tips of the folded wings* (obviously projects in Latham's and Swinhoe's). **In flight:** Almost the entire length of toes projects beyond tail (barely projects in Swinhoe's and Latham's). **In the hand:** Small size with wing length 123–143 mm; tail short, 42–55 mm; tail has 24–28 feathers, of which there are 7 pairs of thin outer tail feathers, more than other snipes, and diagnostic. **Voice:** Alarm call similar to Swinhoe's S. **Range:** Breeds N Asia, winters S Asia to Wallacea and AU.

Swinhoe's Snipe *Gallinago megala* Pl 29
(**Chinese** or **Marsh Snipe**)

27–29 cm. The common NG snipe, a Palearctic winter visitor to short, moist grassland and wetlands from sea level to alpine bogs. Very similar to Latham's S (see preceding). **In the hand:** *Averages smaller than Latham's; wing length 130–150 mm, with shorter tail, 50–63 mm; tail feathers 20–22 [18–26], the outer 5–6 very narrow.* **Voice:** Alarm call, a rasping *squak*. **Habits:** Similar to other snipes. Utilizes dry grassy habitats as well as wetlands. Rarely zigzags when flushed. **Range:** NG, but probably regionwide, 0–3700 m. Breeds NE Asia; winters mainly in Wallacea, Philippines, and NG, but extends from Asia and Sunda Is (rarely) to AU.

Long-billed Dowitcher *Limnodromus scolopaceus*

27–30 cm. Vagrant to coast. Differs from Asian D by smaller size, *all-white back* (not barred), *white trailing edge to secondaries, underwing faintly barred* rather than white, and *bill paler for basal third.* **Voice:** Important to identification. Flight call, *keek* or *keek keek keek*. **Habits:** Similar to Asian D. **Range:** NG: 1 sighting from Port Moresby area. Breeds far NE Asia and N Amer, winters in N Amer.

Asian Dowitcher *Limnodromus semipalmatus* Pl 29
(**Asiatic Dowitcher**)

33–36 cm. Rare Palearctic migrant to coastal mudflats and sandflats. Combines features of a godwit and a snipe; slightly smaller and more compact than a godwit, with *straight, more robust, all-dark bill with blunt tip* and longer face. **Adult:** Nonbreeding plumage brownish grey and faintly marbled above, white below with dark-speckled neck and breast. Breeding plumage rusty, variable, with **Female** duller and frosty; note that *scapulars and wing-coverts have straight white margins*, not pale-notched as in godwits. **Juv:** From Nonbreeding by darker cap and darker scapulars and wing-coverts with sharp, creamy edging; overall buffier. **In flight:** Same *white rump with black barring* as Bar-tailed Godwit. **Similar spp:** Godwits have longer necks and legs, and the bill is upturned and pinkish at the base. See Long-billed D. **Voice:** Flight call a quiet catlike *miau*. Also *chep chep*. **Habits:** Seeks estuaries and mudflats for feeding. Gregarious, associating with other similar-sized waders. Feeds by vertical probing for worms, etc. **Range:** NG: sightings from Trans-Fly and SE Pen, sea level. Breeds in N Asia, winters mainly in Greater Sunda Is, but as widely as SE Asia to AU.

Black-tailed Godwit *Limosa limosa* Pl 29
40–44 cm. Common Palearctic passage and overwintering migrant on mudflats and beaches of sheltered coasts and wetlands. A large, slim, long-billed wader. *Bill only slightly upturned*; pink at the base. **Adult:** Nonbreeding plumage has greyish and *unstreaked upperparts*. Breeding plumage with rusty neck and *heavily barred flanks*; **Female** much paler. **Juv:** From Nonbreeding by buffier plumage and the wings and back feathers with dark, diamond-shaped markings, similar to breeding plumage. **In flight:** *Broad white bar on black tail and flashy black-and-white wings.* Feet and legs project beyond tail. **Subspp (1):** *melanuroides* (breeds NE Asia, winters S Asia to AU). **Similar spp:** See Bar-tailed Godwit. **Voice:** Usually silent. Flight call *tuk* or *kek*. **Habits:** More likely than Bar-tailed G to be found in freshwater lagoons. Sociable, in flocks of its own kind. Probes vertically into mud and sand for worms. Wades deeper than Bar-tailed, up to its belly. **Range:** NG, but probably regionwide. Breeds in N Eurasia, winters Africa, S Asia to AU.

Bar-tailed Godwit *Limosa lapponica* Pl 29
37–39 cm. Uncommon Palearctic migrant, perhaps mostly on passage, but some overwinter; same habitat as Black-tailed G. *Larger and stockier than Black-tailed G, with shorter neck and legs, bill obviously upturned.* **Adult:** Nonbreeding plumage has greyish *upperparts with distinct streaking*. Breeding plumage with rusty head, neck, and *breast, which lacks barring*. **Female:** Much paler and larger on average. **Juv:** From Nonbreeding by overall buffy tinge; darker cap and eye-line that accentuate the white eyebrow; dark scapulars with pale-notched edges. **In flight:** *Finely barred white tail and patch extending up back, and unmarked wings.* Only tips of toes project beyond tail. **Subspp (1 group):** *baueri* (breeds N Asia and Alaska, winters S Asia to NZ). **Similar spp:** Black-tailed G and Asian Dowitcher. **Voice:** Flight call a barking *yik yik yik*. **Habits:** Predominantly coastal, on sandy beaches and mudflats, but visits freshwater lagoons on migration. Associates with knots and whimbrels. Foraging behavior similar to Black-tailed G but keeps to exposed flats and shallows. **Range:** NG, Aru, and SE Is, but probably regionwide. Breeds Eurasia to Alaska, winters Africa, Asia, Australasia.

Little Curlew *Numenius minutus* Pl 30
(Little Whimbrel)
28–31 cm. Locally abundant passage Palearctic migrant on arrival, Sep–Dec; uncommon on return, Apr–Jun. Widespread; many thousands in Trans-Fly. Visits coastal and inland expanses of short grass, such as airfields, playing fields, and burned-over savannah. *The smallest curlew, plover-like and only a bit larger than a golden plover. Daintier than Whimbrel, and bill shorter, slim, and slightly decurved.* **All plumages:** Resembles Whimbrel but buffier, with incomplete dark loral stripe. **Female:** Larger on average. **In flight:** Upperparts evenly colored, lacks white rump. **Similar spp:** Pacific Golden Plover stockier and has larger head with short, straight bill. **Voice:** Flight call a dry, explosive *chu chu chu*. **Habits:** Gregarious—traveling, feeding, and roosting in flocks. Sporadic—numbers vary from one location to the next, and from year to year. Associates with golden plover and other grassland waders. Feeds in short grass, picking and probing for insects. Flock first crouches at approach of danger, blending with the grass. Visits freshwater pools to drink. **Range:** NG (especially Trans-Fly) and Aru Is, but probably regionwide in migration; lowlands. Breeds NE Asia, winters mainly in AU and NG.

Whimbrel *Numenius phaeopus* Pl 30
40–45 cm. Common and widespread Palearctic migrant along and near the coast, also small islands. Present throughout the year, as nonbreeders (Imm?) do not migrate north. Frequents mangrove creeks, estuaries, mudflats, beaches, reef flats, less often fields. *The midsize New Guinea curlew, large with long, downcurved bill.* **Adult:** Mottled brown, with *head prominently striped*. **Female:** Larger on average. **Juv:** Bill shorter; pale spots on the scapulars, tertials, and wing-coverts larger and better defined. **In flight:** *White rump* forming wedge up the back. **Subspp (1):** *variegatus* (breeds NE Asia, winters S Asia to Australasia). **Similar spp:** Little Curlew much smaller, bill shorter. Eastern C much larger, bill longer. Neither has white rump patch. **Voice:** Flight call a tittering, rapidly repeated whistle note, *ti ti ti. . . .* **Habits:** Mainly coastal, but passage birds may stop inland. Solitary or in small groups; associates with godwits and other large waders. Often wary, flushing noisily, scaring up other birds. Feeds at low tide; runs in quick bursts to capture prey, also probes. Takes mainly crabs, other crustaceans, and worms. **Range:** NG Region. Breeds in Arctic, winters along warm coasts worldwide.
Extralimital spp: Bristle-thighed Curlew (*N. tahitiensis*) breeds in Alaska and winters in Oceania, possibly to SE Is. Similar size and shape, but buffier, with pale orange-buff rump, tail, and underwing (these barred grey-and-white in Whimbrel).

Eastern Curlew *Numenius madagascariensis* Pl 30
(Far Eastern Curlew)
60–66 cm. Uncommon Palearctic migrant, mainly in passage, though some overwinter; especially on S coast. Inhabits estuaries, beaches, mangroves, and salt marshes. *Unmistakable: largest curlew in NG Region. A huge, heavyset shorebird with extraordinarily long, downcurved bill.* **Adult:** Nonbreeding plumage buff-grey evenly streaked and speckled; *head without stripes.* Breeding plumage buffier with rufous highlights on the upperparts. **Female:** Larger. **Juv:** Obviously shorter bill; dark scapulars and tertials with sharply defined pale notching. **In flight:** Upperparts evenly colored, lacks white rump. **Similar spp:** Whimbrel. **Voice:** Flight call a loud, ringing, slow *ker leeee* or *ker lew*. Also a bubbling song. **Habits:** Singly or in small, loose parties. Wary, fleeing at the first sign of danger. Foraging movements slow-paced, walks the intertidal flats picking and probing for mollusks and crustaceans. **Range:** NG and Aru Is. Breeds NE Asia, winters mainly in AU.

Common Redshank *Tringa totanus* Pl 30
27–29 cm. Vagrant Palearctic migrant, although possibly more frequent but overlooked. Coastal flats and nearby wetlands. *The only brownish, red- or orange-legged wader in the region. Also, bill orange at the base.* **Adult:** Nonbreeding plumage evenly brownish grey above and dingy white below. Breeding plumage with feathers of upperparts light-and-dark notched; underparts finely streaked and barred. **Juv:** Like breeding plumage but more brightly spotted above, and paler with no barring below. **In flight:** Broad white trailing edge to wing and white rump. **Subspp:** Unknown for NG Region. **Similar spp:** No other sandpipers have such orange legs. **Voice:** Flight call a repeated *teu* note, and variations. **Habits:** Solitary or in small groups, associates with other waders. Wary and excitable, fleeing noisily. Actively forages with quick steps, stopping to pick or probe; wades. **Range:** NG: Sightings from Port Moresby area. Breeds Eurasia, winters from Mediterranean Sea to SE Asia, infrequently to AU. **Extralimital spp: Spotted Redshank** (*T. erythropus*) breeds N Eurasia, winters Africa to SE Asia, vagrant to AU. Also with orange legs, has longer, thinner bill, with slightly drooping tip; no white in wings in flight.

Marsh Sandpiper *Tringa stagnatilis* Pl 30
22–26 cm. Common Palearctic migrant, both a passage and overwintering visitor. Primarily in shallow marshes and freshwater lagoons, also brackish pools around mangroves, but not on the open shore. *A smaller, delicate version of the Common Greenshank, with a straight, thin, needlelike bill.* **Adult:** Nonbreeding plumage plain grey above, *white below and on face and forehead.* Breeding plumage greyish brown above with black diamond and chevron markings; finely speckled whitish head, neck, and breast. **Juv:** Like Nonbreeding, but feathers of upperparts quite dark with broad, buff edges. **In flight:** Dark wings contrast with whitish wedge up the back. Feet project beyond the tail with part of the shanks showing. **Similar spp:** Common Greenshank larger, sturdier, with longer, thicker, upturned bill. **Voice:** Call a short *teu*, sometimes repeated, but not in series like call of Common Greenshank. **Habits:** Singly or in small to large groups, often with other shorebirds. More sociable than Common Greenshank and will gather in parties of more than a hundred where conditions are ideal. Wary and excitable, when flushed it often circles repeatedly before landing. Forages in shallow water, mostly wading and occasionally swimming. Feeds by picking and sweeping; rarely rushes at prey. Eats insects and tiny mollusks. **Range:** NG and Aru Is. Breeds N Eurasia, winters Africa, Asia, Australasia.

Common Greenshank *Tringa nebularia* Pl 30
30–35 cm. Common Palearctic migrant, both a passage and overwintering visitor. Widespread in coastal mudflats, brackish and freshwater lagoons, muddy riverbanks and tidal creeks, and marshes. *A large sandpiper, lanky and large-headed with distinctive upturned, greenish-grey bill, and exaggerated bobbing motions.* **Adult:** Nonbreeding plumage similar to Marsh S. Breeding plumage heavily streaked on the head, neck, and flanks; many scapulars and tertials blackish with white notching on the margins. **Juv:** Similar to Nonbreeding, but scapulars and wing-coverts darker with broad buff or white edges, except at tip. **In flight:** Similar to Marsh S but only feet project beyond tail, not shanks. **Similar spp:** Marsh S similar but obviously smaller, with a very thin, straight bill, and different call. **Voice:** Call a loud, distinctive *tew tew tew*, the notes equally spaced. **Habits:** Singly or in small parties, often with other shorebirds. Wary and noisy. An active wader often seen head-down, rushing at small prey in the shallows. Feeds by picking, probing, lunging, and sweeping techniques, taking insects, crustaceans, mollusks, and fish. **Range:** NG and Aru Is, probably more widespread. Breeds N Eurasia, winters Africa, Asia, Australasia.

Green Sandpiper *Tringa ochropus*
21–24 cm. Vagrant Palearctic migrant. **All plumages:** *Deceptively similar to Wood S but appears almost blackish above (less speckled), especially in flight when all-dark underwing shows* (underwing pale in Wood); white eyebrow does not extend past eye; white tail with 3 thick black bars (Wood has 4 thin bars); *legs greenish* and shorter, only toe tips projecting beyond tail in flight (in Wood most of foot projects). **Voice:** Call *to lu wit* or *weet t weet*, distinctive from Wood. **Range:** NG: 1 record from Trans-Fly. Breeds N Eurasia, winters Africa, Asia to Borneo and Philippines.

Wood Sandpiper *Tringa glareola* Pl 31
19–23 cm. Common Palearctic passage migrant and overwintering visitor. Frequents muddy, freshwater habitats and inundated grassland, never open shore. *A small freshwater wader with dark back noticeably speckled with white, and yellowish legs.* **Adult:** Nonbreeding plumage speckled by *white* notches on wing-coverts and scapulars; *rather smooth grey neck*. Breeding plumage more mottled above, streaked and barred below. **Juv:** From Nonbreeding by darker wing-coverts and scapulars with bold buff notches and streakier neck. **In flight:** White rump patch square, does not extend up back as a wedge. **Similar spp:** See Green S. Marsh S and Common S share Wood's habitat. **Voice:** Call a distinctive *chiff iff iff* or a single *chit*. **Habits:** Solitary or in flocks of its own kind or with other waders. A swift, agile flier with clipped wingbeats. Forages in vegetation, in shallows, or on mudflats, probing and picking for small invertebrates. Expert swimmer. **Range:** NG and NW Is (Waigeo), probably more widespread, 0–500 m, records to 1700 m. Breeds N Eurasia, winters Africa, Asia, Australasia.

Grey-tailed Tattler *Tringa brevipes* Pl 31
(**Siberian** or **Polynesian Tattler**, *Heteroscelis brevipes*)
25 cm. Common Palearctic passage migrant and overwintering visitor. Uses a *variety of habitats on sheltered coasts*, including mudflats, coral rubble, reefs, estuaries, lagoons, rarely beaches or inland. Only the 2 tattler species show such *uniform grey upperparts*, diagnostic in flight. Note the *short, yellowish legs*. The two species are nearly identical. *Best means of distinguishing Grey-tailed T from Wandering T is by the flight call.* Field marks separating Grey-tailed: typically paler coloration, bolder white eyebrow contrasting with dark lores, paler grey tail, whiter flanks in Nonbreeding and Juv plumages, finer barring and more white on belly in breeding plumage, and wing tip in resting posture as long as or only slightly projecting beyond the tail tip (noticeably longer in Wandering). Beware that each of these differences is variable and not entirely reliable; in combination they make a better case. **In the hand:** *Nasal groove extends less than half the length of the bill*; scales on legs overlap. **Adult:** Nonbreeding plumage plain grey, darker above. Breeding plumage finely barred with dark grey crescents; belly to undertail white. **Juv:** Pale-dotted fringes to scapulars and wing-coverts. **Voice:** *Flight call plover-like*, a soft whistle, *too eee*, the second syllable rising. **Habits:** Singly or in loose groups when feeding. Alert, but often allows close approach. Bobs and teeters like Common S, but not as much. Forages quietly on shore or in the shallows, picking and probing for invertebrates. Sometimes actively rushes about in the water pursuing crabs and fish. **Range:** NG Region. Breeds NE Asia, winters Asia, Australasia, Oceania.

Wandering Tattler *Tringa incana* Pl 31
(*Heteroscelis incanus*)
27 cm. Uncommon Palearctic migrant mainly to eastern NG Region. Primarily on *offshore islands and rocky reefs along mainland coast*, often where there is wave activity. *Nearly identical to Grey-tailed T and best separated by flight call. Darker overall*, usually with obscure white eyebrow, darker grey tail, flanks with a thicker obvious grey margin in Nonbreeding and Juv plumages, thicker barring on underparts in breeding plumage, and longer wing tip projection beyond the tail in resting posture. **In the hand:** *Nasal groove extends halfway or greater along the length of the bill*; scales of legs do not overlap. **Female:** Larger on average. **Voice:** *Flight call very different from Grey-tailed T*. A whimbrel-like *li li li . . .* of 4–10 notes. Also gives short calls similar to Grey-tailed. **Habits:** Similar to Grey-tailed. Typically works rocky shorelines, often close to moderate wave action. **Range:** NG Region, mainly in coastal SE Pen and SE Is. Breeds in Alaska and far E Siberia, winters in Oceania, E AU and NZ, and Pacific coast of the Americas.

Terek Sandpiper *Xenus cinereus* Pl 31
(*Tringa terek*)
22–24 cm. Common Palearctic passage migrant and overwintering visitor. Seeks coastal mudflats, beaches, mangroves, and reefs, and freshwater lagoons in passage. Terek's strongly *horizontal stance* is accentuated by the long bill and short legs. The only sandpiper with *conspicuous orange-yellow legs* and *obvious upturned bill*. **Adult:** Nonbreeding plumage plainer (less streaky) and scapular bar obscure. Breeding plumage with head and neck finely streaked, and prominent black bar on scapulars. **Juv:** Similar to Breeding adult but upperparts slightly darker, head and neck less distinctly streaked, scapular bar obscure, and wing-coverts with narrow buff tips (white in Adult). **In flight:** Upperparts appear uniform except for prominent white trailing edge to secondaries. Feet do not project beyond tail. **Voice:** A variety of twittering and whistled calls. **Habits:** Solitary when feeding, but roosts socially. Usually in small numbers among larger assemblages of foraging waders. Behavior diagnostic—hyperactive, often runs through groups of feeding waders with head held low, dashing after prey and abruptly changing direction. Bobs head and jerks tail as it moves. Prey captured by chasing, probing, and sweeping in shallow water. Takes crustaceans and insects. **Range:** NG Region. Breeds N Eurasia, winters Africa, Asia, Australasia.

Common Sandpiper *Actitis hypoleucos* Pl 31
(*Tringa hypoleucos*)
19–21 cm. A widespread Palearctic passage migrant and visitor, this is the *only small sandpiper commonly occurring along streams and rivers* in the interior, also found in coastal mangrove creeks, debris-strewn sandy beaches, and rocky shores; visits large and small islands. *Teeters rear end continually up and down*. Grey-brown above, white below, with diagnostic *white crescent before folded wing*. **Adult:** Nonbreeding plumage has plain, dull brown upperparts with black vermiculation on shoulder. Breeding plumage similar but shows faint dark shaft streaking on crown, face, and neck and black shaft streaking and anchor marks on scapulars and wing-coverts. **Juv:** Similar to Nonbreeding, but has more conspicuous buff fringing on upperparts. **In flight:** Note distinctive, pulsing wing-flutter, with stiff wings held with tips angled downward. Displays white flashes from long wing bar on both upper- and undersurfaces of wing. Brown rump. **Voice:** Flight call a penetrating, high-pitched *swee wee wee*. **Habits:** Typically solitary and territorial on wintering grounds, but sometimes gathers in groups during passage. Usually keeps apart from other shorebirds. Most often forages in tight situations—stream banks, tidal creeks, edge of reed beds. Tends to stay close to vegetation or rocks, seldom venturing far out onto mudflats. Forages mostly out of water, but can wade and swim. Active and agile, scurrying about; mainly picks or dashes after insects and crustaceans. **Range:** NG Region, 0–3500 m. Breeds N Eurasia, winters Africa, Asia, Australasia, Oceania.

Ruddy Turnstone *Arenaria interpres* Pl 32
22–24 cm. Common and widespread Palearctic passage migrant and overwintering visitor. Utilizes a variety of coastal habitats, but prefers rocky shores, reefs, open fields of short grass, and small islands. *A distinctive, stocky shorebird with black breast band and orange legs*. Wedge-shaped bill. **Adult:** Nonbreeding plumage like a much darker, duller version of breeding plumage. Breeding plumage—*pied head and colorful orange-and-black wings and back*; **Female** often darker orange. **Juv:** Difficult to separate from Nonbreeding, has somewhat paler head and more evident pale fringing on dorsal feathers. **In flight:** *Pied wings and tail*. **Subspp (1):** *interpres* (breeds N Eurasia and NA). **Voice:** Flight call diagnostic, a rattling *trik-tuk-tuk-tuk*. . . . Also a short, grating *churr*. Calls lower in tone than most other waders. **Habits:** Usually in small flocks; mixes with other species. Walks slowly along beaches, busily probing, pecking, digging, and turning over small stones and seaweed with its bill. Feeds on a variety of invertebrates and their eggs. **Range:** NG Region. Breeds in the Arctic, winters along warm coasts worldwide.

Great Knot *Calidris tenuirostris* Pl 32
26–28 cm. Abundant on coastal mudflats as a Palearctic passage migrant and overwintering visitor. A medium-sized, robust shorebird. *Proportions differ from the very similar Red K: slightly larger, with longer bill and legs, and smaller head*. More coarsely marked than Red K. Always has *spotting across breast and flanks*. **Adult:** Nonbreeding plumage grey above with *pale scalloping*; head, neck, breast, and flanks with *dark wedge-shaped speckling*. Breeding plumage *densely covered with salt-and-pepper*

markings, especially dark on breast and mantle. Rufous scapulars (many or few). **Juv:** Similar to Nonbreeding but more boldly patterned; mantle and scapulars blackish (not greyish) and head and neck more streaky. **In flight:** White rump. **Similar spp:** Red K very similar, but proportions differ: slightly smaller size, bill and legs shorter, and head larger. Grey Plover has a short stout bill, large eye, and is spangled above. **Voice:** Usually quiet. Flight call similar to Red K, a low double-noted *nyut nyut*. **Habits:** Gregarious, usually in dense flocks. Readily mingles with Red K, godwits, and other shorebirds. Feeds in parties, walking slowly across mud and through shallows, probing deep mainly for small clams. **Range:** NG (mainly Trans-Fly), probably also the satellite islands. Breeds NE Asia, winters Asia to Australasia, mainly in AU.

Red Knot *Calidris canutus* Pl 32

23–25 cm. Status poorly understood—present in numbers during passage on coasts of the Trans-Fly (Oct–Nov and Mar–Apr) with some overwintering; elsewhere rare or absent. Same habitat as Great K, with which it associates. Differs from the more abundant Great K in *more compact shape and proportionally shorter bill*. **Adult:** Nonbreeding plumage *uniformly grey above* (pale scalloping hardly noticeable, unlike for Great K). Breeding plumage unmistakable—*face and underparts bright orange-brown*, upperparts spangled rufous, black, and white. **Juv:** Similar to Nonbreeding, but grey feathers on wings and back distinctly fringed with black and white; breast suffused buff when fresh. **In flight:** Finely barred rump. **Subspp (1):** *rogersi* (breeds NW Asia, winters in Australasia). **Similar spp:** See Great K. Curlew S smaller, with longer, decurved bill. **Voice:** Flight call a low, repeated, single note, *k-nut*. **Habits:** Similar to Great K. Feeding actions quicker than Great K, and diet of invertebrates more varied. **Range:** NG, mainly Indonesian Trans-Fly. Breeds in Arctic, winters Eurasia, S Amer, Africa, Australasia.

Sanderling *Calidris alba* Pl 32

20–21 cm. A rare Palearctic passage migrant and overwintering visitor, sparsely distributed *along beaches*. A small, sturdy sandpiper with large head and short, thick neck. *Bill stout, straight, and black; legs black*. **Adult:** Nonbreeding plumage *silvery grey above*, white below; note prominent dark eye and unique *black shoulder patch*, often covered by white breast feathers. Breeding plumage has *rufous brown, mottled upperparts*. **Juv:** Similar to Nonbreeding, but *wings and back checkered black and white* and crown dark and streaky. **In flight:** Long, white wing-stripe stands out against dark wing—the *most prominent wing-stripe of any small sandpiper*. **Similar spp:** Knots larger, more sluggish. Other small sandpipers have thinner bills and less prominent wing-stripes. **Voice:** Flight call a sharp *cht*, sometimes repeated. **Habits:** Often found as a stray bird, but more typically in groups. Tame, allows close approach. Strictly a beach dweller. Very active, running along sandy beach at the waterline, stopping to pick or probe. Chases waves down the beach at top speed, pausing to capture prey before retreating on the run. Feeds on small crustaceans and other marine invertebrates. **Range:** NG, but probably throughout region. Breeds in the Arctic, winters on warm coasts worldwide.

Stints are tiny sandpipers. They all look much alike and can be difficult to identify to species. Features that help identification are habitat, leg color, bill length, and throat and breast color.

Red-necked Stint *Calidris ruficollis* Pl 32
(Rufous-necked Stint)

13–16 cm. A common, sometimes abundant, Palearctic passage migrant, though a few overwinter. Seeks sheltered coasts with mudflats; sometimes found inland. This is the most often seen NG stint; other stints are rarities. *Note short, tapered bill and short, black legs*. **Adult:** Nonbreeding plumage *pale, clean grey above*, white below; *only stint with white breast*; note *white eyebrow*. Breeding plumage with diagnostic *rufous neck* and faintly dark-spotted breast; *rust-and-black markings on scapulars*. **Female:** Larger on average. **Juv:** Like Nonbreeding, with varying rufous tinge above, dark centers to scapular feathers, and pale fringes to feathers of wings and back, for a scaly effect. **Similar spp:** Little S is nearly identical; not yet confirmed for NG Region. Sanderling larger, with black-and-white wings. **Voice:** Flight call a soft *kreet* or *chirit*, infrequently repeated rapidly to form a short trill. **Habits:** Gregarious, usually in flocks; mingles with other waders. Feeds mostly out of water, but occasionally wades. Rapidly probes and picks for small invertebrates. Works one patch before running to the next. **Range:** NG Region. Breeds in the Asian Arctic and Alaska, winters SE Asia and Australasia.

Little Stint *Calidris minuta* Pl 32
12–14 cm. Hypothetical vagrant with 2 possible records, one of a bird in breeding plumage. A rare Palearctic visitor to AU with numerous records, so almost certainly migrates through NG Region but is probably overlooked. Identify with caution; difficult to separate from Red-necked S. Bill slightly longer; smaller body more compact (Red-necked S body more attenuated); wings a bit shorter. *Call diagnostic.* **Adult:** In Nonbreeding plumage not safely distinguished from Red-necked S; look for size, bill length, and shape. Breeding plumage similar to Red-necked but *throat white, not rufous*; orange-rufous areas of face and neck always finely streaked or spotted (vs unmarked red-orange in Red-necked); sides of breast with rufous and streaky spotting overlapping (vs Red-neck's plain rufous neck separate from dark spotting on white side of breast); wing-coverts and scapulars more evenly brownish rufous (vs greater wing-coverts and tertials tending toward grey, contrasting with rufous on mantle and scapulars). **Juv:** Compared with juv Red-necked, wing-coverts more brownish, showing less contrast with the scapulars and back (greyish and contrasting in Red-necked); prominent white lines running down mantle and back (these obscure or absent in Red-necked). **Voice:** Flight call is a short *tip* or *tit*, vs lower-pitched, longer *kreet* or disyllabic *chirit* of Red-necked. **Habits:** Similar to Red-necked. **Range:** Breeds in Eurasian Arctic, winters Africa to S Asia, a few birds reaching Australasia, but no certain NG records.

Long-toed Stint *Calidris subminuta* Pl 33
13–16 cm. A rare Palearctic migrant, mainly in passage, Sep–Dec. Found in freshwater or brackish wetlands. Differs from other stints in having *pale yellow-olive legs and streaked breast*. The cap is always dark, often rufous. Full-bodied with longish neck; legs long, bill fine-tipped. **Adult:** Nonbreeding plumage pale brownish grey. Breeding plumage with rufous wash on upperparts. **Female:** Larger on average. **Juv:** Brighter rufous of fresher plumage compared with worn feathers of breeding plumage Adult. **Similar spp:** Other stints have black legs. Red-necked Stint paler and has larger head. Sharp-tailed Sandpiper and Pectoral S much larger but otherwise similar with dark, streaky crown, rufous-edged back feathers, and olive legs. **Voice:** Flight call a trilling *trrrt* like a Pectoral Sandpiper, or a *chirrup* suggesting a House Sparrow, but more rolling. **Habits:** Solitary or small groups. Favors areas of soft mud with vegetation cover such as stranded aquatic weeds or inundated grass. Also visits dry, close-cropped grassland. Wary and secretive. Flight is often wild and dashing. Feeds by picking small invertebrates from mud or in shallow water. **Range:** NG, records from Trans-Fly and SE Pen lowlands. Breeds NE Asia, winters Asia to Australasia.

Baird's Sandpiper *Calidris bairdii* Pl 33
14–18 cm. A vagrant N American stint. Favors dry edges of mudflats and short grassland. *Unusually long-winged—folded wings project well beyond tail.* Long, fine-tipped bill. Legs dark. All plumages show *buffy breast band.* **Adult:** Nonbreeding plumage shows scapulars and wing-coverts with long, dark shaft streaks, which are obscure in **Juv.** Breeding adult more heavily mottled above and streaked below. **Female:** Larger on average. **Range:** NG, 2 records: one from Bird's Head (Jefman I, Sorong harbor), the other from Port Moresby. Breeds in N American Arctic, winters in S Amer. **Extralimital spp:** Another N American stint recorded occasionally in AU is **White-rumped Sandpiper** (*C. fuscicollis*), with faint, fine dark streaks along the flanks and an all-white rump; it too shows extensive projection of wing tips beyond the tail.

Pectoral Sandpiper *Calidris melanotos* Pl 33
19–24 cm. A rare Palearctic migrant mainly passage in Oct–Dec to Port Moresby area; few records elsewhere and at other times. Shows distinct preference for grassy verges of wetlands. Similar to but much less common than Sharp-tailed S, from which distinguished by being less rufous, having a longer bill that is paler at the base (Nonbreeding), and *the diagnostic sharp demarcation between the streaked bib and clear white lower breast and belly* giving a heavy, puff-chested look; size larger on average. **Adult:** Nonbreeding plumage has soft, grey-brown tone. Breeding plumage more brownish, especially on the upperparts. **Male:** Larger, bib darker. **Juv:** Similar to Breeding but has a more rufous-capped appearance, closer to Sharp-tailed. **Voice:** Flight call a dry, budgerigar-like trill: *prrrt prrrt . . .* , very different from Sharp-tailed. **Habits:** Similar to Sharp-tailed. Singles or small groups. Will associate with other waders, especially Sharp-tailed and other small sandpipers, but tends to separate by foraging microhabitat. **Range:** NG, mostly Port Moresby area in SE Pen. Breeds in Arctic, winters AU, NZ, and S Amer.

Sharp-tailed Sandpiper *Calidris acuminata* Pl 33
17–22 cm. A common, sometimes abundant, passage and overwintering Palearctic migrant. Widespread; found mostly along the muddy margins of freshwater wetlands, also tidal mudflats and short grass, such as airfields, rarely seashore. Note dark-streaked, *rusty cap* offset by whitish eyebrow (cap obscure and not rusty in Pectoral S) and *buffy wash on breast* with only obscure streaking (unlike well-streaked Pectoral). Facial pattern noticeably different from Pectoral—*dark, shadowy eye-stripe* extends around and behind eye, accentuating the *white eye-ring* (Pectoral has dark lores only, so eye-ring is rather obscure and face looks plainer). Bill dark with only a trace of yellow at the base. **Adult:** Nonbreeding plumage has plain, brownish-grey tone. Breeding plumage has bright rufous tone on back and throat, dark arrow marks on throat and breast (lacking in Pectoral). **Male:** Larger. **Juv:** Has similar rufous-and-buff tone of Breeding, but lacks arrow markings on throat and breast. **In flight:** Very little white in wing and white at either side of dark rump. **Similar spp:** See Pectoral S. **Voice:** Flight call varies, *wheep*, *pleep*, or *chrrt*; may be strung together into swallow-like twittering *sii sii sii* or *cwhe wi wi*. **Habits:** Mostly in the lowlands, but stops in the mountains during migration. Gregarious, usually in flocks, mixing with other shorebirds. Rather tame. Forages at a leisurely pace, picking and probing for small invertebrates and seeds. **Range:** NG Region, islands of all sizes, mainly lowlands (0–3500 m). Breeds in Asian Arctic, winters in Australasia.

Curlew Sandpiper *Calidris ferruginea* Pl 33
18–23 cm. An uncommon, passage Palearctic migrant, mostly Sep–Nov. Frequents margins of freshwater and brackish wetlands, less often on coastal mudflats. The region's *only small shorebird with long, slender bill, decurved throughout its length*. **Adult:** Nonbreeding plumage has grey upperparts, feathers with faint, pale fringing. Breeding plumage with varying amounts of rufous above and below. **Juv:** Like Nonbreeding but feathers of upperparts darker with thicker fringes imparting a stronger scaled appearance, and creamy buff wash on breast. **In flight:** Conspicuous *all-white rump* and feet project beyond tail. **Similar spp:** Broad-billed S shows droop only at tip of bill and is also smaller, with hunched posture and dark rump. **Voice:** Flight call a cheerful purr, *chirrup*, more musical and sweeter than other trilling waders. **Habits:** Sociable; in small groups, often with other waders. Feeds over wet mud and in the shallows, wading in up to its belly. Picks, jabs, and probes deeply, taking an array of small invertebrates. **Range:** NG, mainly S coast. Breeds in Eurasian Arctic, winters Africa, Asia to Australasia. **Extralimital spp: Dunlin** (*C. alpina*) breeds in the Arctic and winters along warm coasts of the N Hemisphere, with a record from Cairns, AU. Has dark rump, shorter legs (feet not projecting beyond tail in flight), and long bill droops less and mainly at the tip. Nonbreeding and juv plumage similar to Curlew S; Breeding very different with pale head and neck, mostly rufous back, and large black blotch on white belly.

Broad-billed Sandpiper *Limicola falcinellus* Pl 33
16–18 cm. An uncommon Palearctic migrant in passage with some birds overwintering. Found locally on mudflats along coast and on shores of freshwater and brackish lagoons, sewage treatment plants, etc. *A tiny, stintlike shorebird with distinctive long, stout bill, drooped at the tip. Reminiscent of a diminutive snipe, with same shape and striped head.* Unique double eyebrow—eyebrow splits, forming a pair of white stripes along the sides of the dark crown. **Adult:** Nonbreeding plumage plain, pale grey above, white below. Breeding plumage has frosted look owing to white fringes of dark brown feathers on wings and back (becoming darker and browner with wear); breast with obscure, dark arrow marks. **Female:** Larger on average. **Juv:** More colorful, neater, and streaky than breeding plumage, with rust- and white-fringed wings and back and brown-tinged breast with dark streaks and dots (rather than arrow marks). **In flight:** *Wing with dark leading edge* and indistinct pale wing bar; rump dark; feet do not project past tail. **Subspp (1):** *sibirica* (breeds E Siberia, winters Asia to Australasia). **Similar spp:** Stints have much shorter bills. Curlew Sandpiper stands taller; has slimmer, more curved bill; and lacks the head pattern. **Voice:** Call a distinctive, buzzy *chrreet*, low in tone, but loud enough to be detected in a flock of calling shorebirds. **Habits:** Singly or in small flocks, often among stints and small sandpipers. Movements and foraging similar to stints, but feeds more slowly with careful, vertical probing. Eats small invertebrates. **Range:** NG coasts, probably widespread in suitable habitat. Breeds in Eurasian Arctic, winters in Africa, Asia, Australasia.

Buff-breasted Sandpiper *Tryngites subruficollis*
18–20 cm. Vagrant Palearctic passage migrant from N America. Favors short, dry grass, but occasionally found on mudflats, swamp edges, beaches. A *plover-like* sandpiper with *uniformly bright, buffy-tan underparts and face;* distinct black spots on side of breast. *Dark, beady eye set in unmarked, buffy face. Legs bright orange-yellow.* **Adult:** Scapulars black with buff fringes. **Male:** Larger. **Juv:** Scapulars more brownish with black subterminal tips and white fringes; belly paler. **Similar spp:** Pacific Golden Plover has spangled, not scaly, back and greyish legs. Sharp-tailed S juv has shorter neck, longer bill, dark eye-line, and pale eyebrow. Ruff female larger and differently shaped, with much longer neck, bill, and legs and smaller head. **Voice:** Usually silent in passage. Flight call low single *tu.* Also a trilling *prrrt.* **Habits:** Solitary in the region, but keeps company with small sandpipers. Forages unobtrusively, picking at the ground with plover-like actions. **Range:** NG: 1 record from coastal SE Pen. Breeds in Arctic of far NE Asia and N Amer, winters S Amer; many records from AU.

Ruff *Philomachus pugnax* Pl 34
(Common name of the species refers to the Male; Female is called Reeve.)
Male 26–32 cm; Female 20–26 cm. A rare, Palearctic passage and overwintering migrant. Found on muddy margins of freshwater swamps and lagoons, less often of brackish or salt water. *A tall, medium-sized wader with characteristic build: small head, long thick neck, robust torso, and long legs.* Bill thin and medium-length. **Male:** Nonbreeding plumage pale grey-brown above, white below obscurely mottled grey on breast; usually with some white on neck. Breeding plumage (never seen in PNG, only on or near breeding grounds) unmistakable with thick neck ruff and puffy crests; color variation beyond words—barred, spotted, or plain; ruff and crests white, grey, black, rufous, and combinations of these. **Female:** Nonbreeding similar to Male. Breeding has dark wings and back with buff fringing and barring; orangish legs. **Juv:** Similar to breeding female but with buffy face, foreneck, and breast and legs brownish. **Similar spp:** Buff-breasted Sandpiper similar to Juv Ruff, but much smaller and more compact. Sharp-tailed S also similar to Juv but smaller, shorter-legged, rusty cap. **Voice:** Silent in NG. **Habits:** Solitary in NG Region, but seeks company of other sandpipers, especially Sharp-tailed. Forages by walking along, bent forward while picking and probing rapidly for small invertebrates and seeds. **Range:** NG, most records from Port Moresby area. Breeds Eurasian Arctic, winters mainly Africa to Asia, rare in AU.

Red-necked Phalarope *Phalaropus lobatus* Pl 34
(Northern Phalarope)
18–19 cm. Common to abundant Palearctic migrant that overwinters on seas NW, N, and NE of NG; rarely on freshwater, with some birds on passage over land to and from AU waters. The only oceanic wader in the region; flocks feed and live at sea during the austral summer. *In flocks far from land; small pied birds bobbing on sea or flying low over water.* Bill black, straight, needle-thin. **Adult:** Nonbreeding plumage grey above, white below with *white face crossed by a thick black eye-mask.* Breeding plumage has slaty grey head and upperparts with contrasting *white throat and spot above eye* and colorful *chestnut stripe* down side of neck. **Female:** Larger on average and more richly colored. **Juv:** Similar to Nonbreeding, but wings and back slaty with buffy fringes on the feathers. **In flight:** Dark wings with white wing-stripe. **Similar spp:** No other waders apart from phalaropes swim on open water or have a black eye-mask. **Voice:** Flight call a repeated *chek* or *twick.* **Habits:** Highly gregarious and tame. Normally seen swimming rapidly on a wandering course, head erect and bobbing. Sits buoyantly on the water. Picks small invertebrates at the surface. **Range:** NG waters from tropical Pacific west to Seram Sea. Breeds in Arctic. Winters in 3 main locations: NW Indian Ocean off Arabian Pen; seas between Sulawesi, Philippines, and NG; and Pacific off western S Amer. **Extralimital spp: Red Phalarope** (**Grey Phalarope**, *P. fulicarius*) breeds in Arctic, winters off western S Amer and W Africa, vagrant to AU and NZ. Larger, stockier, with a shorter, thicker bill. Nonbreeding and juv plumage similar to Red-necked. Summer plumage with chestnut underparts and white eye-patch.

COURSERS AND PRATINCOLES: Glareolidae

Some 17 species of coursers and pratincoles inhabit the warmer regions of the Old World. Only the pratincoles reach New Guinea, where two migratory species spend the winter, one from N Asia, the other from Australia. They are handsome, tern- or swallow-like waders, demurely colored in buffy tans with black-and-white markings. Strong, graceful fliers, they hawk insects from the air, and also run down their prey. Favored habitat is short grass and bare wasteland. Look for them at airfields and open plains.

Australian Pratincole *Stiltia isabella* Pl 34
21–24 cm. A common to abundant AU migrant, mainly south of the Central Ranges. Spends austral winter on open fields of short grass. Likely one of the first birds seen by visitors to PNG upon landing at the Port Moresby airport. *A small, sandy-colored, ternlike wader with short bill and long legs. Dark chestnut flanks look blackish.* **In flight:** Very long, tapered, *scythelike wings*; note the dark M formed by wings and breast that separates sandy head and white undertail. **Adult:** Buffy; bill red with back tip. **Juv:** More greyish; all-dark bill; obscurely mottled; flank patch duller, smaller, smudgy. **Similar spp:** See Oriental P. **Voice:** Often heard before seen in flight. Flight call a sweet 3–4-note whistle. Contact call a plaintive *tsoo wee*. **Habits:** Inhabits dry plains, airfields, burnt savannah. Sociable, forming large parties. Activity mainly crepuscular; also flies at night. Forages mostly on the ground, swiftly running after insects; also springs into flight to take fleeing prey and sometimes hawks. Feeds on insects, spiders, centipedes. **Range:** NG, NW Is (Misool, Salawati, Waigeo), and Aru Is, 0–1700 m. NG: Mainly Trans-Fly and coastal Port Moresby area (SE Pen), less common elsewhere south of Central Ranges and Markham Valley. Breeds AU; winters AU, NG, Wallacea to eastern Greater Sundas.

Oriental Pratincole *Glareola maldivarum* Pl 34
23–25 cm. A scarce Palearctic passage migrant and overwintering visitor during austral summer. Frequents expanses of sparse, short grass. A small, long-winged, short-legged *ternlike wader with pale throat patch, bordered by a thin black line in Adult.* **In flight:** Pale body, darker wings, forked white tail accentuated by a black terminal band. **Adult:** Nonbreeding plumage with neck and breast dull greyish olive; throat patch with streaked border; pale lores and eye-stripe; dark bill. Breeding plumage with buffy tinge on neck and breast; throat patch offset by a black border; black lores; red bill with dark tip. **Juv:** Upperparts and neck mottled, and feathers pale-fringed; lacks black throat border; bill all dark. **Similar spp:** Australian P has much longer legs, blackish flank patch. **Voice:** Alarm call ternlike, *tchik tchik tchik*. **Habits:** Habitat similar to Australian P. Singly or in small parties. Gregarious, associating with Australian P, curlews, and plovers. Feeds on the wing, catching insects in graceful ternlike flight. On the ground, forages with shuffling gait, but also runs well. **Range:** NG, mainly south of Central Ranges, 0–1700 m. Breeds in E Asia; winters SE Asia to AU.

TERNS AND GULLS: Laridae

Of the 102 species of terns and gulls worldwide, 17 terns and 2 gulls are known from New Guinea waters. These birds live along the coasts and out at sea, and a few species visit freshwater lakes, marshes, and rivers. Terns are small seabirds that are strong, swift fliers and very much at home in the air and over the ocean. Gulls are more robustly built, and although they soar and glide effortlessly, their slow flight is no match for the terns and thus gulls keep closer to shore. Terns are the most diverse and abundant seabirds in the region, and a few species breed locally, whereas the gulls are vagrants. They all feed on fish, and some species have broader diets that include crustaceans and other invertebrates; in addition, the gulls scavenge. Terns and gulls are social and often mix with their own and other species. They breed as monogamous pairs; males are larger on average or the same size, depending on species. When nesting, they seek remote islands away from humans and other predators. Unfortunately, few nesting sites are safe in the region, and there are no really large seabird colonies as are found off Australia and in Oceania. Most of the terns we see are visitors. A variety of terns can be found resting on small offshore islets or fishing over baitfish shoals driven to the surface by tuna. Two species are seen on inland rivers and lakes: Whiskered Tern and Gull-billed Tern.

Brown Noddy *Anous stolidus* Pl 10
(**Common Noddy**)
40–45 cm; wingspan 79–86 cm. Widespread and locally common resident over offshore waters.
Adult: A *brown tern* with a conspicuous *white forehead cap* and long, tapered, wedge-shaped tail that shows a shallow notch when fanned. *Upperwing-coverts distinctly paler* than rest of wing; *underwing pale greyish*. **Juv:** Thin white forehead mark. **Subspp (1):** *pileatus* (Pacific and Indian Oceans).
Similar spp: Black N smaller, more compact, and with a proportionately longer bill; appears darker overall because upperwing and underwing both darker than those of Brown N. Bulwer's Petrel has a short, hook-tipped bill and lacks white forehead. **Voice:** Call is a guttural *kar-r-rk* and *ok ok ok*.
Habits: Usually forages offshore, occasionally near coast. Feeds in large flocks at baitfish shoals, often with Black N and other sea terns. Parties rest on the open sea, floating buoyantly or perching on driftwood; gathers on small remote islands to roost in trees. Nests on such islands on the ground or in a shrub or palm; nest a substantial mat of vegetation. Egg (1) white with dark spots.
Range: All seas of NG Region. Also tropical seas worldwide.

Black Noddy *Anous minutus* Pl 10
(**White-capped Noddy**)
35–39 cm; wingspan 66–72 cm. Widespread and locally abundant resident over offshore waters.
Adult: *Smaller than Brown N and more compactly built*, yet with proportionately longer bill. Overall *blackish because of all-dark wings that lack the pale bar across the upperwing and the greyish underwing of Brown N*. Cap seems more gleaming white and extends further back onto the nape. May show pale greyish center to the uppertail (Brown N tail uniform). **Juv:** White cap extends only as far as the crown. **Subspp (1):** *minutus* (equatorial Pacific). **Voice:** Calls include a rattle *kik-krrrr*, screeching, cackling, and other calls. **Habits:** Similar to Brown N, but perhaps more frequent away from the mainland. Nests in trees; nest a pad of vegetation. Egg (1) white with dark spots.
Range: All seas of NG Region. Also tropical Pacific and Atlantic Oceans.

White Tern *Gygis alba* Pl 10
(**Fairy Tern**, **White Noddy**, *Anous albus*)
28–33 cm; wingspan 70–87 cm. Rare oceanic visitor off N coast. Sightings suggest possible breeding sites in Bay Is and SE Is. **Adult:** An easily recognized *all-white tern with moderately forked tail* and buoyant and graceful flight. *Large dark eye and dark slightly upturned bill*. **Juv:** Like Adult but more greyish, has dark splotch behind eye, and mantle, back, and upperwing-coverts with faint brown scalloping, the markings less evident than in other juv terns. **Subspp (1):** *candida* (tropical Pacific and Indian Oceans). **Similar spp:** Black-naped T has tail wires and black nape; coastal, not oceanic. **Voice:** *Heech, heech* and clicking sound. **Habits:** Solitary birds feed far out at sea, typically near their remote nesting islands. Does not build a nest; instead balances egg on a branch. Egg (1) white with speckles and blotches. **Range:** N coast, Bay Is, and SE Is (Budibudi Is); breeds on small islands to the north of the region. Tropical seas worldwide.

Silver Gull *Chroicocephalus novaehollandiae* Pl 10
(*Larus novaehollandiae*)
38–43 cm; wingspan 91–96 cm. Occasional AU visitor to S coast, especially Trans-Fly. **Adult:** *The only gull with all-white head and short bright red bill and legs.* **In flight:** *Wing tip black with white "mirror."* **Juv:** Similarly white-headed, but with dark bill and legs and black markings on the mantle, back, and wing-coverts. **Imm:** Black markings restricted to wing-coverts; bill and legs vary from dark to yellow. **Subspp (1):** *forsteri* (AU). **Similar spp:** See Black-headed G. **Voice:** A loud screaming *keeyow* and *kurr kurr kurr.* **Habits:** Drawn to human settlement to scavenge for discarded food and dead marine life; also captures small prey. Cruises coastline with low, soaring flight. Alights at beaches, towns, and dumpsites and walks with rapid mincing steps. **Range:** S coast, especially Torres Strait. Breeds AU.

Black-headed Gull *Chroicocephalus ridibundus* Pl 10
(**Common Black-headed Gull**, *Larus ridibundus*)
38–43 cm; wingspan 91–94 cm. Palearctic vagrant that turns up in harbors and sewage ponds. **Adult:** Nonbreeding plumage similar to same-size Silver G but with *black smudges around eye and on ear-coverts.* Breeding plumage unique with *blackish-brown head.* **In flight:** Broad white leading edge to the wing; lacks Silver G's black wing tip with white "mirror." **Juv:** Similar to Nonbreeding but with black markings on mantle, back, and wings. **Imm:** Black markings restricted to wing-coverts. **Habits:** Similar to Silver G. **Range:** All NG coasts, especially in W and N. Breeds in Eurasia; winters in S Eurasia, Africa, and Philippine Is.

Gull-billed Tern *Gelochelidon nilotica* Pl 11
(*Sterna nilotica*)
35–43 cm; wingspan 86–103 cm. Uncommon austral winter migrant from AU, mainly to S NG; also vagrant Palearctic winter migrant. Patrols tidal flats, estuaries, other coastal wetlands, and rivers. *Bill diagnostic: stout, black, gull-like.* Tail only slightly forked. **Adult:** Nonbreeding plumage *looks all white* (wings actually pale grey) with *dark trailing edge to wing;* note black smudge across eye and ear-coverts. Breeding plumage with black, crestless cap. **Juv:** Similar to Nonbreeding adult but mantle, back, and wings have black markings. **Imm:** Black markings reduced. **Subspp (2):** *macrotarsa* (also AU); *affinis* (E Asia, vagrant to NG Region) separable in the field—smaller, with proportionately smaller bill, less extensive black eye-patch, and upperparts a shade darker. **Similar spp:** No other tern has stout, black bill. **Voice:** Normally silent away from breeding colony. **Habits:** Solitary or in flocks; will join other terns. Found over fresh and salt water, mudflats, and grasslands. Flies slowly and buoyantly, head tilted downward, searching. Dives in a graceful swoop to pick up prey from the ground. Captures crustaceans, insects, and small vertebrates including fish and lizards. **Range:** All coastal NG. Breeds in Eurasia, Africa, AU, and the Americas.

Caspian Tern *Hydroprogne caspia* Pl 10
(*Sterna caspia*)
48–59 cm; wingspan 127–140 cm. A scarce visitor mainly to coastal Trans-Fly. *The largest tern; robust with thick, red bill—unmistakable.* **Adult:** Nonbreeding plumage with abundant white frosting on the black crown. Breeding plumage with heavy black cap forming a short crest above the nape. **Juv:** Similar to Nonbreeding adult, but with black markings on the mantle, back, and wings; darker primaries; cap darker than other juv terns. **Imm:** Similar to Nonbreeding adult, but retains darker Juv primaries. **Voice:** Normally silent. **Habits:** Solitary or in small parties on coast, lagoons, and large rivers. Flies gracefully with slow beats of long and powerful pointed wings; hovers, then dives, plunging into the water to capture fish. **Range:** S coast only, mainly Trans-Fly; origin of these birds is unknown, presumably AU. Breeds Eurasia, Africa, AU, and N Amer.

Crested Tern *Thalasseus bergii* Pl 11
(**Great Crested** or **Swift Tern**, *Sterna bergii*)
43–48 cm; wingspan 99–109 cm. A widespread and common sea tern that breeds locally. *A very large tern with dull yellow or greenish-yellow bill and black, shaggy crest.* **Adult:** Nonbreeding plumage with white forehead and forecrown merging with black cap. Breeding plumage with a *black cap sharply demarcated from white forehead.* **Juv:** Sooty black cap; black markings on mantle, back, and wings; and dark primaries. **Imm:** Similar to Nonbreeding adult, but retains dark Juv primaries. **Subspp (1):** *cristatus* (Pacific Ocean). **Similar spp:** See Lesser Crested Tern. **Voice:** Calls include

korrkorrkorr or *wep-wep* (AU data). **Habits:** Singly or in gatherings, sociable. Found along the coast and in estuaries, also out at sea; absent from freshwater. Plunges for fish, occasionally squid or prawns. Nests on small, remote islands; nest a shallow depression on the ground. Egg (1) white with dark markings. **Range:** All NG seas. Also Eurasia, Africa, AU, and Oceania.

Lesser Crested Tern *Thalasseus bengalensis* Pl 11
(*Sterna bengalensis*)
38–43 cm; wingspan 89–94 cm. A widespread but typically uncommon visitor to the coast; not known to breed. Plumages similar to Crested T, but Lesser Crested T *smaller, more compact, slightly darker above, and with a smaller, orange bill*. In breeding plumage, *black cap extends to bill* (no white forehead as in Crested T). Juv dorsal markings not as bold. **Subspp (1):** *torresii* (Pacific and Indian Oceans). **Habits:** Similar to Crested T, but usually further from mainland, near islands and reefs. **Range:** All NG seas. Also subtropical waters off Eurasia, Africa, and AU.

Little Tern *Sternula albifrons* Pl 11
(**Least Tern**, *Sterna albifrons*)
20–28 cm; wingspan 50–55 cm. Common visitor during austral summer to inshore coastal waters, lagoons, and swamps not far inland; most birds are probably Palearctic migrants, but some may originate in AU or Bismarck and Solomon Is. Not known to breed in NG. *Smallest NG tern.* Note *dark leading edge to wing* (first few primaries). **Adult:** Nonbreeding plumage with white forehead and cap and black bill. Breeding plumage with *white forehead, short white eyebrow, and yellow bill with black tip*. **Juv:** Similar to Nonbreeding adult but with light-and-dark scalloping on the mantle, back, and wing-coverts (less so). **Imm:** Resembles Nonbreeding adult, but with some Juv dorsal scalloping. **Subspp (1):** *sinensis* (also E Asia to Solomon Is). **Similar spp:** No other small tern has blackish primaries, rapid winnowing wingbeats, and black cap. **Voice:** A chattering *chik chik chik* vaguely recalling a reed-warbler. **Habits:** Solitary or in flocks. Flight diagnostic: flies with rapid wingbeat, suddenly stops and hovers, then dives for small fish. **Range:** Throughout NG Region. Also Eurasia, Africa, and AU.

Grey-backed Tern *Onychoprion lunatu* Pl 10
(**Spectacled Tern**)
35–38 cm; wingspan 73–76 cm. Rare over ocean far from shore. Similar to the much more common Bridled T but *paler grey on mantle, back, and wings* and *less white in the outer tail*. **Juv:** Paler than Bridled juv, but not safely distinguished at sea. **Range:** Sightings from SW of Waigeo, Astrolabe Bay, and Bismarck Sea. Tropical Pacific, away from continents and larger islands.

Bridled Tern *Onychoprion anaethetus* Pl 10
(**Brown-winged Tern**, *Sterna anaethetus*)
35–38 cm; wingspan 76 cm. Common resident and local breeder, usually found near coast and islands. **Adult:** *The grey-brown back and length of eyebrow distinguish it from the black-backed Sooty T.* Diagnostic grey-brown mantle, back, and wings contrast with black cap. White eyebrow extends well past eye, creating black eye-stripe. Long, wiry tail has white outer tail-streamers. Pale grey wash on lower breast and belly. **Juv:** Similar, but tail-streamers shorter; black cap less defined; mantle, back, and wing-coverts with white and buff markings; side and flank grey. White eyebrow and white throat and breast distinguish it from black-hooded juv Sooty T. **Imm:** Intermediate. **Subspp (1):** *anaethetus* (W Pacific). **Similar spp:** See Grey-backed T. **Voice:** Alarm call *wep wep*. High pitched growl and other calls at nesting colony. **Habits:** Singly or in flocks. Flight buoyant with slow wingbeat. Forages well out at sea. Takes fish by swooping and briefly dipping into the water and also by plunging from several meters height. Nests in small, loose colonies on safe offshore islets. Nest in rock cavity or scrape on the ground. Egg (1) white with dark spots and blotches. **Range:** All NG seas; breeds locally on islets off SE Pen, SE Is, and probably elsewhere. Also tropical and subtropical seas of the world, though absent from much of Oceania.

Sooty Tern *Onychoprion fuscatus* Pl 10
43–45 cm; wingspan 86–94 cm. A common visitor that typically lives over the open ocean far from shore, yet occasionally wanders inland. **Adult:** *Black above, white below, with long black-and-white outer tail-streamers.* Similar to Bridled T, but *wholly black above except for a prominent white forehead*

triangle, pure white below (no grey cast to underparts), and white outer web to the long outermost tail-streamers. Nonbreeding birds have some grey scaling on back. **Juv:** *Distinct—black with white breast and belly and white spots on back and wing-coverts. Forked tail lacks streamers.* **Imm:** Intermediate. **Subspp (1):** *serrata* (Moluccas to AU). **Similar spp:** Bridled T. Brown Noddy and Black N differ from the dark juv Sooty T by their pale foreheads, absence of pale spotting on the upperwing, and dark undertails. Juv Sooty T has been confused with a petrel or storm-petrel, but has very different flight, shape, etc. **Voice:** A loud *kirrak rak*. Rather quiet at sea by day; birds call when traveling over land on dark, cloudy nights. **Habits:** Similar to Bridled T, but more gregarious and pelagic. Usually encountered in flocks at sea, foraging on fish shoals with other marine terns. **Range:** All NG seas; a few records of storm-driven birds from the Central Ranges. Tropical seas worldwide; breeds in vast numbers on remote oceanic islands.

Roseate Tern *Sterna dougallii* Pl 11

35–43 cm; wingspan 76–79 cm. Local and uncommon breeder, usually found in vicinity of offshore nesting islands and in association with coral reefs and clear, "blue" water; avoids coastal waters. *Bill unusually long, slender, lance-like.* **Adult:** *Long all-white tail-streamers, whitish upperparts, shallow wingbeat, wholly white underwing, and red legs and feet*. Nonbreeding plumage with white forehead and forecrown; black bill. Breeding plumage with all-black cap, rosy bloom to breast; bill changes to red during nesting. **Juv:** Sooty cap; black scalloping on mantle, back, and wings; tail forked but without streamers; black legs and feet. Leading edge to wing not nearly as dark as in juv Common T, tips of secondaries not particularly dark, and tail lacks dark sides. **Imm:** Intermediate. **Subspp (1):** *bangsi* (Arabian Sea to W Pacific). **Similar spp:** Common T has dark sides to the tail in all plumages, lacking in Roseate. Black-naped T occupies similar habitats but is smaller, lacks tail-streamers, and has black eye-stripe and nape bar. **Voice:** *aaark, aaark*. **Habits:** Singly or in small parties. Joins other terns at fish shoals. Forages within a few kilometers of coral reef. Takes fish by dipping or plunging. Nests on small, sandy islands; nest a shallow scrape. Egg (1) off-white with dark spots and blotches. **Range:** Scattered sightings throughout NG Region; known to breed off SE Pen, likely elsewhere. Tropical to temperate seas near continents worldwide, except E Pacific.

Black-naped Tern *Sterna sumatrana* Pl 11

30–32 cm; wingspan 61 cm. Locally common near breeding islands on outer reefs and atolls, rare or absent elsewhere. **Adult:** A *small sea tern, looks immaculate white* but for a *crisp, crescent-shaped, black nape band* from eye to eye and a thin black leading edge to the outermost primary. *Forked tail lacks streamers.* Bill and legs black. Breeding plumage with rosy blush on breast. **Juv:** Similar but characteristic head pattern somewhat blurred and has sharp black scalloping on mantle, back, and wings and grey outer primaries. **Imm:** Like Adult but retains Juv's darker primaries. **Subspp (1):** *sumatrana* (Pacific and E Indian Oceans). **Similar spp:** Roseate T and Common T adults are larger, have a black cap or nape patch (not band), and the outer 3 primaries are dark; juvs with darker cap and greyer mantle, back, and wings. **Voice:** A sharp *chit, chit, chit*. **Habits:** In small flocks, often with other terns. Essentially pelagic, occasionally along the coast, but never on freshwater. Flight similar to Common T but with quicker wingbeat. Takes fish by swooping and snatching or by diving. Nests on small islands; nest a shallow depression on rock or sand. Eggs (1–2) white with dark spots and blotches. **Range:** All NG seas; breeds off SE Pen and SE Is, probably elsewhere. Tropical W Pacific and Indian Oceans.

Common Tern *Sterna hirundo* Pl 11

32–38 cm; wingspan 79–81 cm. Common to locally abundant Palearctic winter migrant along NG coasts between October and April, with many nonbreeders remaining. **Adult:** *Note the darker outer primaries; white outer tail feather with thin, black edge; and black legs and feet*. Nonbreeding plumage with white forehead and forecrown and well-forked tail but no streamers. Breeding plumage with black cap and long tail-streamers. **Juv:** Similar to Nonbreeding plumage but with darker carpal and forewing and diagnostic darker trailing edge to secondaries. **Imm:** Like Nonbreeding adult, but retains Juv flight-feathers. **Subspp (1):** *longipennis* (breeds NE Asia, winters SE Asia to AU) bill black in breeding plumage unlike red bill with black tip of W Eurasian and American race *hirundo*, which is vagrant to AU and hypothetical for NG. **Similar spp:** Roseate T much paler above, has mostly white primaries, all white tail (no dark edge), and red legs and feet. It also has a slimmer shape and longer bill. **Voice:** A strident *kerr yah* and *kik kik kik* or a subdued *kek* while feeding.

Habits: Mainly coastal, where it is the predominant tern. Also visits freshwater. Sociable, mingling with other terns, often in large flocks. Dives for fish in sea and also picks food off surface in freshwater. **Range:** All NG coastal waters and locally inland. Breeds in N Eurasia and NA, winters in tropical and subtropical coastal seas of the world. **Extralimital spp: Arctic Tern** (*S. paradisaea*) breeds in boreal and Arctic regions and migrates to antarctic waters; difficult to distinguish from Common T in nonbreeding plumage—the best field mark is the paler wing with thin black tips to the primaries forming a crisp black border; distinguished in breeding plumage by all-red bill.

Whiskered Tern *Chlidonias hybrida* Pl 11
(Marsh Tern)
25–26 cm; wingspan 69 cm. The common tern of inland waters. AU migrant during austral winter to shallow bodies of freshwater and occasionally muddy beaches. *A small tern with shallow-forked tail that becomes square or rounded as it is increasingly fanned*; long bill and head shape similar to Common T. Largest of the 3 marsh terns. **Adult:** Nonbreeding plumage *evenly grey across the back to tail* (no contrasting white rump as in White-winged T); white forehead, crown, and underparts; black bill and feet. Breeding plumage unmistakable—*white moustachial stripe separating black cap from dark grey underparts*; red bill and feet. **Juv:** Similar to Nonbreeding adult but for mantle and back heavily marked with black and buff; darker wings. **Imm:** Similar to Nonbreeding adult but retains Juv flight-feathers. **Subspp (1):** *fluviatilis* (breeds AU); *swinhoei* (E Asia) hypothetical, not separable in the field. **Similar spp:** White-winged T and Black T—see those accounts. **Voice:** A dry *krrreck*, like a rasping file. **Habits:** Singly or in small flocks; feeds and roosts socially, often with other species. Forages over rivers, lakes, marshes, sewage ponds. Flies gracefully with slow wingbeats, swooping down to pick food from the surface of the water, occasionally plunging into the water. Hawks insects. Feeds on insects, fish, and other small animal life. **Range:** NG: mainly lakes and marshes of Sepik-Ramu, S Lowlands, and Trans-Fly, 0–500 m. Breeds in Eurasia, N Africa, and AU, winters in the tropics.

White-winged Tern *Chlidonias leucopterus* Pl 11
(White-winged Black Tern)
22–25 cm; wingspan 66 cm. Uncommon Palearctic migrant during austral summer over shallow fresh and salt water. Very similar to Whiskered T, but *smaller and with a shorter, thinner bill*. In all plumages shows a *white rump contrasting with the slightly darker back*. **Adult:** Nonbreeding plumage similar to nonbreeding Whiskered T except has white rump and paler tail and darker head markings. Breeding plumage unique: *black with whitish wings and tail* and black underwing-coverts. **Juv:** From juv Whiskered T by White-winged's mostly black mantle and back. **Imm:** Retains Juv darker flight-feathers. **Similar spp:** Whiskered T and Black T. **Voice:** Noisier and less rasping than Whiskered T; *kreek-kreek* and *keek keek keek*. **Habits:** Similar to Whiskered T. **Range:** All NG Region, but mainly near coast. Breeds in Eurasia; winters in Africa, SE Asia to AU.

Black Tern *Chlidonias niger*
22–24 cm; wingspan 66 cm. Vagrant Palearctic migrant during austral summer. **Adult:** Differs from White-winged T in nonbreeding plumage by *the darker and more extensive black cap and evenly grey back, rump, and tail* (no white rump). In breeding plumage by *evenly dark wings* and black-and-white (not all black) underwing-coverts. **Juv:** Grey rather than white rump. **Subspp:** No information. **Range:** 1 record from Port Moresby (SE Pen). Breeds from Europe to central Asia and in N Amer; winters in Africa and Central and S Amer.

JAEGERS AND SKUAS: Stercorariidae

The jaegers and the larger, more robust skuas comprise a family of 7 species of polar seabirds best known for their predatory and piratical habits. All 3 jaeger species and one skua species have been recorded in New Guinea waters. The jaegers are northern trans-equatorial migrants, breeding in the Arctic and migrating south in the Palearctic autumn through NG waters to reach wintering grounds in subtropical and temperate seas of the Southern Hemisphere. The South Polar Skua migrates in the opposite direction: it breeds around Antarctica and travels to northern oceans for the austral winter. Migrating jaegers and skuas keep to the open ocean, but wintering birds may wander close to the coast. All species are opportunistic predators and scavengers, and they also pursue terns and gulls to steal their prey. Jaegers and skuas forage on both land and sea, usually at sea during migration and winter, when they feed mainly on fish.

Jaegers are difficult to identify because of the similarity between species and the confusing variability in their plumages. Each species has pale and dark color morphs. Jaegers also have breeding and nonbreeding plumages. The breeding plumage is unbarred and sports long tail-streamers. The nonbreeding plumage is similar except for the addition of barring on the flanks and tail-coverts (both upper and lower) and the lack or reduction of the tail-streamers. Finally, young birds pass from a variable, profusely barred juvenile plumage without tail-streamers (look for barring in the underwing, absent in adult), through intermediate immature plumages, to adult plumage. Thus, while identification is aided by attention to plumage details, it often begins with recognizing a bird by its species-specific size, shape, and structure.

South Polar Skua *Stercorarius maccormicki* Pl 9
(*Catharacta maccormicki*)
53 cm; wingspan 127 cm. Rare trans-equatorial, southern migrant during austral spring and autumn. *Larger, more robust, and broader-winged than a jaeger and lacks projecting central retrices.* Only pale morph reported from Australasian waters. **Adult:** Pale morph *with diagnostic pale grey-buff head and underparts contrasting with dark underwing and upperparts.* (*Dark morph* all dark with prominent pale hindneck.) **Juv:** Like Adult pale morph but more greyish, and bill with grey base. **In flight:** Pale underparts contrast with dark underwing-coverts. **Habits:** See family account. **Range:** Several records from Coral, Solomon, and Bismarck Seas. Breeds along the Antarctic coast, winters mainly in N Pacific and Indian Oceans. **Extralimital spp: Brown Skua** (*S. antarcticus*) breeds in sub-antarctic islands and winters in temperate southern seas; adult and juv much darker than South Polar S.

Pomarine Jaeger *Stercorarius pomarinus* Pl 9
(**Pomarine Skua**)
65–78 cm, including tail-streamers that project up to 10.5 cm; wingspan 122–127 cm. A fairly common, overwintering Palearctic migrant during austral summer. *Larger, heavier, and somewhat wider-winged than other jaegers.* Appears small-headed and barrel-chested, with flight like that of a heavy gull. Pale flashes in wings are caused by *white shaft streaks in nearly all primaries* and *a pale panel in the underwing at the base of the primaries.* Both pale and dark morph plumages are difficult to distinguish by plumage from the other, less common jaegers. **Adult:** In Nonbreeding plumage, the *tail-streamers are short and narrow, but are nevertheless wider and more blunt than those of other jaegers*; note barred rump, flanks, and tail-coverts. In breeding plumage, the diagnostic elongated central tail feathers are *broad, twisted, and round-tipped.* **Juv:** Central tail feathers form a mere nubby projection at the tail tip. Both pale and dark morphs barred in the underwing, a characteristic of all juv jaegers. **Imm:** Intermediate between Juv and Adult, difficult to age. **Habits:** See family account. **Range:** Winters on seas regionwide, but especially off N coast. Breeds in Arctic; winters in tropical and subtropical waters.

Arctic Jaeger *Stercorarius parasiticus* Pl 9
(**Parasitic Jaeger, Arctic Skua**)
46–67 cm, including tail-streamers that project up to 10.5 cm; wingspan 96–114 cm. Rare passage migrant from the Arctic during austral spring and summer. *Smaller and more falcon-shaped than Pomarine J, with narrower wings and swift flight.* Pattern of white wing shafts and pale underwing same as in Pomarine. Has pale, intermediate, and dark morphs. **Adult:** In Nonbreeding plumage, the *central tail feathers are much shorter, but still pointed and may project visibly*; heavy light-and-dark scalloping on rump and tail-coverts. In breeding plumage, *tail-streamers narrow, pointed, and of moderate length.* **Juv:** Similar to Nonbreeding adult, but pointed tail-streamers barely project and underwing-coverts heavily barred. **Imm:** Intermediate between Juv and Adult, difficult to age. **Habits:** See family account. **Range:** Passage migrant off N NG coast. Breeds in Arctic; winters in southern subtropical and temperate seas.

Long-tailed Jaeger *Stercorarius longicaudus* Pl 9
(**Long-tailed Skua**)

50–58 cm, including tail-streamers that project up to 18 cm; wingspan 76–84 cm. Rare passage migrant from Arctic. *Smallest jaeger, slender and ternlike, with long, narrow wings.* Upperwing with *only 2 white shafts in outer primaries* (many more shafts white in other jaegers). Underwing has *all-dark primaries*, like rest of underwing (other jaegers have pale panel at the base of the primaries). Only pale morphs reported in the region; dark morph rare in this species. **Adult:** In Nonbreeding plumage the central tail feathers barely project and may not be visible; heavily barred rump, flanks, and tail-coverts; the head typically has a hooded appearance without a dark cap, but varying from dark to pale. *Tail-streamers very long in Breeding plumage.* In both Breeding and Nonbreeding plumage, the pale morph is distinctive for *pale grey upperwings with dark trailing edge.* **Juv and Imm:** Similar to juv Arctic J, but paler, especially on the head. Look for few white wing shafts and lack of underwing panel. **Subspp (2):** No information. **Habits:** See family account. **Range:** Passage migrant off NG coast, mostly in the N. Breeds in Arctic; winters southern temperate and sub-antarctic waters.

PIGEONS AND DOVES: Columbidae

The pigeons and doves constitute a large and distinctive family (338 spp) without close relatives. Strong fliers, they have found their way around the world, yet two regions—South America and Australasia—hold most of the family's diversity and seem to have been its evolutionary homelands. The family attains its greatest diversity of size, form, and ecology in the New Guinea Region, where 52 species reside. Here, pigeons utilize all forested and woodland habitats. Lowland rainforests support the greatest diversity—more than 20 species inhabit some localities. New Guinea's pigeons range in size from the world's largest pigeon, the almost turkey-sized Southern Crowned Pigeon, to perhaps the smallest, the Dwarf Fruit-Dove. Most types of pigeons eat seeds gleaned from the forest floor or in some cases taken as green or ripe fruit, digesting them seeds and all. However, the fruit-doves and imperial pigeons are fruit specialists that digest only the nutritious ripe fruit pulp and pass the seeds intact. Terrestrial species occasionally consume insects, spiders, and snails. Although restricted in diet, pigeons are opportunistic in foraging and breeding. Most appear to either maintain large home ranges or be nomadic, responding to spatial and temporal availability in the abundance of food. The nest is a simple platform, either skimpy and frail (fruit-doves) or substantial but untidy (ground-doves). Both parents share all nesting duties. One or 2 white eggs are laid. Parents feed the chicks a mixture of regurgitated vegetable material or secretions from the lower esophagus ("crop-milk"). In many species, chicks leave the nest well before attaining adult size, and they are cared for by the parents away from the nest.

For convenience we recognize these general types of pigeons and doves in New Guinea:

Feral Pigeon, this worldwide domesticated species is restricted to cities and towns in our region. The **White-throated Pigeon** is related to the preceding species, and it forages in trees for unripe fruits or on the ground for seeds.
Cuckoo-Doves (4 spp) are medium-sized to large, brown or brown and grey, with a long tail, and swift, graceful rowing flight; they forage in trees mainly for unripe fruit.
Emerald Doves and Ground-Doves (6 spp) are small, chunky pigeons with long legs and cryptic terrestrial habits, feeding mainly on fallen seeds and fruits.
Typical Doves (2 spp) are small grey-and-buff long-tailed forms that forage on the ground in open lowland habitats.
Ground-pigeons (4 spp) are a mixed group of large to very large forest-dwelling terrestrial feeders.
Crowned Pigeons or **gouras** (3 spp) are giant terrestrial pigeons that sport spectacular fan-shaped crests.
Fruit-Doves (16 spp) are small to medium-sized arboreal pigeons with cryptic green plumage, often marked with bright colors; they are found most often in fruiting trees, especially figs.
Imperial Pigeons and the **Papuan Mountain-Pigeon** (14 spp) are large arboreal species that consume fruit, and many species are renowned for their nomadic ways and long commutes to colonial nesting sites safe on remote islets or mountain tops.

These groups appear to have had different origins and histories within our region, as indicated by molecular data and biogeography. The White-throated Pigeon, cuckoo-doves, and emerald doves appear to have been late arrivals from Asia via Wallacea, whereas all the other lineages are of greater antiquity and arose within New Guinea and Australia.

In most pigeon species, the sexes share similar plumage and are of nearly the same size, with males larger on average, although a few species (especially the fruit-doves) are sexually dichromatic. Juveniles often resemble their parents but usually have pale tips to the feathers, showing a scalloped effect. Identifying juvenile fruit-doves can be especially problematic. These youngsters abandon their flimsy nest at an amazingly early age, when they are only a fraction of the size of their parents. They live secretly, hidden in the tree foliage, and are almost never seen. Mostly a cryptic dull green, they offer few if any clues to their identity. We have not attempted to describe juvenile fruit-doves.

Rock Dove (Feral Pigeon) *Columba livia*
(**Common** or **Domestic Pigeon**, or just **Pigeon**)
30–34 cm. Introduced to cities and towns, where widespread but local and uncommon. This is *the domesticated pigeon of urban areas*, descended from the wild Rock Dove of Eurasia. Often seen in tight, wheeling flock. Glides briefly with wings held in an upward V. *Grey with iridescent neck and white rump patch, but many color variants exist*. Local populations characteristically of mixed colors: typical grey, white, pied, brown, or black. **Male** noticeably larger and with more iridescence on neck. **Juv:** Duller than Adult; eyes and feet duller, darker. **Similar spp:** No other large ground-feeding pigeon inhabits towns. Torres Strait Imperial Pigeon, which stays in the trees, is larger and has black primaries and tail-band, unlike albino Domestic P. **Voice:** Rolling *oo-rroo coo-coo*. **Habits:** Loiters around markets and town parks. Some people maintain flocks; this has been the source of NG's feral birds. Feeds on seeds and waste food taken from ground. Highly social as pairs or in flocks. Untidy nest on building ledges, cliffs. Eggs (2) white. **Range:** NG, Numfor, Aru Is, and probably other islands, 0–1600 m. Native to S Eurasia and N Africa; introduced around the world.

White-throated Pigeon *Columba vitiensis* Pl 41
(**Metallic Pigeon**)
36–37 cm. Rarely seen, nomadic, and little known. Widespread but sparsely distributed through lowland and montane forest. A *large, blackish pigeon* with a clear *white throat patch* extending to cheek. In good light, note maroon-grey breast and neck, and *glossy green scalloping on wings and rump*. **Male** may be slightly brighter. **Juv:** Duller than Adult, nearly without iridescence. **Subspp (1):** *halmaheira* (Moluccas to Bismarck and Solomon Is). **Similar spp:** Papuan Mountain-Pigeon is slightly smaller and has a white breast and whistling wings.
Voice: Rarely heard; 2 or 3 slow, very low-pitched notes, the first higher and upslurred.

Habits: Solitary or in small flocks in forest; sometimes flushed from the ground. Wings do not whistle in flight. Feeds on fruits and seeds in canopy, subcanopy, and also visits forest floor. As reported elsewhere in range, nest is built 3–6 m up in a tree; a scant stick platform. Egg (1) white. **Range:** NG, NW Is (Batanta, Misool, Salawati, Waigeo), and SE Is (Goodenough, Fergusson, Misima, Tagula, Rossel), 0–2750 m. Also Philippines, E Wallacea, Bismarck Is, and Oceania.

Spotted Dove *Streptopelia chinensis* Pl 35
(**Spotted Turtle-Dove**)
28–30 cm. Introduced to Indonesian NG. A bird of towns and roadsides in agricultural and disturbed habitats. Avoids forest. A medium-sized, *grey-brown dove* with moderately long tail. When flushed, the flared tail shows *black outer feathers with white tips*. **Adult:** *Rosy breast* and *unique black collar with a myriad of small white spots*. **Juv:** Lacks these features. **Subspp (1):** *tigrina* (SE Asia) identified by photos from Sentani. **Similar spp:** Cuckoo-Doves browner and with longer tails. **Voice:** Song *cooo-cooo coo!* **Habits:** Solitary, in pairs, or may gather in numbers, but does not travel in flocks. Feeds mostly on the ground, sometimes in trees, taking seeds and small fruits. Nests on a horizontal branch of tree or shrub. Eggs (2) white. **Range:** So far reported only from Sentani (NW Lowlands) and Biak (Bay Is), but expected to spread. Native to SE Asia (India to China and Greater Sundas), introduced widely in AU and Oceania, locally elsewhere.

Great Cuckoo-Dove *Reinwardtoena reinwardti*   Pl 35
(Giant Cuckoodove, *R. reinwardtsi*)
48–53 cm. Widespread but uncommon in forest from foothills and vicinity up into mountains. Largest cuckoo-dove, with distinctive *elongate shape, including very long tail*. **Adult:** *Pale grey head, neck, and underparts contrast with the dark chestnut mantle, wing, rump, and tail*. Very short hind-crest yielding big-headed profile. **Juv:** Sooty brown, often with some grey feathers on breast. **Subspp (2):** *griseotincta* (NG range except Biak); *brevis* (Biak I) quite different—smaller, with white head and underparts and blackish wing-coverts. **Similar spp:** Other cuckoo-doves are much smaller. **Voice:** Two very different songs: a repeated, upslurred *cookuwook cookuwook cookuwook* . . . very similar to that of Brown CD, but sometimes slower; and a *hoo* followed by a rapidly descending series of *hoos*, suggestive of insane laughter. **Habits:** Passes through the canopy rather than over it; flight swift and graceful, with deep, powerful, rowing wingbeats. A solitary fruit eater favoring umbrella plants (*Schefflera* spp) and related trees in the Ginseng family. Pair or small group may come together in fruiting tree. Usually in the canopy but also descends to ground. Nest placed a few meters above ground in tree or, reputedly, on steep rock face; a platform of sticks, roots, moss, and ferns. Egg (1) white. **Range:** NG, NW Is (Batanta, Misool, Salawati, Waigeo), Bay Is (Biak, Yapen), Manam and Karkar Is, and SE Is (D'Entrecasteaux: Goodenough and Fergusson), 0–1600 m, occasionally to 3000 m. Few records from Trans-Fly. Also Moluccas. **Taxonomy:** The Biak form may be a separate species.

Brown Cuckoo-Dove *Macropygia amboinensis*   Pl 35
(Slender-billed Cuckoo-Dove, Brown Pigeon)
34–37 cm. Widespread from lowlands to mid-mountains; *the common brown pigeon of forest, edge, and gardens. Long, brown, unbarred tail*. **Male:** Rosy grey head and neck; faint greenish iridescence on hindneck; much paler below. **Female:** Entirely brown; head barred and speckled; more heavily barred on breast; lacks neck iridescence. **Juv:** Like Female but wing-coverts with black subterminal bars and rusty fringes to feathers. **Subspp (8):** Male nearly unbarred in East to heavily barred below in West; distinctly barred in Louisiades. **Similar spp:** Black-billed CD smaller, slimmer, darker brown, with shorter black bill and tail barred with black above; male is deep rusty brown, female heavily barred. **Voice:** A series of identical, upslurred disyllabic hoots: *woo-up woo-up woo-up* . . . repeated monotonously at a rate of 6 phrases per 10 sec. One call of Great CD similar from a distance. **Habits:** Frequents forest and regrowth in lowlands, but above ~1500 m confined mainly to disturbed habitats. Solitary or in pairs, gathers in small parties to feed in trees. Feeds on small fruits and seeds; mainly arboreal. Nests in vines or in a small tree; stick nest rather bulky. Egg (1) white. **Range:** All NG and most islands, except Karkar, Trobriands, and Woodlark, 0–1800 m, rarely to 2300 m. Also Sulawesi and Moluccas east to Bismarck Is. **Taxonomy:** Some authorities split this species into Slender-billed CD (*M. amboinensis*) of Wallacea, NG Region, and Bismarck Is and Brown CD (*M. phasianella*) of AU; however, differences are small, and an intermediate population on Cape York, AU, bridges the two taxa.

Black-billed Cuckoo-Dove *Macropygia nigrirostris*   Pl 35
(Bar-tailed Cuckoo-Dove, Rusty Cuckoodove)
29–30 cm. Widespread from lowlands to mid-mountains where locally common. Ecologically similar to Brown Cuckoo-Dove, but in mountains it is the dominant or sole species of cuckoo-dove inhabiting the forest interior. Smaller than Brown CD, with a *short, stout black bill and narrow tail finely barred with black*. **Male:** Entirely rich rusty brown; red orbital skin. **Female:** *Heavily barred with black on back and breast*. **Juv:** Like Female but tail irregularly barred. **Similar spp:** Mackinlay's CD, restricted to Karkar I, is like male Black-billed but with bill more slender and tail not barred. **Voice:** Easily distinguished from that of preceding species by being higher-pitched and rapid. A rapid, descending series of ~12 muted *woi* or *koik* notes, delivered at a rate of 3/sec, decreasing slightly in volume but with no change in rate; like subdued call of Pheasant Coucal but not rising at end. Can be confused with calls of some fruit-doves, from which this species differs in faster delivery, "hard" quality, and lack of acceleration. **Habits:** Similar to Brown CD. Nests on ground on steep slopes, in epiphytic ferns, and in crowns of tree ferns; stick platform. Eggs (2) white. **Range:** NG, Bay Is (Yapen), Karkar I, and SE Is (D'Entrecasteaux: Goodenough and Fergusson), 0–2600 m. All NG except Trans-Fly. Also Bismarck Is.

Mackinlay's Cuckoo-Dove *Macropygia mackinlayi* Pl 35
(Spot-breasted Cuckoo-Dove)
27–30 cm. Karkar I only, mainly in forest interior. *Tail not barred.* Bird overall *bright chestnut without barring* (except Juv). *Neck feathers have unique V-shaped indentations at their tips*, giving a spotted appearance. Individually variable (including grey morph in extralimital populations). **Female** possibly less rufous, more yellowish brown. **Juv:** With distinct barring. **Subspp (1):** *arossi* (Admiralty and Bismarck Is). **Similar spp:** The other small cuckoo-dove on Karkar, the Black-billed CD, is rare; it has a stubbier black bill and a barred tail; female with much barring below.
Voice: A melodious, mellow, repeated, 2-note call *vo ku . . .* , the first syllable higher than second. **Habits:** Similar to Black-billed. Nests in a small tree or palm; stick platform. Egg (1) white.
Range: An insular species reaching NG Region only on Karkar I, with a single sight record on NG, at Bogia (Sepik-Ramu), 0–1200 m. Also Bismarck, Solomon, New Hebrides, and neighboring islands.

New Guinea Bronzewing *Henicophaps albifrons* Pl 36
(White-capped Ground Pigeon, Jungle Bronzewing Pigeon)
33–36 cm. Widespread but uncommon in lowland and lower montane forests. *A large, dark ground-pigeon* with a conspicuously *whitish forehead and disproportionately long bill.* Bronzy iridescence on wing. **Male:** Crown white. **Female:** Crown dingy white to buff. **Juv:** Lacks iridescence. **Subspp (2):** *albifrons* (all of range except Aru); *schlegeli* (Aru Is), underparts dark like back.
Similar spp: Could be confused with emerald doves because of the white forehead and iridescent green wing; however, the bronzewing is much larger and darker and is proportioned differently, with longer tail and bill. Other ground-dwelling pigeons its size lack the prominent whitish forehead.
Voice: A rapid series of 20–60 upslurred whoops suggestive of Brown Cuckoo-Dove but much faster and not disyllabic. Rarely a throaty descending *krrrrrrr*, repeated often. **Habits:** Solitary or in pairs, foraging on the floor of primary forest as well as dry second growth. Quietly flushes from ground and perches on saplings, bobbing its head. One nest was built on a branch 1.5 m over a slow stream; nest was a substantial stick platform. Egg (1) white. **Range:** Endemic. NG, NW Is (Misool?, Salawati, Waigeo), Yapen (Bay Is), and Aru Is, 0–2000 m.

Cinnamon Ground-Dove *Gallicolumba rufigula* Pl 36
(Golden-heart Dove)
22–24 cm. Smallest of the lowland ground-doves, common and widespread in forest and second growth, but difficult to see. Occasionally glimpsed as a brown dove bursting up from the ground, with pale tail tip formed by the outer tail feathers visible in flight. **Adult:** *Clay-brown with bright white underparts* stained yellow. Its *large, dark eye* stands out against the *pale face. Four grey wing bars.* **Juv:** Underparts brown. **Subspp (5):** Subtle differences in grey shading. **Similar spp:** Bronze GD darker, lacks wing bars, has two-toned underparts. **Voice:** Song a faint, upward-inflected, froglike trill: *br r r r r r. . . .* **Habits:** Ground dwelling and solitary in the forest interior. Forages in the leaf litter and on bare earth, taking small seeds. Nests low in the forest, ~1 m off the ground on a firm foundation such as a palm frond or bird's nest fern; nest of flat dead leaves and a few sticks. Egg (1) white. **Range:** Endemic. NG, NW Is (Batanta, Misool, Salawati, Waigeo), Bay Is (Yapen), and Aru Is, 0–1600 m.

White-bibbed Ground-Dove *Alopecoenas jobiensis* Pl 36
(White-breasted Ground-Dove)
24–25 cm. A widespread, nomadic ground-dove of unpredictable occurrence from foothills to mid-elevations. *Distinctive pattern: black with white brow and throat.* In good light, shows purplish gloss on back. Tail longer and more squared off than other ground-doves. **Male:** Bright white brow, throat, and breast. **Female:** Variable—some like Male, others duller with white areas dingy. **Juv:** Dark grey-brown with faint rufous edging to feathers, small purple wing patch, and whitish chin. **Subspp (1):** *jobiensis* (widespread). **Similar spp:** Juv may resemble Bronze GD, which is smaller. **Voice:** Little information. Call a weak, hoarse, froglike note. **Habits:** Solitary or in pairs. Numbers fluctuate locally: often absent, usually scarce when present, but sometimes common. Gathers in forests with mast crops of oak or bamboo seeds. Takes seeds from the forest floor, but also arboreal to some degree, feeding on fruits and seeds there. No nesting information. **Range:** NG, Yapen (Bay Is), Manam and Karkar Is, and SE Is (D'Entrecasteaux: Goodenough and Fergusson), 0–1600 m, sometimes higher. All NG except Trans-Fly and possibly Bird's Head. Also Bismarck and Solomon Is.

Bronze Ground-Dove *Alopecoenas beccarii* Pl 36
(Beccari's Ground Dove)
18–20 cm. Uncommon and widespread in montane forests. *A tiny, compact ground-dove; dark with grey head.* **Male:** Glossy maroon shoulder patch. **Female:** Overall duller, no shoulder patch. **Juv:** Brown edges to feathers, especially on wing. **Subspp (2):** *beccarii* (NG mts) dark olive green; *johannae* (Karkar I, also Bismarck Is) dark brown, note white eye-ring. **Similar spp:** Cinnamon GD has pale wing bars and whitish underparts; lowlands. White-bibbed GD juv similar but larger, usually mottled with grey in the face and throat. **Voice:** Song a series of 5 low, clear notes uttered from the ground or a low perch, *cook cook cook cook cook*. **Habits:** A terrestrial bird of the forest interior. Usually solitary, rarely in pairs. Forages for seeds on the forest floor. Visits terrestrial courts of birds of paradise and bowerbirds to collect regurgitated seeds. Nests not far off ground in young pandan or climbing bamboo; a small stick platform. Egg (1) white. **Range:** NG, 1400–2900 m; Karkar I, 0–1800 m; possibly Fergusson (SE Is). NG: Bird's Head, Central Ranges, Foja Mts, N Coastal Mts, and Huon. Also Admiralty, Bismarck, and Solomon Is. **Taxonomy:** The two subspp *beccarii* and *johannae* may be specifically distinct.

Peaceful Dove *Geopelia placida* Pl 35
(*G. striata*)
18–20 cm. A common, *tiny, grey dove* of savannahs and towns in drier parts of lowland S and E NG. *Barred throat.* **Adult:** Blue eye skin, pale eye, and dark scalloping on upperparts. **Juv:** White "spectacles" around dark eye; pale scalloping on back, rump, and wings. **Subspp (1):** *placida* (also AU). **Similar spp:** Bar-shouldered D much larger and darker, has an unbarred throat, chestnut hindneck, and brownish upperparts. **Voice:** Song is a common suburban sound: a musical, oft-repeated *doodaloo*. Courtship (crouch with tail raised forward and fanned) accompanied by *crOOOuw*. **Habits:** Solitary or in groups, foraging on the ground in short grass and weeds, burned patches, or gravelly sites near towns, roadsides, and riverbeds. Flushes in a steep take-off, wings whistling. Nests on tree branch a few meters up; nest a woven pad of sticks and roots. Eggs (2) white. **Range:** NG and Aru Is, 0–500 m. Trans-Fly to SE Pen, E Sepik-Ramu, S Huon, probably elsewhere locally. Also AU. **Taxonomy:** Sometimes included in Zebra Dove (*G. striata*).

Bar-shouldered Dove *Geopelia humeralis* Pl 35
26–28 cm. A locally common, *medium-sized, grey-brown dove of southern gallery forests and mangroves at savannah edge. Rufous hindneck and unbarred grey throat.* **Adult:** Rufous hindneck and pale eye. **Juv:** White spectacles around dark eye, no rufous neck patch, pale scalloping to upperparts. **In flight:** Bright rufous primaries and secondaries unique; these are grey and black in Peaceful D. **Similar spp:** Peaceful D smaller and paler, no brown or rufous, barred throat. **Voice:** Songs a high-pitched *coolicoo* and an emphatic *hook coo hook coo!* **Habits:** Prefers more wooded habitat than Peaceful D. Singly or in small flocks. Forages on the ground for seeds. Nest built low in a tree; nest similar to Peaceful D. Eggs (2) white. **Range:** NG: Trans-Fly and southern SE Pen, 0–300 m. Also AU.

Thick-billed Ground-Pigeon *Trugon terrestris* Pl 36
32–36 cm. A large and altogether odd-looking terrestrial pigeon, uncommon in lowland forest interior on NG. *Unique face with white-tipped bill, white slash across cheek and ear-coverts, large red eye, and short hind-crest*; this "floating head" seems cut off by the black hindneck. *Bright buff flanks.* The wedge-shaped tail is flicked downward as it walks. **Adult:** Wings evenly grey-brown. **Juv:** Similar, but bill darker and wing feathers faintly edged rufous. **Subspp (3):** Minor. (*mayri* of NW Lowlands has pale grey rather than white on face and paler belly.) **Similar spp:** Might be confused with a scrubfowl. **Voice:** Seldom heard. Song short, low, nasal *whouw* note repeated monotonously at ~5 sec intervals. Calls are various single cooing notes. **Habits:** Singly or in pairs. Unobtrusive and easily missed. Walks about on the forest floor in search of fruit, from which it extracts seeds for consumption. Nests on the ground against a tree trunk between root buttresses; nest a scant stick platform. Egg (1) white. **Range:** Endemic. NG and Salawati (NW Is), 1–600 m. All NG except Trans-Fly, Huon, and northern SE Pen.

Pheasant Pigeon *Otidiphaps nobilis* Pl 37
(Magnificent Ground Pigeon)
44–48 cm. Uncommon in forested hilly country from foothills to lower montane zone. An odd terrestrial pigeon with *large, folded, tentlike tail* that is pumped up and down emphatically as the bird walks. **Adult:** *Generally blackish with bright red-brown wings and mantle and pale grey nape.* The *reddish bill and long yellowish legs* are conspicuous. **Juv** duller: dingy black, dark rufous wings and mantle. **Subspp (4):** *nobilis* (NW Is, Bird's Head to W Ranges, also Foja to Bewani Mts in N Coastal Range) crested, with neck patch iridescent opal, rump and underparts dark purple; *cervicalis* (Border Ranges east to Huon and SE Pens) no crest and with neck patch chalky white, rump and underparts dark green; *insularis* (D'Entrecasteaux Is: Fergusson I) no crest or neck patch, underparts purple; *aruensis* (Aru Is) no crest, neck patch white and large, underparts green and purple. **Similar spp:** Brushturkeys are similar but more robust and have black wings and back, not red-brown. **Voice:** Unmistakable song often heard, repeated at 3-sec intervals; a mournful, tremulous 2-note whistle lasting 2 sec: *wooah woooooooooooo*, the last part descending and fading out. Sometimes only the second note is given. Not loud, but carries a long distance. Easily imitated. **Habits:** Solitary and shy. Runs off or flushes at close range, thrashing off heavily to land on the ground some distance away. Feeds on fallen seeds and fruits. Nests on the ground against a tree trunk between root buttresses; a skimp platform of leaf petioles. Egg (1) white. **Range:** Endemic. NG, NW Is (Batanta, Waigeo), Aru Is, and SE Is (D'Entrecasteaux: Fergusson), 0–1800 m. NG mts (records from all but Cyclops Mts). No recent reports of race *insularis*. **Taxonomy:** Being considered for splitting into 4 species: Crested Pheasant-Pigeon (*O. nobilis*), Greenish Ph-P (*O. cervicalis*), Black-necked Ph-P (*O. insularis*), and White-naped Ph-P (*O. aruensis*).

Western Crowned Pigeon *Goura cristata* Pl 37
(Goura for any crowned pigeon**)**
61–71 cm. Usually rare or hunted out near human settlement; common and tame where undisturbed by people. Inhabitant of undisturbed lowland forest, especially near rivers and streams. *A huge, fan-crested, blue-grey terrestrial pigeon with white patch on wing-coverts.* Larger even than a brushturkey. **Adult:** From other crowned pigeons by its *grey breast (lacking maroon) and maroon mantle (not grey)*. (There is no melanistic morph in this genus; this is an artifact of museum specimens in which oil from the skins has seeped into the feathers.) **Juv:** Duller, markings less pronounced, wing-coverts with grey or buffy edging. **Subspp (2):** Misool and Waigeo Is birds smaller. **Similar spp:** No other crowned pigeon in its range, but see Southern and Victoria CP. **Voice:** Song a loud, deep, drumlike booming *hoom hoom hoom hoom hoom* of 5–6 notes. Contact call a soft *hoom*. **Habits (gouras generally):** Gregarious, in small parties walking the forest floor in search of fallen fruit and seeds. Will flush readily and noisily through vegetation until awkwardly finding a perch on a midstory tree limb. Wings slap loudly (1–3 in a burst) on initial takeoff. Nervously wags tail up and down, rapidly and shallowly. Attracted to preparation sites for sago palm, where it feeds on starchy scraps. Best seen along riverbanks at dawn when birds visit beaches for grit and invertebrate food, or as dusk approaches, when the birds ascend partway up trees before flying across the river. Nest built on a tree branch; nest a bulky platform of sticks and leaves. Egg (1) white. **Range:** Endemic. NG and NW Is (Batanta, Misool, Salawati, Waigeo), 0–300 m. NG: Bird's Head and Neck. Introduced to Moluccas (Seram). **Taxonomy:** Hybridizes with Victoria CP and possibly Southern CP.

Southern Crowned Pigeon *Goura scheepmakeri* Pl 37
71–79 cm. Differs from other crowned pigeons by combination of *all-grey crest, maroon breast, and white wing patch* and by the chin being grey (not black). **Subspp (2):** *scheepmakeri* (southern SE Pen west to Purari R) grey shoulder and foreneck; *sclaterii* (S Lowlands eastward to Fly R) maroon shoulder and foreneck. Identification to race of goura populations between the Fly R and the Purari R has yet to be determined. **Similar spp:** See Western and Victoria CPs. **Voice:** Song a deep, hollow drumming with heartbeat rhythm, *boom-boom, boom-boom*, etc. Reminiscent of a cassowary. **Habits:** See Western CP. **Range:** Endemic. NG: S Lowlands and southern SE Pen, 0–800 m, locally higher. (Possibly meets Western CP in far West.) **Taxonomy:** See Western CP.

Victoria Crowned Pigeon *Goura victoria* Pl 37
58–74 cm. Differs from other crowned pigeons by *white-tipped crest and grey wing patch* (not white). **Subspp (2):** Bay Is birds smaller. **Similar spp:** See Western and Southern CPs. **Voice:** A double-beat booming similar to Southern CP, *uh wuh–uh wuh–uh whu–uh* . . . in 8 pairs. Contact call is a soft grunting, piglike. Another call is a low *oooommm*. **Habits:** See Western CP. **Range:** Endemic. NG and Bay Is (Yapen and Biak, where possibly introduced), 0–300 m. NG: NW Lowlands and Sepik Ramu, also a limited area of N lowlands of SE Pen. **Taxonomy:** See Western CP.

Nicobar Pigeon *Caloenas nicobarica* Pl 40
32–33 cm. A *large, dark, fowl-like pigeon of remote, forested offshore islets*; widespread, but very local, can be common. Seasonally nomadic and able to commute between islands. The Nicobar P appears all dark and tailless, but in good light is a wondrous bird with *iridescent, glossy green plumage* reflecting copper and gold highlights. Resplendent, long, drooping *rooster-like hackles*. Grey, chalky bloom around the head and neck (variable). Bill topped with a black knob. **Male:** Bill knob larger, hackles longer, upperparts greener than **Female**. *Both have immaculate white tail.* **Juv:** Dingier, lacks hackles, and tail black. **Subspp (1):** *nicobarica* (widespread). **Similar spp:** Some local people classify the Nicobar as a type of scrubfowl. **Voice:** Song a disyllable: first part higher and short, second longer and lower, the phrase repeated without pause. Grunting calls. **Habits:** The Nicobar's ecology is focused on pigeon islands—small, offshore islands where imperial pigeons flock in numbers to roost and nest. Here, Nicobars themselves roost at night in the forest interior and at dawn descend to the ground to feed on the abundant fallen seeds passed by the fruit-eating imperials. The Nicobar is wary and difficult to approach when on the ground. It flushes readily—all one sees is the white flash of its tail lifting into the trees. Rising steeply, the bird alights on a branch at midlevel where it stands still and upright, neck craned forward. Many seem to spend their day on the pigeon island, although some will commute to mainland forests to feed. Capable of flying long distances over the ocean, it usually travels in small flocks. Pairs or single birds disperse when foraging on the ground. Largely a crepuscular feeder, including large, hard nuts in its seed diet. Hackles raised in threat or courtship. Nests in trees and shrubs; nest a stick platform. Egg (1) white. **Range:** Locally along coasts of NG (seemingly missing from most of S coast) and many islands (not reported from Gebe, Manam, Goodenough, and Fergusson Is, which it probably visits). Also Indian Ocean to Melanesia.

Common Emerald Dove *Chalcophaps indica* Pl 36
(**Emerald Ground Dove**, **Emerald Dove**)
23–25 cm. Confined to certain small NW and Bay islands. **Adult:** Similar to Pacific ED, but has a *white forehead* (patch smaller in **Female**), grey crown, and smaller white shoulder patch. **Subspp (2):** *indica* (NW Is); *minima* (Bay Is), underparts darker purplish brown. **Similar spp:** Does not co-occur with Pacific ED. **Voice:** Similar to Pacific ED. **Habits:** Presumably similar to Pacific ED. No nesting information for New Guinea. **Range:** NW Is (Gag, Gebe, Kofiau) and Bay Is (Biak, Mios Num, Numfor), 0–800 m. Also India east through SE Asia, Sunda Is, Philippine Is, Wallacea. **Taxonomy:** Formerly included Pacific ED.

Pacific Emerald Dove *Chalcophaps longirostris* Pl 36
(*C. chrysochlora* or *indica*)
23–25 cm. Eastern NG only. Common in forest edge and second growth in lowlands and hills. A chubby ground-dove with *both back and wing-coverts iridescent green or bronze-green* and *white shoulder patch*. Rump marked with two pale grey bands. Orange bill. **Male:** Grey-maroon head and underparts; sooty rump and tail. **Female:** Grey-brown head and underparts; white shoulder patch duller; brownish rump and tail. **Juv:** Breast obscurely barred; tips to flight-feathers brown; lacks white shoulder patch and less green on wing, but has diagnostic green back; rump and tail as in Adults, according to sex. Bill black. **In flight:** Darts across one's path, a meter or two above the ground, directly into the forest. Look for green back and white shoulder patches. **Similar spp:** See Stephan's ED. Common ED does not co-occur. **Voice:** A slow series of up to 7 long, forced *oo* notes, with a peculiar moaning quality, repeated at 1/sec. Each note descends then rises in pitch; successive notes are slightly higher in pitch. **Habits:** A solitary ground feeder but may gather in pairs or groups under fruiting tree. Feeds on fruits and seeds. Nest is a small platform of sticks placed a few meters up in vegetation. Eggs (2) white. **Range:** E NG and all SE Is, 0–1300 m. Locally distributed in Trans-Fly, Sepik-Ramu, Huon, SE Pen, and eastern S Lowlands. Also AU and portions of Wallacea and Melanesia. **Taxonomy:** Formerly with *C. indica* as the Emerald Dove.

Stephan's Emerald Dove *Chalcophaps stephani* Pl 36
(Stephan's Green-winged Pigeon)
24–26 cm. Widespread in lowland and hill forest. Similar to Pacific ED, but rich, darker *reddish brown overall*, with *conspicuous whitish forehead* and *green restricted to wing* (back green in Pacific); *no white at bend of wing*; and 2 rump bands buff (pale grey in Pacific). Bill orange. **Male:** Forehead white; plumage with a purplish cast. **Female:** Forehead grey; plumage more brownish. **Juv:** Lacks the whitish forehead and has a dark bill and brown, not green, secondaries, but shows the diagnostic brown mantle, back, and scapulars, and 2 buff bands on rump. **In flight:** Compared with Pacific ED, shows white forehead and brown mantle. **Subspp (1):** *stephani* (also Kai, Admiralty, and Bismarck Is). **Similar spp:** Pacific ED. Common ED co-occurs in NW Is. **Voice:** A fast, slightly rising series of faint *too* notes lasting as long as 15 sec. Notes are repeated at 3–4/sec. **Habits:** More frequently found in forest interior than Pacific ED. Otherwise ecology, nest, and egg similar. **Range:** NG, NW Is (Batanta, Gag, Misool, Salawati, Waigeo), Bay Is (Yapen), Aru Is, Manam and Karkar Is, SE Is (3 main D'Entrecasteaux Is), 0–1050 m, higher on islands. Also from Sulawesi, Admiralty, Bismarck, and Solomon Is.

Fruit-Doves. As a general rule, fruit-dove species of different body size occur within a given habitat, and species of the same size preferentially occupy different habitats. The most species-rich habitat is lowland rainforest, where one finds 6 fruit-dove species each ~50% heavier than the next smaller: from smallest to largest, Dwarf, Beautiful, Superb, Ornate (seasonally in lowland rainforest), Pink-spotted, and Wompoo Fruit-Doves. Three other species are confined to the mountains as breeders: the small Claret-breasted FD (northern watershed only), the larger Ornate FD (descends to the lowlands as a nonbreeder), and Mountain FD. Coroneted and Beautiful FDs are the same size and have similar calls, but Coroneted is concentrated in lower-rainfall forests, Beautiful in higher-rainfall forests. Orange-bellied FD and Orange-fronted FD are mainly in open habitats and forest edge. Yellow-bibbed, White-bibbed, and Moluccan FDs are confined to small offshore islands in the New Guinea Region. Wallace's and Rose-crowned FDs appear to be only vagrants to the New Guinea mainland.

Fruit-doves are small to medium-sized tree-dwelling species, difficult to observe because of their cryptic green coloration and concealment within foliage. Thus calls are especially important for identification. These can be confusing because most species have 2 quite different vocalizations, and corresponding calls of different species are often similar.

One type of call, termed the "*hoo* series" is shared by most species. The call consists of a series of 7–20 *hoo* notes. They can be distinguished by 7 characteristics: (1) *Pitch*: lowest in Ornate; next lowest in Pink-spotted; medium-low for Mountain, Superb, and Orange-fronted; highest for Orange-bellied, Beautiful, and Coroneted. (2) *Pace*: constant throughout for Ornate, Orange-fronted, and Superb; accelerating somewhat for Orange-bellied, Beautiful, and Pink-spotted; and accelerating greatly for Mountain and Coroneted. (3) Whether note is ascending, descending, or at constant pitch. (4) *Volume*: softest for Ornate and Orange-fronted. (5) Length of each *hoo*: shortest for Pink-spotted. (6) Number of *hoo*s: fewest for Ornate and Superb. (7) Whether series ascends in pitch (Superb), descends (Coroneted), or ascends and then descends (Orange-bellied).

The second type of call, termed "seesaw," is a series of slurred notes on 2 pitches, each alternating with the other. The alternation gives the sense of a seesaw tipping back and forth. Such calls are given by Beautiful, Coroneted, Orange-fronted, Ornate, and Pink-spotted Fruit-Doves. Seesaws of these species differ in (1) *pitch*: Ornate is lowest, Beautiful and Coroneted are highest, (2) number of notes, (3) *speed*, (4) whether an interpolated third short note occurs (only Beautiful), and (5) whether the notes are downslurred, made at a constant pitch, or upslurred then downslurred. See individual accounts for details.

Wompoo Fruit-Dove *Ptilinopus magnificus* Pl 38
(Wompoo Pigeon, **Magnificent Fruit Dove**, *Megaloprepia magnifica*)
29–33 cm. Common in lowland and hill forest. **Adult:** A unique, large, *long-tailed* fruit-dove, with *maroon-purple breast, pale grey head, and band of pale spots on the shoulder*. **Subspp (2):** Minor differences in shading of underparts and shoulder spots (white or yellow). **Similar spp:** All other fruit-doves and imperial pigeons have shorter tail; none has such extensive purple breast but see Claret-breasted FD. **Voice:** Song a low-pitched, throaty, 2-note call, the first disyllabic and upslurred, the second on a constant pitch: *hoowah-hooo*, with quality of a human voice. Easy to recognize. **Habits:** A solitary, subcanopy species foraging at a variety of fruiting trees in the forest interior. Travels through and beneath the forest canopy, and unlike other fruit-doves, does not fly over the forest. Nest placed on a thin branch fork or palm leaf a few meters above ground; a pad of sticks characterized by curly vines. Egg (1) white. **Range:** NG, NW Is (Batanta, Misool, Salawati, Waigeo), Bay Is (Yapen), Manam and Karkar Is, and SE Is (D'Entrecasteaux: Normanby), 1–1400 m. Also AU.

Pink-spotted Fruit-Dove *Ptilinopus perlatus* Pl 38

25–27 cm. Common and associated with fig trees in lowland and hill forest. Often perches conspicuously in tree crowns. **Adult:** A large, long-necked fruit-dove with *mustard-brown neck band, diagnostic pink spotting on shoulder, and terminal tail-band either lacking or indistinct, narrow, and grey*. **Subspp (3, in 2 groups):** *perlatus* (range except that of next race) head capped yellowish green; *plumbeicollis* (Sepik-Ramu and N coast of SE Pen) head all grey. **Similar spp:** The very similar Ornate FD has a purple bar on the shoulder and a broad, yellowish tail-band. Wompoo FD also has grey head and wing spots, but is distinguished by a purple breast and long tail. **Voice:** A *hoo* series and a seesaw, both low-pitched (only Ornate FD is as low). *Hoo* series of Pink-spotted has abbreviated notes, giving a distinctly energetic quality to the series; it shows marked acceleration, the first and second notes sounding almost detached from the rest of the series; notes also initially rise a bit, then fall slightly in pitch through the series. The seesaw consists of 4–6 notes, alternate ones differing slightly in pitch, and each note is initially upslurred then downslurred:

Habits: A gregarious and nomadic species. Nests on tree branch; platform of sticks. Egg (1) white. **Range:** Endemic. NG, NW Is (Batanta, Salawati, Waigeo), Bay Is (Yapen), Aru Is, and SE Is (3 main D'Entrecasteaux Is), 0–1200 m.

Ornate Fruit-Dove *Ptilinopus ornatus* Pl 38

23–26 cm. Common and associated with fig trees. Breeds solitarily in montane cloud forests then wanders in groups to nearby lowlands after breeding. **Adult:** Similar to Pink-spotted FD—with which it co-occurs at lower elevations—but its size averages slightly smaller, and it has a hard-to-see *purple bar on the shoulder, a neck band that is darker and two-toned (mustard above, brown below), and a broad, yellowish terminal tail band*. Note unique Bird's Head race. **Juv:** Shows a much reduced purple wing bar. **Subspp (2):** *ornatus* (Bird's Head) wine-red head; *gestroi* (rest of range) head mustard-yellow. **Similar spp:** Pink-spotted FD. Mountain and Superb FDs are the only species sharing the montane habitat of Ornate. **Voice:** Sings mainly when breeding in the mountains. A *hoo* series and a seesaw, with the lowest pitch and softest volume of any NG fruit-dove. The *hoo* series ascends initially and consists of 7–12 notes. Confusion with the *hoo* series of Superb is possible only if the Ornate series neither descends at the end nor accelerates. The seesaw is similar in pattern to that of Pink-spotted but is lower-pitched and consists of up to 10 notes. **Habits:** Similar to Pink-spotted but breeds in the mountains. Nest found 2–3 m up in a tangle of vines or scrambling bamboo; nest a scant platform of twigs. Egg (1) white. **Range:** Endemic. NG (no records for Kumawa or Wandammen Mts, and much of the S Lowlands and Trans-Fly), 0–2400 m.

Orange-fronted Fruit-Dove *Ptilinopus aurantiifrons* Pl 38

23 cm. Common in disturbed forests, gardens, and towns of lowlands, less common in expanses of primary forest or in foothills. **Adult:** *Unique orange-yellow forecrown*. Diagnostic combination of *white throat (because of shading, looks like white cheek-stripe and grey chin), grey breast, and green belly*. **Similar spp:** From below, Pink-spotted and Ornate FDs both have brown breast band. **Voice:** A soft *hoo* series and a seesaw at medium-low pitch. All other fruit-dove species except Ornate have louder calls. Each note of the *hoo* series is slightly upslurred; rate of delivery may remain constant or accelerate slightly; pitch rises at the start and drops at the end. Seesaw usually consists of 6 notes: the first, third, and fifth upslurred, the others downslurred:

Also a series starting with a single low note followed by a higher series that steadily declines in pitch. **Habits:** Gregarious, often flocking in tops of trees. Frequents fig trees. Nests in isolated trees and those at forest edge; nest a rather substantial stick platform. Egg (1) white. **Range:** Endemic. NG, NW Is (Batanta, Misool?, Salawati), Bay Is (Yapen), Aru Is, and SE Is (D'Entrecasteaux: Fergusson, Normanby), 0–300 m. All NG except Huon and nearby lowlands.

Wallace's Fruit-Dove *Ptilinopus wallacii*   Pl 38

27–29 cm. *Only on islands and along coasts and rivers of Aru Is (1 record in western S Lowlands).* **Adult:** *Striking. Large, with red cap, grey breast and back, and yellow-orange belly.* Often with a whitish throat and breast band. **Similar spp:** Rose-crowned and Beautiful FDs smaller, lacking the white breast band and having some red-purple on the belly. **Voice:** No information from NG Region. **Habits:** On Aru it occupies small patches of forest and savannah. Gregarious, in flocks. Nomadic. **Range:** Aru Is, but also recorded on NG mainland in western S Lowlands (coastal Mimika and Noord R), where possibly vagrant, 0–200 m. Also S Moluccas.

Superb Fruit-Dove *Ptilinopus superbus*   Pl 38
(Purple-crowned Pigeon)

20–23 cm. The most widespread and prevalent fruit-dove, in forest from the lowlands up to middle elevations in the mountains. *The only fruit-dove with white belly and undertail.* **In flight:** *White flank-stripes* conspicuous. Few other fruit-doves have a *grey tail-band that shows well.* **Male:** With *diagnostic orange hindneck and bold black breast band.* **Female:** Obscure, but note unique *all grey-green throat and breast* and dark blue hindcrown. **Subspp (1):** *superbus* (Moluccas to Solomon Is and AU). **Similar spp:** None with Superb's field marks. **Voice:** The *hoo* series is 8–11 notes at medium-low pitch. The series rises in pitch and does not accelerate. This song is distinctive in its simplicity and can be confused only with that of Ornate FD. Second song is more distinctive; a single note with an *mm* sound, followed by a 2-sec pause and then a series of 3–6 *wo-up* upslurs at 1-sec intervals. **Habits:** A solitary, generalized fruit-feeder of the forest interior. Its status as a migrant from AU is possible but undocumented. Nests in the forest midstory on a branch; nest a thin platform of sticks. Egg (1) white. **Range:** Throughout NG Region, 0–1400 m. Also Sulawesi, Moluccas, Bismarck, and Solomon Is, and AU.

Rose-crowned Fruit-Dove *Ptilinopus regina*   Pl 38
(Rose-crowned Pigeon)

20–23 cm. Status unclear, but probably a vagrant AU species. **Adult:** First impression is of a rather pale green fruit-dove. *Grey breast and purple-and-orange belly*; red crown and *yellowish tail tip* are also distinctive. **Juv:** Greenish below; may lack reddish cap, breast patch, or both. **Subspp (1):** *regina* (E AU). **Similar spp:** Coroneted FD lacks the grey throat and breast. Beautiful FD has a white throat and lacks the yellow tail tip. Both are clearly smaller. **Voice:** A low-pitched *hoo* series that starts very slowly, then accelerates greatly. Also a rapidly delivered, low-pitched seesaw, consisting of a downslurred note alternating with a higher note at a constant pitch. **Habits:** Occupies coastal forests. **Range:** NG Trans-Fly: 1 record from Daru I, sea level. Also E Wallacea and AU.

Coroneted Fruit-Dove *Ptilinopus coronulatus*   Pl 38
(Diadem Fruit Dove)

18–19 cm. Common in lowland forest, less so in hills. Prefers drier forest than the similar Beautiful FD. **Adult:** *Small and dark green, with grey head, lilac crown, and yellow undertail.* Lilac crown edged with a gold band—unique but hard to see. **Subspp:** *coronulatus* (S Lowlands to southern SE Pen; Aru Is) crown obviously purple and belly purple and yellow (no orange); *geminus* (Bird's Head, NW Lowlands to northern SE Pen) crown pale grey or purple, and belly purple, orange, and yellow. **Similar spp:** Same-sized Beautiful FD has large white throat patch and pale grey breast that contrast with dark green back. Orange-bellied FD has an all-green head and white undertail. Dwarf FD is smaller, has broad yellow edges to the secondaries, and an all-green head. **Voice:** The *hoo* series is high-pitched, accelerates greatly, and remains at the same pitch or descends slightly. Individual notes are downslurred. The seesaw is high-pitched, simple, and disyllabic: *whoo-oh whoo-oh* . . . , the first note higher. Also a "sad call" *hu-h-h-h-h, hoodle.* **Habits:** Solitary, in pairs, or in groups, mainly in the forest canopy. Nest in a forest sapling 1–5 m up; a shallow stick platform on an accumulation of dead leaves. Egg (1) white. **Range:** Endemic. NG, NW Is (Salawati), Bay Is (Yapen), Aru Is, and Manam I, 0–1200 m.

Beautiful Fruit-Dove *Ptilinopus pulchellus*   Pl 38

18–20 cm. Widespread; common in wet hill forest and lowlands. **Adult:** Small size, red crown, *throat and breast white and pale grey, offset from orange belly by a wine-red band.* In flight, may show faint grey tail-band. **Subspp (2):** *pulchellus* (most of range); *decorus* (NW Lowlands and Sepik-Ramu).

grey breast feathers heavily tipped white. **Similar spp:** See Coroneted FD. Rose-crowned FD larger, with tail tipped yellow. Much larger Wallace's FD lacks purple breast band and has undertail white, not yellow. **Voice:** *Hoo* series is high-pitched and accelerates moderately but not as much as does that of Coroneted FD. The pitch of the series is variable; individual notes are downslurred or unslurred. The seesaw is sad and distinctive, with a low, faint grace note between the 2 main notes, audible only at close range, usually only 2 or 3 in a series:

Habits: Similar to Coroneted FD; occurrence mostly complementary with Coroneted. Nesting similar to Coroneted. **Range:** Endemic. NG and NW Is (Batanta, Misool, Salawati, Waigeo), 0–1200 m, locally higher.

White-bibbed Fruit-Dove *Ptilinopus rivoli*   Pl 39
(White-breasted Fruit-Dove)

20–24 cm. *Common in SE Is and locally in the Bay Is. A small-island species. Bright yellow undertail and lower belly.* Note *yellow-green loral stripe*. **Male:** *White breast band (no yellow) and lacks purple patch on belly*; these lacking in all-green **Female**. **Subspp (2):** *strophium* (SE Is); *miquelii* (Bay Is) smaller. **Similar spp:** Moluccan FD also found in Bay Is but on different islands; it lacks yellow in undertail and male has large purple breast patch. Yellow-bibbed FD, also in Bay Is, shares yellow undertail but is smaller and lacks the obvious yellow line from bill to eye; male has a yellow breast patch and purple belly. Superb FD, the only co-occurring fruit-dove, has white undertail, and its breast pattern is entirely different. **Voice:** The *hoo* series higher-pitched than Mountain FD. Also gives a slurred *hoo-woo* call, the second note at lower pitch (in place of the single *hoo* note of Mountain); this call is repeated after a pause, and the series may continue for a while and end in a single *hoo*. **Habits:** Occupies many small islands and some large ones where other similar-sized fruit-doves are missing. Takes a wide range of fruits, including nutmegs. Solitary or in pairs, foraging at all levels in the forest interior. Nests on a thin branch of a small forest tree; nest a shallow stick platform. No egg information. **Range:** Fragmented range: SE Is (absent from the 3 main D'Entrecasteaux Is and Woodlark I, but present on the many small islands from D'Entrecasteaux and Trobriand Is eastward to and including Misima, Tagula, and Rossel Is) and Bay Is (Mios Num, Yapen), 0–900 m. Also Bismarck Is. **Taxonomy:** See Mountain FD for explanation of split. The NG forms are provisionally classified with *P. rivoli* of the Bismarck Is.

Moluccan Fruit-Dove *Ptilinopus prasinorrhous*   Pl 39
(White-breasted Fruit-Dove, *Ptilinopus rivoli*)

22 cm. A *western small-island specialist*. Similar to Mountain FD but slightly smaller, duller, and not quite as bulky. Note *green undertail* and, in **Male**, the *all-white breast band (no yellow) and large purple breast patch*. **Similar spp:** White-bibbed FD, closely related and also on islands, has yellow undertail, and male lacks large purple breast patch. Yellow-bibbed FD co-occurs in the Bay Is but is smaller and has yellow breast patch and undertail. **Voice:** The *hoo* series similar to Mountain but more high-pitched, on a steady pitch, and increases only slightly in tempo (less rise and fall in pitch, less acceleration). **Habits:** Probably similar to White-bibbed FD. No nesting information. **Range:** Occupies only the smallest of the NW Is, Bay Is (Numfor, others), and Aru Is. Also Moluccas. **Taxonomy:** See Mountain FD.

Mountain Fruit-Dove *Ptilinopus bellus*   Pl 39
(White-breasted Fruit-Dove, *Ptilinopus rivoli*, see below)

23–26 cm. Widespread and common in montane forests; often the only fruit-dove in its habitat. *Large, stolid, dark green fruit-dove*—look for *all-green belly and undertail* (often with yellow edging). The obscure grey tail-band is hidden when bird is perched and doesn't show well even in flight. **Male:** *Broad white-and-yellow breast patch; purple cap and breast spot*. **Female:** *All green*—most fruit-doves have patches of color somewhere. This and the previous 2 species unique in sharing *a bright yellow loral stripe between the greenish yellow bill and eye*, useful in confirming identification of the drab Females. **Similar spp:** See Moluccan and White-bibbed FDs; note that the females of these have all-green heads, whereas Female Mountain has a bluish cast on the crown and face. Superb and Ornate are the only regularly co-occurring FDs; both have different patches of color, except female Superb, which has grey underparts (not green), white undertail, and conspicuous tail-band in flight.

Voice: A familiar sound of mountain forests, the *hoo* series is regretful and haunting, evocative of the cold and wet mountains. It begins slowly, accelerates rapidly, and drops in pitch. Another call is a single *hoo* that first rises, then drops in pitch. Also a series of *hoo* notes without change in tempo or pitch.
Habits: Solitary or in pairs, foraging at all levels in the forest interior for large and small fruits. Nests on a thin branch of a forest sapling or on a rattan frond; nest a shallow stick platform. Eggs (1–2) white. **Range:** Endemic. NG (all mts, but no record yet from Kumawa Mts), Karkar I, and SE Is (D'Entrecasteaux: Goodenough, Fergusson), 1000–3200 m. **Taxonomy:** The Mountain and Moluccan FDs were recently split from the White-bibbed FD (*P. rivoli*). The ranges of all three are generally separate, but Mountain and White-bibbed co-occur in the D'Entrecasteaux Is, and Moluccan and White-bibbed both occupy the Bay Is, although they inhabit different islands.

Yellow-bibbed Fruit-Dove *Ptilinopus solomonensis* Pl 39
18 cm. Only in Bay Is. Co-occurs with larger Moluccan FD from which it differs in smaller size, *bright yellow undertail-coverts, unique bare pale greenish skin around eye*, and **Male** has *unique yellow breast band* (not white) and lacks purple cap, but note purple spot in front of eye.
Subspp (1): *speciosus* (endemic). **Similar spp:** Moluccan and White-bibbed FDs are both larger, show a yellow-green loral stripe, and differ in details of male plumage; Moluccan lacks the Yellow-bibbed's yellow undertail-coverts. **Voice:** The *hoo* series of 7–8 double notes, accelerating slightly: *hoo-woo hoo-woo hoo-woo....* **Habits:** Similar to Moluccan and White-bibbed FDs. Nests in a tree or shrub; nest a meager platform of twigs and vine tendrils. Egg (1) white.
Range: Bay Is (Biak, Numfor, Traitor's, and Marai near Yapen), 0–300 m. Also Admiralty, Bismarck, and Solomon Is.

Claret-breasted Fruit-Dove *Ptilinopus viridis* Pl 39
18–20 cm. Uncommon, with patchy distribution across the northern NG Region. Confined to hills on NG mainland, but lowlands as well on islands. *Small, with grey face and green-and-white striped undertail-coverts.* **Male and usually also Female:** Unique *maroon patch on throat*, variable in size by subspp. **Subspp (4):** Sexes colored differently or not; vary in presence or absence of grey on shoulder and size of breast patch. Race *pectoralis* (NW Is, Bird's Head and Neck) lacks grey shoulder and Male has small purple breast patch, absent in Female (all green); *geelvinkiana* (Bay Is: Biak, Mios Num, and Numfor) Male with grey shoulder and large purple throat-and-breast patch, Female green breast; *salvadorii* (Yapen I, Foja, Cyclops, and N Coastal Mts) both sexes with grey-spotted shoulder and small red-purple breast patch, though smaller patch in Female; *vicinus* (SE Is) grey shoulder and red-purple breast patch in both sexes, smaller in Female. **Similar spp:** Superb FD female shares white in undertail, but has white belly (not green). Female Mountain and Moluccan FDs lack white in undertail and the face is green. Compare with Wompoo FD. **Voice:** Two songs.
(1) a medium- to high-pitched series:

1 ``-_-´-´-´-´-´

(2) a slightly hoarse series:

2   _ `*。   ,*** ***

Habits: Singly, in pairs, or in small flocks in forest canopy. Attracted to figs. Nests on a horizontal tree fork; a platform of twigs. Egg (1) white. **Range:** NG, NW Is (all), Bay Is (all), and SE Is (3 main D'Entrecasteaux Is and Trobriand Is); NG 300–1000 m, islands 0–1000 m. NG: mts of Bird's Head, Bird's Neck (Fakfak, Kumawa, Wandammen), and N ranges (Foja, Cyclops, and N Coastal Mts east to Wewak, plus a sighting from Bogia, near Madang). Also S Moluccas and Solomon Is.

Orange-bellied Fruit-Dove *Ptilinopus iozonus* Pl 39
20–22 cm. A common and widespread fruit-dove that prefers open lowland and foothill habitats.
Adult: A *stocky, stubby-tailed* fruit-dove that is *mostly green with an orange belly* and *white undertail-coverts. Bright eye* stands out on plain, dark green face. **Subspp (5, in 2 groups):** *iozonus* (most of range); *humeralis* (NW Is, Bird's Head south to western S Lowlands) shoulder patch purple, not grey.
Similar spp: Coroneted and Dwarf FDs are obviously smaller, having a purple breast patch and yellow undertail-coverts. **Voice:** A series of high-pitched calls. *Hoo* series begins with a longer upslurred

note, a slight pause, then a rapid series of upslurs that accelerate, rising and then falling. Also a 4-note phrase given without repetition:

1

In East, 2 additional slow calls:

2   or

Habits: In flocks of varying size in open habitats, forest edge, regrowth, mangroves, less frequent in tall forest. Prefers large fig trees. Usually less common where Orange-fronted FD is present. Evidently nests higher in trees than other fruit-doves; nest a scant stick platform. Egg (1) white. **Range:** Endemic. NG, NW Is (Batanta, Salawati, Waigeo), Bay Is (Yapen), Aru Is, and Manam and Karkar Is, 0–800 m. **Extralimital spp: Knob-billed Fruit-Dove** (*P. insolitus*) of Bismarck Is virtually identical but has a large red knob on its forehead. Its nearest location is Long I, and 1 unconfirmed sighting has been reported from Bogia near Madang.

Dwarf Fruit-Dove *Ptilinopus nainus*   Pl 39
(*Ptilinopus nanus*)

13–15 cm. Rare to common in foothill and nearby lowland forest. A tiny dove, *surprisingly small*, with a stubby tail. All green and marked only by *a yellow undertail* and *3 unique yellow wing bars*. **Male:** Small *purple breast patch and unique grey spot in front of wing*, both lacking in **Female**. **Similar spp:** Coroneted FD very similar but larger, with lilac crown, and no obvious yellow edging to the wing-coverts. Might be confused with juv fruit-doves of other species, which early in life are much smaller than their parents and semi-independent. **Voice:** A high-pitched, slow, soft, prolonged upslur repeated 6 or so times at a rate of 1/sec:

  or

Habits: Easily overlooked because of its size. Some local movement. Solitary, paired, or in groups. Usually mixed with other fruit-doves at a fruiting tree. Feeds on small fruits and nectar. Nests in forest interior in midstory; a small stick platform. No egg information. **Range:** Endemic. NG and NW Is (Batanta sighting, Misool, Salawati, Waigeo), 0–1100 m. Distribution along the North poorly known.

Imperial Pigeons. Of the 13 species of this genus in the New Guinea Region, 6 are confined to offshore islands and are absent or virtually absent from the New Guinea mainland (Spectacled, Elegant, Pacific, Spice, Geelvink, and Floury IPs). Three species coexist in lowland forests, where they sort by size: the smaller Purple-tailed IP, the medium-size Zoe's IP, and the large Pinon's IP. The Rufescent IP has mountain forests to itself. Collared IP breeds in mangroves and the 2 sister species Pied and Torresian IPs on small islands and the coast, but all 3 move seasonally up rivers into lowland forests.

Spectacled Imperial Pigeon *Ducula perspicillata*   Pl 40
(**White-eyed Imperial Pigeon**)

44 cm. A large Moluccan species abundant on Kofiau I. **Adult:** A lovely green-and-grey pigeon similar to Elegant IP (does not co-occur), but note the *white eye-ring, darker grey head, green extending partway up the back of the neck* (not sharply demarcated at the shoulders), and the steeper forehead profile. Viewed from below look for *grey undertail-coverts*, not brownish as in all other green-backed IPs. **In flight:** Greyish underwing. **Subspp:** Not determined. **Similar spp:** Spice IP co-occurs and is slightly smaller, has a black bill knob, pink shading on the underparts, and brown undertail-coverts, and lacks the broad white eye-ring. Floury IP seems to be the eastern counterpart of Spectacled. **Voice:** Song a descending series of 7 low, hoarse *coos*, the first note at lower pitch than the second: *hoo—hohohohohoho* (similar to song of Floury IP). Calls include a soft, short, ascending *br-r-r* and a low-pitched *coo* that rises, then falls in pitch. **Habits:** Singly, in pairs, or in small parties. A forest pigeon that can take quite large fruits. No nesting information. **Range:** Kofiau (NW Is). Also Moluccas.

Elegant Imperial Pigeon *Ducula concinna* Pl 40
43–45 cm. A large imperial pigeon resident on small islands within the Aru group and perhaps vagrant on Bird's Neck (S Bomberai Pen). **Adult:** Green wings and *all-grey underparts, except brown undertail-coverts*. Swollen cere gives forehead a flatter profile. *Striking yellow iris*, but no white eye-ring. *Blackish underwing in flight—looks broad-winged*. Slow, deep wingbeats. **Similar spp:** None co-occur. See Spectacled IP. **Voice:** A loud, guttural almost barking *ur aow*. Also a hoarse grunt and a low-pitched upslur. **Habits:** A "super-tramp" species prone to wandering and capable of turning up on small islands or along the mainland coast. Singly, in pairs, or in small parties; associates with other imperial pigeons such as Pied IP. Feeds in the forest canopy on a variety of fruits. Nest built in tree canopy; a rough stick platform. No egg information. **Range:** Small Aru Is, plus 1 record on islets off S Bomberai Pen (Bird's Neck), 0–100 m. Also E Wallacea.

Pacific Imperial Pigeon *Ducula pacifica* Pl 40
34–37 cm. An extremely local, small imperial pigeon so far known from a few small islands off Sepik-Ramu and in SE Is. **Adult:** Green wings and *all underparts rose-grey*; enlarged *black knob atop bill*, especially prominent in **Male**, absent in **Juv**. In flight, *underwing uniformly dark*. **Subspp (2):** Minor. **Similar spp:** Floury IP co-occurs and is larger and has a conspicuous white eye-ring; only the throat is rose-grey; no bill knob. Spice IP has grey throat, seems to be western counterpart to Pacific IP. **Voice:** Song a 5-note series *wooiip—pu pu po po*. Four different 1-note calls: (1) a high-pitched, descending *br-r-r*; (2) a broken note that rises then falls in pitch, with a peculiar throbbing, throaty quality; (3) a low-pitched note that slowly rises in pitch, then quickly drops; (4) an ascending growl. **Habits:** Occupies forest and open habitats with trees and shrubs. Inhabits small coral islands, often in large flocks, moving from island to island. Gregarious. Feeds on a wide variety of fruits. Nest on horizontal leafy branch; a substantial stick platform. Egg (1) white. **Range:** Small islands off Sepik-Ramu (Tarawai and Seleo) and SE Is (Amagusa, Budibudi, Conflict Is, Duchateau, Suau, Teste) and probably more islands in both areas, 0–100 m. Also Admiralty, Bismarck, and Solomon Is, east through Polynesia to Samoa.

Spice Imperial Pigeon *Ducula myristicivora* Pl 40
41–43 cm. A medium-sized imperial pigeon of small islands of *NW Is*. Inhabits coastal forests. **Adult:** Green wings. *Head and neck very pale—almost white—with black bill knob and dark iris*. With closer view note white eye-ring, blood-red iris, pinkish nape. White ring around eye in some populations at least, seemingly lacking in others. Grey throat contrasts with rose-grey breast. **Similar spp:** See co-occurring Spectacled IP. Pacific IP smaller, all rose-grey below. **Voice:** High-pitched *crrruooo* or *urwoow*, upslurred toward the end. **Habits:** This is another nomadic pigeon that crosses between islands and strays to the mainland. Singly, in pairs, or in small groups. On islets and coastal forest of larger NW Is. No nesting information. **Range:** NW Is (all), but mainly on smaller islands. Vagrants reach adjacent Bird's Head coast. Also nearby Moluccas.

Geelvink Imperial Pigeon *Ducula geelvinkiana* Pl 40
(formerly included in Spice Imperial Pigeon, *D. myristicivora*)
38 cm. Restricted to *Bay Is*, large and small. In forests of all types. **Adult:** Similar to Spice Imperial Pigeon but smaller, grey-headed with striking pale yellow iris, and lacking bill knob or white eye-ring. **Similar spp:** None in its range. **Voice:** A medium-pitched, coarse *crwwooo* repeated incessantly, ~1 call/sec. **Habits:** Similar to other island imperial pigeons. No nesting information. **Range:** Endemic. Bay Is (Biak, Mios Num, Numfor), 0–500 m. NG: 1 mainland record from hill forests of Arfak Mts, Bird's Head.

Purple-tailed Imperial Pigeon *Ducula rufigaster* Pl 41
33–39 cm. Common and widespread in lower canopy and midstory of lowland and hill forest. **Adult:** Smallest and most brightly colored of the mainland imperial pigeons: *orange-rufous underparts, green wings, purple tail, white eye-ring*. Conspicuous terminal tail-band. **Subspp (2):** *rufigaster* (NW Is, Bird's Head to S Lowlands and SE Pen) back and rump purplish red; *uropygialis* (NW Lowlands to Huon, plus Yapen I) these parts more green, underparts paler. **Similar spp:** Rufescent IP inhabits mountains, lacks purple rump and tail, lacks the white eye-ring, and the grey on the nape extends to sides of neck and throat forming a "cowl." **Voice:** Diagnostic, mournful; 2 knocks followed by a

medium-pitched, prolonged note that quickly drops in pitch then slowly rises, *buk-buk hooo-oo-hoooooo*.

Habits: A solitary, inconspicuous resident of rainforest interior. Sedentary, not nomadic. Usually noticed flying through the forest (rather than over it). Feeds on fruit from the canopy into the forest midstory. No nesting information. **Range:** Endemic. NG, NW Is (Batanta, Misool, Salawati, Waigeo), and Bay Is (Yapen), 0–600 m, rarely to 1200 m.

Rufescent Imperial Pigeon *Ducula chalconota* Pl 41
37–39 cm. Uncommon and widespread in mountain forest—this is the montane counterpart of the Purple-tailed IP, but much less vocal. **Adult:** From that species by its *green rump* (not purple) and *grey "cowl"* over the head and sides of neck; also lacks a white eye-ring. **Subspp (2):** *chalconota* (Bird's Head and Neck) green upperparts tinged with more purple brown; *smaragdina* (rest of range) upperparts more green. **Similar spp:** Purple-tailed IP. **Voice:** Rarely sings. A slow, deep, mournful, hummed song of 3 notes slurred together:

Also a mournful, low *hoo?*; when alarmed, produces a hollow knock. **Habits:** Similar to Purple-tailed IP. No nesting information. **Range:** Endemic. NG: Bird's Head, Bird's Neck (Wandammen), Central Ranges, and Huon, 1250–2650 m. **Taxonomy:** Sightings from the Foja Mts may be of a very distinct undescribed subspecies or sister-species.

Floury Imperial Pigeon *Ducula pistrinaria* Pl 40
(**Grey** or **Island Imperial Pigeon**)
41–44 cm. Prevalent on small and midsize islands off the Sepik-Ramu coast and in the SE Is. This is the common green-winged imperial pigeon within its range, the smaller Pacific IP being restricted to a few islands. **Adult:** Look for the *absence of a black bill knob*, presence of *white ring of feathers around the eye and at the base of the bill*, and the subtly *pink-tinted throat* contrasting with the grey breast. Another difference is the gradual transition from the grey neck to the green mantle. At close range, the iris, eyelids, and nares of Floury IP are dark ruby red. Birds vary in the degree of green vs grey on the wings. **Subspp (1):** *vanwyckii* (also Admiralty and Bismarck Is). **Similar spp:** Pacific IP smaller, with black bill knob, evenly pink-washed underparts, and sharp division between grey neck and green back. **Voice:** Song a distinctive, high-pitched *ahu ahu ahu ahu ahu*, falling in pitch and volume. Also a peculiar, high-pitched trill. **Habits:** A mobile species, nesting and roosting in numbers on small islets and commuting to coastal forest on larger islands to feed. Possibly also resident on some of the larger islands such as Tagula. Seen singly, in pairs, or in small flocks. Feeds in the forest canopy on a variety of fruits. Nests colonially. Nest built on horizontal branch in tree canopy; a bulk twig platform. Egg (1) white. **Range:** Islands off Sepik-Ramu (Manam, Karkar, others) and SE Is (probably all, but only a local visitor to the 3 large D'Entrecasteaux Is and not yet reported from Trobriands), 0–200 m. Also Admiralty, Bismarck, and Solomon Is.

Pinon's Imperial Pigeon *Ducula pinon* Pl 41
44–48 cm. Common and widespread in lowland and foothill forest. Our largest imperial pigeon, powerfully built; seen flying over the canopy. **Adult:** *Dark grey with black shoulder patch, maroon belly, and narrow white band across the dark tail.* Ruby eye set in a patch of red skin bordered by white ring. **Subspp (3):** *pinon* (all of range except the following) wings deep grey; *jobiensis* (NW Lowlands to Huon, plus Yapen I) wings boldly scalloped (birds on Bird's Neck and SE Pen intermediate); *salvadorii* (SE Is) like *pinon* but lacking the white eye-ring. **Similar spp:** Collared IP is darker and shows a black collar and wider tail-band. **Voice:** A distinctive, resonant, very low-pitched, upslurred disyllable, preceded or followed by a note at the pitch of the lower syllable:

Also, a series of as many as 10 soft, low-pitched upslurs. **Habits:** Gregarious. Singly, in pairs, or in small groups; roosts socially in an emergent forest tree. Feeds on midsize and large fruit in the forest canopy. Nests on a tree branch 10–20 m high; nest a thin platform of sticks. Egg (1) white. **Range:** Endemic.

NG, NW Is (Batanta, Misool, Salawati, Waigeo), Bay Is (Yapen), Aru Is, Manam and Karkar Is, and SE Is (3 main D'Entrecasteaux Is, plus Misima, Tagula, and Rossel), 0–500 m, rarely to 900 m.

Collared Imperial Pigeon *Ducula mullerii* Pl 41
(Mueller's Imperial Pigeon)
38–41 cm. A seasonally common, nomadic imperial pigeon of mangroves and riverine lowland forest. Typically encountered during river travel: a few birds perched on a bare treetop or flying at speed above the canopy. **Adult:** Overall, similar to Pinon's IP but slightly smaller and slimmer. *Appears darker*, and shows a *black collar* strongly contrasting with white throat and neck. *Tail-band broader.* **In flight:** Said to have longer wings and faster wingbeats than Pinon's IP and therefore swifter. **Subspp (2):** *mullerii* (S Lowlands and southern SE Pen, east to Brown R) larger and darker; *aurantia* (NW Lowlands and Sepik-Ramu) smaller and paler. **Similar spp:** Pinon's IP. Papuan Mountain-Pigeon lacks white-black-white collar and has a pale band at tail tip. **Voice:** Song a deep, 3-note series; the first note long, upslurred, followed by a short note, then a longer downslur, *wooo-uh-wooo*. **Habits:** Singly, in pairs, or in small parties. Social, and will mix with Pinon's and Torresian IPs. Feeds in the forest canopy on fruit. Nests on a tree branch a few meters above water; nest a thin stick platform. Egg (1) white. **Range:** Endemic. NG and Aru Is, 0–200 m. NG: S Lowlands to southwestern SE Pen, NW Lowlands, and Sepik-Ramu.

Zoe's Imperial Pigeon *Ducula zoeae* Pl 41
38–41 cm. Common and widespread in lowland and hill forest. A plump, stubby, short-winged imperial pigeon of the forest interior, generally keeping within the leafy canopy. **Adult:** Easily identified within its range by any one of these field marks: *brown shoulders*, two-toned *pink and grey underparts* separated by a *black breast band*, *white-speckled brown "pantaloons,"* bright iridescent *all-green tail*, and *white eye*. **Similar spp:** No other mainland imperial pigeon has these field marks. **Voice:** Distinctive song. A prominent first note, followed by a rolling series starting low and rising up the scale, this series repeated several times: *hoo, h h h h hoo, h h h h hoo, h h h h hoo*, or variations on this. Another song: *buk, buk, buk, buk, woo*. Also a loud, sharp *chsak*. **Habits:** A solitary, canopy-dwelling species, though may gather at a fruiting tree. Feeds in the treetops on a variety of fruit. No nesting information. **Range:** Endemic. NG, NW Is (Salawati), Bay Is (Yapen), Aru Is, and SE Is (3 main D'Entrecasteaux Is), 0–1200 m, locally to 1500 m.

Pied Imperial Pigeon *Ducula bicolor* Pl 40
(Nutmeg Pigeon)
41–43 cm. Common on small islands and coastlines of large ones in far western NG Region. **Adult:** *A creamy white imperial pigeon* with black flight-feathers and tail tip. Some birds with a yellowish tinge. From Torresian IP, its eastern counterpart, Pied IP *lacks black tire-track markings on the undertail-coverts and thighs, and the bill is blue grey with a dark tip*. Some birds appear to be hybrids between the two, showing reduced black markings under the tail and on lower flanks. **Similar spp:** See Torresian IP. **Voice:** Song a low *hoo* series of 10+ notes each upslurred. Also a guttural and hoarse repeated note *hoo-oh!* Or *ka-oh!* One per 4 sec. **Habits:** A nomadic or resident species roosting and nesting in coastal mangroves or on small offshore islands and feeding in the lowland forest of the mainland. Nests alone or colonially. Nest built on a horizontal branch in forest midstory or canopy; stick platform. Egg (1) white. **Range:** NW Is (Misol, Salawati), Bird's Head, W Bird's Neck, and possibly Aru Is (occupied by Torresian IP), 0–300 m. Also Nicobar Is eastward through Sunda Is, Philippines, and Wallacea. Introduced to Bahama Is. **Taxonomy:** Some authors include Torresian IP in this species.

Torresian Imperial Pigeon *Ducula spilorrhoa* Pl 40
(Torres Strait Pigeon)
40–43 cm. Common on small islands and in coastal forests, mangroves, and locally in savannahs across most of NG Region, excluding range of Pied IP. Both resident and migratory. Much of the eastern AU population migrates to NG to spend the austral winter. **Adult:** Differs from Pied in showing *black tire-track markings on the undertail-coverts and by the yellowish bill*. Also, primaries with a greyish cast, rather than pitch black. Some birds with yellowish tinge. Birds resident in Trans-Fly with greyish tinge previously classified as subsp *tarara*. **Juv:** Black markings on the undertail-coverts obscure. **Similar spp:** Pied IP. **Voice:** Slow, hollow, low-pitched, downslurred *hoooo* or *hoo hoooo*

repeated after a pause. Same note maybe be given in a rapid series. A familiar sound on NG islands. **Habits:** Same as Pied IP. Occasionally found far up large rivers. **Range:** NG Region (including Aru Is, where it may overlap with Pied IP), except range of Pied IP (NW Is, Bird's Head, and Bird's Neck), 0–300 m. Also AU. **Taxonomy:** Often treated as a subsp of Pied IP.

Papuan Mountain-Pigeon *Gymnophaps albertisii* Pl 41
(**Mountain**, **Bare-eyed**, or **D'Albertis's Pigeon**)

33–36 cm. Widespread and common, nesting high in the mountains and commuting downslope daily to feed in nearby montane, hill, and lowland forest. This is the greyhound of the NG pigeon world. A sleek, midsize pigeon built for swift, long-distance travel. Social and usually seen in wheeling flocks high over the forest. *The loud whistling of the birds' wings is diagnostic.* **Adult:** Simple color scheme—*dark with a pale breast and a band at the tip of the tail*. Breast patch white in **Male**, dingy in **Female**. Both have *red eye-patch.* **Juv** lacks red eye-patch and is dark and variously mottled with grey and brown. **Subspp (1):** *albertisii* (NG Region and Bismarck Is). **Similar spp:** Imperial pigeons are larger. Collared IP shows a distinct black collar and a pale band in mid-tail. **Voice:** Usually silent. During breeding season it gives a quiet, under-the-breath, low-pitched, upslurred *wooooooo m*. Also a soft, querulous whistle. **Habits:** In small or large flocks, highly mobile, regionally nomadic. Only NG species to swoop down mountain ridges at breakneck speed, wings whooshing. Nests semicolonially on peaks above 2000 m (lower on coastal ranges and islands). Feeds at or below this elevation. Wanders into lowlands at times when not breeding. Cues in on regional abundance of fruit, especially laurels and similar fruits, flocks appearing after months or years of absence. Flock will descend into crown of fruiting tree, birds scrambling about taking fruit, then depart with an explosive burst of whistling wings. Nests on tree branch or tree fern crown; nest a simple twig platform. Egg (1) white. **Range:** NG (all mts, though not reported from Kumawa Mts), Bay Is (Yapen), and SE Is (D'Entrecasteaux: Fergusson, Goodenough), 0–3400 m. Also N Moluccas (Bacan) and Bismarck Is.

COCKATOOS: Cacatuidae

The cockatoos have been recently separated from other parrots and assigned to their own family of 21 species, of which 3 inhabit the New Guinea Region. Originating in Australia and New Guinea, the cockatoos have colonized nearby islands in Wallacea, the Philippines, and Melanesia. They are large parrots, either all black or all white and instantly recognized by their large, curved crest. The crest is folded at rest or snapped open in display when the bird is excited. Their powerful voices are among the loudest of any New Guinea bird. They feed mainly on seeds and nuts, which they chew open with powerful beaks. They form long-term monogamous pairs and nest in tree cavities.

Palm Cockatoo *Probosciger aterrimus* Pl 42

51–63 cm. Uncommon in lowland and hill forest, often depleted near human settlement. *A huge, blackish parrot with high, erectile crest and red cheeks. Massive, hooked beak.* **Adult:** Plumage all black. **Juv:** Beak tipped white; belly with faint pale barring. **In flight:** Long, rounded wings and tail; undertucked beak and flattened crest make for a fist-headed profile; flight is slow, straight, and level with ~4 leisurely flaps, then a short glide. **Subspp (3, in 2 groups):** *aterrimus* (NG minus range of next subsp) exhibits regional variation in body size and crest width; *goliath* (NW Is, Bird's Head and Neck) based on genetic data. **Similar spp:** None really (New Guinea Vulturine Parrot is differently shaped and has red on wings and breast). **Voice:** Distinctive and unlike any parrot or cockatoo. Very loud and far-carrying, yet surprisingly shrill; higher pitched than that of Sulphur-crested C. Calls include a variety of loud, musical whistles, with a large sweep in scale, somewhat squealing—suggestive of the whistles of guinea pig, but at tremendous volume. Alarm calls are jackass-like: *KEEYAANK!* or *EEYOHN!* or *RAAH!!!* **Habits:** Travels singly, in pairs, occasionally in parties of 5–6. Feeds on large, hard seeds of forest trees, such as *Terminalia*, *Canarium*, *Pandanus*, and palm nuts. Nests in a tree hollow; nest lined with splintered twigs. Egg (1) white. **Range:** NG, NW Is (Batanta, Misool, Salawati, Waigeo), Bay Is (Yapen), and Aru Is, 0–750 m, locally to 1350 m. Also AU.

Little Corella *Cacatua sanguinea* Pl 42
(*C. pastinator*)
36–38 cm. Now uncommon in Trans-Fly forest, savannah, and agricultural areas. A medium-sized, white cockatoo with *short crest and pale beak*. Large *bluish eye-patch*. **Adult:** Eye-patch larger and possibly paler than that of **Juv. In flight:** Flight fast, direct, pigeon-like; quick, deep wingbeats; wings narrower than Sulphur-crested Cockatoo. Underwing buff yellow. **Subspp (1):** *transfreta* (endemic). **Similar spp:** Sulphur-crested Cockatoo, with which it sometimes associates, is larger, has a dark beak and a long, yellow crest. **Voice:** Call is a peculiar trisyllabic chuckling cry (AU data). Also a *raah* note shorter than that of Sulphur-crested C. Alarm call is a series of harsh shrieks. **Habits:** Gregarious and noisy; usually in pairs or small flocks, historically in hundreds; now reduced by trapping on Indonesian side. Feeds and roosts communally. Forages mainly on the ground for seeds of grasses and herbs. Visits rice fields. Nests in a tree hollow (AU data). Eggs (2–3) white. **Range:** NG: throughout Trans-Fly, with local movement. Also AU.

Sulphur-crested Cockatoo *Cacatua galerita* Pl 42
38–51 cm. The *often seen white NG cockatoo*. Common in lowland and hill forest and open country, less so near human settlement. A large, white cockatoo with *tall yellow crest and black beak*. S-curved crest thrown forward in flamboyant display. Loudest and most raucous of the NG parrots. **Male:** Iris dark brown. **Female:** Iris red-brown, although some dark brown. **Juv:** Not distinguishable. **In flight:** Heavy bodied; wings rounded, underwing yellow; wingbeats rapid and shallow, followed by a glide. **Subspp (2):** 1 group, *triton* (endemic), blue eye-skin; body size and bill size vary regionally. **Similar spp:** Little Corella co-occurs only in Trans-Fly; see that account. Variable Goshawk white morph can be passed over for a cockatoo. Egrets are similar at a distance. **Voice:** Perhaps the loudest NG bird. Utters a variety of earsplitting notes. Most often a harsh downslurred scream, given singly or repeated (can be confused only with 1 call of the Eclectus Parrot). Contact call: *AH-YAI-YAH*. In flocks: *KAI-YAH!* Individuals when alone give a series of repeated calls, some unexpectedly soft. **Habits:** Gregarious, in pairs or small groups or flocks of 10 or more, but sometimes birds found by themselves. When disturbed it calls loudly and persistently; may break off and drop branches on observer. Sometimes screams at and pursues a hawk or eagle —useful for betraying secretive raptors. Stealthy and quiet when foraging in forest trees for seeds; descends to the ground in savannah country; also raids planted crops like corn, rice, and peanuts. Nests in a tree hollow. Eggs (~2) white (AU data). **Range:** NG and all large and midsize islands, 0–1500 m (2400 m). Also AU; introduced to E Moluccas, Palau Is, and NZ.

NEW GUINEA VULTURINE PARROT and relatives: Psittrichasidae

This relict family includes the New Guinea Vulturine Parrot and the 2 species of vasa parrots in Madagascar. Genetic evidence justifies this association, although the 2 genera are nevertheless somewhat distantly related. The New Guinea Vulturine Parrot has long been recognized for its peculiar morphology; recent molecular studies confirm its distinct position among the parrots.

New Guinea Vulturine Parrot *Psittrichas fulgidus* Pl 42
(**Pesquet's** or **Vulturine Parrot**)
46 cm. Uncommon and confined to primary forest of lower montane zone, hills, and adjacent lowlands; absent from flat lowlands far from hills. Hunted for its red feathers and thus widely extirpated—one of the most vulnerable NG birds. A noisy, crow-sized parrot with *vulture-like head and slaty black plumage* accented by *red on belly, rump, and wings*; unique bare face visible at close range. *Distinctive profile: robust body, long neck tapering to a small, narrow head; hooked beak is long for a parrot.* **Male:** Small red streak behind eye, lacking in **Female. Juv:** Red markings duller. **In flight:** Diagnostic red in the wings and underparts is not always visible. Listen for the voice and look for the profile: *narrow head, broad wings, medium tail.* Rapid, shallow wingbeats followed by short glide. **Similar spp:** None, except at a distance. Against the sky, flying Palm Cockatoo, Eclectus Parrot, Torresian Crow, and this species all look equally black, but each has different shape and voice. **Voice:** Usually seen flying, but heard before it comes into view, so voice is key to discovery. Varies

regionally. A loud series of squawks similar to that of Sulphur-crested Cockatoo but less shrill, deeper, and coarser with an almost honking quality. Or a bellowing roar, like some large animal. Notes repeated in pairs or a short series. Quality suggests a sheet of canvas being torn or a deep, hollow retching—a terrible sound. **Habits:** Singly, in pairs, or in small parties, especially where not hunted. Active and restless, with lory-like hopping gait, wing-flicking, and crane-peering behavior. Specializes on a diet of large figs of only a few species; also reputedly eats fruit or inflorescences of large climbing pandans (*Freycinetia*) and mangos. Nests in a cavity excavated into a dead tree trunk at midstory height. Eggs (2) white. **Range:** Endemic. NG: nearly all mountain ranges, but patchily distributed, 50–1500 m (2000 m). Apparently absent from Fakfak and Kumawa Mts; no information for Cyclops Mts.

AUSTRALASIAN PARROTS: Psittaculidae

The remaining 43 species of New Guinea parrots belong to this large and diverse family (~170 spp) that includes nearly half of all parrots worldwide. The family reaches its greatest evolutionary development in Australasia and from there extends far into Asia, Africa, and Oceania. The 7 groupings within the New Guinea Region are the tiger-parrots, lories and lorikeets, fig-parrots, hanging parrots, king-parrots and allies, Eclectus Parrot and allies, and pygmy parrots. Each group is adapted to feeding on either nectar, fruits and seeds, or fungus. Parrots appear stocky because of the powerful flight muscles that carry them for long distances in search of food, very short legs for scrambling about in trees, and large heads with strongly hooked beaks for chewing their food and excavating a nesting chamber. Australasian parrots range in size from the tiny pygmy parrots (world's smallest) to the cockatoo-sized Eclectus Parrot. This is a colorful family, painted in every of hue of the rainbow, although green predominates. Parrots are highly vocal, with unmusical calls that range from loud whistles and screams to quiet twittering. New Guinea species lay their clutch of white eggs in a cavity chewed into a clump of epiphytes (lorikeets), a tree branch or trunk (lories and most parrots), or termite nest (pygmy parrots). Chicks are tended by both parents, and by helpers in some species. They are often kept as pets, although it is usually difficult to feed them an adequate diet or keep them happy as adults.

Parrots are among the highlights of New Guinea birding. Fortunately, they are easy to find, conspicuous for their loud voices and active flight. Yet actually observing and identifying them is another matter. Upon alighting in a leafy treetop, they instantly vanish—dozens can disappear into a single tree with not one bird visible, all perched perfectly still, cloaked by their protective green plumage and silently alert for signs of danger. Flying in the open the birds are plain enough to see, but their colors don't show against a bright sky. How can so many kinds of parrots all look black? Yet identifying parrots on the wing is a useful skill. Try this from a river or forest clearing where birds are likely to pass over. Note shape and size first, listen to the calls, then carefully look for field marks: the color of the bird's bill, underwing pattern, and tail, as these can vary dramatically. The plate devoted to small parrots in flight will help you (plate 49). Parrots often land in the open on high perches such as dead branches or a clump of foliage; from here they can look around to see what their mates are doing. Scan the treetops; this is the perfect place to view them. As for that invisible flock in the leafy tree—just wait quietly yourself. Though it may take a while, eventually one bird will begin to move about and reveal its location, and the rest will also become active. Finally, realize that a flock of feeding parrots can be remarkably stealthy, not uttering a sound and thereby going undetected. Yet they do make noise—listen for falling bits of fruit as these patter through the leaves and land on the forest floor.

Tiger-Parrots. Tiger-parrots are stubby, short-winged birds endemic to the mountain forests of New Guinea. The 4 quite similar species are recognized by their striped green plumage, red undertail-coverts, and pearl-grey beaks. Residents of the forest interior, they *never* make long flights over the canopy. Found singly, in pairs, or rarely in parties of up to 4, they are inconspicuous and typically located by call and noticed crawling about within leafy branches. Most are rather quiet, calm, and tame.

Brehm's Tiger-Parrot *Psittacella brehmii* Pl 47
21–24 cm. The most often seen tiger-parrot. Common in cloud forest, especially edges and clearings, where it feeds on seeds and fruit in second growth. The *largest* tiger-parrot, with *all-brown head sharply defined* from the green underparts; *red under tail only*. **Male:** Yellow crescent on side of neck. **Female and Juv:** Breast barred. **Subspp (4):** Minor. **Similar spp:** Painted TP, which shares the higher elevations of Brehm's habitat, is slightly smaller and has a red rump (most populations) and/or bluish cheeks or breast; its western race, with yellow-green striped rump, is most similar to Brehm's. The other 2 tiger-parrots are visibly smaller and do not exhibit the sharp contrast between head and breast. **Voice:** Contact call is a quiet and plaintive nasal downslur, *ee-yurr*. **Habits:** Forages at all levels in the forest; at the forest edge, descends into shrubs and onto the ground. Feeds on seeds and fruits; especially fond of bleeding heart (*Homalanthus* spp). Takes grit or clay from road cuts. No nesting information. **Range:** Endemic. NG Central Ranges, Bird's Head, and Huon, 1500–2600 m (1150–3200 m).

Painted Tiger-Parrot *Psittacella picta* Pl 47
17–19 cm. Local and uncommon in high-mountain cloud forest and timberline shrubbery, generally above the elevational range of slightly larger Brehm's TP. *Head two-toned:* brown with *cheeks blue or with bluish cast.* Only tiger-parrot with *rump reddish or yellowish*, barred black. **Male:** *Breast patch blue or bluish;* yellow neck-crescent. **Female and Juv:** Cheeks blue (bluish in adult Male), breast barred. **Subspp (3):** *picta* (SE Pen) head red-brown, rump red, Male with blue breast; *excelsa* (E Ranges) resembles preceding subsp but head dull brown; *lorentzi* (W Ranges and probably Border Ranges) crown dull brown, cheeks and Male's breast blue-green, rump yellow (not red). **Similar spp:** Brehm's TP is larger, has rump green with fine black barring (like back) and evenly brown head, without blue in cheeks or on breast. The other 2 tiger-parrots are smaller, mainly at lower elevation, and show different color patterns. **Voice:** Soft, nasal, and slightly musical. Contact call in East *nhree arehn;* in West *err-eee,* suggesting a streaked honeyeater. **Habits:** Similar to Brehm's TP. No nesting information. **Range:** Endemic. NG Central Ranges, 2500 m to timberline.
Taxonomy: Western *lorentzi* quite different, perhaps deserving species status.

Modest Tiger-Parrot *Psittacella modesta* Pl 47
14–15 cm. A small tiger-parrot locally common in high-mountain cloud forests. Where it co-occurs with same-size Madarasz's TP, the 2 species tend to segregate by elevation, with Modest generally found above 1700 m. Very inconspicuous; look for it in small fruiting trees at forest edge. **Male:** *Brown of head merging into green on breast well below bend in wing,* and usually with *yellow collar on hindneck*. **Female and Juv:** Barred, orange breast. **Subspp (3, in 2 groups):** *modesta* (Bird's Head) Male head all brown; *collaris* (Central Ranges) Male with yellow nape-collar. **Similar spp:** Madarasz's TP is nearly identical to Male Modest, see Madarasz's account; female Madarasz's has blue-green head (not brown) and lacks ventral barring. The other 2 tiger-parrots are much larger. **Voice:** Little information. **Habits:** Frequents forest midstory and small trees at forest edge. A typical tiger-parrot in behavior, sluggish. No nesting information. **Range:** Endemic. NG: Bird's Head to central E Ranges (absent eastward of Mt Hagen), 1200–2800 m.

Madarasz's Tiger-Parrot *Psittacella madaraszi* Pl 47
14–15 cm. Uncommon in mountain forests. Where its range overlaps with similar Modest TP, the 2 species usually segregate by elevation at ~1700 m, with Madarasz's occurring lower. **Male:** Resembles male Modest TP but *brown of head merges into green on breast at bend of wing, yellow hind-collar is lacking*, and *more yellow spotting appears on back of head*. **Female and Juv:** Unique—*blue-green head; barring down the back but little or none on the front*. **Subspp (4, in 2 groups):** *madaraszi* (Central Ranges) Female with red in nape; *huonensis* (Huon) Female lacks red in nape, dorsal barring reduced. **Similar spp:** Modest TP male nearly identical (see above), but female very different, with brown head and barred, reddish breast. Other tiger-parrots larger, with different color patterns. **Voice:** Call is soft

and easily overlooked, a plaintive upslurred phrase, reminiscent of other tiger-parrot calls, but higher pitched and easily distinguished: *huwee hee?* or *ee o oe*. **Habits:** Similar to Modest TP. No nesting information. **Range:** Endemic. NG: Central Ranges, Foja Mts, and Huon, 1200–2500 m (rarely to lowlands in SE Pen).

Lories and Lorikeets. This is a diverse and species-rich group of brush-tongued parrots that feed on flower nectar and pollen, although some species eat fruit and seeds as well. Most species are gregarious and usually seen traveling in flocks flying over the forest or gathered at flowering trees. Their presence at any locality is dependent on the seasonal availability of flowers.

Plum-faced Lorikeet *Oreopsittacus arfaki* Pl 43, 49
(Whiskered Lorikeet)
15–18 cm. Common in cloud forest, especially at higher elevation. A *tiny*, slender, *dark green* lorikeet recognized at a glance by its *red-tipped tail*. *Purple cheek* marked with a *white streak*. *Beak black* (most lorikeets have red or orange beak). **Male:** Red cap, lacking in **Female**. **Juv:** Similar but with dusky scalloping and red forehead in Male. **In flight:** Unique *all-red undertail*; underwing-coverts also red. **Subspp (3, in 2 groups):** *arfaki* (Bird's Head, W Ranges) red patch on lower breast and another on flanks; *grandis* (E Ranges, Huon, SE Pen) all green below. **Similar spp:** None. Other lorikeets lack the red undertail. **Voice:** A soft, short, repeated note, *ts*, weaker than those of other montane lorikeets. A whispered twittering audible at close range, suggestive of a fig-parrot. **Habits:** Travels in small vocal flocks, flying either above or below the forest canopy; flight swift and acrobatic on whirring wings. Forages at canopy and midstory flowers, particularly those of epiphytes. No nesting information, but birds observed exploring and entering masses of epiphytic moss. **Range:** Endemic. NG: Bird's Head, Central Ranges, and Huon, 1700–3650 m.

Red-chinned Lorikeet *Charmosyna rubrigularis* Pl 43, 49
18–20 cm. The *only small lorikeet on Karkar I*, where it occurs throughout, but especially in montane forest. **Adult:** A small, bright green lorikeet with *red chin* and *yellow-tipped tail*. Beak red. **Juv:** Similar, but beak paler with dark base; tail shorter. **In flight:** Green underwing; yellow stripe across wing feathers; yellow undertail. **Similar spp** None on Karkar. **Voice:** A high, shrill note, not as staccato or sharp as those of some other lorikeets. A quiet *seezp* while feeding, repeated in quick succession in flight. Also a harsher note, *see-air*. **Habits:** In small flocks of up to 10, remaining in forest canopy, foraging at flowering trees with other nectar eaters. No nesting information. **Range:** Karkar I, off Sepik-Ramu coast, 0–1500 m. Also mts of Bismarck Is.

Striated Lorikeet *Charmosyna multistriata* Pl 43, 49
(Streaked Lorikeet)
19 cm. In hill and lower montane forest; usually uncommon, although sometimes gathering in large numbers. **Adult:** A *dark green* lorikeet with *yellow streaking from face to belly* and striking *blue-and-orange beak*. Iris red. **Juv:** Markings muted; iris dark. **Similar spp:** Goldie's L is also streaked below, but with reverse pattern—dark streaks on paler yellowish green background; also dark purple face and red cap. Pygmy L is streaked only on the breast and is much smaller, with an all-orange beak and brief, dark tail. **Voice:** Calls are drawn out whistles, in 1, 2, or 3 notes. A calling flock sounds like chiming bells. Also *kss* note similar to other *Charmosyna* lorikeets. **Habits:** Travels in pairs or small groups, foraging at flowers of canopy trees and epiphytes. No nesting information, although has been observed exploring epiphytic moss cushions in the same manner as other *Charmosyna*. **Range:** Endemic. NG, southern slopes from W Ranges to E Ranges (Crater Mt), 80–1800 m.

Pygmy Lorikeet *Charmosyna wilhelminae* Pl 43, 49
11–13 cm. *Smallest lorikeet*. Infrequently encountered, but gathers at flowering trees in hill and montane forest. Difficult to discern color pattern. *Identify by tiny size and the abbreviated, pointed tail.* Breast streaked yellow; look for dark blue patches fore and aft (cap and uppertail-coverts). Beak red. **Male:** *Bright red rump and underwing*; olive nape patch. **Female:** These parts green. **Juv:** Similar to Adult, but markings reduced or absent; beak and iris brown. **In flight:** Obviously smaller than other lorikeets; entire underwing red in Male, green in Female. **Similar spp:** Striated L larger, darker, and entirely streaked below, not just on breast; lacks red underwing. Red-flanked L larger, lacks breast streaking, and has longer, red-tipped tail and different facial pattern. Papuan Hanging Parrot of

similar size and visits flowers, but has compact shape with stubby, rounded tail and red throat.
Voice: Flight calls like those of other small lorikeets, although higher pitched and fainter. Quiet calls given between pair: *ts ts tsee*, reminiscent of call of pygmy parrot. **Habits:** Forages in pairs or flocks in canopies of flowering trees, usually in the company of other lorikeets. Difficult to observe in the treetops as it is so small, active, and easily lost in the foliage. No nesting information. **Range:** Endemic. NG: Bird's Head, Central Ranges, and Huon, 1000–2400 m, occasionally to foothills.

Red-fronted Lorikeet *Charmosyna rubronotata*   Pl 43, 49
15–18 cm. Only in lowlands and hills of Bird's Head and northern NG, where infrequently found. Compare with better-known Red-flanked L. Co-occurs with Red-flanked but generally at higher elevations. Differs by *tail all black above (not green with red tail tip as in Red-flanked) and red rump patch*. Beak red. **Male:** *Red breast patch confined to near bend of the wing* (not a long stripe down flanks); *crown red and chin green* (not red). **Female:** Only traces of yellow streaking behind the eye. **Juv:** No information. **In flight:** Similar to Red-flanked L but lacks red in tail and yellow wing-stripe obscure. **Subspp (2):** Minor. **Similar spp:** Red-flanked L. **Voice:** Similar to Red-flanked L but sharper. **Habits:** Similar to Red-flanked, but said to be mainly a hill species. No nesting information. **Range:** Endemic. NG, NW Is (Salawati), and Bay Is (Biak), 0–900 m. NG: Bird's Head, NW Lowlands, and Sepik-Ramu.

Red-flanked Lorikeet *Charmosyna placentis*   Pl 43, 49
15–18 cm. The widespread green lorikeet of lowland and hill forest and open habitats; seen in villages with flowering coconut palms. Locally common, but may be absent. Either sex shows a unique pattern of red, yellow, and blue. *Tail gren dorsally with red-orange tip* and yellow underside. Beak red. **Male:** Gaudily painted with *long red flank-stripe, red chin highlighting the coral beak, blue ear-coverts, and bright yellow-green forehead*. **Female:** Lacks these colors but shows a spray of *yellow streaking across ear-coverts*. **Juv:** Similar to Adult but duller and color patches reduced. **In flight:** Underwing-coverts red (Male) or green (Female); pale yellow streak across primaries. **Subspp (5, in 2 groups):** *placentis* (NW Is, Bird's Head, NW Lowlands, S Lowlands) blue rump patch; *subplacens* (Sepik-Ramu to SE Pen and SE Is) rump all green. **Similar spp:** See Red-fronted L. Plum-faced L also with red-tipped tail, but has purple face and inhabits cloud forest. Other green lorikeets are streaked below. **Voice:** Call a short, sharp, caustic *kssk-kssk-kssk*. **Habits:** In pairs or small flocks foraging among flowers of tree canopy and palms. Nest is a cavity excavated in an arboreal termite nest, moss clump, or base of a stag-horn fern. Eggs (2) white. **Range:** NG, NW Is (Batanta, Gebe, Kofiau, Misool, Salawati, Waigeo), Aru Is, and SE Is (Woodlark), 0–1200 m. Also Moluccas, Bismarck, and Solomon Is.

Fairy Lorikeet *Charmosyna pulchella*   Pl 44, 49
(Little Red Lorikeet)
16–19 cm. *The small red lorikeet* inhabiting hill and montane forest, locally common. Red head and underparts; *yellow streaking on breast*. Note *multicolored tail grading from green to red to yellow tip*. Beak orange. **Male** rump green, **Female's** with yellow patch on side. **Juv:** Similar to Female but with green replacing much of red; beak dark. **In flight:** Red underwing-coverts, yellowish undertail. **Subspp (2):** *pulchella* (most of NG, excluding range of next subsp) underparts red; *rothschildi* (N slope of W Ranges, plus Foja and Cyclops Mts) larger; ragged green patch on breast. **Similar spp:** Josephine's L larger, has tail red to the base, dark purple shanks and belly patch, and lacks yellow breast streaks. **Voice:** Calls vary geographically. In the West, a nasal *ks* given 2–3 times, and a weak *ss*. In the East, a short high-pitched note, less sharp or staccato than that of Red-flanked L. These calls and those of other small lorikeets are distinguishable only with practice. **Habits:** Similar to other small lorikeets with which it associates. Observation suggests it may nest in a cavity excavated in an epiphytic moss clump. Eggs (2) white (in captivity). **Range:** Endemic. NG mts (records from all but Kumawa and Wandammen Mts), 500–1800 m (0–2100 m).

Josephine's Lorikeet *Charmosyna josefinae*   Pl 44, 49
23–25 cm. Uncommon in hill and lower mid-mountain forest. A western species absent from much of the E Ranges, eastward. Intermediate in size and looks between Fairy and Papuan/Stella's Lorikeets, yet surprisingly difficult to separate from the latter. Look first at the tail. *Tail all red above with yellow tip* (other lorikeets have green tails, at least at the base)*; tail length less than or equal to head-and-body length* (much longer in Papuan/Stella's). Beak orange. **Male:** Lower back red. **Female:** Lower back yellow or greenish yellow. **Juv:** Dusky beak; feathers with dark-scalloped edges; black-and-lilac patches

washed green; Female with little yellow on rump. **In flight:** Tail proportionately longer than that of Fairy L, but much shorter than Papuan/Stella's. **Subspp (3):** Minor, varying mainly in size of belly patch. **Similar spp:** Fairy, Papuan, and Stella's Lorikeets. **Voice:** Flight call, a high-pitched *kris*! When perched, it gives a nasal *engg* note, similar to that of Stella's L. **Habits:** Travels in pairs or small parties, feeding in flowers of canopy trees and in midstory epiphytes and vines, often in company of Fairy L. Inside an epiphytic moss cushion high on a tree limb. No egg information. **Range:** Endemic. NG: Bird's Head to Central Ranges as far east as Purari and Jimi river basins; also Foja, Cyclops, and N Coastal Mts, 700–1800 m (0–2200 m).

Papuan Lorikeet *Charmosyna papou* Pl 44, 49
36–41 cm. Common in cloud forest of Bird's Head. **Adult:** *Easily separated from Stella's L, its counterpart in the rest of NG, by multiple plumage details* including: only red morph (no black morph); blue-and-black crown patch of short feathers positioned above the eye (the blue-and-black patch of Stella's is much longer and positioned behind the eye and connected to it by a black eye-stripe); black eye-stripe absent; presence of a red nape patch, a black crescent on the hindneck, a yellow patch on side of breast at the shoulder, and another on lower flank above the legs (all these absent in Stella's); undertail orange-yellow (yellowish green in Stella's); sexes similar (different in Stella's). **Juv:** As for Stella's. **Similar spp:** See Josephine's L and Fairy L. **Voice and Habits:** Presumably similar to Stella's. **Range:** Endemic. NG: Bird's Head (Arfak and Tamrau Mts), presumably ~1600–2900 m. **Taxonomy:** Split from Stella's L.

Stella's Lorikeet *Charmosyna stellae* Pl 44, 49
(**Papuan Lorikeet**, **Fairy Lory**, *C. papou*)
36–41 cm. Common in cloud forest. One of the most beautiful and enchanting of all parrots—striking in form, movements, color. Unmistakable. Sleek yet compact, with *long neck*, short wings with an exaggerated taper, and *long, drooping tail-streamers*. Flies like a missile propelled on *loudly whirring wings*, followed by a fluttering, yellow-tipped tail; weaves through openings in the treetops. Both red and black morphs show a plumlike, purplish tint. **Adult:** *Red morph* differs from Papuan L by the elongate blue crown patch reaching back to the nape where it joins the black eye-streak and nape patch; lack of yellow in the underparts in most of its range; and plumage difference between the sexes. Also, this species has a *black morph* in which purplish and greenish black replaces red. Beak red. **Male:** Lower back red; **Female's** is bright yellow (or green in black morph). **Juv:** Dusky beak, feathers with dark-scalloped edges, much shorter tail; possesses a pair of erectile occipital tufts, visible at close range when bird is playing. **Subspp (3, in 2 groups):** *stellae* (Central Ranges and Adelbert Mts) breast all red; *wahnesi* (Huon) yellow band across breast (black morph rare). **Similar spp:** Josephine's L rarely visits cloud forest, is smaller, and has a much shorter tail that lacks waggling streamers and is red on the upper surface. **Voice:** Flight call a loud, upslurred and grating nasal downslur, *queea*! A call when feeding or at rest is a nasal *nreeennnggg*! that increases in volume. **Habits:** Singly or in pairs. Aggressive and bullying toward other lorikeets at flowering trees and epiphytes. Fond of flowering umbrella plants (*Schefflera* spp). Black individuals, often paired with red, occur mainly at higher elevations. Observation suggests that these lorikeets nest in a cavity excavated within a large mossy clump of epiphytes. Eggs (2) white. **Range:** Endemic. NG Central Ranges, Adelbert Mts, and Huon, 1800 m to timberline (rarely down to 1400 m). **Taxonomy:** Split from previous species.

Yellow-billed Lorikeet *Neopsittacus musschenbroekii* Pl 43, 49
(**Musschenbroek's Lorikeet**)
19–23 cm. Common in mountain forests and around highland settlements with scattered trees. This and the following species are very similar. Both are small, *mostly greenish, with red breast and yellow-streaked cheeks*. In flight, note bright red underwing. **Adult:** Yellow-billed differs from Orange-billed L as follows: (1) *flanks yellowish green, contrasting with emerald green wings* (vs flanks and wings both emerald green), (2) *undertail orange-yellow* (vs dull green with red), (3) *beak larger and yellow* (vs orange), (4) larger size, and (5) crown and nape with more yellow streaking. **Juv:** Darker beak and eye; less red on breast, in patches. **Subspp (2):** Minor. **Similar spp:** (For both spp) no other lorikeet has the combination of greenish head and red breast. **Voice:** When perched, a high-pitched, slurred, double-noted shriek: *sweet sweew*, the second note at a lower pitch (thereby differing from call of other small lorikeets). Also a trisyllabic, descending and musical trill, *shree-daloo*. Flight call a short, staccato, shrill *ks* or *ts*. Calls are lower pitched, more nasal, and more hollow sounding than those of Orange-billed L. **Habits:** The two *Neopsittacus* species travel separately, in small flocks,

although they may forage together in a flowering tree. Yellow-billed commonly feeds on unripe seeds of forest trees (such as oaks, beeches, podocarps) as well as at flowers. Rarely associates with parrots of other genera. Recorded feeding in casuarinas and eucalyptus in gardens, even feeding on annual weeds at ground level. Altitudinal range overlaps broadly with that of Orange-billed. Observed crawling in and out of tree hollows, where it likely nests. Eggs (2) white (in captivity). **Range:** Endemic. NG mts of Bird's Head, Central Ranges, Foja Mts, and Huon, 1200–2300 m (1100–3000 m).

Orange-billed Lorikeet *Neopsittacus pullicauda* Pl 43, 49
(**Emerald Lorikeet**)
18–20 cm. Common in cloud forest and openings. See previous species. **Adult:** Compared with Yellow-billed L, *flanks dark emerald green, same as wings*, i.e., no contrast (vs flanks paler, yellow green); *beak is smaller, more orange*; and *underside of tail is dull green with red at the base*. **Juv:** As for previous species. **Subspp (3, in 2 groups):** *pullicauda* (E Ranges to Huon and SE Pen) red breast; *alpinus* (W and Border Ranges) orange breast, darker red belly. **Similar spp:** See Yellow-billed L. **Voice:** Similar to Yellow-billed, but higher pitched, more musical, and quieter; recognizable with practice. **Habits:** In small flocks, sometimes with previous species in flowering trees, but generally occupies higher elevations than Yellow-billed L. Feeds primarily on flowers, also fruits, berries, and seeds. Observed entering tree holes, probably to nest. **Range:** Endemic. NG Central Ranges and Huon, 2100–3800 m, rarely lower.

Black-capped Lory *Lorius lory* Pl 44, 49
(**Western Black-capped Lory**, *Domicella lory*)
25–30 cm. *The widespread large red lory with a black cap*; common in forest from lowlands to mid-mountains. This and Purple-bellied L are very similar and in flight share the same stocky profile with head and tail protruding equally. The unusually short wings whir rapidly, flashing yellow from the undersides of the primaries. The species differ in range, calls, and plumage details. **Adult:** Most populations of Black-capped possess a *black hind-collar and in some races a black "vest"* (these are actually dark blue or purple), lacking in Purple-bellied L; this species shows a *dark cere* (skin around the nostril) that is white in Purple-bellied. Beak orange. **Juv:** Dark beak; back patch greenish rather than purple; often with more red in plumage, less black. **Subspp (7, in 5 groups):** *lory* (NW Is, Bird's Head and Neck) has purple-black vest, red underwing-coverts; *erythrothorax* (S Lowlands, SE Pen, except range of next subsp) breast mostly red, hindneck collar blue-black; *somu* (foothills of upper Fly R east to Purari R) similar to preceding, but collar absent or much reduced; *jobiensis* (NW Lowlands to Sepik-Ramu, also Bay Is of Mios Num and Yapen) shows reduced dark vest, and underwing dark blue or black (not red as in all previous races); *cyanauchen* (Biak I) also with dark underwing, but blue nape and black crown merge, and breast mostly red. **Similar spp:** Purple-bellied L. Other large lories show either a pointed tail or longer, rounded tail. **Voice:** Call a short series of varied melodious whistles or squeals, more like that of a wader or Golden Myna than of a parrot: *wheedle wheedle*. "Subsong" is a long series of phrases, each of a few notes repeated over and over before beginning a new phrase. Sometimes gives a monotonous series of identical notes suggesting a goshawk. **Habits:** Usually singly or in pairs, less often in small flocks, foraging in canopy of forest and edge for flowers, also some fruit and small insects. Nest excavated in a tree hollow. Eggs (2) white (in captivity). **Range:** Endemic. NG, NW Is (Batanta, Misool, Salawati, Waigeo), and Bay Is (Biak, Mios Num, Yapen), 0–1500 m.

Purple-bellied Lory *Lorius hypoinochrous* Pl 44, 49
(**Eastern Black-capped Lory**, *Domicella hypoinochroa*)
25–28 cm. A locally common coastal species on the SE Pen mainland; common everywhere on SE Is in forest, coconut palms, and villages, from sea level to mountaintops. *From Black-capped L by its lack of a black hind-collar, its white cere, and a distinctive voice*. Another subtle difference is the shape of the Adult neck feathers: lanceolate textured in Purple-bellied and smooth in Black-capped. **Adult:** Beak orange, back patch maroon. **Juv:** Beak dark, back patch green. **Subspp (3):** Minor. **Similar spp:** On mainland see Black-capped L. None on the islands. **Voice:** Call a loud, drawn-out, nasal wail that rises, then falls—harsh and unmusical, very different from musical squeals of Black-capped. Also a variety of other calls and whistles: *REE!*, or *UWEE!*, or *rayer*, or a repeated tonal whistle, and more. These may include "subsong" in manner of Black-capped. **Habits:** Similar to Black-capped. No nesting information, although seen attending hollows in large trees. **Range:** NG: SE Pen (westward almost to Lae in N and to Cape Rodney in S) and SE Is (all but the smallest, most remote islands); coastal on mainland, 0–2000 m on islands. Also Bismarck Is.

Goldie's Lorikeet *Psitteuteles goldiei*   Pl 43, 49
(*Trichoglossus goldiei*)
19 cm. Inhabits mountain forests and some highland towns; may descend to foothills when foraging. Gregarious, nomadic, and unpredictable—absent most places most of the time, though found regularly in certain highland towns where attracted to tall flowering trees such as eucalypts and silky oaks (*Grevillea*). *Two-toned: dark green above, pale apple-green below.* No other lorikeet has *dark-streaked underparts*. **Adult:** *Purple head with red cap*, note *dark beak and eye*. **Juv:** Similar; red on forehead only. **In flight:** Yellow stripe across flight-feathers. **Similar spp:** None if seen well. **Voice:** Call is a high-pitched shriek typical of other lorikeets, but very dry, less musical and with some vibrato; quieter and less shrill than those of Rainbow L. While feeding, a soft monosyllabic note resembling a hiss or loud whisper. **Habits:** In flocks, sometimes 20 or more. Flight swift and direct; undertakes long flights. Feeds in canopy of tall trees. This species likely capitalizes on the temporary local abundance of flowering trees, moving from one locale to the next in search of bloom. Nests have been found in the crowns of pandanus trees. Eggs (2) white (in captivity). **Range:** Endemic. NG: Central Ranges and Huon, 1500–2300 m (0–2800 m). **Taxonomy:** Placement to genus uncertain.

Dusky Lory *Pseudeos fuscata*   Pl 45, 49
28 cm. Locally common to abundant in forest and near settlements from the lowlands to mid-mountains. Most gregarious of the lories; often in large, high-flying flocks that are exceedingly vocal. **Adult:** A medium-sized, *dusky brown lory with orange markings* and conspicuous, *bright orange beak and facial skin*. Neck heavily scalloped. Note distinctive *white rump patch* and pale cap. Individuals vary in hue: red, orange, or yellow. **Juv:** Darker beak, facial skin, and iris; plumage pattern obscured. **In flight:** Yellow streak across flight-feathers, orange underwing-coverts and undertail. **Similar spp:** The *Chalcopsitta* lories all have black beaks, are larger and longer-tailed. Rainbow L has much green and a pointed tail. These large lories all fly high in flocks and shriek, but can be distinguished by their flight. Dusky flies fast and straight. Rainbow flies fast with abrupt twists of direction. The *Chalcopsitta* lories flap their wings vigorously but progress with low velocity. **Voice:** Loudest of the lories. Flocks call continuously and can be heard at great distance, sounding somewhat musical at first but more irritating on approach. This contact call is a powerful, harsh, grating screech that resembles Rainbow L, but the phrases are shorter, louder, higher pitched, and not upslurred. Alarm call is even more intense, obnoxious, and ultimately intolerable at close range—an effective deterrent. **Habits:** Highly nomadic and perhaps migratory. Makes long daily flights from communal roosts. Forages in canopy of forest, edge, and plantations of coconut, teak, and coffee (attracted to flowering *Leucaena* shade trees). Feeds on flowers, rarely fruit and insect pupae. Birds disperse in pairs into the forest to nest. Nest is in an excavated tree cavity. Eggs (2) white (in captivity). **Range:** Endemic. NG, NW Is (Salawati), and Bay Is (Yapen), 0–1500 m, rarely to 2400 m.

Black Lory *Chalcopsitta atra*   Pl 45, 49
(Rajah Lory)
30 cm. Fairly common in lowland forest edge, swamp forest, mangroves, open country, plantations, and settlements, and presumably rainforest. **All plumages:** *A large, black lory with long, rounded tail.* Note *black beak*. **In flight:** Black with flame-colored undertail; eastern race additionally shows red in face and underwing-coverts. All *Chalcopsitta* species can be distinguished from smaller Dusky and Rainbow Ls by distinctive flight: rapid, shallow wingbeats but slow progress, as though the birds are working hard for little return. **Subspp (3):** *atra* (W Bird's Head, NW Is of Batanta and Salawati) all black, red only in base of tail; *insignis* (E Bird's Head, Bird's Neck) red in face, flanks, thighs; *bernsteini* (Misool I) forehead usually tinged red, thighs red. **Similar spp:** In silhouette, other lories appear black. Identify lories by shape, locality, and in good light, coloration. Ranges of Yellow-streaked and Black Ls meet on Bird's Neck, with evidence of hybridization. In poor light, Yellow-streaked may appear black, so look for green reflection and yellow streaking below. **Voice:** Contact call is a high screech, more shrill than any other lory, but not as loud as that of Dusky or Rainbow. All *Chalcopsitta* species have very similar voices. **Habits:** In pairs, threesomes, and flocks flying with the shallow wingbeat typical of the genus. No nesting information. **Range:** Endemic. NG (Bird's Head and W Bird's Neck) and NW Is (Batanta, Misool, Salawati), 0–150 m.

Brown Lory *Chalcopsitta duivenbodei* Pl 45, 49
30 cm. This is the northern counterpart of the Black and Yellow-streaked Lories. Uncommon in coastal lowlands, generally not venturing far inland; forest, edge, open habitats. **All plumages:** *A large, brown lory with distinctive yellow and violet markings*; black beak framed above by a yellow forehead and below by a yellow crescent around the mouth. Rump, undertail-coverts, and primaries violet. **In flight:** *Yellow flash in wing*, from underwing-coverts; *yellow undertail*. **Subspp (3):** Minor. **Similar spp:** Dusky L is smaller, with an orange beak and two broad yellow or orange bars on the breast. **Voice:** Call nearly identical to Yellow-streaked, a high, thin buzzing screech distinctive to the genus. **Habits:** Similar to Yellow-streaked. No nesting information from the wild. **Range:** Endemic. NG: NW Lowlands and Sepik-Ramu, 0–150 m.

Yellow-streaked Lory *Chalcopsitta scintillata* Pl 45, 49
(**Greater Streaked** or **Yellowish-streaked Lory**)
30 cm. Locally common in lowland forest along rivers, in swamp forest, mangroves, and semi-open habitats including savannah and plantations. **Adult:** A *large dusky green lory with a red forehead*; red patches also on throat and thighs; yellow-and-green streaking often difficult to see. **Juv:** Red forehead reduced; overall somewhat duller. **In flight:** Blackish with red forehead and flame-colored undertail; western race additionally shows red underwing-coverts. **Subspp (3):** *scintillata* (Bird's Neck and S Lowlands to lower Fly R) dark green with red underwing-coverts; *chloroptera* (eastern S Lowlands from upper Fly R to southern SE Pen) similar, but underwing-coverts green with or without some red; *rubrifrons* (Aru Is) broad orange streaking on breast. **Similar spp:** See Black L, which meets Yellow-streaked on the Bird's Neck. Dusky L is also dark, but has more compact shape, an orange beak, and orange belly and bands across the breast. **Voice:** Flight calls shrill (weaker than those of Dusky and Rainbow Lories) and include a series of several buzzing screeches in descending series and a burst of musical twittering. Also, when perched, a variety of typical lory calls. **Habits:** Usually in pairs or small parties, feeding on nectar and fruit in treetops. Observed inspecting tree hollows, presumably for nesting. **Range:** Endemic. NG and Aru Is. NG: Bird's Neck, S Lowlands, and Trans-Fly to southern SE Pen, 0–800 m, but usually flat lowlands.

Violet-necked Lory *Eos squamata* Pl 45, 49
(**Moluccan Red Lory**)
28 cm. The *red-and-black lory of the NW Is*. Primarily a tramp species found on tiny islets and small islands. Favors coastal habitats, including villages and coconut plantations. Fairly common. **Adult:** Much individual variation in extent of the *smudgy purple collar*. Beak orange. **Juv:** Beak dark; dark-scalloped appearance. **Subspp (1):** *squamata* (endemic). **Similar spp:** None within its restricted island range. Black-winged L occurs elsewhere and possesses a blue ear patch and black wings and back. **Voice:** Flight call quite similar to that of Rainbow L but said to be less harsh, more musical. **Habits:** A flocking, wide-ranging island species that visits flowering and fruiting trees in small flocks. No nesting information. **Range:** NW Is (Batanta, Gebe, Kofiau, Misool, Waigeo, and smaller associated islands), 0–500 m. Also Moluccas.

Black-winged Lory *Eos cyanogenia* Pl 45, 49
(**Biak Red Lory**)
30 cm. The *red-and-black lory of the Bay Is* (minus Yapen). Locally common and conspicuous throughout in forest and open habitats; found along the coast near villages. **Adult:** Note black saddle on wings and back, and vivid blue ear patch. Beak orange. **Juv:** Beak dark; red plumage with dark scaling. **Similar spp:** None within its restricted island range, but compare with local races of Black-capped and Rainbow Ls. Violet-necked L occurs elsewhere and lacks the black wings and back and violet ear patch. **Voice:** Contact call is a high screech like that of Rainbow L, but stronger, more abrupt, and less prolonged. **Habits:** Similar to Violet-necked L. Gregarious and seemingly nomadic. Observed tending a possible nest in a tree hollow. **Range:** Endemic. Bay Is: Biak, Manim, and Numfor (possibly Mios Num), 0–500 m (likely higher).

Rainbow Lorikeet *Trichoglossus haematodus* Pl 45, 49
(For synonyms, see Subspp groups below)
25–30 cm. *The most often seen NG parrot*; common to abundant in towns, gardens, savannah, and other open country, less so in forest, from the coast to mid-mountains. **Adult:** Unmistakable: a noisy

green-and-red parrot with pointed tail and yellow hind-collar contrasting with the dark blue head. Beak orange. **Juv:** Duller; beak dark. **In flight:** Darting and unpredictable, with fluttering wingbeats; yellow wing-stripe, red breast. *Subspp (4, in 3 groups): haematodus*, "Coconut L" (Moluccas to Melanesia; throughout NG Region, except range of next 2 subspp) maroon head with blue face, barring on breast thin or heavy, belly green; *nigrogularis* (Trans-Fly and Aru Is, likely part of *moluccanus* group, "Rainbow L", in AU) all-blue head, breast barring obscure, blackish belly; *rosenbergii*, "Biak L" (Biak I) wide yellow-orange hind-collar, heavily barred breast, blackish belly. **Similar spp:** This familiar lory may be confused with others only at a distance. Dusky L is larger and has a rounded tail; flight more direct with regular wingbeats. *Chalcopsitta* lories larger, with long, rounded tails. *Neopsittacus* lories are quite similar but smaller and have green heads. **Voice:** Shrill, upslurred screeches alternating with musical notes. **Habits:** Usually in flocks of 5–20 or more. Large night roosts in eucalypts. Forages in treetops, taking flowers, soft fruit, sap, and rarely insects or seeds. Nests in excavated tree hollow. Eggs (probably 2) white. **Range:** NG, NW Is (Batanta, Misool, Salawati, Waigeo), Bay Is (Biak, Numfor, Yapen), Aru Is, Manam and Karkar Is, and SE Is (Budibudi Is and Misima, but no recent records from the latter), 0–1800 m, rarely to 2500 m. Also Wallacea east to Melanesia and AU. **Taxonomy:** This species, which ranges far outside the NG Region, is ripe for breakup, but a molecular analysis is needed to shed more light on species' boundaries.

Large Fig-Parrot *Psittaculirostris desmarestii* Pl 47, 49
(Desmarest's Fig Parrot)
18–19 cm. The large fig-parrot of southern and far western NG. Uncommon in forest and edge of lowlands, hills, and mid-mountain basins. This and the next 2 species are big-headed, chunky parrots with brief spiky tails. **Adult:** Unique for its *fiery orange crown*; cheeks green, yellow, or orange. Sexes alike except in race *godmani*. **Juv:** Crown not as bright, either green or with orange restricted to forehead, depending on race. **In flight:** Yellow across underside of flight-feathers; blue underwing-coverts. *Subspp (5, in 4 groups): desmarestii* (Bird's Head and Neck, Batanta and Salawati Is) blue spot below eye, cheeks green or yellow, brown band on lower neck; *blythii* (Misool I) head all orange; *godmani* (western S Lowlands) yellow cheeks contrast with orange crown, Male with yellow hindneck; *cervicalis* (eastern S Lowlands to SE Pen) face golden orange with blue-and-purple nape and collar. **Similar spp:** No other parrot has the crown entirely orange. Compare with Double-eyed FP, which sometimes associates with Large FP. **Voice:** A sharp, broken note. Also, a high, thin, downslurred flight call, and a small clicking noise. Calls of this and other *Psittaculirostris* species are stronger than those of smaller fig-parrots and small lorikeets, but weaker than those of Red-cheeked Parrot. At nest, birds call excitedly: *chet chet*, repeated continually—a beacon to investigate. **Habits:** Singly, in pairs, or in small flocks, usually quite vocal. Flies rapidly just above or through the forest canopy. Feeds mainly on fig seeds. This species appears to nest colonially in cavities in upper tree branches, with many pairs in noisy attendance. **Range:** Endemic. NG: Bird's Head and Neck, NW Is (Batanta, Misool, Salawati), S Lowlands to SE Pen, 0–1650 m. **Taxonomy:** Plumage differences among the subspecies seem extreme, warranting investigation.

Edwards's Fig-Parrot *Psittaculirostris edwardsii* Pl 47
18–19 cm. The large fig-parrot representative in the Sepik-Ramu and Huon. Locally common to abundant in lowland and hill forest, edge, and villages. Unique head pattern: *bright red cheek bounded by blackish eye-line and collar. Conspicuous yellow ear patch*. **Male:** *Red breast*; green in **Female**. **Juv:** Similar to Female, but ear-coverts greenish yellow, not golden yellow. **In flight:** Underwing similar to Large FP, except underwing-coverts green. **Voice:** Flight call is louder, harsher, than similar calls of Double-eyed FP, *screet screet* or *screet-a-lut*. Also a short, metallic *ks*. **Habits:** Similar to Large FP, feeding mainly on figs, and also casuarina seeds and nectar. Often many birds active and noisy around the nest, but whether this indicates cooperative or colonial breeding is unknown. Nests in a high tree hollow. **Range:** Endemic. NG: Sepik-Ramu and Huon, from Jayapura east to Huon Gulf, 0–800 m.

Salvadori's Fig-Parrot *Psittaculirostris salvadorii* Pl 47
18–19 cm. The large fig-parrot representative in the NW Lowlands. Local and uncommon in lowland and hill forest, edge, and villages. *Green parrot with turquoise highlights and yellow-streaked cheeks*. Note *turquoise-green crown* and blue mark behind the eye. **Male:** *Orange breast band*. **Female:** *Turquoise-blue breast band*. **Juv:** Similar to Female but iris dark (Adult's is reddish) and male has blue breast band with some orange. **In flight:** Underwing same as Edwards's FP.

Similar spp: Blue-collared Parrot is larger, has different shape including a squared tail, and shows much blue in the underwing. **Voice:** Similar to calls of Large and Edwards's FPs. **Habits:** Similar to Large and Edwards's FPs, including multiple pairs at a single nest hole. **Range:** Endemic. NG: NW Lowlands from E shore of Cenderawasih Bay to Sentani and Cyclops Mts, 0–400 m. This and Edwards's FP may meet in the Jayapura area.

Orange-breasted Fig-Parrot *Cyclopsitta gulielmitertii*   Pl 48
(*Opopsitta gulielmi III*)

11–13 cm. Common in the South, rare and local in the North; forests and edges from lowlands to foothills. A very small, stubby-tailed parrot. *Black-and-white facial pattern* unique, with forecrown regionally black or blackish blue. *Often with orange on the breast.* **Male:** White or yellowish face, without orange; large orange breast patch (lacking in *amabilis* group of NE). **Female:** Ear-coverts orange or pale orange, in contrast to white cheek; breast generally green (orange in *amabilis* group). **Juv:** Generally similar to Female. **In flight:** Lacks yellow in flight-feathers; underwing-coverts yellowish. **Subspp (6, in 3 groups):** *gulielmitertii* (Bird's Head, NW Lowlands to Sepik R) Male with yellowish cheeks and narrow comma-shaped cheek patch, Female upper breast obscurely orange compared with other races; *amabilis* (Ramu R to northern SE Pen) Male with ivory-white cheeks, throat, and breast (no black cheek patch, no orange on breast), Female shows orange extending far down onto breast, as in males of other races; *melanogenia* (S Lowlands to southern SE Pen; Aru Is) Male with ivory-white cheek and thick, black cheek patch, Female with yellowish or orange restricted to upper breast. **Voice:** Similar but higher pitched and sweeter than call of Double-eyed FP. In flight, a shrill, penetrating *tseet* repeated at frequent intervals. Also an explosive *chit chit chit*. A subdued chittering is given when feeding and is useful for locating hidden birds. **Habits:** Noisy and gregarious. Usually in small parties. Forages on figs, other fruits with small seeds, and unripe palm fruits. Often feeds quite low and is readily observed at figs that bear their fruits like bead strings or in clumps against the trunk. Roosts and nests colonially in arboreal termite nests. **Range:** Endemic. NG, Salawati (NW Is), and Aru Is, 0–300 m (rarely to 1100 m). **Taxonomy:** Perhaps more than 1 species is involved; research needed.

Double-eyed Fig-Parrot *Cyclopsitta diophthalma*   Pl 48, 49
(*Opopsitta diophthalma*)

14–15 cm. Mainly a foothill and lower montane species; locally common in forest and edge. Tends to be absent or rare where Orange-breasted FP is common, and the other way around. A small, compact, stub-tailed parrot with diagnostic *yellow flank-stripe*; colorful face of blue, red, and yellow varies with sex and geography. Mainland races *have blue spot in front of eye—the "double eye."* **Male:** Generally has red cheek, that in **Female** is yellow, blue, or grey. **Juv:** Resembles Female. **In flight:** Underwing pattern mostly yellow; green-and-yellow underwing-coverts. **Subspp (4):** *diophthalma* (N NG, Bird's Head to SE Pen) Male has red face with blue "double eye," Female with red crown and eye-stripe, yellowish cheek; *aruensis* (S Lowlands east to Fly R and Aru Is) Male similar to preceding, Female with blue-and-grey face (no red or yellow); *virago* (D'Entrecasteaux Is) Male with red face lacking "double eye," Female resembles next race but with a hint of blue on the cheek; *inseparabilis* (Tagula I) sexes alike, showing only a red-and-blue forehead spot. **Similar spp:** No other parrot has yellow flank-stripe. Compare with the large fig-parrots and Orange-breasted FP, with which Double-eyed associates at fruiting trees. **Voice:** Flight call is a lorikeet-like *tseet* rapidly repeated a few times, but with stronger attack than a lorikeet, weaker than those of the large fig-parrots; has been likened to the jingling of coins. **Habits:** Singly, in pairs, or in small parties. Forages in fig trees in the canopy or in midstory. Also consumes flower nectar and insects. Observed taking gravel or mud from road cuts. Appears to nest in solitary pairs, with no evidence for colonial nesting as in other fig-parrots. Nests in a cavity chewed in a tall, dead stump. Eggs (probably 2) white. **Range:** NG, NW Is (Kofiau, Misool, Salawati, Waigeo), Aru Is, and SE Is (3 main D'Entrecasteaux Is, plus Tagula I), 0–1600 m. NG: throughout except eastern S Lowlands and southern SE Pen. Also AU. **Taxonomy:** Eligible for splitting into multiple species.

Orange-fronted Hanging Parrot *Loriculus aurantiifrons*   Pl 48
(Papuan Hanging Parrot)

9–10 cm. An uncommon and frequently overlooked tiny parrot of lowland and hill forest. Listen for calls as it flies over the forest. *Fast-flying and bullet-shaped*, with small beak and brief, rounded tail. Smooth bright green, with *red rump and gorget*. **Male:** Yellow cap; white iris. **Female:** Face bluish (missing yellow cap); dark iris. **Juv:** Like Female, but no red gorget. **In flight:** Bright blue underwing. **Subspp (3):** Minor. **Similar spp:** Even the smallest lorikeets are larger, and they have longer, pointed tails. Pygmy parrots are duller and creep on bark. Fig-parrots larger. **Voice:** Call, a harsh, sibilant, high-pitched, rapid *tseo tseo tseo* (2–4 notes). Song reported as a series of 3–5 buzzing, warbler-like *chzee* notes, delivered by a bird perched in midstory. **Habits:** Singly, in pairs, or in small parties of 4–5. Feeds on flowers, fruit, and probably insects in canopy and forest interior. Noted feeding in casuarina and albizia. Roosting habits in the wild unknown, but in captivity sleeps hanging upside down. Reported entering a hole in an arboreal termite nest and also nesting in a cavity at midstory height in slender tree. Hanging parrots carry leaf strips and other nesting material tucked into their contour feathers; as yet unreported for this species. **Range:** Endemic. NG, NW Is (Misool, Waigeo), Karkar I, and SE Is (D'Entrecasteaux Is: Fergusson, Goodenough), 0–1200 m (1600 m).

Moluccan King-Parrot *Alisterus amboinensis* Pl 46
38 cm. The far-western counterpart of the Papuan KP. **Adult:** *Wing entirely dark green* (Papuan has yellow-green stripe across wing-coverts). *Mantle blue* (not black or green). **Juv and Imm:** Sequence similar to Papuan. **Subspp (1):** *dorsalis* (endemic). **Voice:** Similar to Papuan KP but shorter and not as high pitched and ringing. **Habits:** Similar to Papuan. **Range:** NG (Bird's Head and Neck to Weyland Mts in W Ranges) and NW Is (Batanta, Salawati, Waigeo), 0–1200 m. Also E Wallacea.

Papuan King-Parrot *Alisterus chloropterus* Pl 46
(Green-winged King Parrot)
38 cm. The widespread king-parrot, locally common in forest interior and regrowth from foothills to mid-mountains. Usually noticed gracefully flying with rowing wingbeats through or below the forest canopy; alights with mothlike fluttering. *Green with long, blackish tail and red underparts.* **Male:** Red head, *glowing yellow-green stripe across wing-coverts.* **Female:** Either mostly green and showing red only from breast to undertail, or resembles Male in one race. **Juv:** Similar to Female but tail feathers narrower and with pointed, pink tips; darker beak and iris. **Imm:** Intermediate between Juv and Adult; takes several years to attain Adult plumage. **Subspp (3):** Female very different from Male in first 2 races; *chloropterus* (SE Pen and Huon) Male with blue nape as well as mantle, Female green with red breast; *callopterus* (S slopes of W, Border, and E Ranges) Male with red nape, Female similar to previous subsp; *moskowskii* (N slopes of W Ranges and Border Ranges, east to Aitape) Male similar to previous, but Female resembles Male except that the mantle and back are green, as is side of breast. **Similar spp:** See Moluccan KP. **Voice:** Call is a sharp *shhk!* or *keech!* repeated in flight. When perched gives a series of shrill whistles gradually descending in pitch in mournful cadence: *eree eree eree eree....* **Habits:** Singly, in pairs, or in small groups of up to 10. A relatively sedentary inhabitant of the forest subcanopy and midstory, feeding on seeds. No nesting information. **Range:** Endemic. NG: all mts except range of Moluccan KP, 0–2300 m (rarely to 2800 m). **Taxonomy:** Relationships among the taxa of king-parrots is in need of study.

Red-winged Parrot *Aprosmictus erythropterus* Pl 46
30 cm. Trans-Fly savannah and open woodlands, where locally common and perhaps nomadic. *Brilliant yellow-green with red patch on wing.* Iris reddish. **Male:** Possesses a large, oval, scarlet patch on wing; black back. **Female:** Overall green with red restricted to lower edge of wing-coverts. **Juv:** Similar to Female but iris dark brown and beak yellow. **Subspp (1):** *coccineopterus* (also N AU). **Similar spp:** None. **Voice:** Contact call is a metallic *crillik-crillik* or *ching ching.* Alarm call *chik-chik-chik.* **Habits:** In pairs or flocks. Flight buoyant and undulating on a weaving course. Arboreal, feeding mainly on seeds and fruits, also nectar and insects. Nests in a tree cavity, often high in eucalypts (AU data). Eggs (4–6) white. **Range:** NG: Trans-Fly between mouths of the Digul and Fly Rs, 0–100 m. Also AU.

Eclectus Parrot *Eclectus roratus* Pl 42
38 cm. The most often seen large, lowland parrot; common to abundant in forest and openings such as rivers; visits the outskirts of towns. Male and Female plumages utterly different. **Male:** *Emerald green with orange beak*; note red flank-stripe and orange iris. **Female:** *Mostly red with black beak*; blue breast, collar, and flight-feathers; white iris. **Juv:** Resembles Adult but with brown-and-orange beak and dark iris. **In flight:** Both sexes usually appear all dark against the sky; note long, rounded, steadily beating wings; short, squared tail; and rather slow flight; Male with red underwing-coverts, Female blue.

Subspp (1): *polychloros* (also Kai Is of E Wallacea). **Similar spp:** None, but see Great-billed P for NW Is. **Voice:** Noisy. Contact call is a loud, harsh, grating squawk: *graAAH!* In flight a rhythmic *kedek kedek kedek. . . .* Female gives a musical *keleng*. **Habits:** Alone, in pairs, or in small gatherings. Eats seeds and fruit of trees and vines, also crops of bananas, corn, and papaya. Night roosts may consist of dozens of birds. Social life complex: the Female owns and ferociously defends a nest in a high tree cavity or hollow spout, spending much time inside or perched at the entrance with only her colorful red head and blue shoulders showing—a warning to all. Here she waits and watches with those glaring white eyes. More than one Male attends not only to her but also to other nesting Females over a wide expanse of forest. Kwila trees (*Intsia*) are favored for nesting and worth looking over for nesting eclectus. Eggs (2) white. **Range:** NG, NW Is (all), Bay Is (all), Aru Is, and SE Is (3 main D'Entrecasteaux Is, Trobriand Is, and Tagula I), 0–1000 m, rarely higher. Also Wallacea, NE AU, and the Admiralty, Bismarck, and Solomon Is.

Red-cheeked Parrot *Geoffroyus geoffroyi* Pl 46

20–25 cm. The common, medium-sized, green parrot throughout the lowlands, usually seen flying high overhead calling noisily. Inhabits all forest types, mangroves, regrowth, open country with trees, savannahs, and town environs. *Head pattern distinct, otherwise generally bright green.* **Male:** *Purple head with red cheeks.* **Female:** *Dull brown head.* **Juv:** Green crown, olive cheeks. **In flight:** Swift flight on tapered wings; sky-blue underwing-coverts. **Subspp (7):** 3 southern ones with green rumps: *aruensis* (Aru Is, S Lowlands, Trans-Fly to Huon, SE Pen, and D'Entrecasteaux Is) reddish shoulder patch; *sudestiensis* (SE Is: Misima and Tagula) similar, lacking red shoulder patch; *cyanicarpus* (SE Is: Rossel) similar, with leading edge of wing blue. 4 northern races with reddish-brown rumps: *minor* (NE Lowlands to Sepik-Ramu) obscure bronze shoulder patch, underwing-coverts pale blue; *pucherani* (NW Is, Bird's Head and Neck) similar, underwing-coverts dark blue; *jobiensis* (Yapen Is) similar to previous, turquoise collar; *mysoriensis* (NW Is: Biak and Numfor) dark blue underwing-coverts, shoulder patch red-brown, and head pattern extends onto neck. **Similar spp:** None in the lowlands. Blue-collared P lives in the mountains and has a green head; the 2 species overlap in hills. **Voice:** Varies geographically. In SE, a metallic *kee! kee! kee! . . .* , with piercing, ringing quality as of a metal surface being struck. In Sepik, *kee-keer! kee-keer! kee-keer! . . .* **Habits:** Singly, in pairs, or small parties, occasionally larger flocks. Its wheeling flight is conspicuous over open spaces, appearing singly or in small groups chasing each other. Perches on exposed branches. Feeds in the treetops and forest interior on seeds, fruits, flowers, and nectar. Swoops into villages to raid mango trees. Nests in a cavity excavated in a dead tree trunk at midstory height. Eggs (2–4) white. **Range:** NG, NW Is (all), Bay Is (all), Aru Is, and SE Is (all except Woodlark), 0–800 m, rarely higher. Also Wallacea and NE AU.

Blue-collared Parrot *Geoffroyus simplex* Pl 46
(Lilac-collared Parrot)

23–24 cm. The only parrot, other than lorikeets, making long flights high over mountain forest. A vocal, flocking species. Heard often, seen flying regularly, but rarely observed perched. Highly nomadic—abundance varies dramatically with season and year depending on food supply. *Entirely leaf-green without distinctive field marks.* **Adult:** At very close range note pale iris and **Male's** obscure bluish collar. **Juv:** Collarless like Female, but iris dark, bill somewhat pale when young; bluish cast on crown and cheek; inner margin of flight-feathers whitish (as in Adult). **In flight:** Streamlined, short-tailed silhouette; twisting flight; bright blue underwing. Forms high-flying, strung-out lines. **Subspp (2):** Minor. **Similar spp:** Red-cheeked P, confined to lower elevations, has a different voice and shows a colored head; juv with olive cheek and dark (not whitish) inner margin on flight-feathers. **Voice:** Flight call tinkling at a distance, likened to the sound of sleigh bells; harsher at close range, a loud, rhythmic *kree-kro kree-kro kree-kro. . . .* **Habits:** Inhabits mid-mountain forests, especially dry ridge crests with oaks *Castanopsis* and *Lithocarpus*, the nuts of which are its favorite food. Also eats other seeds, fruits, and flowers. Nest similar to Red-cheeked P. Eggs (~3), presumably white. **Range:** Endemic. NG mts (records from all but Wandammen, Cyclops, and Huon Mts), 600–2000 m, occasionally descends to foothills and lowlands.

Great-billed Parrot *Tanygnathus megalorynchos* Pl 46
38 cm. A large parrot of the NW Is. *All green with tapered body shape, long wings, and pointed tail. Big head and beak.* **Adult:** *Orange-red beak; shoulders marked gold-and-black.* **Juv:** Beak smaller, yellowish; shoulder mostly green, the black-and-gold pattern nearly lacking. **In flight:** Distinctive shape (see above); *wings turquoise above, yellow below*. **Subspp (1):** *megalorynchos* (also N Moluccas). **Similar spp:** Eclectus P male has different shape with wider wings and shorter, squared tail; shows red flanks and lacks black-and-gold shoulders. **Voice:** Flight calls include a whine like a squeaky door, *kyahn!*, and a high-pitched, harsh, repeated *eeyaak-eeyaak!* unlike other large parrots. **Habits:** Singly or in small parties in forests, including mangroves, foraging on nuts and fruit. Prefers small islands and travels between them. Nests in a high tree hollow. **Range:** All NW Is, especially small ones, 0–1000 m. Mainland sightings southeast of Sorong (Bird's Head). Also Wallacea.

> **Pygmy Parrots.** These tiny "microparrots" are no bigger than a man's thumb. Gregarious at all times, they live in small flocks that are probably families. When foraging, they creep about with animated stop-and-go movements on trunks and branches, picking at the bark for food. Their diet is not well understood—variously reported to include lichen, fungi, small seeds, and insects—it is in much need of careful study and should prove most interesting. The high-pitched voice is key to their discovery. In flight they seem too small to be recognized as a parrot. Their stubby shape and whirring pulsed wingbeats are like those of a mannikin.

Yellow-capped Pygmy Parrot *Micropsitta keiensis* Pl 48
9 cm. Common in lowland and hill forest, but inconspicuous and difficult to locate, even when calling. **Adult:** Dark green with *dark brown face and dingy yellowish cap*; beak dark. **Juv:** Pale yellowish beak with dark tip. **Subspp (2):** *keiensis* (Aru Is and S Lowlands) entirely green underparts; *chloroxantha* (NW Is, Bird's Head and Neck) Male with red streak down center of lower breast and belly. **Similar spp:** Buff-faced and female Red-breasted PPs have pale buff face and blue crown. **Voice:** Thin, high-pitched, and almost inaudible. Flight call, an incessant *ssii ssii*. Contact calls are weak, high, tinkling notes, *tseeet*, repeated every second or so, often uttered by several members of the foraging party. Feeding birds are generally silent but call in response to others. **Habits:** In small parties. These forage at any height on trees but mostly in the canopy and midstory. Flocks can be seen flying both over the forest as well as through it. Appears to be a cooperative breeder with as many as 6 Adults in the nest at once. Nests and roosts in a tunnel chewed into an active, arboreal termite mound; the round hole has a wide oval entrance with a platform at the front. Eggs (2) white. **Range:** NG, NW Is (all), and Aru Is. NG: Bird's Head and Neck, S Lowlands east to Kikori R basin, 0–800 m. Also Kai Is in E Wallacea.

Geelvink Pygmy Parrot *Micropsitta geelvinkiana* Pl 48
9 cm. The only pygmy parrot on the Bay Is of Numfor and Biak. *Blue and brown head.* **Male:** *Yellow patch down breast and belly*; face usually darker; may have yellow nape spot. **Female:** Face usually paler and more mottled; lacks pale nape spot. **Juv:** Pale face set with dark eye; beak horn-colored. **Subspp (2):** *geelvinkiana* (Numfor I) and *misoriensis* (Biak I), difficult to distinguish. **Similar spp:** None in its range. No other pygmy parrot has as much blue on the head and a yellow breast patch as well. **Voice:** Similar to other pygmy parrots. **Habits:** Apparently identical to other pygmy parrots. Also nests in a tunnel excavated into an arboreal termite nest. **Range:** Endemic. Bay Is of Numfor and Biak, 0–300 m (higher?).

Buff-faced Pygmy Parrot *Micropsitta pusio* Pl 48
8–9 cm. Common to abundant in lowland and hill forest and tall regrowth, including eucalypts and coconuts. **Adult:** *Face buff or brown, with blue cap.* Dark beak. **Juv:** Greenish-blue cap and bicolored beak. **Subspp (4, in 2 groups):** *pusio* (southern SE Pen, SE Is, and Bismarck Is) face buffy brown; *beccarii* (NW Lowlands to northern SE Pen) face brown. **Similar spp:** Red-breasted PP occurs in the mountains; female is nearly identical but has a much narrower pale eyebrow (Buff-faced has broad, buff eyebrow), and the crown is dull blue, not azure blue; juv difficult to separate. Yellow-capped PP has an obvious yellow forehead and crown. **Voice:** Similar to that of Yellow-capped PP. **Habits:** As for Yellow-capped PP. Nest and nesting behavior similar. **Range:** NG, Manam and Karkar Is, SE Is (Fergusson, Misima, and Tagula Is), 0–800 m. NG mainly in N from E Bird's Head and Neck to NW Lowlands and to SE Pen; in South, from eastern S Lowlands (Crater Mt and Purari R) to SE Pen. Also Bismarck Is.

Red-breasted Pygmy Parrot *Micropsitta bruijnii*   Pl 48
9 cm. The only pygmy parrot in montane forest above 1000 m elevation. Usually uncommon and local. **Male:** Unique, with *red underparts and blue nape, collar, and tail*, and bright buffy face and crown. **Female and Juv:** Mostly green with *blue cap and pale face*; note that the *pale eyebrow is narrow*, not nearly so broad as Buff-faced PP. **Subspp (1):** *bruijnii* (endemic). **Similar spp:** See Buff-faced PP. **Voice:** Call, a repeated, high-pitched tinkling note that is louder, harsher, and more disyllabic than that of Buff-faced PP. **Habits:** Typical for a pygmy parrot. Forages creeperlike on the trunks and larger branches of trees. Occasionally in large flocks (50+). Nest cavity is excavated into a stump, usually only a few meters up; entrance identified by its sloping, triangular landing chewed at the foot of the hole. Eggs (2) white. **Range:** NG mts (records from all but Fakfak, Kumawa, and Cyclops Mts), 1000–2300 m, strays much lower. Also S Moluccas and Bismarck and Solomon Is.

COUCALS: Centropodidae

This family contains but a single genus of 28 species of large, black cuckoos extending from Africa and Asia to Australasia, with 4 species in the New Guinea Region. Formerly classified with Old World Parasitic Cuckoos in the family Cuculidae, coucals differ from them greatly in their morphology, behavior, and DNA. Coucals are unmistakable: their somber plumage, hunched posture, drooped wings, and floppy tail give them a downtrodden look, enhanced by their awkward, scurrying gait. They rarely fly and instead scramble around in thick vegetation, foraging for arthropods and small vertebrates. In contrast to parasitic cuckoos, coucals live in pairs, build and tend a nest, and raise their own young. Females are larger than males. Social organization and breeding are known only for the Pheasant Coucal and reveal an odd family life: the male incubates and does most of the feeding of the chicks. Usually shy and keeping to cover, they often emerge in early mornings and late afternoons, a good time to look for them. The loud hooting or booming calls are often heard. Coucals have a reputation for being bad tasting to humans, and perhaps as a consequence they can be found unmolested in the vicinity of villages and gardens.

Greater Black Coucal *Centropus menbeki*   Pl 50
(Ivory-billed Coucal)
56–69 cm. The widespread forest-dwelling coucal. Keeps to thick cover such as vine tangles from the ground up to lower levels of the canopy. *Large black bird, with an oversized tail. Bill ivory with a dark grey base.* **Adult:** All oily black with green, blue, and purple sheen; iris red. **Juv:** Similar but tail with indistinct rufous bars and iris pale tan. **Subspp (2):** *menbeki* (species' range excluding Aru Is) blue-green gloss; *aruensis* (Aru Is) purple gloss. **Similar spp:** Pheasant C has a shorter tail, dark bill, variable brown barring, and occupies scrub and grassland, not forest. Lesser Black C considerably smaller, with dark bill and iris, and inhabits scrub at the forest edge. Eastern Koel male smaller still and inhabits forest canopy. **Voice:** Low-pitched, resonant, far-carrying, booming hoots, given in a slow and cadenced series that descends in pitch: *uh—oo-oo-oo—oo—-ooo-oh*; resembles call of the New Guinea Harpy-Eagle, and like that species also heard at night. Sometimes gives a single note: *oodle*. Or may give a peculiar and totally unmusical call consisting of a staccato upslurred grunt, followed by a dry rattled series rising in pitch. Calls not as rapid or high pitched as Pheasant C, more complex and longer than those of Lesser Black C. Duets frequently. **Habits:** Hops and scrambles clumsily through vegetation but difficult to observe; flies infrequently. Solitary or as a separated, vocalizing pair. One nest reported in a pandanus; nest a large mass of leaves. Egg white. **Range:** Endemic. NG, NW Is (Batanta, Misool, Salawati), Bay Is (Yapen), and Aru Is, 0–800 m (rarely to 1275 m).

Biak Coucal *Centropus chalybeus*   Pl 50
44–46 cm. Biak Island's only coucal. Common and noisy in lowland forest and second growth. **Adult:** The only NG coucal with a *yellow iris* and black bill. **Juv:** Washed with rufous, not barred; iris color unknown. **Voice:** Varied loud notes: a broken, very harsh rasp; a descending, slightly accelerating series of upslurred hoots; a repeated *bup*. **Habits:** Lives from ground to treetops. Behavior much like other coucals. No nesting information. **Range:** Endemic. Biak (Bay Is), 0–500? m.

Lesser Black Coucal *Centropus bernsteini* Pl 50
(Black-billed Coucal)
41–51 cm. Common in grassland, scrub, and forest edge in western and central NG, overlapping range of very similar Pheasant C from about the Indonesian border eastward in the North to Bulolo Valley and locally in the South to the Purari R. **Adult:** Differs from Pheasant C by *all-black wings, smaller size, and shorter tail* (in some races of Pheasant C the barring may be difficult to see). *Iris dark brown* (not red). **Juv:** Black with faint rufous or buff barring; head buff-spotted, not streaked as in Pheasant C. **Voice:** Three hoots, *woop woop woop*, slowly delivered, on a descending scale; occasionally a longer series is given. Pairs often duet. Hoots suggest the call of a Brown Cuckoo-Dove or a weaker, flatter version of the Greater Black C. **Habits:** Similar to Pheasant C. Nests in grass; nest a smooth, irregularly shaped ball made of grass and with side opening. Eggs (2) white. **Range:** Endemic. NG, NW Is (Salawati?), and Manam Is. On NG, east to western SE Pen (Lae, Bulolo, and Purari R), 0–500 m (rarely 900 m).

Pheasant Coucal *Centropus phasianinus* Pl 50
43–60 cm. The coucal of grassland and scrub in eastern NG. **Adult:** Variable. A medium-sized blackish coucal with *brown barring in the wing*. Seasonal and age differences in plumages not well understood. May be more streaked and barred outside of breeding season? *Iris dark red*. **Juv:** Head and underparts more heavily pale-streaked than Adult and with pale buff throat and upper breast; *iris dark brown*. **Subspp (3, in 2 groups):** *nigricans* (most of range) nearly all black, with obscure barring; *thierfelderi* (S Lowlands and Trans-Fly) rufous bars on wings and tail. **Similar spp:** Lesser Black C replaces Pheasant in similar habitats in West and is all black. Greater Black C inhabits forest and is larger, blacker, and has whitish bill. **Voice:** A conspicuous call of open habitats: a slightly musical, rapid, booming series of *hooh* notes, falling then rising in pitch, accelerating, then decelerating. One series may contain 50 or more notes. **Habits:** Clambers about in thick vegetation, often perching atop greenery with wings and tail spread or seen scurrying across the road. Spends much time on the ground. Singly or in pairs. Nests low in herbaceous or grass cover; nest a globular structure with side entrance. Eggs (2) white. **Range:** NG and SE Is (3 main D'Entrecasteaux Is), 0–1600 m. NG: E half of NG, westward to NW Lowlands and Trans-Fly; has colonized E Ranges. Also AU.

OLD WORLD PARASITIC CUCKOOS: Cuculidae

This peculiar family of 57 species extends from Eurasia and Africa to Australasia. Parasitic cuckoos are diverse in New Guinea (18 spp), and the family seems to have had a long history in Australasia, a region particularly rich in passerines, their only hosts. The parasitic cuckoos are slender birds of sleek plumage, upright posture, poised appearance, and graceful flight. They dwell mainly in the treetops and lay their eggs in the nests of other birds, abandoning the care of their eggs and young to foster parents. Features that distinguish cuckoos from any passerines they might resemble are their short legs with a foot with 2 toes turned back, the circular nostril, and underparts usually barred (many birds of paradise, trillers, and some cuckooshrikes also have ventral barring). The smallest species of parasitic cuckoos are flycatcher-like in size and behavior, whereas the largest species are jay- and crowlike and eat mainly fruit. All but the largest species have a taste for caterpillars and in fact eat prickly, toxic ones avoided by other birds. They are drawn to caterpillar infestations. Parasitic cuckoos are usually solitary, although they may join mixed-species flocks. Perhaps because their food fluctuates seasonally, many cuckoo species are migratory, with overwintering or passage migrants in the NG Region breeding in nearby Australia (9 spp), New Zealand (1), and Asia (2).

The most intriguing aspect of cuckoos is their parasitic breeding behavior. Most species are thought to be promiscuous, without a lasting pair bond. Males, which are generally the larger sex, do a lot of singing and displaying, whereas female cuckoos are quiet. Females have all sorts of tricks to deceive and outmaneuver the small songbirds that are the hosts for their eggs. Each cuckoo species has a range of host species but tends to prefer just a few. The hosts themselves are often well prepared to thwart the would-be parasite, but in due course many fail. Very little is known about the breeding of NG parasitic cuckoos, and the hosts of 4 species remain to be discovered. In this guide, we have paid particular attention to describing juvenile plumages and would encourage the reporting of host species feeding juvenile cuckoos.

Juveniles undergo complete molt into adultlike immature plumage, retaining none or very few juvenile flight-feathers. Be aware, too, that plumage coloration of female cuckoos varies individually. Identification of females and juveniles of some species can be difficult!

Dwarf Koel *Microdynamis parva* Pl 52

20–22 cm. An uncommon canopy-dwelling fruit eater of lowland and hill forest. Shaped like miniature koel, with a short, thick, slightly hooked bill and rounded tail. Note distinctive *white streak below eye*. **Male:** Has a *metallic blue-black crown and black malar streak*; also note the *red eye*. **Female:** Like a much duller version of Male and without the dark crown; shows traces of barring in underparts; iris hazel to red. **Juv:** Like Female but with indistinct dark barring in the wings, tail, and underparts. **Subspp (2):** *parva* (range except that of next subsp) back brown, underparts buff; *grisescens* (Sepik-Ramu to northern SE Pen) upperparts greyer, underparts rufous brown. **Similar spp:** Long-billed Cuckoo female much like smaller version of female Dwarf K, but with a longer, thinner bill and darker iris. Eastern K female much larger and spotted above. Tawny Honeyeater with longer, narrower curved bill and bare patch around eye. **Voice:** Two similar, far-carrying songs. One is medium to high pitched with a frantic, resonant quality: a series of identical upslurred notes (1/sec), *whuri-whuri-whuri . . .* , lasting 30 sec or more:

1 ᴗ ᴗ ᴗ ᴗ ᴗ ᴗ ᴗ ᴗ

(Eastern K gives a similar *koel* song that is lower pitched, slower, and shorter in duration.) Second song is a rapid series of downslurred notes, *wur-dr*, the series ascending in pitch then reaching a plateau and proceeding on a steady pitch, lasting 2.5–5 sec:

2 ⌐⌐ ⌐⌐ ⌐⌐ ⌐ ⌐ ⌐ ⌐

Habits: A fruit eater, feeding in crowns of trees in association with fruit-doves and other birds. Will sing from an open perch. Often seen in pairs. Hosts and egg unknown. **Range:** Endemic. NG (minus Trans-Fly), plus audio records from the 3 main D'Entrecasteaux Is (SE Is), 0–1450 m.

Eastern Koel *Eudynamys orientalis* Pl 51
(**Common**, **Indian**, or **Pacific Koel**, *E. scolopacea*)

38–42 cm. Uncommon fruit-eating cuckoo of lowland and lower montane forest canopy. Shape distinctive: large, slender, and with long, rounded tail. *Red iris and pale bill in Adult.* **Male:** A *large, glossy black cuckoo reminiscent of a coucal or manucode*, but differs by *long shape* and combination of *red iris* and *pale bill*. **Female:** *Individually variable*, but with *white malar streak, dark upperparts boldly spotted, and many-barred tail*; pale underparts with *dark barring*. Female plumages differ between the resident NG and migrant Australian races. Darker Females have head black rather than brown and streaked. **Juv:** Compared with Female has *black eye-patch framed by pale crown and white malar streak*; *wing-coverts barred* as well as spotted; *dark iris*. Juv of migratory race should be molting to Adult plumage on NG wintering grounds. **Subspp (2):** *rufiventer* (NG resident) Female with upperparts chestnut-spotted and underparts with ochre wash; *subcyanocephalus* (AU migrant to S NG) Female with whitish dorsal spotting and paler underparts; some have a black head with the usual white malar streak. **Similar spp:** Manucodes have a black bill and nervous, jerky posturing. Coucals are larger and have black or ivory bill. Dwarf K much smaller than Female Eastern K. Long-tailed Cuckoo is streaked below. **Voice:** The two songs are loud, far-carrying, deep whistles, delivered most frequently at dawn and dusk, sometimes predawn. (1) The *koel* or *cooee* song: a slow-paced series of slurred notes, each note with a sense of forward momentum. Series (1 note per 1.5 sec) may ascend and swell somewhat:

1 ⌣ ⌣ ⌣ ⌣ ⌣

(2) A rapid series of 2 or 3 notes, *whurdur* or *whurdurdur*, that rises and then levels in pitch (This may be the same as the *wurroo* song of Australian koels, given in the presence of other koels.):

2 ⌒⌒⌒ ⌒⌒⌒ ⌒⌒⌒

(Similar to the songs of the Dwarf K, which gives higher-pitched, disyllabic notes.) Also gives a loud, staccato, rapidly repeated *week week week WEEK . . .* ; this could be the *keek* call of Female AU birds.

AU Male also gives a *whik* call. AU koels duet with Female *keek* notes mingled with Male *whik* notes and *wurroo* song. **Habits:** Rarely seen except at fruiting trees, where it behaves warily but aggressively. Male will encamp at a productive tree and chase away smaller songbirds. Takes insects in typical cuckoo manner. Hosts in AU include friarbirds, Blue-faced Honeyeater, Magpielark, and Australasian Figbird. Egg reddish orange to pinkish buff, sparingly spotted and blotched. **Range:** NG, NW Is (Batanta, Misool, Salawati, Waigeo), Aru Is, Manam I, and Karkar I, 0–1500 m. Also Wallacea eastward to Bismarck Is and AU. **Taxonomy:** Formerly included with other forms in the Common K, *E. scolopacea.*

Long-tailed Cuckoo *Urodynamis taitensis* Pl 51
(**Long-tailed Koel**, *Eudynamys taitensis*)
36–39 cm. A vagrant during austral winter to small islands off northern NG and SE Is; no records from mainland. *A brown, hawklike koel with streaked underparts. Distinctive slender profile with long tail. Note bicolored bill and conspicuous pale eyebrow.* Probably only Adult birds in NG Region. **Adult:** *Upperparts entirely barred; underparts white with sharp, dark streaking; iris yellow.* Some birds, particularly Females, with buffy underparts. **Juv:** Upperparts white-spotted like female AU race Eastern K; underparts buffy; iris dark. Probably all Juv have molted into mostly Adult plumage by the time they reach the NG Region. **In flight:** Flight rapid, straight, and powerful, with continuous wingbeats; note long tail. **Similar spp:** Eastern K female and juv with spotted upperparts and barred, not streaked below; note differences in eye and bill color and absence of pale eyebrow. Oriental Cuckoo brown morph barred below and lacks eyebrow. Small hawks have a shorter, more square-tipped tail. **Voice:** Probably quiet in region. Twittering call by migrating birds. Two advertisement calls: shriek call, a penetrating, buzzing, upslurred whistle; and chatter call, a series of *wheet* notes. **Habits:** Passage dates unknown, but possibly northward in Feb–Apr, southward in Aug–Nov. In migration and on wintering grounds remains concealed in canopy of coastal scrub, coconut plantations, and occasionally lowland forest. Feeding behavior little known, cryptic, and wary. *Forages (mainly?) nocturnally* for large invertebrates, reptiles, and small bird eggs, nestlings, and adults; occasionally fruit and seeds. **Range:** SE Is, where known from a single record from Kimuta I near Misima. Expected elsewhere in SE Is and along north coast of NG. Breeds in NZ. Winters in vast arc of small islands extending from Micronesia to Polynesia.

Channel-billed Cuckoo *Scythrops novaehollandiae* Pl 50
(**Stormbird**)
51–58 cm. An uncommon austral winter migrant to forest and river edge and other open habitats in lowlands. An unforgettable, giant, *crow-sized* cuckoo. Shape unique—*massive bill, long pointed wings, and long tail with black-and-white tip.* **Adult:** Grey, with red eye and orbital ring; but Female has smaller bill with longer cream tip (>30%) and underparts with heavier ventral barring. **Juv:** Head and underparts buff rather than grey; upperparts spotted buff. Migrant young birds in region molting from Juv into Adult plumage. **In flight:** Unmistakable shape, with long bill, wings, and tail; distinctive flight with slow and regular wingbeat. **Similar spp:** Grey Crow has much shorter wings and tail and no markings in plumage. Flying Blyth's Hornbill in silhouette has broader wings with noisy flapping. **Voice:** Can be very vocal day or night, perched or flying, and especially when breeding. A wide variety of repeated screeching calls. A loud, unpleasant, guttural *gaak* or *graah-graah*, distinctive. **Habits:** Migrates N in Feb–Apr, S in Aug–Sep. A few birds present in summer, so breeding is possible but not proven. Slow-moving. Feeds in crowns of trees on fruit, especially figs, and on insects and eggs and nestling birds. In AU parasitizes Torresian Crow, butcherbirds, Australian Magpie, and Magpielark. These hosts present in NG Region, where Channel-billed observed being chased by Torresian and Grey Crows. Egg buff, dark-spotted and blotched. **Range:** Migrant throughout most of NG Region (no records from Gebe in NW Is, nor Bay Is, nor Rossel and Woodlark in SE Is), 0–650 m. Possibly breeds on D'Entrecasteaux Is (3 main islands), where it is common and vocally active in austral summer. Breeds in AU, Sulawesi, and New Britain I (and possibly nearby islands).

Long-billed Cuckoo *Rhamphomantis megarhynchus* Pl 52
(*Chrysococcyx megarhynchus*)
18–19 cm. New Guinea's rarest and most mysterious cuckoo, known from a few scattered locales but may be expected anywhere in lowland forest on mainland NG and land-bridge islands. *Long, deep bill with droop tip and short, squared tail suggestive of a honeyeater. Tail with indistinct dark barring on the underside.* **Male:** *Sooty black head* and *striking red-brown iris emphasized by red orbital ring;*

rest of upperparts brown; underparts pale greyish brown. Some Males have darker wings, tail, and breast. **Female:** Similar but with *head grey-brown* and *iris dark brown with narrow cream outer ring and dark orbital ring*; rest of upperparts brighter (cinnamon); and underparts more rufous buff. Some Adults with vague dark barring or vermiculation on the breast. **Juv:** *Resembles Tawny-breasted Honeyeater*—head variably patterned brown, crown and ear-coverts darkest, contrasting with *a broad greyish-white zone around dark eye, on the throat, and also sometimes on the nape or line behind the eye*; iris and orbital ring blackish; remainder of upperparts rufous brown to brown; upper breast brownish or greyish, rest of underparts pale rufous; may have greyish vermiculations on the breast and belly. **Similar spp:** Dwarf Koel larger, with much shorter bill and a white streak below eye. Tawny Honeyeater more active and does not perch upright, bill more gradually curved, and facial pattern sharply defined with yellow or orange markings and bare orbital patch; up close note slit nostril and songbird feet. Both koel and honeyeater lack vague dark barring in undertail. **Voice:** Seldom heard. Song reminiscent of Malay Bronze Cuckoo, but a louder, more distinct trill—a descending series of constantly spaced notes, the trill lasting 4 sec, and repeated at 5-sec intervals; sibilant, soft, but far-carrying. Also a 2- or 3-note whistle. **Habits:** Seen singly. Behavior typical of a small, insectivorous cuckoo. Once located, rather easily viewed, confiding. Perches motionless with an upright posture for long periods, sometimes on an open branch. Forages in subcanopy. Visits caterpillar infestations, where it associates with other birds, including other cuckoos and Tawny Honeyeaters. Seen removing spittlebug nymphs from their foamy mass. Flies in the direct manner of a bronze cuckoo, not undulating. Hosts and egg unknown. **Range:** Endemic. NG, NW Is (Misol, Waigeo), and Aru Is, 0–500? m.

Horsfield's Bronze Cuckoo *Chalcites basalis* Pl 52
(*Chrysococcyx basalis*)

15–20 cm. Rare austral winter migrant to Trans-Fly and Aru Is savannah, open country, and forest edge. Only bronze cuckoo with *prominent pale eyebrow, sooty ear patch curving down side of neck, and scaled feathers on upperparts*. Upper tail with rufous edge basally. *Iris dark* in all plumages. **Adult:** Iridescent bronze upperparts rather dull and greyish. Underparts with *incomplete bold, black barring*. Some Females duller, more brownish above, but most not safely distinguished from Male. **Juv:** Head grey to grey-brown; pale eyebrow and dark eye-line both vague but still diagnostic; rufous edge to base of tail also present; scaling on upperparts rarely visible; underparts without barring. Overwintering young birds probably molting from Juv to adultlike Imm plumage. **Similar spp:** Other bronze cuckoos lack the distinctive Horsfield's facial pattern, dorsal scalloping, and rufous edging to base of tail. Shining and Little BCs overlap in habitat. **Voice:** Probably quiet in NG. Song is a slow, monotonous series of 15–30 steeply descending notes, *seeuuu*, given without change in pitch or rate at ~0.7 notes/sec. **Habits:** Feeds in foliage of trees and bushes, perches on wires and fences, occasionally on the ground far from cover. May dive to the ground for prey. Flight swift, direct, and slightly undulating. **Range:** NG and Aru Is, 0–500? m. NG: Bensbach region (potentially throughout the Trans-Fly) and vagrant to Port Moresby. Could turn up elsewhere. Breeds in AU.

Black-eared Cuckoo *Chalcites osculans* Pl 52
(*Chrysococcyx* or *Misocalius osculans*)

19–22 cm. Austral winter vagrant to savannah and other open habitats. *A small, pale grey cuckoo of desert colors with distinctive head pattern: a thick white eyebrow contrasting with black eye-patch extending from bill, through the eye and ear-coverts to side of neck. No barring* anywhere. **Adult:** *Note unique buffy underparts* in fresh plumage, ultimately fading to white. **Juv:** White-and-black facial markings less distinct and rump greyer than Adult. Should be molting into adultlike Imm plumage when in NG. **In flight:** *Note pale rump.* **Similar spp:** None. **Voice:** Probably quiet in NG. Song a monotonous series of descending *peeer* notes repeated at same pitch and speed, ~10 notes/15 sec. **Habits:** Forages from treetops and often flies to the ground to take prey. Quiet and shy. **Range:** A few records from Aru Is, 0–100 m. Breeds in AU dry country.

Rufous-throated Bronze Cuckoo *Chalcites ruficollis* Pl 52
(*Chrysococcyx ruficollis*)

15–17 cm. An uncommon resident of montane forest. Unmistakable within its habitat. **Adult:** *Extensive rufous brown wash on forehead, face, and throat*; upperparts with iridescent bronze, casting green and rufous highlights. Sexual difference in iris color: **Male** red brown, **Female** brown. **Juv:** Upperparts greyish green; faintly barred underparts grey fading to white belly.

367

Similar spp: White-eared BC at lower elevations lacks rufous in the sides of the face and throat and instead adult has a white ear patch and rufous at base of primaries; juv lacks barring on flanks and instead has chestnut base to flight-feathers. **Voice:** Song a series of 8–9 identical, high-pitched downslurred whistles (2/sec) similar to but higher pitched than those of White-eared: *feer feer feer.* . . . This note may be repeated at irregular intervals. **Habits:** A typical bronze cuckoo of the forest canopy. Only 1 host record so far: Papuan Scrubwren. Egg undescribed. **Range:** Endemic. NG: Bird's Head, Bird's Neck (Wandammen), Central Ranges, Foja Mts, and Huon, 1800–2600 m (1130–3230 m).

Shining Bronze Cuckoo *Chalcites lucidus*   Pl 52
(Golden Bronze-Cuckoo, *Chrysococcyx lucidus*)

15–17 cm. Austral winter migrant, uncommon but widespread in savannah, gardens, plantations, second growth, and rarely forest canopy. Male located by whistled song, commonly heard on winter territories. **Adult:** *A glittering, emerald cuckoo with white face speckled black; underparts banded by thick, dark, iridescent barring on white*; little or no rufous in tail. *Brown cap* (except Male of race "*lucidus,*" mainly of SE Is). **Male** *iris pale.* **Female** iris dark brown; plumage similar, although usually duller, with slightly reduced speckling and barring and usually with obscure rufous spotting in spread undertail, where Male typically has none. **Juv:** Duller, with barring much reduced in face and throat, and ventral barring neither as extensive nor as prominent, nevertheless more so than other juv bronze cuckoos; iris dark. On wintering grounds, Juv undergoes molt to adultlike Imm plumage. **Subspp:** Only 1 race, *lucidus,* now recognized for all NG birds; however, Males often separable by plumage, Females not, into 2 forms: "*plagosus*" (breeds in AU) Male with more brown in plumage, particularly on crown, nape, and mantle; "*lucidus*" (breeds in NZ) Male evenly green from head to back. A better character is the bill, which in the hand is noticeably wider in the NZ birds. **Similar spp:** Little BC usually has rufous in plumage, especially in the tail, and male has a red eye. Horsfield's BC has a dark eye-line. White-eared BC has a green face. **Voice:** Song a series of about a dozen identical, upslurred whistled notes; series declines slightly in pitch but not in rate, which is ~1.6 notes/sec, twice as fast as Horsfield's, faster and shorter than White-eared, and much slower than Little's. Series sometimes followed by a second series of downslurred notes (AU data). **Habits:** Usually remains hidden in treetops. Male sings regularly during winter on what may be his territory. **Range:** Population "*plagosus*" breeds in AU and winters throughout the NG Region; "*lucidus*" breeds mainly in NZ and winters in the NG Region mainly in the SE Is (recorded from Woodlark and Misima); 0–1900 m. Also SE Asia to AU.

White-eared Bronze Cuckoo *Chalcites meyerii*   Pl 52
(*Chrysococcyx meyeri*)

14–15 cm. The resident forest bronze cuckoo of hills and mid-mountains, where uncommon. Rare in adjacent lowland forest and absent away from hills. *Rufous patch at base of primaries* at all ages. **Adult:** *Boldly colored.* Of the bronze cuckoos, White-eared has the *darkest face* with contrasting "*white ear*"; upperparts vary individually from deep iridescent emerald to coppery green; underparts white with thick, dark barring. **Male:** Eye and eye-ring red. **Female:** Dark eye and *chestnut forehead,* variable in extent. **Juv:** Rufous patch at base of primaries (some juv Little BC have this); otherwise plain greyish brown, much paler below and without barring. **In flight:** Bright rufous wing patch. **Similar spp:** Compare with co-occurring bronze cuckoos, Little (lowlands), Shining (second growth), and Rufous-throated (high mountains), but these are white or rufous in the face rather than green, and they have less pronounced barring below. **Voice:** Song is a diagnostic series of 5–8 high-pitched, bright, drawn-out, slightly plaintive, clear notes at the rate of 1/sec: *peer peer peer* . . . The series drops slightly in pitch, and there is a brief pause before the final note, which is at a lower pitch than others. (Similar call of Black Cicadabird is louder, richer, and lower pitched. One song of Brush Cuckoo is similar but louder, faster, with notes shorter and no pause before the last note.) Also a complex song, sounding somewhat like a cross between the calls of an accipiter and a cuckoo: 4 pairs of downslurred notes, the pairs rising and falling. **Habits:** The sole nonpasserine species regularly found in mixed-species foraging flocks of hill forest. Hosts and egg unknown. **Range:** Endemic. NG mts (records from all but Wandammen and Kumawa Mts on Bird's Neck) and NW Is (Batanta), 0–1800 m.

Little Bronze Cuckoo *Chalcites minutillus* Pl 52
(Malay Bronze-Cuckoo, *Chrysococcyx malayanus*. See Taxonomy, below.)
14–17 cm. A widespread resident of lowland forest and mangroves, where it is the most common bronze cuckoo. Some winter birds may be AU migrants. *Smallest lowland bronze cuckoo.* Typically a *brownish-green* cuckoo with underparts *densely barred* white and black. Variable: iridescence of upperparts variable in brightness of green; some birds have white forehead and eyebrow; breast with or without a rufous wash. Shows rufous in the spread undertail. **Male** eye and eye-ring red, these dark brown in **Female**. **Juv:** Dull version of the Adult with all-white underparts lacking barring or with faint, incomplete barring; some without rufous in the undertail; most lack rufous edges to the wing feathers (a trait of White-eared BC); eye dark. **Subspp (3):** *poecilurus* (NG resident, some possibly AU migrants, includes *russatus*) forehead and crown variably brown or green, upperparts bronze to bronze-green, breast often with rufous tinge, barred; *misoriensis* (Biak resident) upperparts bronze-green; *barnardi* and/or *minutillus* (AU migrant) forehead white, crown green, upperparts bronze green, underparts clean white barred black. **Similar spp:** Shining BC has thicker, wider-spaced barring in underparts and lacks red eye in male. Horsfield's BC has black eye-line. White-eared BC has rufous patch on flight-feathers only and on crown of female; also note white ear. **Voice:** Two songs, both distinctive. (1) A very fast, steady, long trill of dozens of notes ~4 sec duration, increasing then decreasing in volume, easily overlooked because of its soft delivery. (2) A lilting, lightly whistled, descending series of 5–7 notes that slows and drops in pitch. **Habits:** Difficult to observe in treetops. In NG Region, hosts are mainly Large-billed Gerygone; a single additional record for Mangrove G. Egg glossy buff-olive, speckled especially at the large end. **Range:** NG, NW Is (Batanta, Gag, Gebe, Kofiau, Misool, Waigeo), Bay Is (Biak), Aru Is, Manam and Karkar Is, and SE Is (D'Entrecasteaux and Trobriand Is), 0–500 m (to 1400 m). SE Asia through Sunda Is, Philippine Is, Wallacea, and AU. **Taxonomy:** Geographic variation in this cuckoo is a most complex issue. In AU, 2 forms, sometimes treated as species, occur: Little Bronze Cuckoo (*C. m. minutillus* and *barnardi*) and Gould's BC (*C. m. poecilurus/russatus*). However, birds in intermediate plumage could be (1) hybrids, or (2) plumage variants of Gould's, or (3) just variants in a larger population combining both *minutillus* and *russatus*.

Pallid Cuckoo *Heteroscenes pallidus* Pl 51
(*Cacomantis* or *Cuculus pallidus*)
31–32 cm. Rare austral winter migrant to Trans-Fly savannah. *Same size as Oriental C, but paler and with dark line behind the eye* extending downward to side of the neck, *dark iris,* and a pale nape spot. **Male:** Variable, with whitish, unmarked breast and rather evenly grey mantle, although some birds have the crown and upperparts slightly darker and browner and have a small rufous patch on the back of the neck, below the white spot. White notching on mantle feathers wears with age, and some birds can appear dull grey. **Female:** Also variable, browner than Male with upperparts more heavily marked with brown and buff, rufous nape patch more conspicuous, and breast obscurely barred, heaviest on upper breast, although some birds are paler, approaching Male in appearance. **Juv:** By winter most have molted out of distinctively *mottled white-and-black* juvenile plumage and resemble Adults. **Similar spp:** Oriental C barred below. Brush and Fan-tailed Cs much smaller. **Voice:** Song a series of tonal notes rising in pitch. **Habits:** Feeds in trees or on the ground for caterpillars and large insects. Flight swift, undulating; profile falcon-like. **Range:** NG in Trans-Fly, plus 1 record from Bird's Head (Momi). Also AU.

White-crowned Cuckoo *Caliechthrus leucolophus* Pl 51
(White-crowned Koel, *Cacomantis leucolophus*)
30–36 cm. An uncommon, medium-large, black cuckoo in hills and lower mountains, extending into nearby lowlands. A conspicuous voice in the hill forest canopy, but the bird itself is difficult to see. **Adult:** Unmistakable if seen well—distinguished from other black birds by *white crown-stripe.* When viewed from below note the (usually) *white-tipped undertail-coverts and tail.* Bill and iris dark. Plumage variation not understood—some birds faintly barred below. **Juv:** All black, with virtually no scaling. **Similar spp:** Eastern Koel male larger, with pale bill, red eye, crown all black, and no white feather tips. **Voice:** Distinctive—a loud, mournful song repeated over and over, consisting of 3 (or 4) descending whistled notes *whir wher whur*:

⁻ ⁻ ＿   or   ⁻ ⁻ ⁻ ⁻

Alternate song is 3 or 4 clear notes, delivered rapidly on the same pitch, with a ringing, laughing quality: *wher-wher-wher-wher*. Also a single burry, downslurred *whurr!* **Habits:** Little known.

Feeds on invertebrates and fruit. Hosts and egg unknown. **Range:** Endemic. NG and NW Is (Salawati and possibly Waigeo), 0–1900 m.

Chestnut-breasted Cuckoo *Cacomantis castaneiventris* Pl 51
(*Cuculus castaneiventris*)
20–23 cm. Common, widespread resident of forest interior, usually on hilly terrain, from lowlands to mid-mountains. **Adult:** *Dark, blue-grey upperparts* and *bright rufous or chestnut underparts.* Note bright yellow eye-ring and feet. **Juv:** *Plain and unmarked*; brownish grey above; white, buff, or brownish below, palest on belly; *no barring*. **Subspp (3):** Vary slightly in size and intensity of color, with SE birds darker and nearly identical to resident race of Fan-tailed C. **Similar spp:** Confusingly similar Fan-tailed C occurs mostly above the elevational range of Chestnut-breasted; see account for that species. **Voice:** Two songs, both similar to Fan-tailed. (1) A musical but mournful, descending, whistled trill of about a dozen notes, often repeated ~1/sec; very like song of Mountain and Yellow-billed Kingfishers, but last two are richer, longer, and end in a rising trill. (2) A mournful, slow, 3-note series in the same pattern as the 3-note series of the Brush C, but the triplets are much slower and usually not immediately repeated.

Habits: Perches inconspicuously in the forest, taking food from any level. Hosts are scrubwrens (Grey-green S and Tropical S reported). **Range:** NG, NW Is (Batanta, Misool, Salawati), Bay Is (Yapen), and Aru Is, 0–1500 m, rarely to 2500 m. Also AU.

Fan-tailed Cuckoo *Cacomantis flabelliformis* Pl 51
(*C. pyrrhophanus* or *Cuculus pyrrhophanus*)
26–28 cm. Uncommon resident of higher montane forests. Also rare austral winter migrant to southern lowlands and disturbed areas of intermontane valleys. **Adult:** *Resident Adult similar to smaller Chestnut-breasted C* from lower elevations, especially in SE, but usually darker and the underparts dull brown with extremely faint, fine barring on the breast and belly visible at close range. *Migrant race much paler*, with upperparts grey and *underparts pale tan-brown.* Tail feathers with *upper surface notched on the outer edge with white*, faintly in outermost tail feathers of resident race (hard to see), obvious in migratory race. **Female:** Noticeably paler than Male in each race, with lower flanks and breast finely barred and speckled with dark grey. **Juv:** *Resident race variable. Upperparts uniform greyish brown; underparts grey or brownish grey, barred and speckled black.* (Chestnut-breasted juv plain below, no barring or speckling.) *Migrant race Juv paler and greyer, also faintly barred and speckled below, with overwintering birds molting into adultlike plumage.* In both races the *eye-ring is dull yellow*, helpful for distinguishing them from barred female and juv Brush C and from smaller juvenile bronze cuckoos. **Subspp (2):** *flabelliformis* (AU migrant); *excitus* (NG resident). **Similar spp:** Chestnut-breasted Cuckoo. Brush Cuckoo smaller and paler and with inconspicuous eye-ring; some females and juvenile more heavily barred, especially on back. **Voice:** Two songs very similar to Chestnut-breasted but slower and first song of longer duration. **Habits:** More conspicuous but similar to Chestnut-breasted, which it replaces at higher elevations. Residents' recorded hosts are White-shouldered Fairywren and Large Scrubwren. Egg is white, speckled with brown and grey all over (AU data). **Range:** Resident race in NG mts only (Bird's Head, Central Ranges, and Huon), 1500–3000 m (1200–3900 m). Migratory AU race winters in NG (S Lowlands and E Ranges, but probably more widespread) and Aru Is, 0–2000 m. Also AU, New Caledonia, Solomon Is, to Fiji.

Brush Cuckoo *Cacomantis variolosus* Pl 51, 52
(**Grey-breasted Brush-Cuckoo**, *Cuculus variolosus*)
19–23 cm. A common, widespread resident mainly of gardens and forest edge in lowlands and mid-mountains, less common in forest interior. Also an austral winter migrant. *Voice is the single distinctive feature of this dully plumed, medium-sized cuckoo.* Eye-ring and feet dull yellowish. **Male:** *Plain brownish grey head and upperparts; grey throat and breast; buffy belly and undertail-coverts.* **Female:** Polymorphic (1) *plain morph*, like Male with unbarred underparts, (2) *intermediate morph, underparts faintly dark-barred*, (3) *barred morph*, underparts more heavily barred and *upperparts with feathers pale-notched and barred.* **Juv:** *Unmistakable as heavily notched and barred buff throughout.* **Subspp (3):** Confusingly variable in depth of color; resident and migratory races not distinguishable in the field.

Similar spp: Chestnut-breasted and Fan-tailed Cs have the eye-ring and feet bright yellow and the underparts chestnut; juv has plain dark brown back (no markings). **Voice:** Two distinct songs. (1) A series of similarly clear, high, whistled notes (3–16 notes;1 note/sec), each slightly downslurred, the series dropping in pitch slightly *peer peer peer peeyr peeeyr . . .* (2) A loud excited series of clear, fairly high-pitched whistled notes; notes come in groups of 3, *sea to SEA*, and are repeated 5–10 times and gradually rise up the scale, gain in speed and volume, and become more excited:

Habits: Perches in the treetops, where it sings or forages for caterpillars and other invertebrates. Recorded hosts in NG are White-shouldered Fairywren, Yellow-tinted Honeyeater, Brown-backed H, and Lemon-breasted Flycatcher; mainly small honeyeaters and flycatchers in AU (AU data). Egg white, or sometimes cream or pale brown, with band of grey to brown spots and blotches at large end (AU data). **Range:** Throughout NG Region except outer SE Is (Trobriand, Woodlark, Misima, Tagula, Rossel), 0–1300 m, rarely to 1800 m. Also SE Asia to AU and Solomon Is.

Himalayan Cuckoo *Cuculus saturatus* Pl 51
(Oriental Cuckoo)

Apparent vagrant to NW. Probably inseparable from Oriental C in NG Region, except for smaller size on average. See account for Oriental C. Expected to be silent on wintering range, but song differs from that of Oriental: a high note followed by 3 lower notes, *hoop hoop* or *tun-tadun*; Female call *quick quick quick*. **Range:** Distribution in NG Region unclear, but records so far only from NW Is (Waigeo) and Bird's Head. Breeds in Himalayas and SE Asia; winters from Malay Pen reportedly east to NG.

Oriental Cuckoo *Cuculus optatus* Pl 51
(**Horsfield's Cuckoo**, *C. saturatus* or *horsfieldi*)

29–34 cm. A regular Palearctic winter migrant during austral summer, widespread to open habitats such as gardens, savannah, and forest edge of lowlands and intermontane valleys. A *medium-large cuckoo*, unobtrusive and wary, *usually with grey upperparts and white breast barred with black*, but *some Females brown and barred throughout*; iris pale to brown. Most often seen in flight. **Male:** Clean grey upperparts and white breast with black barring. **Female:** Common *grey morph* resembles Male, but with faint rufous wash on neck and breast; rare "*hepatic morph*" with upperparts rufous brown and heavily barred throughout. **Juv:** Upperparts with white-edged feathers, but probably molting to adultlike plumage when in NG; iris dark. **Similar spp:** Himalayan C differs by song and smaller average size. Yellow-eyed Cuckooshrike has darker upperparts, is more closely barred below, and has shorter, squared, less graduated tail. See rare Pallid C. **Voice:** Silent in NG; but just in case, song is 4 low, *hoop* whistles, the first one soft; also a harsh call, *gaak-gaak-gak-ak-ak-ak*. **Habits:** Solitary. Often seen at forest edge and along rivers. Forages in canopy and from tree trunks for invertebrates, especially caterpillars. Near Port Moresby it gathers in teak plantations to feed on caterpillars. Swift flight; falcon-like profile. **Range:** Specimen records throughout NG mainland and from NW Is (Gebe, Misool, Salawati), Biak (Bay Is), Manam I, and Karkar I, 0–1600 m (1 record at 4400 m). Breeds in NE Asia, wintering from Malay Pen to AU. **Taxonomy:** Oriental Cuckoo (*C. saturatus*) split into Oriental C (*C. optatus*) of northern Asia and Himalayan C (*H. saturatus*) of the Himalayas and China; both in the books as wintering in NG Region, but this assumption should be reexamined. Only *C. optatus* regularly reaches AU.

BARN-OWLS: Tytonidae

The barn-owl family is distributed worldwide with 18 species and is composed mostly of a single genus, *Tyto*, suggesting a recent radiation of these successful birds. The 4 New Guinea *Tyto* species are all shared with Australia. In New Guinea, barn-owls occupy virtually all forest and nonforest habitats. Barn-owls are recognized by their heart-shaped face, dark eyes, speckled plumage lacking bars and streaks, and long legs. They forage in twilight and at night for rodents and small marsupials which they catch with their powerful taloned feet. Barn-owls are typically found alone, but they live in pairs and defend a breeding territory. Females are the larger sex, and parents share nesting duties. They roost and nest in tree hollows, among rocks, or in grass, depending on species. These sites are strewn with an accumulation of regurgitated pellets containing the fur and bones of prey—a gold mine of dietary information. Barn-owls most often reveal their presence by piercing screams, whistles, or hisses, but may sometimes be encountered at dusk or nighttime at the forest edge or other open habitat.

Sooty Owl *Tyto tenebricosa* Pl 53
(Greater Sooty Owl)
33–38 cm. This is the forest barn-owl of NG. Uncommon in forest and subalpine grasslands from the lowlands near the ranges to well above timberline. **All plumages:** *The only sooty-colored NG owl.* Some individuals darker than others. Looks deceptively pale when caught in the light of a torch or automobile headlight. **Female:** Larger. **Downy chick:** Also sooty (AU data). **Subspp (1):** *arfaki* (endemic). **Similar spp:** Other barn-owls have a whitish breast. Should not be confused with much smaller Barred Owlet-Nightjar, also sooty. Papuan Boobook has yellow eyes. **Voice:** Song is a far-carrying, tremulous, descending whistle that has been likened to a falling bomb, whinnying horse, or whistling teakettle. When birds are interacting, they produce bizarre whinnies, cacklings, and sputterings. **Habits:** Little known in NG. Strictly nocturnal and seldom seen. Feeds on marsupials, rodents, and bats. Prey taken from the ground or in trees. Roosts in tree hollow, rock crevice, or under tree roots or log. Nests in tree hollow. Eggs (2) white (AU data). **Range:** NG and Yapen (Bay Is), 0–4000 m. Also AU.

Australian Masked Owl *Tyto novaehollandiae* Pl 53
(Masked Owl)
38–43 cm. A little-known savannah owl from the Trans-Fly, where it is the only *Tyto* species. Largest NG barn-owl; powerfully built. **All plumages:** Similar to more widespread Australian Barn-Owl, but with separate range. *Much larger than Australian BO, back and wings marked with dark grey, underparts heavily speckled black, and facial mask outlined with a thick dark border.* Plumage individually variable, ranging from darker with buffy underparts, to paler with white underparts. **Female:** Larger. **Downy chick:** Cream (AU data). **Subspp (1):** *calabyi* (endemic). **Similar spp:** Australian BO. Eastern Grass-Owl with mantle and wings dark brown and less patterned; mask with dark border barely evident. **Voice:** Song is a loud rasping screech, *ha-a-a-a-a* (AU data). Calls include a rhythmic cackling, like the sound of a handle being wound (AU data). **Habits:** Similar to Australian BO but takes larger prey. Can be found roosting at the same site year after year. Roosts and nests in hollow of a large tree (especially eucalyptus) or cave (AU data). Eggs (2–3) white. **Range:** NG: Trans-Fly, 0–200 m. Also AU.

Australian Barn-Owl *Tyto delicatula* Pl 53
(Barn Owl or **Eastern Barn Owl,** *Tyto alba* or *javanica*)
30–33 cm. Frequents disturbed habitats close to people—roadsides, gardens, plantations, and forest edge. Found along N coast, in mid-mountain valleys, and in SE Pen. Uncommon. **All plumages:** *Rather slender. Very pale, with white breast* (faintly spotted at most). White facial disc with only a hint of a black border. Bird looks all white by auto headlight. **Female:** Larger on average and tends to be more buffy. **Downy chick:** White to pale grey (AU data). **Subspp (1):** *meeki* (endemic). **Similar spp:** Eastern Grass-Owl has dark mantle and wings; Australian BO often seen on ground, leading to confusion with grass-owl. Australian Masked Owl does not co-occur. **Voice:** Thin reedy screech *sk-air* or *skee-air*, in flight or perched (AU data). **Habits:** Seen singly. Forages in open and takes small rodents. Roosts and nests in tree hollow, old building, cave, or rock crevice (AU data). Eggs (3–6) white. **Range:** NG (Sepik-Ramu, E Ranges, Huon, and SE Pen) and Manam and Karkar Is, 0–1600 m. Also Wallacea to AU and Oceania. **Taxonomy:** Split from the formerly worldwide Barn-Owl (*T. alba*).

Eastern Grass-Owl *Tyto longimembris* Pl 53
(**Grass Owl**, *Tyto capensis*)
30–36 cm. Patchily distributed in mid-montane grasslands; common where present.
All plumages: *A lanky, two-toned owl—dark grey-brown above with buffy white face and underparts.* Upperparts heavily smudged brown on buff, plain-looking, with white speckling hardly visible. Long-faced and small-eyed compared with other barn-owls. **Female:** Larger on average.
Downy chick: Buffy to tawny rufous (AU data). **Subspp (2):** *papuensis* (endemic).
Similar spp: Australian Barn-Owl smaller and much paler above. **Voice:** A harsh, rasping screech, *scairr*. Contact call a soft, high-pitched cricket-like trilling (AU data). **Habits:** Somewhat more social than other barn-owls, with gatherings of a few birds. Forages at dusk and dawn over open grassland, taking mainly small rodents. Roosts and nests on the ground under tussocks of grasses and sedges (AU data). Eggs (3–8) white. **Range:** NG: W Ranges and E Ranges (no records from Border Ranges) and Huon, 1000–2500 m. Also India and E Asia to AU.

OWLS: Strigidae

Of the 211 species of typical owls known worldwide, only 6 inhabit New Guinea. The family has a long fossil record in Eurasia, from where it seems to have spread rather late to the Australasian region. Only 2 lineages are represented: the speciose hawk-owls (*Ninox* and *Uroglaux*) and the scops-owl (*Otus*), with 1 species. Owls are nocturnal predatory birds with slow flight and deep wingbeats. New Guinea species differ from barn-owls in having a smaller, rounder facial disc and yellow eyes. In New Guinea, owls are found primarily in the lowlands, although the Rufous Owl and Papuan Boobook range well up into the mountains. The smaller species prey on arthropods; larger species take small mammals and birds. Typical owls occur singly or in pairs, and they lay white eggs in a tree-hollow nest. Both parents share nesting duties. Most species are vocal and perhaps territorial. Sometimes owls can be found at their day roost in tree foliage or other shelter, and their location may be given away by the scolding of small forest birds. But more typically, owls are heard at night or seen at dusk flying at the forest edge or near a road.

Biak Scops-Owl *Otus beccarii* Pl 54
(**Moluccan Scops-Owl**, *Otus magicus*)
25 cm. Endemic to Biak forests, where it is the only owl. At night in the beam of a torch it appears to be a small, unpatterned owl with square head and inconspicuous ear tufts. **Adult:** *Seemingly without pattern except for line of whitish spots above the shoulder.* (Actually finely vermiculated all over, without shaft streaking.) Color morphs either tawny or dark grey-brown. **Juv:** No information, but probably small differences in pattern. **Similar spp:** Papuan Frogmouth and Large-tailed Nightjar are the only other night birds on Biak. **Voice:** Male song a series of hoarse croaking notes. Female call *rick* or *ren ren ren . . .* for long periods. **Habits:** Little known. Roosts in tree foliage. Eats insects. Nests probably in tree hole; no nesting material. Eggs (no.?) white. **Range:** Endemic. Biak (Bay Is), 0–1000 m.
Taxonomy: Split from Moluccan Scops-Owl (*O. magicus*).

Rufous Owl *Ninox rufa* Pl 54
(**Rufous Boobook**)
41–51 cm. Rare and little known in lowland, hill, and mid-mountain forest. **Adult:** *A large, robust owl, finely barred below as well as above.* Individuals vary in shade of rufous brown. **Male:** Larger on average (AU data). **Juv:** Similar to Adult, but downy young have white head and underparts and a dark mask. **Subspp (2):** *humeralis* (NG); *aruensis* (Aru Is) smaller. **Similar spp:** Papuan Hawk-Owl, the only other barred owl, has streaked underparts. **Voice:** Generally quiet. Song is 2 low notes, softer and less clear than that of Barking O, transcribed as *woo-hoo* or *mumph-mumph*. **Habits:** Little known in NG. Best discovered during the day by listening for the agitated twittering of small birds scolding an owl at its tree crown roost. In AU, preys mainly on mammals, including flying foxes. Nests in a large tree cavity; no nesting material (AU data). Eggs (2) white. **Range:** NG, Waigeo I (NW Is), and Aru Is, 0–1800 m. Also AU.

Barking Owl *Ninox connivens* Pl 54
33–36 cm. Locally common in open habitats in eastern lowlands. **All plumages:** More slender and small-headed than boobook owls. *Upperparts brown; underparts white with neat brown streaking* (each feather with a straight brown line). **Male:** Larger on average (AU data). **Subspp (1):** *assimilis* (endemic). **Similar spp:** Southern Boobook is smaller, with rusty upperparts, chin, legs, and tail; underparts white-and-brown mottled. Papuan Hawk-Owl slimmer and with heavily barred mantle, wings, and tail. **Voice:** May call during the daytime, as well as at night. Song is a doglike barking: 2 quick, short notes of medium pitch, the second note lower. Male, *wuf-wuf*; Female, higher pitched *wok-wok*. **Habits:** Inhabits forest openings, edge, plantations, and scrub. Usually in pairs or small groups. Feeds mostly on insects. Nests in tree hollow; no nesting material (AU data). Eggs (2–3) white. **Range:** NG and Manam and Karkar Is, 0–200 m. NG: Sepik-Ramu, S Lowlands, and Trans-Fly eastward. Also AU and Moluccas.

Southern Boobook *Ninox novaeseelandiae* Pl 54
(**Common** or **Australian Boobook**, *N. boobook*)
25–28 cm. Trans-Fly savannahs only. **All plumages:** *A small, reddish-brown owl; underparts with messy white and red-brown mottling; face with dark mask.* Breast-feather pattern differs from that of Barking O: the dark markings on each feather being hourglass-shaped, creating a coarse, white-spotted effect. **Subspp (1):** *pusilla* (endemic). **Similar spp:** Barking O is larger, dark brown above, and white underparts have neat, long, dark streaking. **Voice:** Song *boo-book* (AU data). **Habits:** Little known in NG. Roosts during the day high in trees, sometimes in small groups that are mobbed by honeyeaters and other small birds. In AU feeds mainly on insects. Nesting similar to Barking O (AU data). **Range:** NG Trans-Fly, 0–100 m. Also AU and NZ. **Taxonomy:** Some authorities split the AU and NG populations (*N. boobook*) from the Morepork (*N. novaeseelandiae*) of NZ.

Papuan Boobook *Ninox theomacha* Pl 54
(**Jungle Boobook** or **Jungle Hawk Owl**)
25–28 cm. The common forest owl, from sea level to cloud forest. **All plumages:** *Chocolate brown* all over. **Subspp (4, in 2 groups):** *theomacha* (NG, NW Is) no markings; *goldii* (SE Is.) larger, pale spotting on shoulders and breast. **Similar spp:** Other owls either streaked or barred below. Sooty O much larger, blackish, and has dark eyes. **Voice:** Vocal. Song is an emphatically uttered pair of identical, slightly hoarse notes at medium pitch, each strongly downslurred: *hyu-hyu* or *kyo-kyo*, repeated every 3–4 secs. **Habits:** A familiar nocturnal inhabitant of forest and gardens. Often close to human habitation and may be seen taking insects attracted to lights. Solitary or in pairs. Feeds primarily on large insects. Nests in a tree hollow; no nesting material. Eggs (2) white. **Range:** Endemic. NG, NW Is (Misool and Waigeo), and SE Is (3 main D'Entrecasteaux Is, plus Tagula and Rossel Is), 0–2500 m. **Taxonomy:** Classification of *goldii* uncertain—may instead be its own species or possibly a race of Southern Boobook.

Papuan Hawk-Owl *Uroglaux dimorpha* Pl 54
30–33 cm. A rare and little-known owl of lowland and hill forest. Note distinctive hawklike shape with small head and long tail that is finely barred. **Adult:** *Underparts white or buff with fine dark streaking; upperparts brown and barred.* Some birds paler, others darker. **Juv:** Similar to Adult, but downy Juv with white head and underparts, and lacks black mask of Rufous O. **Similar spp:** Barking O has thicker streaking below and is not barred on the back. Rufous O also barred above but is much larger and darker below with fine barring. Doria's Goshawk has similar color pattern but is obviously a hawk and is much larger and has a longer tail. **Voice:** Song a series of slow-paced, double *hoo* notes of medium pitch (higher than Rufous Owl, lower than Papuan Boobook); notes ~1 sec apart, phrases ~2 sec apart. **Habits:** A mysterious owl of the forest interior and edge. Feeds on insects, birds, and rodents. No nesting information. **Range:** Endemic. NG and Yapen I (Bay Is), 0–1500 m. Spottily reported but probably throughout NG mainland.

FROGMOUTHS: Podargidae

This small family of tropical night birds (16 spp) ranges from India through SE Asia to Australasia. The 2 New Guinea species are shared with Australia, and these are sometimes classified as a family separate from the Asian birds. Frogmouths are large nocturnal birds with an oversized and wide head, heavy bill, short weak legs, and a long, pointed tail. They perch motionless on a thick branch and appear to be no more than another piece of wood. The head is characteristically held beak upward while the bird watches its observer obliquely through half-closed eyes. The 2 species broadly co-occur and are much alike, differing mainly in size and voice. They feed mostly on insects taken from the ground or trees. Their diet also includes small vertebrates. The often-heard songs are repetitive series of humming notes. The small, flat nest is built on a tree branch. Frogmouths can be readily seen in towns and at the forest edge, either by asking residents for the location of roosting or nesting birds or by searching at night along a road, river, or forest edge.

Marbled Frogmouth *Podargus ocellatus* Pl 53
33–38 cm. Common in lowland and hill forest and edge. *Distinguished from Papuan F by voice and size (Marbled two-thirds the size of Papuan).* Complex plumage pattern resembles Papuan, including much plumage variation and some overlap between the sexes. **Male:** Usually overall grey or brown, with much coarse white spotting, especially on throat and breast. **Female:** Usually overall rufous, with finer buffy spotting (or scarcely any)—a plainer, smoother-looking bird; pale eyebrow and scapular streak often less prominent. **Juv:** Similar to Adult, but look for fragments of down clinging to feather tips.
Subspp (3): *ocellatus* (all NG range except SE Is); *meeki* (SE Is: Tagula) no obvious sexual dimorphism in plumage, both sexes coarsely marked, plumages tending toward grey and brown rather than rufous; *intermedius* (SE Is: D'Entrecasteaux and Trobriand Is) larger, plumages similar to mainland race.
Voice: Readily distinguished from song of Papuan F by its fewer notes and higher pitch. Song a haunting series of 4–6 drawn-out, slightly upslurred, hummed notes with an insistent quality; *hoooa hoooa hoooa hoooa*. Also an excited song that starts high and descends the scale rapidly, a rolling series of trilled notes, ending with a distinct clap made by the snapped bill. D'Entrecasteaux frogmouth gives typical song, but the excited series is a rapid *hoo-woo, clap!* **Habits:** Typical for a frogmouth. Scant stick nest. Egg (1) white. **Range:** NG, NW Is (Misool, Salawati, Waigeo), Bay Is (Mios Num, Yapen), Aru Is, and SE Is (3 main D'Entrecasteaux Is, Trobriand Is, and Tagula I), 0–1500 m. Also AU.

Papuan Frogmouth *Podargus papuensis* Pl 53
46–53 cm. Common in lowland and hill forest and openings. More apt than Marbled F to occur in open habitats such as towns and parks. *Identified by voice and large size*, otherwise resembles Marbled F. Much plumage variation and some overlap between the sexes. **Male:** Grey or brown, always with underparts marbled with white blotches highlighted by black borders; scapulars patterned with white spots and black markings; more massive bill. **Female:** Usually greyish brown, underparts normally lacking marbling, or with only a few blotches on belly; scapulars smooth-textured, unpatterned, or obscurely patterned buffy. **Juv:** As with Marbled. **Voice:** A very low, resonant, slightly rising, and monotonously repeated series of 9–15 notes: *ooom ooom ooom . . .* sometimes repeated for long periods. **Habits:** Typical for a frogmouth. Nest and egg similar to Marbled. **Range:** NG, NW Is (Batanta, Misool, Salawati, Waigeo), Bay Is (Biak, Numfor, Yapen), and Aru Is, 0–1000 m, locally to 2000 m. Also AU.

NIGHTJARS: Caprimulgidae

The nightjars include 95 species of cryptic, nocturnal birds. The family is found throughout the world. Three species reside in the New Guinea Region, another is an uncommon Australian migrant, and 2 others are vagrants. Two genera are present: *Eurostopodus*, mostly in Australasia, and *Caprimulgus*, represented here by mainly Asian species. Nightjars in the New Guinea Region occur in lowland forests, second growth, and savannah; in the mountains they inhabit only man-made grasslands, cloud forest, and subalpine heath and moors. With long wings and tail and short, weak legs, a resting nightjar holds its body in a characteristic horizontal posture. During the day, nightjars roost on the ground or on a tree branch, substrates so closely matched by their camouflaged plumage that a bird is usually overlooked until it is flushed. Nightjars forage by sitting motionless on an exposed perch (often on a path or road) and flying swiftly upward to ambush passing insects. They also course back and forth over forest and clearings and may take a break by alighting on a branch. Nightjars are found alone or in pairs. All species lay their eggs on the ground without preparing any sort of nest. They sing at night, exclusively or mainly when breeding.

Nightjar species look much alike and are difficult to identify. Particular attention should be paid to shape (does the tip of the folded wing reach the tip of the tail?), the presence or absence of white signaling patches in the wing or tail, and the details of their beautiful and wonderfully cryptic plumage (but individual variation in plumage can be expected). The *Eurostopodus* nightjars lack rictal bristles fringing their mouth and an obvious, pale moustache-stripe. All species are uncommon and noteworthy. In flight, these dark nightjars lack white patches evident on the primaries and outer tail feathers; the wing is narrow and pointed. The *Caprimulgus* nightjars show prominent rictal bristles and white moustache-stripe and white wing and tail patches. The wing shape is shorter and more rounded. One *Caprimulgus* species, the Large-tailed Nightjar, is the common species to which all other New Guinea nightjars are compared.

Look for nightjars hawking insects at dusk or dawn. Flying nightjars silhouetted in the twilight can be recognized by their flight of flaps and glides on wings held upward in a V. During the day, nightjars are most often encountered when they are flushed from the ground. Nightjars perched in trees have been discovered when betrayed by the scolding of small birds.

Spotted Nightjar *Eurostopodus argus*   Pl 55
(*E. guttatus*)
27–35 cm. Vagrant to Aru Is but could turn up elsewhere, especially in the Trans-Fly. An austral winter migrant from AU to open habitats in lowlands. **Adult:** *Comparatively pale nightjar with buffy band across scapulars made up of sharp streaking* (suggesting dried grass); buffy spotting on wing-coverts. White patch on the primaries, but not on tips of outer tail feathers. **Juv:** Pale rufous brown, finely and uniformly patterned, without streaked scapulars; primaries with obvious pale tips. **Imm:** Like grey version of Adult; has streaky mantle, but retains Juv primaries with pale tips. Migrant young birds in NG would be expected to be in Imm plumage, or in transition from Juv to Imm plumage. **Similar spp:** Other nightjars lack sharp scapular streaking and combination of white patch on primaries but not on tail. **Voice:** Probably silent in NG. Song, a weird laugh beginning with accelerating cawing and ending in gobbled notes (AU data). **Habits:** In AU, inhabits dry woodland and savannah. Foraging behavior similar to White-throated N. Roosts on the ground. **Range:** Aru Is, plus a possible sighting from Trans-Fly (at Obo). Breeds AU.

White-throated Nightjar *Eurostopodus mystacalis*   Pl 55
(*E. albogularis*)
30–35 cm. An uncommon austral winter migrant from AU to open habitats at low to mid-elevations, including intermontane valleys, same habitats as Large-tailed N. **Adult:** *Large, grey nightjar* said to resemble discarded eucalyptus bark. Typically has a thick, grey band along the scapulars, and another across wing-coverts. *Long falcon-like wings, which when folded nearly reach tail tip. Primaries and tail without white markings.* (The few small white spots on outer primaries do not show.) **Juv:** Rufous brown, more finely and uniformly patterned than Adult; primaries with obvious brown tips. **Imm:** Similar to Adult, but retains Juv primaries with pale tips. Migrant young bird in NG would be expected to be in Imm plumage, or in transition from Juv to Imm. **Similar spp:** Large-tailed N

with white patches in primaries and tail; tip of folded wing falls well short of tail tip. Papuan N smaller, brownish, and without grey band across scapulars. Spotted N with narrow buff band across scapulars and with white patch on primaries. **Voice:** Generally silent in NG? Song, 3–4 mellow ascending whistles followed by a rapid, accelerating chuckle not ending in gobble notes (AU data). **Habits:** Poorly known in NG. Solitary. Hawks night-flying insects over clearings, rivers, and forest. Mostly keeps level with or above treetops in straight runs or broad, open circuits, higher and less erratic than Large-tailed N. Flaps, then glides with wings held in a dihedral V. Roosts on the ground or in thickets. **Range:** Eastern and central NG, 0–1650 m. Breeds in AU.

Papuan Nightjar *Eurostopodus papuensis* Pl 55

25–28 cm. Uncommon lowland forest nightjar of gaps and edge, such as river courses. **Adult:** *Dark, marked with grey tertials and buff scapular patch that contrast with the dark wing. Wings and tail without white markings.* Breast with horizontal rows of buff spots. **Juv:** Very different. Milk-chocolate colored with black markings much reduced; round black spots on center of crown. **Similar spp:** Large-tailed N has vermiculated, greyish wing-coverts, with pale spots more evident; tail proportionately longer; and white markings in primaries and tail. White-throated N larger and has more greyish back and tail. **Voice:** Sings perched or on the wing. Song, an infrequently heard but distinct series of 4 notes, *coo coo coo coo*, rapidly delivered. Call described as a low throaty chattering note. **Habits:** Poorly known. Solitary or in pairs. Active after dark. Forages over forest and breaks such as rivers. Hawks for insects. Perches in trees and on ground. Nests under thicket at edge of clearing; egg laid on ground. Egg (1) pale brown, heavily blotched in shades of brown. **Range:** Endemic. NG and Salawati (NW Is), 0–400 m.

Archbold's Nightjar *Eurostopodus archboldi* Pl 55
(Mountain Nightjar)

26–30 cm. *The only high-mountain nightjar.* Rare inhabitant of cloud forest gaps and edges and alpine heaths and moors. **Adult:** *Blackish, appearing lichen-encrusted, with mottling of silvery grey blotches and golden buff spots.* Prominent grey eyebrow. *Underparts black, conspicuously spotted buff or white.* No white patches in primaries or tail. **Juv:** Very different. *Ashy grey throughout*; black eye-patch; black spots and diamonds on upperparts; *line of large black spots running along scapulars*; sooty-barred underparts with a few scattered rufous feathers. **Similar spp:** None in habitat. **Voice:** Mostly silent. Voice poorly known (odd, as the species is occasionally seen by birders). Reported to be a quiet guttural flight-note. **Habits:** Solitary or in pairs. Hawks insects in twilight or darkness, low over forest and in clearings; also sallies from branch or fallen log, often returning to the same perch. Roosts on branch or ground. Nests in forest clearing or edge; egg laid on ground. Egg (1) white. **Range:** Endemic. NG: Bird's Head (Arfak Mts), Central Ranges, and Huon, 2200–3260 m.

Grey Nightjar *Caprimulgus jotaka* Pl 55
(Jungle Nightjar, *C. indicus*)

25–28 cm. Vagrant; 1 record from Bird's Head, but could turn up elsewhere in West. A Palearctic migrant during austral summer; presumably of open habitats in NG. **Adult:** Dark greyish to brownish grey nightjar with *plumage rather obscurely and evenly textured, without bold markings.* Folded wings nearly reach tip of tail. **Male:** White tips to outer tail feathers and small white patch on longest primaries. **Female and Juv:** These markings narrower and buff, obscure. **Subspp (1):** *jotaka* (breeds E Asia, winters SE Asia, Greater Sunda Is, Philippine Is). **Similar spp:** Large-tailed N has more prominent white markings in wing and tail, boldly patterned brown plumage, and folded wings far shorter than long tail. **Voice:** Presumably mostly silent in NG, perhaps sings prior to northward migration. Song, a long rapid series, *tuck tuck tuck* . . . delivered at 3–4 notes/sec, faster than Large-tailed N. **Habits:** Similar to Large-tailed N. **Range:** Recorded once from NG: Bird's Head (Manokwari). Breeds in Asia. **Taxonomy:** Split from Jungle Nightjar (*C. indicus*) of India.

Large-tailed Nightjar *Caprimulgus macrurus* Pl 55

25–28 cm. The common open-country, roadside nightjar from lowlands to mid-mountain valleys. Announces its presence nightly during breeding season with *loud knocking song.* **Adult:** *White patches on the primaries and tail* usually evident when perched or in flight. Perched birds rather boldly patterned but individually variable; *shoulders with 2 rows of buff spots* formed by the pale-tipped wing-coverts; obvious, *white-based rictal bristles and white gape streak* (lacking in *Eurostopodus* nightjars). *Folded wings do not reach tip of long tail.* **Male:** Wing and tail patches large and white.

Female: Patches smaller, buff. **Juv:** Poorly known; wing-covert spots much less prominent, primaries tipped rufous. **In flight:** Shorter, rounder wings than *Eurostopodus* nightjars; white patches show in wings and tail. **Subspp (1):** *schlegelii* (Australasia and E Wallacea). **Similar spp:** All other nightjars, including White-throated and Papuan that overlap in habitat, lack white patches in both tail and wings, but see vagrant Grey N. **Voice:** Song, a loud series of around 10 staccato notes, slightly less than 2/sec, *tok tok tok . . .* , a monotonous chopping or knocking. Calls are very different and often mistaken for another night bird, a series of soft croaking or grunting notes. **Habits:** Most often seen flushed up from roads at night. Singly or in pairs. Sallies from perch for insects; also hawks, usually below treetop level in the confined space of a clearing. Flight maneuverable, involves fluttering, gliding, and swooping on a twisting or circular route. Roosts on ground or often in trees. Nests at edge of forest; egg laid on bare ground. Eggs (2) whitish, indistinctly blotched brown and grey. **Range:** NG, NW Is (Batanta, Gag, Kofiau, Misool, Salawati, Waigeo), Bay Is (Biak, Yapen), Aru Is, Manam and Karkar Is, and SE Is (3 main D'Entrecasteaux Is, Tagula I), 0–1740 m. Also S Asia to N AU.

OWLET-NIGHTJARS: Aegothelidae

Few birds are as special to the New Guinea Region as the owlet-nightjars. Seven of the 10 species included in this family are found only in our region, and an eighth species is shared with Australia; 2 other species occur elsewhere, in the nearby N Moluccas and New Caledonia. Surprisingly, the owlet-nightjars are most closely related to the diurnal swifts, treeswifts, and hummingbirds, an archaic group of families that long ago split off from nightjars and other night birds. In shape and ground-feeding habits, owlet-nightjars resemble tiny frogmouths. Unlike frogmouths, the owlet-nightjars have a long, rounded tail, and the bill is framed by their signature feature, a complex array of whiskers. In the dark, whiskers may help the bird detect objects near its face in the same way a cat uses its whiskers. The whiskers and the large, owlish eyes give owlet-nightjars an appealing mammal-like face. Active only at night, owlet-nightjars feed on diverse insects and other invertebrates taken on the ground and from tree trunks, branches, and vegetation. Their usual mode of prey capture is by pouncing, either from a perch or by hopping along the ground. They also sally for flying insects.

Studies of the Australian Owlet-nightjar reveal this species to be highly territorial, defending its patch of forest by calling. Presumably the NG species behave similarly—they are certainly vocal and can be heard throughout the night. Songs of NG owlet-nightjars are usually a simple series of repeated notes; depending on species, these comprise hoarse whistles, muffled yapping, or squeaking; however, study is needed before vocalizations can be used as a reliable means of identifying species. During the day, the smaller owlet-nightjars roost in tree hollows, whereas the larger ones (Feline and perhaps Starry) roost in vegetation at least occasionally. Owlet-nightjars occupy one roost for a period of nights before moving to another. It is at tree hollows where they are most often seen, sitting in the entrance, resting motionless, staring outward. They also nest inside such hideaways, often choosing a vertical hollow inside the trunk of a dead tree.

Although lacking the glamour of other New Guinea novelties, owlet-nightjars are much sought by birders. These night birds are less elusive than one might think, living at the outskirts of towns and villages and on the grounds of lodges and bush-camps. Calling birds are often not evident until after the human world has settled down for the night. Owlet-nightjars can be quite tame, yet they are hard to spot with a headlamp, for they sit motionless among the foliage and do not show eye-shine. During the day, one can locate owlet-nightjars by scanning tree hollows for a small possum-like face. Identifying an owlet-nightjar to species may be difficult, especially because the plumages of some species are so individually variable. Owlet-nightjar species can be assigned to 3 groups: the large species with white striping (Feline and Starry), the small species with white flecks on the wing-coverts (Wallace's, Mountain, and Archbold's), and the small, barred species showing no white flecks on the wings (Mimic, Barred, and Australian). Much remains to be learned about owlet-nightjar distribution and all aspects of their lives in the wild.

Feline Owlet-nightjar *Aegotheles insignis* Pl 55
(**Large Owlet-nightjar**, *Euaegotheles insignis*)
25–30 cm. Uncommon in montane forests. *Largest owlet-nightjar, strikingly patterned with white or buff stripes on the face and underparts.* **Adult:** Ranges from rufous to brown. All show a distinctive facial pattern with a thick white eyebrow and a broad white stripe from chin to throat, and underparts marked by 3 broad white stripes, 1 down the middle and 1 on each flank. Short, rounded feather "horns" above eyes usually folded away. Adult plumage with some dark, stippled barring, at least on wings. **Juv:** Rufous only, body plumage and wings lacks barring. White markings brighter, enhanced by upraised feather tips. **Subspp:** Birds from N Coastal and Adelbert Mts smaller, but not racially distinct. **Similar spp:** Mountain ONJ much smaller, lacks such pronounced white markings. Starry ONJ in lowlands, not mountains. **Voice:** Song, a moaning and haunting series of 3 (2–5) unpleasant, hoarse whistles of medium pitch, each note usually a half tone higher than the preceding, the last notes coarser: *whor . . . whor, whor.* Can sound human—a startled, gasping scream, *ah . . . ah, ah!* **Habits:** Little known. Solitary. Inhabits forest interior and edge (Juv). Diet of insects (beetles often noted) and other arthropods, presumably from vegetation and ground. Usually seen by day when flushed from its daytime roost in cluster of dead leaves or vine tangle in the forest midstory. Possibly roosts/nests in tree hollows. No nesting information. **Range:** Endemic. NG mts (records from all but Fakfak, Kumawa, and Cyclops Mts), 1200–2800 m.

Starry Owlet-nightjar *Aegotheles tatei* Pl 55
(**Spangled Owlet-nightjar**, *Euaegotheles tatei*)
23 cm. Recently rediscovered and so far known from only 2 localities. Rare along rivers winding through low, hilly country covered in thick, tall forest. *A slender owlet-nightjar*, with short, tight plumage. *Always bright rufous*, so far as known. *Feather horns curve above the eye and taper to a point.* Smaller than Feline ONJ. **Adult:** Similar to rufous morph Feline but differs in having *tail with white dorsal bars* and rufous lores (not white). Adult (vs Juv) may show frosty pale barring (buff or white) on the breast and belly and tiny starry spots on back and scapulars. **Juv:** Lacks barring and star-spots. **In the hand:** Unlike other owlet-nightjars, lacks filamentous tips to feathers of face, chin, throat, and breast; ear-coverts stiff and bristly, not soft; rictal semibristles have short tips; tarsus short. **Similar spp:** None in its habitat. **Voice:** Song similar to Feline ONJ but notes higher pitched, usually 4, *whor . . . whor . . . whor, whor.* Readily mimicked. Call, given after singing in response to playback, a sharp medium-pitched squeak, *peh!* **Habits:** Solitary and little known. Keeps to the upper midstory and canopy. Its structural features suggest a departure from the typical owlet-nightjar mode of prey capture by pouncing. There are a few records of single birds being flushed from a daytime roost on a perch a few meters up inside the forest. **Range:** Endemic. NG: central S Lowlands (upper reaches of Fly R near Kiunga) and southern SE Pen (inland of Amazon Bay), 30–80 m (probably much broader). **Taxonomy:** Split from Feline ONJ.

Wallace's Owlet-nightjar *Aegotheles wallacii* Pl 56
20–23 cm. Rare and sparsely distributed in hill and lower montane forest, plus flat lowland forest near foothills. This is the hill forest counterpart of the Mountain ONJ. **All plumages:** *Dark above, charcoal grey or blackish*, obscurely barred giving finely mottled texture without distinct pattern; no pale collar (some have an indistinct collar). To best separate from Barred ONJ, look for *fine, white, streaky spots on scapulars and wing-coverts*, lacking in Barred. Underparts vary from pale with few markings to dark with heavy chevron marks. *Buffy chin and throat contrasting with dark sides of neck.* **Subspp:** Local populations vary by size. **Similar spp:** Barred and Allied ONJs are distinctly barred, usually not as dark above, and lack white spots on wing-coverts. Mountain ONJ more coarsely patterned. **Voice:** Song, a pretty series of rapid, squeaky whistles, *per pew-pew*, or *peer pew*, or variation. **Habits:** Little known. Solitary, in forest and edge. Behavior presumably like Mountain ONJ. No nesting information. **Range:** Endemic. NG and Aru Is, plus sightings from Waigeo (NW Is). Most of NG, but absent from Huon and SE Pen and no record yet for Bird's Neck, NW Lowlands, and E Sepik-Ramu; 80–1500 m.

Mountain Owlet-nightjar *Aegotheles albertisi* Pl 56

18–20 cm. Mountain forests. Common and ubiquitous from mid-mountains to timberline, except in the Western and Border Ranges, where poorly known and seemingly patchy and in mid-mountains only. *The small montane owlet-nightjar, lacking white facial and breast streaks* of the larger Feline ONJ. **All plumages:** Amazingly variable. Brown, grey, rufous, or shades and combinations thereof. Dark or pale. Variously mottled and barred; *breast with irregular dark spots and streaks*. Usually with a distinct pale collar. Note white, streaky flecks on wing-coverts. **Subspp:** Patterns of geographic variation are obscured by the large degree of individual variation. **Similar spp:** See Archbold's ONJ. Wallace's ONJ occurs at lower elevations and is more finely patterned. **Voice:** Call (or song?) a steady series of explosive squeaks, each note a second or more apart, *ee! . . . ee! . . . ee!. . . .* Sometimes double-noted. Also a faster series of the same notes, rapid at first, then slowing. **Habits:** Solitary. Usually seen by day perched motionless at the entrance of a tree hollow with only head and foreparts visible. At night, forages for insects within the forest and in small openings, taking food from vegetation and the ground. Diet includes insects, especially beetles, and earthworms. **Range:** Endemic. Most NG mts: Bird's Head, Kumawa Mts, Wandammen Mts, Central Ranges, Foja Mts, and Huon (no information for Fakfak and Cyclops Mts), 1200–2800 m (1000–3300 m). **Taxonomy:** Likely includes Archbold's ONJ as a series of color morphs.

Archbold's Owlet-nightjar *Aegotheles archboldi* Pl 56

18–20 cm. Counterpart of Mountain ONJ in high mountains of Western and Border Ranges. Usually ranked as a species, but intermediate birds exist between it and Mountain, and the 2 species are not safely separable. Archbold's is more *richly colored and heavily mottled*, especially on the breast; it shows *reduced barring on body*; and the *tail bars are usually solid*. As both forms are highly variable, and as birds of one form exist in the range of the other, it is possible that they are little more than variants along a continuum of plumage patterns (irrespective of color), with boldly patterned *archboldi* prevalent in the high-elevation cloud forests of the West. Mountain ONJ ranges throughout, though locally missing or scarce in the W Ranges, where it is found together with Archbold's but at the lower end of the elevational range. It is more finely patterned, with definite barring on sides of neck, and is often less color-saturated; the tail bars are usually broken. **Range:** Endemic. NG: W Ranges and Border Ranges, although a few high-elevation birds of E Ranges and SE Pen fit the description, 2200–3600 m.

Barred Owlet-nightjar *Aegotheles bennettii* Pl 56

20–23 cm. The common owlet-nightjar of lowland and hill forest and edge. **All plumages:** *Dark charcoal grey, finely barred,* and paler below. (A few mainland birds with brownish cast; D'Entrecasteaux Is birds always with buff cast.) *Usually lacks brown on cheeks* (compare with Australian ONJ), unless plumage is overall brownish. Face has an indistinct black eyebrow or mask and usually an obvious whitish hind-collar. **In the hand:** Primary no. 7 nearly always longest. **Subspp (4, in 2 groups):** *bennettii* (lowlands of mainland NG) nearly always charcoal grey and barred across breast (no dark streaking), folded wing (flattened) 117–139 mm; *terborghi* (E Ranges: Karimui) similar to *bennettii* in plumage, but much larger in size, folded wing 155 mm; *plumifer* (D'Entrecasteaux Is) washed buff, especially on head and breast, and with dark shaft streaking as well as barring on breast, folded wing 111–124 mm. **Similar spp:** Australian ONJ, co-occurring in savannah country, is very similar but paler, showing buff or rust on cheeks; see that account. Allied ONJ in mountain forests of Bird's Head fluffier and with brownish cast. Wallace's ONJ has buffy streak down throat and shows white flecks on wing-coverts. **Voice:** Song, an unevenly paced series of muffled yapping barks, suggesting a small dog, *ap . . . ap . . . ap . . . ap. . . .* Sometimes ~20 squeaky notes, *chyek . . . chyek. . . .* **Habits:** Little information. Solitary, usually seen perched at entrance of tree hollow roost. Foraging behavior not described but likely similar to Australian ONJ and Mountain ONJ. Nest, no information, but presumably in tree hollow. Eggs (clutch?) white. **Range:** Endemic. NG and SE Is (3 main D'Entrecasteaux Is); *bennettii* 0–800 m; *terborghi* 1100 m; *plumifer* 800–1100 m, probably broader. Eastern NG, westward to eastern NW Lowlands and Trans-Fly. **Taxonomy:** Subsp *terborghi* is huge compared with other Barred and Australian ONJs, and a DNA study grouped it with Allied ONJ rather than Barred. Alternatively, *terborghi* may be a distinct species living in the lower montane forests along the southern escarpment of the Central Ranges; more sampling is needed before the issue can be settled. Subsp *plumifer* has distinct plumage color and pattern, but a DNA study placed it within *bennettii* rather than as a separate species.

Allied Owlet-nightjar *Aegotheles affinis* Pl 56
(Vogelkop Owlet-nightjar)
23 cm. Lower montane forests of Bird's Head. **All plumages:** Barred pattern like Barred ONJ but with *brownish cast* and bird *thickly feathered, producing plump shape with large head*. Washed buffy brown, with buff cheeks and whitish belly. Head coarsely patterned. Most birds show a distinct pale collar. **Similar spp:** Barred ONJ absent from Bird's Head; inhabits lowlands; usually lacks brown in plumage and possesses shorter, tighter feathering for a slimmer shape with smaller head; white on belly usually either more restricted or lacking. Wallace's ONJ may co-occur but is darker and has pale streak down center of throat and small whitish flecks on the wing-coverts. **Voice:** No information. **Habits:** No information. **Range:** Endemic. NG: Bird's Head (Arfak Mts), elevational range not yet defined, somewhere within 800–1500 m and perhaps lower. **Taxonomy:** Split from Barred ONJ or Australian ONJ. May include *A. bennettii terborghi*, see account above.

Australian Owlet-nightjar *Aegotheles cristatus* Pl 56
23–24 cm. The pale owlet-nightjar of southern savannahs and open country; apparently common around Port Moresby. **All plumages:** *Medium grey above* (paler than Barred ONJ), with *mostly white breast and belly* (these usually more darkly barred in Barred ONJ). Some birds with brown wash. *Always with faint buff or rust color on cheeks*. Distinct pale hind-collar bordered with a dark line above and below. **In the hand:** Primary no. 8 nearly always longest. **Subspp:** Birds of Port Moresby area are same size as Barred ONJ (wing arc 122–131 mm). Trans-Fly birds larger (wing 134–146 mm). **Similar spp:** Barred ONJ darker, often much darker; lacks any brown on cheek unless overall plumage has brownish cast (rare). **Voice:** Port Moresby birds described as giving a repeated shriek. Song in AU, a single- or double-noted repeated series, the second note at lower pitch, *chur chur*; also a call, *kair!* **Habits:** In territorial pairs, but usually encountered alone. Feeds on insects and other invertebrates taken by sallying or pouncing. Seen sitting on the ground beside the road. Roosts and nests in tree hollows; nest unlined or scantly lined with leaves, grass, bark, or fur (AU data). Eggs (~4) white. **Range:** NG: Trans-Fly and Port Moresby area of SE Pen, 0–100 m. Also AU.

TREESWIFTS: Hemiprocnidae

A small family (4 spp) ranging from India to the Solomon Is, with 1 species in the New Guinea Region. A medium-sized aerial bird with a small body, long slim wings, broad head, and a tail that is long and forked. Much larger than the swiftlets or swallows. Unlike true swifts, treeswifts have strong enough legs and feet to perch upright on a branch. Flight often consists of graceful circuits starting from an exposed, high perch in a forest clearing or garden edge. Sometimes makes long flights hundreds of feet above the ground. Feeds on insects captured in flight. Somewhat vocal and sociable, occasionally in large flocks. One white egg is laid in a tiny cup nest on a tree limb.

Moustached Treeswift *Hemiprocne mystacea* Pl 57
28–30 cm. Common in lowlands and foothills, favoring forest edge at clearings, roadways, rivers. *A very large swift with long, curved, pointed wings and long, deeply forked tail*. When folded, tail looks like a long spike. **Adult:** Sooty grey with glossy blue wings and crown and *white eyebrow and moustache plumes*. **Male** has a small chestnut patch on ear-coverts, lacking in **Female**. **Juv:** Mottled black, rufous, buff, and white. **Similar spp:** Fork-tailed Swift has a white rump and throat, less prominently forked tail, no "moustache," and flies very differently. **Voice:** Diagnostic. Call a loud, clear, drawn-out, single downslurred note, *kyee*; somewhat hawklike but shorter and usually only repeated 2–3 times at ~1 sec intervals. Also a series of muted, upslurred, softly whistled notes *owee wee wee wee*. **Habits:** Usually occurs singly or in pairs, but occasionally gathers in flocks, sometimes of hundreds. During day, perches on high snag overlooking gardens or other open habitat. Lights out after flying prey, then circles back to perch. At dawn and dusk becomes more active in the twilight, energetically hawking flying insects with dashing flight. Nest built on a bare, horizontal branch; nest a small saucer of feathers and plant material cemented with saliva, hidden under sitting Adult. Egg (1) white. **Range:** NG, NW Is (Batanta, Gebe, Misool, Salawati, Waigeo), Bay Is (Biak, Numfor, Yapen), and Aru Is, 0–1000 m, rarely higher. Also Moluccas, Bismarck, and Solomon Is.

SWIFTS: Apodidae

There are some 105 species of swifts worldwide, with 8 species known from New Guinea, 2 of which are northern migrants. These, the most specialized of aerial birds, have cylindrical bodies and long, tapered wings, with the bend of the wing very close to the body. Flight is fast, with stiff and rapid wingbeats. Swifts spend most of the time in the air. They are rarely seen at rest and then only at hidden roosting and nesting sites. Although generally abroad by day, many species are actually most active in dim light, either at dawn or at dusk, or before or after rain. This comfort in dim light hints at their ancestry, for their relatives are nocturnal birds, the owlet-nightjars and nightjars. Swifts feed exclusively on insects, caught on the wing. The sexes are alike in plumage and size, and juveniles show small plumage differences, if any. Swifts occupy virtually all New Guinea habitats. They nest in caves, overhanging boulders, or tree hollows. Both parents attend the nest, usually a small cupped structure of fine plant material, often cemented to its substrate with saliva.

In New Guinea there are 2 groups, swiftlets and swifts. The swiftlets (5 spp) are common and nearly ubiquitous residents, although their cave-roosting and nesting habits keep them close to hills and mountains. Certain *Aerodramus* swiftlets have a remarkable ability of navigating in the total darkness of the caves in which they roost and nest. They manage this by continually uttering clicking calls and then listening for the echoes bouncing off the walls of their hideaway, a process of navigation called echolocation. At least the Mountain and Uniform Swiftlets are known to echolocate, and other New Guinea species may as well. (Bats navigate in a similar manner, but the echolocation calls of most bats can't be heard by the human ear.) Swiftlets use saliva as glue to hold their nests together and plaster them to vertical substrate. Nests of 2 swiftlet species in Asia are made entirely or mostly of hardened saliva, and it is these nests that are harvested to make "bird's nest" soup. The nests of swiftlets in New Guinea contain so much moss and other plant material that they are not edible.

The 3 species of swifts are larger, and each is easy to recognize by its shape alone. They do not nest in caves and cannot echolocate.

Swiftlets lack distinctive field marks, and separating them in the field may be difficult, often impossible. The effort is worth a try, however, particularly because a careful search may turn up either of the 2 rarest and least known species, Bare-legged and Three-toed Swiftlets. All swifts and swiftlets are gregarious, so look for the rarer species among flocks of more common ones. Note that in the hand the presence of white fringes on the flight-feathers often distinguishes birds in juvenile plumage.

Glossy Swiftlet *Collocalia esculenta* — Pl 57

9 cm. Common, from the foot of the ranges to above timberline, over forest and open habitats; avoids extensive flat country and savannah. Smallest NG swift. *Tail moderately long* and slightly notched. **All plumages:** *The only swiftlet with glossy blue upperparts and whitish belly.* Underwing all dark. **Subspp:** In need of review. **Similar spp:** Papuan Spinetailed Swift similarly glossy, but heftier with much shorter, square tail and pale streak across underwing; usually flies higher. Other swiftlets grey-brown without noticeable gloss and with sooty underparts. **Voice:** Usually silent. Call a sharp twitter. Does not echolocate. **Habits:** Forages singly or in small flocks low over the forest and in forest gaps, especially beside streams and over roads. Flight slow; glides with frequent changes in direction; rapid winnowing of wings followed by short glides with wings held downward. Nests alone or in colonies in the twilight zone of caves, overhanging road cuts, sheltered rock faces; nest a small cup of moss, lichen, and fiber glued with saliva to rock. Eggs (2) white. **Range:** All NG Region, except much of S Lowlands and Trans-Fly, nor any records from Gebe I or Manam I, 0–4500 m. Also SE Asia to Solomon Is.

Mountain Swiftlet *Aerodramus hirundinaceus* — Pl 57
(*Collocalia hirundinacea*)

11–13 cm. Common in the mountains from foothills to tree line, over forest and open habitats. *The grey-brown swiftlet most often seen in the highlands.* **All plumages:** A uniformly grey-brown swiftlet with notched tail. **In the hand:** From Uniform S by somewhat larger size; and *tarsus usually feathered* (a tract of short feathers running down the leg). **Similar spp:** See Uniform, Bare-legged, and

Three-toed Ss. Bare-legged and Three-toed are larger but difficult to distinguish from Mountain S in flight. **Voice:** Flight call is a rapid twittering that repeats its pattern. Echolocates with clicking call. **Habits:** Gregarious, usually in small or large flocks, mixing with other swiftlets. Flocks come and go during the day, their movements a mystery. Found especially over open country and less commonly over forest. Flies both high and low. Nests colonially in caves, sinkholes, and mine shafts, usually well into the dark zone; nest a cup of moss and other plant material without saliva adhesive built on a niche on rock surface. Egg (1) white. **Range:** Endemic. NG, Yapen (Bay Is), Karkar I, and SE Is (D'Entrecasteaux Is: Goodenough I, possibly others), 0–4000 m.

Bare-legged Swiftlet *Aerodramus nuditarsus* Pl 57
(Whitehead's Swiftlet, *Collocalia whiteheadi*)
14 cm. A rare montane swiftlet. *Largest NG swiftlet, with large-headed profile.* Known from very few specimens. **All plumages:** Grey-brown above, uniform lead grey below. **In the hand:** Tarsus bare. Extensive pale fringes on the eyebrow distinctive and unique. **Similar spp:** Mountain and Three-toed Ss. **Voice:** No information. **Habits:** Difficult to separate from other swiftlets in the field and therefore little known. Over both forests and open habitats? Roosts in rocks overhanging mountain streams. No nesting information. **Range:** Endemic. NG Central Ranges, with fewer than 10 records, 1500–1850 m (1 record at 30 m). **Taxonomy:** Split from Whitehead's Swiftlet (*A. whiteheadi*) of the Philippine Is and Mayr's Swiftlet (*A. orientalis*) of the Bismarck and Solomon Is.

Uniform Swiftlet *Aerodramus vanikorensis* Pl 57
(Lowland Swiftlet, *Collocalia vanikorensis*)
13 cm. Common in lowlands and foothills over forest and open habitats, including coasts and small islands. *The grey-brown swiftlet most often seen at low elevations*, although not separable from Mountain S in the field. **All plumages:** Uniformly grey-brown swiftlet with notched tail. **In the hand:** From Mountain S by smaller size and tarsus usually not feathered. **Subspp:** In need of review. **Similar spp:** See Mountain, Bare-legged, and Three-toed Ss. **Voice:** Flight call a repeated twittering. Echolocates with clicking call. **Habits:** Similar to Mountain S. Nesting similar to Mountain S but includes saliva as glue. **Range:** NG, NW Is (Batanta, Kofiau, Misool, Salawati, Waigeo), Bay Is (Biak, Numfor, Yapen), Aru Is, and SE Is (all except Rossel), 0–1400 m. Also Philippines and Sulawesi to New Hebrides Is. **Extralimital spp:** White-rumped Swiftlet (*Aerodramus spodiopygius*) on islands to the N and E (Admiralty, Bismarcks, Solomons to Tonga Is) has been reported in error from Manam I but nevertheless may wander to the eastern NG Region; distinguished by its dingy white rump patch.

Three-toed Swiftlet *Aerodramus papuensis* Pl 57
(*Collocalia papuensis*)
14 cm. Uncommon and local in foothills and mountains. Unique among swifts for having only *3 toes instead of 4. Large size, closest to Bare-legged S*, but slightly smaller and more slender. **All plumages:** Similar to other all-brown swiftlets. One supposed field mark does not work: the throat is often, but not always, slightly paler (impossible to see in the field because it is ever so obscure and is confounded by counter-shading); this character is shared with other swiftlets. **Similar spp:** Mountain, Uniform, and Bare-legged Ss. **Voice:** Flight call, no information. Echolocates with clicking calls. **Habits:** Recorded foraging in flocks over lagoons. Flies at night outside roosting caves giving echolocation calls. No nesting information, but probably nests in roosting caves. **Range:** Endemic. NG: only a few records, from W Ranges to E Ranges; possible sightings as far east as SE Pen, 0–2400 m. **Taxonomy:** At present, molecular data point to a relationship closest to the Waterfall Swiftlet (*Hydrochous gigas*) of Malay Pen and Greater Sunda Is.

Papuan Spinetailed Swift *Mearnsia novaeguineae* Pl 57
(New Guinea Spine-tailed Swift, *Chaetura novaeguineae*)
11 cm. An uncommon lowland swift over forests and open habitats. *Unique stubby shape accentuated by short, square tail.* Body thicker and more oval than the slender, long-bodied swiftlets. Tail very brief compared with longer-tailed swiftlets. **All plumages:** Resembles Glossy Swiftlet in dark, glossy upperparts and pale belly; however, note *pale underwing-stripe across secondaries* (obscure in northern race). **In the hand:** "Spinetailed" refers to the projecting shafts of the tail feathers. **Subspp (2):** *novaeguineae* (S Lowlands and SE Pen) throat grey; *buergersi* (NW Lowlands and Sepik-Ramu) throat dark, like back. **Similar spp:** Glossy S has slower, more erratic, gliding flight.

Voice: Flight call distinctive: 3–4 squeaks. **Habits:** Forages singly or in small groups; often associates with swiftlets. Fast flight with rapid wingbeats. Makes long, straight runs between changes in direction; similarly follows river courses, dropping low to the water. Roosts in hollow trees, but occasionally perches on vertical spires in the forest canopy. Seen carrying nesting material (palm fiber) into tree hollows; no other nesting information. **Range:** Endemic. NG (only sightings from Bird's Head and Salawati I), 0–500 m (to 1200 m).

White-throated Needletail *Hirundapus caudacutus* Pl 57
(*Chaetura caudacuta*)
19 cm. An uncommon Palearctic migrant, both in passage and overwintering during austral summer. Over lowland forests and open country. *Largest NG swift; short-tailed with long, thin wings.* Flight can be amazingly fast. **Adult:** Unique combination of *white throat and undertail*, otherwise dark below. Darker above, but back is paler than head and tail. White spot on upper scapulars. **Juv:** Noticeably smaller white patches on forehead and scapulars; upperparts less glossy. **Subspp (1):** *caudacutus* (breeds NE Asia, winters S NG and E AU). **Similar spp:** No other swift or swiftlet is as large or shows a white undertail. **Voice:** Flight call, a rapid, high-pitched twittering; also *chur* notes. **Habits:** In wandering flocks, sometimes with Fork-tailed S. One of the fastest flying of all birds. Flight usually direct and powerful, but may also be more leisurely and circling with pulses of flapping and glides. Forages at both high and low altitudes. **Range:** Breeds E Asia, winters S Asia to NG Region and AU.

Fork-tailed Swift *Apus pacificus* Pl 57
(**White-rumped** or **Pacific Swift**)
18 cm. A Palearctic migrant, both on passage and overwintering during austral summer. Common over Trans-Fly, rare and occasional elsewhere. Mainly over savannah, but also forest and other habitats in passage. Note long, thin, curved wings and long tail. **All plumages:** *White rump and long, forked tail diagnostic.* Beware that the fork disappears when the tail is folded. Chin and throat white; rest of *underparts scaled white*. **Subspp (1):** *pacificus* (breeds NE Asia, winters SE Asia to AU). **Similar spp:** No other swift has forked tail, but some swallows do. **Voice:** Can be heard even when bird is far off. Flight call, a shrill, buzzing scream, *skree-ee-ee*. **Habits:** In flocks, sometimes with White-throated N, swiftlets, swallows. Typically flies very high. Flight slower and more changeable than needletail. **Range:** NG Region. Breeds E Asia, winters S Asia to AU.

ROLLERS: Coraciidae

Twelve species of rollers inhabit Africa and Eurasia, with 1 species reaching New Guinea as a breeder and widespread migrant from Australia. Rollers are long-winged, graceful fliers. They spend much time perched on bare branches or power lines at the forest edge or clearings, sailing out to capture flying insects or snatching small vertebrates on the ground. They nest in a tree cavity.

Oriental Dollarbird *Eurystomus orientalis* Pl 60
(**Broad-billed Roller**)
25–28 cm. Inhabits savannah, forest edge, gardens, and roadsides in lowlands and hills. Seen perched in or sallying from a tall isolated tree or the forest edge. Both races widespread, but the resident race is uncommon, whereas the migratory AU race is seasonally common during austral winter. **Adult:** Uniquely shaped—*chunky with large head and broad bill; appears all dark with red bill and legs.* In good light note brown head and blue body. **In flight:** Wing with *pale "dollar coin" mark on primaries.* Floppy but graceful wingbeats distinctive. **Juv:** Like a darker, blackish version of Adult; bill blackish. **Subspp (2):** Distinguished with care; both are widespread; *waigiouensis* (endemic) brighter blue; *pacificus* (migratory, breeds AU) paler grey-blue. **Similar spp:** Mynas or Eclectus Parrot may look similar in flight. **Voice:** Calls are distinctive coughlike croaks, *kak* or *chak*, sometimes repeated in a series. **Habits:** Solitary or in pairs when not migrating. On migration, passes southward in long straggling lines of hundreds of birds. Has 2 modes of foraging: from a high, exposed perch sallies out to capture passing insects, or flies continuously high in the sky, sometimes in flocks, often just after dawn. Roosts socially on passage, but solitary migrants may roost in a tree hollow. Nests in a tree

hollow or spout. Eggs (3–4?) white. **Range:** All NG region, although resident race not reported from Salawati, Biak, Mios Num, Numfor, Manam, Aru, and Trobriand Is; both races 0–1600 m. Also E Asia to AU and Solomon Is.

WOODLAND KINGFISHERS: Halcyonidae

Of the 62 woodland kingfisher species worldwide, a substantial proportion—21 species—inhabit the New Guinea Region. Kingfishers are a distinctive and old lineage of birds believed to have originated on the northern continents. The presence of kingfishers across the seas in Australasia is not surprising given their attraction to rivers and coasts and an ability to disperse long distances over water.

All kingfishers have at least some blue feathering and are easily recognized by shape: the large head wields a long, powerful bill, and the body is supported on feet so small the bird appears glued legless to its perch. Kingfishers sit motionless for a time looking around for prey, then dive down to snatch it with the bill. Other characteristic and comical behaviors include bobbing the head and sometimes the whole body (the smaller species do this) and cocking the tail upward, especially when calling. Kingfishers would often be overlooked if it were not for their calls, and fortunately most species are noisy.

Woodland kingfishers, in contrast to the family of river kingfishers, obtain nearly all their prey from land and typically have no affiliation with water. They have a heavier bill and sport a tail of normal length, or one of extraordinary length in the paradise-kingfishers. A few species—Blue-black, Collared, and Beach Kingfishers—actually fish on occasion. Prey taken on land include large insects, worms, crabs, lizards, frogs, rodents, and even small birds. This group includes the celebrated kookaburras and the spectacular paradise-kingfishers, both groups especially diverse in New Guinea. Two nonconforming New Guinea species are the nocturnal Hook-billed Kingfisher and the fossicking Shovel-billed Kookaburra.

Woodland kingfishers are mainly lowland birds, with only a few species being found in the mountains. Even though they usually forage alone, they live in territorial pairs or, depending upon species, in families of parents and their offspring. They do not build a nest but instead excavate a nest chamber in an earth bank, termite mound, or tree hollow. All species in the region are breeding residents, except one, the Sacred Kingfisher. That species, the Forest Kingfisher, and the Buff-breasted Paradise-Kingfisher nest in Australia and migrate to New Guinea to spend the winter, but the latter two also have resident populations on mainland New Guinea.

Although some woodland kingfishers are easily spotted when perched in the open, most forest-dwelling species are best located by their distinctive vocalizations. Another way to find them is to be alert to the scolding of small songbirds mobbing a kingfisher, a potential predator. Identifying kingfishers is rarely a problem when a bird is seen well. Juvenile plumage differs from adult plumage and can also be identified to species. The juvenile undergoes a partial molt (retaining flight-feathers), after which it looks like the adult.

Common Paradise-Kingfisher *Tanysiptera galatea* Pl 59
33–43 cm. The widespread paradise-kingfisher. Generally common in forest interior, but may be locally scarce or absent. Very inconspicuous, but vocal. Look for the white breast and movement of the *flaglike tail being pumped up and down* as the bird sits quietly—a habit of all paradise-kingfishers. **Adult:** *Striking pure white underparts and rump; bright blue crown and wing-coverts; very long streamer tail with spatula tip* (tail usually longer in Male). Only paradise-kingfisher on mainland NG with *mostly white outer tail feathers. Bill red.* **Juv:** Mostly brown above with crown showing at least some feathers edged blue; dirty buff-white below with dark scalloping; central tail feathers only slightly lengthened; bill all or partly black. Molts in a few months to Imm. **Imm:** Adultlike plumage but with some retained Juv flight and tail feathers. Some birds with long, all-white tail without spatulate tips may be in imm plumage? **In flight:** White streamers obvious. **Subspp (4, in 1 group):** *galatea* (all NG Region). **Similar spp:** Little PK (Trans-Fly and Aru) most similar; see that account.

Voice: The soft, clear, musical, whistled trill is one of the characteristic songs of lowland forest. Consists of a short series of downslurred notes, slowly rising in pitch and speeding up, usually preceded by 1–4 sharper initial notes:

When excited, the bird repeats the series rapidly, running the notes together, and adding louder, higher-pitched scolds. Call is a single, slow, mournful, whistled downslur or upslur of 2 sec duration. **Habits:** Occasionally in mangroves; migrates from monsoon forest to rainforest in dry season. Solitary or in pairs, territorial and aggressive. Perches in forest understory slowly pumping tail and occasionally sallying out to take prey from foliage or ground, even digging for earthworms. Feeds on invertebrates. Nest chamber excavated in arboreal, active termitarium, 2–6 m above ground. Eggs (~5) white? **Range:** NG, NW Is (Batanta, Gebe, Salawati, Waigeo), and islands off N coast (Walis, Manam, Karkar, Bagabag), 0–500 m. All NG lowlands, but may be absent from Purari-Kikori area, and there are few records from western S Lowlands (Etna Bay to Digul R) and N coast of SE Pen. Also Moluccas.

Rossel Paradise-Kingfisher *Tanysiptera rosseliana* Pl 59
(formerly under Common Paradise-Kingfisher, *Tanysiptera galatea***)**
35–36 cm. Rossel I (SE Is), where it is the only paradise-kingfisher; common in forest.
Adult: Differs from Common PK in *darker crown and bluer back* yielding much less color contrast over the upperparts. Further, tail shorter and all white (not blue at the base). **Similar spp:** None on Rossel. **Habits:** Similar to Common PK. Nest and egg unknown. **Range:** Endemic. Rossel (SE Is), 1–300 m. **Taxonomy:** Formerly a subspecies of Common PK.

Biak Paradise-Kingfisher *Tanysiptera riedelii* Pl 59
35–36 cm. Biak I (Bay Is), where it is the only paradise-kingfisher; common in forest interior, edge, and degraded forest. **Adult:** *Pale blue head, dark blue back, and long, spatulate, nearly all-white tail.* Bill red. **Juv:** Head and upperparts similar to Adult but duller; underparts buff with dark scalloping; tail somewhat elongate. Bill dark. **Voice:** Song, a series of ~20 squeaky notes, begins slowly then accelerates and ascends in pitch. **Habits:** Similar to Common PK but often perches in the upper midstory. No nesting information. **Range:** Endemic. Biak (Bay Is), 0–600 m.

Kofiau Paradise-Kingfisher *Tanysiptera ellioti* Pl 59
33 cm. Kofiau I (NW Is), where it is the only paradise-kingfisher; common in lowland forest interior.
Adult: *Similar to Common PK, but tail shorter, thicker, and all white.* Bill red. **Juv:** Head and upperparts similar to Adult but duller; underparts buff with some dark scalloping and streaking; somewhat elongated tail blue above. Bill dark. **Voice:** Song is a trill similar to that of Common PK but shorter and more plaintive. Call is 3 mournful high-pitched downslurs. **Habits:** Generally similar to Common PK. No nesting information. **Range:** Endemic. Kofiau (NW Is).

Numfor Paradise-Kingfisher *Tanysiptera carolinae* Pl 59
36–38 cm. Numfor I (Bay Is), where it is the only paradise-kingfisher; common in forest, second growth, and gardens. **Adult:** *Dark blue* with white rump and *long, white tail with spatulate tip.* Bill red, feet yellowish. **Juv:** Dark blue above, whitish or rufous marked with black below, and tail somewhat elongated, black. Bill dark. **Voice:** Song begins very slowly with plaintive *yak* notes then speeds up to a rapid trill, *k-k-k-k-k* . . . all on the same pitch. **Habits:** Feeds on insects and nests in termitaria. **Range:** Endemic. Numfor (Bay Is). Lowlands.

Little Paradise-Kingfisher *Tanysiptera hydrocharis* Pl 59
31 cm. Uncommon and local in riverine forests of S Lowlands and Trans-Fly on NG, and the only paradise-kingfisher on the Aru Is. *Nearly identical to the Common PK* in appearance, habitat, and behavior, but slightly smaller and with subtle color differences. **Adult:** *Outer tail feathers entirely blue*, only thin white edging to central pair of feathers (adult Common has tail feathers white or blue-and-white); *crown dark blue* (Common has pale blue crown); and *wing-coverts dark blue* (Common has a contrasting pale blue patch on the wing-coverts, same color as crown). Bill red. **Juv:** Distinguished from juv Common by size and darker blue crown, if present.
Similar spp: Common PK. **Voice:** A sure means of identification and separation from Common.

Song is an 8-note series that drops slightly in pitch, *tr-tr-tr-tr-tr-tr-tr-tr*, each of the downslurred notes delivered slowly and separately; by comparison, the song of the Common PK is a faster and longer trill that rises in pitch.

Habits: Co-occurs with Common and habits apparently similar. No nesting information. **Range:** Endemic. NG and Aru Is, 0–300 m. NG: Trans-Fly north to central S Lowlands (Kiunga area). Not found recently on Aru.

Buff-breasted Paradise-Kingfisher *Tanysiptera sylvia* Pl 59
(Australian or White-tailed Paradise Kingfisher)

37 cm. A rare and very local resident in monsoonal lowland and hill forests of SE Pen; also an uncommon austral winter migrant to much of lowland NG, particularly S Lowlands, in monsoonal forest and rainforest. In all plumages, note *pale patch on dark back*. **Adult:** Look for *buffy orange breast and contrasting red bill*. *Diagnostic whitish patch on center of upper back*, merging with white rump and tail-streamers; *tail pointed, no spatulate tip*. **Female:** Differs from Male by shorter tail, blue margins on central tail feathers, and variable dusky streaking on white back and rump. **Juv:** Like a much duller, shorter-tailed version of Adult, but feathers of upperparts edged brown and underparts edged sooty; from other juv paradise-kingfishers by *buffy triangular patch on back*; bill dark. **Subspp (2):** Not safely distinguished; *salvadoriana* (resident) greenish tinge to blue of head and wing and paler below; *sylvia* (AU migrant). **Similar spp:** Brown-headed and Red-breasted PKs are deep salmon below (not buff-orange) and have all-dark back, pinkish brown rump, and blue tail. **Voice:** Vocal when breeding. Song in NG is a soft, purring trill and differs from that of Common PK by lacking introductory notes, being more rapid and at a lower pitch, and decelerating rather than accelerating. In AU, the trill is 1 sec long and descends slightly. A 4-note call is thought to be the territorial call in Australia, where successive notes rise in pitch: *chuga, chuga, chuga, chuga*. Little known about vocalizations of AU birds wintering in NG. **Habits:** Similar to those of other paradise-kingfishers. Occupies higher levels of the forest when sympatric with Common. Feeds on insects and other invertebrates. Nest chamber excavated into ground-level termitarium. Eggs (3) white (AU data). **Range:** NG: resident in SE Pen (Angabanga R to Kemp Welsh R, including Port Moresby area); migrant AU birds overwinter in Trans-Fly, S Lowlands, and Sepik-Ramu, 0–500 m. Also AU.

Red-breasted Paradise-Kingfisher *Tanysiptera nympha* Pl 59
(Pink-breasted Paradise Kingfisher)

32 cm. Note disjunct and patchy distribution. Locally common in foothill forest interior and second growth, particularly ravines, also lowlands and mangroves. Generally above elevational range of Common PK and west of range of Brown-headed PK. In all plumages note *red-orange breast and blue crown*. **Adult:** Blue crown, blackish mantle, red-orange underparts, and pink rump. Tail with blue, white-tipped streamers, short for a paradise-kingfisher. **Juv:** From other juv paradise-kingfishers by combination of blue feathers in crown and red-orange wash on breast and rump; bill orange, black on culmen. **Similar spp:** Common PK, which shares its lowland range, has white underparts and rump. Buff-breasted PK, a rare winter visitor, has buff-yellow underparts and white back and tail-streamers. Brown-headed PK, further east, most similar but has brown head and mantle. Juv Common and Buff-breasted have underparts whitish or buffy, without red-orange wash; juv Brown-headed very similar but lacks blue in crown. **Voice:** Very similar to that of Brown-headed PK. **Habits:** Usually perches in an understory sapling, ~5–6 m up, in shaded forest interior. No nesting information. **Range:** Endemic. NG: patchily distributed in Bird's Head and Neck, Sepik-Ramu, Adelbert Mts, Huon, and northwestern SE Pen (Upper Watut, Wau, and Waria R), 500–900 m (0–1400 m).

Brown-headed Paradise-Kingfisher *Tanysiptera danae* Pl 59
(Brown-backed Paradise Kingfisher)

28–30 cm. Only in SE NG. Locally common in forest interior mainly in foothills, occasionally nearby lowlands; generally above the range of Common PK and east of range of Red-breasted PK. In all plumages note striking *red-orange breast and all-brown head*. Search suitable habitat for silent birds by scanning the forest at the ~5 m height looking for the bright orange breast of a bird facing and watching you. **Adult:** The only paradise-kingfisher with head entirely buff-brown and back dark brown.

Underparts rich red-orange; rump pink; tail short for a paradise-kingfisher, blue with white tip. Bill red. **Juv:** From other juv paradise-kingfishers by combination of all-brown head, red-orange wash on breast and pink rump; bill orange and black. **Voice:** Song a slow, plaintive, slightly nasal, descending, and decelerating trill, suggestive of call of Yellow-billed Kingfisher, but slower, softer, and decelerating. **Habits:** Sometimes in same forest as Common, but perches higher in vegetation. No nesting information. **Range:** Endemic. NG: SE Pen (westward to Waria R and Aroa R, including Port Moresby area), 300–1000 m (locally to sea level).

Hook-billed Kingfisher *Melidora macrorrhina* Pl 60
(Hook-billed Kookaburra)
25 cm. Often heard, but nearly impossible to see. Widespread and adaptable; common in lowland and hill forest interior, but can be any place where there is dense tree cover, even in towns and tree plantations. A *loud and inveterate caller at dawn and dusk.* Otherwise perches inconspicuously and silently in the middle story of the forest during the day. *A white-breasted, brown-backed kingfisher with a bluish crown.* Note *unique facial pattern* with black eye-stripe and moustachial stripe; heavy and flattened, hooked bill; and buff-scalloped back; underparts white or with ochraceous wash and faint, dark scaling. **Male:** Crown scalloped blue. **Female:** Crown mainly black, scalloped bluish green; blue nape band. **Juv:** Similar to Adult, but ochraceous underparts with feathers distinctly dark-edged (faintly so in Adult). **Subspp (2):** Minor. **Similar spp:** None, but vaguely recalls a juv Common PK. **Voice:** One of the iconic voices of the NG bush, though song varies regionally. Sings before sunrise and after sunset, often well into the night; rarely during day except when breeding. The musical song begins slowly with one or a few hesitant, rising, whistled notes of medium pitch, immediately followed by a rapid, accelerating, falling trill: *keeer keeer keer ker-ker-ker . . .* (trilled). Tone reminiscent of a paradise-kingfisher. **Habits:** Mainly crepuscular and nocturnal; also active during the daytime when breeding. Solitary. Usually perches within a few meters of the ground when foraging, but ascends into trees to sing. Feeds on insects, other invertebrates, and frogs. Most prey is taken from ground; digs for food as well. Nesting chamber excavated in an active, arboreal termitarium, >3 m up a tree. Eggs (2) white. **Range:** Endemic. NG, NW Is (Batanta, Misool, Salawati, Waigeo), and Bay Is (Yapen), 0–750 m (rarely to 1200 m).

Shovel-billed Kookaburra *Clytoceyx rex* Pl 60
(Shovel-billed Kingfisher)
32 cm. Rare and patchily distributed in forest interior from lowlands to mid-mountains. Appears to prefer areas of wet climate—damp, rainy, foggy. Often flushed from the ground, where this odd kingfisher digs with its massive, stubby bill. *Bill unique, bizarre: broad, thick, and rounded at the tip like a pair of salad tongs, often mud-caked. Bird large, all brown, and with a pale blue rump.* Plumage pattern recalls a dull, brown version of Rufous-bellied K. Note white throat and buff collar. **Male:** Tail blue. **Female:** Tail brown. **Juv:** Similar to Female, but has rusty tips to feathers of upperparts; underparts with feathers dark-scalloped. **Subspp (2):** Minor. **Similar spp:** Rufous-bellied K is smaller, with black cap, deep chestnut underparts, and a long whitish bill. **Voice:** Song, often given predawn and dusk, is a ringing, far-carrying, monotonous series of insistent, plaintive whistling notes, 1/sec: *ru? ru? ru? . . .* , very similar to the monotone call of the Dwarf Koel, but with notes not upslurred, more resonant and deeper. Alarm call is a loud, kingfisher-like, guttural chatter. **Habits:** Seen singly or in pairs. Forages on ground devoid of leaf litter, such as stream banks, land slips, boggy places, adjacent to tree buttresses, and flat alluvial forest. Uses bill as a plow or shovel for digging up arthropods, grubs, earthworms, small vertebrates, and apparently in mangroves, small crabs. Plowed-up leaf litter and soil may be a clue to its presence. Flushes to a perch 3–4 m up, where it sits motionless except for the slowly moving tail; then typically flies off, giving its rattling alarm call. Nest excavated in the rotten trunk of a still-standing tree. Egg undescribed. **Range:** Endemic. NG, Sea level –2400 m.

Blue-winged Kookaburra *Dacelo leachii* Pl 60
38–40 cm. Savannah and edge of monsoon forest. *Largest NG kingfisher. Streaked head; whitish collar and underparts*; note pale eye. Wings and rump brilliant blue; back is black. **Male:** Tail blue. **Female:** Tail barred brown and blue. **Juv:** Differs subtly from Adult—more buffy and with heavier barring in underparts. Young male also has some rufous barring near tip of tail. **In flight:** Impressively large, with much blue, and white outer tail feathers. **Subspp (1):** *intermedia* (endemic).

Similar spp: Spangled K is smaller, dark-eyed, and lacks the all-black mantle, pale collar, and white outer tail feathers. **Voice:** Song is a series of repeated downslurred barks similar to those of other kookaburras. Also, demoniacal, screeching laughter in duets. **Habits:** In vociferous, territorial pairs and family groups. Feeds on large insects, other invertebrates, and small vertebrates. Nests in tree hollows, or excavates chamber in arboreal termitarium. Eggs (2–3) white. **Range:** NG: Trans-Fly (Mimika R to mouth of the Fly R) and southern SE Pen (Angabanga R to Amazon Bay, including Port Moresby area), 0–600 m. Also AU.

Spangled Kookaburra *Dacelo tyro* Pl 60
(*Sauromarptis tyro*)
33 cm. Restricted to Trans-Fly and Aru Is. Common, inhabiting monsoon forest and adjacent savannah. *Smaller and prettier than similar and more widespread Blue-winged K. Head and mantle darker and spangled with pale crescents; iris dark; tail all blue, lacking white in outer tail feathers.* **Adult:** Heavily spangled; pale underparts with or without feathers faintly margined with black. **Female:** Blue parts duller and greenish; tail blue, unlike other female kookaburras. **Juv:** Head and back appear darker because of reduced spangling; underparts more noticeably dark-scaled. **Subspp (2):** *archboldi* (Trans-Fly) whitish spangling and underparts, *tyro* (Aru Is) these buff. **Similar spp:** Blue-winged K. **Voice:** Song begins with introductory gurgling followed by a rattled laugh almost on 1 pitch, repeated at length; lacks the variations of those of Blue-winged. Also described as a monotonous series of identical, short, dry coughs or doglike barks, ~10 notes per 6 sec. Sometimes calls in answer to Blue-winged. **Habits:** Shares its range with Rufous-bellied and Blue-winged Ks, the former inhabiting rainforest and monsoon forest, whereas the latter is confined to monsoon forest and savannah. Occurs in vocal pairs and small family groups, generally keeping to cover and not perching in the open. Found in treetops or forages from perch a few meters up, diving to the ground for prey. Recorded taking mainly insects. No nesting information. **Range:** Endemic. NG southern Trans-Fly and Aru Is, 0–300 m.

Rufous-bellied Kookaburra *Dacelo gaudichaud* Pl 60
(**Gaudichaud's Kookaburra**, *Sauromarptis gaudichaud*)
28 cm. The common, lowland forest kookaburra. Its voice is one of the familiar sounds of lowland forests. *A large kingfisher*, though small for a kookaburra; *black head with whitish bill, white collar, and deep rufous underparts.* Mantle black with blue on wings and rump. **Male:** Blue tail. **Female:** Brown tail. **Juv:** Similar to Female but with blackish bill and dark edging to feathers of the collar and underparts. **Similar spp:** Shovel-billed K is larger and brownish, has shovel-bill. Blue-black Kingfisher, rufous-breasted race male, is smaller, black-billed, lacks white collar. **Voice:** Song consists of loud barking or chopping notes, *tok* or *chok*, repeated many times, at a rate of ~1/ sec. Also a loud descending "laugh" similar to, but drier than, the rattle of Blue-winged K. Two birds often call synchronously, with tails cocked upward. **Habits:** Noisy territorial pairs inhabit the lower forest canopy, foraging for arthropods and small vertebrates (especially lizards) taken from foliage and branches, also ground, where it finds earthworms. Nest chamber excavated in an active, arboreal termitarium. Eggs (2) white. **Range:** Endemic. NG, NW Is (Batanta, Misool, Salawati, Waigeo), Bay Is (Mios Num, Yapen), and Aru Is, 0–750 m (rarely to 1300 m).

Blue-black Kingfisher *Todiramphus nigrocyaneus* Pl 61
(**Black-sided Kingfisher**, *Halcyon nigrocyanea*)
23 cm. A stunning *deep blue* kingfisher of lowland forest interior, often near streams; also mangroves. Rare and poorly known. *Diagnostic black mask* from bill to nape, and *black back;* glowing blue crown, wings, rump, and tail. Note white chin and throat (dim in Males of one race) framed by the *dark blue breast.* **Male:** Underparts vary with race, but belly dark blue or rufous, not white as in Female. **Female:** Underparts white in all races, broken by the broad blue breast band. **Juv:** Upperparts same as Adult but duller; underparts dingy white with rufous breast band, faintly scaled. **Subspp (3):** Males differ: *nigrocyaneus* (NW Is, Bird's Head to S Lowlands east as far as Princess Marianne Straits) breast and belly dark blue crossed by a white band; *stictolaemus* (Kurik and Fly R in central S Lowlands to southern SE Pen) entire underparts black with deep blue cast; *quadricolor* (Yapen I, NW Lowlands, and Sepik-Ramu) with thin white breast band and rufous belly. **Similar spp:** No other NG kingfisher has such strikingly patterned underparts. Other kingfishers of similar size have blue or green on back. Rufous-bellied K is larger and much more conspicuous, with a

white bill and collar. Juv paradise-kingfishers differ from juv Blue-black by having some red in the bill and a longer tail. **Voice:** Often sings before dawn. Song is high, musical, and clear, 4–5 notes, the first note longest, followed by 2 shorter, higher notes (all 3 are upslurred, each higher than the preceding), the final 1–2 notes are lower, downslurred or unslurred: *wee wheh di—wu-wu* or *yaaah wheh wheh wu*. Reminiscent of Hook-billed K song, but easily distinguished. **Habits:** Perches in understory (3–5 m up), often by a small, slow-moving stream, but also away from water. Takes small fish, crabs, and lizards. No nesting information. **Range:** Endemic. NG, NW Is (Batanta, Salawati), and Bay Is (Yapen), 0–600 m. NG: absent from most of SE Pen (present eastward in N to Markham R; in S to Mt Cameron). **Taxonomy:** The species may eventually be split up.

Forest Kingfisher *Todiramphus macleayii* Pl 61
(*Halcyon macleayii*)
20–21 cm. Uncommon resident of SE Pen and fairly common austral winter migrant at forest edge, in gardens, towns, savannah, and other open habitats (not forest) in lowlands and hills. *Underparts clean* white (may have ochre wash on flanks) and *crown deep blue*. Also note *white forehead spot*. **Male:** White collar encircles hindneck. **Female:** Collar interrupted on nape. **Juv:** Feathers of upperparts edged buff, those of underparts edged dusky; has buffy loral spot and underparts; molts into adultlike **Imm** plumage while overwintering. **In flight:** The only small kingfisher showing *white "silver dollar" wing patch*. **Subspp (2):** *macleayii* (resident; also Northern Territory, AU) bright blue back; *incinctus* (migratory from E AU) greenish-blue back. **Similar spp:** Sacred K more common; has underparts and forehead spot tinged buff, back more greenish, and could especially be confused with buffy-breasted juv Forest. Collared K usually larger, has more greenish back, confined to different habitat. Both species lack the white wing patch. **Voice:** Resident race gives a short, rapid, musical trill very similar to song of Sacred K, *ki-ki-ki-ki-ki-ki*. Migrants are usually silent. **Habits:** Solitary or in pairs; breeds in family groups of up to 5 in AU, probably also NG; migrating groups along coast. Conspicuous on exposed perches in open habitats, e.g., bare branches, roadside wires. Dives to the ground for invertebrates, frogs, and reptiles. Vocal. Nest chamber excavated in arboreal termitarium (AU data). Eggs (4–5) white. **Range:** Resident race breeds on NG only: SE Pen and Huon, 0–700 m. Migrants visit NG (S Lowlands, E Sepik-Ramu eastward), Aru Is, and SE Is, 0–1800 m. Also AU, Bismarcks, and Solomon Is.

Collared Kingfisher *Todiramphus chloris* Pl 61
(**Mangrove** or **White-collared Kingfisher**, *Halcyon chloris*)
20–25 cm. Uncommon in mangroves and other coastal habitats throughout region, except N coast of NG and nearby islands, where replaced by Beach K. **Adult:** *A heavy-billed, white-collared kingfisher, with a blue-green crown and back*. Underparts white (occasionally with buff). Female often duller, but not reliably so. **Juv:** Faint buff scaling in upperparts; obscure dark scaling on underparts and collar. **Subspp (3):** *chloris* (NW Is, Bird's Head and Neck, also Wallacea) a coastal race, bright bluish green; *sordidus* (NG S coast and Aru Is, also AU) a mangrove specialist, dusky olive green; *colonus* (SE Is) a small-island tramp, darker above and smaller than *sordidus* and Sacred K. **Similar spp:** Sacred K has smaller bill, buffy collar and underparts. Forest K is pure blue above, with a white "silver dollar" wing patch. **Voice:** Song of form *colonus* like that of Sacred K, but slower and louder: a series of deliberate, 2- or 3-note phrases of nasal *kek* notes, with the first note highest. **Habits:** Solitary or in pairs. Forages for crabs and fish in saltwater habitats, insects and lizards on land. Excavates nest chamber in tree hollow or termitarium. Eggs (3) white. **Range:** NG, NW Is (Gag, Misool, Salawati, Waigeo), Aru Is, and SE Is (mostly on small islands, but also Misima, Tagula, and Rossel), 0–100 m. NG: Bird's Head and Neck along S coast to Milne Bay. Also Red Sea to AU, Solomon Is, and Oceania. **Taxonomy:** This species will likely be split up.

Beach Kingfisher *Todiramphus saurophagus* Pl 61
(**White-headed Kingfisher**, *Halcyon saurophaga*)
28 cm. Seacoasts, especially on small, offshore islands; in strand, mangroves, coconut groves. *The only white-headed kingfisher* in the NG Region. Note *large size and massive bill*. **Adult:** All-white head and underparts; back blue-green, wings and tail blue. **Female** mantle sometimes more greenish. **Juv:** Head and underparts buffy; upperparts buff-scaled. **Subspp (1):** *saurophagus* (Moluccas to Solomon Is). **Similar spp:** Collared K smaller; green cap. **Voice:** Song of repeated notes reminiscent of related species. A strident *kio kio kio kio* or *kee kee kee*. **Habits:** Solitary or in pairs. Typically seen

alone on exposed perch overlooking shore. Dives on crabs, insects, lizards, and fish. Nests in a tree hollow. Eggs (2–3) white. **Range:** NG, NW Is, Bay Is, and SE Is, sea level. NG: Bird's Head, along N coast to SE Pen, where recorded locally from small islets on both N and S coasts. Also Moluccas, Bismarck, and Solomon Is.

Sacred Kingfisher *Todiramphus sanctus* Pl 61
(*Halcyon sancta*)

21 cm. Common austral winter migrant to forest edge, gardens, towns, and other open habitats at low to mid-elevations, where it is the most common kingfisher. **Adult:** *Underparts and collar buffy white; cap and back greenish.* Underparts range from dirty white to pale ochre, with variable amounts of faint, dark edging. Female sometimes duller. **Juv:** Buff scaling in upperparts; molts to adultlike **Imm** plumage while overwintering. **Subspp (1):** *sanctus* (breeds in AU). **Similar spp:** Forest K is same size and shares habitat, but cap and back are pure blue, underparts usually pure white, and wings with white "silver dollar" patch. Collared K has larger bill, whitish collar and underparts (usually), and different call. **Voice:** Song, a slightly musical series of 3–7 nasal *kenkk* notes, at a rate of 5 notes/sec, at constant rate and pitch. Call is soft rasped notes. **Habits:** Solitary or in pairs. Occupies same winter territory from year to year. Forages from exposed perches by swooping down on large arthropods, frogs, and reptiles. **Range:** Entire NG Region, 0–2400 m. Also AU and NZ; migrates as far as Wallacea and Bismarck Is.

Yellow-billed Kingfisher *Syma torotoro* Pl 62
(**Lowland** or **Lesser Yellow-billed Kingfisher**, *Halcyon torotoro*)

18–20 cm. Widespread and common in lowland and hill forest. Its song is one of the distinctive sounds of the forest edge. *A small kingfisher with yellow-tan head and yellow-orange bill that lacks black markings.* **Male:** Plain yellow-tan crown; nape marked with 2 black patches. **Female:** Black patch atop head; often with black nape patches joined. **Juv:** Like Adult but bill black or partly black; also, with white patch between the 2 black nape patches and underparts with faint dusky scalloping. **Subspp (2):** *torotoro* (all except that of the next race); *ochracea* (D'Entrecasteaux Is) much larger and underparts more ochre. **Similar spp:** See Mountain K. **Voice:** Song is a whistled trill with the quality of a referee whistle. The usual song is 3–4 downslurred trills, each separated by a pause and ending with a longer trill that drops, rises, and drops. When the distinctive final trill is not heard, the descending trills are very similar to trills of Chestnut-breasted Cuckoo but are longer and richer, with longer intervals between successive trills.

Another song is a mournful, slow, monotonous series. Race *ochracea* song unlike Yellow-billed or Mountain K of mainland NG; a deeper-pitched string of 3-sec phrases repeated over and over; each phrase begins with 3 trills so brief they act as single notes and ends with a short, jumbled trill; phrase thus, *dee, dee, dee, deedrdrdrdr*. **Habits:** A solitary, vocal inhabitant of the midstory and canopy. Snatches arthropods and small vertebrates from foliage and larger branches, or dives to the forest floor. Nest chamber excavated in an active termitarium >3 m up a tree. Egg (1) white. **Range:** NG, NW Is (Batanta, Misool, Waigeo, Salawati), Aru Is, Bay Is (Yapen), and SE Is (3 main D'Entrecasteaux Is), 0–700 m (rarely to 1200 m). Also AU. **Taxonomy:** Hybrids between this and Mountain may occur rarely. Race *ochracea* possibly a separate species.

Mountain Kingfisher *Syma megarhyncha* Pl 62
(**Mountain** or **Greater Yellow-billed Kingfisher**, *Halcyon megarhyncha*)

21–23 cm. Uncommon, montane counterpart to Yellow-billed K. Mountain K nearly identical, but *noticeably larger with proportionately longer bill. Note blackish marking atop yellowish bill* (bill unmarked in Huon Pen birds). Beware that imm Yellow-billed can have yellow bill with black patches. For a bird in hand: folded wing measure (flat against rule) is 70–86 mm for Yellow-billed; 83–91 mm for Mountain. Bill measure (to juncture with skull) is 34–44 vs 42–53 mm. **Male:** Crown yellow.

Female: Crown with black patch; nape patches always form a bar. **Juv:** Ear-coverts mottled black; bill all or mostly black. **Subspp (3, in 2 groups):** *megarhyncha* (Central Ranges) bill with black tip; *sellamontis* (Huon) bill all yellow. **Voice:** Song similar to that of preceding species mainland race but deeper pitched. (Song of Chestnut-breasted Cuckoo considerably higher pitched.)
Habits: Similar to preceding species. Nest chamber dug into earthen bank and dry timber. Eggs (2) white. **Range:** Endemic. NG: Central Ranges and Huon (possible records from Foja and Adelbert Mts), 1100–2100 m (rarely down to 700 m).

RIVER KINGFISHERS: Alcedinidae

The 39 river kingfisher species occur throughout the Old World, with 4 species inhabiting the New Guinea Region. These are small kingfishers with very long, lance-shaped bills and short, stubby tails. Nearly all species are found close to water, into which they plunge for fish, crustaceans, insects, and spiders. One can occasionally spot a river kingfisher on its fishing perch by scanning the vegetation just above the waterline, one can occasionally spot a river kingfisher on its fishing perch. More often, however, they are seen zipping along a waterway or through the forest but may be overlooked unless one is aware of the high-pitched flight calls. The Papuan Dwarf-Kingfisher, an exception, has forsaken the waterside for the forest.

Common Kingfisher *Alcedo atthis* Pl 62
(**River Kingfisher**, the Kingfisher)
14–15 cm. The familiar kingfisher of Eurasia is local and habitat-specific in the NG Region. Along coasts, mangroves, and rivers; rarely inland, where displaced by Azure K. **Adult:** Like the slightly larger and darker Azure K, but *sky blue streak up the back and bands of sequined highlighting on crown and wing-coverts*; underparts rusty orange; whitish streak on ear-coverts sometimes hidden. **Female:** Red in lower bill. **Juv:** Duller, more greenish, with some dusky-edged feathers on breast. **In flight:** *Brilliant pale blue dorsal streak as bird flies off.* **Subspp (1):** *hispidoides* (Sulawesi to Bismarck Is). **Similar spp:** Papuan Dwarf K is much smaller, has a proportionately longer and thinner bill, is darker blue above, and does not fish. **Voice:** Call, shrill high-pitched notes uttered singly or in a series. **Habits:** Prefers open streams more than other kingfisher species. Solitary. Patrols a length of riverbank, with a number of regular perches, and dives into water for small fish, crustaceans, and insects. Nest chamber dug in a stream bank; nesting little known in NG Region. **Range:** NG and all SE Is, 0–100 m. (Records from NW Is need confirmation.) NG: Sepik-Ramu to SE Pen (both N and S coasts; apparently absent westward of Aroa R in the S). Also Eurasia to Solomon Is.

Papuan Dwarf Kingfisher *Ceyx solitarius* Pl 62
(**Variable Dwarf Kingfisher**, *Ceyx lepidus*, *Alcedo lepida*)
12 cm. A widespread, common, tiny kingfisher of lowland and hill forest interior, often away from water. **Adult:** Fluorescent deep blue upperparts; honey golden underparts; *crown and wing-coverts with glittering, sequined feathering; back and rump paler blue*. Bill black; feet yellow-orange. Female duller, but not reliably separable. **Juv:** Duller and may have some dark scaling on breast. **Similar spp:** Azure K is clearly larger, with darker, more uniformly blue upperparts (including crown and back). Common K also larger, with upperparts paler blue and more heavily sequined. Both are found strictly near water. **Voice:** Call is a single, high-pitched *tseet* or *ss*, given while perched, but especially in flight—a moving *ss* sound. Azure call similar. Pygmy parrots give series of similar notes delivered higher in the trees. **Habits:** Found zipping through the understory, calling as it goes. Solitary and inconspicuous, perching at low to middle heights in forest interior. Forages for arthropods taken at various levels (occasionally into canopy) and beside small pools or slow streams in forest. Does not dive into the water for food. Nest chamber dug in earth bank or tree root tip-up mound. Eggs (2) white. **Range:** Endemic. NG, NW Is (Batanta, Misool, Salawati, Waigeo), Aru Is, Bay Is (Biak, Yapen), Karkar I, and SE Is (D'Entrecasteaux Is: Fergusson, Normanby, and possibly Goodenough), 0–1200 m. **Taxonomy:** A recent split, one of many from the former Variable Dwarf Kingfisher (*C. lepidus*).

Azure Kingfisher *Ceyx azureus* Pl 62
(*Alcyone* or *Alcedo azurea*)
16 cm. Widespread and fairly common along any body of freshwater or tidal creek with adequate cover, including mangroves. **Adult:** *Small, deep blue kingfisher with evenly dark blue crown and upperparts, and black bill with pale tip*; tawny below, to almost white in some birds. Female duller, but not reliably so. **Juv:** Blackish forehead and sides of breast; underparts paler; bill with more pronounced pale tip. **Subspp (2):** *lessonii* (all range except that of next race) underparts rich ochre-tan; *ochrogaster* (NW Lowlands to Sepik-Ramu) underparts paler. **Similar spp:** Papuan Dwarf K much smaller, with sequined crown and paler blue back and rump contrasting with darker wings and tail. Common K has paler blue upperparts with fluorescent blue midline on back and sequined crown feathers. **Voice:** Similar to, but higher pitched and less trilled than call of Papuan Dwarf K. **Habits:** Solitary. Perches close to water's edge at forest margins and mangroves, taking small fish, crustaceans, insects, etc. Hovers over water (AU data). Excavates tunnel in stream bank. Eggs (5–7) white. **Range:** NG, NW Is (Batanta, Misool, Waigeo), Bay Is (Biak, Numfor, Yapen), Aru Is, Karkar I, and SE Is (D'Entrecasteaux: Fergusson and Normanby), 0–1500 m. Also Moluccas and AU.

Little Kingfisher *Ceyx pusillus* Pl 62
(*Alcyone* or *Alcedo pusilla*)
11 cm. Widespread and common in mangroves and small, lowland creeks and pools. *Smallest NG kingfisher; the size of a fairywren.* **Adult:** The only small kingfisher with *white underparts*; dark blue upperparts. Bill and feet black. **Juv:** Greenish-blue crown appears barred; some buff in underparts and black edging. **Subspp (2):** *pusillus* (all range except that of next race); *laetior* (NW Lowlands to Sepik-Ramu) lighter blue. **Similar spp:** Other small kingfishers have orange-buff underparts. **Voice:** Call an insect-like *ts!* brief and very soft. Similar to, but higher pitched, than call of Variable Dwarf K; difficult to distinguish. **Habits:** Perches solitarily low over water, diving for tiny fish, crustaceans, and insects. Excavates nest chamber in an earth bank or rotted tree stump (AU data). Eggs (4–5) white. **Range:** NG, NW Is (Batanta, Gag, Kofiau, Misool, Waigeo), Aru Is, and SE Is (3 main D'Entrecasteaux Is), 0–500 m (rarely 800 m). Also AU, Moluccas, and Solomon Is.

BEE-EATERS: Meropidae

The 27 species of bee-eater are distributed mainly in Africa and Eurasia, with only 2 species in Australasia. Both occupy the New Guinea Region—one shared with Asia, the other with Australia. They are blue-green birds with long, pointed bill and pintail. In New Guinea, bee-eaters occupy open habitats, where they perch on exposed branches or wires and dart out after flying prey including wasps, bees, flies, butterflies, beetles, and dragonflies. After each foray, the bee-eater will gracefully swoop back to its perch to consume the captured prey. The bird will whack its prey against the perch, and if there is a stinger, it will be carefully removed. Bee-eaters are gregarious and vocal and live in family groups. They nest colonially, each pair, often with related helpers, in a tunnel it has dug in flat, bare ground or in an earthen bank. Bee-eaters are easy to find in suitable habitat and often noticed because of their characteristic calls and flight pattern.

Blue-tailed Bee-eater *Merops philippinus* Pl 60
27–29 cm. Resident and patchily distributed in lowland savannah, grassland, airstrips, and other open habitats with suitable perches; common where found. **Adult:** A greenish bee-eater with impressively long central tail feathers. Compared with more widespread Rainbow BE, *tail longer, more tapered, and blue*, rather than black, and *brown throat blending into breast*, not separated by a blackish bar, and body slightly larger. **Juv:** Similar to Adult, but duller and lacks tail wires. Differs from juv Rainbow by having the breast green rather than blue, and the brownish throat blends with the breast. **In flight:** *Wings green*, without Rainbow's chestnut patch; *tail blue*. **Voice:** Flight call a double note, separable from that of Rainbow, and a rolling trill when alarmed. **Habits:** In pairs or flocks. Foraging behavior and diet typical of a bee-eater. Nests colonially, sometimes in hundreds of pairs; helpers not yet reported. Digs a burrow into the ground or earth bank (Asian data). Eggs (5–7) white. **Range:** NG: Coastal in the North from Lake Sentani (NW Lowlands) eastward to Lae; Trans-Fly and adjacent S Lowlands, 0–200 m. Also India, SE Asia, Wallacea, the Philippines, and Bismarck Is.

Rainbow Bee-eater *Merops ornatus*   Pl 60

25 cm. Widespread and common AU migrant during austral winter (March–October) and very local resident breeder. Occupies open habitats, but migrates over forest and mountain passes. **Adult:** *Has black throat bar, rufous in wing, and a distinctly squared, black tail with thin central tail wires.* **Juv:** Duller and lacks black throat bar and tail wires. **In flight:** *Wing with bright rufous flight-feathers; black tail contrasts with blue rump* (Blue-tailed BE also has blue rump). **Similar spp:** Blue-tailed BE, see that account. **Voice:** Call notes herald birds migrating overhead or foraging nearby; call a *prr, prrip,* or *preee* note repeated in quick succession. **Habits:** In pairs or small parties that coalesce during migration into large flocks. Roosts in flocks. Foraging behavior and diet typical of a bee-eater. Nests solitarily or in small colonies (tens of pairs); helpers frequent. Digs a burrow into the ground or earth bank. Eggs (4–5) white (AU data). **Range:** Australian migrants spread across all NG Region, 0–1500 m (to 4000 m when migrating); resident on NG mainland in Port Moresby region and possibly in the Sepik-Ramu, 0–200 m. Breeds in AU; migrates from Wallacea to Bismarck Is.

HORNBILLS: Bucerotidae

The 59 species of hornbills are a tropical Afro-Asian family that ranges eastward to the Solomon Islands. One species inhabits New Guinea. Hornbills are large to very large birds with oversized, pointed bills frequently adorned with a casque. They feed mainly on fruit and small animals. The nest is in a tree hollow, into which the female is sealed for the duration of the incubation period, being fed by the male through a small slitlike opening.

Blyth's Hornbill *Rhyticeros plicatus*   Pl 50
(**Papuan Hornbill**, *Aceros plicatus*)

76–91 cm. Widespread in lowland and hill forest, common where not heavily hunted. Presence announced by *boisterous grunting and loud whooshing wingbeats. A huge black, white-tailed bird* of the forest canopy seen flying across forest openings. Note grotesquely large pale bill. **Male:** Yellow-orange head, varying from rufous to buff. **Female:** Head black. **Juv:** Like Male but casque with just 1 wrinkle; casque grows, adding a new wrinkle each year until ~6 wrinkles, when the older wrinkles begin breaking off. **Similar spp:** Palm Cockatoo, which is black-tailed, is the only other really large black bird cruising over the forest. Channel-billed Cuckoo is the only other bird with a hornbill-like bill. **Voice:** Deep, resonant, far-carrying honks and grunts, singly or in a series of notes rising and falling slightly in pitch. The beating wings make a terrific roaring sound, and when gliding, a high, long-continued ripping sound. **Habits:** Singly, in pairs, but usually in small parties. Gregarious and highly vocal. Commutes above the forest over a large area. Forages in the forest canopy mainly for fruits, but also for large insects, reptiles, frogs, mammals, and nesting birds. Nest in a high tree hollow. Eggs (2?) white. **Range:** NG, NW Is (Batanta, Misool, Salawati, Waigo), Bay Is (Yapen), and SE Is (3 main D'Entrecasteaux Is), 0–1400 m. Also Moluccas, Bismarck, and Solomon Is.

PITTAS: Pittidae

This tropical Old World family of 33 species ranges from Africa to Australia, with its center of distribution in SE Asia. This is the only family of suboscine passerines inhabiting New Guinea, with 3 species in the region. Pittas are brilliantly colored, solitary songbirds that inhabit the tropical forest floor. They have a plump appearance, with short wings and tail, long legs, and a large head. The New Guinea species all show a white wing patch in flight. Sexes are alike. Pittas frequent heavily forested habitats in lowland and hills, occasionally to 1200 m. Often heard but rarely seen, they are very shy and remain out of view. On the rare occasion when a pitta is seen, the bird is typically standing with an upright posture and hopping about, occasionally bobbing its head and flicking the wings or tail. They forage by turning over leaves and decayed wood in search of insects, worms, and snails, the latter being broken open on a rock. The loud, hoarsely whistled song is sometimes delivered from a perch in a tree, mostly during the day but occasionally at night. The nest is dome-shaped, made of twigs and leaves, and built on the ground.

Red-bellied Pitta *Erythropitta erythrogaster* Pl 63
(Blue-breasted Pitta)
15–17 cm. A common but secretive resident mainly in hill forest, extending from lowlands to lower montane zone. Some may be migrants from N AU. **Adult:** *A brown-headed pitta with bright red belly.* Breast blue; wing blue, without bright shoulder patch. **Juv:** Brown and mottled, with pale or white throat patch and hint of red on the belly. **Subspp (6):** *bernsteinii* (NW Is: Gebe): blue back, rufous brown head; *macklotii* (NG range, except the following) with greenish wings and back, dark brown head, and rufous nape; *habenichti* (NW Lowlands and Sepik-Ramu) similar to preceding but nape more orange; *loriae* (SE Pen, east of Kumusi R and Port Moresby area) similar to *macklotii* but head all dark brown; *finschii* (SE Is: D'Entrecasteaux Is) blue wings and back, dark brown head; *meeki* (SE Is: Rossel) green back, pale brown head. **Similar spp:** Other pittas quite different. **Voice:** Song a weird, low, tremulous, double-noted whistle, somewhat hoarse and upslurred. First note rises, second note rises then falls. Alarm call, a short scream usually uttered twice (1–3 times). **Habits:** Solitary, but lives in pairs. Vocal and best located by song. Hops about on the forest floor, feeding on invertebrates. Nests at the base of a tree; domed nest made of leaves, etc. on a platform of sticks. Eggs (2) white, with dark spots and splotches. **Range:** NG, NW Is (Batanta, Gebe, Misool, Salawati, Waigeo), Bay Is (Yapen), Aru Is, SE Is (Goodenough, Fergusson, Rossel), 0–1200 m. Also Wallacea, Philippines, AU. **Taxonomy:** Recently proposed for a split into 5 species for the NG Region, but the genetic divergence of these is rather shallow, and 2 are not reliably identifiable by appearance or voice (*macklotii* and *habenichti*).

Hooded Pitta *Pitta sordida* Pl 63
15–17 cm. Common but secretive resident in lowland and hill forest. **Adult:** *A black-hooded, green-bodied pitta.* Shining *turquoise shoulder patch* and throat band at the base of the hood. Glittering blue rump patch in **Male**, reduced or absent in **Female**. A small red patch under the tail. **Juv:** Mostly sooty with suggestion of Adult color pattern. **Subspp (4, in 2 groups):** *novaeguineae* (NG range except Bay Is) blue-green flanks, glittering throat-band, and lower breast black; *rosenbergii* (Bay Is: Biak, Numfor) deep blue flanks, no glittering throat-band, and lower breast red like belly. **Similar spp:** Other pittas quite different. **Voice:** Song, a loud, musical, somewhat hoarse pair of whistled notes: the first note is flat, the second an interrogative disyllable at higher pitch: *kuhwih kuwee?* Sings day or night. Alarm call, a repeated *kiaw*. **Habits:** Similar to Red-bellied P, with which it broadly co-occurs. At least somewhat nomadic when not breeding. Nest similar to preceding species, roofed over with moss. Eggs (4) white with brown spots. **Range:** NG and most satellite islands, including all NW Is, Bay Is (Biak, Numfor), Aru Is, and Karkar I, 0–500 (1200) m. Also S Asia through Wallacea and Philippines. **Taxonomy:** Subsp group *rosenbergii* perhaps merits species rank.

Noisy Pitta *Pitta versicolor* Pl 63
17–18 cm. Resident in monsoon forest and mangroves of Trans-Fly. Some birds may be migrants from AU. **Adult:** *Yellow-tan breast and flanks with black blotch in center.* Black head with *chestnut crown.* Shining, pale shoulder patch. **Juv:** Similar but duller and with simpler pattern, lacking dark chin and throat, dark belly patch, and pale shoulder patch. **Subspp (1):** *simillima* (also AU). **Similar spp:** Other pittas lack yellowish underparts and chestnut crown. **Voice:** Song resembles Hooded P, but lower pitched: *walk-to-work!* Alarm call similar to that of Hooded P. **Habits:** Similar to other pittas. Nests on the ground at the base of a tree trunk or rarely on a stump; nest a bulky structure with side entrance, made of sticks, leaves, bark, etc. (AU data). Eggs (3–4) white with dark spots and blotches. **Range:** NG: Trans-Fly and vagrant to Port Moresby area in SE Pen, 0–100 m. Also AU.

BOWERBIRDS AND CATBIRDS: Ptilonorhynchidae

The bowerbirds are an old and distinctive family of 21 species confined to Australia and the New Guinea Region (with 13 spp). Their namesake feature is the *bower*, a complex structure of sticks and other materials built on the ground by the male and used by him as a display stage for attracting females with which to mate. All but the catbirds are bower builders and polygynous. Each species builds a bower unique to it, with some variation. There are 2 broad types of bowers that define the 2 main branches of the bowerbird family tree. *Maypole bowers* are typically constructed around a central pole, usually the trunk of a sapling. *Avenue bowers* are platforms supporting 2 or more walls of upright sticks that form an avenue between them. The NG bowerbirds are found mainly in the hills and mountains, although 3 species are primarily lowland dwelling. All are highly frugivorous, and some are fig specialists.

White-eared Catbird *Ailuroedus buccoides* Pl 108
(**White-throated Catbird**)
24–25 cm. A common and vocal but retiring forest dweller of *lowlands and foothills*; difficult to observe. Frequents midstory of forest interior. Note bright green back, wings, and tail, *pale buff breast with profuse small black spots*, and *white ear-patch*. Lacks white tail tip of Black-eared CB. **Adult:** Bill whitish, iris deep red. **Juv:** Bill dusky, iris grey. **Subspp (3, in 2 groups):** *buccoides* (widespread, exclusive of next subsp) crown dark brown or blackish; *geislerorum* (NW Lowlands to northern SE Pen) crown pale brown. **Similar spp:** Black-eared CB generally does not co-occur, has a black ear, scalloped or mottled breast, and a white tail tip. **Voice:** Song, given from cover of vegetation, is a very harsh, drawn-out, rasping or grating *shh n shh n shh n . . .* of 3–4 sec duration. Reminiscent of song of Stout-billed Cuckooshrike. Call is a very high-pitched, short, metallic *chink!* Also a short, piercing *ss*. Wings produce a rustling noise. **Habits:** Wary, a master at concealment, but will approach "spishing" or song imitations. Generally observed only by waiting quietly in the forest interior and allowing the bird to approach. Encountered singly, in pairs, or in family groups of 3. Feeds on fruit and large insects. Witnessed attacking small birds; consumes small birds trapped in mist-nets; probably depredates nests as well. Monogamous. Nest a cup of sticks with a few interior leaves placed in a pandanus or sapling 2–3 m up. Egg (1) pale coffee-colored with no markings. **Range:** Endemic. NG, NW Is (Batanta, Salawati, Waigeo), and Bay Is (Yapen), 0–800 m (locally to 1200 m). Absent from Trans-Fly.

Black-eared Catbird *Ailuroedus melanotis* Pl 108
(**Spotted** or **Green Catbird**, *A. crassirostris*)
28–29 cm. The catbird of *lower mountains and remnant rainforest and monsoon forest in the Trans-Fly.* Distribution patchy, but often common where found. Mostly above elevational range of White-eared CB, where their ranges overlap. A hefty, green-backed forest dweller with a *dark ear-patch, underparts variously scalloped rather than black-spotted, and tail conspicuously white-tipped.* **Adult:** Red iris. **Juv:** Grey iris; finer spotting and streaking on head. **Subspp (6):** Geographic patterns complex and showing variation mainly in darkness of head pattern, the amount of green vs brown on the head, and markings on breast. **Similar spp:** White-eared CB exhibits a white ear-patch, has black-spotted underparts, and lacks the white tail tip. **In flight:** Catbirds sail through forest on their short rounded

wings, much like a small green goshawk; look for the white tail tip of Black-eared. **Voice:** A weird, nasal, wavering 3- or 4-part song: *nranh a ranh a ranh*, repeated every 15 sec or so—a distinctive vocalization of the hill forest, reminiscent of a growling cat. Similar to vocalization of Stout-billed Cuckooshrike, but the latter lacks the peculiar catlike quality. Call note easily overlooked—a loud, high-pitched *chink!* **Habits:** Wary and difficult to observe, even at favored fruiting trees. Inhabits forest canopy, but regularly visits the ground. Diet is mainly fruit (especially figs) and arthropods. Habitually depredates bird nests, eating the eggs and young, and with stealthy flight will opportunistically attack small birds, presumably to capture and eat. Monogamous. Nest is placed in a low sapling or pandanus; nest is a large, bulky, open cup. Eggs (1–2) creamy white or buff. **Range:** Nearly all NG mts (no record from the Cyclops Mts), 900–1800 m (rarely up to 2250 m); lowland populations occur on Misool I, Aru Is, and the Trans-Fly. Also NE AU.

Streaked Bowerbird *Amblyornis subalaris* Pl 66

22 cm. *SE Pen*, where it is a locally common inhabitant of mid-mountain forests at elevations mostly lower than those of the very similar Macgregor's BB, from which separated by its *distinctly streaked throat and breast and smaller size*. **Male:** Exhibits a short, broad orange crest, typically obscured except in display. **Female, Juv, and Imm:** Lack crest. **Similar spp:** The very similar Macgregor's BB inhabits higher elevations and lacks mottling ventrally; male exhibits a longer crest. Black Pitohui female also lacks throat mottling. **Voice:** Loud, unmusical notes, similar to those of Macgregor's BB. Calls from inside bower. **Habits:** Generally similar to other gardener bowerbirds. Male builds a hutlike bower of sticks, with 2 entrances and a vertical front wall decorated with flowers and fruit; bower is sited only on slopes off the ridge crest. Nest a cup of leaves, sticks, and twigs. Egg (1) yellowish white. **Range:** Endemic. NG: SE Pen (S watershed from the Angabanga R northwest of Port Moresby to Milne Bay, around the East Cape, and on the N scarp west to Mt Suckling), 1000–1400 m (extremes 700–1800 m).

Macgregor's Bowerbird *Amblyornis macgregoriae* Pl 66

26 cm. This is the *widespread gardener bowerbird*, occupying upper mid-mountain and cloud forests of the Central Ranges and Adelbert Mts. Common. Male is predictably found at or near his bower, which is set atop a ridge crest; otherwise, the species is difficult to observe. *A chunky, uniformly olive-brown bird with a large head and rounded wings and tail. Lacks any streaking on underparts.* **Male:** Head shape influenced by the *long, orange nuchal crest*, which extends down back (orange plumes often folded in and not seen). **Female, Juv, and Imm:** Entirely olive brown, with blunt blackish bill. **Subspp** (2): Minor. **Similar spp:** Female Black Pitohui similar but longer-tailed, more greyish olive, and with hooked bill. Female Streaked BB has visible mottling on throat and breast. Female Crested Satinbird has olive-green plumage and diagnostic forehead groove. Female Loria's Satinbird is olive green. **Voice:** Call, a conspicuous, loud, harsh note repeated twice: *urschweet urschweet*, rapidly delivered. Advertisement calls at the bower imitate other birds and environmental sounds, mixed with odd scraping, rasping, and clicking sounds. **Habits:** Usually seen singly. A shy canopy dweller that forages for fruit and large insects in forest interior, rarely at edge. Male displays at his maypole bower of small twigs built up around a sapling pole surrounded by a circular mossy base of ~1 m diameter with an elevated rim. Various bower decorations may include charcoal, berries, insect frass, fungi. Bower placed atop a ridge crest. Nest typically built in the head of a pandanus 2–3 m above ground; nest a cup of leaves and sticks. Egg (1) creamy white. **Range:** Endemic. NG Central Ranges, Adelbert Mts, and Mt Bosavi, 1600–3300 m (occasionally as low as 1200 m). **Taxonomy:** The Huon Peninsula form has been separated as a distinct species.

Huon Bowerbird *Amblyornis germana* Pl 66
(formerly considered a subspecies of **Macgregor's Bowerbird**)

25 cm. Confined to the mountains of the Huon Peninsula, where it is the only gardener bowerbird. **Male:** *Short-crested* and short-winged compared with Macgregor's BB. **Female, Juv, and Imm:** Similar to Macgregor's. **Similar spp:** Black Pitohui female. **Voice:** Apparently similar to that of Macgregor's BB. **Habits:** Bower situated off the ridge crest. Bower is mushroom-like, with a head of twigs and a thick "stem" of more twigs; the circular base is carpeted with moss that rises onto the stem where it is hung with small decorations; this decorated wall faces downslope; the perimeter of the bower is a flat layer of sticks and leaves. Nest unknown. **Range:** Endemic. NG: Huon Pen, 1700–2400 m. **Taxonomy:** Recently split from Macgregor's B because of distinct bower and different bower location.

Golden-fronted Bowerbird *Amblyornis flavifrons* Pl 66
(**Yellow-fronted Bowerbird**)
24 cm. Endemic to the Foja Mts, where it is the only gardener bowerbird. Common in cloud forest. **Male:** Exhibits an *orangish-yellow forehead*, crown, and nuchal crest, and rich olive-brown body plumage. **Female, Juv, and Imm:** Lack crest; similar to others of the genus. **Similar spp:** Black Pitohui female. **Voice:** Male gives repeated harsh notes and also a repeated *kuk kuk kuk kuk*. Advertisement calls similar to those of other *Amblyornis* spp. **Habits:** Similar to Macgregor's BB. Relatively tame at the bower; otherwise difficult to observe, but vocal. Polygynous. Builds a maypole similar to that of Macgregor's BB but with ornamentation of mainly blue fruit and perimeter ridge absent. Bower set on the ridge crest. Nest unknown. **Range:** Endemic. NG: Foja Mts, 1000–2000 m.

Vogelkop Bowerbird *Amblyornis inornata* Pl 66
25 cm. *The only maypole bowerbird in the Bird's Head and Bird's Neck regions*, where common in mid-mountain and lower cloud forest. **All plumages:** Plain, drab, olive brown, large-headed. *No crest.* **Similar spp:** Not separable from female and imm of Macgregor's or Golden-fronted BB. Female Black Pitohui is very similar but has a heavier, blackish, hooked bill and a smaller head. **Voice:** Call, a strident *chree chree* or *chuck chuck chuck*, as well as other weird metallic sounds, and the *scheeurr* note similar to that of other gardener bowerbirds. Advertisement calls at the bower include catlike *meows*, drumming, rapping, ticking, and creaking noises, plus mimicry of songs of other bird species. **Habits:** Generally solitary. Vocal and active. Feeding behavior and ecology presumably similar to Macgregor's BB. Polygynous. Builder of the largest and most elaborate bower in NG—a conical hut 1 m high supported by a pole in the center and 1.6 m in diameter, with entrance and front lawn, both decorated with discrete piles of flowers, fungi, and fruits. Bower is built on a flat patch of forest floor on a ridgetop or flank. Nest 2.5 m up in sapling; a stick cup with leaf lining. Egg (1) plain whitish. **Range:** Endemic. NG: Bird's Head and Neck (Arfak, Tamrau, Wandammen, Fakfak, and Kumawa Mts), 1000–2000 m. **Taxonomy:** Populations from the Kumawa and Fakfak Mts are apparently referable to this species, but their bower is a maypole without a conical stick hut.

Archbold's Bowerbird *Archboldia papuensis* Pl 66
(**Sanford's** or **Tomba Bowerbird**, *A. sanfordi*)
35–37 cm. *An elusive, blackish canopy dweller restricted to a few high portions of the Central Ranges. Note longish tail.* Usually rare, in high cloud forest and at timberline. **Male:** Long, narrow, *yellow crest*, otherwise dull black; crown flat; tail forked. **Female:** Entirely dull blackish but for *ochre mark on base of primaries* visible at rest and in flight. **Imm Male:** Lacks the yellow crest. **Subspp (2):** Minor, differing in measurements. Race *papuensis* known from W Ranges (upper Baliem, Ilaga, and near Paniai Lakes). Race *sanfordi* known only from the E Ranges (SW slopes of Mt Hagen and NE slopes of Mt Giluwe westward to Doma Peaks). **Similar spp:** The astrapias have longer, narrow, unforked tails. Short-tailed Paradigalla has a very brief tail. The blackish *Melidectes* and *Melipotes* honeyeaters are smaller and exhibit facial wattles. **Voice:** An unpleasant 1- or 2-note song; harsh, downslurred *sherrd sherrd* or catlike *gnaad!* Various unmusical notes and mimicry are delivered by Male at the bower. **Habits:** Apparently inhabits only open, frost-disturbed, high plateau forest with pandanus, podocarp trees, and scrambling bamboo; absent from steep, ridge and ravine habitat predominating in most montane regions. To date, known only from a few localities. Seen singly or in Female-offspring pairs. Wary while foraging at fruiting trees. Polygynous. Male displays at a most unusual bower, with 2 structural components: (1) a broad stage of matted ferns decorated with snail shells, King of Saxony Bird of Paradise plumes, and other small objects and (2) over the stage, the low branches shrouded in trailing stems of epiphytic orchids. Nest situated in a low sapling; nest is a cup of sticks, leaves, and twiglets. Egg (1) pale buff with slight gloss and sparse, tiny spots. **Range:** Endemic. NG: W and E Ranges, 2300–2900 m (extremes 1800–3660 m).

Masked Bowerbird *Sericulus aureus* Pl 67
(**Golden Bowerbird**, *Xanthomelus aureus*)
24–25 cm. An elusive canopy dweller of lower mountain forests of the North and West. Uncommon. **Male:** A spectacular *orange, yellow, and black bird with a black face*, yellow iris, and black-tipped grey bill. **Female and presumably Juv:** Olive brown dorsally, *yellow below with fine dark scalloping on chin and throat.* **Imm Male:** Sequence presumably as for Flame BB. **Similar spp:** Flame BB male lacks the black face; female has yellow chin and throat streak. Golden Myna has a black back and wings.

Voice: Apparently typical of the genus—harsh rasping notes and faint *ksh* notes. **Habits:** Active but shy. Small avenue stick bower constructed in forest interior on a low, flat ridge. Decorates bower with leaves and blue and purple fruits. Nest and egg unknown. **Range:** Endemic. NG: Bird's Head and Bird's Neck (Wandammen Pen) and northern portion of the Central Ranges (Weyland Mts eastward to Jimi R) and the Foja and N Coastal Mts, 850–1400 m. **Taxonomy:** Formerly included Flame BB.

Flame Bowerbird *Sericulus ardens* Pl 67

25–26 cm. Confined to lowland and foothill forest of the Southern Lowlands. Uncommon.
Male: *Bright yellow with black wing edgings and tail, and a flaming reddish-orange head, nape, and mantle.* Bill ivory and iris white. **Female and presumably Juv:** Olive brown above and *yellow below*, with a dark face and *yellow chin and throat; lacks mottling and scalloping on the throat*; iris dark.
Imm Male: Resembles Female but iris white and bill yellowish at the base; Adult plumage attained first on the head. **Similar spp:** Male unique in lacking black face of other golden bowerbirds. Female similar to female Masked BB but shows a clear yellow chin and throat streak (these dark-scaled in Masked). **Voice:** Harsh rasping and hissing notes typical of the genus; also an oft-repeated churring *shh* or faint *ksh* note. **Habits:** Solitary except at fruiting trees and at bower, where many congregate. Inhabits lowland rainforest and monsoon forest. Especially attracted to patches of regrowth and village rubber plantations. A canopy-dwelling frugivore partial to figs; diet includes some insects. Makes midmorning flights over the canopy. Polygynous. Builds a small avenue bower consisting of a parallel row of vertically woven sticks. Bower sited on flat ground in a small opening in the forest, at forest edge, or in the forest itself. Nest and egg unknown. **Range:** Endemic. NG: S Lowlands (Mimika and Wataikwa Rs eastward to Nomad R, Mt Bosavi, and very locally in the Trans-Fly), 0–750 m.
Taxonomy: Formerly treated as conspecific with the Masked BB. The 2 species presumably meet at the Mimika and Wataikwa Rs, where males with intermediate plumage occur.

Fire-maned Bowerbird *Sericulus bakeri* Pl 67
(**Adelbert** or **Beck's Bowerbird**, *Xanthomelus bakeri*)

26–27 cm. Endemic to the lower montane forest of the Adelbert Mts. Uncommon.
Male: Unmistakable: *velvety black with an orange-red cape and broad, yellow wing patch.*
Female and presumably Juv: Grey-brown dorsally and *pale grey below with fine scalloping*, reminiscent of a female King of Saxony Bird of Paradise. **Imm Male:** Sequence presumably as for Flame BB. **Similar spp:** Female is duller and more lightly barred than female birds of paradise.
Voice: Call, *k-skgg!*, similar to the sharp hissing and rasping notes of the Yellow-breasted BB. This call is the best means of locating it. **Habits:** A shy inhabitant of forest canopy where it forages for figs, other fruits, and arthropods. Small avenue bower of sticks is placed on a gentle slope below a ridge and decorated with blue fruit and a few leaves. Nest situated in an *Asplenium* fern, 15 m up in a tree. Egg and nest structure unknown. **Range:** Endemic. NG: Adelbert Mts, 900–1400 m.

Yellow-breasted Bowerbird *Chlamydera lauterbachi* Pl 67
(**Lauterbach's Bowerbird**)

25–28 cm. A streaky bowerbird with a yellow cast that inhabits montane forest edge, margins of grassland, scrubby regrowth, abandoned gardens, and cultivation, where locally common; also patchy in like habitats of Sepik-Ramu lowlands. Shy and nervous, keeping to cover, but also curious and will mount a high perch to watch human intruders. Most often seen flying through scrub openings, with its peculiar head-up undulating flight. Note square-shaped head with flat crown. **Adult:** Greybrown dorsally with pale feather-edgings; throat heavily streaked; *breast and belly washed with yellow* (Female and Imm duller). *Crown plain and unstreaked*, greenish or coppery. **Juv:** Crown with white shaft streaking; upperparts with less obvious white spotting. **Subspp** (2): *lauterbachi* (Sepik-Ramu and Huon) orange-washed crown; *uniformis* (remainder of range but includes parts of the Sepik) dull greenish crown. **Similar spp:** Fawn-breasted BB has an orange-buff breast and belly, otherwise very similar. The two co-occur in some localities. **Voice:** A sharp *chilp chilp chilp*, and rasping and hissing calls, probably not distinguishable from those of the Fawn-breasted BB. Alarm call is like a rap on a cardboard box. **Habits:** Solitary or in small groups. Takes fruit and arthropods from low trees and shrubs. Polygynous. The bower, an impressively substantial stick structure unique for having 4 walls, is situated in thickets near grassland and decorated with blue, red, or green berries and pebbles. Nests in shrub or tall grass, ~2 m up; nest is a neat cup of twigs and grass on a loose foundation of twigs. Egg (1) pearl grey marked with grey and black. **Range:** Endemic. NG: patchily distributed in mid-mountains from W Ranges to E Ranges, north through the lowlands to N Coastal Mts and east to head of Ramu R and N slopes of the Huon Pen, 0–1750 m. **Taxonomy:** The relationship of the 2 forms, which both occur in the Sepik-Ramu, needs to be clarified.

Fawn-breasted Bowerbird *Chlamydera cerviniventris*   Pl 67

28–30 cm. A pale greyish bowerbird mainly of lowland scrub, preferring the transitional thickets between grassland and forest, also edges of mangroves and towns; has colonized areas of habitat clearing in mid-mountain valleys. Locally common. In flight, note short blunt bill, flat crown, and its distinctive head-up, undulating flight pattern. **All plumages:** A heavy bird, *grey-tan above, orange-buff below. Upperparts with fine, pale scalloping.* Crown, face, and throat with profuse pale streaks. **Similar spp:** Yellow-breasted BB exhibits yellow breast and belly and unstreaked cap of yellowish green or orange. Australasian Figbird and other orioles lack the fawn breast. **Voice:** Many rasping, churring, and harsh notes strung together. Also bizarre mechanical sounds. As with other bowerbirds, an accomplished mimic. Sometimes ventriloquial. **Habits:** Solitary, sometimes gregarious. Noisy, but shy and wary. Often perches on an open dead branch overlooking grasslands. Diet is mainly fruit and some insects. Polygnynous. Builds a substantial, 2-walled avenue bower of sticks decorated with clusters of green berries on the bower platform, situated in a thicket. Nest built in a tree or shrub, ~4 m up; nest similar to that of Yellow-breasted BB. Egg (1) creamy with hint of green, heavily marked. **Range:** NG: Largely confined to coastal areas of eastern NG, westward in North to Jayapura, westward in South to the Trans-Fly, 0–500 m; also on the Bird's Head (Ransiki, Kebar Valley) and locally in the mts of SE Pen, to 1800 m. Also NE AU.

AUSTRALASIAN TREECREEPERS: Climacteridae

A small family of 7 species endemic to Australia and the mountains of New Guinea (1 sp). In shape and habits, the Australasian treecreepers resemble the creepers of Eurasia and North America, except that they do not use the tail as a prop when creeping. They are unrelated to the creepers, instead being an ancient family closest to bowerbirds. Treecreepers are small brown birds with enormously elongated toes and claws. They are the only songbirds in the region that habitually climb up the trunks of trees, searching for small arthropods hiding in the bark. Although usually silent when foraging, during certain seasons they deliver an attractive whistled song throughout the day. The New Guinea species probably breeds in simple pairs and nests in tree hollows, like its closest relative in Australia.

Papuan Treecreeper *Cormobates placens*   Pl 83
(*Climacteris placens*)

14–15 cm. Locally common in montane forest. Treecreepers and pygmy parrots are the only birds in the NG Region that *creep up tree trunks* or larger branches. *A small olive bird with scaly markings on the crown, belly, and flanks.* **Male:** Lacks rufous facial stripe. **Female:** Shows rufous moustachial stripe and usually more markings on belly. **Juv:** Breast indistinctly mottled. **In flight:** Pale buff stripe in wings. **Subspp (3, in 2 groups):** *placens* (Bird's Head and Border Ranges east to Tari, E Ranges) greyish olive; *meridionalis* (SE Pen) brownish olive. **Similar spp:** None. **Voice:** Song, a beautiful, mellow series of repeated flutelike notes slowly increasing in volume and emphasis, the last note held longer, *du du du du du du duu.* When paired birds call together, one answers the other with a call at lower pitch. Another song similar to trill of Fan-tailed Cuckoo. Another call is a faint nasal triplet, *chu chee chu.* **Habits:** Solitary or in pairs. Usually on vertical trunks, hitching its way upward until reaching some limit, then swooping to the base of the next tree and beginning to climb again. Feeds on insects and spiders. (Its closest relative in AU takes mostly ants.) While foraging often accompanied by a fantail, which snatches flushed insects. Vocalizations and occasional aggressive behavior suggest territorial habits. No nesting information. **Range:** Endemic. NG: Bird's Head and Central Ranges, although oddly absent from most of the E Ranges, but present in SE Pen, 1200–3000 m. **Taxonomy:** The 2 races are quite isolated and different; should be investigated for degree of phylogenetic divergence.

FAIRYWRENS AND ALLIES: Maluridae

The fairywrens, emuwrens, and grasswrens are a small family of 29 species confined to Australia and New Guinea. Only the fairywrens are represented in the New Guinea Region, where there are 6 species. Fairywrens are small perky insectivores with a broad bill and long tail that is held erect. Many species have strikingly beautiful plumage, vibrant blue and black being their hallmark colors. They are mainly birds of the lowlands and hill country, and only the Orange-crowned Fairywren dwells solely in the mountains. Of the New Guinea species, Wallace's Fairywren is the most distinct in form and habitat, occupying the interior of primary forest and foraging both high and low in the trees. The other species prefer dense thickets in openings and live close to the ground. Fairywrens forage quickly and keep to cover in their search for insects and spiders, which they take by gleaning. They live in small family groups composed of parents and older offspring, the latter helping to raise the young of the year. These groups are territorial and vocal, with high-pitched songs and calls. Despite their colorful plumage and social ways, fairywrens are somewhat secretive, so the best way to locate them is to listen for their high-pitched scolding. There are 2 groups of species: a group that inhabits forest, with broad bills and long rictal bristles (Wallace's, Orange-crowned, Broad-billed, and Campbell's), and a group that dwells in scrub and thickets (Emperor and White-shouldered), with narrow bills and short rictal bristles.

Wallace's Fairywren *Sipodotus wallacii*   Pl 68
(*Todopsis wallacii*)
11–12 cm. Uncommon and inconspicuous, but widespread in hill forest and second growth. In groups, often with other small insectivores, calling incessantly while foraging in midstory. Oddly for a fairywren, the species holds its body horizontal as though stretched, and the tail is not cocked. **Adult:** *A tiny, uniquely patterned, warbler-like bird with all-white underparts and wing bars*, rufous back, and black crown streaked with blue. Face marked with a partial eye-ring and white cheek tuft. Throat white in **Male**, buff white in **Female**. **Juv:** Duller; crown dark grey and spotted buff, wing bars buff. **Similar spp:** Gerygones that show white underparts lack wing bars, have a shorter bill, and upperparts patterned differently. Female Emperor FW similarly patterned but larger and has blue throat and no wing bars. **Voice:** A rapid series of ~10 *tss* notes lasting 5 sec, or some variation; notes so high pitched as to be barely audible. Also a distinctive buzzy note. **Habits:** Forages from the shrub layer to canopy, mainly higher than 3 m, in forest tangles and thickets. Lives in pairs or small groups, probably families; some evidence of cooperative breeding with immature helpers. Joins foraging flocks of warblers, monarchs, whistlers, and fantails. Takes insects and spiders by gleaning. Nests in midstory vine tangles; nest domed and pendent with side entrance, built of grass and palm leaf strips. Eggs (2) white, a few blotches. **Range:** Endemic. NG, NW Is (Misool), Bay Is (Yapen), and Aru Is, 0–1100 m. All NG except eastern Sepik-Ramu, Huon, and most of northern SE Pen.

Orange-crowned Fairywren *Clytomyias insignis*   Pl 68
14–16 cm. A rare and local fairywren of cloud forest thickets. **All plumages:** Unique *orange-rufous cap*. **Subspp (2):** Minor. **Similar spp:** None. **Voice:** Song a sibilant chattering. An active vocalist with a variety of high, scolding notes and squeaks *jiyub! jiyub!* **Habits:** Lives in small groups occupying large territories that encompass forest openings or edge overgrown with shrubbery and climbing bamboo. Active, quick moving, difficult to observe. Cocks tail. Nest built in vine or shrub ~1 m up; nest domed with hooded side entrance, constructed mainly of live moss. Eggs (2) white with sparse spotting. **Range:** Endemic. NG: Bird's Head, Central Ranges, and Huon, 1700–2800 m.

Broad-billed Fairywren *Chenorhamphus grayi*   Pl 68
(*Malurus grayi*)
12–14 cm. A rare, *soft blue* fairywren fond of small patches of regrowth appearing after tree falls and landslides within primary forest of hills and nearby lowlands. *Both sexes with blue mantle.* **Male:** *Crown mottled blue*, belly blue. **Female:** Crown brown, belly white. **Juv:** Mostly brown with little blue; look for brown in the crown and underparts. **Similar spp:** See Campbell's FW. Emperor FW prefers more disturbed habitat—forest edge and second growth; has deep, brilliant purplish blue plumage; breast and belly deep blue in male and white in female. **Voice:** Song a 3-sec series of

many rapid-fire *ts* notes. High-pitched *ss* calls similar to other fairywrens and a high-pitched upslur. **Habits:** Although it favors breaks in tall forest, this species differs from Emperor FW by avoiding regrowth in man-made clearings and forest edge, such as along rivers; thus the 2 species are rarely found together. In pairs or small family groups; territories far apart. Forages similarly to other fairywrens on the ground and in the undergrowth to ~5 m. One nest reported 0.5 m up in a cavity in a moss clump attached to a sapling. No egg information. **Range:** Endemic. NG and NW Is (Salawati), 0–1000 m (rarely 1500 m). NG: Bird's Head and Neck, NW Lowlands, and W Sepik-Ramu.

Campbell's Fairywren *Chenorhamphus campbelli* Pl 68
(formerly included in **Broad-billed Fairywren**, *C. grayi*)
11–12 cm. Replaces Broad-billed FW in S Lowlands. Distinguished from that species by smaller size, *crown black in both sexes, and mantle brown*. Sexes alike. **Voice:** Presumably similar to Broad-billed FW. **Habits:** Similar to Broad-billed FW. Nest and egg: no information. **Range:** Endemic. NG, where its range in S Lowlands is known only from Mt Bosavi and sightings near Kiunga (presumably this form), 100–800 m. **Taxonomy:** Arguably a form of Broad-billed FW, although distinct in plumage and DNA.

Emperor Fairywren *Malurus cyanocephalus* Pl 68
(*Todopsis cyanocephala*)
13–16 cm. The common fairywren of dense thickets in abandoned gardens and at the forest edge. **Male:** *Intense dark blue*—an electric spark when it flies! **Female:** Boldly patterned: blue head, chestnut back, white underparts, and white-tipped tail. **Juv:** Similar to Female, but duller; head dusky black, not blue. **Subspp (3):** Minor. **Similar spp:** Wallace's FW has white throat and wing bars. Broad-billed FW has blue breast and is not as brightly patterned. Fantails might be confused with female Emperor, but lack blue head. **Voice:** Song unusual for a fairywren yet easily recognized although it varies locally—a long, medium-paced series of burry warbles and brief rapid trills suggestive of a thrush. Scolding call a rapid chitting: *tst, tst, tst. . . .* **Habits:** Lives in noisy family groups foraging within 2 m of the ground in dense growth. Tends not to join mixed flocks. Territorial and probably breeds cooperatively. Cocks tail like most fairywrens, but also partially fans it. Nest 1 m off the ground in a bush; gourd-shaped with side entrance, built of strips of fern and other leaves and mosses. Egg, no information. **Range:** Endemic. NG, NW Is (Salawati), Bay Is (Biak, Yapen), and Aru Is, 0–1000 m. Throughout NG lowlands except Huon and northern SE Pen.

White-shouldered Fairywren *Malurus alboscapulatus* Pl 68, 105
10–13 cm. A common and widespread inhabitant of grasslands and low shrubby regrowth; likely seen near human settlement and along road verges. **Male:** *A tiny, jet-black bird with white shoulder patch and cocked tail.* **Female:** Color varies geographically, see Subspp. All differ from Male in having white partial eye-ring or brow. **Juv:** Similar to Imm, but duller, i.e., greyer or browner. **Imm:** Mostly similar to Female, but all-black subspecies have some white on chin and abdomen or not. **Subspp (6):** 2 groups by color of Female. Black Group, *alboscapulatus*: either pied as in *alboscapulatus* (Bird's Head) with mostly white underparts and *naimii* (Sepik-Ramu, eastern E Ranges, western SE Pen) more black on flanks; or all black like Male but duller, wings brownish as in *aida* (NW Lowlands, probably Border Ranges), *kutubu* (northeastern S Lowlands, southwestern E Ranges), and *moretoni* (Huon and SE Pen, eastward in South to Port Moresby area). Brown Group, *lorentzi* (S Lowlands, southern W Ranges) Female pale brown, white brow and throat. **Similar spp:** Pied Bushchat shares habitat, but much larger, chunkier, does not cock tail. Cisticolas and grassbirds could be confused with brown female White-shouldered FW in S Lowlands. Other FWs occupy forest, not grass and scrub. **Voice:** Song a high-pitched, rapid-fire, twittering reel lasting ~5 sec, typical of a fairywren. Calls are an incessant high churring and chattering. **Habits:** One of the most familiar grassland birds. Usually shy and remains hidden in thick shrubs or tall grass, but groups are noisy, and inquisitive birds eventually ascend to an elevated perch. Lives in small, territorial family groups; breeds cooperatively. Gleans small insects and spiders in dense cover. Nests in shrub or tall grass within 2 m of ground; nest domed with side entrance, composed of grass and leaves. Eggs (2–4) pinkish white, speckled. **Range:** Endemic. NG, although no records from Bomberai Pen (Bird's Neck), 0–2000 m, occasionally to 3000 m. (A sighting from Fergusson I in SE Is lacks confirmation.) **Taxonomy:** Females quite distinct and may indicate more than 1 species.

HONEYEATERS: Meliphagidae

The honeyeaters, with 184 species altogether and 65 inhabiting NG, are the largest Australasian songbird family. Oddly, while this important group extends into Oceania, it has advanced toward Asia only as far as Bali. Most honeyeaters are easily recognized by their curved bills and sloping foreheads, yet bill shape and size vary considerably, and different genera approximate sunbirds, warblers, flycatchers, and jays. Some possess bare wattles and facial skin of various bright colors. Most produce unmelodious vocalizations, and some are noisy with conspicuous calls. Many are unusually pugnacious. Honeyeaters form an important component of the birds in NG that visit flowers for nectar and insects (with parrots and fruit-bats), glean foliage, twigs, and branches (together with warblers), and visit fruiting trees (with pigeons and birds of paradise). Their habitats include savannah, all forests from treetops down to undergrowth, and also shrublands at timberline. Pair formation is the rule, with the male sharing some nesting duties. For most species, the nest is cup-shaped and hangs suspended by the rim from a forked branch; the eggs are spotted.

Myzomelas. Of the 9 species in the New Guinea Region, 6 are small, while 3 are larger. Among the 6 small species, 3 occupy forests as an altitudinal sequence with considerable overlap: Red-collared Myzomela at higher elevations, Red M at middle elevations, and Papuan Black at low elevations. Elfin M lives in the mid-mountains and is very uncommon in forests, commoner in gardens. Red-headed M is confined to south-coast mangroves, Sclater's M to Karkar Island. Among the 3 related and similar larger myzomela species, Ruby-throated is widespread in lowland forests, Dusky occupies open habitats of southern New Guinea, and White-chinned is the only myzomela of some SE Is.

Red-collared Myzomela *Myzomela rosenbergii* Pl 69
(Red-collared Honeyeater)
11–12 cm. Male is a small, jewel-like bird of the cloud forest; ranges into subalpine shrubland and downslope into mid-mountain forest. The only myzomela above 2000 m, though it overlaps seasonally with Elfin and Red Ms in mid-mountain forest. Common at flowering trees. **Male:** *Striking, black and red.* **Female:** Mottled sooty and rich brown, with *black face and red breast* and rump. **Juv:** Mottled brown all over, lacks black face, and the red of the rump is reduced or absent. **Subspp (2):** *rosenbergii* (mainland NG) Female brownish and mottled, with red confined to breast and rump; *longirostris* (Goodenough I) larger, bill much longer, and Female distinctly greenish with red collar and back, while Juv male similar to Female mainland race. **Similar spp:** Elfin M much smaller, not easily mistaken. **Voice:** Song is an energetic, breathless, high-pitched sibilant trill or rapid alternation between 2 notes on different pitches. Call a high-pitched *tswi* or a high *ts ts*. **Habits:** Vocal, active, and pugnacious. Solitary, but often in congregations at flowering trees, where Males predominate. A bird of the treetops that may descend lower in search of food. Obtains nectar and insects by gleaning. One nest found in a shrub, hung from a forked branch; bulky, cup-shaped, built of roots, and lined with rootlets. Egg (clutch at least 2), no description. **Range:** Endemic. Mountains of NG (records from all but Wandammen Mts) and Goodenough I (SE Is: D'Entrecasteaux Is), 1200–3700 m (600–3950 m).

White-chinned Myzomela *Myzomela albigula* Pl 69
(White-chinned Honeyeater)
13–14 cm. *The drab, grey-brown myzomela of select SE Is.* Common where found, but missing from many seemingly suitable islands. Most likely encountered on islands supporting other small-island specialists and lacking other myzomelas or sunbirds. Inhabits all types of forest. The *diagnostic white throat* is absent in Males of the Rossel I race. *Obscurely streaked below.* **Male:** Larger, exhibits more distinct red throat-streak (hard to see). **Subspp (2):** *albigula* (Rossel I) white throat patch missing in Male; *pallidior* (other SE Is in the Louisiades) much paler and with white throat present in both sexes. **Similar spp:** Probably none on its islands. Papuan Black M elsewhere in the Louisiades; its female has a rose forehead and chin. Dusky M on NG mainland lacks red throat-streak and streaked breast. **Voice:** Male call *tink tink tink.* **Habits:** Much like Dusky M. Nest and eggs undescribed. **Range:** Endemic. Eastern SE Is: Louisiade Archipelago (Alcester/Nasikwabu I, Bonvouloir Is, Conflict Group, Deboyne Is, Kimuta I, Rossel I), 0–300 m.

Dusky Myzomela *Myzomela obscura* Pl 69
(Dusky Honeyeater)
13–14 cm. The common drab myzomela of southern coastal towns, gardens, and other open-wooded habitats. *Plain dull brown and relatively large*; this is the dullest-plumaged myzomela.
Male: Larger, more greyish. **Female:** Browner. **Juv:** Like Female but paler and face with reddish tinge.
Subspp (2): *fumata* (NG and Aru Is) drab brown; *rubrobrunnea* (Biak I) washed with red.
Similar spp: Ruby-throated M prefers forest and shows a red throat-stripe except when young. White-chinned M of SE Is does not co-occur, has streaky underparts and white chin. **Voice:** Song is a jumble of squeaky high-pitched notes. Dawn song in AU is *tip-tip-eeee-chip*. Call is a weak, high-pitched *si-si-si-si*; also an upslurred squeaky note and mournful whistle. **Habits:** Solitary or in pairs. Attracted to garden flowers. Also frequents mangroves, coastal scrub, savannah edge, second growth, and forest edge. Gleans for insects and takes flower nectar. Nest slung on a thin, forked branch in a small tree; a small basket of woven plant material. Eggs (2) pinkish white, finely spotted (AU data).
Range: Entire S NG coast and northern SE Pen west to Popondetta, Aru Is, Biak (Bay Is), and Pulau Adi off Triton Bay, 0–100 m (rarely to 600 m). Also AU and Moluccas.

Ruby-throated Myzomela *Myzomela eques* Pl 69
(Red-throated Honeyeater or **Myzomela)**
14–15 cm. Uncommon in lowland and hill forest and edge, though gathers at flowering trees. Striking in its own way: *a large, dark brown myzomela adorned with a ruby-red throat-stripe*.
Male: Much larger; throat-stripe more prominent; may show reddish forehead spot, lacking in **Female**.
Juv: Generally duller. **Subspp (4):** *eques* (NW Is, Bird's Head and Neck); *nymani* (S Lowlands, SE Pen, Huon) overall greyer, and red of throat more extensive; *primitiva* (NW Lowlands, Sepik-Ramu) small red throat patch; *karimuiensis* (Karimui in E Ranges) plumage darker and richer. **Similar spp:** In its rainforest habitat, only the drab females of the smaller Papuan Black and Red Ms are similar; these show pinkish red on chin and forehead, but lack the throat-stripe. Dusky M difficult to distinguish from juv Ruby-throated. **Voice:** Calls include a typical myzomela high-pitched *chip* and a series of chirps and twittering. **Habits:** Feeds alone or in groups in canopy flowers; Males especially seen at flowering trees. Frequents canopy and midstory, descending lower to join mixed feeding flocks of small birds. Slower moving than other myzomelas. Gleans mainly nectar and insects, rarely fruit. Nests in a forest sapling, ~3 m up or more; nest, slung on a thin forked branch, is a scant basket made of black lichen strands. Eggs (clutch?) pinkish with fine brownish spots. **Range:** Endemic. NG and NW Is (Misol, Salawati, Waigeo), 0–500 m, rarely to 1200 m.

Red Myzomela *Myzomela cruentata* Pl 69
(Red or **Red-tinted Honeyeater)**
10–11 cm. Mainly in hill and mid-mountain forest and edge, frequenting the canopy and usually noticed only at flowering trees; less frequent in lowlands at the foot of the ranges. Uncommon except at gatherings of nectar-feeding birds. **Male:** *Uniquely all red*. **Female:** Drab brown marked with red forehead and chin; *diagnostic reddish rump and tail*. **Juv male:** Resembles Female, but often with reddish wash on the upperparts. **Subspp (1):** *cruentata* (endemic). **Similar spp:** Papuan Black M female lacks the red rump and tail. Elfin M female also lacks the red tail and is smaller, shorter-tailed, lacks red forehead, and inhabits higher elevations. **Voice:** Call is a high-pitched, insect-like *tseet* repeated at intervals of 5–10 sec. **Habits:** Seen singly or in small groups. Quiet and inconspicuous. Partly nomadic. Feeds on nectar and insects. Nesting habits undescribed. **Range:** Virtually all mts of NG (no records from Kumawa and Wandammen Mts), plus Yapen I (Bay Is), 600–1500 m (extremes: 0–1600 m). Also Bismarck Is.

Papuan Black Myzomela *Myzomela nigrita* Pl 69
(Black or **Carbon Honeyeater)**
12–13 cm. Hill and lower montane forest canopy, savannah woodland, and regrowth; common at flowering trees such as eucalypts and albizias. Also Trans-Fly. Locally common, but nomadic.
Male: *Lacquer black with white underwing-coverts* (latter rarely show). Looks matte black in the field.
Female: Smaller; *drab brown with combination of reddish forehead and throat but not tail*. Some show sooty black face with no red. In the Southeast many Females are all black, but duller than the Male.
Juv: Variably resembles grey-brown Female or is darker. **Subspp (4, in 3 groups):** *nigrita* (species' range minus next 3 races) Male black; *steini* (Waigeo I) both sexes brown-plumaged, like typical Female; *forbesi* (D'Entrecasteaux Is) larger and Male black with red cap. **Similar spp:** Red M female has

red tail. Elfin M female smaller, lacks red on forehead. Black Sunbird male flashes iridescent blue cap, shoulder, and tail. Black Berrypecker male has short bill. **Voice:** Song is formless series of high-pitched notes. Call frequently, a dry, high-pitched *sit* or *zeet*. **Habits:** Singly, in pairs, or gathers with other nectar feeders at favored blooming trees, where Males outnumber Females. Feeds on nectar, insects. Nesting undescribed. **Range:** Endemic. NG, NW Is (Waigeo), Bay Is (Mios Num, Yapen), Aru Is, and SE Is (3 main D'Entrecasteaux Is and Louisiade Is: Misima, Tagula, and Woodlark), 0–1250 m. NG: throughout lower mts (although no records from Fakfak, Kumawa, Wandammen, and Cyclops Mts), plus lowlands of Trans-Fly.

Red-headed Myzomela *Myzomela erythrocephala* Pl 69
(**Red-headed Honeyeater**)
11–12 cm. Southern mangroves only; locally common. **Male:** *Diagnostic red hood and rump.*
Female: Similar to other drab female myzomelas but shows *red on chin but not forehead* (trace at most). **Juv male:** Resembles Female, but may have a few red feathers on the rump.
Subspp (2): *erythrocephala* (SE Pen and AU) smaller and paler; *infuscata* (S NG and Aru Is) larger and Male much darker. **Similar spp:** Papuan Black M female, absent from mangroves, always with red forehead. Elfin M smaller; mountains only. **Voice:** Song a jingling trill of metallic notes (AU data). Call in NG, *zit*, sharper than other myzomelas. **Habits:** Usually seen in pairs, in mangrove canopy and midstory. Nest suspended from a forked branch in mangrove foliage; a cup of bark strips and rootlets (AU data). Eggs (2) white with speckles (nest and eggs). **Range:** NG (S coast from Triton Bay on Bird's Neck east to Cape Rodney) and Aru Is, sea level. Also islands of Torres Strait and coastal N AU.

Elfin Myzomela *Myzomela adolphinae* Pl 69
(**Mountain Red-headed Honeyeater**)
9 cm. Mid-mountain forest canopy, edge, and second growth, but most easily seen in highland towns and rural villages where it seeks flowers of eucalypts and other trees and shrubs. Generally common. A tiny mite of a bird; the *smallest myzomela*, with short, curved bill. **Male:** *Red head and rump.*
Female: *Plain grey with pink only on chin and below eye.* **Juv male:** Resembles Female but may have reddish wash on rump. **Similar spp:** Papuan Black M and Red M females are larger, with pink wash on forehead. Red-headed M larger; mangroves only. Flowerpeckers have stubby bills.
Voice: Song, given from the treetops, is of 2 high-pitched notes: *tyink*, followed by a slightly musical trill at lower pitch, slightly downslurred. Calls various, composed of 1 or a few high-pitched notes.
Habits: Generally solitary. Active and vocal. Its small size does not deter it from visiting flowering trees occupied by aggressive nectar feeders. Takes nectar and insects. One nest found in a suburban garden built in a shrub with comingled dead fern leaves; it was made of fine plant materials and lichen. Eggs undescribed, but clutch size at least 2. **Range:** Endemic. NG: Bird's Head, E Ranges (east from Telefomin) to SE Pen, N Coastal Mts, possibly Adelbert Mts (sighting), Huon, and SE Pen, 800–2000 m, rarely lower.

Sclater's Myzomela *Myzomela sclateri* Pl 69
(**Scarlet-throated** or **Scarlet-bibbed Honeyeater**)
11–12 cm. In the NG Region found on Karkar I only, where it is the sole myzomela species.
Resides in montane forest, but visits flowering trees down to sea level. **Male:** *Prominent red throat.*
Female and Juv: Drab, with a streaked throat and red chin-spot. **Similar spp:** Females of some other myzomelas are similar, but do not occur on Karkar. Compare with Black and Olive-backed Sunbirds, both on Karkar. **Voice:** Song is a pair of short, hoarse, buzzy notes, the second note lower and trilled, suggestive of Elfin M. Calls various and high pitched. **Habits:** Solitary or in small gatherings at food sources. A lively, noisy species. Forages at all levels, from treetops to understory, taking insects, nectar, and fruit. Gleans and hover-gleans. Nest built in shrub, tree, palm, or cane, 2–10 m up; a very thin-walled, suspended basket. Egg (no clutch size data) white or pinkish with reddish spots.
Range: Karkar I, 0–1850 m. Absent from NG mainland, but look for it on other islands off N coast.
Also small islands off New Britain.

Green-backed Honeyeater *Glycichaera fallax* Pl 70
10–12 cm. Lowland and hill forest canopy and midstory, especially seen at forest openings and edge. Uncommon. *A warbler-like honeyeater with straight bill.* Petite, nondescript except for *whitish iris and narrow pale eye-ring*. Iris pale grey in **Adult**, brown in **Juv**. **Male** larger. **Subspp (2):** *fallax* (all of range except next race) greener and underparts yellowish; *pallida* (NW Is: Batanta, Salawati, Waigeo)

paler and greyish. **Similar spp:** Spectacled and Pygmy Longbills are dark-eyed and have prominently decurved bills. Gerygones are smaller-billed, iris dark. White-eyes have a pronounced white eye-ring, and the throat is bright yellow. **Voice:** Usually quiet. Call is a nondescript, repetitive *whit–whit–whit . . .*, the notes in 1-sec intervals. In AU, also gives a chicken-like *peep* and busy twittering. **Habits:** Usually seen alone, but sometimes in pairs or small groups; regularly joins mixed feeding flocks of warblers and flycatchers. Actively gleans and hover-gleans insects from foliage. No nesting information. **Range:** NG, NW Is (Batanta, Misool, Salawati, Waigeo), Bay Is (Yapen), and Aru Is, 0–850 m, rarely to 1300 m. Also Cape York Pen, AU.

Streaked Honeyeaters (*Ptiloprora* spp.). Of the 6 species of this genus endemic to New Guinea mountains, 5 (all except Leaden Honeyeater) are larger and segregate partly or wholly spatially. Grey-streaked, Rufous-backed, and Mayr's Streaked Hs are high-mountain birds that segregate entirely spatially: Grey-streaked inhabits a wide range of elevations in western NG; the slightly smaller Rufous-backed lives at elevations below Grey-streaked in the mountains of eastern NG and extends to higher elevations as well locally in the absence of Grey-streaked; Mayr's Streaked is the only *Ptiloprora* of 3 mountain ranges of the north coast. Rufous-sided H is ecologically similar to those species and is the only *Ptiloprora* in the Bird's Head and Neck but co-occurs with the larger Grey-streaked in the W Ranges, mostly at lower elevations. Yellow-streaked H also appears to be ecologically equivalent but patchily distributed and rare. Grey-streaked, Rufous-sided, and Yellow-streaked have been found at similar elevations in the mountains of western NG, where it is unknown how they segregate ecologically. Among these 5 species, all except Yellow-streaked give essentially the same plaintive high-pitched slurred calls. The sixth species, Leaden H, is smaller and very rare.

Leaden Honeyeater *Ptiloprora plumbea* Pl 70

14–15 cm. Patchily distributed and rarely encountered in mid-mountain forest and second growth. This species has the shape and typical behavior of a *Ptiloprora*, but is the smallest, drabbest, and rarest of the group. Cocks its tail. **All plumages:** *A small grey bird with a thin, slightly curved bill, and a silver-grey iris.* Evenly grey below, but note frosting on the chin. Pale brown iris in some birds may denote Juv status. **Male** larger. **Similar spp:** Marbled H is also grey but has white-scaled underparts and white cheek-stripe. **Voice:** Calls include a whistled, asthmatic *wshee wshee wshee . . .* and a weak *teu teu* or *tswee*, about every 3 sec. **Habits:** Usually solitary. Forages actively at all levels in the forest, gleaning insects and taking nectar from flowers. No nesting information. **Range:** Endemic. NG: Central Ranges (absent from much of the E Ranges), 1000–2100 m.

Yellow-streaked Honeyeater *Ptiloprora meekiana* Pl 70
(**Olive-streaked** or **Meek's Honeyeater**)

16–17 cm. Patchily distributed and rare. Found at transition between mid-mountain and cloud forests in SE, but inhabits cloud forest in W Ranges. **All plumages:** *A medium-sized, yellowish, streaked honeyeater.* Iris pale grey; birds with greenish-grey iris are possibly Juv. **Male** larger. **Subspp (2):** Minor and needing study. **Similar spp:** Juveniles of Grey-streaked and Rufous-backed Hs have an olive-yellow wash below, but these are darker and without yellowish above. **Voice:** Call note *cht* or *chip!*, repeated frequently. **Habits:** Difficult to observe, little known. Singly or in pairs. Mainly in the canopy, where often found at flowering trees, but ranges to all levels. Gleans insects and probes flowers for nectar. No nesting information. **Range:** Endemic. NG: N slope of W Ranges (2200–2800 m), Huon, and SE Pen (1500–2100 m, extremes 1300–2300 m), plus a few records from E Ranges (Mt Michael at 2400 m, Tauna Gap, and Tari Gap).

Rufous-sided Honeyeater *Ptiloprora erythropleura* Pl 70

16–17 cm. An uncommon western *Ptiloprora* of mid-mountain and cloud forests.
All plumages: Note *russet flanks, all-grey back, and reddish iris.* (Reports of birds from W Ranges and Bird's Neck with green iris and yellowish wash below are possibly Juv or may represent local variation.) **Male** larger. **Subspp (2):** Minor. **Similar spp:** Grey-streaked H is usually more common and generally occurs above 1500 m where the 2 species co-occur; it is larger, with bold streaking on flanks, and green iris. **Voice:** Call, a single, plaintive, high-pitched slur similar to that of Grey-streaked and Rufous-backed Hs. **Habits:** Often seen in pairs or small groups. Forages at all levels of the forest, gleaning insects and probing flowers for nectar; rarely takes fruit. No nesting information.
Range: Endemic. NG: Bird's Head and Neck (Fakfak and Kumawa Mts) and W Ranges, 1300–2800 m.

Taxonomy: Birds in the Fakfak and Kumawa Mts have not been formally described and are thought to be a race or races of this species or possibly a new species.

Grey-streaked Honeyeater *Ptiloprora perstriata* Pl 70
(**Black-backed** or **Black-backed Streaked Honeyeater**)
19–20 cm. Common in cloud forest and subalpine scrub. In the E Ranges and SE Pen, this species is confined to the highest elevations and related Rufous-backed H occurs below. Active, conspicuous, and familiar. Cocks tail. **Adult:** *A big* Ptiloprora *with blackish back and upperparts* (no russet on the back) and bold streaking on underparts. Iris green. **Juv:** Duller, with greenish wash to underparts. Iris brown. **Male** larger. **Similar spp:** Rufous-backed H is very similar but for back color and smaller size. **Voice:** A varied repertoire of high, plaintive slurs, similar to vocalizations of Rufous-backed H: *deeyur whit, wee dyu,* or *peeyu seeyu,* repeated every few sec. **Habits:** Solitary or in pairs. Forages at all levels in the forest, but typically below the canopy and at forest edge. Seeks flowering shrubs and vines, also gleans insects and fruit. This and the next 2 species are flower trap-liners, attuned to searching out the small patches of brightly colored flowers of epiphytes, vines, shrubs, and understory plants—rare points of light in the cloud forest gloom. These include the red, pink, or yellow flowers of various rhododendrons and others in the heath family, gingers, and innumerable species of small orchids. Nest suspended from a forked branch in a forest sapling; a bulky cup of grass, leaves, and twigs, live moss outside and fine plant fibers inside. Egg (1) pale pink with dark spots and blotches. **Range:** Endemic. NG: Central Ranges and Wandammen Mts, 1700–3750 m (in E Ranges and high mts of SE Pen, only above 2400 m or missing locally).

Mayr's Streaked Honeyeater *Ptiloprora mayri* Pl 70
(**Mayr's Honeyeater**)
19–20 cm. A large *Ptiloprora* confined to cloud forests of the N coastal ranges, where it is the only streaked honeyeater. **Adult:** *Larger and duller* than Rufous-backed H. **Subspp (2):** Minor. **Voice:** Plaintive whistles not distinguishable from those of Rufous-backed H. Song a high plaintive slur repeated once per 3 sec. Call: *cheep.* **Habits:** Similar to Rufous-backed. No nesting information. **Range:** Endemic. NG: Foja, Cyclops, and N Coastal Mts, 1200–2150 m. **Taxonomy:** Formerly classified as a subspecies of the Rufous-backed H, from which it is thinly differentiated.

Rufous-backed Honeyeater *Ptiloprora guisei* Pl 70
(**Guise's** or **Rufous-backed Streaked Honeyeater**)
16–18 cm. The abundant streaked honeyeater of E mountains. Inhabits mid-mountain and cloud forest, mostly below the elevational range of Grey-streaked H where the geographic ranges of the two overlap.*Rufous mantle diagnostic.* Tail often cocked. **Adult:** Clear, grey streaking below. Iris green. **Male** larger. **Juv:** Duller, has greenish-yellow wash on breast. Iris dull green. **Similar spp:** Grey-streaked and Rufous-sided Hs have a grey mantle. Mayr's Streaked H does not co-occur, has similar plumage but is larger and duller. **Voice:** Calls are plaintive, 1-, 2-, or 3-noted, slurred whistles that vary regionally; these may be repeated monotonously. Also a staccato *chip!* and a short buzzy note. **Habits:** Similar to Grey-streaked H. An unwary, curious, active forager. Comes readily to "spishing." Loud wingbeat. Nest not well described but similar to Grey-streaked as far as known. Eggs, no information. **Range:** Endemic. NG: E Ranges (east of Strickland Gorge), Huon, and SE Pen, 1750–2400 m (in absence of Grey-streaked H, as on Huon, extends up to the tree line at ~3000 m).

Plain Honeyeater *Pycnopygius ixoides* Pl 71
18 cm. Inhabits forest and edge from lowlands to lower mid-mountains. Inconspicuous, but may be locally common. **All plumages:** *A uniquely nondescript, drab brown honeyeater.* Face without any of the usual field marks such as ear-spot, rictal streak, or wattle. At most, crown slightly greyer and faintly scaled; faint, pale rictal streak; and an obscure pale streak extending from behind the dark brown eye—all these nearly impossible to see in the field. *Note small head and bill and rather upright posture.* **Male** larger. **Subspp (4, in 2 groups):** *ixoides* (most of range, except the following) brown; *finschi* (SE Pen, east of Kumusi R in N and Port Moresby in S) variably rufous brown. **Similar spp:** The highly variable Tawny-breasted H is bulkier, longer-billed, and usually shows an ear-spot and other facial pattern; beware the dark, plain form occupying NW Lowlands to Sepik-Ramu. Meyer's Friarbird is larger, longer-billed, longer-tailed, with a blackish face. Streak-headed H juv also entirely plain brown except for a whitish cheek-stripe. Tawny Straightbill has a sharper, brown bill (not black); red iris; and

pinkish brown legs (not bluish grey). **Voice:** Typically quiet. Song a soft *pu-dee, pu-dee, pu-dee, pu-dee*. Call note *petert* or *peetrt*, similar to that of Streak-headed and Marbled Hs. **Habits:** Solitary, in pairs, or small gathering in a food tree. Forages mainly in the canopy, visiting flowering and fruiting trees. Feeds on small fruits, nectar, and probably insects. No nesting information. **Range:** Endemic. NG (apparently absent from Trans-Fly), 0–1150 m (rarely to 1400 m).

Streak-headed Honeyeater *Pycnopygius stictocephalus* Pl 71
20–22 cm. A familiar voice of lowland forest edge, gardens, regrowth, and riverine cuts through tall forest. Uncommon and keeps to the treetops. **Adult:** *A dull-brown honeyeater suggesting a friarbird, but smaller and with a small head and a much smaller bill.* Look for the diagnostic *white cheek-stripe and whitish-speckled crown.* **Male** larger. **Juv:** Plain brown—the only clue is the indistinct white cheek-stripe. **Similar spp:** Streak-headed H is a visual mimic of the much larger Helmeted Friarbird. Helmeted FB is considerably larger, longer-billed, with no white streak on face. Meyer's FB similar at a distance, but lacks head markings and has a longer bill. Tawny-breasted and Plain Hs lack crown markings. Brown Oriole has longer, thicker bill (red in adults) and streaked underparts. **Voice:** Sings loudly and often from a treetop perch; also has a song-flight. Song distinctive, musical, medium pitched, and liquid. This catchy tune often sounds like *wheeta wheeta wheer* or *whita-tee, whita-teer* repeated every 2 sec; there are many variations, but the notes and basic theme are easily recognized. Suggestive of songs of Brown Oriole or Meyer's Friarbird. **Habits:** Usually solitary, but joins its own kind and other species in fruiting and flowering trees. Feeds on insects, fruit, nectar. Nest hung from a branch fork in the foliage of a tree crown; nest is a woven basket of fibers or stems. Egg undescribed. **Range:** Endemic. NG, Salawati (NW Is), and Aru Is, 0–300 m, locally to 1000 m.

Marbled Honeyeater *Pycnopygius cinereus* Pl 71
22–23 cm. Mid-mountains, where it is an uncommon and unobtrusive inhabitant of oak forests and adjacent second growth. A medium-sized, grey honeyeater with white-scaled underparts. Seemingly nondescript, but easily recognized by its *short bill, unique white crescent below the eye, and white cheek-stripe.* **Adult:** *Greenish edges to primaries and tail feathers*; iris pale grey (brown in W Ranges). **Male** larger. **Juv:** Wings tinged brown; iris color unknown. **Subspp (2):** *cinereus* (Bird's Head to Weyland Mts of W Ranges) faint scaling below; *marmoratus* (most of range, exclusive of previous) pronounced scaling on breast. **Voice:** *Tyert!* repeated every 2 sec. **Similar spp:** Leaden H smaller, lacks facial markings and white-scaled underparts, cocks tail. **Habits:** Mostly solitary. An inconspicuous canopy dweller, usually seen foliage-gleaning for insects, or coming to fruiting or flowering trees. Nest undescribed. Egg (clutch?) is pale pink with spots. **Range:** Endemic. NG: Bird's Head, Central Ranges, Adelbert Mts, and Huon, 1000–2000 m (sometimes as low as 500 m).

Olive Honeyeater *Lichmera argentauris* Pl 72
14–15 cm. A Moluccan species of westernmost NW Is. Coastal forest and gardens. A greenish relative of the more familiar Brown H. **Adult:** Small, pale, olive green, meliphaga-like. Note diagnostic *white-and-yellow ear-spot.* **Male** larger on average. **Juv:** Undescribed. **Similar spp:** Within its range, no other honeyeater has the combination of pale olive dorsal plumage, dark iris, and obscure, two-tone cheek patch. Green-backed H lacks ear-patch; iris color differs. **Voice:** Song a squeaky series of 4–5 notes, bright and upbeat. Also a chattering. Call a harsh *zhip*. **Habits:** Singly, pairs, or small parties. Feeds on insects, flowers, and nectar. Nest and egg, no information. **Range:** NW Is: Gebe, Misool, Schildpad Is, and Waigeo, at sea level. Also Moluccas.

Brown Honeyeater *Lichmera indistincta* Pl 72
13–14 cm. Savannah and other open, wooded, lowland habitats near water, including mangroves. Locally common and conspicuous. *A small, olive-brown honeyeater with loud, cheery song.* The only markings are *a small yellow spot behind eye, yellow gape (becoming black in breeding male), and bright yellow-green edges of wings and tail.* **Male:** Larger, with greyish crown contrasting with grey-brown back. **Female:** Smaller, with crown same grey-brown color as back. **Juv:** Similar to Female but with a yellowish tinge to the head; eye-spot absent. **Subspp (2):** *ocularis* (Trans-Fly and E AU); *nupta* (Aru Is) darker, lacks yellow eye-spot, and has larger area of bare skin around the eye and yellowish wash to chin. **Similar spp:** Dusky Myzomela darker and more nondescript. **Voice:** Song of both sexes is surprisingly loud, liquid, but choppy and reminiscent of a reed-warbler; a note or pair of notes is repeated before switching to another, e.g., *sweet-sweet-quarty-quarty* (AU data). Calls include

ke-ke and a loud *plik*. **Habits:** Singly or in pairs, but more may gather in a flowering tree. Forages in all levels of the vegetation, particularly at flowers, but also gleans insects and fruit. Nests in dense foliage at various heights; a small cup, slung from a forked branch, of woven grass and bark, bound with spiderwebs, often lined with plant down (AU data). Eggs (2) white with spots. **Range:** NG (locally in Trans-Fly) and Aru Is, sea level. Also AU.

Silver-eared Honeyeater *Lichmera alboauricularis* Pl 72

15–16 cm. Patchy distribution. Open coastal and riverine habitats: cane grass with shrubs and trees along river and lake edges, coconut plantations, settlements, and scrub. Gregarious and noisy, but often hard to see, as the bird hides within vegetation and moves rapidly. **Adult:** Looks like a small, grey-brown meliphaga with *silvery ear-spot and dark-spotted pale breast*. **Male** larger on average. **Juv:** Obscure ventral spotting and ear-spot. **Subspp (2):** Minor (N birds darker).
Similar spp: Meliphagas larger and lack prominent dark spotting below. **Voice:** Call (or song?) is a loud, harsh, descending, accelerating series of upslurred notes: *shree shree shree . . .* rapidly delivered, usually from the cover of vegetation. Often several birds call together. Dawn call, a repeated *jit*.
Habits: Singly, in pairs, or in groups. Feeds on insects and nectar. The nest is built low in a tree or shrub, slung from a forked branch in the foliage; a woven cup of dead leaves, fibers, small pieces of bark, and dried grass, bound with spider-webs and lined with plant down. Eggs (2) pinkish white with some speckling. **Range:** Endemic. NG: NW Lowlands, Sepik-Ramu, and SE Pen (Popondetta and Port Moresby area east to Milne Bay), 0–100 m.

Tawny-breasted Honeyeater *Xanthotis flaviventer* Pl 72
(*Xanthotis chrysotis, Meliphaga flaviventer* or *chrysotis*)
18–21 cm. Common and widespread. Inhabits forests of all sorts and second growth from lowlands to mid-mountains. Forages in midstory and canopy, keeping within the cover of vegetation. Much geographic variation. Distinctive shape—a *medium-sized, robust honeyeater with medium-long and sharply pointed bill*. Generally drab, but with *facial pattern showing bare eye and small, flamelike ear-stripe*. **All plumages:** Some shade of brown, with a grey throat, pale bare skin under and behind the eye, and a small yellow (or orange) ear-stripe. Male larger on average.
Subspp (7): In 4 quite different groups.
Green group: *fusciventris* (Batanta and Waigeo Is)—with greenish cast.
Race *flaviventer* (Bird's Head; Misool and Salawati Is) is intermediate with next.
Tawny group: *visi, saturatior, madaraszi* (Bird's Neck, Aru Is, S Lowlands, Huon, SE Pen)— tawny brown, grey throat, pale bare eye with dark border, and yellow ear-stripe.
Dusky group: *meyeri* (Yapen, NW Lowlands to Sepik-Ramu)—all dark grey-brown, indistinctly spotted below, dark eye-skin, yellow or flame-orange ear-stripe.
Spotted group: *spilogaster* (SE Is)—grey-brown, distinctly pale-spotted below, white spectacles, yellow ear-stripe.
Similar spp: Plain H is smaller and slimmer, with shorter bill and no ear-stripe or other facial features. Meyer's Friarbird especially resembles dusky group of Tawny-breasted but is larger with a longer bill. Long-billed H has a much longer bill and lacks ear-stripe. Long-billed Cuckoo had straighter bill and barring on underside of tail. **Voice:** Song is a cheerful, whistled, 3-note phrase, *wheet whoot whit*, repeated 3–10 times; delivery is ~1 phrase/sec in the S and twice as fast in the N. Call is a weak *whuk* or *tup* like that of a meliphaga. **Habits:** Usually alone, less often in pairs or groups. Joins mixed flocks of babblers and pitohuis, gleaning in vines. Forages for fruit, nectar, and insects. Gleans insects from foliage and inspects dead leaves. Nests in the forest midstory; a hanging cup of leaves and strips of plant matter. Eggs (2) pinkish white, sparsely spotted. **Range:** NG, NW Is (Batanta, Misool, Salawati, Waigeo), Aru Is, Bay Is (Yapen), and SE Is (3 main D'Entrecasteaux Is and Trobriand Is), 0–1500 m. Also Cape York, AU. **Taxonomy:** Likely to represent more than 1 species, perhaps split along the lines of the 4 groups described above, but species limits unclear at present. The AU race *filiger* may belong to one of these groups or its own group.

Spotted Honeyeater *Xanthotis polygrammus* Pl 73
(*Meliphaga polygramma*)
16–17 cm. Hill and lower montane forests; also monsoon forest and well-wooded savannah in Trans-Fly. Uncommon. Most often seen in flowering trees. Forages in canopy and midstory.
Adult: *Resembles a meliphaga, but paler below and with profuse, black, triangular spotting*; bright

greenish above; face marked with a *big, dark eye surrounded by blush pink or yellow bare skin* and yellow ear-streak (most races). **Male** larger. **Juv:** Similar, but buff tips on upperwing-coverts, breast spots more diffuse, ear-streak reduced or absent. **Subspp (5):** Minor; yellow ear-streak is absent on Waigeo I and NW Lowlands to Sepik-Ramu. **Voice:** Song (?) is a musical, pleasant, but indistinct and upslurred *wudee wudee wudee. . . .* A repeated trisyllable is given by a group. Calls include a single *tyert!* **Habits:** Solitary, pairs, rarely small groups; joins other birds at flowering trees and vines. Inconspicuous, though sometimes noisy. Mainly a nectar feeder, but also gleans foliage for invertebrates, paying special attention to curled dead leaves. No nesting information. **Range:** Endemic. NG and NW Is (Misol, Salawati, Waigeo, plus a sighting from Batanta), 500–1500 m; lowlands locally at foot of the ranges and Trans-Fly.

Meyer's Friarbird *Philemon meyeri* Pl 71

21–22 cm. Lowland and hill forests, including openings and second growth. Uncommon and usually inconspicuous. Mainly in the canopy; best seen when it ascends an exposed perch to sing. *A plain, dark, sooty-brown friarbird, much smaller than the more familiar Helmeted F.* Typical friarbird shape and behavior. Heard more often than seen, and best identified by song. When seen well, note the long, pointed, slightly curved bill. **Adult:** Plain dark brown. **Juv:** Similar, but upperparts scaled whitish, underparts tinged and scaled with yellow. **Imm:** Like Adult but for bronzy yellow wash across upper breast. **Similar spp:** Tawny-breasted H, particularly the dusky northern form, is smaller with a shorter bill and may show a yellow or orange ear-streak. Plain and Streak-headed Hs have much shorter bills. Helmeted F much larger, noisier. Long-billed H slimmer, with much longer bill, eye orange or with pale ring. **Voice:** Song is a loud, brisk alternation between whistled notes repeated over and over, ~5 notes/sec. Call is a single, loud, clear, explosive *tyew*. **Habits:** Alone or in a pair. Less sociable than other friarbirds. Pugnacious at fruiting trees, but often inconspicuous, especially when foraging for arthropods in tree foliage. Diet also includes nectar and fruit. The nest is undescribed. Eggs orange-red with some darker smears and clouding. **Range:** Endemic. NG except for Bird's Head and Trans-Fly, 0–500 m (rarely to 1200 m).

Brass's Friarbird *Philemon brassi* Pl 71

21–22 cm. Known from a few localities in the NW Lowlands. Found along rivers and associated lagoons, in dense second-growth trees and cane grass. Locally common. A small, slender friarbird, similar to Meyer's F, but much *paler below and with a slightly shorter bill*; at close range, *whitish chin contrasts with black face and dark grey neck*. **Adult:** Wings and back plain greyish brown. **Juv:** Similar but with rufous tinge on head, wings, and back; olive edges to flight and tail feathers; more brownish breast. **Imm:** Similar to Adult but with white-tipped feathers on the back and scapulars and retained Juv flight and tail feathers. **Similar spp:** Meyer's F much darker overall and with plain face. **Voice:** Little information. Upslurred call. **Habits:** Singly, in pairs, or small groups. Gathers in flowering and fruiting trees. Diet includes nectar, fruit, and insects. Nest and egg unknown. **Range:** Endemic. NG: NW Lowlands (Mamberamo, Taritatu, and Wapoga Rs), 0–300 m.

Little Friarbird *Philemon citreogularis* Pl 71

24–25 cm. Savannah and monsoon forest of Trans-Fly. Common, sometimes abundant. *Smallest of the 3 savannah friarbirds, distinctive for lacking a knob at the base of the bill.* Note the *white upper throat* and nape and *blue-grey skin on the face*. **Adult:** Plain grey-brown, lacking pale dorsal scalloping or yellowish markings. **Male** larger on average. **Juv:** Similar to Adult but with yellowish throat and greenish edges to flight-feathers. **Imm:** Similar to Juv but with white-tipped feathers on the back and scapulars, and sides of breast with yellow scalloping. **Subspp (1):** *papuensis* (endemic) smaller and darker than AU races. **Similar spp:** Helmeted and Noisy Fs are larger and show a knob at the base of the bill and have a blackish face. **Voice:** Song is a series of oft-repeated, mellow double-whistles, *tiew-tiew*. A chatter call known from AU. **Habits:** Often gregarious. Pugnacious toward smaller birds. Forages in the treetops, mainly probing flowers for nectar and gleaning for invertebrates; sometimes sallies after flying insects. Some fruit taken. Nest suspended from a forked branch in tree foliage, often not far from a nest of another bird species; nest is a large, deep cup of loosely woven bark strips, grass, and plant stems bound with spiderweb (AU data). Eggs (2–3) pale pinkish to orange-red, spotted and streaked. **Range:** NG: Trans-Fly (Kurik to Oriomo R), sea level. Also AU.

Helmeted Friarbird *Philemon buceroides* Pl 71
(**New Guinea Friarbird, Leatherhead,** *Philemon novaeguineae*)
33–35 cm. One of the most conspicuous birds of town, garden, and forest edge. Most common in disturbed habitats. Has followed settlement from the lowlands into the mid-mountain valleys. Its natural habitat includes lowland forest edge, second growth, mangroves, monsoon forest, and savannah. Mainly in the treetops. *A large, slender-necked, drab bird with pugnacious habits and a raucous call. Face shows bare black skin, with or without a knob at base of the large bill.*
Adult: Grey-brown without pale scalloping; iris red-brown. **Male** larger. **Juv:** Resembles Adult, but bill shorter, knob rudimentary, and iris grey or dull brown. **Imm:** Similar to Adult but bill-knob smaller and feathers of upperparts with pale edges. **Subspp (5, in 2 groups):** *jobiensis* (NW Lowlands and Yapen I to Huon) bill-knob small or missing; *novaeguineae* (remainder of range) prominent bill-knob, though SE Is birds intermediate. **Similar spp:** Birds of similar size, plumage, and dark face: Fawn-breasted and Yellow-breasted Bowerbirds, Brown Oriole juv, and Australasian Figbird female, but none of these have the large bill and thin neck of a friarbird. Little and Noisy Fs are smaller and restricted to Trans-Fly, where Helmeted also occurs. Meyer's F co-occurs broadly with Helmeted but is obviously smaller and darker. **Voice:** A prominent element of the dawn chorus in most NG lowland habitats. Song is a varied series of repetitive, loud, harshly musical slurred notes repeated over and over; duets and group calling are common. Calls composed of loud single or double notes, with much geographic variation. **Habits:** Conspicuous and gregarious, often in pairs or groups. Often perches atop dead branches calling loudly and incessantly. Engages in aggressive chases with conspecifics, as well as with other birds. Diet is fruit, nectar, insects. Nests in the crown of a leafy tree, the nest suspended from a forked branch. May nest close to another species (or vice versa), such as Spangled Drongo or Hooded Butcherbird; neighbor species in AU are frequently drongos and figbirds, but also a range of pigeons and large to midsize passerines, including another Helmeted F. The purpose may be mutual defense from predators. The nest is a deep cup of loosely woven plant stems and bark-strips, lined with fine semi-woody stems. Eggs (2–3), pale pink with brown spots. **Range:** NG, NW Is (Batanta, Kofiau, Misool, Salawati, Waigeo), Bay Is (Yapen), Aru Is, and SE Is (D'Entrecasteaux Is, Trobriand Is, and Tagula I), 0–1000 m (rarely to 1500 m in disturbed habitats). Also Lesser Sundas and AU.

Noisy Friarbird *Philemon corniculatus* Pl 71
30 cm. A common friarbird of the Trans-Fly savannahs, and a familiar Australian species. The only friarbird with a *bald, black head*. Note bill-knob and *silvery-feathered chin, neck ruff, and breast*. A good field mark when the front end can't be seen is the *obvious white tail tip*.
Adult: Head entirely bald, bill-knob prominent, upperparts plain grey-brown, lanceolate breast feathers, iris reddish. **Male** larger. **Juv:** Presence of feathering above the eye and on nape, bill-knob rudimentary, breast feathers not lanceolate, dark iris. **Imm:** Similar to Adult but with white-tipped feathers on back and scapulars, yellowish bib, bill-knob smaller. **Subspp (1):** *corniculatus* (also NE AU). **Similar spp:** Helmeted F has a feathered head with only the face bare, and the tail lacks a pale tip (may have a narrow, obscurely pale tip at most). **Voice:** Song (AU data) a repeated 3-note phrase, second note at higher pitch: *cockalock!* Vocalizations in NG reported as varied and including a single, high, nasal note like a weak version of the call of the Grey Crow; also an antiphonal duet. **Habits:** Similar to Helmeted F, differing mainly in habitat preference. Nests in outer branches in tree crown; nest suspended from a branch fork, a large, sturdy cup or basket of bark and dry grasses, bound with spiderwebs (AU data). Eggs (2–4) orange-pink and spotted. In AU, other songbirds often nest nearby, presumably for protection. **Range:** NG: Trans-Fly, sea level. Also AU.

Blue-faced Honeyeater *Entomyzon cyanotis* Pl 72
30 cm. Trans-Fly savannah only, where common. *Unmistakable—large and conspicuous with a green back, black cap, white underparts, and wings flashing white while flying.* **Adult:** *Bright blue face patch.* **Male** larger. **Juv:** Shows Adult pattern, but head grey, upperparts brown, and face patch yellowish. **Imm:** Similar to Adult but face patch yellowish green. **Subspp (1):** *griseigularis* (also Cape York, AU). **Similar spp:** White-throated H shares the same plumage pattern but is very much smaller, lacks black bib. **Voice:** Piping call is a far-carrying, repeated, querulous *woik . . . woik . . . woik . . .* or *queet . . .* etc. (AU data). **Habits:** In small parties. Active and noisy. Perhaps locally nomadic. Feeds on insects, nectar, and fruit. Commonly visits coconut palms. Probes under bark in search of insects. A cooperative breeder at least occasionally (AU data). Nests typically in old nest of another species, particularly Grey-crowned Babbler, but also friarbird or magpie; or may build its own nest, a cup of bark strips and

grass wedged into a branch fork in tree crown (AU data). Eggs (2) buff-pink with bold, dark blotches. **Range:** NG: Trans-Fly, sea level. Also AU.

White-throated Honeyeater *Melithreptus albogularis* — Pl 72
12–13 cm. Associated with eucalyptus and paperbark trees in Trans-Fly and SE Pen. Common. A noisy, gregarious species, typically heard before seen. *A small honeyeater with black cap, green back, and all-white underparts.* Note white nape patch and blue skin above eye. **Adult:** Black cap, all-green upperparts. **Juv:** Cap brownish black; mantle, back, wing-coverts tan-brown. **Male** larger. **Subspp (1):** *albogularis* (also AU). **Similar spp:** Blue-faced H much larger and with black bib. **Voice:** Song a rather loud, high, musical series, *see see see see see see. . . .* Also, *tserp tserp.* **Habits:** Lives in pairs and small groups. Mainly in the treetops but will descend to forage lower. Active and acrobatic when feeding. Takes nectar from flowers of eucalypts, paperbarks, mistletoes, and banksias. Gleans insects from foliage. Some evidence for cooperative breeding (AU data). Nests in foliage, usually in eucalypt or paperbark crown; a suspended cup nest of bark strips bound with spiderwebs. Eggs (2–3) buff with dark blotches (nest and eggs, AU data). **Range:** NG: Trans-Fly and SE Pen (Port Moresby area, Musa Valley and Oro Bay, and upper Markham Valley), 0–800 m. Also N AU.

Giant Wattled Honeyeater *Macgregoria pulchra* — Pl 73
(**Macgregor's Bird of Paradise** or **Honeyeater**)
38–40 cm. Highest mountains of the Central Ranges, though patchily distributed. Inhabits mossy podocarp forests at timberline. Conspicuous, unwary, and vulnerable to hunting, hence local and typically uncommon. *A strange, black, crow-sized, arboreal bird of fowl-like shape.* Immediately recognized by its *floppy, egg-yolk-colored facial wattles and bright ochre wing patches prominent in flight.* The flapping wings make a *loud rustling noise*; wings when held in a glide produce a *ripping sound: zzzinng*—these sounds are the best means of detection. Takes a horizontal stance with tail cocked and wings drooped, nervously flicking. Runs and bounds through the trees. **Adult:** All black; iris red or red-brown. **Male:** Larger, with bigger wattle. **Juv:** Duller, slightly brownish; iris brown. **Subspp (2):** *pulchra* (SE Pen); *carolinae* (W and Border Ranges) considerably larger, but shorter-tailed. **Similar spp:** Smoky Honeyeaters have yellow face patch but are much smaller. **Voice:** Calls incessantly when active. Contact note a rapidly repeated *jeet jeet jeet . . .* , pathetically weak for so grand a bird; also a repeated *peer* and other small notes. **Habits:** In pairs or small groups. Probably mostly sedentary, but periodic absences suggest local movement. Diet is predominantly fruit, especially those of the podocarp *Dacrycarpus compactus.* Also observed probing moss cushions with its bill, presumably for invertebrates. This is the behaviorally dominant and largest fruit eater in subalpine forests. Establishes an all-purpose territory when nesting, intolerant of other species in its food trees, drives them away with noisy chasing. Monogamous. Nest built in tree crown; a large, bulky cup of moss, plant stems, lichens, and sticks, lined with slender stems and small leaves. Egg (1) pink and spotted. **Range:** Endemic. NG: W Ranges, Border Ranges, and SE Pen, 3200–3500 m (rarely 2700–4000 m). **Taxonomy:** Long thought to be a bird of paradise before DNA studies revealed its relationship with the smoky honeyeaters.

Western Smoky Honeyeater *Melipotes gymnops* — Pl 73
(**Arfak Honeyeater** or **Melipotes**)
21–22 cm. The smoky honeyeater of the Bird's Head and Wandammen Mts. Abundant in montane forest and edge, including village gardens. **All plumages:** Distinguished from other smoky honeyeaters by *black chin and white streaks on the black breast and belly*; also different are the white-spotted wing-coverts and ochre lower belly and undertail. **Male** larger on average. **Similar spp:** Within its restricted range, only Cinnamon-browed and Vogelkop Melidectes are similar, but they are longer-billed. **Voice:** More vocal than other *Melipotes* species: *wee wee wee . . .* given while perched or in flight. **Habits:** Solitary, although gathers at fruiting trees, together with other species. Feeds on fruit at any height, but mainly in canopy. No nesting information. **Range:** Endemic. NG: Bird's Head and Bird's Neck (Wandammen Mts), 1200–2700 m. **Taxonomy:** A *Melipotes* population in the Fakfak Mts has not been described.

Common Smoky Honeyeater *Melipotes fumigatus*   Pl 73
(Smoky Honeyeater, Common Melipotes)
21–22 cm. This is the widespread smoky honeyeater—a ubiquitous and common montane forest bird. *Smoky honeyeaters are characterized by sooty plumage, yellow facial skin and wattles (can flush red), and short, straight bills.* They eat mainly fruit and are not as active as other honeyeaters, though just as scrappy. Each species has a range separate from the others. **All plumages:** Only markings are the grey chin and grey-edged breast feathers. **Male** larger on average. **Subspp (3, in 2 groups):** *fumigatus* (all range except that of next race) appearance typical for the species, but varies geographically in size and darkness; *kumawa* (Kumawa Mts) white-scaled lower breast and belly, white-spotted wing-coverts, and black chin and throat. **Similar spp:** Can be momentarily mistaken for a melidectes. High-elevation Giant Wattled H is huge and bulky in comparison. First-time birders in the upland forest may mistake this species for a bird of paradise. **Voice:** Usually silent. Song is a sibilant, weak *swit swit swit . . .* repeated monotonously. Has several small, weak calls. **Habits:** Solitary, in pairs, or small, quarrelsome aggregations at food trees. Feeds on small fruit in middle and upper stories of the forest. Gleans foliage and flowers for insects; also sallies. Nest suspended from a forked branch of a small tree inside the forest; nest is a bulky cup mostly of green moss bound with black fungal fiber. Egg (only 1?) pinkish with spots. **Range:** Endemic. NG: Central Ranges, Bird's Neck (Kumawa Mts), Cyclops Mts, and N Coastal Mts, 1000–4200 m. **Taxonomy:** Subsp *kumawa* might be better assigned to Western Smoky H or separated as its own species.

Wattled Smoky Honeyeater *Melipotes carolae*   Pl 73
21–22 cm. Restricted to the Foja Mts. Common. **All plumages:** Closest in appearance to Common Smoky H and separated by its *darker chin and underparts, and pendent wattle below the eye.* **Male** larger on average. **Similar spp:** Within its range, only Cinnamon-browed M is similar (but longer-billed). **Voice:** Usually silent. **Habits:** As for Common Smoky H. The cup nest is a bundle of debris 20 cm in diameter and woven around the trunk of a sapling, 9 m up. Eggs, no information. **Range:** Endemic. NG: Foja Mts, 1050–2200 m.

Spangled Honeyeater *Melipotes ater*   Pl 73
31–32 cm. Common in montane forests of the Huon Pen. Much larger than other smoky honeyeaters, being the size of a bird of paradise, which it further resembles by its long tail. **All plumages:** Black, with white spangles down the front. **Male** larger. **Similar spp:** In Huon mountains, no other large black bird has the yellow face. Wahnes's Parotia has a wedge-shaped tail and loud calls. Huon Astrapia female is longer-tailed and black-faced. Huon Melidectes has a longer bill and facial striping. **Voice:** More vocal than other smoky honeyeaters. Calls are a soft *hoo-ee*. Scold is *sht sht sht chh chh*. **Habits:** Similar to other smoky honeyeaters, but more social and lives mainly in the canopy. Sometimes travels through canopy in considerable numbers. No nesting information. **Range:** Endemic. NG: Huon Pen, 1200–3300 m.

Long-billed Honeyeater *Melilestes megarhynchus*   Pl 71
22–23 cm. A furtive denizen of forest interior and thickets from lowlands to mid-mountains. Frequents lower and midstories; common. Seen flying through forest interior and across garden clearings. Stops to probe bark and vines, sometimes moving creeper-fashion on a trunk. *A midsize, olive-brown honeyeater with a long, curved bill.* Much individual variation in the amount of streaking. **Adult:** Plain, dark underparts; *orange eye*. **Male** larger. **Juv:** Paler, streaky breast (variable); dark brown eye with pale, yellow-feathered eye-ring. **Subspp(2):** Minor. **Similar spp:** Meyer's Friarbird has shorter, stouter bill and is duller grey-brown. Tawny-breasted H has shorter bill and an ear-stripe. **Voice:** Three quite different calls. One, a short sneeze. Also a single, short, harsh downslur reminiscent of call note of Magnificent Bird of Paradise: *chur-r-r*. In flight, a distinctive, small, sucked-in, staccato note: *chhkk!* **Habits:** Solitary, sometimes seen in pairs; occasionally joins mixed-species flocks of insectivores. Sedentary and perhaps territorial. Forages mostly by deep probing, taking arthropods and lizards from crevices in bark, leaf bases of climbing pandans, and other hiding places. Searches out large tubular flowers of vines, gingers, bananas, etc., and blooming trees with smaller flowers. Takes some fruit. Nest, unusual for a honeyeater, is not suspended but built on a shelf, for instance at the base of pandan leaves, or on an accumulation of dead leaves, or in a crevice on the underside of a thatched roof. Nest is a thick-walled cup of green moss, lined with thin stems. Eggs (2) white with fine spotting. **Range:** Endemic. NG, NW Is (Batanta, Misool, Salawati, Waigeo), Bay Is (Yapen), and Aru Is, 0–1500 m (locally to 2100 m).

Olive Straightbill *Timeliopsis fulvigula* Pl 70

13–14 cm. Interior of mid-mountain forest and lower cloud forest, frequenting midstory and dense undergrowth. While uncommon and inconspicuous, it attracts attention with its social foraging and incessant calling. A moderately small, *buff-brown, warbler-like bird with bright green wing edgings; a sharp, straight bill; and an orange eye*. Posture wrenlike, with head held horizontal and tail slightly raised. Utters scolding call while foraging. **Adult:** Iris orange. **Juv:** Similar to Adult but greyer below; iris grey or brown. **Subspp (2):** Minor. **Similar spp:** Scrubwrens also forage socially, gleaning and inspecting dead leaves, but they are smaller and lack the sharp bill and green wing edges (except Grey-green SW). **Voice:** Call an oft-repeated *sreed-sreed-sreed* . . . sounding like the begging of a baby bird. **Habits:** In small parties of 2–4 birds foraging together, leaf-gleaning and often probing dry, rolled leaves for insects and spiders. Occasionally joins insectivorous mixed-species feeding parties. Exclusively insectivorous; does not visit flowering trees. Nest is an open cup of fine roots and fibers, externally covered with dry grass blades and lined with feathers. Egg (1) white, flecked with pale red. **Range:** Endemic. NG: Bird's Head, Bird's Neck (Fakfak), Central Ranges, Foja Mts, N Coastal Mts, and Huon, 1400–2200 m (extremes 750–2800 m).

Tawny Straightbill *Timeliopsis griseigula* Pl 70

18 cm. A rare bird of lowland forest (sparingly in lower hills). Seemingly with patchy distribution. Met in mixed-species flocks of other rusty brown birds, such as Rusty Pitohui, Papuan Babbler, and others. Not obviously a honeyeater. A rather nondescript songbird with *rusty plumage; short, straight, pointed bill; bright iris (tan or reddish); and pinkish grey legs*. Behavior distinctive: often hangs upside down to probe leaves; quick moving, constantly on the go; flares tail. **Adult:** Iris usually reddish. **Male** larger. **Juv:** Plumage with a yellow wash, especially on the head; iris grey or dark brown. **Subspp (2):** *griseigula* (Bird's Head to NW Lowlands, also Gogol R in Sepik-Ramu) olive brown above; *fulviventris* (SE Pen) more rufous brown above, paler below. **Similar spp:** Easily confused with the array of rusty-plumaged, flock-forming birds. Little Shrikethrush has heavier bill with a hooked tip; dark eye; and brownish-grey legs. Tawny-breasted H is slightly larger and has eye-patch, but is behaviorally similar. The monarchs are more active, flying more often, and with bluish bills. Other species are significantly larger. **Voice:** Song (call?) a rapid descending series of upslurred whistles, *swee-swee-swee-swee-swee-swee-swee*, speeding up toward the end and fading away; quality somewhat similar to that of Golden Cuckooshrike. Also mimics call of Papuan Babbler. **Habits:** Alone or in small family parties of up to 4 birds. Forages at all levels in the manner of a Papuan Babbler. In constant motion while searching vines and leaves, gleaning, probing, and often hanging upside down. Feeds mostly on insects, also fruit and flowers. No nesting information. **Range:** Endemic. NG: Bird's Head and Neck, NW Lowlands, Sepik-Ramu (records only from the lower Gogol R and the lower Watut R), S Lowland (Kiunga sightings), and SE Pen (east from Kumusi R in the North and Lakekamu R in the South), 0–800 m.

Rufous-banded Honeyeater *Conopophila albogularis* Pl 72

13 cm. In open habitats near coast and rivers: towns, roadsides, denser savannah woodland, reed beds, and mangroves. Locally distributed, but common where found. **Adult:** A small honeyeater with *pale rufous breast band, white throat, grey hood, and golden green wing edges*. **Male** larger on average. **Juv:** Nondescript, but note whitish throat and may have yellowish eye-ring. **Voice:** Song consists of a 10-note series, beginning with 2 wheezy upslurs and continuing with alternating upslurs and downslurs in a seesaw pattern. Calls include small, high-pitched, squeaky notes, e.g., *szweeit*, and a chattering, *kwee-kwee-kwee*. . . . **Habits:** Sociable, quick-moving, and vocal; usually in isolated trees and flowering plants. Forages at all levels for insects and flower nectar. Nest built in low tree or shrub, suspended from fork in foliage, or in reeds near or over water; sometimes nests near the nest of a figbird, Willie Wagtail, or its own species. Nest is a deep cup or pouch woven from strips of bark, grass, and plant fiber, bound with spiderwebs and lined with finer material. Eggs (2) white and spotted. **Range:** NG and Aru Is, 0–600 m. Patchily along NG S coast from Bird's Head (Sorong) and Bird's Neck (Triton Bay) to tip of SE Pen; on the N coast found on Sepik R, in Markham Valley (Nadzab), and in the Musa Valley of SE Pen. Also AU.

Brown-backed Honeyeater *Ramsayornis modestus* Pl 72
11 cm. Another coastal species, common near water in open, lowland habitats such as roadsides and other clearings, savannah, and mangroves; locally present in some towns. **Adult:** Small and rather nondescript. Faded brown above and white below with *indistinct fine brown barring on breast and pinkish bill and legs*. No other has *white throat* and dark ventral barring. **Juv:** Similar, but streaky below. **Similar Species:** Silver-eared H has white ear-spot and is streaked below. **Voice:** Song is a series of rapidly repeated, faint, dry notes (~4/sec) all on the same pitch. Call is a dry *chh chh*, second note slightly higher. **Habits:** Usually in pairs or small groups. Active, conspicuous, and unwary. Flits about in shrubs at edges of watercourses and scrubby clearings. Gleans or sallies for insects and also visits flowers. Frequently flies high, chirruping noisily. Flight swift and undulating. Domed nest made of bark strips is often seen in vegetation over water. Sometimes nests colonially, a few nests within close proximity. Eggs (2) white, speckled. **Range:** NG, NW Is (Batanta, Salawati, and Waigeo), Aru Is, and SE Is (3 main D'Entrecasteaux Is), 0–600 m. NG: S coast; along N coast only at Mamberamo R, Wewak and Kairiru I, Madang, and eastern SE Pen. Also AU.

Obscure Honeyeater *Caligavis obscura* Pl 73
(*Lichenostomus* or *Oreornis obscurus, Meliphaga obscura*)
18–19 cm. In hill forest and adjacent lowlands. Mostly gleans insects in the midstory, ascending to feed in the canopy of flowering or fruiting trees. Generally uncommon and unobtrusive but for its song. Resembles a meliphaga, though the face is more colorful. **Adult:** A medium-sized, *dark olive* honeyeater with a diagnostic *bicolored ear-spot and yellow-washed throat*. Ornate facial pattern also includes a pale spot in front of eye and another behind, and a black moustachial stripe. **Male** larger. **Juv:** Similar, but rump brownish. **Similar spp:** In Obscure's habitat, meliphagas could be confusing, but these exhibit a simple facial pattern showing only a unicolored ear-spot and a rictal streak. Black-throated H is the most similar but lives at much higher elevation and has sooty underparts. **Voice:** Similar to song of Black-throated H. A typical song begins with 2 lower notes (second higher than first), then 4 rapidly delivered higher notes in a descending scale, repeated once or twice. Suggestive of laughter. Also a bubbling series of repeated *chearp* notes in a rising and falling series. Foraging note is a light *ssit* repeated every 3–4 sec. **Habits:** Singly or in pairs; joins other birds feeding in flowering and fruiting trees. Diet includes insects, flower nectar, and fruit (such as figs). One nest was suspended from a fork low to the ground; a cup of moss and twigs, lined with finer material. Eggs (2) pinkish with spots. **Range:** Endemic. NG (no records from Fakfak, Kumawa, and Wandammen Mts; absent from Adelbert Mts, Huon, and northwestern SE Pen to Morobe), 100–1100 m (rarely to 1400 m).

Black-throated Honeyeater *Caligavis subfrenata* Pl 73
(*Lichenostomus* or *Oreornis subfrenatus, Meliphaga subfrenata*)
20–21 cm. Inhabits cloud forests and subalpine shrubbery. A vocal, active, dusky olive honeyeater with *brightly marked face*; easy to hear but often difficult to observe. Uncommon, except where gathered in flowering trees and shrubs. **All plumages:** Suggests a miniature melidectes; however, the combination of *bicolored (yellow/white) ear-spot and orange legs* is diagnostic. **Male** larger. **Similar spp:** The sooty underparts and green wings suggest a melidectes; these have grey legs, brown vent, and different facial patterns. Meliphagas inhabit lower elevations, as does the Obscure H. Orange-cheeked H is much larger and has a broad orange cheek patch. **Voice:** Song is a rapid, loud, bubbling, and cheerful series of notes that rises then falls in pitch, progressively decelerating. Often given as a duet. Similar to song of Varied H. Call note *whik whik whik . . .* , 6 per 10 sec. **Habits:** Forages alone except when gathered in flowering trees popular with honeyeaters and lorikeets. Actively displaces other nectar feeders at sources of abundant nectar. Forages in all levels of the forest for insects (gleans and sallies), nectar, and some fruit. No nesting information. **Range:** Endemic. NG: Bird's Head, Central Ranges, and Huon, 2000–3500 m (1070–3700 m).

Orange-cheeked Honeyeater *Oreornis chrysogenys* Pl 73
(*Lichenostomus* or *Meliphaga chrysogenys*)
24–26 cm. Confined to the highest peaks of the W Ranges, where locally common. Timberline forest and shrubbery bordering alpine grasslands. Often forages in low shrubs well away from closed forest. Confiding and somewhat stolid. **All plumages:** *A big, olive-green honeyeater with a conspicuous goldenorange cheek patch and yellowish legs.* **Male** larger. **Similar spp:** Black-throated H smaller, darker, and exhibits a white-and-yellow ear-spot. **Voice:** One call is a single bubbly note, rising in pitch, *chilwip*. Also a wheezy, downslurred, nasal *nnyau* and a squeaky *weechit*. **Habits:** Presumably mainly solitary or in pairs. Forages in trees, shrubs, and on the ground. Diet is composed of fruit, insects, and sedge seeds. One nest was in the crown of a tree fern in alpine grassland; it was a thick cup of moss and lichens on the outside with an inner wall of grass stems and fern material, and lined with an insulating layer of down from the leaves of tree ferns and feathers. Egg (1) pale pink with spots. **Range:** Endemic. NG: W Ranges, 3250–4000 m, rarely lower.

Sooty Honeyeater *Melionyx fuscus* Pl 74
(Sooty Melidectes, *Melidectes fuscus*)
23–24 cm. High mountain cloud forest and shrubland. Common in the East, rare in the West. Forages with tail cocked, frequently giving a soft call note. *A slender, sooty black honeyeater with pale blue tear-shaped wattle at the corner of its eye and thin black bill.* Bare skin behind the wattle is normally red, but sometimes orange or yellow. **Adult:** Black. **Male** larger on average. **Juv:** Browner. **Similar spp:** Long-bearded and Short-bearded Hs are larger and have a prominent white "beard." Melidectes are larger and have greenish wings and complex facial patterns. **Voice:** Song is a long, repetitive series of paired, high-pitched whistles repeated tirelessly: *chirp-chip, chirp-chip, chirp-chip.* . . . Call is a pair of upslurred notes: *schweep schweep*. **Habits:** Solitary, less often as a pair. Rather passive, avoiding encounters with larger honeyeaters. At all levels in vegetation. Gleans twigs and foliage for arthropods and takes nectar and occasionally small fruit. Often seen feeding at flowers of rhododendrons and others in the heath family. The nest is built into an upright fork low in a shrub or small tree; it is a bulky cup composed of moss and other materials and lined with feathers or tree-fern scales. Egg (1) pale pink and blotched. **Range:** Endemic. NG: Central Ranges, 2200–3600 m (extremes 1700–3720 m). **Taxonomy:** Traditionally assigned to the genus *Melidectes*, the *Melionyx* honeyeaters are quite distinct in shape, voice, and behavior.

Long-bearded Honeyeater *Melionyx princeps* Pl 74
(Long-bearded Melidectes, *Melidectes princeps*)
26–27 cm. Mossy cloud forest at timberline and in scattered clumps of shrubbery in alpine grassland. *A large, blackish honeyeater of the highest peaks in the E Ranges*, where fairly common away from people. Note *unusually long, slim, curved bill and conspicuous, wispy, white "beard."* Yellow spot behind eye. **Adult:** Gleaming white beard. **Male** larger on average. **Juv:** Yellowish white beard. **Similar spp:** Sooty H is smaller, has shorter bill, and lacks beard. The various high-elevation melidectes have shorter white throat-stripes, blue facial skin, and shorter bill. Short-bearded H occurs separately to the West; has shorter bill and shorter, bushier beard. **Voice:** Unrecorded. **Habits:** Typical of genus. Forages in shrubbery flowers for nectar and also presumably for arthropods. No nesting information. **Range:** Endemic. NG: E Ranges, 3000–3800 m (extremes 2450–4200 m). Ranges west to Kaijende Highlands and presumably to highlands just east of the Strickland Gorge.

Short-bearded Honeyeater *Melionyx nouhuysi* Pl 74
(Short-bearded Melidectes, *Melidectes nouhuysi*)
27–28 cm. Inhabits subalpine forest and shrublands of W and Border Ranges. This is the better-known W counterpart of the Long-bearded H. Plumages similar but bird differs by shorter bill and short pair of bushy, white beards. Yellow spot behind eye. Cocks tail. **Similar spp:** See comparisons for Long-bearded H. Belford's M has a similar pair of white "beards." **Voice:** Song unrecorded. Call is a metallic *pwik*. Scold is a low buzzy *chsh* . . . repeated rapidly 4–6 times, sounds like a much smaller bird. When chasing each other, a sweet scolding *swee* is repeated. **Habits:** Forages in low shrubbery and on ground; flies with a jerky flight from bush to bush. Diet is fruit, sedge seeds, arthropods, and nectar. One nest was concealed in an isolated alpine shrub; nest, wedged into upright branch forks, was a bulky cup mainly of mosses and lined with grass stems, other soft material, and feathers. Egg (1) pink with spots. **Range:** Endemic. NG: Western and Border Ranges (eastward to Star Mts, PNG), 3050–4500 m.

Melidectes. This is a genus of large, vocal honeyeaters endemic to the mountains of New Guinea. Of the 6 species, all except Ornate M, a small species of mid-mountain forests, are ecologically equivalent, similar vocally, and separated spatially by geography or elevation. On the Central Ranges, the 2 large melidectes are Belford's, which lives throughout these ranges, and Yellow-browed, confined to east-central New Guinea. Belford's M lives at elevations above those of Yellow-browed in localities where both co-occur, but the 2 species hybridize, sometimes extensively. They are replaced on the Bird's Head and Neck by the Vogelkop M and on the Huon Pen by the Huon M. Cinnamon-browed M is restricted to elevations below those of the larger melidectes throughout most of its wide range but lives by itself in the Wandammen and Foja Mts.

Ornate Melidectes *Melidectes torquatus*   Pl 74
(Ornate Honeyeater)

22–23 cm. Mid-mountains in forests, disturbed areas with trees, and highland towns. Common. A medium-sized, *colorful* melidectes with *banded throat and breast.* **Adult:** Unmistakable— *yellow-orange eye-skin, white throat patch, and buffy breast.* **Male** larger. **Juv:** Duller markings and paler below. **Subspp (4, in 2 groups):** *torquatus* (most of range, exclusive of next) large white throat patch, throat wattles small or absent; *emilii* (SE Pen, except Herzog Mts occupied by previous group) small white throat patch, throat wattles large. **Similar spp:** Other melidectes have dark underparts. **Voice:** Loud, raucous, and similar to some calls of Yellow-browed M. Pattern consists of complex disyllabic gurgles, with the first a higher-pitched *stook!* Bird's Head version prettier and more bubbly. **Habits:** Solitary or in pairs, but congregates in small quarrelsome groups at flowering and fruiting trees. Easily observed in towns and gardens but more difficult in forest interior. A pugnacious canopy dweller. Gleans foliage for insects, sips flower nectar, and eats small fruit, more so than other melidectes. Nests in tree foliage, suspended from a forked branch; nest cup-shaped. Egg, no information. **Range:** Endemic. NG: Bird's Head, Central Ranges, Adelbert Mts, and Huon, 1100–1800 m (extremes 950–2200 m).

Cinnamon-browed Melidectes *Melidectes ochromelas*   Pl 74

24–25 cm. Occupies mid-mountain forest and edge, mainly in the canopy. Locally common. Note that geographic range is disjunct and lies outside that of Yellow-browed M. **All plumages:** A midsize, dark melidectes with *pale greenish-yellow orbital skin and lacking a white throat-stripe.* Ruby-red wattles stand out alone against the dark throat. **Male** larger. **Similar spp:** Belford's, Yellow-browed, and Huon Ms are larger, greyer below, and show a white moustachial stripe. **Voice:** Typical of the genus; often a series of 4–5 nasal slurs *nraa nraa nraa . . . ;* also a loud descending *whee oo wooo;* another call is like the bell notes of the Blue Bird of Paradise. This species often duets. **Habits:** Found alone, in pairs, or in small groups drawn to flowering trees. Noisy and pugnacious, like other melidectes. Nest undescribed. Egg (clutch?) dark buff with a few very faint blotches. **Range:** Endemic. NG, with interrupted distribution: present on Bird's Head and Neck (Wandammen), in western W Ranges, Foja Mts, Huon, and SE Pen, 1600–2000 m. Absent in the middle Central Ranges.

Vogelkop Melidectes *Melidectes leucostephes*   Pl 74

26 cm. Mid-mountain forest and edge. Common but shy and difficult to observe. The *large,* blackish melidectes on the *Bird's Head and Neck;* note the prominent, pearly white bill. **All plumages:** Differs from other melidectes by its *heavily white-spotted breast and exaggerated white facial markings.* Neck wattle either orange or red. **Male** larger. **Similar spp:** Co-occurs with two other melidectes, the smaller Cinnamon-browed M, which lacks the white facial markings and breast spots, and the very different Ornate M. **Voice:** A series of 3–9 nasal slurs, like those of other melidectes. Vocal, but not to the extent of Belford's M. **Habits:** Similar to other large melidectes, as far as known. No nesting information. **Range:** Endemic. NG: Bird's Head and Neck (Fakfak, and Kumawa Mts), 1000–1800 m.

Huon Melidectes *Melidectes foersteri*   Pl 74
(Huon Wattled Honeyeater)

28–29 cm. The large melidectes of *Huon Peninsula,* occupying mid-mountain and cloud forest. *Largest of all melidectes.* **All plumages:** Resembles Yellow-browed M, except for the blue eye-skin, white brow and ear-stripe, and *unique red gape wattles* in addition to red throat wattles. **Similar spp:** Cinnamon-browed M co-occurs, but is much smaller and lacks white on the face.

Ornate M also inhabits Huon Mts. **Voice:** Similar to that of Belford's and Yellow-browed Ms. **Habits:** Similar to Belford's H. No nesting information. **Range:** Endemic. NG: Huon, 1600–3700 m.

Belford's Melidectes *Melidectes belfordi* Pl 74
(Belford's Honeyeater)

27–28 cm. The cloud forest melidectes of the Central Ranges, living at mostly higher elevations than Yellow-browed M, but locally hybridizing with it where they meet. Common, noisy, and conspicuous, a familiar bird of the high mountains. Resembles the bearded honeyeaters in its white moustachial tufts and dark bill and legs. **Adult:** "Pure" Belford's distinguished by the *blackish bill and legs, black forehead, white brow and ear-streak, blue eye-skin, a small white gape wattle (may be absent), and absence of throat wattles.* The bill is not as long as that of Yellow-browed. This is the only melidectes lacking a pair of throat wattles. **Male** larger. **Juv:** Similar but duller, with reduced eye-skin and white facial markings. **Subspp (4):** Minor; probably most are stable hybrid populations from crosses with the Yellow-browed M. **Similar spp:** See Yellow-browed M. Cinnamon-browed M has a grey bill and lacks black throat-stripe. **Voice:** Loud repeated calls, including hoarse rasps, nasal *caw* notes, and clear, piping, high notes. Duets, staccato notes, and gurgles are common. An inveterate caller; one of the first birds heard in montane habitat. **Habits:** Singly or in pairs. Frequents mountain forest and disturbed habitats, usually in the canopy, especially at epiphytic flowers and canopy blossoms. Noisy and aggressive. Attacks other species at flowering trees and shrubs. Feeds on nectar and arthropods. Often sails across openings with wings held widely spread. Nests on a branch fork at various heights; nest is a deep cup of moss, woolly fern scales, and twigs. Egg undescribed. **Range:** Endemic. NG: Central Ranges, 1600–3800 m. **Taxonomy:** Belford's and Yellow-browed Ms have historically formed hybrid swarms where they meet in eastern NG. Some populations seem to be stable products of this introgression (e.g., *M. rufocrissalis stresemanni* on S slopes of Wau Valley).

Yellow-browed Melidectes *Melidectes rufocrissalis* Pl 74
(Yellow-browed Honeyeater)

26–27 cm. The large, mid-mountain melidectes of the Eastern and Border Ranges. Locally common. Originally a forest bird, but now most often seen in the casuarina groves and scrub of extensive agricultural landscapes in the intermontane valleys; this habitat lies below the cloud forest–covered ridges and peaks occupied by Belford's M. The 2 species hybridize, so identification of individuals to species sometimes is not possible. **Adult:** "Pure" Yellow-browed has *pearly grey bill and legs; white forehead; yellow brow and ear-streak; extensive, pale, greenish-yellow facial skin; and presence of a large, pale gape wattle and red throat wattles on each side of the face.* **Male** larger. **Subspp (3):** Two possibly of hybrid origin. **Similar spp:** See Belford's M. Cinnamon-browed M occupies the same habitat outside the range of Yellow-browed. Vogelkop and Huon Ms are geographically isolated and have blue eye-skin. **Voice:** A series of hoarse *caw* notes on the same pitch, similar to those of the Superb Bird of Paradise, but with no deceleration; also a series of 2 kinds of notes, strictly alternating with each other, 1 note high and clear, the other garbled. Also antiphonal duets similar to those of Belford's. **Habits:** Similar to Belford's. Nest hangs from a branch fork a few meters up; a cup woven of grass and fern stems, externally decorated with moss. Eggs, no information. **Range:** Endemic. NG: Border Ranges east to NW section of SE Pen, 1100–2450 m. Actual distribution uncertain owing to massive hybridization with Belford's M (hybrid birds occur east at least to Wau).

Varied Honeyeater *Gavicalis versicolor* Pl 73
(*Lichenostomus* or *Meliphaga versicolor*)

19–22 cm. Strictly coastal. Strand forests, mangroves, small islands, village shade trees, coconut plantations, and many seaside towns, especially on the N coast (Jayapura, Vanimo, Wewak, Madang, Lae). Common, even locally abundant, but shy. A familiar, melodious voice in the right habitat, yet the bird is wary and elusive. **Adult:** *Streaky underparts, black mask, and yellow-and-white ear-patch.* **Male** larger on average. **Juv:** Paler and less strongly patterned. **Subspp (2):** *sonoroides* (N coast: Bird's Head to Milne Bay, plus NW Is and SE Is) green above, strong yellow wash below; *versicolor* (S coast: Merauke to Milne Bay; also AU) grey and green above, whitish below; birds of N coast SE Pen are intermediate. **Similar spp:** In its habitat, no other bird has the streaking and ear-spot, except for the very much smaller Silver-eared H. **Voice:** Song, a loud, mellow, bubbling series of jumbled notes, rising and falling in pitch. Musical and bright, often given as a duet. Also harsh scolding. **Habits:** Singly, in pairs, or in small gatherings. Active and belligerent, chases other birds. Forages in foliage at all levels, feeding on insects and nectar. Nests in a shrub, mangrove, or small tree; the nest, suspended from a fork a few meters up, is a cup of stems or rootlets bound with spiderwebs and lined with finer rootlets and fibers. Eggs (2) pale pinkish buff. **Range:** NG, NW Is (Batanta, Misool, Salawati, Schildpad Is, Waigeo), Bay Is (Yapen), SE Is (D'Entrecasteaux Is: Fergusson, Normanby), and many small islets on the coastal shelf, sea level. NG: patchily distributed along coast. Also AU.

Yellow-tinted Honeyeater *Ptilotula flavescens* Pl 72
(*Lichenostomus* or *Meliphaga flavescens*)

12–13 cm. Eucalypt savannahs and nearby town gardens, where locally common. Unique in its very restricted range. *A small, pale, yellowish honeyeater with black-and-gold ear-stripe.* Dark beady eye set against yellow face. **Adult:** Bill black. **Male** larger on average. **Juv:** Bill pale with dark tip. **Subspp (1):** *flavescens* (also N AU). **Voice:** Song in NG described as a short series of bubbling notes; in AU, as *porra-cheu, porra cheu, porra cheu cheu cheu*. Various calls, including a high-pitched, repeated *wi wi wi* . . . of a dozen or more notes, particularly at dawn. **Habits:** Usually in small groups. Sprightly, tame, and feisty. Forages in vegetation at all levels; also gleans from branches. Takes nectar and arthropods. Nest suspended from a forked branch in a low bush or tree; nest sack-shaped and finely woven of bark strips and rootlets bound with spiderwebs and lined with feathers and other soft material (AU data). Eggs (2) pinkish with small spots. **Range:** NG: Port Moresby area and savannahs of SE Pen, 0–450 m. Also AU.

> **Meliphagas.** In the region there are 10 species of these plain honeyeaters, all confusingly similar. Any one site may support 2–4 species. Many individual meliphagas seen in the field will be either very difficult or impossible to identify with certainty. A combination of location, elevation, and habitat, in addition to plumage, size, and shape can aid in identification. Field marks to study are ear-spot color, size, and shape, bill length and thickness, gape flange prominence and color, rictal streak size and color, plumage hue (from olive to brown), and pattern-texture of underparts. Learn the Mimic Meliphaga, the most common species, first and then use it as a standard to compare with the rest. Some calls are shared among the species with slight differences, whereas other calls are diagnostic of one species.
>
> Of the 10 species, the 2 widespread species in forests of the lowlands and foothills are Mimic M and the stout-billed Puff-backed M. The Yellow-gaped M is widespread but rare in these habitats and may be a flowering-tree specialist. The Graceful and Elegant Ms are confined to the lowlands of the South and Southeast, respectively. The Scrub M is widespread and restricted to second growth. Three meliphaga species are montane: the larger White-eared M of N hill forests and Mottled M of S hill forests, and the smaller and more slender-billed Mountain M of mid-mountain forests, which is ecologically equivalent to the Mimic M that lives at lower elevations. Finally, the Tagula M, the sole meliphaga of Tagula Island, is endemic to that island.

Puff-backed Meliphaga *Meliphaga aruensis* Pl 75
(Puff-backed Honeyeater)

16–18 cm. Common in midstory and undergrowth of lowland and hill forest and edge. *Stout-billed*, chunky. **All plumages:** *Thick, olive rump tuft shows white fringe on the sides*—white part usually hidden under the wing and hard to see, but diagnostic. *Ear-spot pale yellow, varies from large and squarish (S Lowlands) to elongated and tear-shaped*, may connect with the yellow rictal streak. Iris dark brown. **Male** larger on average. **Subspp:** None, but shows some plumage variation that is difficult to define. See Taxomony. **Similar spp:** Mimic M exhibits a small ear-spot, slimmer bill, and often a grey iris. Mottled M exhibits a mottled breast. Yellow-gaped M more delicate and shows more pronounced yellow gape, small round ear-spot, and yellow-olive legs in the South. **Voice:** Song of 2 types—a querulous piping series, ~2 notes/sec, suggestive of a bronze-cuckoo, and an emphatic, rapid, decelerating trill, probably diagnostic. An additional song from S watershed of eastern NG, and diagnostic, consists of a loud, medium-pitched, ascending and decelerating series of notes. Calls include a short, generic *chup* and a single low whistled note. **Habits:** Solitary or in pairs. Shy: usually hidden in vegetation and difficult to see well. Diet is mainly fruit and arthropods, less often flower nectar. Nests in the forest on a branch fork about a meter from the ground; nest is a neat, hanging cup of fibers and leaves and often decorated with bark. Eggs (1–2) pinkish white with sparse brown spots. **Range:** Endemic. NG, NW Is (Batanta, Misool, Salawati, Waigeo), Bay Is (Yapen), Aru Is, and SE Is (3 main D'Entrecasteaux Is; Trobriand Is), 0–1200 (1580) m. **Taxonomy:** DNA studies indicate what may be a cryptic species in the S watershed of the SE Pen.

Yellow-gaped Meliphaga *Meliphaga flavirictus* Pl 75
(**Yellow-gaped Honeyeater**)
15–16 cm. A rare and little-known canopy-dwelling species of lowland and hill forest and edge. Small and thin-billed. **All plumages:** *Has the most prominent yellow gape flange and rictal streak of any meliphaga*. *Small, pale yellow ear-patch* (paler than rictal streak). Yellowish-tinged legs in the South also diagnostic (grey in other meliphagas). **Male** larger. **Subspp (2):** *flavirictus* (S catchment from Trans-Fly to SE Pen) paler, bill dark to pale grey-brown, legs yellowish olive, ear-spot almost whitish; *crockettorum* (W and N NG) darker and brighter yellow-green, bill dark, legs greyish. **Similar spp:** Scrub M is larger, has white ear-spot, and does not have such a prominent rictal streak. Mimic and Puff-backed Ms have less prominent gape flange and rictal streak; rictal streak the same color or paler than the ear-spot, the reverse of Yellow-gaped. Puff-backed M has heavier bill, larger and often longer ear-spot, somewhat less prominent gape; otherwise confusingly similar to N race of Yellow-gaped. **Voice:** Song a bubbling, descending series of chips, suggestive of song of Obscure H, but without the mix of rising and falling series. Call is a distinctive whinny *kweea-hoo!* **Habits:** Solitary, but gathers at canopy of fruiting or flowering trees. Feeds on nectar, insects, and fruit. No nesting information. **Range:** Endemic. NG, though no records from Sepik-Ramu and the far SE Pen (eastward from Morobe and Port Moresby), 0–1400 m.

White-eared Meliphaga *Meliphaga montana* Pl 75
(**Forest White-eared Meliphaga** or **Forest Honeyeater**)
16–17 cm. Restricted to hill forests of the northern catchment of NG. Keeps to the forest interior, frequenting midstory and undergrowth, where common but seldom seen. **Adult:** Only meliphaga in which the *gape skin, rictal streak, and ear-spot are all white*. This is also the *most brownish of the meliphagas*. **Male** larger. **Juv:** Ear-spot yellowish white. **Subspp (3):** Minor. **Similar spp:** Scrub M has whitish ear-spot but is greener overall, with orange gape. **Voice:** Call is diagnostic—a slightly hoarse, loud, upslurred *wheep*. **Habits:** Solitary, secretive, and shy. Diet is fruit and arthropods obtained by gleaning and hover-gleaning; does not visit flowering trees. Has a loud wingbeat, unlike other meliphagas. No nesting information. **Range:** Endemic. NG, Batanta (NW Is), and Yapen (Bay Is), 500–1150 m (rarely to 1500 m). NG: Bird's Head and Neck (no records from Wandammen Mts), to Central Ranges (N catchment only), and all peripheral mts.

Mottled Meliphaga *Meliphaga mimikae* Pl 75
(**Spot-breasted Meliphaga** or **Honeyeater**)
16–18 cm. The southern, yellow-eared counterpart of White-eared M. Lives in similar habitat and also common. Large size, large bill, and fine rictal streak. **All plumages:** Recognized by its *distinctly mottled breast*. Note *conspicuous eye-ring* that is narrow, pale olive, more prominent than in other meliphagas. Iris brown. Yellow gape skin, rictal streak, and ear-spot. **Male** larger. **Similar spp:** Puff-backed and Mimic Ms share its habitat but lack mottling below. Mountain and White-eared Ms are less mottled below. Mountain M is smaller, has a smaller bill, and lives at mostly higher elevations. White-eared occurs in N catchment. **Voice:** Song (?) is a repeated mournful piping, similar to that of Mimic M. The only member of the group to give a soft nasal note like that of the Eurasian Bullfinch. Also gives a typical meliphaga *tuck* or *chip*. **Habits:** Similar to White-eared. Nest 1–2 m up on a horizontal fork of a forest sapling branch; suspended open cup made of fine plant fibers, twigs, and vines, with live moss and fern on the outside and lined inside with plant down. Eggs (2) no description. **Range:** Endemic. NG: S slopes of the Central Ranges and N slopes of SE Pen (Hydrographer Range), 500–1150 m (extremes 0–1800 m).

Elegant Meliphaga *Meliphaga cinereifrons* Pl 75
(formerly included in **Graceful Meliphaga**, *m. gracilis*)
14–16 cm. Restricted to lowlands and hills of SE Pen, mainly in second growth and edge, also other open wooded habitats and mangroves. Forages at all levels in the vegetation. Rather small; note long, slender bill. **All plumages:** Look for *dark sides of face and often blackish hind edge of the yellow ear-spot*. Ear-spot large and round, often with *a projection reaching upward behind the eye*. Gape skin orange; thin yellowish-white rictal streak. Underwing-covert with a dilute ochraceous wash. Variable pale eye-ring. Iris dark brown to grey. **Male** larger on average. **Subspp (2):** *cinereifrons* (S watershed); *stevensi* (N watershed: Markham R east to Hydrographer Range) has slightly darker and greyer underparts, and rump tuft edged whitish—see Taxonomy. **Similar spp:** See Graceful M. Mimic M shares its habitat and is slightly larger, paler on the side of the face, and has a round ear-spot. Mountain M is darker, has

a faintly mottled breast, and lives at higher elevations. **Voice:** Song is a thin trill ending in an upwardly inflected note. Call a sharp *tchik*, *tik*, or *tup*. Also, a high *ki-ki-ki-ki-ki . . .* in alarm. **Habits:** Singly or in pairs. Diet is fruit, insects, and nectar. Nest slung from a thin fork or vine, 1–6 m up in forest; a deep cup of grass, leaves, and spiderweb, lined with plant down. Eggs (2) pinkish with darker red spots. **Range:** Endemic. NG: SE Pen, 0–900 m (1200 m in North). **Taxonomy:** Split from Graceful M by molecular analysis and plumage. Population *stevensi* contains individuals assigned to either *M. analoga* or *M. cinereifrons* by a DNA study, indicating apparent hybridization between species-populations.

Graceful Meliphaga *Meliphaga gracilis* Pl 75
(**Graceful Honeyeater**)
14–16 cm. Common in lowland forests of all types, but predominantly in second growth, edge, mangroves, swamp forest, and savannah. Frequents all levels in vegetation. *Smallest lowland meliphaga, with slender bill.* **All plumages:** *A small, thinner-billed version of Mimic M*; lacks distinctive field marks. In the hand, note large, fluffy, and loose rump tuft; underwing is pale yellowish to white. Iris brown to greyish. **Male** larger on average. **Subspp (1):** *gracilis* (also AU). **Similar spp:** Mimic M larger, with heavier bill on average, and the rump tuft is denser. Elegant M has different facial pattern and ochraceous underwing-coverts. **Voice:** Song is a series of piping notes. Song on Aru: 8 repeated, whipped upslurs, whistler-like but less explosive. Also a short series of snapped notes. The common call is a sharp *tick* or *tuk*. **Habits:** Feeds on nectar and arthropods, rarely fruit. Nest slung from a thin forked branch 1–20 m up; a deep cup of fibers, grass, bark shreds, etc., and decorated with moss and lichen (AU data). Eggs (2) pale orange-pink, spotted. **Range:** NG (S Lowlands and Trans-Fly, east to Purari Delta) and Aru Is, 0–300 m (rarely to 800 m). Also AU.

Mountain Meliphaga *Meliphaga orientalis* Pl 75
(**Hill Forest** or **Mountain Yellow-eared Honeyeater**)
14–16 cm. Exclusively montane, *this meliphaga lives at higher elevations than any other*, but co-occurs with some at the lower half of its elevational range. Common in mid-mountain forest, mainly in the canopy; uncommon in scrub. *Small, with elongated, slender bill.* **All plumages:** Variable, see Subspp. Undistinctive—best identified by habitat and size. Small yellow ear-spot; underparts dark and mottled. **Male** larger. **Subspp (3):** *orientalis* (SE Pen, east of Angabanga and Mambare Rs) dark side to face and ventral mottling pronounced; *citreola* (N slopes of W Ranges to N Coastal Mts) side of face green and underparts yellower with obscure mottling; *facialis* (Waigeo I east along S slopes of Central Ranges to Huon and northeastern SE Pen) less dusky face, reduced mottling. **Similar spp:** Mottled M resembles race *orientalis* but is larger and more robust. Mimic M is larger, with heavier bill. **Voice:** Common call note *tuck* or *chup* (latter somewhat musical and resembling House Sparrow). Also *cheeyur!*; *weet!*; *weeyurt*; *chureep!* Calls from canopy. **Habits:** Solitary or in pairs. Feeds on insects and nectar (occasionally fruit), small and agile; visits flowering trees. Nest a tight cup of fiber, twigs, and vines decorated on the outside with moss and bamboo leaves, lined inside with plant down. Eggs (2), no description. **Range:** Endemic. NG (records from all mts except the Cyclops Mts) and Waigeo I (NW Is), 800–1750 m (common above 1300 m).

Mimic Meliphaga *Meliphaga analoga* Pl 75
(**Mimic Honeyeater**)
16–19 cm. This is the standard meliphaga—ordinary in plumage and shape, present everywhere in lowland and hill forests of all sorts, and easy to observe. Size medium to large; long slender bill. **All plumages:** *Greenish upperparts; unmottled underparts. Note sides of head not so dark. Ear-spot yellow, large, and elongate*; gape skin orange; rictal streak pale yellow. Iris varies from grey to brown; legs grey; bill blackish. **Male** larger on average. **Subspp:** None recognized now, but see plate. **Similar spp:** Graceful M, especially the larger male, can be difficult to distinguish from females of this species; Graceful usually smaller with a thinner bill. Puff-backed M is same size, but stouter with a stubby bill and white-edged and dark-tipped rump. **Voice:** The common and characteristic call, shared with other meliphaga species, is a single *tup!* In both Mimic and Scrub Ms, the *tup* call is louder than that of other meliphagas. Also a disyllabic call and a querulous piping. **Habits:** Singly, in pairs, or in small parties. Mainly in canopy, midstory, and edge. Feeds on fruit and arthropods; also visits flowering trees for nectar. As yet no nesting information, owing to possible confusion with other meliphagas. **Range:** Endemic. NG (everywhere except N coast of SE Pen), NW Is (Batanta, Misool, Salawati, Waigeo), Bay Is (Mios Num and Yapen Is), and Aru Is, 0–1250 m (rarely to 1500 m).

Scrub Meliphaga *Meliphaga albonotata* Pl 75
(Scrub White-eared Meliphaga, Scrub Honeyeater)
16–17 cm. A scrub and garden specialist. Numerous in mid-mountains, less so in lowlands. Large size. **Adult:** *White ear-spot. Upperparts distinctly green, underparts unmottled olive grey.* Gape bright yellow to orange-yellow; rictal streak thin and obscure, yellowish or whitish. **Male** larger. **Juv:** Yellow-tinted, whitish ear-spot; often has an irregular darker olive band on the breast. **Similar spp:** White-eared M brown and darker, without indication of green on wings or tail; prefers forest interior. Yellow-gaped M has very pale yellowish ear-spot and green upperparts, but is smaller, paler, and has a conspicuous yellow rictal streak, and yellowish-olive legs. Mottled-breasted M resembles juv Scrub, but has mottled underparts. **Voice:** Song is a series of 10 or more thin notes at a rate of ~5/sec, rising first, then dropping in pitch. Calls include the typical meliphaga *tup!*; also a bright, cheery call: *cherrup!* or *cheereeo*; and a distinctive piping. **Habits:** Singles or pairs. Diet includes insects, nectar, and some fruit taken mostly by gleaning. Nests 1–5 m up on a forked branch of tree, shrub, or bamboo; cup nest hung by its rim, composed of various plant materials. Eggs (1–2) pinkish white, small spots and blotches. **Range:** Endemic. NG (except NW Lowlands, Sepik-Ramu, and Trans-Fly), 500–1500 m (0–1950 m).

Tagula Meliphaga *Meliphaga vicina* Pl 75
(Louisiades Honeyeater or **Meliphaga)**
17–18 cm. The only meliphaga on Tagula I, where it is common in all wooded habitats. Possibly the largest in the group, with a *heavy, longish bill.* **Adult:** Resembles Mimic M, but underwing-coverts ochraceous (whitish with olive wash in Mimic). Iris grey to dark brown. **Male** larger. **Voice:** Usually gives calls in pairs or triplets, *cheerie! . . . cheerie!* **Habits:** Resemble those of Mimic M. Frequents mainly canopy and midstory. No nesting information. **Range:** Endemic. Tagula I (SE Is), 0–800 m.

AUSTRALASIAN WARBLERS: Acanthizidae

The Australasian warblers comprise a family of 63 species of small insectivores, 20 of which occur in New Guinea. This family of Australian and New Guinean origin is allied to the fairywrens and honeyeaters rather than to other warblers in New Guinea. The latter are more closely related to one another and evolved in Asia and Africa. Despite the numerous genera of Australasian warblers, they share many features besides being mainly small, insect-gleaning birds. Common features include foraging in a small, noisy party; a song composed of repeated phrases; and a domed nest with a side entrance. Included are the Goldenface, mouse-warblers, scrubwrens, thornbills, and gerygones, as well as some strictly Australian genera. The gerygones, thornbills, and Goldenface tend to frequent the forest canopy and midstory where they mostly glean insects from foliage and twigs. The scrubwrens prefer the forest interior from midstory to understory, the larger species often foraging on branches. The mouse-warblers inhabit the forest floor. Most gerygone species are lowland birds, whereas most scrubwren species are montane, replacing one another by elevation. These warblers often join mixed-species flocks with other small insectivores, such as monarchs. Australasian warblers are among the most familiar New Guinea birds, and it is usually not difficult to find them.

Goldenface *Pachycare flavogriseum* Pl 86
(Dwarf Whistler)
13 cm. An unusual, yellow-faced warbler with an explosive, whistler-like song. Uncommon and reclusive in the canopy foliage of hill and lower montane forest. **Adult:** Strikingly patterned. *Entire underparts golden yellow or orange,* depending on location; *dark, beady eye framed by yellow face; upperparts blue-grey* with white spotting on the tertials. **Female and possibly Juv:** Sooty ear patch. **Subspp (4):** *flavogriseum* (Bird's Head and Neck) and *subpallidum* (eastern E Ranges, Huon, SE Pen), face and underparts bright yellow; *subaurantium* (W, Border, and western E Ranges) more orange-yellow; *lecroyae* (N Coastal Mts) blaze-orange face and underparts. **Similar spp:** None close. Sclater's Whistler male with black head and white throat. **Voice:** Song and calls of explosive, whistled notes. Song composed of a few phrases. Each phrase a note (e.g., *tink!, wheep!,* or *chew!*) repeated 2–3 times and getting louder; the next phrase may repeat the same note or a different one. Song often ends in a loud, sneezing *CHEW!* or *CHEW! CHEW!* **Habits:** Usually solitary, sometimes as a pair; frequently

joins mixed flocks of small insectivores. Forages actively in the canopy and midstory for arthropods which it takes by gleaning in a horizontal posture like a gerygone. Nest built on sloping ground against the downward side of a small tree-trunk; a complexly constructed large globe of woven leaf strips capped with dead leaves; downward-facing side entrance and small vestibule. Eggs (2) pink with fine reddish or maroon spots and blotches. **Range:** Endemic. NG: most mts within its elevational range (no records yet from Fakfak, Kumawa, Cyclops, or Adelbert Mts; hypothetical for the Foja Mts), 800–1600 m, rarely higher. **Taxonomy:** Formerly classified as a whistler.

Rusty Mouse-Warbler *Crateroscelis murina* Pl 65

12 cm. The mouse-warbler of *lowland and hill forest*; common. Dark brown above with *contrasting white throat*; breast rusty-buff, except in the central S Lowlands and Aru Is where breast is mostly white. **Male:** Crown usually darker than back. **Female and Juv:** Crown same as back. Iris color in Males varies from red to brown, in Females from red-brown to brown; possibly iris changes with age from brown to red or red-brown. **Subspp (4, in 3 groups):** *murina* (species range excluding next races) white throat, cinnamon breast; *pallida* (S Lowlands) flanks and belly very pale cinnamon; *monacha* (Aru Is) underparts mostly white. **Similar spp:** Bicoloured MW lacks white throat, co-occurs at top of Rusty MW's elevational range. **Voice:** Song is a characteristic sound of lowland forest. Beautiful, clear, high-pitched, plaintive whistled tunes of up to a dozen notes are slowly delivered. Typical song is a simple 3- to 5-note series, often with some brief introductory notes, and sometimes with a trill in the middle. Series are repeated over and over before switching to a different tune. Much variation and may include mimicry of other bird species. Alarm call is a harsh buzzing note. **Habits:** Secretive and difficult to observe. Hops along ground and ascends into understory up ~1 m searching for arthropods. Singly or as a pair. Nest built on ground; globular with side entrance. Eggs (2) variously buffy or brownish, may have smudgy dark spotting. **Range:** Endemic. NG, NW Is (Batanta, Misool, Salawati, Waigeo), Bay Is (Yapen), and Aru Is, 0–1700 m.

Bicoloured Mouse-Warbler *Crateroscelis nigrorufa* Pl 65

12 cm. Rare, local, and often overlooked. A *gregarious, black-and-rufous* warbler of the *mid-mountain forest* interior within a narrow elevational range. **Adult:** Black above, bright chestnut below. Pale feather shafts on the back impart a coarse texture. Iris varies from red to brown. **Juv:** Extent of chestnut varies from throat only to all of underparts. **Similar spp:** Other MWs and scrubwrens lack such a contrasting black-and-tan pattern. Rusty MW can be cinnamon below but is brown above and has a white throat. **Voice:** Song is 2 short whistles followed by a longer third with a slight upslur. Call note is a nasal chip. **Habits:** Elevational range generally lies above that of Rusty MW and overlaps the lower range limits of Mountain MW. Forages for arthropods on the forest floor but also in the understory more frequently than other MWs. Gregarious, usually in active parties of 5 or more. Nest built ~0.5 m up in a sapling—a globe of dark rootlets with side entrance. Eggs (2) white with fine dark speckling. **Range:** Endemic. NG: Central Ranges and Huon (possible sighting in Foja Mts), 1300–2000 (1220–2500) m.

Mountain Mouse-Warbler *Crateroscelis robusta* Pl 65

12 cm. The mouse-warbler of *high mountain forest*; common. Much geographic variation in color and pattern. **Subspp (7):** Assigned to 3 groups by plumage.

1. WHITE-THROATED DICHROMATIC GROUP: White throat and belly divided by a dark breast band in Male, lacking or faint in Female. Iris of Male white, grey, yellow, orange, red-brown, or brown; of Females mostly red-brown or brown. *robusta* (eastern E Ranges, Huon, SE Pen east to Owen Stanley Mts); *pratti* (far eastern SE Pen: Mt Simpson) much smaller, darker, stronger pattern; *diamondi* (Foja Mts) also dark with strong pattern, but large. Birds from western and central E Ranges less patterned, approaching buff-breasted group.
2. PALE-WASHED MONOMORPHIC GROUP: All plumages with underparts uniformly pale. Subspp: *peninsularis*, Bird's Head; *deficiens*, Cyclops Mts.
3. BUFF-BREASTED MONOMORPHIC GROUP: All plumages with underparts uniformly rusty brown. Iris similarly variable. Subspp: *sanfordi*, W and Border Ranges; *bastille*, N Coastal Mts.

Similar spp: Bicoloured MW black above with contrasting rufous below. In E Ranges where Mountain MW can have pale brown throat and rusty brown breast and belly, Rusty MW has a distinctly white throat and paler belly. Large Scrubwren smaller, has rusty face, forages mostly in low vegetation. **Voice:** Songs of the various forms are alike: a plaintive, whistled series reminiscent of

Rusty MW, but sometimes slightly trilled or more throaty and coarse; a phrase of ~4 notes repeated over and over. Mimicry may be included in song. Buzzy scold notes *zig zig zig*.... **Habits:** Similar to Rusty MW. Nest placed just above ground level among tree roots, low branches, or embankment; globular with side entrance; made of fine roots, fiber, and moss. Egg undescribed. **Range:** Endemic. NG, all mts except Adelberts, 1750 (or lower)–3600 m. **Taxonomy:** Likely constitutes more than 1 species. Future molecular systematic analysis will help clarify species boundaries.

Scrubwrens. The 8 New Guinea species include 2 altitudinal series. The Tropical and Large SWs are larger branch-foragers that sort out by habitat with minimal overlap—Tropical in the lowlands and foothills, Mountain in mid-mountain and cloud forest. Another elevational series consists of 3 small leaf-gleaning warblers: Pale-billed SW at lower elevations, Grey-green SW at middle elevations, and Buff-faced SW at higher elevations. Vogelkop SW is the sister species of Buff-faced and replaces it on the Bird's Head. Papuan SW is a gleaning warbler slightly larger than Buff-faced and overlaps elevationally with it and Large SW.

Pale-billed Scrubwren *Sericornis spilodera* Pl 76

10–12 cm. Common in hill forest, ranging locally to lowlands. This species has the lowest elevational range of the smaller scrubwrens. **Adult:** *Pale bill diagnostic.* A rather dark, greenish scrubwren with contrasting *pale throat with black spots* (all-white throat on Aru Is). Iris red. **Juv:** Lacks dark spots below. **Subspp (6, in 5 groups):** *ferrugineus* (NW Is: Batanta and Waigeo) face rufous, overall paler than mainland races; *aruensis* (Aru Is) very pale overall, nearly spotless below; *spilodera* (N NG from Bird's Head to Sepik/Ramu) black crown, overall dark, and heavily spotted below with clinal variation in darkness, hue, and strength of pattern, merging with next races; *granti* (S Lowlands) brown crown; and *guttatus* (SE Pen) green crown. **Similar spp:** All other scrubwrens have a dark bill. Tropical and Grey-green Ss overlap in habitat. The only other pale-billed, lowland and hill forest, warbler-like birds are Yellow-bellied Gerygone on Bird's Head and Obscure Berrypecker. **Voice:** Song a phrase of 2 or 3 notes repeated over and over; varies geographically, can include trills and twittering. Call *chit chit*. **Habits:** In pairs or small parties, sometimes with other small insectivores. Forages for arthropods in foliage and branches of understory and midstory. Flicks tail. No nesting information. **Range:** Endemic. NG, NW Is (Batanta, Waigeo), Bay Is (Yapen), and Aru Is, 200–1200 (1650) m. NG: Likely all mts (no records yet from Kumawa and Cyclops Mts) and locally in Trans-Fly.

Tropical Scrubwren *Sericornis beccarii* Pl 76
(**Beccari's Scrubwren**; includes **Perplexing Scrubwren**, *S. virgatus*)

10–13 cm. This species forms a confusing and controversial complex with the Large Scrubwren. Dwelling in lowland and hill forest, the Tropical S lives immediately below the elevational range of the montane Large S but locally hybridizes with it. Moreover, the populations of Tropical S from the Bird's Head to northern NG—those races formerly classified as Perplexing Scrubwren—lack the bright markings of southern populations of Tropical S and resemble more the plain-looking Large S. **All plumages:** A *large* scrubwren showing one or both of these plumage features: *an incomplete white eye-ring and dark wing-coverts edged white.* **Subspp (8):** Arranged in 3 groups for convenience: (1) "Typical Tropical" *beccarii* and *randi* (Trans-Fly and Aru Is; lowlands) brightly patterned with olive upperparts, whitish throat, pale bill, white eye-ring, and dark wing-coverts edged white. Next, the following 2 groups of hill forest forms are distributed in a checkerboard pattern with respect to each other, a pattern that makes no sense biogeographically and is better explained as the outcome of hybridization in varying degrees: (2) "Perplexing 1" group is less patterned but tends toward Tropical S, illustrated by *wondiwoi* (Wandammen and Foja Mts) and *idenburgi* (N slope of W Ranges); (3) "Perplexing 2" group shows little or no pattern and is more rufous, tending toward Large S, illustrated by *virgatus* (Middle Sepik and N Coastal Mts). Birds resembling both Perplexing 1 and 2 have been found in the southern foothills of the Border Ranges (Mts Bosavi and Sisa); these are highly variable and appear to also be hybrids between Tropical and Large Ss. **Similar spp:** Large S co-occurs on Bird's Head, northern Western and Border Ranges, and Foja Mts; generally more rusty and lacks pattern on wing-coverts. High elevation Large S and lower elevation Tropical S replace each other somewhere at 1200–1800 m, depending upon locality. **Voice:** Song of short, high-pitched phrases repeated rapidly many times suggesting Green-backed Gerygone or Large S (perhaps higher pitched and less run together). Call a repeated rasp, like Fantailed Monarch. **Habits:** Similar to Large S but forages in the understory and midstory, searching mainly on twigs and branches. Nest (AU data) an untidy dome suspended from vines a meter or so above ground. Eggs (2, AU data) similar to Large S. **Range:** NG, NW Is (sightings from Waigeo), Bay Is (Yapen) and Aru Is, 0–1800 m. NG: Bird's Head and Neck (Fakfak, Kumawa, Wandammen Mts), N foothills of Western and Border Ranges, and Van Rees, Foja, Cyclops, and N Coastal Mts. Also AU. **Taxonomy:** Some authorities treat Tropical S as conspecific with the Large-billed Scrubwren (*S. magnirostris*) of AU. Others treat the Perplexing forms as specifically distinct (*S. virgatus*) but disagree about which populations should be included in Perplexing S.

Large Scrubwren *Sericornis nouhuysi* Pl 76

12–14 cm. The *large, common scrubwren of the higher montane forests* and timberline.
All plumages: *Olive or rufous brown* with a contrast between the upperparts and underparts, especially evident at the eye line. *Face rufous* and eye brown to red. Tail lacks black band.
Subspp (5): *nouhuysi* (W Ranges) more rufous brown; *stresemanni* (Border and E Ranges) intermediate between the previous and next races; *oorti* (Huon, SE Pen) more olive above and yellowish below; *adelberti* (Adelbert Mts) even more olive overall and pale below; *cantans* (Bird's Head) more olive with dark spot at wrist, likely of hybrid origin with Tropical S.
Similar spp: Papuan S frequently co-occurs and is deceptively similar, but is smaller, base of primaries whitish behind dark alula, tail with black subterminal band, iris grey or brown. Buff-faced S has buffy face contrasting with greyish cap. Tropical S inhabits lower elevations and not as rufous. Mountain Mouse-Warbler is essentially ground dwelling. **Voice:** Song is a pretty, descending trill: *weesee, weesee, weesee, weesee* . . . or *sit tu tuwit*, etc., each phrase at a lower pitch. Scold: *skz-skz-skz-skz*. Call note *spit!* **Habits:** In pairs or small parties, often with other scrubwrens and small insectivores. Forages in understory for arthropods, creeping mouselike through epiphytes and mossy branches. Nest a bulky dome of dead leaves and bryophytes built into clumps of similar substrate a short height off the ground. Eggs (2) white washed brown, with dark markings.
Range: Endemic. NG: Bird's Head, Central Ranges, Foja Mts, Adelbert Mts, and Huon, 2100–2500 (1200–3750) m. **Taxonomy:** See Tropical S; form *cantans* has been treated as belonging to Perplexing S.

Vogelkop Scrubwren *Sericornis rufescens* Pl 76

9–10 cm. Endemic to the Bird's Head, where it replaces the Buff-faced and Papuan Ss.
All plumages: *Pronounced buff eye-ring*. Plumage otherwise drab olive brown. Dark tail band.
Similar spp: Large and Tropical Ss co-occur but are larger, lack tail band, and differ in habits and voice. Grey-green S co-occurs at generally lower elevations; is much more greenish, with a whitish eye-ring. Papuan S most similar but does not co-occur; larger, darker below, pale band above bill. **Voice:** Song is an energetic outburst, consisting of several groups of 2 or 3 notes, each group repeated several times before proceeding to the next: *weesee weesee weesee tue tue tue tue*. . . . Resembles song of Fairy Gerygone in pattern and a weak version of song of Papuan S in quality. Call is a dry *chip* or a nasal *chee chee* when flocking. **Habits:** Similar to Buff-faced S. No nesting information. **Range:** Endemic. NG: Bird's Head and Neck (Kumawa Mts), 1300–1800 m.

Buff-faced Scrubwren *Sericornis perspicillatus* Pl 76

10 cm. Common in montane forest. Its elevational range lies between those of the Grey-green S below and the Papuan S above. **All plumages:** *Buff face framing a dark, beady eye*. Note *contrasting olive-grey crown*. Dark tail band. **Similar spp:** Vogelkop S very similar but does not co-occur; lacks grey crown. Papuan S with higher elevational range has a plain grey-brown face with orange buff restricted to a narrow eye-ring and forehead band. Buff-faced, Papuan, and Large Ss can be found together in the same forest patch, so do not identify solely by elevation. **Voice:** Song is a pleasing series of high-pitched warbles rising and falling in pitch. Call a series of musical chips ascending the scale or rising then falling. Also, soft scolding notes, including a sweet *jib!* **Habits:** Travels in small noisy parties, often mixed with other species. Searches for arthropods in foliage in the forest mid- and lower stories. Nest a bulky dome with side entrance and slight porch or hood; suspended from a shrub or sapling. Egg undescribed. **Range:** Endemic. NG mts, except Bird's Head and Neck and Cyclops Mts, mainly 1700–2600 m (850–2800 m).

Papuan Scrubwren *Sericornis papuensis* Pl 76
10–11 cm. Common in montane forest, mainly at elevations above that of Buff-faced S.
Adult: A small, nondescript brown or olive-brown scrubwren. Note *pronounced buffy orange eye-ring and narrow forehead band* at the base of the bill. There is also a unique *pale spot at the base of the primaries that contrasts with the black alula.* Obscure dark subterminal tail-band. Iris brown, hazel, or grey. **Juv:** Olive green with pale forehead, yellowish below. **Subspp (3):** *papuensis* (SE Pen) and *meeki* (W Ranges) more olive than brownish *buergersii* (E Ranges). **Similar spp:** Large S is larger and more rufescent on the forehead and face, lacks pale spot at base of primaries and dark tail band, often with red iris. Buff-faced S has a buff face and greyish crown, usually lacks the Papuan's rusty forehead band at the base of the bill. New Guinea Thornbill at higher elevations lacks buffy orange in the face and is more olive-colored overall and has a pale iris. **Voice:** Song a distinctive, loud, chipping phrase of ~8 notes (e.g., *tootsee tootseetoo tootseetoo weetsee*) repeated over and over; a characteristic dawn or dusk song of the high mountain forests. Call, dry chip notes. **Habits:** Similar to Buff-faced S. Nest suspended ~1.5 m above ground, reported as a teardrop structure coated with green moss, featuring a side entrance and hood or porch. Eggs (2) fawn with fine dark spots and squiggles. **Range:** Endemic. NG: Central Ranges, Foja Mts, and Huon, 2100–3500 m, in some places lower.

Grey-green Scrubwren *Sericornis arfakianus* Pl 76
9–10 cm. Common in interior of lower montane forest throughout mainland NG. Its elevational range is rather narrow and fits between that of the Pale-billed S below and the Buff-faced S above.
Adult: *A small, nondescript grey-green scrubwren.* Tail has blackish subterminal band. **Juv:** Slightly more brownish. **Similar spp:** Only Pale-billed and Tropical Ss regularly share its range; Buff-faced S at higher elevation but may overlap. Black Berrypecker female grey-green with short, thicker bill and longer tail. **Voice:** Song a repeated phrase of 2 notes: *weedu weedu weedu.* . . . Call a dry scolding typical of a scrubwren. **Habits:** Inconspicuous but otherwise similar to Buff-faced S. No nesting information. **Range:** Endemic. NG, throughout (no record yet from Kumawa Mts), 1100–1700 m, occasionally to 700 m.

New Guinea Thornbill *Acanthiza murina* Pl 76
(**Mountain** or **Papuan Thornbill**)
9–10 cm. *Highest elevational range of any warbler.* Common in subalpine forests in the Central Ranges.
All plumages: A small, grey-olive warbler, plain but for *dingy white underparts* and *banded tail.* *Short tail* has a dark subterminal bar contrasting with pale tip. Diagnostic *whitish iris* and *grizzled cheek* visible at close range. **Similar spp:** Grey T, clean white below, grey above, dark eye. Papuan S more evenly brown above and below, subterminal bar not nearly as pronounced. **Voice:** Song a series of paired notes, the first note at a higher pitch. Call when flocking diagnostic—a continually delivered sweet note like those often given by baby birds: *teeyuk, teeyuk, teeyuk.* . . . **Habits:** Usually in a mobile and vocal flock busily moving through the forest canopy. May associate with scrubwrens. Forages in foliage for tiny arthropods. Does not sally and hover as do other warblers. Domed nest with side entrance located in a forest tree. Egg undescribed. **Range:** Endemic. NG: Central Ranges, 2500 m to timberline (rarely down to 2150 m).

Grey Thornbill *Acanthiza cinerea* Pl 77
(**Ashy**, **Grey**, or **Mountain Gerygone**, *Gerygone cinerea*)
9 cm. Uncommon in montane forest below the elevational range of New Guinea T.
All plumages: *Clear grey above, white below.* Tail with black terminal band. *Iris brown.*
Similar spp: New Guinea T underparts not clean white; iris white. Brown-breasted G brown above and not so white below; white spots in tail. **Voice:** Song given infrequently, a slow thin, descending musical series. Call when flocking, a high-pitched *tee tee tee* or *tee taa tee*. Scolding is a soft, myzomela-like note. **Habits:** Typically very active and busy foraging in small flocks of its own species. Works the forest canopy and midstory. Gleans foliage and hovers to take tiny arthropods. No nesting information. **Range:** Endemic. NG: Bird's Head and Neck (Fakfak and Wandammen Mts), Central Ranges, N Coastal Mts, and possible sighting in Foja Mts; 1750–2700 m, rarely descending lower. **Taxonomy:** Recently recognized as a thornbill based on DNA evidence.

Gerygones. Of the 7 New Guinea species, 3 are habitat specialists: Mangrove G in mangroves of south NG, White-throated G in southeast NG savannahs, and Large-billed G along rivers and in mangroves throughout NG mainland (in forest as well as on islands). Of the other 4 species, 1 is confined to mountains—the Brown-breasted G of open habitats and forest. The remaining 3 species share lowland and hill forests: Green-backed G in the canopy, Yellow-bellied G at any forest level in mixed flocks, and Fairy G in mixed flocks in the midstory and canopy of hill forests.

Yellow-bellied Gerygone *Gerygone chrysogaster* Pl 77
10 cm. Common in lowland forest interior. **Adult:** Olive-brown upperparts with *contrasting brown tail* (hard to see). *Underparts two-toned—white on throat and upper breast changes abruptly to pale yellow on lower breast and flanks.* **Juv:** Ear-coverts greyer. **Subspp (4, in 3 groups):** *chrysogaster* (most of NG, from NW Lowlands, Bay Is, Aru Is, and S Lowlands eastward) upperparts olive, breast white, bill dark; *notata* (Bird's Head; NW Is of Salawati and Misool) upperparts greener, bill pale (*chrysogaster* and *notata* intergrade at Bird's Neck); *neglecta* (NW Is: Waigeo and Batanta) breast and belly with yellow wash, bill pale. **Similar spp:** Green-backed G broadly co-occurs, but its head is grey, back green, and underparts clean white, with yellow confined to the flanks. Fairy G is more a hill forest species but female very similar; underparts more extensively yellow, all but white chin and throat; note white loral spot. White-throated G in savannah only. **Voice:** Song commonly heard—persistent, thin, plaintive, and musical, repeated over and over: a few rising notes, followed by a repeated falling pattern. Quality and pattern vary geographically.

Totally different from much faster Green-backed G song. Contact call while foraging is a series of dry, nasal notes. **Habits:** Usually in small flocks; often mixes with other small insectivores. Gleans leaves and hovers for arthropods mainly in midlevels of the forest. Nest suspended in midstory; oval with downward spout serving as side entrance, woven of fiber and other plant material and featuring a long, dangling tail. Eggs (3) undescribed. **Range:** Endemic. NG, NW Is (Batanta, Misool, Salawati, Waigeo), Bay Is (Mios Num, Yapen), and Aru Is, 0–800 m, but mainly in lowlands.

White-throated Gerygone *Gerygone olivacea* Pl 77
9–10 cm. Restricted to the savannahs and suburban gardens of SE Pen (Port Moresby and environs), where locally common. **Adult:** *White throat contrasting with bright yellow belly.* Note pale edges to wing feathers. Iris red. **Juv:** Underparts entirely pale yellow; pale yellow eyebrow. Plumage pattern muted. Iris brown. **Subspp (1):** *cinerascens* (also Cape York, AU). **Similar spp:** No other gerygone in its habitat. Large-billed and Green-backed Gs in nearby forest lack striking yellow underparts. Fairy G female similar in that the underparts are yellow, but less brightly so, and the tail lacks white spots and the back is green. **Voice:** Song a medium-speed, hesitating, sweet series of weak whistled notes ending in a trill. The series said to be repeated twice. **Habits:** Singly or in pairs, sometimes small parties. Joins mixed flocks of insectivorous birds. Forages in eucalyptus and other trees for arthropods. Rarely in coconut trees and mangroves. Nest a suspended globe with tail, high in a savannah tree. Egg (AU data) white or pink spotted with red or purple. **Range:** NG: southern SE Pen (approximately, Malalaua east to Rigo), 0–650 m. Also AU.

Green-backed Gerygone *Gerygone chloronota* Pl 77, 105
8–9 cm. Keeps to the canopy of lowland and hill forest and edge. Locally common. Often heard but rarely seen. Tiny, short-tailed, and stubby. **Adult:** Distinctive contrasting, 3–colored plumage pattern with *white underparts, grey head, and green back*. The yellowish wash on the flanks is difficult to see—in the field this bird appears all white below. Lacks an eye-ring. Iris red. **Juv:** Similar, but pattern duller and does have a broken eye-ring and brown iris. **Subspp (1):** *cinereiceps* (endemic). **Similar spp:** Yellow-bellied G has yellow breast and less contrasting pattern overall. **Voice:** Song is a very rapid musical series of high notes rolling up and down the scale, over and over. Varies regionally but always much faster than songs of other gerygones. Call is a dry chip note. **Habits:** Singly, in pairs, or in small groups usually high in treetops and difficult to observe. Forages in the canopy and midstory, gleaning and hovering for arthropods. Nest (AU data) a suspended globe with brief tail; entered via a short spout. Eggs (2–3) white with dark spots. **Range:** NG, NW Is (Batanta, Salawati, Waigeo), and Aru Is, 0–1500 m, thought to be mainly a hill forest bird. NG: east to Port Moresby, but perhaps mostly absent from SE Pen. Also AU.

Fairy Gerygone *Gerygone palpebrosa* Pl 77
(Black-throated Warbler)
10 cm. Common in forest of foothills and lower mountains, less frequent in the lowlands. The only gerygone with sexually dichromatic plumage. **Male:** *Unique black face marked with white loral spot and cheek-stripe.* **Female:** Rather plain, *green above and yellow below with white throat* and loral spot. **Juv:** Similar to Female but underparts all yellowish and iris brown, not red. **Subspp (3, in 2 groups):** *palpebrosa* (S NG, from NW Is and Bird's Head to SE Pen), Male with green crown and nape; *wahnesi* (N NG, from NW Lowlands to SE Pen), Male with black crown and nape. **Similar spp:** Yellow-bellied G similar to Female, but back olive (not as green) and tail brown. Green-backed G mainly white below, head grey. **Voice:** Song is an undulating musical warble of sweet, whistled notes, each series repeated a few times before starting another series. Recognizable as a gerygone song, but not so high pitched, and also slower and more rambling than songs of other gerygones. **Habits:** In small parties, often in mixed flocks, gleaning and hovering among the foliage in canopy and midstory. Also monsoon forest of Trans-Fly. Nest suspended from a tree, shrub, or rattan; an elongate dome with a spout side entrance and elongated tail, decorated with lichen, etc. Egg white or pink, spotted red-brown or purple (AU data). **Range:** NG, NW Is (Misool, Waigeo), Bay Is (Yapen), and Aru Is, 0–1450 m. Also AU.

Large-billed Gerygone *Gerygone magnirostris* Pl 77
10 cm. Common and widespread along lowland rivers, creeks, the margins of swamps, and in mangroves. On islands it ranges into forest away from water. **Adult:** *Plain brown or olive-brown upperparts and face marked by a broken white eye-ring* and sometimes a white loral spot. Underparts dingy white. *Tail pattern distinctive: pale terminal band bordered by a dark subterminal band.* Iris red. **Juv:** Eye-ring yellowish; chin to upper breast also yellowish. Iris dark brown. **Subspp (5, in 2 groups):** *magnirostris* (all NG range, excluding that of next race) underparts variable whitish to pale buffy; *hypoxantha* (Bay Is: Biak) underparts washed pale yellow. **Similar spp:** Within its habitat only the Mangrove G is similar, differing in its conspicuous white eyebrow and cheeks and white in outer tail feathers. **Voice:** Songs are slow and cadenced 6- to 9-noted patterns of rising and falling musical slurs, with a somewhat drunken effect, especially when compared with other gerygone songs; reminiscent of call of Black-browed Triller. **Habits:** Found as singles, pairs, or small parties. Forages from canopy to ground level, but often low. Feeding behaviors similar to other gerygones. Nest a suspended, untidy mass with long tail; frequently positioned over water and resembles dangling flood debris. Eggs (2–3) pinkish with red-brown spotting. **Range:** NG, NW Is (Batanta, Kofiau, Salawati, Waigeo), Bay Is (Biak, Yapen), Aru Is, Manam and Karkar Is, and SE Is (3 main D'Entrecasteaux Is, plus Misima, Tagula, Rossel), 0–1200 m. Reaches highest elevational range in broad montane valleys. Also AU. **Taxonomy:** The rare Biak G (*G. m. hypoxantha*) sometimes treated as separate species, but differences are slight.

Mangrove Gerygone *Gerygone levigaster* Pl 77
9–10 cm. An uncommon mangrove specialist of S NG. **Adult:** *Diagnostic facial pattern with white eyebrow and cheeks.* Grey-brown above, white below. *Tail with white spots* toward tips of the feathers. Iris red. **Juv:** Yellowish eyebrow, broken eye-ring, and throat. Iris brown. **Subspp (1):** *pallida* (endemic). **Similar spp:** Large-billed G co-occurs, but lacks white eyebrow and cheek, instead has a brown face with broken white eye-ring and a white loral spot, not always visible. Brown-breasted G, in different habitat, lacks white eyebrow. **Voice:** Song attractive, soft, rising and descending cadence; similar to White-throated G. Call, quiet chatter. **Habits:** Singly, pairs, or small flocks. Gleans foliage of low mangroves and adjacent shrubbery. Nest a suspended pear-shaped mass of woven plant material decorated with cocoons and spider's egg sacs (AU data). Eggs (2–3) white to buff densely speckled with red-brown. **Range:** NG S coast from Bird's Neck (Bintuni Bay) to SE Pen (Rigo). Also AU. **Taxonomy:** Formerly included in Western G (*G. fusca*).

Brown-breasted Gerygone *Gerygone ruficollis* Pl 77
(Treefern Gerygone or **Warbler)**
9–10 cm. A common bird of highland villages, gardens, and second growth wherever there are tall trees. Also at forest edge, less common in primary forest. **Adult:** *Brown above and white below*, it lacks the grey, olive, or yellow of other gerygones. Buffy brown wash on throat gives it its name. White band in outer tail feathers evident in flight. Iris red. **Juv:** Underparts with yellowish wash. Iris brown? **Subspp (2):** Minor. **Similar spp:** Grey Thornbill co-occurs, but is grey above, all white below. Co-occurring scrubwrens and thornbill have underparts brownish or greyish, lack white tips to outer

tail feathers. Large-billed and Mangrove G most similar but occupy different habitats and lack pale edging on the tertials. **Voice:** One of the distinctive songs of montane habitats: a very high-pitched, long, breathless series of mournful whistled notes, seemingly in a minor key, descending the scale unevenly, then monotonously leveling out at a lower pitch.

Call—twittering. **Habits:** Single bird or flock energetically forages at midlevel and treetop. May join with white-eyes and other foraging small birds. Favors casuarinas near settlements. An unusual habitat in SE Pen is tree ferns in alpine grassland. Gleans arthropods from leaves, also sallies and hovers. Nest suspended, globular and lichen decorated, with sided entrance. Egg undescribed. **Range:** Endemic. NG: Bird's Head, Bird's Neck (Fakfak), Central Ranges, Foja Mts, and Huon, 1450–3300 m, rarely down to 1100 m.

AUSTRALASIAN BABBLERS: Pomatostomidae

A small Australasian group of 5 species (2 in New Guinea) that until recently was believed to be a part of the huge and poorly defined babbler family (Timaliidae) of Asia and Africa. Recent studies in molecular systematics have demonstrated that the Australasian babblers show no close affinities to the Asian family in spite of the superficial resemblance and instead are an old and separate lineage. The two Papuan species inhabit lowland forest (Papuan Babbler) and savannah (Grey-crowned Babbler). Both live in family groups that often join mixed-species flocks. They forage with characteristic bounding movements, working their way up through a tree before making a short flap-and-glide flight to the next tree. Australasian babblers build large, covered nests that are also used for roosting.

Papuan Babbler *Garritornis isidorei* Pl 85
(**Rufous Babbler**, *Pomatostomus isidorei*)
25 cm. Common in tangles in lowland and hill forest, though may be locally absent. Always found in small to large noisy flocks. **All plumages:** *A rufous brown bird with curved, yellow-orange bill.* Diagnostic shape: long bill, small head, slim body, and long, rounded tail. Many birds have yellow eyes. Some or all dark-eyed birds are Juvs. **Subspp (2):** Minor—underparts of N birds darker. **Similar spp:** Often associates with a confusing variety of all-brown insectivores: Rusty Pitohui is larger, rangier, black-billed, and white-eyed; Piping Bellbird chunkier, with full crest and black bill; Black Cicadabird female is smaller, with a short, thick bill. **Voice:** Continuously vocal, usually with soft contact notes: *ker—ker—ker. . . .* Also a low, rasping *kowoit?* And a *kss—kss—kss. . . .* Papuan Babbler mimics Rusty P.'s call when flocking with it. **Habits:** Permanent flocks travel together through the forest, from the midstory to the canopy. Very active. Shy but inquisitive. Diet includes mainly large arthropods. The birds roost communally in their nest; this roosting site is replaced each year near the old nest. The pendent nest dangles at midstory height in an open space in the forest interior and is attached to a thin branch or vine, often the spiny whip of a climbing rattan. Built of dried stems of vines and fiber, the nest is a marvelous, large, shaggy, purselike structure with a long, trailing tail. Egg, no information. **Range:** Endemic. NG and NW Is (Misool, Waigeo), 0–300 m. No records from Huon. **Taxonomy:** Separated as a monotypic genus based on degree of molecular genetic divergence.

Grey-crowned Babbler *Pomatostomus temporalis* Pl 85
24 cm. Locally common in Trans-Fly savannahs. *A dark bird with a whitish head and sooty eyestripe.* Cocks *black tail with white tip.* Note curved bill. **Adult:** Pale iris. **Juv:** Dark iris; wing-coverts edged rufous. **Subspp (1):** *temporalis* (shared with AU). **Similar spp:** None. **Voice:** AU data. Noisy. Antiphonal call, *ya-hoo*; Female gives the first syllable, Male the second. A great many other calls, including chuckling, single *chuck, ooi-ai, wee-oo,* etc. **Habits:** Inhabits open savannah woodland, particularly where *Banksia* trees grow. Always in permanent, tight flocks of 6–10 birds. These forage in shrubs and trees, sometimes descending to the ground. The flock builds a group of nests used for roosting as well as breeding. Nests in treetop branches; nest globular in shape with side entrance, built of twigs and bark. Eggs (2–3) pale brown marked with dark squiggly lines. **Range:** NG: Trans-Fly, 0–100 m. Also AU.

LOGRUNNERS: Orthonychidae

This small family of 3 species includes 2 species endemic to Australia and 1 to New Guinea. Logrunners are an old lineage of songbirds without close relatives. They are robust ground birds with drab, concealing plumage and a unique spine-tipped tail that is used as a prop when scratching in leaf litter. Comical, they scurry about the forest floor with semierect posture like windup toys, but stop and hunch forward attentively when searching for food in their own peculiar manner. This involves pushing aside dead leaves and other debris, first with one foot then the other, using a unique, sideways raking motion that clears small patches of ground. Such activity exposes insects and other prey items. Logrunners live in pairs and small groups, but usually only a pair attends the dome-shaped nest built against the base of a tree. Group members of the Australian species sing in a loud chorus to advertise their territory, but this behavior is unknown in NG.

Papuan Logrunner *Orthonyx novaeguineae* Pl 64
(**Logrunner**, *O. temminckii*)
18 cm. Patchily distributed along the main ranges and rarely encountered, but can be locally common. A quail-shaped ground dweller of wet mid-mountain and cloud forest. Mottled brown and black above with *prominent wing bars*. Note *grey face* and frayed-looking, *spiny tail*. **Male:** White throat. **Female:** Orange throat. **Juv:** Mottled black and rufous all over, except belly, which is olive with random white markings. **Subspp (3, in 2 groups):** *novaeguineae* (Bird's Head) paler dorsally, black border on breast patch absent or much reduced in Female; *victorianus* (Central Ranges) darker, and breast border present in Female. **Similar spp:** Forest-rails are rufous. Other forest ground dwellers less compact (and quail not in forest). **Voice:** Usually quiet. Song entirely unlike Australian relatives, a lovely series of 4–6 clear, whistled notes, reminiscent of a mouse-warbler or ground-robin; series descends, as *tee too too too too too* or *tsee too tee toooo*; numerous variations. Alarm call, *eee*. **Habits:** In pairs or small parties. Forages in typical logrunner manner raking with its feet. Reported to hide under leaves and other debris or in holes. Nest, at the base of a tree or multiple stems, a large domed structure outwardly concealed with sticks. Egg (1) white. **Range:** Endemic. NG: Bird's Head and locally in the Central Ranges, but poorly known and seemingly patchy; 1200–3500 m, perhaps mainly in the lower and middle cloud forest zone. Apparently missing from the central and northern portions of the E Ranges. **Taxonomy:** Split from the Australian Logrunner (*O. temminckii*).

SATINBIRDS: Cnemophilidae

The 3 species of this small and enigmatic lineage can be found only in the central ranges of New Guinea. Formerly considered a subfamily of the birds of paradise, the satinbirds have been shown to be only distantly related to them, in spite of their superficial similarity. Molecular studies indicate that cnemophilines are sister to the New Zealand wattlebirds (Calleatidae) and that these two are sister to the berrypeckers and longbills (Melanocharitidae). The 3 sexually dimorphic satinbird species are shy inhabitants of mossy forests, where they feed on a diet largely or entirely of fruit. As far as known, they are polygynous breeders, the males intent on attracting females but leaving nesting duties to the care of their mates.

Loria's Satinbird *Cnemophilus loriae* Pl 66
(**Loria's Bird of Paradise**, *Loria loriae*)
20–23 cm. A common canopy fruit eater of cloud forest and edge throughout the Central Ranges.
Male: *Compact and satiny black with bluish, scaly iridescence on forehead and pale yellow gape.*
Female and Imm: Olive green with brownish-olive wings. Lacks forehead groove. **Juv:** Dull grey.
Similar spp: Crested SB female is larger and more olive brown dorsally and shows a forehead groove. Yellow-breasted SB female and juv are distinctly two-toned (dark above, pale below). *Amblyornis* bowerbirds are much more brown, and larger. **Voice:** Male song is a powerful, ventriloqual, bell-like toll repeated at intervals, delivered 1 per 5 sec. Female utters soft and rasping calls.
Habits: Inconspicuous; seen singly or occasionally in small groups at fruiting trees. The singing male

frequents a favorite exposed perch in the canopy. Little is known of mating behavior other than male displays hanging upside down, wings open, while giving a clicking note. Nest is a cryptic, globular mass of moss and ferns with a side entrance, usually placed on a rock face, but also found on the side of a mossy tree trunk, 1.5 m up. Egg (1) pale pink-buff, spotted with blotched markings. **Range:** Endemic. NG Central Ranges, from the W Ranges to SE Pen, 1800–2400 m (rarely 1450–3000 m).

Crested Satinbird *Cnemophilus macgregorii* Pl 66
(Crested Bird of Paradise)
25 cm. A quiet, plump, and retiring fruit eater, uncommon in high mountain forest and tangled subalpine thickets. **Male:** *Orange or yellow-orange dorsally, and blackish brown ventrally*; the bizarre crest feathers resemble moth antennae and rise from a groove in the forehead.
Female and Imm: Brownish olive, palest on belly (with buff wash). Note diagnostic *forehead groove* and unusual *grey eye-color* (brownish in Imm). **Juv:** Dull grey. **Subspp (2):** *sanguineus* (E Ranges) Male deep orange dorsally; *macgregorii* (SE Pen) Male yellow-orange dorsally. Populations from W and Border Ranges undescribed. **Similar spp:** Male unique. Loria's SB female is smaller and greener and lacks the forehead groove. Macgregor's Bowerbird female is browner, thicker-billed, and has a dark iris. **Voice:** Usually silent. Male song is an explosive muffled bark similar to call of Macgregor's Bowerbird and repeated after a long interval of several minutes. His wings make a whirring noise in flight. Female: at nest, *wark wark* and a soft, sharp *whit*; when disturbed, a soft, churring growl.
Habits: Solitary and rather inactive, occasionally 2–3 birds together. Inhabits forest interior and edge, feeding entirely on fruit. Mating habits and courtship display unknown, but presumably polygynous breeding. Nest is a mossy globe with a side entrance set on a stump or aside a trunk, 2.6 m up. Egg (1) pinkish buff with broad streaks and blotches. **Range:** Endemic. NG: W Ranges (Lake Habbema above Baliem Valley) to SE Pen (Owen Stanley Range), but possibly absent from many sites in the West; 2300–3650 m, usually above 2600 m.

Yellow-breasted Satinbird *Loboparadisea sericea* Pl 67
(Yellow-breasted, Wattled, Wattle-billed, or Shield-billed Bird of Paradise)
16–18 cm. Rare, patchily distributed, enigmatic. A retiring and seldom-seen fruit eater of mid-mountain forest canopy of the Central Ranges. **Male:** *Rich brown dorsally*, darker on crown; *rump and underparts silky yellow*; a unique 2-lobed, bulbous, *blue-green wattle sits atop the bill*.
Female: Duller than Male and lacks wattles; note slightly larger size and obscurely mottled underparts. **Juv:** Like the Female, but underparts buffy with darker brown edging to breast and belly feathers, most concentrated in throat and upper breast. **Subspp (2):** Minor. **Similar spp:** Male and Female unique. Juv resembles a Streaked Bowerbird (which occurs further east of this species). **Voice:** Song is a series of loud, harsh, grating notes slightly upslurred: *sssh sssh sssh . . .* ; the series slows and becomes more emphatic or described as 2 notes followed by a brief pause then another 2 notes; not unlike that for Superb Bird of Paradise. **Habits:** An unobtrusive, fruit-eating inhabitant of forest, being found both low and high in the vegetation. Mating behavior unknown. Nest and egg, no information. A local informant noted the species lays a single egg in an open moss structure. **Range:** Endemic. NG: W Ranges to northeastern SE Pen (Herzog and Kuper Ranges), 625–2000 m, mainly above 1200 m.

BERRYPECKERS AND LONGBILLS: Melanocharitidae

This family of 10 species is endemic to New Guinea. The relationships of the family are not as yet resolved. The most recent molecular studies either aligned the Melanocharitidae within the core Corvoidea or were unable to resolve their placement. These small forest-dwelling songbirds are divided into 2 groups, the berrypeckers (6 spp) and the longbills (4 spp). The berrypeckers were traditionally placed with the flowerpeckers in the Dicaeidae, but it is now clear that the two are unrelated. Similarly, the longbills were thought to be either honeyeaters or sunbirds. Beginning in the 1990s, molecular studies established that the berrypeckers and longbills are each other's closest relatives, and the 2 groups were joined in family Melanocharitidae. In general they are small, very active, forest-dwelling songbirds, olive green in color. The females are the larger sex in 2 species of berrypecker. Berrypeckers and longbills feed by gleaning and hovering for fruit or nectar, respectively; both take insects and spiders. They build small, compact, cup-shaped nests that are carefully woven and lined with plant material or lichen.

Berrypeckers. Of the 6 species, 3 are common and form an altitudinal sequence in forests: Black B in the lowlands, Mid-mountain B at middle elevations, and Fan-tailed B at higher elevations. The other 3 species, Obscure, Streaked, and Spotted Bs, are rare species of hill and montane forests, with poorly understood ecological requirements; Streaked seems to be nomadic and congregates where there is abundant fruit.

Obscure Berrypecker *Melanocharis arfakiana* Pl 78
11–12 cm. One of New Guinea's rarest birds. Known from infrequent sightings in hill forests. **Adult:** A tiny, stubby, greenish bird with grey underparts, *pale tan-horn bill, prominent orange gape, and lemon-yellow pectoral tuft.* Head *greyish.* Easily recognized by the pale bill and bright gape. **Similar spp:** Female Black B has pectoral tuft white or the palest yellow, bill dark, and head green like back rather than greyish green. Female Mid-mountain B with tail much longer and belly with a yellow wash. Pale-billed Scrubwren also has a pale bill, but is usually mottled below. **Voice:** Song said to be a rapid *swizzly swizzly swizzly swit*, faster and higher pitched than Black B; also described as a rapid, 3-part jiggly trill on a single pitch, the first note separated by a minor pause. Call described as *zit zit zit.* **Habits:** Active in the lower forest canopy and midlevels. Takes small fruits and arthropods. Habits presumably similar to other berrypeckers. No nesting information. **Range:** Endemic. Known from one specimen from Bird's Head, another from SE Pen, and 2 more from S Border Ranges (Tabubil). The range map shows these locations plus sightings from the Adelbert Mts, Huon, and near Port Moresby (SE Pen), 640–1100 m. **Taxonomy:** Whether this species even existed was at one time in doubt, but recent sightings and collected specimens have revealed the bird in life.

Black Berrypecker *Melanocharis nigra* Pl 78
11–12 cm. The common berrypecker of lowland and hill forest. *A small and stubby bird with white pectoral tuft* (yellow in N Huon). **Male:** *Black upperparts,* but wing-coverts and underparts vary regionally (see Subspp). **Female and Juv:** Green above, grey below. **Subspp (4, in 3 groups):** *nigra* (NW Is, Bird's Head and Neck) Male black above, grey below; *chloroptera* (S Lowlands and Aru Is) Male similar, but wing green; *unicolor* (NW Lowlands to SE Pen) Male all black. **Similar spp:** Mid-mountain B has a longer tail, yellow pectoral tuft, and in the female a yellow wash on the belly. Obscure B has shorter tail, pale bill, and bright gape. Flowerpeckers have shorter tail; most plumages have some red. **Voice:** High pitched, twittering, variable song with rolling phrases repeated over and over. Notes high pitched and scratchy. **Habits:** Solitary or in pairs. Regularly with mixed flocks of small birds. Active and quick in its movements. Feeds on small fruits and arthropods in forest, from understory to canopy. Drawn to fruits of weedy *Piper* shrubs and vines at forest edge. Hovers when taking fruit and insects. Nest a neat, compact cup decorated with pale flakes of lichen and bound to a thin branch in the forest midstory. Eggs (1–2) creamy with dark streaks and blotches. **Range:** Endemic. NG, NW Is (Misool, Salawati, Waigeo), Bay Is (Mios Num and Yapen I), and Aru Is, 0–1200 m, occasionally higher.

Mid-mountain Berrypecker *Melanocharis longicauda*   Pl 78
13 cm. The common berrypecker of mid-mountain forest. *Tail longer than Black B's and with white outer margin. Belly with yellow wash, and pectoral tuft yellow (not white).* **Male:** Upperparts glossy blue-black; note the white basal half of the outer tail feathers. **Female and Juv:** Upperparts olive. **Subspp (4):** Slight variation in amount of yellow or shading in the underparts; SE Pen birds have a large white spot on the inner web of the tail feathers. **Similar spp:** Black B. Obscure B with shorter tail and pale bill. Fan-tailed B with much longer tail; male clean grey below without yellow in the underparts and pectoral tuft; female with white nearer base of tail. **Voice:** Song a rolling series of several repeated phrases, e.g., *jjeteleet jjeteleet jjeteleet.* Call is a downslurred *bzhsst!* **Habits:** Behavior, nest, and egg similar to those of Black B. **Range:** Endemic. NG mts (no records yet from Cyclops, Kumawa, or Adelbert Mts), 700–1900 m.

Fan-tailed Berrypecker *Melanocharis versteri*   Pl 78
(*Pristorhamphus versteri*)
14–15 cm. Common in montane cloud forests. The brilliant and sprightly Male is eye-catching. Both sexes have a *long, graduated tail* flashing white when fanned. *Pectoral tuft white.* **Male:** *Iridescent blue-black with contrasting clear grey underparts.* Smaller than Female, with proportionately longer tail. **Female and Juv:** Olive with underparts varying from unpatterned to obscurely streaked. Note yellow gape line. Juv male shrinks slightly as he matures and molts into Adult plumage. **Subspp (3):** Slight regional variation in darkness of the underparts (darkest in W Ranges) and size of the white tail patch (largest in SE Pen). **Similar spp:** Mid-mountain B with shorter, more evenly tipped tail and yellow pectoral tuft. Garnet Robin female smaller with white patches in wing. Papuan Whipbird female larger, exceedingly rare, forages on or near the ground (berrypecker sometimes does, too), but lacks whitish belly and flanks; no yellow gape mark. **Voice:** A harsh, buzzy *shwet, dee dee dee dee dee*, and other high-pitched notes, squeaks, and nasal scolds. **Habits:** Solitary or in pairs. Foraging behavior similar to Black and Mid-Mountain Bs. Hover-gleans conspicuously. Often encountered in dense thickets at forest edge where it may descend to the ground. Nest and egg similar to those of Black B but larger; much fern fiber recorded in nest. **Range:** Endemic. NG mts (absent from Fakfak, Kumawa, and Adelbert Mts), 1700 m to tree line, 3680 m maximum. Wanders to lower elevations, e.g., 1450 m.

Streaked Berrypecker *Melanocharis striativentris*   Pl 78
(*Neneba striativentris*)
13–14 cm. Inhabits mid-mountain forests where it is usually rare but may be locally common or abundant. Always inconspicuous and probably often overlooked. **Adult:** Olive with *streaky breast and belly* and *broad buff eye-ring and orange gape.* Bulkier than other berrypeckers. Adult Male averages smaller than Female but with longer tail; Juv similarly proportioned to Female. **Subspp (3):** Minor; Male in E Ranges and SE Pen with obscure white base of outer tail feathers; this feature is much reduced or absent in Female and Juv. **Similar spp:** Other berrypeckers lack buff eye-ring and streaking. **Voice:** A scratchy *schree schree schree schree*, harsher than other berrypeckers. **Habits:** Secretive and poorly known. Solitary, pairs, or small groups. Sometimes absent and perhaps nomadic. Forages for small fruits in the lower and middle stories of forest interior. Fond of *Piper* fruit. Nest similar to Black B but larger. Eggs (2) larger and with darker blotches. **Range:** Endemic. NG: Central Ranges and Huon, 1150–2300 m.

Spotted Berrypecker *Rhamphocharis crassirostris*   Pl 78
11–12 cm. A rare berrypecker of mid-mountain forests. Irregular occurrence suggests nomadic movements. From other berrypeckers by the *longer bill*, suggestive of a honeyeater's. Bill differs from honeyeater's in being *straight, rather blunt-tipped.* Also contributing to this species' odd shape is the narrow, squared-off tail. **Male:** Except for the long bill, Male resembles the females of other berrypeckers—olive green above, grey below, with yellow pectoral tuft. **Female and Juv:** Dark brown and spotted throughout. **Subspp (3, in 2 distinct groups):** *crassirostris* (Bird's Head to Border Ranges) much shorter bill (approaching the proportions of other berrypeckers) and Female with a pale, dark-spotted breast; *piperata* (E Ranges, Huon, SE Pen) longer bill and Female dark below with white spots. **Similar spp:** Spotted Honeyeater is larger with dark spotting on the underparts and yellow ear patch. **Voice:** For *piperata*, a rapidly delivered series of downslurred notes, on 1 pitch: *seeu seeu seeu seeu . . .* or *tseeut tseeut . . .*, suggestive of an insect, but otherwise typical of a berrypecker. Also a sibilant, sucked-in outburst, like that of a Red-collared Myzomela. **Habits:** Poorly known. Generally solitary, but does gather at popular fruiting trees, such as figs. Forages for small fruits and insects in

flowering trees, frequenting the lower canopy and midstory. No nesting information. **Range:** Endemic. NG: Bird's Head, Central Ranges, Foja Mts, and Huon, 1500–2500 m, though wanders lower and higher. **Taxonomy:** The 2 forms are being considered for species status as the Spotted Berrypecker (*R. crassirostris*) and the Long-billed Berrypecker (*R. piperata*).

Spectacled Longbill *Oedistoma iliolophus* Pl 79
(**Dwarf** or **Grey-bellied Longbill, Dwarf Honeyeater,** *Toxorhamphus iliolophus*)
11 cm. An inconspicuous, small, plain-colored longbill with yellow eye-ring inhabiting forests from the lowlands to mid-elevation in mountains. **All plumages:** Identified by *olive-grey color, short tail, and somewhat long bill*. The distinctive yellow, fleshy eye-ring is difficult to see in the field. Female smaller. **Subspp (4, in 3 groups):** *iliolophus* (NG) with olive crown; *cinerascens* (Waigeo I) with grey crown; *fergussonis* (D'Entrecasteaux Is) like *iliolophus* but much larger. **Similar spp:** Pygmy L nearly identical but appears nearly tailless, is much smaller, flocks more, has a different call, and lacks a yellow eye-ring. Yellow-bellied and Slaty-headed Ls darker with yellow breast and longer bill. See also female sunbirds. **Voice:** No song reported. Call a rapid series of spitting notes, often delivered in flight; harsher on Bird's Head, like 2 pebbles struck together. **Habits:** Forages singly or in pairs for insects; sometimes joins mixed feeding flocks of small insectivores. Gleans insects from leaves, often hovering, and from bark while scaling branches. Visits flowering trees for nectar. Ranges from understory to canopy; forages higher than the larger longbills. Nest a small cup covered with dried leaves and lined with plant silk. Egg (1) white with dark spots. **Range:** Endemic. NG, NW Is (Waigeo), Bay Is (Mios Num and Yapen), and SE Is (3 main D'Entrecasteaux Is), 0–1750 m.

Pygmy Longbill *Oedistoma pygmaeum* Pl 79
(**Pygmy Honeyeater**)
7 cm. *Smallest New Guinea bird.* This tiny mite forages in active flocks that call incessantly from the treetops in lowland and hill forest. Common. **All plumages:** *Short tail* and slender, curved bill on an otherwise *plain, olive-grey, insect-sized bird*. Female smaller on average. **Subspp (2):** *pygmaeum* (NG and NW Is) greenish head, yellowish underparts; *meeki* (D'Entrecasteaux Is) larger with grey head, whitish underparts. **Similar spp:** Spectacled L is similarly colored but larger, with proportionately longer tail and bill and a yellow eye-ring. **Voice:** No song reported. Call a repeated, weak, rapid train of sputtered notes easily distinguishable from that of Spectacled L in being higher pitched and sweeter, with series of notes slurred together. **Habits:** In pairs or small flocks. Gleans outer shell of vegetation, not hidden in shaded forest like Spectacled L. Forages for arthropods and also visits flowering trees for nectar. Hovers. No nesting information. **Range:** Endemic. NG, NW Is (Misool, Waigeo), and SE Is (3 main D'Entrecasteaux Is), 0–700 (1300) m.

Yellow-bellied Longbill *Toxorhamphus novaeguineae* Pl 79
12–13 cm. A common, hyperactive nectar feeder that darts between flowering plants mainly in the understory and midstory of lowland and hill forest. Note the *very long, curved bill* and short tail. **All plumages:** *Olive green with thin, pale eye-ring.* White terminal spots in tail may be visible. Female smaller. **Subspp (2):** *novaeguineae* (northern lowlands from NW Is and Bird's Head to Sepik-Ramu) with mostly olive underparts; *flaviventris* (S Lowlands) with yellow breast and belly. **Similar spp:** Slaty-headed L, with dark grey head and wing, is usually segregated at higher elevations. Olive-backed Sunbird female has yellow eyebrow and favors open habitat. Spectacled L smaller and has a much shorter, finer bill and is pale yellowish grey below. **Voice:** Sings and calls often. Song, similar to Slaty-headed L, is a few downslurred whistles followed by jumbled notes. Various call notes: a high *pee*, weak sneezelike notes, a sharp *ks*, a rapid *ks ks ks*, and a repeated piping note like that of a meliphaga. **Habits:** Forages for insects and nectar. Solitary or in pairs. No nesting information. **Range:** Endemic. NG, NW Is (Batanta, Misool, Salawati, Waigeo), Bay Is (Yapen), and Aru Is. Absent from SE Pen (east of Huon Gulf and Lakekamu Basin). Usually 0–500 m in the South, extending higher in the North, occasionally to 1200 m.

Slaty-headed Longbill *Toxorhamphus poliopterus* Pl 79
(**Slaty-chinned Longbill**)
13 cm. Replaces Yellow-bellied L in montane forests. **All plumages alike:** Look for the combination of *grey head and wings and yellow throat patch*. White tail tips flash as the bird flits past. Female smaller and paler on average. **Subspp (2):** Minor. **Similar spp:** Yellow-bellied L olive overall; lowlands.

Black Sunbird female (lowlands and hill forest) with pale grey throat and breast; prefers open habitat and forest canopy. Spectacled L pale yellowish grey below, yellow eye-ring, and shorter bill. **Voice:** Song is a bubbling profusion of slurs, disyllabic notes, and other repeated notes. One call is a musical, downslurred *cheep!* or *tsip*; also *tsee tsee tsee*; scolding; other high-pitched notes. **Habits:** Similar to Yellow-bellied L. Nest a neat cup of fine, green vegetation and fibers woven onto a thin sapling or branch in the understory or midstory; lined on the outside with spider silk and egg sacs, on the inside with white plant material. Egg (1) white. **Range:** Endemic. NG Central Ranges, Adelbert Mts, and Huon, mostly 500–2000 m (extremes 300–2450 m).

PAINTED BERRYPECKERS: Paramythiidae

This family of 3 colorful species is endemic to the mountains of New Guinea. Although the crested and tit berrypeckers share certain structural features, and based on molecular genetics also seem to be each other's closest relatives, their relationships with other oscine songbirds have not been resolved and await further study. The crested berrypeckers occupy an elevational range mostly above that of the Tit Berrypecker. All species feed mainly on berries and are quite social. Neither species is known to sing, but their calls are distinctive. Their nests are mossy cups.

Tit Berrypecker *Oreocharis arfaki*   Pl 78
(Painted Berrypecker)
12–14 cm. A flocking berrypecker usually high in the treetops of montane forests, common throughout New Guinea. Diagnostic call frequently heard in flight. Look for the orange-and-yellow bellied Male. *Yellow tertial spots* in both sexes diagnostic when seen. **Male:** *Underparts yellow with burnt orange stain down the center; black head with golden cheek.* **Female and Juv:** Field marks difficult to see when high overhead. Combination of green upperparts, grey throat, and *dark scalloped yellowish breast and belly* is diagnostic. Note white-spotted cheek. **Similar spp:** None, but Female must be seen well to separate from other small, cryptic-plumaged birds. **Voice:** A high-pitched, thin contact note, *sss* or *zzz*, frequently repeated. **Habits:** Usually paired and in flocks actively foraging for small fruit. Spends most of its time in the forest canopy. A strong flier through or over the canopy. Nest a mossy cup. Egg undescribed. **Range:** Endemic. NG: Bird's Head, Bird's Neck (Wandammen Mts), Central Ranges, Foja Mts, N Coastal Ranges, and Huon, 1700–3000 m (descends to 1450 m).

Eastern Crested Berrypecker *Paramythia montium*   Pl 78
(Crested Berrypecker)
20–22 cm. A large, elegant, and confiding berrypecker heralded by its weak, scratchy voice. Common in the high montane forests and shrublands in the E sector of the Central Ranges and Huon Pen. This and the following species are immediately identified by the *distinctive black mask, pointed crest, and white eyebrow*. Otherwise a blue and green bird with *yellow undertail-coverts*. Female averages smaller. **Adult:** Distinguished from Western CB by its shorter, drooping crest; long, greyish-white eyebrow; bright green back; yellowish flank streak (often hidden under the wing); and greenish-yellow undertail. **Juv:** Duller; pattern obscured; black mask and blue plumage much reduced, replaced by green. **Subspp (2):** *montium* (Central Ranges); *brevicauda* (Huon) tail noticeably shorter. **Similar spp:** See Western CB. **Voice:** Faint squeaking, kissing, and rasping sounds. Wings rustle. **Habits:** In pairs or small groups. Often in mixed-species flocks with insectivores. Feeds on small fruits at all levels in the forest. Nest a shaggy, mossy cup built a few meters above ground in a shrub or small tree. Egg (1) white with dark spots. **Range:** Endemic. NG: Border and E Ranges, SE Pen, and Huon, 2500 m to timberline (rarely down to 2100 m). **Taxonomy:** No longer includes populations now classified as Western CB.

Western Crested Berrypecker *Paramythia olivacea*   Pl 78
(formerly classified as part of an inclusive **Crested Berrypecker**, *P. montium*)
20–22 cm. Western CB and Eastern CB approach each other somewhere in the W Border Ranges. Western is distinguished by its longer, upward-curved crest; short, all-white eyebrow; black (vs white) nape; olive back; absence of a yellow flank streak; and pure yellow undertail. **Voice and Habits:** Similar to Eastern CB, as far as known. **Range:** Endemic. NG W Ranges to W Border Ranges.

WHIPBIRDS: Psophodidae

Often classified with the jewel-babblers and quail-thrushes, the whipbirds now belong in their own family according to molecular studies. This small family of 5 species of ground birds lives mainly in Australia, with only 1 New Guinea representative, the infamously rare and elusive Papuan Whipbird.

Papuan Whipbird *Androphobus viridis* Pl 64
16–17 cm. A little-known ground bird of cloud forests. A lot smaller than Australian whipbirds. *Dark olive green.* **Male:** *White moustachial stripe, black throat.* **Female:** Looks all moss green; lacks Male's white facial stripe and black throat. **Juv or Imm (presumed):** Darker green with sooty face and underparts, lacks white facial stripe. **Similar spp:** Female Fan-tailed Berrypecker resembles female whipbird, will descend to the ground, and is common and more likely to be seen; however, it is really an arboreal bird and smaller and less robust, and it has paler, greyish underparts (whipbird evenly green-colored); note also the pale yellow gape mark, lacking in whipbird. Juv Rufous-naped Bellbird is also mostly or entirely green, is the same size and lives on the ground, but is stockier and often has some brown patches in the plumage. **Voice:** Song unknown. Calling between a pair is a rapidly repeated, strident scolding, like a person squeaking between pursed lips to attract birds. **Habits:** Perhaps usually in pairs. Forages for insects and other invertebrates on the ground or in thickets while keeping to low cover along trails and roads. The genus name translates to "afraid of people"—how true. Extreme patience and quiet are required to see this bird. No nesting information. **Range:** Endemic. NG: known from a few localities in W Ranges and E Ranges (Tari area), 1400–2700 m. **Taxonomy:** Its relationship within the family has not been studied.

JEWEL-BABBLERS AND QUAIL-THRUSHES: Cinclosomatidae

This family of 11 species of medium-sized, insectivorous ground birds is endemic to Australia and New Guinea (5 spp). As with the preceding family, the jewel-babblers and quail-thrushes were once classified with the Asian and African babblers but are now placed in their own family of Australasian songbirds. In the New Guinea Region, these are shy, skulking birds of the forest interior that sing and call often but avoid showing themselves. Seen by chance, particularly when a bird is distracted by territorial activity, including response to sound playback.
Each species has numerous songs and calls, the signature elements of which are high-pitched tonal whistles and explosive whip-cracks. All species find their prey of invertebrates and small vertebrates by walking and searching, stopping occasionally to turn fallen leaves. They live in pairs or small groups and nest on the forest floor.

Spotted Jewel-babbler *Ptilorrhoa leucosticta* Pl 64
(*Eupetes leucostictus*)
20 cm. The jewel-babbler of cloud forest; uncommon. **Adult:** *Large, white cheek patch* on black face. *Rows of white spots on shoulder.* Tail with broad, white tips to the outer feathers. Male larger.
Juv: Heavily washed olive; wing spots whitish or rufous. **Subspp (7):** Much regional variation in amount of green and blue on the breast and hue of chestnut cap, but can be divided into 4 groups based on dorsal color and breast spotting: *leucosticta* (Bird's Head) mantle, back, and rump olive green, bib black with white speckling; *centralis* (Bird's Neck to Border Ranges) mantle, back, and rump brown, little or no spotting on small bib; *sibilans* (Cyclops Mts) green-backed with green breast band; *loriae* (E Ranges, N Coastal Mts, Huon, SE Pen) mantle, back, and rump olive green, no spotting on upper breast. **Similar spp:** Chestnut-backed JB lives at lower elevation and is chestnut and blue with all-white throat. **Voice:** Less vocal than other jewel-babblers. Song, a high, clear, ringing note repeated rapidly for up to 10 sec, easily confused with song of Black-capped Robin. Various calls: a short, clear, piercing, and slightly downslurred *siuk!* or *seeya!*; a rapidly repeated, unmusical rattled note; a soft growl; a short *chup.* **Habits:** In pairs or small parties on the forest floor. Only moderately shy. Walks along ground, turning leaves and searching for invertebrates and small frogs. Occasionally eats fruit. Nest located within a meter of the ground in a cavity at the base of a tree or tree fern; a cup of rootlets, moss, and

dead leaves built on a platform of sticks and leaves. Eggs (2) pink or pale brown with spots. **Range:** Endemic. NG: nearly all mts (no records from Fakfak, Kumawa, or Adelbert Mts), 1750–2400 m.

Blue Jewel-babbler *Ptilorrhoa caerulescens* Pl 64
(**Lowland Jewel-babbler**, *Eupetes caerulescens*)

22 cm. The widespread jewel-babbler of lowland forest; common. *All blue-grey, with clear white throat and black eye-stripe.* **Adult:** Sexes alike, except Female in South with white eyebrow and in far West (Bird's Head and Misool I) with thin black eye-stripe. **Juv:** Browner underparts. **Subspp (3, in 2 groups):** *caerulescens* (Bird's Head, Misool I, NW Lowlands, Sepik-Ramu) sexes alike (Female from Bird's Head and Misool has white of throat extending up to eye); *nigricrissus* (S Lowlands, SE Pen) blackish undertail and Female has white eyebrow. **Similar spp:** Dimorphic JB range narrowly overlaps in the Adelbert Mts (E Sepik-Ramu) and Huon; male has brownish cap, female is all brown. Chestnut-backed JB has chestnut back. **Voice:** Striking geographic variation. Song is a high, drawn-out whistle, *eeeee-eeeee-eeeee* . . . , lasting many seconds and broken with brief pauses; it sometimes ends in a loud, short, explosive note. Also variations of this thin, whistled song. Only from NW Lowlands is a loud unmusical crescendoing note. Another song is a pulsing, repeated, whirring trill that increases in volume but remains monotone, *tsip fish-fish-fish* . . . (rising to) . . . *FISH-FISH-FISH.* . . . Alarm call is a loud *CHOP ti-ti-t-ti.* . . . Other calls are loud *chink* notes in various short series, suggestive of a Sooty Thicket-Fantail. **Habits:** A vocal ground dweller in territorial pairs. Shy. Often in thickets; when undisturbed, it forages in open forest floor. Searches for insects by scanning ahead, peering under logs and other objects, and tossing leaves. Wings whir when flushed. Nest placed at the base of a tree or sapling; built as a cup sunk into the forest floor and lined with dead leaves and rootlets. Eggs (2) brownish white with spots and streaks. **Range:** Endemic. NG and NW Is (Misool and Salawati Is), 0–300 (to 800) m. Lowlands throughout NG, except Huon and northern SE Pen, and possibly Trans-Fly.

Dimorphic Jewel-babbler *Ptilorrhoa geislerorum* Pl 64
(**Brown-capped Jewel-babbler**)

22–23 cm. Common in lowland and hill forest in its limited range; seems confined to hills only where it meets Blue JB, although both seen near each other in the Adelbert Mts. **Male:** Similar to male Blue JB, but has *crown with a brownish tinge.* **Female and Juv:** *Brown with white throat and black eye-stripe.* **Similar spp:** Blue JB lacks brownish cap. Chestnut-backed JB is two-toned, with brown back and blue breast. **Voice:** Song, a long series of connected, disyllabic, whistled notes with a slightly bouncing rhythm; the notes begin short and faint and progressively become longer, higher in pitch, and increase greatly in volume. Another song is a quiet *peepeepee* . . . , the notes very rapid. Also a variety of other whistled or explosive calls. **Habits:** Similar to Blue and Chestnut-backed JBs. Coexists with the latter in the Adelbert Mts. Nest undescribed. Egg (clutch?) similar to Chestnut-backed JB. **Range:** Endemic. NG: Adelbert Mts, eastern Huon, and northern SE Pen (to Collingwood Bay), 0–1200 m. **Taxonomy:** Split from Blue JB.

Chestnut-backed Jewel-babbler *Ptilorrhoa castanonota* Pl 64
(**Mid-mountain Jewel-babbler**, *Eupetes castanonotus*)

23 cm. The jewel-babbler of mid-mountain forests; common. *White throat, chestnut back, and blue underparts.* **Male:** Rump blue, eyebrow blue. **Female:** Rump chestnut, like back (blue like Male in race *uropygialis*); some E races have cream or pale blue eyebrow. **Juv:** Dark brownish grey with white throat. **Subspp (7):** Complex variation in darkness and hue of brown upperparts, blue underparts, and details thereof, but lumped here into 4 groups: *castanonota* (Bird's Head and Batanta I) underparts blue, Female with blue eyebrow; *saturata* (W Ranges) dark overall, underparts purple-blue, Female with blue eyebrow; *pulchra* (eastern E Ranges, Huon, SE Pen) underparts blue, Female with eyebrow white, tan, or pale blue; *uropygialis* (N slope of W Ranges and N Coastal Mts) nearly sexually monochromatic, Female with lower back blue as in Male. **Similar spp:** Dimorphic JB is either all blue (male) or all brown (female), not two-toned brown and blue. Spotted JB is greenish and has white spots on wing-coverts and black throat. **Voice:** Song is a series of drawn-out whistles building into a complex jumble of loud, sputtering, explosive *CHEW* notes. Shorter versions give only a few whistles before the CHEW notes. Call when flushed, a loud *CHEW-CHEW!* **Habits:** Similar to Blue JB. Nest built on the ground at the base of a tree, structure similar to Blue JB. Eggs (1–2) buff or yellowish, marked with dark spots or blotches. **Range:** Endemic. NG, Batanta (NW Is), and Yapen (Bay Is), 900–1450 m (300–1580 m). Nearly all mts of NG (no records for Cyclops Mts).

Painted Quail-thrush *Cinclosoma ajax*   Pl 64
23 cm. Locally distributed in foothill forests of the southern Central Ranges and gallery forest in the Trans-Fly, uncommon where found. A grey-brown bird of the forest floor that shows *a white facial stripe or throat*. Note *white tips of outer tail feathers* when it flushes. Eye orange. **Male:** Boldly patterned, with black face and foreparts marked with a white moustachial stripe and rufous or buff flanks. **Female:** White eyebrow and throat, rufous breast. **Juv:** Like Female, but iris paler. **Subspp (3):** Minor, becoming paler from W to E. **Similar spp:** Papuan Scrub-Robin is also brown and has white tail tips when flushed, but is all white below and has a black diagonal stripe through the eye. Blue Jewel-babbler has similar voice and shape, but is blue-grey and does not show white in tail when it flies. **Voice:** Song begins with 1–3 short notes, then a thin, drawn-out whistle that increases in volume then trails off, similar to a jewel-babbler but of shorter duration. A high-pitched whistle-call is like that of the Papuan Scrub-Robin, but each drawn-out note drops in pitch. Another call is an upslurred, explosive *FWIP* repeated every 2 sec in a long series, far-carrying although the volume is not great. Contact call is a high-pitched *tsst*, given while the bird walks. **Habits:** A shy, insectivorous ground dweller. Quiet, slow-moving, and found in pairs or small parties. Forages by walking along, stopping to investigate or turn aside leaves. No nesting information. **Range:** Endemic. NG mainland: Bird's Neck, S Lowlands, Trans-Fly, and southern SE Pen, 0–800 m.

BOATBILLS: Machaerirhynchidae

This family of 2 species inhabits Australia and New Guinea. Molecular systematic research and morphological studies now point to family-level designation for the boatbills within the corvoid songbirds. Boatbills are small rainforest flycatchers that resemble monarchs. Their defining feature is a wide, flat bill. They build a small nest suspended from a branch fork.

Yellow-breasted Boatbill *Machaerirhynchus flaviventer*   Pl 90
11–13 cm. The boatbill of lowland and hill forest; uncommon. *A tiny, yellow-breasted bird that twitches from side to side* on its perch and flicks its *small, cocked tail*. Note *white chin*. **Male:** Upperparts black, often mixed with green; bright yellow below; black on lores only. **Female:** Dusky olive above, paler yellow below with faint dusky markings on throat and flank; dark eye-stripe passes through eye onto ear-coverts. **Juv:** Resembles the Adult of its sex but shows reduced wing bars; Male's underparts with faint dusky markings, Female's often heavily dark-scalloped. **Subspp (4, in 2 groups):** *albifrons* (NW Is, Bird's Head to Sepik-Ramu) forehead and eyebrow white; *xanthogenys* (S Lowlands, Huon, SE Pen) these parts yellow. **Similar spp:** Black-breasted B has black spot on breast, confined to mountains. Golden Monarch male orange-yellow, has black throat. **Voice:** Song a cheerful phrase repeated over and over; quiet and easily overlooked but diagnostic—weak, musical, and trilled, at first descending (and with vibrato), then rising: *teee-yoo o o o-wheet!*, as if a small teakettle were boiling over. Also: *twi twi twit—twe tw twet*. Call notes are spitting or grating, reminiscent of weak version of call of Fantailed Monarch. **Habits:** Singly or as a pair, joins mixed foraging flocks of small insectivores, particularly Yellow-bellied Gerygone. Forages in the canopy and midstory. Actively gleans foliage and sallies. In AU, nest is built in midstory, in fork at end of thin branch; nest a suspended, frail cup. Eggs (2) white with spotting. **Range:** NG, NW Is (Misool, Salawati, Waigeo), and Aru Is, 0–800 m (rarely to 1300 m). Also AU.

Black-breasted Boatbill *Machaerirhynchus nigripectus*   Pl 90
13–14 cm. The highland boatbill; common in forest and edge. Resembles Yellow-breasted B but larger and less warbler–like and has a *black patch in the middle of the breast* and yellow (not white) throat. **Male:** Glossy black above, deep yellow below, yellow face with black lores. **Female:** Duller greenish black, medium yellow below, breast patch smaller and more ragged, black eye-stripe extends behind eye. Black breast patch with more ragged edge. **Juv:** Female duller and may have scalloped breast. **Subspp (3):** Minor. **Similar spp:** Yellow-breasted B and *Microeca* robins lack black breast spot. **Voice:** Song, a scratchy, buzzy, strident but musical descending series: *swee bzzbzzbzzbzzbzz*, first note upslurred, the trill of *bzz* notes descending in scale, and often repeated once or twice. Sometimes mellow notes are inserted. Call a metallic *tsk*. **Habits:** More overt than Yellow-breasted B. Nest built on a thin, midstory branch; a neat cup with thick walls, made of moss and fiber and decorated with lichen. Eggs (2) white with fine dark speckling and blotches. **Range:** Endemic. NG: Bird's Head and Neck (Fakfak Mts), Central Ranges, Foja Mts, Adelbert Mts, and Huon, 1300–3000 m.

BUTCHERBIRDS AND ALLIES: Cracticidae

The butcherbirds, Australian magpie, and 2 peltops belong to an Australasian songbird family of 12 species, 7 of which inhabit New Guinea. They vary in size from medium to very large, are black or black and white, and have a heavy bill. They inhabit lowland forest and savannah and penetrate the highlands in disturbed habitats and openings, although the Mountain Peltops is strictly a montane bird. Their food is insects and small vertebrates, less often fruit. All species are conspicuously vocal. They are either solitary or live in small groups. The nest is a bulky, cup-shaped structure placed in a branch fork in the crown of a tree. Depending on species, both parents, and sometimes additional individuals, attend the nest. Three groups can be recognized: Peltopses (2 spp) are aberrant, fly-catching species of the treetops. Butcherbirds (4 spp) have a disproportionately large head and short legs; mainly arboreal. The magpie, actually a form of terrestrial butcherbird, possesses a more robust body and longer legs; it spends much time in trees but forages on the ground.

Lowland Peltops *Peltops blainvillii* Pl 80
(Clicking Shieldbill)
18–20 cm. Uncommon and patchy in lowland and hill forest, where it perches upright on high, open snags and sallies for insects. The 2 peltopses are NG's most spectacular aerial salliers. They inhabit forest openings—tree falls, road cuts, river edges. Peltopses are unmistakable for their shape and *black Adult plumage with white and bright red markings*. **Adult:** This species is best separated from Mountain P by its *clicking song*. Lowland P is smaller (wing 90–102 mm) and differs by the proportionately larger bill and smaller white back patch. The *white facial patch is somewhat tear-shaped, dropping below the level of the eye*. **Juv:** A sooty version of Adult with more white on face. Wing-coverts tipped white. **Similar spp:** Mountain P nearly identical, but song differs. The 2 may coexist and possibly hybridize where their elevational ranges meet, e.g., the Sogeri Plateau near Port Moresby. **Voice:** Song easily overlooked yet unlike any other bird—a burst of 2 or 3 mechanical clicks like a clock being wound: *PT-TIK!, PT-TIK!, PT-TIK!* . . . or *CLICK!-CLICK!-CLICK!, CLICK!-CLICK!-CLICK!, CLICK!-CLICK!-CLICK!*. . . . The series is repeated ~4–5 times, with a brief interval between each series, duration 4–5 sec or more. Call is a monarch-like *wheeeeit*. A less common, twittering call is a repeated series of 4 short notes, high-pitched and whistled. **Habits:** Usually in pairs or family groups that perch and call for long intervals from high open perches and sally forth in long swoops for flying insects. Nest built in tree crown on a fork on outer branches; a small, compact, cryptic cup of twigs and rootlets. Eggs (clutch?) yellow-white with a ring of dots at the large end. **Range:** Endemic. NG (but absent from Trans-Fly) and NW Is (Misool, Salawati, Waigeo), 0–550 (750) m.

Mountain Peltops *Peltops montanus* Pl 80
(Highland Peltops, Singing Shieldbill)
20 cm. Common but patchy in montane forest, favoring tall trees overlooking a clearing or valley. **Adult:** Nearly identical to Lowland P, and *best separated by song*. Larger than Lowland P, with wing 101–119 mm. Mountain P has a proportionately smaller bill and has a larger, though variable, white patch on the back. The *white facial patch is fan-shaped, spreading above as well as below the level of the eye*. **Juv:** Like juv Lowland P. **Voice:** Song (or call) is a twittering musical series of 7–8 notes, the series descending: *tit tit tit tit tit tit tit*, rapidly delivered. It resembles the sound of a fingernail running along the teeth of a comb. Also a single upslurred *schweep!*, similar to Lowland P. Also a rare twittering call, like Lowland P. **Habits:** Similar to Lowland P. Multiple Adult birds in nest tree suggests cooperative breeding. Nest nearly invisible under the sitting bird. Egg not described. **Range:** Endemic. Nearly all NG mts (no records from Cyclops Mts), 550–2400 (3000) m.

Black Butcherbird *Cracticus quoyi* Pl 80
33–36 cm. Uncommon in mangroves and forest interior from lowlands to foothills, higher locally. Quite shy, best located by its loud, ringing voice. **Adult:** A large, *all-black* butcherbird, with *contrasting pale grey bill*. Male larger on average. **Juv:** Nearly identical but duller, iris paler, bill darker. **Subspp (2):** *quoyi* (all NG range except next race) smaller; *alecto* (coastal Trans-Fly and Aru Is) larger and bill more slender, said to mainly inhabit mangroves. **Similar spp:** Crows and manucodes have a black bill. Manucodes have jerky, flap-and-glide, roller-coaster flight, whereas butcherbird flight is level with steady wingbeats. Male Eastern Koel has a shorter pale bill and much longer tail.

Voice: Song is a complex and striking series of notes. Voice is loud and varies geographically, but the mixture of high ringing notes and low cawing notes is diagnostic. For example, a rapidly delivered *kwa wen kwa*, middle note highest; or *kwa kwa caur*; or *kokwoiee koi*; or *tolk-kwingwing*! Long vocalization is similar but slower and more slurred than that of the Hooded B. The brief, 2-note vocalization, *kwing-wing*, is unique to the species. **Habits:** Solitary, often wary and difficult to observe. Forages from the ground to lower canopy and feeds on a variety of small animal life and fruits taken by gleaning. Breeds in pairs, rarely with a helper. Nest is placed in a tree crotch at midstory; the cup nest is built of sticks and twigs. Eggs (2) pale green with dark flecks. **Range:** NG, NW Is (Misool, Salawati, Waigeo), Bay Is (Yapen), and Aru Is, 0–750 m (rarely to 2200 m). Also AU. **Taxonomy:** Molecular studies have revealed 2 species of Black B—one in the NG Region (*C. quoyi*), the other in AU (*C. spaldingi*). As it is not yet known to which species the Trans-Fly/Aru race *alecto* belongs, nor is it known how to separate the 2 species in the field apart from bill size and shape, we continue to recognize only 1 species of Black B.

Black-backed Butcherbird *Cracticus mentalis*   Pl 80
25–28 cm. Common in savannah woodland and nearby towns, where easily observed. Look for *white throat* in all plumages. **Adult:** *Pied with black head.* White hind-collar. **Juv:** Same pattern as Adult but black replaced by messy brown; bill darker. **Subspp (1):** *mentalis* (endemic). **Similar spp:** Hooded B, usually absent from savannah, has black throat and is larger. **Voice:** Noisy. Song, a slower version of Hooded B, distinguishable. **Habits:** Solitary or in pairs. Forages for lizards, arthropods, and small birds in tree crowns; also snatches prey off trunks of trees and building walls, and from the ground. Perhaps breeds as a simple pair. Nest built in a tree; a cup of sticks and twigs. Eggs (3) grey brown, spotted and blotched. **Range:** NG: Trans-Fly and Port Moresby area (SE Pen), 0–600 m. Also AU.

Hooded Butcherbird *Cracticus cassicus*   Pl 80
32–35 cm. Common and conspicuous along forest edges, river courses, and in tall second growth in lowlands and hills. Seen perched on an exposed high branch at forest edge or in a garden. **Adult:** *Black-headed, with a black throat and upper breast.* Pied, with much white on back and underparts. Large, prominent bill is pale blue with dark tip. **Juv:** Same pattern, but black feathers tipped brown and bill darker. **Subspp (2):** *cassicus* (mainland and W islands); *hercules* (SE Is) much larger. **Similar spp:** Black-backed B, a savannah dweller, is smaller and white-throated. **Voice:** A fine songster, author of one of the most characteristic sounds of NG's lowlands. Song, a loud, jumbled bugling and yodeling, which combines bell-like notes and liquid rollicking phrases with hoarse notes, gurgles, musical croaks, and duets. Occasionally mimics other species. **Habits:** Usually in pairs or small parties. Raises and lowers wings while calling. Feeds on large insects, spiders, and fruit. Reports of more than 2 birds attending a nest suggest cooperative breeding. Nests in outer branches in treetop; nest a cup of sticks and twigs. Eggs (2–3) olive with dark spots and blotches. **Range:** Endemic. NG, NW Is (Batanta, Gebe, Misool, Salawati, Waigeo), Bay Is (Biak, Numfor, Yapen), Aru Is, and SE Is (3 main D'Entrecasteaux Is, plus Trobriands), 0–1450 m.

Tagula Butcherbird *Cracticus louisiadensis*   Pl 80
(Louisiades Butcherbird)
27–30 cm. The only butcherbird on Tagula I, where uncommon in forest interior and edge. **Adult:** *Almost entirely black, with white marking on the wing and tail.* **Juv:** Similar to Adult but duller, the black feathers faintly pale-tipped. **Similar spp:** None on Tagula. **Voice:** Song easily recognized as a butcherbird's. Besides typical daytime singing, will also sing at night, hours before daybreak. **Habits:** Similar to Hooded B. No nesting information. **Range:** Endemic. SE Is of Tagula, Sabara, and probably other small, nearby islands, 0–500 m.

Australian Magpie *Cracticus tibicen*   Pl 80
41 cm. The crow-sized, stocky, black-and-white songbird of the Trans-Fly savannah, where uncommon. *Black face and underparts.* **Male:** Brilliant white dorsally from nape to tail. **Female, Imm:** Black above with white nape and rump. **Juv:** Similar to Female, but pattern subdued, greyish. **Subspp (1):** *papuanus* (endemic). **Similar spp:** Black-backed and Hooded Bs have white on breast. Magpielark much smaller, has horizontal posture and white cheek and breast. **Voice:** Song, a series of whistled and yodeled notes, suggestive of a butcherbird. **Habits:** A familiar Australian bird, yet poorly known in NG, where encountered singly, in pairs, or in small parties. Wary, flying off at the first sign of danger. Seen perched in a tree or on the ground. Forages mainly on the ground, walking slowly with an erect posture; takes arthropods and small animals. Social system is complex and

involves cooperative breeding in AU, but unknown for NG. Nest built in tree crown, a cup of sticks and twigs lined with grass (AU data). Eggs (~3) bluish or greenish often with dark spots and blotches. **Range:** NG: Trans-Fly, 0–100 m. Also AU.

WOODSWALLOWS: Artamidae

The woodswallows (11 spp) are mainly Australasian. One species ranges as far as India, and another is widespread in the southwest Pacific. Three inhabit New Guinea. Woodswallows are medium-sized (18–20 cm), compact, aerial birds with a heavy, almost conical bill, characteristically triangular wings, and a square-tipped tail. Their close relatives, the butcherbirds and allies, are sometimes merged into the Artamidae. In New Guinea, species range throughout the lowlands to as high as 2800 m. Plumage is generally sooty grey, usually patterned with white. Note pale blue-grey bill in all species. Woodswallows prefer open sites—mountain ridges, forest clearings, towns, and savannahs—where there are exposed perches from which they sally out after flying insects. The birds often soar on rising thermals. All species are highly social and vocal. Parties have the charming habit of perching side-by-side on a branch, bodies pressed together. New Guinea species are cooperative breeders. Nest is cup-shaped, built of twigs in fork of a branch.

White-breasted Woodswallow *Artamus leucorynchus* Pl 80
18 cm. Conspicuous and common in lowland towns, open habitats, and coasts. Perches on electrical wires or high on bare branches, at times clustered in tight little groups. Sallies out with twinkling wingbeats and slow, graceful glides. **Adult:** *Ashy grey head and upperparts with contrasting white rump and underparts.* Grey hood stops on throat above the level of the shoulder. Lacks white shoulder spot in flight. **Juv:** Pale chin and throat; upperparts somewhat brownish with fine, pale scalloping. **Imm:** Like Adult, but retains Juv flight-feathers with pale tips. **Subspp (1):** *leucopygialis* (Moluccas to AU). **Similar spp:** Great WS is larger, more blackish, with bib extending farther onto breast (to the bend of the wing or lower). Black-faced WS has all-grey underparts. **Voice:** Song quiet, includes mimicry of other species. Contact call a soft, pleasant *dirt*. Alarm call harsh and strident.
Habits: Lives in family groups and often forages in flocks. Hawks insects. Leisurely flight with much gliding and soaring. Communal roosts include hundreds of birds. Cooperative breeder. Nest built in tree crown in deep fork or against trunk, or on power-pole cross arms; cup nest of twigs and grass. Eggs (3) white with dark spots and blotches. **Range:** NG, NW Is (Batanta, Gag, Gebe, Misool, Salawati, Waigeo), Aru Is, and SE Is (D'Entrecasteaux Is), 0–800 m. Also SE Asia (Malay Pen) to AU.

Great Woodswallow *Artamus maximus* Pl 80
(New Guinea Woodswallow)
20 cm. Montane counterpart of White-breasted WS, usually common. *A larger, darker version of White-breasted WS.* **Adult:** *Sooty black* with white rump and underparts. Hood extends a short way down onto breast below the shoulder. Shows *white shoulder spot in flight.* **Juv and Imm:** Same as for White-breasted WS, but darker. **Voice:** Song a rapid, prolonged, soft jumble of squawks, chirps, and short trills, including mimicry of other species. Calls similar to White-breasted WS. **Habits:** Similar to White-breasted WS. Soars on updrafts. Cooperative breeder. Nest similar to preceding species, but no egg information. **Range:** Endemic. NG, nearly all mts (no record from Cyclops and Kumawa Mts), 800–2800 m.

Black-faced Woodswallow *Artamus cinereus* Pl 80
18 cm. Rare in Trans-Fly, where its status is unclear as a resident or an AU visitor. Occupies savannah and open plains studded with patches of pandanus and scrubby trees. **Adult:** A small, *uniformly grey woodswallow with black face mask and white-tipped, black tail.* **Juv:** Back and wings spotted buff. **Subspp (1):** *normani* (also AU). **Similar spp:** Other NG woodswallows have white breasts. **Voice:** Song animated, includes mimicry. Call soft, scratchy chirps, *chiff chiff* or *chap chap*. Also twittering in flight. **Habits:** Singly or in small parties. Habits similar to other woodswallows, but also takes insects from the ground. Usually breeds as simple pair, sometimes cooperatively (AU data). Nest built in tree; a cup of twigs, grass, and other plant material (AU data). Eggs (~3) whitish, marked with spots and blotches. **Range:** NG: Trans-Fly, 1–100 m. Also AU.

CUCKOOSHRIKES: Campephagidae

Ninety-two species; 16 in New Guinea. The sleek, streamlined cuckooshrikes are strong fliers and highly mobile. They have spread across the Old World tropics, from Australasia and Oceania westward through Wallacea and SE Asia to India and Africa. They are a well-defined family that appears to have arisen in Australasia. In New Guinea, they are most abundant in the lowlands and foothills. Cuckooshrikes are characteristically birds of the forest canopy, but 3 species occupy open habitats. Insects and other arthropods form the bulk of their diets, while fruit is preferred by certain species. Many cuckooshrikes shuffle their folded wings upon alighting; the function of this peculiar behavior is unknown. Most are conspicuous vocalists, and several species even perform male-female duets. Cuckooshrikes live in pairs or family groups, and it seems likely that many species could be cooperative breeders; however, the breeding behavior of most species is essentially unknown. All cuckooshrikes typically build the same sort of cryptic nest: an amazingly small, tightly woven cup that is carefully decorated on the outside so as to blend in with the branch upon which it is built. The nest and its clutch of 1–2 eggs fit snugly beneath the sitting parent. Cuckooshrikes are usually easy to find but hard to see well. Rather noisy birds, a pair or flock sing and call loudly to stay in contact while traveling through the treetops. They are best observed at the forest edge, at a fruiting tree, or among a mixed-species flock. Note that assignments to genus have recently changed.

Stout-billed Cuckooshrike *Coracina caeruleogrisea* Pl 81
(Large-billed Cuckooshrike)
33–37 cm. *The large, elongate cuckooshrike of the forest interior from lowlands at the foot of the ranges up into the lower mountains*; uncommon. *Uniform grey with long, stout bill and long tail.* Cinnamon wing linings. **Male:** Blue-grey with black lores. **Female:** Duller grey than Male. Large black, beady eye set in all-grey face. **Juv:** Grey, some birds mottled with white; tail white-tipped and secondaries pale-edged. **Subspp (3):** Minor, W birds smaller and darker. **Similar spp:** Other cuckooshrikes in its habitat are smaller and with shorter bills. Boyer's CS similarly colored but much smaller and more compact, with short bill; it is the only other cuckoo-shrike with cinnamon underwing. **Voice:** Song a loud, slightly musical but harsh, forced catlike cry, very similar to that of the White-eared Catbird. Also a short chirp, a soft mewing slur, a soft rasp, and a harsh buzzy note. **Habits:** Singly or in pairs. Shy, sluggish, moving inconspicuously through forest subcanopy searching for large arthropods and lizards, especially on the larger limbs. Nest a small, compact, lichen-covered cup molded onto a branch fork in the canopy. Egg, no information. **Range:** Endemic. NG, Bay Is (Yapen), and Aru Is, 0–1600 (2100) m.

Hooded Cuckooshrike *Coracina longicauda* Pl 81
33 cm. *The large, long-tailed cuckooshrike of cloud forest high in the mountains;* uncommon. Sociable and vocal. *Grey with black wings and tail.* **Male:** *Black hood.* **Female, Imm:** *Black face.* **Juv:** Like Female but with messy, white scalloping and edges to wing and tail feathers. **Similar spp:** Black-bellied Cicadabird is the only other cuckooshrike to share its habitat; it is smaller, shorter-tailed, with different plumage pattern. Black-faced CS, an open-country species, is never found in cloud forest habitat of the Hooded; has similar size, shape, and plumage pattern as female Hooded, but black face lacks crisp edge, and flight and tail feathers are mostly grey, not all black. **Voice:** Song is a loud, medium-pitched, descending series of squealed notes: *jeer-jeer-jeer*, suggestive of a White-bellied CS, often given by several birds in a foraging party. Can be confused with call of young male King of Saxony Bird of Paradise, which is briefer. Also a variety of other calls. **Habits:** Forages through the forest canopy in family parties seeking arthropods and lizards, sometimes fruit. The birds travel slowly, as a group, calling frequently. Group may include more than 1 adult Male, suggesting delayed dispersal and cooperative breeding. Nest built on a horizontal branch fork in a tree crown; nest a small, shallow cup of rootlets on a foundation of moss, decorated on top with lichens. Egg (1) bluish, with dark spots or blotches. **Range:** Endemic. NG Central Ranges and Huon, 1800–3600 m.

Barred Cuckooshrike *Coracina lineata* Pl 82
(Yellow-eyed Cuckooshrike)
23 cm. Uncommon and sparingly distributed in hill forest, edge, and partly cleared areas. Encountered more often in outlying ranges than on the Central Ranges. Drawn to fig trees. The only cuckooshrike with *pale yellowish eye and barred underparts.* **Male:** Common race all grey with black

tail and barred underwing. **Female and Imm:** All races neatly *barred white-and-grey from breast to undertail-coverts* and on underwing-coverts. **Juv:** Like Female but with obscure barring on chin and throat and with white-tipped tail feathers; iris ranging dark to yellow. **Subspp (3):** *axillaris* (NG range, minus next 2 races) Male all grey, not barred; *maforensis* (Numfor) Male darkly barred below; *lineata* (Trans-Fly, shared with AU; either a vagrant or rare resident) larger, both sexes barred, undertail-coverts with white bars wider than black ones (in *axillaris*, black bars usually wider or the same width). **Similar spp:** No other grey cuckooshrike has pale iris or barred underparts. **Voice:** Song, a high-pitched whistle descending slightly in pitch: *wheeuuu* or *whee*, often given as a duet in flight. Also medium, high-pitched, catlike notes reminiscent of those of Black-faced CS. **Habits:** Gregarious in small, active parties foraging in canopy for fruit, especially figs, and some arthropods. Nest on a horizontal branch fork; small cup decorated on the outside with green moss (AU data). Eggs (2) dull white or greenish with dark blotches. **Range:** NG, Waigeo (NW Is), Numfor (Bay Is), and Aru Is, mainly 600–1450 m. NG: Bird's Head, N ranges from Foja Mts to Huon (no records from Cyclops Mts), SE Pen, S Lowlands (Tabubil/Kiunga), and Trans-Fly (Merauke). Also E AU, Bismarck and Solomon Is.

Boyer's Cuckooshrike *Coracina boyeri* Pl 82

22 cm. Common in lowland forest canopy. Gregarious, vocal. *Small size, short bill, paler grey crown* contrasting with darker, blue-grey back and wings. *Cinnamon underwing linings* visible in flight and when bird flicks it wings. **Male:** *Black lores and chin.* **Female:** Grey lores, uniquely *pale and contrasting* in N race. **Juv:** Resembles Female but with irregular light and dark markings throughout, and wing feathers white-edged and tipped. **Subspp (2):** *boyeri* (N NG, from NW Is and Bird's Head to SE Pen) underwing rufous, Female with contrasting pale lores; *subalaris* (S NG, from S Lowlands to SE Pen) underwing buff, Female with plain grey face. **Similar spp:** Other species of similar shape and size lack pale crown and cinnamon underwing. Stout-billed CS with similar plumage pattern, but has different, elongate shape. **Voice:** Song, a sweet, upslurred or downslurred *speeyu*, as single or paired notes, often repeated by several birds. Other quieter calls are a low clucking, a catlike mew, and a repeated chirping *chuck*. A duet that varies regionally begins with harsh notes by the Female, followed by the Male's downslurred whistles, for example: *ticky ticky, seeyu seeyu seeyu*. **Habits:** In small parties in treetops foraging for fruit and arthropods taken from bark. Flies from tree to tree with bouncing flight. Nest built on canopy branch; structure typical of a cuckooshrike. Egg, no information. **Range:** Endemic. NG, NW Is (Misool, Salawati), and Bay Is (Yapen), 0–1100 m, rarely higher.

Black-faced Cuckooshrike *Coracina novaehollandiae* Pl 81

32–35 cm. The *large, elongate cuckooshrike of open country and towns*; common in Trans-Fly, S Lowlands, and SE Pen, less so in cleared highland valleys. Breeds in Port Moresby area, but most birds are migrants from AU during austral winter. **Adult:** *Underparts grade from black face and throat to grey breast to whitish belly.* Note white-tipped tail in flight. **Juv:** Small, black mask that extends to ear-coverts; may show obscure barring in underparts. **Subspp (2):** *melanops* (breeds in Port Moresby area, also austral winter migrant from AU); *novaehollandiae* (vagrant from Tasmania) larger, with shorter bill. **Similar spp:** White-bellied CS is obviously smaller and has black mask that does not extend past eye. Hooded CS (female) of the high mountains has crisply edged mask and all-black flight and tail feathers. **Voice:** Song, a cheerful, rolling trill, *cher-reer cher-reer cher-reer* usually given in flight. Also, a plaintive *plee-urk*. **Habits:** In small groups or migratory flocks foraging through savannah, coastal vegetation, plantations, urban areas. Peculiar undulating flight. Large communal roosts. Forages in tree crowns by sallying, hovering and snatching, and gleaning for insects and spiders, occasionally fruit. Nest placed on a horizontal limb in tree crown; nest a small, shallow cup built on a branch fork. Eggs (clutch?; 2–3 in AU) green, variously spotted or blotched. **Range:** NG and satellite islands as a widespread migrant and local breeder in Port Moresby area, 0–500 (1850) m. Breeds and winters in AU; winters outside AU mainly in the NG Region, but extends from N Moluccas across to Bismarck and Solomon Is.

White-bellied Cuckooshrike *Coracina papuensis* Pl 81

25–26 cm. The *midsize cuckooshrike of open habitats and towns* from lowlands to settled highland valleys. Common and conspicuous. Forest clearings, gardens, savannahs, coasts, mangroves. *Conspicuous black lores with thick, cushiony texture and sharp border.* In S and SE, *uniquely pearly grey above and white below*. In N and NW, evenly grey and more easily confused with other, same-size cuckooshrikes. **Adult:** *Note thick, black lores with crisp border.* **Juv:** Lores less distinct, breast obscurely scalloped. **Subspp (5, in 2 groups):** *papuensis* (N NG from Bird's Head to Huon) throat and breast

grey; *hypoleuca* (S NG from S Lowlands to SE Pen, plus Aru Is) throat and breast white.
Similar spp: Black-faced CS is larger, with black mask extending behind eye. In N, no other cuckooshrike has thick lores with sharp margins. **Voice:** Song, a weak, squealing, slurred *whee-eeyu* or *wee-yeer*, often repeated. Also a longer jumble of notes of similar quality, suggestive of the sound of wind chimes. **Habits:** Solitary, in pairs, or in small parties, gleaning insects and spiders from tree foliage, occasionally sallying. Also eats small fruits. Most often seen in flight, which is distinctly undulating: the bird flaps to gain altitude, then glides, holding its wings stiffly downward. Nest on a horizontal branch fork; a small, shallow cup of twigs and vine stems, decorated with lichen. Eggs (1–2) pale blue-green with dark markings. **Range:** NG, NW Is (all except Waigeo), Bay Is (Yapen), Aru Is, and SE Is (Tagula), 0–1500 m. Also E Wallacea to Bismarck and Solomon Is, south to AU.

Golden Cuckooshrike *Campochaera sloetii*   Pl 81

20 cm. Locally common in forests of the foothills and adjacent lowlands. *A striking golden orange and black bird* of the treetops. Slender and rather small-headed, with a dark face. **Male:** Black face. **Female:** Grey face. **Juv:** Like Female but the face is yellow, leaving grey only on the breast; greenish central tail feathers. **Subspp (2):** *sloetii* (N NG from Bird's Head to W Sepik-Ramu) crown grey; *flaviceps* (S NG from S Lowlands to SE Pen) crown olive. **Similar spp:** Compare with Golden Monarch and Golden Myna, both black-and-gold birds living in the same lowland forests but otherwise quite different. **Voice:** In S, the song is an excited, musical, high-pitched whistle: *tut u tu duit!* Or *teeto teeto tu tuweet* repeated in rapid succession, often with more than 1 bird contributing. Birds from NW Lowlands lack the musical slurs and commence with explosive clicking. **Habits:** In small, active, and noisy pairs or small parties. Gleans and sallies-then-hovers, taking fruit and small insects from the canopy foliage. Nest quite small, compact, and cryptic, mostly hidden beneath the sitting bird and built on a horizontal fork in outer branch of a high forest tree. Egg, no information. **Range:** Endemic. NG: Bird's Head and N lowlands eastward as far as Wewak while keeping north of Sepik R; S Lowlands eastward to SE Pen as far as Port Moresby area, 0–1100 m.

White-winged Triller *Lalage tricolor*   Pl 81
(Australian Trilller, *Lalage sueurii***)**

16–18 cm. The triller of *savannah and other open habitats*; coastal locally (Port Moresby area). Distribution patchy. Generally scarce, sometimes common. Nomadic. Birds in Trans-Fly likely AU migrants, elsewhere probably resident. Male vocal and easily noticed, especially in song-flight displays; Female usually inconspicuous. **Male:** Breeding—black above, white below; note *large white patch across wing, grey rump*. Nonbreeding—difficult to distinguish from Female, but primaries and tail black (not dark brown) and rump grey (not pale brownish grey). **Female:** Drab and rather nondescript, yet with distinctive markings. *Pale brown*. Note *pale eyebrow, wings with pale edges to coverts and secondaries,* and *greyish rump*. **Juv:** Similar to Female but obscurely mottled and scalloped throughout, and throat diffusely streaky. **Imm male:** Breeding—shows Adult body coloring while retaining brown-edged Juv primaries and tail. **Similar spp:** Other male trillers have less white on wing and are strictly tree dwellers. Common Cicadabird female is much larger and shows barring below. **Voice:** Male sings from perch or in flight. Song, a clear, sustained *chiff-chiff-chiff joey-joey-joey* or *deet-deet-deet-dip-dip-drr*. **Habits:** Singly, in pairs, or in small groups; forms flocks after breeding. Perches in trees and shrubs, often low. Actively gleans foliage and forages on the ground, favoring bare areas or short grass. Takes arthropods, especially insects. Flight swift and undulating. Sometimes nests colonially. Nests on horizontal branch of a small tree; nest a tight, shallow cup of grass, twigs, and bark. Eggs (2) green, dark streaked, and blotched. **Range:** NG: Mainly in Port Moresby area, but also other widely scattered dry localities in SE Pen (Musa Valley, Bartle Bay), Sepik-Ramu (upper Markham Valley, upper Ramu Valley, and central Sepik), and possibly Trans-Fly (questionable sight record, Bensbach), 0–700 m. Also AU. **Taxonomy:** Split from White-shouldered Triller (*L. sueurii*) of Wallacea; either way, the English name does not change.

Black-browed Triller *Lalage atrovirens*   Pl 81

18–19 cm. The northern forest triller. Common in rainforest canopy and openings in the lowlands and hills. *"Black-browed"—cap all black, without white eyebrow* of Varied T and *vent white*, not buff. *Rump white*, not grey or all barred. *White shoulder bar*. **Male:** Glossy black with all-white underparts, no barring. **Female and Juv:** Duller above and barred below. **Similar spp:** White-winged T co-occurs; male's shoulder is nearly all white; female entirely different. Varied T has a white eyebrow and buff or cinnamon vent. **Voice:** Song a sweet, rapidly repeated, upslurred whistle, *twee twee twee. . . .* Another, *see o seet weo*, repeated 5–6 times. Another song gerygone-like, a rapidly cycling 3-note phrase, the third note higher pitched. Paired birds counter-sing sweet notes: *tewhit tewheet wheetu*, rapidly delivered. Call, *whick!-whick!, whick!-whick!*. . . . Also, a few hard *chak* notes, accelerating and with the cadence of laughter. **Habits:** Similar to the better-known Varied T. No nesting information. **Range:** Endemic. NG and NW Is (Kofiau, Misool, Salawati, Waigeo), 0–1400 m. Northern NG from Bird's Head to northern Huon.

Biak Triller *Lalage leucoptera*   Pl 81
18–19 cm. The only triller on Biak I; inhabits forest and edge. Closely related to Black-browed T of mainland NG. **Male:** Compared with male Black-browed, has the *black cap extending well below eye* (vs just below eye), *a single large white patch across shoulder* (vs white-edged secondaries), and *all-black back* (no white rump patch). **Female:** Duller than Male and shows the same differences from female Black-browed (but has a trace of white on the rump), and in addition, not barred below. **Voice:** Song is a series of loud, medium-pitched, clear 2-note whistles at a rate of ~1 phrase/min: *whir-whee! whir-whee!*. . . . As the bird becomes more excited, the song achieves a bubbling quality. **Habits:** Presumably similar to Black-browed. **Range:** Endemic. Biak (Bay Is). **Taxonomy:** Formerly a subspecies of Black-browed T.

Varied Triller *Lalage leucomela*   Pl 81
18 cm. The forest triller of the South and Southeast. Common in rainforest canopy and openings of the lowlands and hills. Pied with *white eyebrow* and *buff or cinnamon vent*. Back black or black barred with white. Geographically variable in color of underparts. **Male:** Black above with white shoulder bars; underparts unbarred or with some barring. **Female:** Duller and less boldly patterned; brownish black or slaty above; underparts barred from throat to belly. **Juv:** Like Female but somewhat pale-mottled above. **Subspp (4):** *polygrammica* (mainland NG minus Trans-Fly; Aru Is) Male pale grey below, finely barred; *yorki* (Trans-Fly and AU) paler and Male underparts white, barring obscure or absent; SE Is races, Male heavily barred over mostly white underparts in Louisiade birds (*pallescens*) and over mostly buff in D'Entrecasteaux birds (*obscurior*), with Trobriand birds intermediate. **Similar spp:** Other trillers lack white brow and cinnamon vent. The 3 trillers segregate by habitat and range. **Voice:** Song, a rolling, churring trill, *breeer-breeer-breeer-breeer* or *cherwee, cherwee, breer-breer-breer* . . . , each note swelling then diminishing. **Habits:** Solitary or in pairs, sometimes small parties. Inconspicuous for such an active singer. Forages for fruit and caterpillars in the canopy of forest, edge, regrowth, mangroves. Nests in the canopy on a horizontal branch or fork; nest is a small, tightly woven cup. Egg (1) greenish to brownish with dark markings. **Range:** NG, Aru Is, and SE Is (3 main D'Entrecasteaux Is; Trobriand Is; Louisiade Is of Misima, Tagula), 0–1500 m. NG: S Lowlands, Trans-Fly, SE Pen, and southern Huon. Also AU, Kai Is, and Bismarck Is.

Black-bellied Cicadabird *Edolisoma montanum*   Pl 82
(*Coracina montana*)

24 cm. A common, vocal canopy dweller of montane forest, from mid-mountains to lower limits of the cloud forest. **Male:** Unique—*blue-grey above, black below*. **Female and Imm:** Resemble other midsize grey cuckooshrikes but distinguished by the *all-black tail and wing feathers*. Forehead, lores, and chin black. **Juv:** Obscure barring and streaking (heavy in some birds); white-edged secondaries and tail tip. **Subspp (2):** Minor. **Similar spp:** No other cuckooshrike has the all-black underparts of male Black-bellied. Female resembles males of Papuan and Common CBs, but these have shorter, grey-washed tails and mostly grey wings; female Papuan in addition has a grey face. **Voice:** Song is a conspicuous duet. Male alternates between rising and falling musical, nasal, whistled phrases, *tituwi, teeuw,* or *twei*, while the Female interjects after each phrase a series of 2–3 rapidly repeated identical notes: *ch ch ch*. So, *tituwi, ch ch ch, teeuw, ch ch ch,* repeat. . . . Duet is repeated for 8–10 sec. Call, *kek*. **Habits:** In pairs or small groups that move through the canopy singing and calling as they traverse their large territory. Gleans or sally-hovers for insects from foliage and branches, seldom takes fruit. No nesting information. **Range:** Endemic. NG: nearly all mts (no records from Wandammen Mts), 800–2400 (2800) m.

Papuan Cicadabird *Edolisoma incertum*   Pl 82
(**Black-shouldered, Sharpe's,** or **Müller's Cuckooshrike;** *Coracina incerta* or *morio*)

21–22 cm. Common in hill and lower montane forest and edge. Most similar to Common CB, but more compact because of the shorter wings and tail. **Male:** *Overall darker than Common CB, especially the*

445

head (often has all-black face and throat). *Wing-coverts edged dark grey, like mantle, and less patterned, poorly contrasting.* **Female:** Like Male but paler, and black only on lores. **Juv:** Like Female, but face flecked with white and underparts obscurely pale-barred. **Similar spp:** Common CB avoids forest proper; male is larger, slimmer, has strongly contrasting pale edges to wing-coverts and secondaries. Male Grey-headed CB also similar but has different song and paler, all-grey wing-coverts and secondaries. Black-bellied CB female has all-black tail. **Voice:** Song is a series of loud, repeated notes at a rate of 2/sec for several seconds, either buzzy or clear and mellow, depending on locality. In some areas, the notes are paired: *chur-chur chur-chur chur-chur.* . . . Songs with repeated single notes resemble variations of Common CB song from Trans-Fly and have yet to be sorted out. Male and Female also duet, flying from tree to tree: *du du du whit . . . du du du whit.* . . . **Habits:** Similar to Common CB. No nesting information. **Range:** Endemic. NG (all mts), NW Is (Batanta, Waigeo), and Bay Is (Mios Num, Yapen), 450–1450 m. **Taxonomy:** Split from Sulawesi Cicadabird (*C. morio*).

Common Cicadabird *Edolisoma tenuirostre* Pl 82
(Cicadabird, Slender-billed Cicadabird, *Coracina* or *Lalage tenuirostris*)
24–27 cm. Uncommon in treetops at forest edge and in second growth, savannah, and mangroves on the mainland; also inhabits forest on islands; lowlands and hills throughout. Status complex and not entirely clear. There are basically 3 types: (1) a migratory AU subsp, *tenuirostre*, that visits the mainland and islands in austral winter, (2) a resident, southern mainland subsp, *aruensis*, and (3) many endemic island subspp. *Male difficult to distinguish from either sex of Papuan CB and from male Grey-headed CB.* Common CB is larger and more elongate. To help identify the male of one of these 3 cuckooshrikes, *look for the accompanying female—very different in each species.* **Male:** *Slightly paler grey* than the other 2 species. *Wing-coverts have black centers and pale edges, forming rows of dark spots and pale wing bars,* whereas the wings of the other 2 species are more evenly colored. *The secondaries show strongly contrasting silver-grey edges*—diagnostic. **Female and Imm:** Brown and drab, but easily recognized. *Barred below. Conspicuous, crisp black eye-stripe in line with bill.* **Juv:** Above brown mottled with darker brown and pale scalloping, below off-white with brown streaking. **Subspp (7):** *tenuirostre* (= *muellerii*, widespread AU migrant; likely also resident on SE Is of Fergusson and Woodlark); *aruense* (resident in S Lowlands, Aru Is, and Trans-Fly) smaller on average; *nehrkorni* (Waigeo I) Male with ear-coverts and throat black; *numforanum* (Numfor I) Female with barring reduced to spots or absent; *meyerii* (Biak I) Male dark grey with chin and throat blackish, Female underparts buff/russet with few markings; *tagulanum* (Misima and Tagula Is) long bill; *rostratum* (Rossel I) large, and Female with reduced barring. **Similar spp:** Papuan CB male is darker, often has nearly black chin and throat, and has less contrasting pattern in the wing; female similar to Male Common CB but has less black in the face (black lores only). Male Grey-headed CB has even, pale grey wing-coverts and secondaries, without pattern. Other barred, medium-sized birds that could be confused with Female cicadabird are female birds of paradise (dark crown and longer bill) or various cuckoos (longer tail). **Voice:** Song of migratory Australian birds is a rapid series of dry, cicada-like, energetic, pulsing notes, *jj jj jj* . . . and variations. Song of resident race in Trans-Fly, *aruense*, is a slow, drawn-out cicada-like buzz, recalling song of Papuan CB. Songs of island subspecies sound quite different, but the pattern of repeated notes is similar. Contact call between pair, *tchuit* or *tchuit-tchuit*. **Habits:** Often in pairs—singing Male usually spotted first, but Female more easily identified, so listen for her call notes. Most of the time quiet and easily passed over. Noticed when singing or calling as it flies from tree to tree. Seen perched and looking about for insects, which it takes from foliage or branches by gleaning or sally-hovering. Nest on a high, horizontal branch; a small, tight, shallow cup (AU data). Egg (1) bluish or greenish with dark spots and blotches. **Range:** NG and probably all satellite islands, 0–1300 m. Also N Moluccas to AU and Oceania. The AU population winters mainly in the NG Region. **Taxonomy:** This is a racially complex species with a great many forms throughout the Western Pacific; some will likely be raised to species level.

Grey-headed Cicadabird *Edolisoma schisticeps* Pl 82
(Black-tipped Cuckooshrike, *Coracina schisticeps*)
21–22 cm. Hills and nearby lowlands, where locally common in forest and edge. Lives below the elevational range of Black-bellied CB, but overlaps with Papuan CB. Normally occurs in vocal parties of both sexes, often in groups of 3, remaining high in the treetops. Shares compact shape of Papuan CB. **Male:** Similar to Papuan and Common CB males, but *shoulders are a smooth, unpatterned, paler grey*, same color as neck and mantle; *secondaries solid grey* when wing is folded, not showing pale edges against dark centers, as with the other 2 species. **Female and Juv:** Rufous brown with dark lores and lacking rufous eyebrow of female Black CB. **Subspp (4):** Females differ. Race *schisticeps* (NW Is,

Bird's Head and Neck) Female with grey crown; *poliopsa* (S Lowlands to southern SE Pen) Female with grey crown, ear-coverts, and chin; *vittatum* (SE Is: D'Entrecasteaux Is) like the preceding subsp, but Female underparts barred (Male with grey face and black lores, no black on ear-coverts); *reichenowi* (NW Lowlands and Sepik-Ramu) most different, with Female showing all-brown head and Male with a darker chin. **Similar spp:** Papuan and Common CB males. Black CB female. **Voice:** Among the distinctive calls of the hill forest. Duet resembles that of Black-bellied CB. Male starts with a musical *chickity-choo*, Female answers with a double, nasal upslur, *wee wee*. Or rapidly repeated nasal, slurred whistles by Male followed by chatter of other group members, *chuk-chuk-chuk*. The later notes also used as call notes in flight. **Habits:** Forages in pairs or small parties in hilly forest, taking mostly fruit, also insects in typical cuckooshrike fashion. No nesting information. **Range:** Endemic. NG, NW Is (Misool, Salawati), and SE Is (3 main D'Entrecasteaux Is), 150–1200 m. NG mainland: All mts except N watershed from E Ranges to SE Pen.

Black Cicadabird *Edolisoma melas* Pl 82
(**Black**, **Papuan Black**, or **New Guinea Cuckooshrike**, *Coracina melas* or *melaena*)
23 cm. Common in lowland and hill forest. Forages below the canopy and usually found in mixed-species flocks with other black or brown birds such as babblers and pitohuis. **Male:** *Entirely black*. **Female and Imm:** *All brown*, paler below; subtle but distinctive facial pattern—*pale eyebrow contrasts with dark eye-stripe and cheeks*. **Juv:** No information. **Subspp (4–6):** Difficult to sort out; Females vary in shade of brown. **Similar spp:** Other male cuckooshrikes and cicadabirds are grey, but may look black when backlit. Manucodes and starlings also black, but have different shape and red eyes. Rusty Pitohui and Papuan Babbler differ from Female cicadabird in shape and more active foraging behavior. **Voice:** Song, a loud series of ~10 repeated, liquid, downslurred whistles, ~1 note/sec, *deer-deer-deer-deer-deer...* (deeper and slower than White-eared Bronze Cuckoo). Flocking call is a short nasal *whit*, repeated 1–4 times. Also a chattering call in feeding flocks is a repeated *wutchay*. **Habits:** In pairs or small parties. Joins mixed flocks, moving through the midstory with drongos, pitohuis, and babblers. Actively shifts from perch to perch, gleaning foliage or branches for insects; sometimes sallies in pursuit of fleeing prey. Takes some fruit. Nests on a branch fork in a tree crown. Egg, no information. **Range:** Endemic. NG, NW Is (Batanta, Salawati, Waigeo), Bay Is (Yapen), and Aru Is, 0–750 (1200) m.

SITTELLAS: Neosittidae

This distinctive Australasian family includes only 2 species, both of which inhabit New Guinea. Sittellas are small, nuthatch-like birds that live in tight flocks and forage for insects and spiders over branches and limbs of forest trees, continually uttering high twittering notes. They move rapidly from one tree crown to the next and are instantly recognized by their foraging behavior. They inhabit mountain forest in New Guinea. Breeding habits in New Guinea are unknown.

Varied Sittella *Daphoenositta chrysoptera* Pl 83
(**Papuan Sittella**, *D. papuensis*, *Neositta chrysoptera* or *papuensis*)
11–12 cm. Seen infrequently in mid-mountain and lower cloud forest zone. *In twittering flocks* that move hurriedly through the canopy. *A very small, compact, streaked bird that busily forages on branches and trunks*. Note *white rump*. **Male:** Head usually darker than Female's, see Subspp. **Female:** Head pale, either white or grey, and unblemished or faintly streaked. **Juv:** Dark upperparts with white tips to feathers. **In flight:** White wing bar. **Subspp (5, in 4 groups):** *papuensis* (Bird's Head) top of Male's head black; *alba* (northern W Ranges) Male's head white; *toxopeusi* (southern W Ranges to E Ranges) Male's head streaked; *albifrons* (SE Pen) similar but top of head darker. **Similar spp:** Black S entirely blackish with pink face and tail tip; occurs at higher elevation. Red-breasted Pygmy Parrot has a similar size, shape, and habits, but shows obvious parrot features and green plumage. Papuan Treecreeper is solitary, more elongate, and creeps up tree trunks and branches. **Voice:** Flock call a rapid *sipsipsip...*, repeated incessantly like a baby bird (Eastern Ranges) or slower *zzip zzip zzip...* (Bird's Head). **Habits:** In open canopy. Only in flocks of 4–10 birds that move and call continually. Searches bark and wood for hidden prey, often moving downward (treecreeper moves upward). Gleans and excavates. Probably a cooperative breeder, but nest and nesting habits in NG undescribed. Nest built on an upright branch fork in treetops; a small cup, tightly woven and decorated to blend in with branch (AU data). Eggs (~3) bluish, greenish, or greyish, dark spotted and blotched.

Range: NG: Bird's Head, Central Ranges, and Huon, 1440–2200 m (1075–2600 m). Also AU.
Taxonomy: Relationship of the New Guinea forms to the Australian ones is in need of study.

Black Sittella *Daphoenositta miranda*   Pl 83
12 cm. Rare in cloud forest at high elevation, above the range of the previous species.
Adult: *Small, sooty-black sittella with pink face* and tail tip. **Male's** iris and legs dark; these yellow in
Female. Juv: Blackish with rusty spotting and face. **In flight:** White wing bar noticeable in underwing.
Similar spp: Variable S is paler and streaked, lives mostly at lower elevations. Red-breasted Pygmy
Parrot. **Voice:** Flock call is a faint, sucked-in, and slightly squeaky, *sweek, sweek, sweek*. . . .
Habits: Similar to previous species. Usually in canopy, but in gaps it may descend to forage on low
trees and fallen logs. No nesting information. **Range:** Endemic. NG Central Ranges, 2000–3600 m.

BERRYHUNTERS: Rhagologidae

Recent studies have shown that the Mottled Berryhunter is not a whistler (family Pachycephalidae) and it has recently been removed to its own family. *Rhagologus* is an odd, mysterious songbird unlike any other in NG. It possesses a beautiful and memorable song of slurred, musical notes.

Mottled Berryhunter *Rhagologus leucostigma*   Pl 87
(Mottled Whistler)
15–16 cm. An elusive fruit eater of the montane forest interior. Locally common and sings often,
yet difficult to see. Female is the more ornate sex. **Male:** In far West similar to Female but somewhat
duller; elsewhere *plain drab olive*. **Female and Juv:** *Dark olive with rusty face. Streaked and spotted
with white and rufous.* Adult both sexes have olive or yellow-brown iris, Juv dark brown.
Subspp (2): *leucostigma* (Bird's Head to W Ranges) Male patterned like Female but markings less
conspicuous and face not as rufous; *obscurus* (Border Ranges eastward to SE Pen) Male lacks Female
pattern, instead is dingy greyish olive, faintly mottled grey on face and throat, and Female bright and
heavily spotted. **Similar spp:** Berrypeckers are all much smaller. Imm of true whistlers have rufous
in plumage, suggestive of female Mottled B, but their plumage is not streaky, nor mottled in the same
way. Wattled Ploughbill female is yellow-green below, has different bill. Little Shrikethrush is larger,
stockier, not as mottled. **Voice:** Song is unique—a loud, slowly delivered, cheerfully whistled tune of
5–13 slurred notes. **Habits:** Alone or less often in pairs. Usually quiet and inactive. Inhabits middle
story of montane forest and less commonly second growth. Diet includes much fruit, yet the bird is
rarely noticed at fruiting trees. Perhaps takes fruit mostly from shrubs, vines, and small trees.
Nest 2–3 m up in the fork of a small tree; a woven cup covered on the outside with moss and
liverworts. Egg (1) cream white with speckles. **Range:** Endemic. NG: Bird's Head to Central Ranges
and Huon, 1500–2300 m (800–2900 m).

AUSTRALO-PAPUAN BELLBIRDS: Oreoicidae

Also once placed in Pachycephalidae, these two genera plus the Crested Bellbird (*Oreoica gutturalis*) of Australia appear to have no close relatives besides each other. All 3 live on the ground and will ascend partway up trees to forage along trunks and large branches. They are stocky birds with a full, bushy crest and distinctive songs.

Rufous-naped Bellbird *Aleadryas rufinucha*   Pl 83
(Rufous-naped Whistler, *Pachycephala rufinucha*)
16–18 cm. Common but skulking in thickets and lower story of mid-mountain and cloud forests.
Adult: *A chunky, olive-green bird with white eye and forehead, yellow throat, and rufous nape.* **Juv:**
Variably olive brown above, chestnut below, with green wings and tail (some have Adult pattern but
with brown, soiled appearance); pale bill; dark iris. **Imm:** Resembles Adult but with reduced markings

and darkish iris. **Subspp** (**4, in 2 groups**): *rufinucha* (Bird's Head) lacks white forehead, rufous nape patch smaller; *gamblei* (rest of range) white forehead. **Similar spp:** Papuan Logrunner female shows barred wings, bristle-tipped tail, and different behavior. Papuan Whipbird female resembles a green juv bellbird but has thinner bill and longer tail. **Voice:** Song variable, a series of repeated, high, clear, ringing, whistled notes or upslurs, at times alternating between notes or repeated monotonously for prolonged periods. Examples include a robin-like series of repeated pure, tonal whistles, at a rate of 3/2 sec. Another is a repeated pair of whistler-like notes, the first rising, the second falling in pitch. Call (alarm?) is a conspicuous, loud, harsh, disagreeable rasp or hiss, *bshh SHHHT*. **Habits:** Solitary and often secretive, but vocal. Forages mostly by hopping along the ground searching, but also follows trunks and branches up into the forest midstory and even into the canopy. Takes arthropods and some fruit. Nest is built in the forest understory or midstory, in the fork of a vertical stem; nest a bulky cup of moss, fiber, and rootlets. Eggs (2) white with dark spots. **Range:** Endemic. NG: nearly all mts (no records from Cyclops and Kumawa Mts), 1750–2600 m (1450–3600 m).

Piping Bellbird *Ornorectes cristatus*   Pl 84
(**Crested Pitohui,** *Pitohui cristatus*)
24–25 cm. Locally common in forest interior of foothills and nearby lowlands. *A bushy-crested, stocky, brown bird* living on the ground and as high up as the midstory. *Extremely shy. Amazing song is key to discovery.* **Adult:** Brown above, paler rusty brown below. Rather variable. Some birds have a blackish face, of unknown significance. **Juv:** Rufous edges to wing-coverts. **Subspp** (3): Minor. **Similar spp:** Rusty P is lankier, larger, with pale iris. Little Shrikethrush smaller and slimmer. Neither spends much time on the ground. **Voice:** Song an incredibly long series of identical notes delivered uninterrupted for as much as 15 min. The rate of delivery gradually increases from several to many per sec, and the pitch gradually descends. The notes have a musical but throbbing quality, like that produced by blowing across the mouth of a large open jar. From afar, sounds like *ko ko ko . . .*, but at close range, like *ur-kohwahn, ur-kohwahn, ur-kohwahn. . . .* Song carries long distances but is ventriloquial, and the singer hides from view. Bell-like calls are very different; 2–4 clear notes can be produced by more than 1 bird. Alarm call, a harsh *chhrr*. **Habits:** Elusive, staying hidden in the undergrowth and almost impossible to observe. Solitary, in pairs, or small parties. Occasionally joins mixed-species flocks with pitohuis and birds of paradise. Forages for invertebrates while hopping along the ground or up into branches. No nesting information. **Range:** Endemic. NG, except most of Sepik-Ramu, Huon, and northern SE Pen, 400–1000 m.

PLOUGHBILLS: Eulacestomatidae

Unlike the previous two families that have been removed from Pachycephalidae, the Wattled Ploughbill does actually appear to be related to the whistlers and allies, but it is very different from that family.

Wattled Ploughbill *Eulacestoma nigropectus*   Pl 83
(**Wattled Shriketit**)
13–14 cm. Uncommon in cloud forest, especially thickets of climbing bamboo. *A small, chubby bird that creeps on tree branches and in bamboo,* stopping to excavate or strip stems with its *unique, deep-biting bill.* The bill has a reinforcing ridge on the pointed upper half for chiseling; the lower half is upward curved for biting. **Male:** *Black breast patch and wings. Large, pink gape wattles.* **Female:** Entirely olive green (wings, tail with brownish wash; faintly scalloped below) and lacks wattles. **Juv:** Grey-green with chestnut brown on back, rump, wings. **Imm:** Like Female but with retained Juv chestnut wing-coverts and flight-feathers. **Similar spp:** Male unmistakable. Mottled Berryhunter male darker, with some mottling and thin bill. **Voice:** Several songs or calls: a barely audible, high, thin, even whistle like a jewel-babbler but emanating from treetops or midstory; a series of ~6 rising and falling whistled notes; and a buzzy, repeated note. **Habits:** Solitary, in pairs, or in groups of 3–4; joins mixed flocks of small insectivores. In humid mossy forest, especially thick, regenerating vegetation. Eats insects. Vigorously digs and hammers on bark, dead wood, and bamboo; hops about, occasionally hanging upside down to check undersides of limbs. No nesting information. **Range:** Endemic. NG: Central Ranges, in the zone of climbing bamboo at ~2000–2500 m (1300–2800 m).

WHISTLERS AND ALLIES: Pachycephalidae

This is a family of some 50+ species made up of a core group of whistlers, shrikethrushes, and the pitohuis. Also traditionally placed in the family is a mix of other species whose affinities lie elsewhere according to recent molecular systematic studies. The Goldenface has been moved to the Australasian Warblers. Four other oddballs are the aberrant Mottled Berryhunter, Rufous-naped and Piping Bellbirds, and Wattled Ploughbill; these we have classified into new families, which we have placed in front of the whistler family for convenience. The pitohuis were once all thought closely related, but that is not the case now—most still belong in the whistler family, but at least Piping Bellbird does not. The 4 species in the genus *Pitohui*, which a molecular study indicates are most closely related to orioles, are for the time being still classified with the whistlers.

Whistlers and allies are widespread in Australasia, with outlying species ranging west through Wallacea to the Philippines and S Asia, and eastward into Oceania. The New Guinea Region is the family's center of distribution and radiation, with 22 species, and the family is represented everywhere, from small offshore islands through all types of mainland forest up to timberline. In general, the whistlers and allies are compact and robust birds, with powerful, hooked bills. Most species are insectivorous, taking arthropods from bark or leaves in a deliberate fashion. Some eat fruit, and certain of the larger species may prey upon small vertebrates. Most are behaviorally unglamorous but conspicuously vocal, with loud, varied, and attractive whistled songs. The pitohuis have become famous for being among the few birds poisonous to eat, thanks to a type of small beetle in their diet from which a toxin is taken up and sequestered in the bird's tissues and feathers.

Little Shrikethrush *Colluricincla megarhyncha* Pl 85, 70
(**Rufous Shrikethrush**, *Myiolestes megarhynchus*)
17–19 cm. Ubiquitous in forests and scrub of all sorts, from lowlands to mid-mountains. Great regional variation, with more subspecies than any other NG bird. **Adult:** *A stocky, medium-sized, nondescript songbird modestly colored in shades of brown* (or olive green in SE Is). **Juv:** Reddish-brown edges to wing-coverts and secondaries. **Subspp (16):** *Following are 8 candidate species, each corresponding to an identified genetic lineage*: (1) *affinis* (NW Is: Waigeo) above olive greyish brown, below pale buffy grey obscurely streaked and finely barred, bill grey-brown. (2) *megarhyncha* (Bird's Head and Neck, S Lowlands, Aru Is) either rufous with pale brown bill (West) or like *despecta* but more evenly pale grey-rufous below, with throat mottling very obscure (East). (3) *normani* (Trans-Fly, N AU) like *despecta* but has cream throat and is evenly bright pale rufous below, with little or no streaking or darker neck; pale bill. (4) *despecta* (SE Pen) above olive brown, below greyish rufous with pale greyish chin and throat mottled and streaked dark grey-olive, bill greyish brown. (5) *fortis* (SE Is: D'Entrecasteaux and Trobriand Is) above greenish olive with grey head, below grey with obscure streaking and yellowish undertail, bill black. (6) *discolor* (SE Is: Tagula I) above yellowish green with grey head, below dingy white with pronounced streaking. (7) *tappenbecki* (Sepik-Ramu, N Huon) inseparable from *despecta* and intergrades with *obscura* in West. (8) *obscura* (NW Lowlands) above olive or brownish grey, below greyish and streaked, bill black. *Three island forms were not included in DNA analysis, of which misoliensis* (NW Is: Misool) is like *megarhyncha, and 2 are distinct and possibly separate species*: (9) *batantae* (NW Is: Batanta) dark olive brown, paler below with heavy throat streaking, bill grey-brown. (10) *melanorhyncha* (Bay Is: Biak) above dark olive brown with rufous brown wings; below, including cheeks and ear-coverts, pale buff with darker breast band but no streaking; bill brownish grey; song suggests a whistler, not a shrikethrush. **Similar spp:** Sooty ST of mid-mountains is of similar size and shape but much darker. Tawny Straightbill has pointed bill, pinkish legs, and reddish iris. Grey Whistler has white eyebrow and throat. Rusty Whistler has grey cap, white throat. White-bellied Pitohui has restricted range and is larger and paler below. Other brown pitohuis are larger, lankier, and have specific field marks. **Voice:** Song (SE Pen) a series of 3–5 notes whistled, mellow on a moderate pitch: *who-WHI-oo* or *uwih weeteeyou* or *hoot hootuWEEtu*. Call is a loud, musical, liquid, downslurred note: *dyoong* or *tchuck*. Often answers loud, sudden noises. **Habits:** Solitary or in pairs. Sedentary and probably territorial. Usually in understory but occasionally into canopy, foraging for arthropods in branches and leaves, or rummaging about in clumps of dead leaves. Feeds on insects and other invertebrates, rarely fruit. Nest 1–2 m up in branches of a sapling or vine tangle; nest a large, messy bowl of dead leaves and stems sparingly bound together with fiber. Eggs (2) whitish with plentiful dark spots

and blotches. **Range:** NG, NW Is (Batanta, Misool, Salawati, Waigeo), Bay Is (Biak, Yapen), Aru Is, and SE Is (3 main D'Entrecasteaux Is, Trobriand Is, Tagula I), 0–1800 (2300) m. Also AU. **Taxonomy:** Sorely in need of taxonomic revision. This is actually a complex of species, as convincingly demonstrated by a molecular systematic study that identified 8 lineages so divergent they would represent separate species or genera in other birds. These look-alike shrikethrush lineages are distributed in a mosaic pattern across the region; however, a few cannot as yet be separated by plumage or song, and some seem to intergrade. Research is needed to corroborate and define any proposed breakup of the Little Shrikethrush. Candidate species are given in the Subspp section.

Sooty Shrikethrush *Colluricincla tenebrosa* Pl 83
(Sooty Whistler, *Pachycephala tenebrosa* or *umbrina*)

18–19 cm. Rare and local in mid-mountain forest. Very poorly known. Shy, elusive; skulks in shrubbery and thickets. Above the elevational range of Little ST. **Adult:** *A robust, dark bird, entirely brownish black.* Same shape and size as Little ST, but tail longer. **Similar spp:** Black Pitohui larger, hooked stout bill, plumages either sooty black or grey-brown. Little ST not nearly so dark. **Voice:** Little known. Alarm call a loud, metallic *huija*. **Habits:** Solitary or in pairs. Perches and watches for prey for long periods. Takes insects. No nesting information. **Range:** Endemic. NG: W, Border, and E Ranges as far as Schrader Mts, 1450–2150 m.

Grey Shrikethrush *Colluricincla harmonica* Pl 85

23 cm. A bird of savannahs, towns, and open lowland country, following human settlement into highlands. Best known for its powerful and musical song. *Medium-sized, grey, with white face and belly*; note distinctive white highlighting on lores, eyebrow, and throat, and contrasting dark eye. **Male:** Dark bill, immaculate underparts. **Female:** Bill greyish, underparts obscurely streaked. **Juv:** More heavily streaked underparts, rufous wash on wings. **Subspp (1):** *superciliosa* (also AU). **Similar spp:** The larger, open-country bowerbirds and figbirds are brownish, and with brownish/yellowish or streaked underparts, respectively. Cuckooshrikes are differently shaped, usually have black in the face, and typically are seen in the canopy. **Voice:** A noted songster. Song highly varied, consists of loud, ringing notes; song pattern is whistler-like, slowly delivered with strong emphasis on certain notes, often the last note, e.g., *du du du du du du DYO!* **Habits:** Singly or in pairs. Forages in deliberate whistler fashion for invertebrates, small vertebrates, and fruit; visits all levels in the trees and descends to the ground. Breeds in pairs. Nest placed a few meters up in a small tree; nest a bulky cup of twigs, grass, and other materials. Eggs (2–3) white with dark spots and blotches. **Range:** E NG, westward to Trans-Fly and Sepik-Ramu, and locally in interior valleys of E Ranges, 0–1700 m. Also AU.

Rusty Pitohui *Pseudorectes ferrugineus* Pl 85
(*Pitohui* or *Colluricincla ferrugineus*)

25–28 cm. Common in lowland and hill forest. In small groups, often associated with mixed-species flocks. **Adult:** *A large, rusty brown pitohui with staring white eye.* **Juv:** Eye dark. **Subspp (5, in 3 groups):** *ferrugineus* (W and N NG) black bill, plumage uniform rusty; *clarus* (eastern S Lowlands, SE Pen, Huon) black bill, pale ochre breast; *leucorhynchus* (NW Is: Batanta, Waigeo) yellowish bill, uniform dark rusty. **Similar spp:** Associates with other rufous insectivores: Papuan Babbler has smaller head, longer yellow bill; Piping Bellbird has dark eye, bushy crest; shrikethrushes are smaller, dark-eyed; Black Cicadabird female has fine bill, dark eye. **Voice:** Contact calls range from simple *tieu* notes to various loud, strongly ascending, repeated musical phrases, e.g., *hoodleee* followed by hollow repeated notes *kyo kyo kyo*. . . . Alarm call, a repeated, harsh, rasping note. **Habits:** In pairs or groups, members usually scattered but keeping in contact by calling. Conspicuous, but wary. Foraging birds range from understory into the canopy. Bird pauses to look around before actively gleaning insects and fruit. Nest built a few meters up in a forked sapling; nest a bulky cup of sticks and stems. Egg (1) pinkish grey marked with a some darker spots. **Range:** Endemic. NG, NW Is (Batanta, Misool, Salawati, Waigeo), Bay Is (Yapen), and Aru Is, 0–1000 m.

White-bellied Pitohui *Pseudorectes incertus* Pl 85
(*Pitohui* or *Colluricincla incertus*)

22–23 cm. A local inhabitant of the riverine floodplain forest that winds through high-rainfall areas of the S Lowlands; common in this restricted habitat and absent from adjoining lowland forest. In shy, noisy groups, often joined by other species. **Adult:** *Pale and tan*. Belly dingy white. *Dark beady eye prominent against the pale face. Pale bill.* Larger than a Little Shrikethrush but smaller than other pitohuis. (Eye color variable, from dark to medium brown, and may relate to age and sex.) **Juv:** No information. **Similar spp:** Little Shrikethrush darker and more uniformly colored; solitary rather than gregarious. Rusty P juv also has dark eye, but note black bill and lanky shape. Grey Shrikethrush similar yet grey, does not co-occur. **Voice:** Song is a lovely, bubbling duet, the leader singing a loud, rapid, mellow, whistled phrase at medium pitch, similar in quality to the voice of the Little Shikethrush and repeated several times. The follower sings a shorter phrase, high pitched, louder, and more piercing. Call resembles that of King Bird of Paradise: 5 or more downslurred notes, repeated 2/sec, the series is constant in pitch or ascends. **Habits:** Travels in noisy parties of 3–6 in midstory where the canopy is open and viny tangles are dense. May be joined by other species to form mixed-species flocks. Actively forages for insects and fruit. No nesting information. **Range:** Endemic. NG: S Lowlands, from northwest (Lorentz) R eastward to Fly R, 0–100 m.

Black Pitohui *Melanorectes nigrescens* Pl 83, 66
(*Pitohui nigrescens*)
22–23 cm. Uncommon in the mountains in a rather narrow elevational band straddling the lower limit of the cloud forest. Note large, heavy, *strongly hooked, black bill*. In shape, posture, and movements, the bird *suggests a giant whistler* rather than a pitohui or shrikethrush. **Male:** *All sooty black.* **Female:** *Brown, varying from greyish brown to rufous brown.* **Juv:** Brown? **Subspp (5, in 2 groups):** *nigrescens* (Bird's Head) Female with greyish cap; *schistaceus* (rest of range) uniform rufous- or grey-brown, depending on locality. **Similar spp:** *Amblyornis* bowerbird females are similarly colored to Female Black P, but stockier and with shorter tails and thicker, more tapered, brownish bills without the hooked tip; they never sally. Sooty Shrikethrush is stockier, browner, and smaller than Male Black P. Pygmy Drongo perches upright and is largely confined to lower elevations. **Voice:** Songs vary geographically, and each individual has several songs. These can be a series of breezy upslurs or downslurs; a rising series of 10 staccato, medium-pitched *kwik* notes; and a low mellow whistle that sweeps upscale and is repeated every 4 sec. Call a buzzy, inflected *whurr?* or *wheet?* Flies with audible wingbeats. **Habits:** Forages singly or in pairs, ranges broadly from the ground to lower canopy, usually in the midstory. Not shy. Sluggish; looks about from its perch before moving on. Gleans and sallies insects and spiders. Nest a cup of fern leaves and rootlets in a tree. Eggs (1–2) maroon or pink, spotted and blotched. **Range:** Endemic. NG: Bird's Head and Neck (Fakfak and Wandammen Mts), Central Ranges, Foja Mts, and Huon, 1600–2200 m (1000–2600 m). **Taxonomy:** Allied with whistlers rather than other pitohuis.

The Whistlers. The 14 species of the New Guinea Region fall into 7 groups in their ecological distributions. Six are forest species that approximately replace each other with altitude. From the lowlands to high elevations, they are Grey, Rusty, Sclater's, Regent, and Lorentz's and Brown-backed Whistlers. Of the 2 species of highest elevation, Lorentz's is confined to the West, Brown-backed to the East. The 6 species also differ in vertical distribution within the forest, especially where they overlap elevationally: Lorentz's, Brown-backed, and Grey forage high in the canopy; Rusty, Sclater's, and Regent from the undergrowth to the canopy. The last mainland forest species, Vogelkop Whistler, is small and confined to the Bird's Head (and possibly Foja Mts). There are 2 nonforest mainland species: Golden-backed Whistler of low second growth, shrubs, and cane grass in the East, and Black-headed and White-bellied Whistlers in savannah and open woodland where other whistler species are absent. The last mainland species is the extremely range-restricted Baliem Whistler, known only from the Baliem Valley in the Western Ranges; it inhabits mid-mountain and lower cloud forest and second growth below the elevational range of Regent W. The remaining 3 species are virtually confined to islands—Island Whistler on small islands in the West, Mangrove Golden Whistler in mangroves on small islands in the East, and Louisiade Whistler on midsize SE Is.

Regent Whistler *Pachycephala schlegelii* Pl 86
(Schlegel's Whistler)
15–16 cm. Common in cloud forest and subalpine shrubbery. Mostly above the elevational range of Sclater's W, and in West shares habitat with Lorentz's W and, locally, with Baliem W. **Male:** *Broad, black breast band, orange-brown breast patch, large yellow nape patch.* White throat patch relatively small. **Female:** *Grey head; chin speckled pale grey-and-white and separated by grey border from olive breast band.* **Juv:** Mainly rufous brown. **Imm:** Like Female but retaining some rufous, especially in wing. **Subspp (2):** Minor. **Similar spp:** Lorentz's W resembles Female Regent and shares habitat, but it lacks olive-green border between grey bib and yellow belly. Baliem and Sclater's Ws females lack grey head and bib. **Voice:** An inveterate singer. Song similar to other whistlers but distinguished

by characteristic slurred notes and unmusical sounds resembling lip-smacking, kissing, or meowing. These strange notes and whistles are repeated in loud swelling series. Common song: *whi-whi-whi, tu, tu, tu, tu, tu*. But much geographic variation. Call is a faint *ss*. **Habits:** Solitary or in pairs, alone or in mixed-species flocks. Forages in middle and upper stories of forest and in shrubbery at timberline. Hops slowly along limbs searching for stationary insect prey. Often 2 or more Males will confront each other and counter-sing, expanding the bright yellow nape patch. Nest, no information. Eggs (2) white with dark spots. **Range:** Endemic. NG: nearly all mts (no records for Kumawa and Adelbert Mts), 1700–3650 m (not as high in West where Regent overlaps with Lorentz's W).

Sclater's Whistler *Pachycephala soror* Pl 86
(Hill Golden Whistler)

15 cm. Common in mid-mountain forest. Overlaps with Grey and Rusty Ws (these at lower elevations) and Regent W (higher). In W Ranges, lives at lower elevations than Baliem W. **Male:** Resembles male golden whistlers but *lacks yellow hind-collar*. **Female:** Duller version of Male, with grey-brown cap and *broad olive breast band* that merges with yellow breast; *lacks extensive grey on the head*. **Juv:** Mostly chestnut brown. **Imm:** Similar to Female but shows some rusty Juv feathering in wings and tail. **Subspp (4, in 2 groups):** *soror* (all of range except that of next race) Male head black with thick, black breast band, and tail color varies from black or mostly black to mostly green in SE Pen; *octogenarii* (Bird's Neck: Fakfak and possibly Kumawa Mts) Male head grey with thin breast band. **Similar spp:** Males of Mangrove Golden and Baliem Ws have yellow hind-collars; females not safely separable except by habitat and locale. Regent W female has a grey head and grey-mottled throat patch bordered by a grey collar and thick olive breast band. **Voice:** Song a loud and ringing whistled series, varied and usually ending in a loud crescendo; a series of 4–15 notes, such as *chink-chink-chink-chink-CHINK* or *wu-weet wu-weet wu-weet twee-twee-twet-tweet-weet weet—WEET*, increasing in volume but remaining on the same pitch. Also a piping whistle. Contact call a weak *zweet?* **Habits:** Singly or in pairs. Often in mixed-species flocks with scrubwrens and other small insectivores. Moderately active but inconspicuous except when singing. Gleans insects from medium-sized branches in forest midstory. Nests a few meters up on a sapling branch; nest a small cup of stems, leaves, moss. Eggs (1–2) pinkish white with dark blotches. **Range:** Endemic. NG and Goodenough I (SE Is: D'Entrecasteaux Is), 1100–1900 m (600–2200 m). NG: all ranges except Wandammen, Foja, Cyclops, and N Coastal Mts.

Louisiade Whistler *Pachycephala collaris* Pl 86
(formerly under **Golden Whistler**, *P. pectoralis*, or **Bismarck Whistler**, *P. citreogaster*)

15–16 cm. Coastal forest and scrub on certain small to midsize SE Is, where scarce. Louisiade, Baliem, and Mangrove Golden Whistlers are nearly identical but are separated by habitat and range. The Males share yellow nape and breast, and the drab Females, a whitish throat, tan breast band, and yellow breast. *Louisiade W differs by its much longer bill and green tail*. **Subspp (2):** *collaris* (species range except next); *rosseliana* (Rossel I) Female bill brown, underparts somewhat buffy. **Similar spp:** Mangrove Golden W, also in SE Is, has not been found on the same islands as Louisiade W; it has silvery margins to the wings and a blackish tail. **Voice:** Similar to other golden whistlers. **Habits:** Typical of other golden whistlers. Nest is a cup of small twigs decorated with dead leaves. Eggs (1–2) pale pinkish grey with dark spots. **Range:** Endemic. SE Is, including Alcester (Nasikwabu), Egum, Kimuta, and Rossel Is; plus Bonvouloir, Calavados, Conflict, Deboyne, and possibly Engineer groups, 0–300 m. **Taxonomy:** Molecular systematics has shown the Louisiade Whistler to be genetically divergent from the rest of the golden whistler group.

Baliem Whistler *Pachycephala balim* Pl 86
(formerly under **Common Golden Whistler**, *Pachycephala pectoralis*)

15–16 cm. Known only from the Baliem Valley (W Ranges) where it is found in mid-montane and cloud forest and in pockets of second-growth forest and stands of casuarina in a widely cleared landscape. Apparently lives above the elevational range of Sclater's W and below that of Regent W. *Both sexes of Baliem W have green edges to the wing feathers* (these are silvery grey in Mangrove GW) and a *short bill*. **Male:** Differs from co-occuring male Sclater's by *yellow hind-collar*. Thick black breast band and black tail (tail same for Mangrove GW, but green in Louisiade W). **Female:** Similar to female Sclater's and Louisiade Ws. **Juv:** Much rufous brown. **Imm:** Similar to Female but may retain brownish Juv wing feathers. **Similar spp:** Louisiade W occupies SE Is; Mangrove Golden W inhabits both NG mangroves and SE Is forest. Sclater's W also in W Ranges but at lower elevations; male lacks

yellow collar; female may not be distinguishable. **Voice:** Not known, but probably similar to Mangrove Golden and Sclater's Ws. **Habits:** Similar to other golden whistlers, but in different habitat. Nest, no information. **Range:** Endemic. NG: W Ranges (Baliem Valley), 1600–2400 m.
Taxonomy: Molecular systematics indicate this relictual NG population is a distinct species.

Mangrove Golden Whistler *Pachycephala melanura*   Pl 86
(Black-tailed Whistler)
16–17 cm. Locally common in mangroves and other types of inundated coastal forest on mainland NG and in forest and scrub on small SE Is. Differs from Louisiade W in *silvery grey (not green) margins to secondaries*. **Male:** *Tail always black (not green)*. **Female:** *Grey-tan breast band* and *tail partly black*. Underparts vary from bright yellow like Male's to quite pale (along S coast). **Subspp (2):** Minor. **Similar spp:** Louisiade W (green tail) occurs only on large islands not occupied by Mangrove Golden. Baliem W lives only in mountains. Sclater's W lives in mountains, not coast; male lacks yellow hind-collar; female has olive-brown (not grey-tan) breast band. **Voice:** Song is a series of loud, clear disyllables, the first note much softer and shorter than the second, which is louder, longer, whipped, and upslurred. Series comprise 2–3 disyllables that are repeated irregularly. **Habits:** Usually solitary. Stolid insect gleaner on midstory limbs and deep in thickets. Keeps to cover when near ground and then difficult to observe. Nests in an upright branch fork; nest an untidy cup of twigs, vine stems, leaves, grass. Eggs (2) white or buff, spotted with red-brown. **Range:** NG (Bintuni Bay; coastal Trans-Fly to SE Pen and Huon to Madang) and SE Is (Fergusson, Normanby, and small, offshore islands), sea level. Also AU, Bismarck Is, and Solomon Is. **Taxonomy:** DNA studies have confirmed this species as separate from Louisiade Whistler.

Golden-backed Whistler *Pachycephala aurea*   Pl 86
15–16 cm. Specialist of second growth and tall cane bordering rivers and lakes. Patchily distributed in foothills and nearby lowlands, seldom found in mountains. Common where found. *Golden back*. **Adult:** Unique among whistlers for its *black-and-gold pattern*. White throat patch reduced, nearly absent in some birds. **Imm:** Similar to Adult but with olive wash. **Similar spp:** Other, similar whistlers, such as Sclater's W or Mangrove Golden W, have a green back. Also compare with Golden Cuckooshrike and Golden Monarch. **Voice:** Song an accelerating series of short, high-pitched, whipped, upslurred, whistled notes at a rate of 2 notes/sec. **Habits:** Often in pairs. Inhabits large-leaved shrubs and small trees, 3–7 m above ground. Sometimes in tall riverine cane grass. At Garaina, in albizia shade trees in tea plantations. Similar to golden whistlers in behavior. No nesting information. **Range:** Endemic. NG: S Lowlands, SE Pen, and Sepik-Ramu, 0–1460 m.

Vogelkop Whistler *Pachycephala meyeri*   Pl 86
14–15 cm. Restricted montane range: Bird's Head, plus a single sight record from the Foja Mts. Inhabits forest and thickets. **Adult:** *Pale, buff-grey ear-coverts; same color patch on sides of breast, merges with pale yellowish abdomen.* Grey cap, whitish throat. *Small bill.* **Juv:** No information. **Similar spp:** Other yellow whistlers are larger in size and have larger bills. Sclater's W female has a dark grey-olive breast band. Regent W female has a dark green breast band; occurs mainly at higher elevations. **Voice:** Song a descending series of clear whistles, usually followed by an explosive upslurred note:

Habits: No information. **Range:** Endemic. NG: Bird's Head and hypothetical for Foja Mts (possible sight record; further documentation needed), 970–1450 m. **Taxonomy:** Foja Mt population undescribed.

Lorentz's Whistler *Pachycephala lorentzi*   Pl 86
15–16 cm. Common in western cloud forest and subalpine scrub. Regent W also lives in cloud forest but does not extend as high in elevation where Lorentz's is present. In habitat occupied by both, Lorentz's lives in the canopy, Regent at lower heights. **Adult:** Like female Regent W, but *lacks speckling on chin* and *lacks olive-green band between grey bib and dull yellow belly*. Bill slightly smaller. **Juv:** Mostly rusty, inseparable from juv Regent. **Imm:** Similar to Adult or with white belly. **Similar spp:** Regent W female has olive-green border between grey bib and yellow belly. Baliem and Sclater's Ws females have olive-brown breast bands, not grey. **Voice:** Song is a modest whistle, *swit swit wit tu wit teo wit* or *whit weet tyuu* repeated 3–5 times, each series increasing in volume.

Habits: A typical whistler. Quite active, flicking tail and hopping from limb to limb in shrubbery. No nesting information. **Range:** Endemic. NG: W and Border Ranges (Victor Emanuel Range), 1750–2700 m.

Brown-backed Whistler *Pachycephala modesta* Pl 87
14 cm. Common in canopy of eastern cloud forest and subalpine scrub. Note small size. **Adult:** *Dark brown upperparts, white throat*; breast ranges from off-white (E) to pale grey (W); note *grey crown*, visible in good light. **Juv:** Rufous brown. **Subspp (2):** Minor. **Similar spp:** Grey Gerygone much smaller. No other canopy insectivore in montane forest has entirely clear white or pale grey underparts. **Voice:** Song typical of a whistler, soft, pleasant, and more complex than that of co-occurring Regent W; consists of some 20 notes, lasting 8 or more sec. Begins with rising, whistled *wheet* notes, the first 4 successively lower pitched, subsequent syllables alternate in pitch, the phrase ending with 2 explosive notes, *CHIEW! CHIEW!* **Habits:** Solitary or in pairs. Inconspicuous if not for its song. Forages in the treetops by hopping slowly but steadily on upper limbs and branches; takes insects from branches and foliage by gleaning or sallying and hovering. No nesting information. **Range:** Endemic. NG Border Ranges (Hindenberg Range) eastward to SE Pen and Huon, 1750–3500 m (descends lower locally).

Island Whistler *Pachycephala phaionota* Pl 87
16–17 cm. *Only on tiny islets of Indonesia* where it occupies all wooded habitats. **Adult:** *A brown whistler with greyish head, white throat, and dull white underparts* washed greyish tan. **Juv:** Top and sides of head olive brown. **Similar spp:** None in its habitat. **Voice:** Song is a loud *weet-chut-wweeeee*. **Habits:** Frequents understory. Feeds on insects. No nesting information. **Range:** NW Is (Kofiau, Misool, Salawati, Wai, Schildpad, Sorong, Waigeo—probably absent from large islands and instead found on nearby small offshore islets), Bay Is (Rani), and Aru Is (Pulau Babi and smaller islands), sea level. Also E Wallacea.

Rusty Whistler *Pachycephala hyperythra* Pl 87
15 cm. Hill forest midstory; locally distributed, often common and conspicuously vocal where present. **Adult:** *Brown with tan or orange-tan underparts.* Grey cap, white or pale grey throat. **Juv:** Largely rufous. **Subspp (4, in 3 groups):** *hyperythra* (Bird's Head to western S Lowlands) rufous brown above, pale tan below, indistinct throat patch; *sepikiana* (NW Lowlands to Sepik-Ramu) olive brown above, bright orange-tan below; *salvadorii* (eastern S Lowlands, Huon, SE Pen) grey-brown above, tan below. **Similar spp:** Little Shrikethrush larger and lacks contrasting white throat patch. **Voice:** Song diagnostic, most musical and varied of the whistlers. A typical whistler song, beginning with 1–8 clear, faint, bell-like or whistled notes in a crescendo terminating in explosive, usually repeated slurs: *whik! whik!* Contact call is a pair of mournful notes on same or different pitch: *tooi* or *tu-ee*. **Habits:** Solitary or in pairs. Hops along branches and gleans in typical whistler fashion. Feeds on insects. Nest (location?) bowl-shaped, made of leaves of small climbing ferns, moss, and tendrils. Eggs (2) pinkish grey with dark spots. **Range:** Endemic. NG, lower slopes nearly throughout (no records Fakfak, Kumawa, Cyclops, and Adelbert Mts, and N slope of SE Pen), 600–1300 m.

Grey Whistler *Pachycephala simplex* Pl 87
(**Brown Whistler** for *simplex* group and **Grey-headed Whistler** for *griseiceps* group)
14–15 cm. The only whistler in lowland forest and regrowth, where common. Extends up into the hills and lower mountains where it overlaps with Rusty W and at the top of its elevational range meets Sclater's W. Inconspicuous, but vocal and unwary. A small, dainty whistler. **Adult:** Plain, except for *diagnostic pale eyebrow* and pale throat. **Juv:** Largely rufous (but showing the pale eyebrow?). **Subspp (6, in 2 groups):** *griseiceps* (most of NG range and E AU) olive with pale yellowish belly and flanks, greyish cap; *simplex* (includes *dubia* of SE Pen and *simplex* of N AU) brown above, white washed with brown below. **Similar spp:** Gerygone warblers similarly colored, but much smaller and more active. Rusty Whistler has well-defined white throat patch, lacks white eyebrow, and is evenly tan or yellowish orange below. **Voice:** Habitually sings in the middle of the day when most other birds are quiet. The song is a pleasant, varied warble, distinctive for its flexibility of tempo. Tempo shifts back and forth between faster and slower, and volume between louder and softer, with notes rising and falling below a standard pitch; *hoo hee ho hoo, TYU! TYU!* Also a melodious chirping unmistakable once learned. The pattern consists of alternation between a monosyllabic, slightly upslurred chirp and a disyllabic chip with flexibility of tempo and volume. **Habits:** Solitary or less often in pairs. Regularly joins mixed-species groups of gerygones, fantails, and monarchs. Forages from the

midstory to lower canopy. Stolid, casually searches on small limbs and in leaves for insects and spiders. Nest built in branches of midstory tree; nest is a cup of various plant materials (AU data). Eggs (2) whitish with dark spotting. **Range:** NG, NW Is (Batanta, Gag, Gebe, Misool, Salawati, Waigeo), Bay Is (Mios Num, Yapen), Aru Is, and SE Is (3 main D'Entrecasteaux Is and Tagula I), 0–1450 m. Also AU and Kai Is. **Taxonomy:** Some authorities split the species into 2, the Brown and Grey-headed Whistlers.

Black-headed Whistler *Pachycephala monacha*   Pl 87

15–16 cm. In widely scattered populations associated with human settlement. A loud songster in tall trees of open habitats, such as casuarinas, typically in hills and mid-mountain valleys, rarely in lowlands and along coast. Hybridizes with White-bellied W, producing birds with intermediate plumage. **Male:** *Black with white lower breast and belly.* **Female:** *Same pattern in dark grey and white.* **Juv:** Rufous brown above, dingy white below. **Subspp (2):** *monacha* (Aru Is) upperparts brownish black; *lugubris* (NG) upperparts sooty black. **Similar spp:** Satin Flycatcher male is more slender and vibrates tail up and down. Willie Wagtail larger, with longer tail and different behavior and voice. Hooded Monarch in different habitat (forest interior), has grey bill and longer tail. **Voice:** Song a typical loud, whistler repertoire, clear and whistled with explosive notes, for example: *hoo-hoo-hoo-hoo-hoo-WHIT*; *wi teur teur wee TYUR*. **Habits:** Usually in pairs. Active, vocal, and conspicuous in upper branches and foliage where it gleans insects from twigs and foliage. Nest built on branch in tree crown; a substantial cup of plant material. Eggs, no information. **Range:** Endemic. NG and Aru Is. NG: Central Ranges and Huon; also lowlands at foot of N Coastal Mts and coastal western S Lowlands, 0–1750 m.

White-bellied Whistler *Pachycephala leucogastra*   Pl 87
(**Rufous Whistler**, *P. rufiventris*)

14–15 cm. Scarce in its localized distribution in E NG but common on Rossel I (SE Is). Occurs in open-wooded habitats on the mainland and also in closed forest on Rossel. Hybridizes with Black-headed W in vicinity of Port Moresby. **Male:** *Unique—black hood, white throat and breast, grey back.* **Female:** *Grey with whitish or dingy grey throat*; *grey breast band*, whitish breast and belly. **Juv:** Grey above, white below with distinct dark streaking, white eye-ring. **Imm male:** Resembles Female but often with indistinct black breast band. **Subspp (2):** *leucogastra* (NG mainland) Male upperparts paler grey, side of breast white, and Female with whitish breast, Juv pale grey; *meeki* (SE Is: Rossel) Male upperparts darker, side of breast grey, Female with breast washed buff, Juv washed with buff. **Similar spp:** Mangrove Robin similar to Female whistler but lacks grey breast band. **Voice:** Song a series of clear whistles followed by a sharp, explosive whip-crack; for example, *pee pee pee, joey joey joey, eeeeCHONG*. **Habits:** Solitary or in pairs. Vocal resident of crown foliage in scrub, mangroves, eucalypt savannah, rubber plantations, and other open habitats of coastal lowlands; also primary forest on Rossel. Nests high in a tree in disturbed habitat; nest a cup of plant material. Eggs (2) undescribed. **Range:** Endemic. NG and Rossel (SE Is), 0–1200 m. NG: Sepik-Ramu, SE Pen (Purari Delta and Port Moresby area). **Taxonomy:** Has been classified as a subsp of Rufous Whistler (*P. rufiventris*) of AU or possibly a race of Black-headed W, even including the Wallacean Whistler (*P. arctitorquis*).

Northern Variable Pitohui *Pitohui kirhocephalus*   Pl 84

20–25 cm. Lowland and hill forest, common at edge and in second growth, less so in primary forest. Striking geographic variation; usually separable from Southern Variable P either by *grey head, wings, and tail* or by *all rusty-brown plumage*. However, the local Wandammen form is black and chestnut and resembles Southern Variable and Hooded P, see details below. This species possesses toxic feathers and skin. *Bill grey-brown* (black in race *dohertyi*). Eye color variable, from dark red to dark brown, likely an indicator of age and sex. **Adults** usually alike, although in some subspecies the Female is duller, plainer. **Juv** differences, if any, unknown. **Subspp (9, in 4 groups):** *kirhocephalus* (eastern Bird's Head and Bird's Neck except Wandammen Pen) grey head, wings, tail; *dohertyi* (Wandammen Pen) black head, wings, tail; *jobiensis* (NW Lowlands east to Vanimo, plus Yapen I) all rufous brown; *brunneicaudus* (Sepik-Ramu) like western grey-headed group but underparts paler. **Similar spp:** Southern Variable P range meets Northern Variable's on Bird's Head and Bird's Neck; here Southern is always black and chestnut, whereas Northern is grey and chestnut (black-and-chestnut Wandammen subsp not in contact with Southern). Hooded P's black-and-rufous pattern resembles only the Wandammen subsp of Northern Variable, which has brown uppertail-coverts and is a near-perfect match but larger, and some individuals show grey instead of black. Rusty P very similar to NW form of Northern Variable but has pale eye. **Voice:** Songs (or calls?) are variable, loud, bubbly, jumbled series of gurgling whistles, interspersed with scratchy catlike sounds, often delivered by 2–3 birds together. Sound confusingly like Hooded P, but often more scratchy. In Sepik-Ramu, these begin with softer, shorter phrases and end with longer, louder phrases, suggestive of a Helmeted Friarbird, oriole, or Hooded P. Dawn duet: one bird calls a whistled upslur, the second replies with a *tup*. **Habits:** In pairs or small groups, often joining mixed-species flocks. Shy, difficult to view well. Habits similar to Hooded P. No nesting information. **Range:** Endemic. NG and Yapen (Bay Is), 0–1100 (1500) m. NG: Bird's Head and Neck, NW Lowlands to N Huon. **Taxonomy:** The original Variable Pitohui is now 3 species, this one and the next two.

Raja Ampat Pitohui *Pitohui cerviniventris* — Pl 84

20–21 cm. Waigeo and Batanta Is (NW Is) only. A small, insular pitohui. **Adult:** Unique pattern— *pale brownish-grey head and upperparts; underparts tan-brown*. **Juv:** No information. **Subspp (2):** Minor. **Similar spp:** Color pattern differs from Hooded and both variable pitohuis, neither of which occurs on these islands. **Voice:** No information. **Habits:** Presumably similar to related pitohuis. **Range:** Endemic. NW Is (Batanta, Gam, Waigeo), presumably throughout. **Taxonomy:** Split from Variable Pitohui (*P. kirhocephalus*).

Southern Variable Pitohui *Pitohui uropygialis* — Pl 84

22–25 cm. Southern counterpart of Northern Variable P. Note disjunct distribution: W Bird's Head and S Lowlands, divided at the Bird's Neck by Northern Variable P. This species also with striking geographic variation; usually separable from Northern Variable P by either *black head, wings, and tail, or these parts dusky brown*. (One subsp of Northern is black and chestnut but does not co-occur with Southern.) Black-headed forms of Southern always have *black uppertail-coverts*, unlike the black-headed Northern Variable and Hooded Ps. *Bill black.* Eye color variable as in other pitohuis, likely an indicator of age and sex. **Subspp (6):** 4 groups. *uropygialis* (Misool, Salawati, W Bird's Head) black with chestnut back, breast, belly; *aruensis* (western S Lowlands, plus Aru Is) black with chestnut back, flanks, belly; *brunneiceps* (eastern S Lowlands) dusky brown head, wings, and tail, brown body; *meridionalis* (southern SE Pen) black head, wings, and tail, body rufous tan. **Similar spp:** See Northern Variable P and Hooded P. **Voice:** Resembles Northern Variable P, but the means to distinguish the two has not yet been worked out. **Habits:** Similar to Northern Variable P. Also toxic. No nesting information. **Range:** Endemic. NG, NW Is (Misool, Salawati) and Aru Is, 1–1000 m. NG: western Bird's Head (absent from Bird's Neck), S Lowlands, and southern SE Pen to Milne Bay. **Taxonomy:** Split from Variable Pitohui (*P. kirhocephalus*).

Hooded Pitohui *Pitohui dichrous* — Pl 84

22–23 cm. Common and adaptable in mid-mountain forests, second growth, and gardens, from understory to canopy. Its bold plumage serves as warning that the bird is poisonous to eat. Eye color variable, from dark red to dark brown, likely related to age and sex. **Adult:** A strikingly patterned, *orange-brown bird with black head, wings, and tail*. Note brown uppertail-coverts, to separate from other black-and-brown pitohuis, see Similar spp. **Juv:** Same except for brownish edges to black throat feathers. **Similar spp:** Certain races of Southern Variable P and Northern Variable P are nearly identical but usually have black on uppertail-coverts and are slightly larger, and only the race *dohertyi* of the Northern Variable always has brown uppertail-coverts. The mimic races of Southern occur on SE Pen, W Bird's Head, and the NW Is of Misool, and Salawati (where Hooded is absent). **Voice:** Bird's name is a rough transcription of the call, *pi-to-hui?* A variety of soft and melodious calls, for instance a low mournful *koo-koo oh* often followed by a guttural *schk ggh*; or a higher *kuukuu koh WEEoh WEEoh weeii?*, the last notes on an ascending scale. When several birds assemble, the calls may be jumbled together. Usually delivered from thick vegetation, out of view. Voice similar but less scratchy than that of variable pitohuis. Wings make distinctive fluttering sound in flight. **Habits:** In pairs or small groups; known to be a cooperative breeder. A generalist forager gleaning small fruits, insects, and spiders. Nest built in tree is a cup woven of vine tendrils. Eggs (1–2) creamy with grey patches marked by dark spots and blotches. **Range:** Endemic. Probably all NG mts (no record from Kumawas) and Yapen I (Bay Is), mainly 600–1700 m, but descends to sea level in absence of variable pitohuis.

SHRIKES: Laniidae

The 33 species of shrikes are predominantly African and Eurasian. Two species have been recorded from New Guinea—one a resident, the other a vagrant from SE Asia. Shrikes are predatory songbirds with strong, hooked bills. They typically wait for long periods on exposed perches in open habitat, swooping down to capture large insects and small vertebrates. Shrikes are often vocal and usually found alone or in pairs. The nest is a deep woven cup placed in reeds or shrubs. Both parents attend the nest.

Brown Shrike *Lanius cristatus* Pl 104
18 cm. Palearctic vagrant to open habitats during austral summer. **Adult:** *A pale brown-and-grey bird with striking black mask* framed above by *white eyebrow* and below by white throat. **Juv:** Similar but duller and greyer; breast and flanks marked with fine, dark scaling. **Subspp (1):** *lucionensis* (breeds far E Asia, winters SE China to Philippines). **Similar spp:** Long-tailed S is larger, and adult has black cap, wings, and tail; juv lacks white eyebrow; all ages show pale wing patch—white in adult, buff in juv. **Habits:** Similar to Long-tailed S. **Range:** NG: 1 record from Bird's Head. Breeds over much of Asia, winters in southern Asia, Sunda Is, and Philippines, rarely in eastern Wallacea.

Long-tailed Shrike *Lanius schach* Pl 104
(**Black-capped** or **Schach Shrike**)
25 cm. Common resident in mid-mountain and alpine grassy clearings and gardens. **Adult:** A *black-capped* shrike with russet back; *wings and tail are patterned black and white.* **Juv:** Paler and duller with black mask and fine, dark scaling above and below. **Subspp (1):** *stresemanni* (NG). **Similar spp:** See Brown S, a vagrant. **Voice:** Song is beautiful and melodious, a series of whistled notes interspersed with grating ones, reminiscent of a reed-warbler. Mimics other birds. Call is a harsh, repeated buzzing note, *tchick, tchick, tchick . . .* or *grennh, grennh, grennh. . . .* **Habits:** Conspicuous, but wary and easily flushed. Singly or in pairs. Frequents open perches from which it drops down to capture prey. Nest built ~1.5 m up in cane grass or shrub branches among cane; nest a cup of fine grass stems. Eggs (2) pale blue with dark spots. **Range:** Central and eastern NG: Border and E Ranges, Huon, and northern SE Pen (Telefomin to Woitape), 800–2700 m. Also India to China, to Sunda Is and Philippines.

ORIOLES AND FIGBIRDS: Oriolidae

The oriole family probably originated in Australasia, yet the majority of its 34 species inhabit Asia and Africa, with only 4 in the New Guinea Region. (Recent molecular systematic studies indicate that the genus *Pitohui* may also be a close relative of the orioles.) Orioles are medium-large, arboreal fruit eaters with throaty, melodious songs. They inhabit forests and open wooded country from the lowlands to mid-mountains. Orioles forage singly or in pairs in treetops for fruit and some insects, and they join feeding assemblages at popular fruit trees. The nest is a pendent cup, often far out on a branch, or a simple cup positioned in a tree fork.

Australasian Figbird *Sphecotheres vieilloti* Pl 108
(**Figbird**, *S. viridis*)
27 cm. Locally common in open wooded habitats and towns. A noisy and social, short-billed oriole with bare skin around the eye. **Male:** *Unique—yellow-green with black head and reddish facial skin.* White outer tail feathers conspicuous in flight. **Female and Juv:** *Dull olive grey above, whitish below and heavily streaked; dark beady eye surrounded by grey facial skin.* **Imm male:** Like Adult but with streaked breast. **Subspp (1):** *salvadorii* (endemic). **Similar spp:** Brown Oriole has a large blackish ear patch, streaked head, buff (not white) belly, and red bill (black in Imm). Fawn-breasted Bowerbird is buff-brown without the heavy streaking below. **Voice:** Song is a series of a few whistles followed by a louder downslur. Call is a squeaky, emphatic, nasal upslur: *chyer!* suggestive of a Singing Starling. **Habits:** Usually in groups or flocks; roosts communally. Flocks shift about locally. Often perches in the open; conspicuous. Forages for fruit in tree canopy, occasionally lower; particularly fond of figs. Also takes nectar and insects. Breeds as solitary pairs or in communities with 2–10 active nests in

a single tree (AU data). Usually nests in simple pairs in AU, but sometimes with more Adults at the nest. Nest is an open cup suspended from a forked branch in a tree crown. Eggs (2–3) dull olive with dark spots (AU data). **Range:** NG: mainly SE Pen (Port Moresby area from Bereina to Kupiano in the South; also Alotau and Collingwood Bay); very locally in S Lowlands (Balimo), 0–50 m. Also AU and Moluccas (Kai Is). **Taxonomy:** Split from Green FB (*S. viridis*) and Wetar FB (*S. hypoleucos*).

Olive-backed Oriole *Oriolus sagittatus* Pl 108

25–28 cm. Locally common resident of Trans-Fly savannah, where it replaces the Brown O. (Some may be winter visitors from AU, but this needs confirmation.) Differs from Brown O by *upperparts grey-olive and underparts white with sharp, dark streaking*. **Adult:** Red iris and bill. **Juv:** Greyer; dark iris and bill, pale eyebrow, and buff edges to wing feathers. **Subspp (1):** *magnirostris* (endemic). **Similar spp:** Brown O has buff underparts (never white), black facial patch, and brown upperparts. Green O usually entirely washed greenish yellow. **Voice:** Songs are mellow, quick, and oft-repeated variations on the word "oriole," e.g., *oree oree ole* or *olio* (AU data). Also mimics. **Habits:** Similar to Brown O but prefers eucalypt savannah and savannah/forest edge. Nests in a tree on a thin, forked branch; nest a woven basket of bark and fiber. Eggs (2–3) white with dark spots and blotches (AU data). **Range:** NG Trans-Fly, 0–100 m. Also AU.

Brown Oriole *Oriolus szalayi* Pl 108

25–28 cm. Common in forest canopy and regrowth, from lowlands into mid-mountains. Vocal but retiring—can be difficult to observe well. **Adult:** Grey-brown, with *blackish face patch*, much fine dark streaking on head and underparts, and a *blood-red iris and bill*. **Juv:** *Prominent white eyebrow; bold black face*; blackish iris and bill. **Similar spp:** Other orioles are greenish. Brown O, particularly the Juv, is deceptively similar to friarbirds, which lack streaking and have a more pointed, curved bill. Streak-headed Honeyeater has a shorter bill and unstreaked underparts, and a prominent malar stripe. **Voice:** A familiar voice of the lowland chorus. Song includes a variety of loud, musical, liquid warbles, for example (1) a gurgling series of descending notes ending in an ascending disyllable and (2) a slur that drops and rises in pitch while increasing then decreasing in volume. Subsong includes mimicry of other species. **Habits:** Singly or in pairs, less often in groups. Keeps high in the forest canopy, usually out of view. Consumes fruit and insects. Not a colonial breeder, but will nest in the same tree as a friarbird, presumably for protection. Nest suspended from a tree fork in the midstory or canopy; a cup made of bark strips. Eggs (1–2) pale brown with dark spots. **Range:** Endemic. NG and NW Is (Batanta, Misool, Salawati, Waigeo), 0–1500 m, rarely higher.

Green Oriole *Oriolus flavocinctus* Pl 108

(**Yellow Oriole** is the AU name, but a Neotropical species is also named Yellow Oriole.) 25–28 cm. Common in all wooded habitats within its limited Trans-Fly range. **Adult:** *Yellow-green; dark wings and tail tipped white*. Iris and bill red. (Rare grey morph, perhaps just a variant individual, is washed-out grey without any green or brown, yet is identical in plumage pattern.) **Juv:** Duller olive with heavier streaking; pale eyebrow; dark iris and bill. **Subspp (1):** *muelleri* (endemic, but doubtfully distinct). **Similar spp:** Olive-backed O is white below. **Voice:** Vocal all day long. Song is unusually loud, an oft-repeated, yodeling or bubbling series of 3–4 notes: *yok yok yoddle*; also harsh notes and a subsong with softly warbled notes (AU data). **Habits:** Similar to other orioles. May prefer denser habitats (monsoon forest, mangroves) than Olive-backed O. Nests on a thin, forked tree branch; a basket woven of bark, leaves, fiber. Eggs (2–3) whitish with dark spots and blotches (AU data). **Range:** NG (Trans-Fly) and Aru Is. Also AU and eastern Lesser Sunda Is.

MONARCHS: Monarchidae

This large family of about 100 species is widely distributed from Africa, through South and East Asia to the Philippines and Wallacea, to Australasia and Oceania. It is in the latter two regions where their center of diversity lies, particularly in New Guinea, with 24 species. The monarchs of New Guinea are a uniform group of small fly-catching insectivores colored in combinations of black, grey, brown, yellow, and white, typically with blue-grey bills. The *Myiagra* flycatchers can be quickly recognized by their nervously vibrating tails, but several of these species pose a significant problem in field identification. Monarchs are confined to forest habitats in lowland and lower montane forest, with only 1 inhabiting cloud forest. Most frequent the lower and middle stories of the forest. A few prefer mangroves or other open habitats. Three are Australian migrants. All spend much of each day searching for arthropod prey, which they capture by gleaning from their perch (all spp), hover-gleaning, or sallying (the flycatchers). Many species of monarchs join mixed-insectivore feeding parties. Monarch voices are conspicuous, with whistled songs and scratchy or buzzing call notes. Most breed as monogamous pairs and build tidy cup-shaped nests, often beautifully camouflaged with moss and lichen.

Biak Black Flycatcher *Myiagra atra* Pl 88
(Biak Flycatcher)
13–14 cm. In all types of forest, though most common in hilly country. **Male:** *A small and entirely blue-black flycatcher.* **Female and Juv:** *Slate grey above; pale, dirty grey below.* **Similar spp:** Shining F male is noticeably larger, shinier, and more square-headed; prefers scrub and mangroves, and the voice is distinctive; note orange mouth when it sings. Males of the two are superficially similar, females very different. **Voice:** Song has ventriloquial quality and begins with a short burst of typical *Myiagra* grating, *zit-zit*, followed by a tuneful *sitseeu sitseeu sitseeu*. Also *tiu tiu tiu. . . .* Call is a high note followed by a trill at slightly lower pitch. **Habits:** Singly or in pairs, often in mixed-species flocks. Forages in midstory and canopy. Nervous, flicking wings and shivering tail; erects small crest; changes perch continually. No nesting information. **Range:** Endemic. Bay Is (Biak and Numfor), 0–400 m.

Leaden Flycatcher *Myiagra rubecula* Pl 88
13–16 cm. Both a year-round NG resident and an AU migrant during austral winter. Locally common in open lowland habitats of all sorts: savannah, mangroves, forest clearings, settlements. Leaden is 1 of 3 flycatcher species that are difficult to separate, the other 2 being Broad-billed and Satin. Besides appearance, note differences in habitat and foraging height. **Male:** *Dull, slaty blue-grey* with *white belly*; note that the grey-white division cuts across from breast to wing *in a slightly bulging line and meets the wing at a right angle.* **Female and Imm:** *Plain grey above, with pale cinnamon throat and breast*; from Satin F, Female Leaden is paler with *pale margins to the secondaries*, noticeable on the folded wing; from Broad-billed F by the narrower bill. **Juv:** Grey-brown above with pale eyebrow and white streak across the scapulars; whitish below. **Subspp (3):** Races indistinguishable in the field; *rubecula* (breeds E AU; winters NG) larger, greyer; *papuana* (breeds mainland NG) smaller with paler underparts; *sciurorum* (SE Is) smallest and darker and more glossy. **Similar spp:** Satin F male is shiny black above, with concave separation of black and white on the breast, and the underside of tail feathers is blackish (not grey); female is darker with solidly colored wing (edges to secondaries not conspicuously paler) and underside of tail is dark brown-grey (not pale grey). Broad-billed F best separated by broader bill and more graduated tail. **Voice:** Song is a diagnostic *wheeyu wheeyu wheeyu wheeyu*, sweet and breezy. Call is a deep, slightly harsh, guttural *zhirrp*, singly or repeated. **Habits:** Singly or in territorial pairs. Very active. Forages in tree canopy and middle levels. Sallies from an open perch, taking both flying and stationary prey. In AU, often nests near Noisy Friarbird, presumably for protection from predators. Nest built in a tree crown; a tight cup of bark and other plant material bound with spiderwebs and decorated with a few pieces of lichen. Eggs (2–3) whitish with a ring of spots around the middle. **Range:** Breeds in NG (S Lowlands, Trans-Fly, SE Pen) and SE Is (D'Entrecasteaux Is of Fergusson and Normanby; Conflict Group; Misima, Tagula, and Rossel Is); AU migrants overwinter in Trans-Fly and probably elsewhere, 0–650 m. Also AU.

Broad-billed Flycatcher *Myiagra ruficollis* Pl 88
15–16 cm. A rare mangrove specialist; also locally in thickets along river oxbows; usually under cover and only a few meters above the ground or water, though forages into canopy.

Adult: Resembles female Leaden F, less so female Satin F; distinguished *by broad bill, tips of tail feathers not overlapping completely (graduated), glossy crown, and in Female more white in the outer 2 tail feathers*. **Female** differs by paler throat and lores than either male Broad-billed or female Leaden. **Juv:** Plumage similar to juv Leaden; identify by structural differences and association with parents. **Subspp (1):** *mimikae* (NG Region, AU). **Similar spp:** Leaden F female is duller (not glossy) grey on head and upperparts, has narrower bill, and the tail tip is squared and completely overlapping (not graduated). Satin F female is darker with same structural differences. **Voice:** Song, *hrinney hrinney hrinny*, clear, far-carrying. Calls include *theeoooo-uuu*, and various churring and scratching sounds, similar to Leaden (AU data). **Habits:** Singly or in pairs. Foraging behavior similar to Leaden. Nest built a few meters up on a fork in mangroves; structure and eggs similar to Leaden. **Range:** NG (S coast from Mimika R to Port Moresby; Fly River) and Aru Is, sea level. Also AU and E Wallacea.

Satin Flycatcher *Myiagra cyanoleuca*   Pl 88
15 cm. A rare, overwintering AU migrant to open country, forest edge, and other disturbed habitat, from lowlands to mid-mountain valleys. Perhaps mainly in passage in the South, overwintering in the North and mountains. **Male:** *Glossy blue-black* with clear *white belly*; note that the black-white division on the breast is *concave, running under the wing at a sharp angle*. Quivers its tail. **Female:** Resembles female Leaden F but is *darker, somewhat glossy above, with solidly dark wings, and grey-brown undertail*. **Imm male:** Similar to Female but many birds with a dark, smudgy breast band. (Juv plumage not seen in NG Region.) **Similar spp:** Leaden F male is dull grey, has a convex separation between the slaty throat and white breast; female is paler, not glossy above, and has a pale panel on the secondaries of the folded wing; in both sexes the underside of the tail is greyish (not blackish or dark brownish). Broad-billed is separated from Female Satin by structural differences—see that account. Hooded Monach has black "bib" and white border alongside the folded wing. Black-headed Whistler is more sluggish, chunkier. **Voice:** Call, *zurp*. **Habits:** Usually single. Fly-catches actively from tree at the forest edge at midstory level. Raises feathers of crown to form a peak. **Range:** NG (Jayapura area, Sepik-Ramu, and S Lowlands to SE Pen) and SE Is, 0–1400 m. Breeds AU, winters rarely as far as Bismarck Is.

Shining Flycatcher *Myiagra alecto*   Pl 88
(**Common Shining Flycatcher**, *Monarcha* or *Piezorhynchus alecto*)
16–18 cm. Common in thickets near water in swamp forest, mangroves, and lowland second growth; in mountains, only in broad valleys; also occupies rainforest on islands, but does so less often on mainland. Forages from or just above ground. Reclusive but noisy; conspicuous when it flies.
Male: *Entirely shiny blue-black*. Note angular head. Bright *orange-red mouth* shows when bird sings and displays. **Female:** *Bright orange-rufous with black cap and white underparts* (may be tinged). **Imm male:** Similar to Female but often with black sides of neck. **Juv:** Similar to Female, but cap dull brown-black. **Subspp (5):** Minor, Manam I and SE Is birds with paler upperparts.
Similar spp: For Male, none on the mainland and most islands; Biak Black Flycatcher resembles co-occurring Shining but is smaller and tends to forage higher. For Female, Frilled Monarch female similar but has conspicuous blue eye-ring and rufous throat (not white). **Voice:** Song is a rapid trill of clear, whistled notes, swelling somewhat, all on 1 pitch or slightly rising; the slight crescendo is distinctive. Call is a rather short, low, buzzing *yeannnNNN!* **Habits:** Singly or in territorial pairs; occasionally joins mixed-species flocks. Actively gleans and hovers-and-snatches in the understory or shrubs along riverbanks. Raises crest; fans and jerks tail. Nests in a shady location on a thin branch 1–2 m above ground; nest a deep cup of tightly woven plant material, bound with spiderwebs and decorated with small pieces of lichen and bark. Eggs (2) pale green with dark spots. **Range:** Entire NG Region, except Rossel (SE Is) and no record from Mios Num (Bay Is), 0–1200 m. Also Moluccas and AU.

Restless Flycatcher *Myiagra inquieta*   Pl 88
(**Paperbark Flycatcher**, *M. nana*)
16–18 cm. Trans-Fly only, where common in savannah, edge of gallery forest, and scrub bordering rivers, entering sedge beds of riverside. Active, noisy, conspicuous. Hovers near ground while hunting, uttering distinctive call. Sweeps tail from side to side. **Adult:** *Glossy blue-black above, white below, including throat*. **Juv:** Similar but duller and shows white wing bars and streak across scapulars.
Subspp (1): *nana* (also N AU). **Similar spp:** Willie Wagtail has a black throat. Satin and Leaden Fs also have black throats, are much smaller, and forage from trees. **Voice:** Song (AU data) is a clear, repeated whistle *chewee chewee chewee*, each phrase rising at end. Call in NG is drawn out *switch*. (Does not give the "grinding" or "zap" calls of the AU population.) **Habits:** Singly or in territorial pairs;

will forage with other species. Forages mainly near the ground, but also up into trees. Takes insect prey by fly-catching and hovering-and-snatching or -pouncing. Nests on a thin branch, alive or dead, in a tree or shrub, often near water (AU data). (In AU, often nests near Willie Wagtail or Magpielark.) Nest is a cup of grass and bark strips matted together and bound with spiderwebs, lined with rootlets, feathers, or fur. Eggs (2–3) whitish with dark spots and blotches (AU data). **Range:** NG: Trans-Fly and adjoining S Lowlands, sea level. Also N AU. **Taxonomy:** Some authorities treat this race as a separate species, Paperbark F (*M. nana*).

Frilled Monarch *Arses telescopthalmus* Pl 89
(*A. telescophthalmus*)
15–16 cm. The most common and widespread forest flycatcher in lowlands and foothills. This is the southern member of a distinctive pair of monarch species—the other is Ochre-collared M—sporting a *blue, fleshy eye-ring* and expandable *neck ruff*. **Male:** Boldly patterned black and white with erectile *snow-white ruff* and white patch on back. **Female and Juv:** From Ochre-collared by *collar same color as back and wing*; belly varies from white to buff. **Subspp (5, in 2 groups):** *telescopthalmus* (excluding range of next subsp) Female belly white; *henkei* (SE Pen) Female belly buff in N, white or buff in S. **Similar spp:** See Ochre-Collared M. Other black-and-white flycatchers and monarchs lack the white collar and patch on the back. **Voice:** Songs include a dry, slow cicada-like rattling; a trill that progressively drops in pitch; and a ringing trill. Also a buzzy and monotonous series of 6–8 notes in rapid succession: *schweet schweet schweet. . . .* Call is a grinding *schweeit*. **Habits:** Singly or in a pair, often associated with mixed feeding flocks of monarchs and other small insectivores. Active. Forages mainly in the midstory, Males exploring trunks, vines, and understory, Females often higher, gleaning and fly-catching in the lower canopy. Takes insects and spiders. Nests at midlevels in the forest, the nest woven between two slender, dangling vines and often unconcealed in an open space; nest is a thin-walled cup of fine stems and roots, bound with spiderwebs and decorated with a few pieces of lichen. Eggs (2) whitish with brown dots and markings. **Range:** Endemic. NG, NW Is (Batanta, Misool, Salawati, Waigeo), and Aru Is, 0–1200 m, locally higher. Throughout western and southern NG, but excluding NW Lowlands and Sepik-Ramu. **Taxonomy:** Formerly included Ochre-collared M and Frilled-necked M (*A. lorealis*) of Cape York, AU.

Ochre-collared Monarch *Arses insularis* Pl 89
(**Frilled Monarch**)
15–16 cm. This is the northern counterpart of the Frilled M. Common in forest interior and edge in lowlands and hills. Like Frilled, shows a *blue, fleshy eye-skin*. **Male:** *Black and white with buffy orange throat and ruff*. **Female and Juv:** *Rufous neck and breast, white belly*; from female Frilled M by *grey-brown saddle*. **Similar spp:** Frilled M female is variable, but always has an orange-brown mantle and wings, same as the collar. Rufous M resembles Female but is larger, without dark cap or blue eye-skin, and is all rufous below. **Voice:** Song is a series of 15–20 rapid, buzzing, cicada-like notes; series increase in volume, 2.5–4 sec; longer, faster, higher pitched than that of Frilled. Alarm call, *bzzzt*, shorter and less emphatic than that of Frilled. **Habits:** Somewhat shy. Habits similar to Frilled M, as far as known, although Males seem to forage more on foliage and branches, less so on trunks. Where the two co-occur on Huon Pen, this species is found in foothills and mountains rather than in lowlands occupied by Frilled. Nest location and construction also similar to Frilled. Egg, no information. **Range:** Endemic. NG and Yapen (Bay Is), 0–1500 m. NG: NW Lowlands to Sepik-Ramu and N coast of Huon. **Taxonomy:** Split from Frilled M (*A. telescopthalmus*).

Torrentlark *Grallina bruijnii* Pl 101
(*Pomareopsis bruijnii*)
20 cm. Scarce in noisy pairs and family groups along forest-lined, fast-flowing creeks and rivers in the mountains. Prefers primary forest and steep terrain. *A medium-sized, black-and-white terrestrial songbird that bobs its body and wags its tail*. **Male:** Black face and underparts. **Female:** White brow and breast. **Juv:** Like Female but with black forehead and white eyebrow. **Similar spp:** Torrent Flycatcher perches on branches or rocks, is smaller, and has a pale grey back and white throat. Other black-and-white birds are arboreal forest birds. **Voice:** A penetrating, harsh, buzzy, upslurred call rising above the roar of streams and carrying a great distance: *jjirrreeee . . .* or *bzzzeeee*, repeated irregularly. **Habits:** Highly conspicuous and vocal but shy, flying around a stream bend to avoid disturbance. Walks along gravel bars and among rocks, and may wade into the water, searching for insects. Rhythmically bobs, wags tail side to side, and fans wings. Sometimes forages on wet gravel roads.

Travels through forest between streams. Nest built on a branch near water; nest a mud cup. Egg (1) undescribed. **Range:** Endemic. NG, probably all mts (no information for Fakfak, Kumawa, Wandammen, and Cyclops Mts), 400–2400 m.

Magpielark *Grallina cyanoleuca* Pl 101
(**Mudlark**)
26 cm. Restricted to southern Trans-Fly where resident locally. Common in wide open grassland with scattered trees and along river courses. *A medium-large, pied, terrestrial songbird that struts about on long legs wagging its tail.* Note rounded black wings in flight. Distinguished from Australian Magpie and the arboreal butcherbirds by the *smaller bill and white markings on its face.* Adult has distinctive white eye. **Male:** White eyebrow and black throat. **Female:** White forehead and throat. **Juv:** White eyebrow and throat and dark eye. **Subspp:** No information. **Similar spp:** Torrentlark lives along mountain streams and so does not co-occur; it is smaller, has a dark eye and somewhat different plumage pattern. **Voice:** AU data. Songs of Male and Female the same, either given solo or in a duet. Solo songs are several repetitions of the same phrase, e.g., a liquid mellow *cloop cloop cloop* or *peewit peewit peewit*. Duets contain 2 alternating phrases, one given by each bird, e.g., *peewit cloop peewit cloop*. Each bird rhythmically opens and raises its wings and spreads its tail. Alarm call is a strident *pee! pee! pee!* **Habits:** Singly, in pairs, or in small parties. Takes insects from the ground. Builds nest on tree branch near watercourse; nest a substantial, deep cup of mud and grass (AU data). Eggs (3–5) pinkish with dark spots and blotches. **Range:** NG: S Trans-Fly, 0–200 m. Also AU and Timor I.

Fantailed Monarch *Symposiachrus axillaris* Pl 89, Pl 92
(**Black Monarch**, *Monarcha axillaris*)
15–16 cm. Uncommon in interior of mid-mountain forest, frequenting the understory and midstory. A small, *black, fantail-like* monarch with *small white tuft at the bend of the wing.* **Male:** Satiny black. **Female:** Dull grey-black. **Juv:** Similar to Female but with browner flight-feathers, lacks white tufts under the wing. **Subspp (2):** Minor. **Similar spp:** Black Fantail male is velvety black, with a small white eyebrow, black-and-orange bill, and no pectoral tufts; note that tail of Fantailed M is not as open-fanned as Black Fantail's, and feather tips of the monarch are slightly pointed, not rounded. Black Berrypecker is satiny black but has a shorter, square-tipped tail and blunt black bill. **Voice:** Calls include a very harsh, buzzing rasp, rapidly repeated 3 or 4 times, or longer; usually a single note with a short pause, followed by 3 notes in quick succession: *schweg—zsh zsh zsh*. **Habits:** Singly, in pairs, or in small groups (3–4), sometimes forages with fantails. Crouches and holds tail cocked and partially fanned. Fly-catches in shaded forest. Nest site, no information; nest a cup of moss lined with black rootlike fiber. Eggs (1–2) whitish with darker spots. **Range:** Endemic. NG, nearly all ranges (no records from Kumawa and Cyclops Mts) and Goodenough I (SE Is), 800–2300 m.

Rufous Monarch *Symposiachrus rubiensis* Pl 89
(*Monarcha rubiensis*)
18 cm. Uncommon and patchily distributed in forests of W and N lowlands and foothills, frequenting lower and middle stories of the forest interior. *A large* monarch, *entirely rich rufous brown.* **Male:** Black face. **Female and Juv:** Rufous face. **Similar spp:** Females of Frilled and Ochre-collared M have a blue eye-ring and dark cap. **Voice:** Song, 4–6 buzzing notes, similar to but louder than those of Frilled M: *chwe chwe chwe chwe*, often followed by a whistler-like upslur. Call, *chhh chhh chhh*. . . . **Habits:** Usually solitary; joins mixed-species flocks. Gleans and hovers for insects and spiders on foliage. Nest a mossy cup 3 m up in fork of a small sapling in forest interior. **Range:** Endemic. NG: Bird's Head eastward in N to Sepik-Ramu and western S Lowlands (with isolated record from Nomad R), 0–300 m.

Hooded Monarch *Symposiachrus manadensis* Pl 89
(*Monarcha manadensis*)
15–16 cm. Uncommon and local in forest interior of foothills and adjacent lowlands; in understory and midstory. **Adult:** *Shiny blue-black above, white below* with characteristic *black breast-bib framed by white sides.* **Juv and Imm:** Pale ochraceous or white breast, and grey replaces black of Adult. **Similar spp:** Spot-winged M is grey above, with white wing spots. No other black-and-white forest bird has the black bib. **Voice:** Songs are a series of long, swelling, monotone whistles, each note ~1 sec in length; a sweet series of whistles on a monotone, like that of a treecreeper, except last notes become slurred into 1 tone, each series lasts ~1.5 sec; and high-pitched whistle, *dee-yuu—dewi dewi*. Calls are a sweet upslurred set of notes: *swee dee dee* and harsh notes like a Spot-winged M.

Habits: Singly, in pairs, or in family parties; joins mixed-species flocks. Generally quiet and inconspicuous. Gleans and flycatches for arthropods. Behavior resembles Spot-winged M, which is smaller and with which it may associate. Nests ~1 m off the ground in fork of a small sapling; nest a deep cup of green moss lined with dark rootlets. Eggs (2) white with dark spots. **Range:** Endemic. NG, 0–400 m, rarely to 1200 m.

Spot-winged Monarch *Symposiarchus guttula* Pl 89
(*Monarcha guttula*)
14–15 cm. The common and widespread monarch of lowland and hill forest interior. Usually quiet and easily missed; forages in understory and midstory. Small. *White tail spots* flash when tail is fanned. **Adult:** *Grey and white with black face*; row of *white spots on the wing* (diagnostic). **Juv and Imm:** Slaty face, buff or white breast, wing spots reduced or absent. **Similar spp:** Hooded M has black back and bib. Spectacled M juv resembles Juv Spot-winged but lacks the blackish chin. **Voice:** Vocally diverse. (1) A faint, tremulous whistled note repeated several times; (2) a dry rasping scolding resembling that of Frilled M or Slaty-headed Longbill; (3) a harsh, downslurred *shee shee*; and others. **Habits:** Usually alone, sometimes in pairs, often with mixed flocks. Actively gleans, hovers-and-snatches, and flycatches for insects and spiders, taken mostly from leaves. May assume fantail-like posture. Nests ~1–3 m off the ground in fork of a small sapling; nest a deep cup, decorated with moss and globs of silk and lined with dark rootlets. Eggs (2) whitish with dark spots. **Range:** Endemic. NG, NW Is (Gag, Gebe, Batanta, Misool, Salawati, Waigeo, and some small ones), Bay Is (Mios Num, Yapen), Aru Is, and SE Is (3 main D'Entrecasteaux Is; Misima and Tagula Is), 0–800 m, locally to 1200 m.

Kofiau Monarch *Symposiachrus julianae* Pl 89
(*Monarcha julianae*)
15 cm. Kofiau I (NW Is) only, sister species to Spot-winged M, which is absent on Kofiau. Common in lowland forest interior. **Adult:** Similar to Spot-winged M but *upperparts blackish and lacks wing spots* (or these obscure). **Juv (?):** Brown above, dingy white below; mask dark, but not black; pattern similar to Adult. **Imm (?):** Retains brown Juv wings and tail; rest of upperparts grey, breast orange, belly white, dark mask. **Similar spp:** Spot-winged M. **Voice:** Similar to Spot-winged M. **Habits:** Similar to Spot-winged M. Forages at all levels. No nesting information. **Range:** Endemic. Kofiau I (NW Is).

Spectacled Monarch *Symposiachrus trivirgatus* Pl 89
(*Monarcha trivirgatus*)
15 cm. Uncommon resident on SE Is in lowland and hill forest. Also, a rare AU migrant to Trans-Fly monsoon forest, where most birds are in imm plumage. This is mainly an island bird in the NG Region, and it co-occurs with closely related Spot-winged on the larger islands in its range. Also shows *white in tail*. **Adult:** Resembles Spot-winged but has a *cinnamon breast*, *lacks spots on the wing*, and the shape of the face mask is different. **Juv and Imm:** Similar to juv Spot-winged but only chin grey (not including throat). **Subspp (2):** *melanopterus* (SE Is) shoulders black (some with white spots), broad white tail tip, breast pale rufous; *gouldii* (presumably, breeds E AU) shoulders grey (same as back), breast rich rufous. **Similar spp:** Spot-winged M. *Myiagra* flycatcher females quiver the tail. **Voice:** (AU data) Song is a drawn-out, high-pitched whistle with a slight buzzy quality, repeated immediately 2–4 times or at irregular intervals. A similar song is shorter and mixed with rapid high-pitched chatter. Also a short squeaky warble. Scolding calls are harsh and rasping, *zzit zzit*. **Habits:** Similar to Spot-winged M. Nest placed in a sapling fork; a deep cup of woven grass and fiber, decorated outside with bark, leaves, and globs of silk. Egg (1) white with dark spots. **Range:** NG (S Trans-Fly) and SE Is (D'Entrecasteaux—Normanby; Louisiades—Alcester, Misima, Rossel, Tagula, and some smaller islands, e.g., Conflict group), 0–750 m. (Record from Salawati I an error.) Also AU, Lesser Sundas, Moluccas. **Taxonomy:** May be composed of more than one species.

Biak Monarch *Symposiachrus brehmii* Pl 89
(*Monarcha brehmii*)
15–16 cm. Biak I only. Rare in lowland forest interior; poorly known. **Adult:** Unique *bold pattern of creamy yellow and black*, the yellow of an unusual washed-out hue; the *black head* is marked with a distinctive, *pale, tear-shaped patch on the ear-coverts*. **Juv (from photo):** Resembles juv Spectacled M but with mostly whitish outer tail feathers. **Similar spp:** Golden M, also on Biak, is smaller, deeper orange-yellow, and has yellow crown and white eye-spot. **Voice:** Song is a series of the same phrase repeated 7–8 times: *whHUEEP wh-wh-wh-wh-wh*—the first part a loud, mellow, swelling whistle,

the second a rapidly repeated, soft, breathless, tonal whistle. Grating call. **Habits:** Similar to Spot-winged, so far as known. No nesting information. **Range:** Endemic. Biak I (Bay Is).

Golden Monarch *Carterornis chrysomela* Pl 90
(Black and Gold Monarch, *Monarcha chrysomela*)
13–14 cm. The common treetop monarch in lowland and foothill forest. Often out of view in the canopy; listen for the easily recognized song. Sexes differ, but both show *unique, white tear-spot below eye*. **Male:** Striking *black-and-yellow pattern*; note the *yellow wing patch*. **Female:** Generally yellowish olive; *look for white tear-spot conspicuously set against dark cheek*. **Juv:** Similar to Female, but bill black (not bluish) with yellowish base. **Subspp (5, in 2 groups):** *chrysomela* (all of range except for Biak I) Female mostly all green; *kordensis* (Bay Is: Biak) Female has bright orange-yellow head and underparts and patch on the wing, Male deep orange head and underparts. **Similar spp:** Golden Cuckooshrike male resembles a giant Male Golden M, but it lacks the white eye-spot and yellow on the wing. Biak M is pale creamy yellow, has black head with pale crescent on ear-coverts. **Voice:** Song is a loud, cheery, rambling sequence of varied phrases, many notes repeated 2–3 times; phrases composed of mellow notes, scratchy notes, sputtering notes, cheeps, and squawks, accompanied by frequent snapping of the bill, e.g., *ditoweet duweet duweer teerteertee*. **Habits:** Usually in pairs, often in mixed-species flocks of small insectivores. Lives in the forest canopy, and best viewed in openings created by tree falls or trail cuts. Active, nervous, and noisy. Gleans and sally-gleans, taking insects from leaves. Nests higher than other monarchs, in the midstory and perhaps lower canopy; nest, on a horizontal branch, is a cup covered with moss. Egg, no information. **Range:** NG, NW Is (Batanta, Misool, Salawati, Waigeo), Bay Is (Biak), Aru Is, and SE Is (3 main D'Entrecasteaux Is), 0–700 m, rarely to 1200 m. Also Bismarck Is (New Ireland and islands to the north).

Islet Monarch *Monarcha cinerascens* Pl 90
17–19 cm. Confined to islets off the coast; mainly littoral. Occupies forest and edge, especially strand vegetation; common. This and the next 2 species are large, sluggish monarchs, *grey with rufous breast and belly*. **Adult:** Dull grey with cinnamon breast; note *dark, beady eye*. **Juv:** Differs in buffy wash from throat to breast, brownish edges to wings, and dark bill, paler at the base in the youngest birds. **Subspp (6):** Minor. **Similar spp:** Sister species Black-faced M co-occurs mainly in SE Is; juv/imm of that species is similar to Adult Islet M, but its face is paler grey around the eye; note that in SE Is, Adult Islet M may show some black at the base of the bill and paleness around the eye. Spectacled M may co-occur but has a black eye-patch, paler rufous underparts, and much white in the tail. **Voice:** Generally similar to that of the better-known Black-faced M; what differences there are have not been reported. **Habits:** Similar to Black-faced M, differing mainly in its ecology as a tramp species of small islands and willingness to forage in low strand vegetation. Nest 6–17 m up on a horizontal tree branch; nest a bulky cup of grass, fiber, and spiderwebs lined with fine black stems. Eggs (1–2) whitish with red-brown spots and band at wide end. **Range:** Islands off the N coast of NG (e.g., Karkar, Manam), but also locally on coast of Bird's Head; NW Is (Kofiau, Misool, Waigeo, and likely others); nearly all Bay Is (no record from Mios Num); Aru Is; and all SE Is, 0–1400 m. Also Wallacea to Solomon Is. **Taxonomy:** Islet and Black-faced Ms are closely related, and the presence in the SE Is of birds that appear to be intermediate is confusing.

Black-faced Monarch *Monarcha melanopsis* Pl 90
17–19 cm. Breeds AU, migrates to eastern NG and SE Is during austral winter; some Imm birds remain over summer. Seasonally common in all forest types and edge in lowlands mainly, less frequent in foothills; forages in canopy and midstory. Sister species of Islet M; the two overlap in the SE Is. **Adult:** *Black forehead and throat*; prominence of *beady black eye* enhanced by *pale grey eye-patch*. **Juv:** May not be present in NG Region; resembles Adult but the grey plumage is brownish grey and the face, instead of black, has orange-rufous wash and pale loral spot. **Imm:** Shows blue-grey plumage of Adult, but retrains brown-grey Juv wings, and the grey face has pale loral spot. **Similar spp:** See Island M. Black-winged M has black wings and tail, and the black on face usually extends to the eye; only at higher elevations, does not co-occur. **Voice:** Song is a cheerful, mellow whistle: *wyy-oo ich-ee-ou*. Calls include a *quick* note and various monarchine chatterings and grindings. **Habits:** Usually singly, sometimes in mixed flocks. A slow-moving gleaner, rarely sallying, taking most prey from twigs and foliage. **Range:** NG (mainly in the East, westward to Huon and Trans-Fly), Aru Is, and SE Is, 0–800 m. Breeds E AU.

Black-winged Monarch *Monarcha frater* Pl 90

18 cm. The common monarch of treetops and edge of hill and mid-mountain forests. Forages from the canopy down into midstory. A resident species, though the AU race must overwinter in NG somewhere (no record as yet). Easily identified in its habitat: *note striking contrast between pearl-grey body and black face, wings, and tail; rufous breast and belly*. **Adult:** Well-defined plush, black mask; blue bill. **Juv:** No information. **Imm:** Black mask small and poorly defined; wings brownish grey (not black); bill dark. **Subspp (3 or 4):** Minor; black mask does not reach eye in *frater* (Bird's Head, N slope of W Ranges) and *canescens* (migratory AU race); does in *kunupi* (Weyland Mts) and *periophthalmicus* (remainder of NG). AU race not separable in the field from nominate *frater*. **Similar spp:** Black-faced and Islet Ms, at lower elevation, have grey wings and tail. **Voice:** Varied, reminiscent of a cuckooshrike or Black-faced M. Song sweet, a whistled series of identical notes (1/sec) that rapidly rise then fall in pitch: *witchewu* or *wichew* or upslurred *whuree*. Call is a nasal chattering. **Habits:** Singly or in pairs, often in mixed-species flocks. Actively gleans and sallies. Nests on a tree branch or fork in midstory or canopy; a cup of plant material with covering of moss and mats of silk. Eggs (2–3) white with reddish spots. **Range:** NG, nearly all mts (no records for Kumawa, Wandammen, and Cyclops Mts), 400–1600 m, rarely to sea level. Also AU.

IFRITS: Ifritidae

Previously classified as a babbler or with the logrunners and allies, the Blue-capped Ifrit according to DNA studies belongs with a group of families including monarchs. Recently assigned to its own family.

Blue-capped Ifrit *Ifrita kowaldi* Pl 83

16–17 cm. Common in upper cloud forest. A tame, active, vocal bird with creeping habits. Chunky. Buffy brown with *distinctive blue-and-black crown and pale eye-stripe*. **Male:** Eye-stripe white. **Female:** Eye-stripe buff. **Juv:** Resembles Female, but forecrown brown, not blue. (Buff tips on greater and median wing-coverts variably present in Juv and Adult.) **Subspp (2):** Minor. **Similar spp:** Wattled Ploughbill and Large Scrubwren, with which it associates, also creep, but lack the blue crown and are smaller. Papuan Treecreeper slimmer, darker, no blue crown. **Voice:** Song is a loud, squeaky, musical series of exuberant notes, rising and falling in pitch with the quality of a baby's squeeze-toy. Call buzzy but musical, *zig-zig-zig-zig-zig-zig*, rising and falling. Flocking call is a 3-note *jit—jit-jit*. **Habits:** In pairs or small foraging parties, sometimes with other insect eaters. Forages at all levels in the forest. Creeps on mossy trunks and limbs, probing or excavating. Eats insects, spiders, rarely fruit. Nest, built in the fork of a sapling, is a bulky cup of moss and plant fiber. Egg (1) white with a few dark spots and blotches. **Range:** Endemic. NG: Central Ranges and Huon, 2100–2800 m (1450–2900 m).

DRONGOS: Dicruridae

The drongos (25 spp) are primarily an African and Asian family, with a few species that reach as far east as New Guinea, the Bismarck Is, and Australia. There is just 1 species in New Guinea now that the Pygmy Drongo has been moved to the fantails. Characteristics of drongos are their black plumage (most species), upright posture when perched, and conspicuous and demonstrative flycatcher-like foraging behavior.

Spangled Drongo *Dicrurus bracteatus* Pl 107
(*D. hottentottus*)

25–28 cm. Ubiquitous resident in lowland forest and edge, up into the mid-mountains in disturbed habitat. The AU form migrates to NG savannah country in austral winter. *A noisy, conspicuous black bird that sits upright* and is easily recognized by its *forked "fish tail"* with the tips turned outward and upward. Often flicks tail upward. **Adult:** Glossy, metallic, blue-black with spangled breast; red iris. **Juv:** Duller, sooty black; dark iris. **Imm:** Plumage mostly adultlike, but in AU form retains some white-tipped undertail-coverts; iris red. **Subspp (3):** Minor, although the migratory AU form *bracteatus* (breeds SE AU north to central Queensland, wintering in Trans-Fly) is larger, tail shorter and less forked, bill longer, Juv with white tips to undertail-coverts, and Adult with somewhat different calls than resident *carbonarius* (said to have gloss blue-and-green vs blue-and-purple, but this seems questionable). **Similar spp:** Manucodes and drongos can be seen together in the same feeding flock. Manucodes are larger, with rounded tails, horizontal posture, and very different vocalizations. Black Cicadabird male also in same mixed flock lacks fish tail. **Voice:** Song, associated with a flight display, is a variety of creaky and nasal notes jumbled together. Calls are varied, all of which are loud, ringing, and conspicuous, often repeated 2 or more times. Sometimes duets. **Habits:** Solitary or in pairs, often mixed in with flocks of babblers, pitohuis, and birds of paradise, the drongos acting as sentinel species, calling loudly and taking insects flushed by other foragers. When feeding, sallies out to capture insects in flight or from tree foliage, branches, or trunks. Residents live in territorial pairs. The AU migratory birds migrate in flocks. Aggressive toward other birds, even swooping down on flying raptors. Breeds as a monogamous pair. Nest built in midstory, slung from a forked twig; nest is a thin-walled basket of twigs and vine stems. Eggs (2–3) whitish with dark spots or blotches. **Range:** NG, NW Is (Batanta, Gebe, Kofiau, Misool, Salawati, Waigeo), Bay Is (Biak, Numfor, Yapen), Aru Is, and SE Is (3 main D'Entrecasteaux Is, Trobriand Is, Tagula Is), 0–1450 m. Migratory AU birds widespread throughout the Trans-Fly and adjacent S Lowlands. **Taxonomy:** Formerly included within Hair-crested Drongo (*D. hottentottus*) of Asia and Wallacea. (Proposal to elevate race *carbonarius* of NG to species status not accepted given intermediate forms in N AU.)

FANTAILS: Rhipiduridae

The 46 species of fantails range from India to New Zealand and Oceania; 14 fantail species inhabit the New Guinea Region. The high diversity of fantails in New Guinea demonstrates the central role this region has played in the evolution of the family. Recent genetic research has identified the Pygmy Drongo (and Silktail, *Lamprolia victoriae*, of Fiji) as sister species to the fantails, and it is provisionally placed in this family. Except for the Pygmy Drongo, the New Guinea species present a rather homogeneous group. Fantails are small and delicate flycatching birds, 13–20 cm long. While foraging, the long tail is often fanned and the wings are drooped in a characteristic crouching posture. In some species the tail is cocked upward. The smaller fantails are fidgety in movements, twisting this way and that. Most fantails favor closed forest, although one (Willie Wagtail) inhabits lawns and gardens, and others dwell in mangroves or scrub. Fantails build a neat, cup-shaped nest usually with a pendent "tail" attached to the base. Many species readily join mixed bird flocks. Most are unwary and easy to find. Thicket-fantails have loud, explosive songs but are typically much more difficult to observe.

Pygmy Drongo *Chaetorhynchus papuensis* Pl 107
(Mountain Drongo)
20 cm. Common in the interior of hill and mid-mountain forests, where it is the only *black, flycatcher-like bird with upright posture and unforked tail*. Looks like a large, black monarch or a small drongo. Note short bill with prominent hooked tip. **Adult:** Metallic blue-black plumage, short nuchal crest, and concealed white patch at bend of wing. **Juv:** Duller; white gape. **Similar spp:** Spangled D much larger, with "fish tail." Black Pitohui male is similar in shape and posture, but is larger, duller, and confined to higher elevations. Fantailed Monarch is much smaller and has horizontal posture and a pearl grey bill. **Voice:** Song is a loud, long, brilliant, jumbled, musical mixture of chips, squeaks, whistles, and warbles. Call is a strong, explosive *pik* or *peep*, given either once or repeated in rapid succession. Also a downslurred, whistled *pi-yew*. **Habits:** Solitary or paired. A noisy bird of forest midstory; perches on horizontal branches and sallies for insects. Foraging behavior and posture reminiscent of Northern Fantail. Frequently joins mixed foraging flocks, stationing itself in open spaces beneath the canopy where it can pursue insects disturbed by other birds foraging above. Nest a small, shallow basket slung from a tree fork. **Range:** Endemic. NG: All mts (no record yet for Cyclops Mts), 600–2000 m, rarely lower. **Taxonomy:** Recently transferred from the Dicruridae as a result of molecular studies.

Fantails. The 13 species of the New Guinea Region fall into 6 ecological groups. The largest fantail species, Willie Wagtail, is met almost everywhere in human settlements and other open habitats, often on the ground. Three species of very similar thicket-fantails are strictly confined to the dense lowland understory: Sooty in forests, Black at the edge and locally in forests, and White-bellied strictly in thickets. Five species of the forest interior forage from the understory into the canopy but differ in preferred foraging height, altitudinal range, and foraging method. From the lowlands to high elevations, they are Rufous-backed, Chestnut-bellied, Black, and Friendly and Dimorphic Wagtails. Of these species, Black is most concentrated in the understory, Rufous-backed is the only species found in flat lowland forests, and Chestnut-bellied and Friendly have mutually exclusive altitudinal ranges (Chestnut-bellied below Friendly). The remaining mainland forest species, Northern Fantail, lives in open areas of the forest and at the forest edge. The Mangrove Fantail is confined to mangroves of the South and East. Two species similar to each other and often considered conspecific are Arafura Fantail of NW Is forests and southern mangroves, and Rufous Fantail resident on SE Is and also occurring as an Australian winter visitor to forests of the Trans-Fly.

Willie Wagtail *Rhipidura leucophrys* Pl 91

17–20 cm. A familiar bird of towns, villages, and gardens; originally an inhabitant of savannahs, beaches, lagoons, and riverside openings in forested country. Conspicuous and active. *Wags entire body, calling persistently.* Tail cocked and waved but not fanned. Distinctive silvery flash in the wings when it flies. **Adult:** *Black except for white belly and eyebrow.* **Juv:** Duller, with bigger white eyebrow and buff tips to wing-coverts. **Subspp (1):** *melaleuca* (Moluccas to Solomon Is). **Similar spp:** Restless Flycatcher (Trans-Fly) has an entirely white throat and underparts. **Voice:** Song (both sexes) a loud, jerky, musical, squeaky whistling best likened to "sweet pretty little creature." Often sings at night. Alarm call is a harsh rattle, *tikka tikka tikka. . . .* **Habits:** Alone or in pairs. Feeds actively on the ground, walking and running and chasing insects; follows livestock. Also forages from low shrubs and trees, picking from branches or sallying. Territorial, aggressive, and bold, attacking intruders near the nest. Will chase raptors. Breeds in pairs. Nests on a variety of structures: low shrub or tree (usually) and rarely post, power pole, or coastal piling or boat in the water; nest is a tightly woven cup (without tail) of fine plant material often lined with hair or feathers. Eggs (2–3) white with spots. **Range:** NG, NW Is (all), Bay Is (Biak only?), Aru Is, Karkar I, and SE Is (3 main D'Entrecasteaux Is), 0–2000 m, higher locally. Also AU and Moluccas, Admiralty, Bismarck, and Solomon Is.

Black Thicket-Fantail *Rhipidura maculipectus* Pl 91

18–19 cm. Local and uncommon, secretive in lowland forest edge, swamp forest, mangroves, and second growth, particularly near water and where the forest is sometimes flooded. Closely resembles the more widespread White-bellied TF, except *overall more blackish and the belly is black*; also the *central tail feathers lack white tips.* **Female:** Duller, with reduced wing-spotting. **Juv:** Sooty black with much reduced white markings; tail pattern same. **Similar spp:** White-bellied TF. Sooty TF lacks white in tail and is brown above, not black, and lacks black throat-stripe. **Voice:** Song similar to that of White-bellied but more rapid and ends with the next-to-last note loudest and downslurred, *WHIT-chee.* Alarm call *tchep* or *wheck.* **Habits:** Similar to White-bellied TF but usually close to water and occupies shady cover; may co-occur with other thicket-fantails. No nesting information. **Range:** Endemic. NG, NW Is (Batanta, Salawati), and Aru Is, 0–300 m. NG: Bird's Head to S Lowlands and southern SE Pen to Laloki R.

White-bellied Thicket-Fantail *Rhipidura leucothorax* Pl 91

18 cm. A thicket specialist. Forages near the ground in a broad range of disturbed lowland and hill forest habitats: gardens, second growth, forest edge, streamside thickets, and mangroves, less common in forest interior where associated with light gaps. Locally common and often heard, but keeps to cover and difficult to observe. **Adult:** Blackish brown; *black face marked with a prominent white eyebrow and white submoustachial patch; belly white; all* tail feathers white-tipped. **Juv:** Greyer; all-dark bill; black breast band spotted white. **Subspp (2):** Minor. **Similar spp:** Sooty TF of forest interior has all-black tail and all-white throat. Black TF, also in second growth, is darker, has black belly and dark tips on central tail feathers. **Voice:** Conspicuous. Song is a loud pair of explosive notes, *chirPING!* or *teyuPINK!,* sometimes preceded by a few soft introductory notes. **Habits:** Singly or in pairs. Slightly easier to observe than other thicket-fantails—makes brief forays into the open. Fans tail exaggeratedly. Gleans and flycatches for insects, staying close to the ground, rarely ascending higher. Nest, built in a thicket and ~1 m up on a thin stem, is a small, woven, tailed cup coated with cobwebs. Eggs (2) white with blurry markings. **Range:** Endemic. NG mainland except Trans-Fly, 0–750 m, rarely to 1350 m.

Sooty Thicket-Fantail *Rhipidura threnothorax* Pl 91
17–18 cm. Common but secretive in thickets within forest interior in lowlands. A long-tailed, sooty brown bird, always near the ground, usually in dense cover. *Tail all black, lacking white tips.* Also look for *all-white throat.* Plumage of the small head usually unkempt-looking. **Male** darker; **Female** browner. **Juv:** Paler, with breast and belly grey marked by pale shaft streaks. **Subspp (2):** Minor. **Similar spp:** The other 2 thicket-fantails are blacker above, have white tail tip, dark forehead, and dark chin streak. **Voice:** Song, characteristic of lowland forest, is loud 3-note *chew chew chew* (could be mistaken for song of a jewel-babbler, but fantail almost always delivers 3 notes, jewel-babbler only 1 or 2). Calls include a repeated *teek, pink!* or *kraaank-pink!* Also scolding notes when foraging. **Habits:** Singly or in pairs. Often follows babblers or ground-feeding birds, some much larger than itself. While foraging, assumes a horizontal posture with tail fanned. Gleans insects from the ground or flycatches at ground level or from low perch. Nest built on a forked branch in undergrowth within 1 m of the ground; a small, tight cup bound with spiderwebs and possessing a long tail. Egg, no information. **Range:** Endemic. NG (throughout, except Trans-Fly), NW Is (Misool, Salawati, Waigeo), Yapen (Bay Is), and Aru Is, 0–1100 m.

Rufous-backed Fantail *Rhipidura rufidorsa* Pl 92
(Chestnut-backed Fantail)
13–14 cm. Common in lowland and hill forest interior. The plainest and smallest lowland fantail. **Adult:** *Head grey and underparts dingy white*; *tail tipped white*; back and rump rufous. **Juv:** Similar, tail shorter. **Subspp (3):** Minor. **Similar spp:** Rufous and Arafura Fs have a prominent black "necklace." Female Black F has rufous underparts and tail, without white tips. Chestnut-bellied F, broadly co-occurring, is grey above, rufous below. **Voice:** Song is a series of 3–5 thin, sweetly whistled upslurs and varies geographically; quality resembles voice of Rusty Mouse-Warbler. Call is *tseet*. **Habits:** Singly or in pairs. Joins mixed flocks of warblers and flycatchers. Forages actively in foliage, from ground to canopy, occasionally at edge. Takes insects by gleaning and flycatching. Nest is built within a meter of the ground on a thin, forked twig; a small, tailed cup tightly bound with cobwebs. Eggs (2) pinkish white with small spots. **Range:** Endemic. NG (minus Trans-Fly), Misool (NW Is), and Yapen (Bay Is), 0–650 (900) m.

Dimorphic Fantail *Rhipidura brachyrhyncha* Pl 92
15–16 cm. Common in cloud forest and subalpine shrubland. **Adult:** Small, pale brown fantail. *Contrasting rufous rump and lower back against dull buff-brown head and upper back.* 2 morphs: *pale morph* has all ashy-grey tail and pale grey-brown underparts; *dark morph* has blackish tail with broad rufous tip and dull ochraceous underparts. **Juv:** No information. **Similar spp:** Black F female is larger and brighter rufous, and shows different tail pattern with dark central feathers. **Voice:** Song has a tinkling quality when heard nearby, but sounds squeaky at a distance. It begins with a couple of detached, higher-pitched, slower notes, then descends in a run-together cascade of 7 or more fast notes. Calls include scolding *didlit didilit didlit . . .* or nasal *weedint weedint.* **Habits:** Tame and active. Often in pairs, of one or both morphs. Forages with typical fantail behavior in undergrowth and midstory. Fans tail more frequently than Black F. No nesting information. **Range:** Endemic. NG: Bird's Head, Central Ranges, Foja Mts, and Huon, 1700–3900 m, rarely lower.

Rufous Fantail *Rhipidura rufifrons* Pl 92
14–15 cm. Uncommon AU migrant to Trans-Fly; also resident on small SE Is; both populations inhabit forest. This species and Arafura F are very similar but have different ranges. They share a *rufous patch at the base of the tail, a black throat "necklace," and a rufous forehead* (missing in one race of Arafura F). **Adult:** Broad rufous patch at base of tail and on rump. **Juv:** Duller, browner, with rufous tips to wing-coverts, necklace brown and reduced. **Subspp (3, in 2 groups):** *rufifrons* (migrates to Trans-Fly, breeds AU) grey tips to tail feathers; *louisiadensis* (resident SE Is, endemic) white tips to tail feathers. **Similar spp:** Rufous-backed F has plain grey head and lacks black necklace. **Voice:** Song of AU race begins with a high-pitched squeak and accelerates into a series of descending seesaw notes; in SE Is it is a long, brilliant, tinkling, descending series. Calls include a squeaky *chip* and a high-pitched *ps*. **Habits:** Singly or in pairs. Works the understory and midstory in typical fantail fashion—very active and constantly fans tail. Nest and eggs (AU data) similar to that of Rufous-backed F. **Range:** NG (Trans-Fly) and SE Is (found on small islands, but includes Goodenough, Fergusson, and Rossel), 0–300 m. Also AU, Melanesia, Micronesia. **Taxonomy:** Formerly included Arafura F.

Arafura Fantail *Rhipidura dryas*   Pl 92
(formerly under **Rufous Fantail**, *R. rufifrons*)
14–15 cm. Western counterpart to Rufous F. There are 2 very different resident subspecies. Differs from Rufous F in *small* rufous patch at base of tail, and *tail tip is always white*. Also, *the island races lack the rufous forehead and have a white eyebrow instead*. **Subspp (2)**: *squamata* (forests of NW Is and Aru Is; also Banda Is) black forehead, white eyebrow; *streptophora* (mangroves of Mimika R; birds in Gulf of Papua mangroves undescribed and possibly belong to this form) extensive rufous forehead and difficult to separate from Rufous F. **Similar spp**: See Rufous F. **Voice**: NG forms, no information. **Habits**: Presumably similar to Rufous F. **Range**: NG, NW Is (Misool, Salawati, Schildpad Is, Waigeo, and Kofiau with subspp undet.), and Aru Is, sea level. NG: S coast at Mimika R (western S Lowlands) and Gulf of Papua (Purari R east to Hisiu), the latter population presumably belonging to this species. Also E Wallacea and AU. **Taxonomy**: The NG mangrove populations are in need of review.

Black Fantail *Rhipidura atra*   Pl 92
16–17 cm. Common in undergrowth of montane forest. **Male**: *Velvety black*; the small white eyebrow can be hard to see. **Female and Juv**: *Rufous with all-black central tail feathers*. **Subspp (2)**: Minor. **Similar spp**: Fantailed Monarch is glossy blue-black with white axillaries, no white eyebrow; its bill is pearl grey rather than black and yellow; and voice is very different. Dimorphic F is smaller, duller below, and lacks black central tail feathers. **Voice**: Varied. Songs include a rising series of ~10 thin, sweet notes, each becoming shorter and more closely spaced at end, lasting 2 sec; a series of ~10 sharp syllables, *tyeek*, rising and falling slightly in pitch for ~3 sec; and a sweet, high-pitched, piercing, tinkling jumble with an abrupt change of quality on the final slur. Notes include a single, soft *tseep* recalling voice of Northern F and a loud metallic *chink!* like coins struck together. **Habits**: Often in pairs. Easily observed—approaches at first, then flees. Active. Sallies after flying insects and gleans off foliage. Elevational range overlaps Dimorphic F and Fantailed Monarch. Nest built on a forked twig near the ground; a tight cup of fine plant material, with a tail. Egg (1) cream with dark speckles. **Range**: Endemic. NG (all mts) and Waigeo (NW Is), 1000–2200 m (700–3200 m).

Friendly Fantail *Rhipidura albolimbata*   Pl 91
14–15 cm. Common in cloud forest interior, edge, and openings. **Adult**: Generally *sooty-colored with white facial markings, wing bars, and tail tips*. **Juv**: Upperparts mottled rufous; wing bars and pale portions of underparts buffy. **Similar spp**: Northern F is larger, more upright; lives at lower elevation. Chestnut-bellied F has chestnut belly; also at lower elevation. **Voice**: Song a high, squeaky, brightly whistled series of 3–7 staccato notes, sometimes followed by a rapid jumble:

Also other more complex variations made up of rising and falling, pleasantly whistled notes, repeated over and over. **Habits**: Singly or in pairs. Tame, inquisitive, active in the lower and middle levels of the forest and edge; occasionally canopy. Sallies after insects and gleans from foliage and branches. Behavior similar to Chestnut-bellied F. Nest built in midlevel in forest on sloping vine; a tightly woven cup of plant fiber, with tail, bound with cobwebs. Egg, no information. **Range**: Endemic. NG: nearly all mts (no records from Adelbert and Wandammen Mts), 1600 m to timberline.

Chestnut-bellied Fantail *Rhipidura hyperythra*   Pl 91
14–15 cm. Common in forest midstory, from lowlands at foot of the ranges, up to mid-mountains. **Adult**: Pattern diagnostic—*slaty grey above, rich rufous below*. Wing-coverts with 2 rows of white or buff spots. **Juv**: Paler, duller below; wing-covert spots rufous. **Subspp (2)**: Minor. **Similar spp**: Northern F larger, paler below, less active. Friendly F grey and white below, inhabits cloud forest. **Voice**: Song a variety of soft and musical whistled notes: a descending series of 3 notes; an ascending series of 5 notes spaced at half-sec intervals and ending abruptly; and a rapid series of rising and falling notes, similar to those given by Friendly and Northern Fs. Song may closely resemble a gerygone. Call is 2 faint, clear, high-pitched short notes on different pitches. **Habits**: Singly or in pairs. Fans tail frequently. Actively gleans from branches and leaves; sometimes sallies. Often in mixed flocks with gerygones, etc. Behavior similar to Friendly F. Nest is built in midstory on a thin, forked, horizontal twig; a tightly woven cup of plant fiber, with tail, bound with cobwebs. Eggs (2) cream with small spots. **Range**: Endemic. NG, Yapen (Bay Is), and Aru Is, 0–1600 (1750) m. NG: hills, mts, and adjacent lowlands (no records for Fakfak, Kumawa, and Wandammen Mts); also lowlands around Oriomo, Morehead, and Tarara Rs.

Mangrove Fantail *Rhipidura phasiana* Pl 91
(Mangrove Grey Fantail)
14–15 cm. In mangroves only; locally common. **Adult:** A small, *pale grey* fantail with *white eyebrow and white wing bars*. **Juv:** Paler still, with upperparts suffused with buff. **Similar spp:** Northern F, which may share its habitat, is larger, more upright, rarely fans tail, lacks wing bars. Friendly F much darker; only in high mountains. **Voice:** Song a tinkling, rising series of soft, high-pitched notes. **Habits:** Similar to other small fantails; fans tail often. Generally forages in midlevels, higher than co-occurring Rufous/Arafura Fs. Nest (AU data) is on a low, thin branch; a tailed cup. Eggs (2) white with spots and blotches. **Range:** NG: Spotty distribution from Trans-Fly (Dolak I to Merauke) east to southern SE Pen (Lea Lea); Aru Is. Also N AU. **Taxonomy:** Formerly treated as a race of Grey Fantail (*R. albiscapa* or *fuliginosa*) of AU. Identity of Aru and Trans-Fly birds needs confirmation from specimens, to separate from Grey F. **Extralimital spp:** Grey F may occur in S NG as migrant from AU; none reported yet. It is slightly larger, darker, and not confined to mangroves.

Northern Fantail *Rhipidura rufiventris* Pl 91
17–18 cm. The widespread and common forest fantail of lowlands, hills, and mid-mountains. Perches upright and is relatively inactive and rarely fans tail. **Adult:** *A grey fantail lacking white markings on wings*. Regionally variable: either grey or black with *whitish underparts and dark breast band*. Unlike other grey fantails, *rarely shows its white eyebrow*. **Juv:** Duller, browner; wing-coverts tipped buff. **Subspp (4):** *gularis* (widespread NG form) grey with buffy belly; *vidua* (NW Is: Gag, Gebe, Kofiau) breast band pale-spotted, belly white; *kordensis* (Bay Is: Biak) black with white belly and strong white eyebrow; *nigromentalis* (SE Is: Misima, Tagula) black chin, white tail tips larger. **Similar spp:** Mangrove F smaller, has a pronounced white eyebrow and white wing bars and actively fans tail. **Voice:** The tune of one song, a 4–7-note series, sounds like a mouse-warbler, yet with a softer, faster, more halting delivery, and descending in pitch. Another song, given in excitement, is composed of 2 alternating, high, piping notes in a long repetitious series, *tiptuptiptup* . . . , ending in a stuttered jumble. **Habits:** Singly or in pairs, often with a mixed-species flock of warblers, monarchs, and small honeyeaters. Perches conspicuously in an open space in the forest interior and sallies for small insects. Also frequents the forest edge. Nest is built in the midstory on a thin horizontal or sloping stem; a tight cup with a tail, made of fine plant material bound with spiderwebs. Eggs (2) yellowish white with spots. **Range:** NG Region, all NW Is, Bay Is (except Numfor), Manam I, and SE Is (3 main D'Entrecasteaux Is; Misima and Tagula), 0–1600 m. Also Wallacea, AU, to Admiralty and Bismarck Is. **Taxonomy:** A highly variable species that may someday be broken up.

CROWS AND ALLIES: Corvidae

The crows have a worldwide distribution, with 130 species in the family. The greatest diversity of forms inhabits tropical Asia. The family includes jays and others, but only true crows are found in New Guinea (3 spp). These are large and either black or grey, with a heavy bill and powerful legs; all have the iris white or pale blue, like the Australian crows. The New Guinea species inhabit forest, savannah, and coastal areas, where they forage for fruits, insects, spiders, crabs, frogs, lizards, and other animal foods. Crows are vocal and sociable, and some species have helpers at the nest.

Brown-headed Crow *Corvus fuscicapillus* Pl 93
46 cm. Known from only 4 localities in lowland forests of the West. A *robust* crow with *large, high-arched bill* and short, square tail (barely extends beyond folded wing tip). *Brownish-black* head and breast. Adults have a striking, pale blue eye. **Male:** Bill black. **Female:** Bill color uncertain; reported as yellow with a dark tip, but this could indicate a young bird; needs confirmation. **Juv:** Paler and more brownish than Adult, with much white showing from bases of feathers; bill all yellow. **Similar spp:** Torresian C has the same shape but lighter build, has a smaller, tapered bill, and is glossy black. Grey C has a long tail and pink face. **Voice:** Deeper than Torresian C. Call, a loud *ow ow ow . . . aahhh* and variations; also *gakock gakock* (possibly a duet). **Habits:** Singly or in pairs, rarely in flocks of as many as a dozen; shy and retiring, normally staying within the forest. A fruit eater of the forest canopy in primary and tall secondary lowland forest; while reported from mangroves, it usually avoids the coast and open habitats used by Torresian C. No nesting information. **Range:** Endemic. NG

(NW Lowlands: mouth of Mamberamo R and Nimbokrang forest west of Cyclops Mts), NW Is (Waigeo), and Aru Is, 0–500 m.

Grey Crow *Corvus tristis* Pl 93
(Bare-eyed Crow, *Gymnocorvus tristis*)
51–56 cm. Common in vocal family groups in lowland and hill forest. A lanky, *long-tailed* crow, unusual for its *pale grey or sooty-brown plumage*. Note *bare, pink face*. **Adult:** Sooty brown; bill slaty grey. **Juv:** Paler and greyish; bill pink. **Imm:** Intermediate. **Similar spp:** Other crows are darker and have much shorter tails. Channel-billed Cuckoo is similar, but has a curved bill, longer pointed wings and tail, and shows some barring in the flanks and tail. **Voice:** Call is a short, nasal, plaintive, high-pitched *wenh?* repeated every few seconds. Flock calling at a distance sounds like yelping. **Habits:** In small, vocal, wary, but curious family parties, foraging for fruit and animal prey in forest canopy and edge. Often seen along rivers. Flies mainly at canopy level. Nest built in crown of a forest tree; a platform of thin sticks. Eggs, no information; 4 young observed in the only nest reported. **Range:** Endemic. NG (all except Trans-Fly), NW Is (Batanta, Salawati), Bay Is (Yapen), SE Is (3 main D'Entrecasteaux Is), 0–1400 m (rarely to 2000 m).

Torresian Crow *Corvus orru* Pl 93
46 cm. The widespread, common black crow of coasts, settlements, and open lowland country. Compact shape, with tail barely projecting beyond tip of folded wing. *Tapered, medium-sized bill.* **Adult:** *Shiny black*. Iris whitish or pale blue. **Juv and Imm:** Duller, iris matures from grey, to brown, to hazel, to white. **Subspp (1):** *orru* (NG Region and Moluccas). **Similar spp:** Brown-headed C, in forests of the West, is the same size and shape, but is more robust, and the bill is larger and arched, head and breast brownish black. Grey C has much longer tail and is grey or sooty brown. **Voice:** Call an annoying, oft-repeated, nasal caw: *cah cah caaaaaaaooooowww*, and variations of this. **Habits:** Singly, in pairs, or in small gatherings. Conspicuous but wary. Forages on the ground, scavenging along roads and shorelines; also takes fruit from trees. Feeds on a variety of animal life, dead or alive, and fruit. Nest, built in a tree crown, is a bulky cup of sticks and other material. Eggs (2–4) pale green with spots. **Range:** All NG Region except Aru Is and Rossel I (SE Is), but no record from Mios Num (Bay Is), 0–650 m. Also AU and Moluccas.

MELAMPITTAS: Melampittidae

> The genus *Melampitta* has recently been classified in its own family. The 2 species are black, long-legged, short-winged ground birds. The Lesser M is rather pitta-like, but more noteworthy is the subterranean roosting Greater M. DNA studies have placed the melampittas closest to birds of paradise and related families.

Lesser Melampitta *Melampitta lugubris* Pl 63
18 cm. A widespread but reclusive ground dweller of cloud forests up to timberline. Difficult to see, so listen for its double-click call. *An entirely velvet-black pitta-shaped bird*, hopping on stilt legs. Distinctive domed forehead. **Male's** eye red, **Female's** dark brown. **Juv:** Dull, sooty black with brownish cast in underparts. **Similar spp:** None in its habitat. Greater M much larger, bulkier yet less compact, with bigger bill, longer tail; not in cloud forest. **Voice:** Song *tzee-sisisisisisi*, the last series a descending, insect-like trill. Contact call is a sputtered, loud, double clicking or spitting, like the clicking or grinding of 2 pebbles: *SPITjup* or *SHLIK* or *SHLIP*; varies regionally. **Habits:** Singly or in pairs, concealed in the undergrowth. Hops or runs. Seeks insects, worms, snails by searching or flipping leaves and gleaning, probing, or digging. Nest built near the ground on the side of a tree fern; bulky nest of live moss and rootlets, with side entrance. Egg (1) white with dark spots and blotches. **Range:** Endemic. NG: Bird's Head, Central Ranges, Huon, 2100–2800 m (1200–3500 m).

Greater Melampitta *Megalampitta gigantea* Pl 84
(*Melampitta gigantea*)
29 cm. Habits unique: associated primarily with forested, hilly, limestone terrain where it shelters in sinkholes. Very local. Shy and difficult to see; best located by its loud, distinctive song. *The only large,*

all-dark, terrestrial songbird. Pitohui-like, but strictly terrestrial. Often with unkempt appearance, particularly the worn, frayed tail. Many birds have white feather-mite egg cases encrusted behind the eye. **Adult:** All black. **Juv and Imm:** Head black, body mostly or all dark brown, although age sequence not understood. **Similar spp:** Piping Bellbird also terrestrial but not as dark, more rufous, smaller. The hooded juv Greater M shares the black-and-brown plumage pattern with Southern Variable, Northern Variable, and Hooded Pitohuis, but is much larger, and there is no distinct division between the black-and-brown plumage. **Voice:** Song is loud, carrying, and musical—a connected pair of medium- to high-pitched notes repeated over and over at slightly irregular intervals for several minutes. The second note of each pair is either higher or lower in pitch; phrases are spaced 2–4 sec apart. **Habits:** Poorly known; no information on its social system, although birds are typically encountered alone. Presumably forages for arthropods on the forest floor in the vicinity of subterranean roosting sites. Progresses in long, bounding leaps or a scurrying run, like a small mammal. Seldom flies. Nest reported by local people to be a large suspended basket built down in a sinkhole. Egg, no information. **Range:** Endemic. NG: Known from 7 scattered localities (may actually be widespread in limestone areas) including Bird's Head (Arfak Mts), Bird's Neck (Fakfak and Kumawa Mts), S slopes of Western Ranges (Otakwa and Setekwa Rs), S slope of Border Ranges (Tabubil and P`nyang Camp), N Coastal Ranges (Torricelli Mts), and SE Pen (Mt Mura, NW of Mt Simpson), 650–1400 m.

BIRDS OF PARADISE: Paradisaeidae

Birds of paradise (BoP) are a principally New Guinean bird family, with 37 of the 41 known species inhabiting the NG Region. The family ranges from northern Moluccas through NG and thence south along the coast of eastern Australia. In NG, the species inhabit forests at all elevations, but the greatest concentration of species can be found in the mid-mountain zone, ~1500–2100 m. As currently circumscribed, the Paradisaeidae no longer includes the satinbirds (now in their own family) and the genus *Macgregoria* (Giant Wattled Honeyeater).

Although the birds of paradise exhibit a diverse array of feeding habits, bill shapes, body forms, and plumages, they nevertheless constitute a taxonomically compact group, famous for the males' remarkable nuptial plumages and elaborate courtship displays. The bird of paradise body plan is somewhat crowlike, with a powerful bill and feet, and all species are adapted to a diet of fruit and insects. This is an important group of seed dispersers.

There are 2 branches to the family, and they differ in breeding habits. The glossy black manucodes (5 spp), in which the sexes look alike, are socially monogamous, meaning that the male and female form a pair bond and share in nesting duties. While they may lack ornamental plumes, manucodes are nevertheless remarkable as the only songbirds in the world with an elongated trachea (windpipe) that coils (in *Phonygammus*) or loops (*Manucodia*) beneath the breast skin and, like a trumpet, serves to amplify the bird's voice.

In the polygynous "true" birds of paradise, the colorful adult males sport ornamental plumage and advertise themselves with loud songs and calls. The cryptic females are brown or black plumaged, often with barring ventrally. Young males look exactly like females but are somewhat larger, and it takes them many years to eventually acquire adult plumage. Once fully adult, the male faithfully attends a display court or perch, where he attracts and mates with females. Depending on species, a male displays by himself away from other males or with a group of males called a "lek." Females of polygynous species alone build the nest and raise the offspring.

One outcome of these promiscuous matings deserves mention: hybrids—the offspring of parents of different species. Museum collections contain a surprising diversity of bird of paradise hybrids, including those between different genera, such as *Paradigalla* × *Astrapia* or Magnificent × King BoP crosses. The hybrids may look intermediate between the 2 parental species, but just as often their appearance can be quite bewildering. Intergeneric hybrids are very rarely encountered in the field. The only commonly seen hybrids are between Ribbon-tailed and Stephanie's Astrapias and between Greater, Lesser, and Raggiana BoPs.

Birds of paradise are much admired for their beauty and their entertaining displays. The birds feature prominently in local folk traditions, and their plumes are used in dress on festive and ceremonial occasions. Despite the hunting of adult males for plumes, many birds of paradise seem able to maintain populations close to rural villages. Whether near human settlement or in the most remote wilderness, the brown-plumaged females and immature birds predominate by far, so learning to identify them is key. Be aware that males of many iridescent species appear partly to completely black depending on light conditions, obscuring what would otherwise be colorful plumage and possibly confusing someone trying to identify them. For best results finding and observing male birds of paradise, knowledge of the specific location of display courts and perches is required (though not for manucodes, which lack a fixed display site). Visitors should employ a local guide to show them display sites. To find displaying males on one's own, first learn about the courtship behavior of individual species, then seek out the songs of the males in appropriate habitat—the advertising males may lead you to their display site. When not at his display site, a male can sometimes be spotted resting on an open perch in a tall tree, particularly after rain, so scanning ridge crests with a telescope is a productive technique. Another method is to wait patiently at a favored fruiting tree; over time a selection of species may visit the tree to feed. Even chance encounters are frequent, particularly of female and immature birds, which are often seen hopping slowly along a branch carefully searching for prey. Both males and females can be found in mixed flocks in the lowlands.

Trumpet Manucode *Phonygammus keraudrenii* Pl 93
(**Trumpetbird**, *Manucodia keraudrenii*)
28–33 cm. Frequents fruiting trees (especially figs) of lower montane forest, and lowlands in the South. Shy and stealthy, difficult to observe. Much nervous wing-flicking and hesitant posturing. Smaller than other manucodes. **Adult:** A glossy black canopy dweller with *diagnostic, loose feather hackles adorning the nape, neck, and throat*; spiky crest; iris red. **Male:** Famous for an outlandish anatomical feature—the extraordinarily long trachea (to 800 mm) that lies like a tightly coiled garden hose beneath the skin of its breast. **Female:** Smaller; trachea shows as a shorter loop or coil. **Juv and Imm:** Dull blackish with reduced gloss and virtually no hackles; iris grey-brown, slowly changing to red with age. No tracheal loop or coil. **Subspp (6):** *keraudrenii* (Bird's Head and Neck, and far western W Ranges) a montane form of average size and plume length, bluish back, purplish wings; *neumanni* (N slopes of W and Border Ranges, plus Foja and N Coastal Mts) similar but with spiky crest short; *adelberti* (Adelbert Mts) small size, short crest and hackles, greenish gloss, shrill voice, short tracheal coil; *jamesii* (S Lowlands and Aru to southern SE Pen) a lowland form, upperparts greenish blue, produces a distinctive *ooo-uh* vocalization not given by adjacent upland forms; *purpureoviolaceus* (E Ranges and SE Pen) a montane form of large size with long crest and hackles, purplish upperparts, deep voice; *hunsteini* (D'Entrecasteaux Is) a rare, quiet, and inconspicuous insular form that is very large, less glossy, and shows slightly twisted central tail feathers. **Similar spp:** Other manucodes are larger. No other manucode exhibits the loose mane of head feathers; all others show a "crew-cut" instead. **Voice:** Display song and contact calls not well understood among populations, yet differences seem significant. Subsp *jamesii* from the S Lowlands gives a distinctive *ooo-uh*, as well as the more typical calls of the species. A putative Male song is a low, tremulous, prolonged, fluttering *wodldldldld* accompanying a display with plumage erected and wings fanned forward. Call notes include: *kauaugh* (like clearing a sore throat—perhaps Female response), *kowp*, or *kyawk*. Upland birds have been heard giving Male-Female duets (comprising sounds listed above). **Habits:** In pairs or in small parties at fruiting trees, where it aggressively displaces other birds. A fig-eating specialist. Flight distinctly floppy and undulating. Nest built at middle-story height suspended from a fork in the branch of a small tree; a shallow cup of vine tendrils without leaves or wood. Eggs (1–2) pink-buff, spotted and blotched. **Range:** NG, Aru Is, and SE Is (3 main D'Entrecasteaux Is), 0–2000 m. Lowland-dwelling populations: S Lowlands and Aru Is. Montane-dwelling forms: Bird's Head and Neck (Fakfak), Central Ranges, most outlying ranges (Van Rees, Foja, N Coastal, and Adelbert Mts), but no records from Kumawa, Wandammen, Cyclops, and Huon Mts. Also Cape York, AU. **Taxonomy:** May comprise 2 or more distinct species.

Curl-crested Manucode *Manucodia comrii* Pl 93
43 cm. Common canopy fruit eater in forests, gardens, and strand throughout the D'Entrecasteaux and Trobriand Is. Its musical, far-carrying voice is an ever-present and otherworldly sound of the islands. **Adult:** *A large, crowlike bird with curly feathers atop crown and twisted central tail feathers*; iris red-brown. Note undulating, floppy flight typical of manucodes. **Male:** The trachea forms a J-shaped loop on the breast, and in full development the bottom of the J extends around the end of the body and up onto

the back. **Female:** Smaller; no tracheal loop. **Juv and Imm:** Blacker, duller, and lacking crinkly or curly feathering; iris grey-brown changing to red-brown with age; no tracheal loop. **Subspp (2):** Minor. **Similar spp:** Trumpet M shares habitat but is smaller, thinner, and exhibits shaggy feathering of neck and nape. Spangled Drongo, with curved outer tail feathers, is much smaller and sits upright. **Voice:** Typical song is a beautiful, haunting, low, rolling series, descending slightly in pitch and dropping in volume, *woodloodloodloodloodl*; another version is higher-pitched, more musical and xylophone-like, rapidly descending; another *wodldldldld* is a deep exhalation and resembles the display song of Crinkle-collared and Trumpet Ms. These songs are delivered throughout the day and even before dawn. Duet is composed of different songs given in sequence. Calls are a variety of single notes: a metallic croak, a scolding *ench*, and a low clacking. **Habits:** The most conspicuous bird species in its habitat; Males often perch and vocalize from an exposed branch. Unwary and inquisitive. Solitary, in pairs, or small foraging parties. Eats mainly fruit. Nest built in a midstory tree, suspended from a branch fork; a loosely made cup of vinelets and twigs, decorated outside with leaves, on a base of rotten wood. Eggs (1–2) pale pinkish buff with dark spots and blotches. **Range:** Endemic. SE Is: 3 main D'Entrecasteaux Is and Trobriand Is, 0–2200 m.

Crinkle-collared Manucode *Manucodia chalybatus* Pl 93
(*Manucodia chalybata*)

33–37 cm. Common in midstory and canopy of *forest in hills and lower mountains,* including disturbed scrubby habitat of highland valleys. Stealthy, wary, and difficult to observe except at favored fruiting trees. **Adult:** A large, blue-black, crowlike bird with an evident *"bump" above each eye,* formed by a ridge of erectile feathers. Hard to see are the *crinkled feathers of breast and mantle, showing in good light as alternating bands of matte black and glittering iridescence.* Neck and upper breast usually appear black, but in direct lighting display yellowish-green iridescence; lower breast and belly dark blue with purple gloss. Iris red. **Male:** Larger. The tracheal loop passes down over the pectoral muscles and extends slightly over the edge of the sternum (lacking in Female and Juv). **Female:** Smaller; tends to be less purplish, more green-blue. **Juv and Imm:** Blacker, duller, with feather-crinkling absent; iris brown. **Similar spp:** Glossy M, in lowlands, is essentially the same shape and color but lacks the "bump" over the eye and the crinkling of feathers of breast and mantle; best separated by song. Jobi M, also lowland dwelling, is nearly identical but it too lacks a bump above the eye and is smaller, shorter-billed, and shorter-tailed, and looks stockier. Trumpet M has shaggy neck hackles. **Voice:** Song is a slow series of up to 8 hollow, haunting, low-pitched *hoo* notes on the same pitch; can be followed by response from another bird that is *woo owoo owoo owoo owoo owoo* on a descending scale. Display song of Male is a deep pigeon-like *ummmh* or similar single or double note; in display the bird leans forward and expands its feathers and wings. Also a deep, hollow *hoouw.* Common call note like that of other manucodes, *tuck* or *chook.* Makes a variety of other calls. **Habits:** Seen singly or in pairs. Diet is almost entirely fruit, especially figs; also a few insects. Nest of vine stems similar to other manucodes, includes leaves and is lined with finer stems and fibers. Eggs (1–2) cream-colored with dark markings. **Range:** Endemic. All NG mts and Misool (NW Is), mostly 600–1500 m (0–1750 m); also a local population on the lower Fly R.

Jobi Manucode *Manucodia jobiensis* Pl 93

30–36 cm. A secretive, little-known, and difficult-to-identify manucode of *lowland and hill forest of the North.* Shares its habitat with Glossy M, and replaced at higher elevations by the closely related Crinkle-collared M. Jobi M merits attention in the field to better delineate field characters in plumage, behavior, and voice. **Adult:** All but identical to Crinkle-collared, though *lacks eyebrow "bumps"* and is smaller on average, more compact, and *tail is relatively short and bill marginally smaller.* Also, the upper breast is purplish blue (not yellowish green) and the lower breast and belly greenish blue, rather than purplish blue. Iris red. **Male:** Larger; tracheal loop as in Crinkle-collared. **Juv and Imm:** As with other manucodes. **Similar spp:** Glossy M significantly larger and lacks glittery crinkled feathering on breast and mantle, and the dorsal color is generally bluish or greenish, not purplish blue. **Voice:** Poorly documented. Song is a series of hollow *hoo* notes, much like that of Crinkle-collared. What may be the display song is described as a drawn-out moan. Call is a harsh *chig* or *bcheg.* **Habits:** As for other manucode species; mainly solitary and shy; probably best observed at fruiting fig trees. Nest and egg similar to Crinkle-collared. **Range:** Endemic. NG and Bay Is (Yapen, where it is the only manucode), 0–750 m. NG: NW Lowlands and Sepik-Ramu; also western S Lowlands to the Mimika and Setekwa Rs. No records from Bird's Head and Neck.

Glossy Manucode *Manucodia ater* Pl 93
(**Glossy-mantled Manucode**, *Manucodia atra*)
Male 38–42 cm; Female 33–37 cm. *Widespread in lowlands*, where common but shy. *Prefers forest edge and open woods*, i.e., swamp forest, riverine scrub, monsoon woodland, gallery forest, heavy savannah, and mangroves; generally absent from deep forest. The *largest mainland manucode, with proportionately long tail. Sleek, lanky, and thin-necked. Breast and mantle glossy, lacking crinkled feathering.* Neck feathers short with pointed tips, rather than crinkled. Best identified by its song, a *high whistled tone*. Like all manucodes, exhibits an exaggeratedly floppy and undulating flight pattern. Wings produce a swishing sound. **Adult:** Iridescent; iris red. **Male:** Larger; tracheal loop of mainland birds short, but on Tagula I it is long, reaching over the hind edge of the sternum. **Juv and Imm:** Dull blackish without prominent iridescence. **Subspp (2):** *ater* (NG mainland, NW Is, and Aru Is) dorsal color either greenish or bluish, or most often purplish on islands and SE Pen, high-pitched song; *alter* (SE Is: Tagula) larger with more massive bill, purplish color, and deep song. **Similar spp:** Jobi M, also in lowlands, is smaller, more compact, and stubbier-billed, generally purplish, not bluish or greenish, and shows crinkled breast and mantle. Crinkle-collared M co-occurs at the foot of the ranges and exhibits crinkled feathering on breast and mantle and bump over the eye. Eastern Koel male and Black Butcherbird have whitish bills and steady, level flight. **Voice:** Song is unique and sounds like an electronic tone: a drawn-out, high-pitched, penetrating whistle on a single pitch, *eeeEEEEee*, that swells in volume then tapers off, and lasts several sec. Song accompanied by a display with erected feathers and partially fanned wings. On Tagula I, the song is a deep hum, *MMM....* Call of all races is a *chook* note typical of the genus. **Habits:** Singly, in pairs, or in small parties. A shy fruit eater of canopy fruiting trees (mainly figs), but regularly sings from a high, open perch at forest edge. Nest as for other manucodes. Eggs (1–3) cream-colored with blotches. **Range:** Endemic. NG, NW Is (Batanta, Gebe, Gam, Kri, Misool, Salawati, Waigeo), Aru Is, and SE Is (some small ones near the mainland, plus Tagula), 0–300 m (rarely to 1000 m). **Taxonomy:** The Tagula I form may qualify as a species.

King of Saxony Bird of Paradise *Pteridophora alberti* Pl 97
Male 22 cm; Female 20 cm. A small canopy dweller of cloud forest in the Central Ranges. Uncommon and seldom encountered except at song perches or at favored fruiting trees. Listen for the weird, insect-like, buzzing song. **Male:** Unique. *Black with a pale ochre breast and belly*, and two remarkable erectile and enameled, *sky blue head plumes that far exceed the length of the bird*. The stiff head plumes can be moved around in all directions, like the antennae of a long-horned beetle. Broad orange wing band in flight. **Female, Juv, and Imm:** Obscure. The *small bill, grey plumage, and fine, scalloped barring below* suggest something other than a bird of paradise, perhaps a whistler or bowerbird. Note *ochre undertail*. **Similar spp:** Other female birds of paradise are brownish, larger-billed, and show heavier barring. Female bowerbirds and satinbirds lack barring below. **Voice:** The Male's song possibly the weirdest of any NG bird. Sounds like a cicada or electrical malfunction, a series of sputtered and jumbled notes poured out at machine-gun pace and suggestive of bad radio static. Lasts 3–4 sec, and gradually swelling in volume finally rising to a twittering climax. Young Males monotonously repeat a jeering call reminiscent of that of the Hooded Cuckooshrike, each note harsh, rolling, and downslurred: *chweer chweer chweer....* **Habits:** Solitary. Diet mainly fruit with some arthropods, gleaned or excavated from bark and moss. Especially attracted to false fig (an epiphytic *Timonius* shrub). The territorial Male sings in early morning and late afternoon from a high, open perch, usually a dead branch, but subsequently displays to Female on a vine in the understory. Nest, built in a midstory tree, is a shallow cup of orchid and fern stems. Egg (1 only?) pale buff with numerous dark, longitudinal blotches. **Range:** Endemic. NG Central Ranges minus SE Pen, 1800–2500 m (1400–2850 m).

Western Parotia *Parotia sefilata* Pl 94
(**Arfak Six-wired Bird of Paradise**)
Male 33 cm; Female 30 cm. Bird's Head and Wandammen Pen only. Common in mid-mountain forests. *Tail of medium length* compared with other parotias (see Lawes's P). **Male:** *Diagnostic combination of medium tail and white forehead.* **Female, Juv, and Imm:** *Whitish rather than buff below compared with other parotias; upperparts darker, less brownish and showing less contrast with black head.* Note whitish malar stripe. **Similar spp:** Superb BoP female is smaller, less robust, and smaller-headed and lacks the pale malar stripe. Long-tailed Paradigalla has a longer, pointed tail and longer, narrower bill, and facial wattles. Magnificent Riflebird inhabits lower elevations and is longer-billed. **Voice:** Typical of the genus. Male "song" is 1 or more very loud, cockatoo-like shrieks: *gnaad-gnaad!* Also musical piping or mewing calls near display court. Note the varied calls of the Vogelkop Bowerbird could be confused for

those of this species. **Habits:** Social behavior, feeding ecology, courts, and displays similar to Lawes's P. Nest and egg unknown. **Range:** Endemic. NG: Bird's Head and Wandammen Pen, 1100–1900 m. Parotias should be looked for in the Fakfak and Kumawa Mts of the Bird's Neck.

Wahnes's Parotia *Parotia wahnesi* Pl 94
(Wahnes' Six-wired Bird of Paradise)
Male 43 cm; Female 36 cm. Confined to Huon Pen (locally common) and Adelbert Mts (nearly extinct). In mid-mountain forest. *Tail longest of all parotias.* **Male:** *Golden-bronze frontal crest.*
Female, Juv, and Imm: *Russet dorsally and with a black head offsetting a pale eye-stripe and moustachial streak.* **Similar spp:** Unmistakable within its range, but compare with Huon Astrapia and Spangled Honeyeater, the 2 other large, long-tailed, black passerines of the Huon mountains. **Voice:** Male gives a loud and rather harsh but musical shriek, *yeah-yeah* or *yack-yack* or *khh kaaakkk* with a cockatoo-like quality. Musical, minor squeaks, whines, and twittering notes given around display sites. **Habits:** Social behavior, feeding ecology, courts, and displays similar to Lawes's P, as far as known. Nest, no information. Eggs (2) laid in captivity, cream with heavy streaking. **Range:** Endemic. NG: Huon Pen (1100–1700 m) and Adelbert Mts (1300 m to summit at 1600 m). In the Adelberts, much of its primary forest habitat has been felled for village gardening, and recent searches have failed to find it.

Lawes's Parotia *Parotia lawesii* Pl 94
(Lawes's Six-wired Bird of Paradise)
Male 27 cm; Female 25 cm. The parotia of eastern mid-mountain forests, where common. *Distinctive shape: a moderately large, robust, and stolid* bird of paradise with *broad, rounded wings in flight.* Striking *blue-and-yellow iris* in both sexes at any age (as with Western, Wahnes's, and Eastern Ps, but not Carola's and Bronze Ps). **Male:** *A chunky, stub-tailed, all-black bird with a long, silvery white patch above bill,* head wires with black flag tips (sometimes hard to see), and small, shining, green breast shield. **Female, Juv, and Imm:** *Black-headed and deep brown body, with fine dark barring ventrally.* Note pale moustachial streak. **Similar spp:** See Eastern P. Female could be confused with other female BoPs when not seen well. Female Superb BoP smaller, slimmer, and smaller-headed. Female Carola's P without black on head and with twin facial stripes. Female riflebirds longer billed. **Voice:** A loud *shaak!* and other harsh shrieking notes, delivered once or a few times. **Habits:** Mostly solitary, but gathers in small numbers at fruiting trees; a bit more socially tolerant than other birds of paradise. Diet is mainly fruit with some animal prey (arthropods, lizards, and probably small frogs) taken in forest canopy and edge. An accomplished bark gleaner of limbs and trunks. Male dances at a terrestrial display court (1–2 m across) established in thicket in tree-fall opening. Each Male has his own court, and courts of different Males are often grouped near each other. The court is fastidiously cleared of leaves (but not carpeted like that of Carola's P) and features a horizontal perch above for Female viewing. The dance is elaborately choreographed and varies in its sequence among parotia species. Nest placed on a midstory branch or in a vine tangle; a shallow cup of fern tendrils and vines. Egg (1) pale buff with broad dark streaks and some small spots and blotches. **Range:** Endemic. NG: E Ranges and SE Pen (west to Oksapmin), 1200–1900 m (500–2300 m). Where the ranges of Lawes's and Carola's Ps overlap, Lawes's lives at higher elevations than Carola's.

Eastern Parotia *Parotia helenae* Pl 94
(Previously classified as a subspecies of **Lawes's Parotia**, *P. lawesi*)
Male 27 cm; Female 25 cm. Replaces Lawes's P on *N watershed of SE Pen*. Both sexes of Eastern P nearly identical to Lawes's but show a *steeper, more concave forehead profile* (vs forehead sloped and extending forward over the bill in Lawes's). Eastern thus has *more rounded, less tapered head shape.* **Male:** Identical to Lawes's, but *feathered patch over the base of bill is small and bronze-colored* (not elongate and gleaming white). **Female, Juv, and Imm:** Differ from Lawes's by reduced feathering over the bill that leaves more of bill tip exposed (at least half the bill or more), and *the culmen (dorsal ridge of the bill) is sharply keeled* (angular or slightly rounded in Lawes's). **Voice:** Similar to Lawes's. **Habits:** In need of study. Similar to Lawes's as far as known. Nest and egg similar to Lawes's. **Range:** Endemic. NG: N catchment of SE Pen, westward in N to the Waria R, in S probably only to Keveri Hills just west of Mt Suckling, 1100–1500 m (1 record at 500 m). Contact with Lawes's P has yet to be found. **Taxonomy:** Previously treated as a subsp of Lawes's; however, structure and color of the frontal tuft, bill shape, molecular genetics, and preliminary data on display behavior together tip the scales toward species status.

Carola's Parotia *Parotia carolae*   Pl 94
(Queen Carola's Parotia or **Six-wired Bird of Paradise)**
Male 26 cm; Female 25 cm. A boxy and big-headed canopy dweller in western and central portions of the Central Ranges; uncommon. Overlaps with Lawes's P in the E Ranges. Distinctive *pale yellow eye* in all plumages, unlike blue eye of other parotias. **Male:** Black with diagnostic *white flank plumes, buffy face and throat,* and pinkish-gold breast shield. **Female, Juv, and Imm:** Brown, with breast and belly finely barred black and *brown face marked by pale stripes above and below eye.* In flight note distinctive *russet wing patches.* **Subspp (4):** Minor. **Similar spp:** All other male parotias, besides this and Bronze P, lack the white flank-stripe. Lawes's P female has a black head and blue iris and lacks russet wing patches. The smaller Superb BoP female is smaller-billed, smaller-headed, but otherwise a close match in plumage, including russet wing patches in parts of its range (W Ranges). Bronze P most similar, but has isolated range (see that account). **Voice:** Vocalization is best means of locating this elusive species, but knowledge of local dialect needed. Male's shriek is higher pitched than that of other parotias and varies geographically, e.g., *kwoi-kwoi-eeeng* (Mt Stolle, Border Ranges) or *kwa-a-a-a-ng* (Crater Mt in E Ranges) or *whee o weet* and other versions. Calls given at courts are small chips, squeaks, and chortles. **Habits:** Solitary. Birds forage for fruit and animal prey in forest canopy. Gleans insects from bark of branches creeper-fashion. Male dances at a terrestrial court (1–5 m across) cleared in thicket of a tree fall in forest interior. Display courts are often clustered. The Male carpets the court with a mat of rootlike fungi. A horizontal perch above is used by Females for viewing. The Male's dance is highly complex. Nest and egg unknown. **Range:** Endemic. Central Ranges, eastward to the Bismarck Range and Crater Mt, 1200–1800 m.

Bronze Parotia *Parotia berlepschi*   Pl 94
(Foja Parotia)
Male 26 cm. The Foja Mts sister-form of Carola's P; uncommon in its isolated range. Differs from Carola's in *the Male's blackish face and throat and overall bronzed cast, and the iris of both sexes is grey-blue, rather than overall yellowish.* **Male:** Bronzed blackish brown with white flank plumes; iris blue-grey with orange outer ring. **Female, Juv, and Imm:** Grey-brown dorsally, with rufous brown wing edges. Finely barred below dark on buffy background; forehead frosted white, extending to pale eyebrow. Iris plain grey. **Similar spp:** No other similar bird of paradise in its range. **Voice:** Male's high, sweet call notes are Carola-like, but differ: *whee-deent* or a rapid *di-di-di* with the quality of a child's plastic squeeze-toy. Quiet musical notes, musical squeals, and low scolding near display court. **Habits:** As for Carola's P. A reclusive inhabitant of forest canopy. Nest and egg unknown. **Range:** Endemic. NG: Foja Mts, 1200–1600 m.

Twelve-wired Bird of Paradise *Seleucidis melanoleucus*   Pl 98
Male 33 cm; Female 35 cm. Inhabits swampy lowland forest and regrowth, especially along rivers. Uncommon. Best seen from a boat. The wary Male may be tracked down by his song and hissing wings, or in early morning at his display site. The more approachable Female is commonly found in mixed flocks. **Male:** Shows a *bicolored pattern: black in front with rich yellow below and behind.* The 12 bent flank-wires are difficult to see. The bird is often observed in undulating flight crossing a lowland river, long bill protruding and wires trailing. **Female, Juv, and Imm:** Strikingly plumaged with *reddish-brown back and tail, black cap,* and finely barred underparts. The combination of *red eye and pink legs* is unique for a BoP. *Long, black chisel-bill* is prominent. **Similar spp:** Male riflebirds' broad and rounded wings rustle, whereas the Male Twelve-wire's wings hiss. The smaller female riflebirds are brown-capped with a pale eyebrow. **Voice:** Male gives 3 songs or calls: a *harnh* or *hahn* note produced intermittently; a powerful, far-carrying series of 3–8 resonant notes, *hahr—haw haw haw*; a higher-intensity song is a series of 5 or more *koi* notes, with a pause after the first note. These series have quality reminiscent of the paradisaeas. **Habits:** Solitary. Prefers seasonally inundated lowland forest, especially where pandanus and water-loving palms like sago are common. Unplumed individuals seem to wander farther from this ideal habitat into adjacent flat lowland forest and karst forest (Sepik region). Forages in the canopy and midstory for arthropods, fruit, and flower nectar. Especially fond of red pandanus fruit. Seeks arthropods hidden under bark or in pandanus leaves. Males are widely dispersed and presumably territorial. Male sings and displays from atop an exposed vertical spire protruding above the vegetation. Nest situated in the head of a pandanus or sago palm; a bulky cup of pandanus leaf strips, vines, and sticks. Egg (1) off-white with dark longitudinal streaks. **Range:** Endemic. NG and Salawati (NW Is), 0–200 m. Absent from Trans-Fly and hilly coasts, including Huon and northern SE Pen.

Magnificent Riflebird *Ptiloris magnificus*   Pl 98
Male 34 cm; Female 28 cm. An often-heard but seldom-seen inhabitant of the forest canopy in lowlands and hills through the western two-thirds of NG. Listen for the double-whistle song. Common in rainforest, less so in monsoon forest and swamp forest. Male terribly shy, Female and young birds less so. Female often seen in mixed flocks with other rufous birds. *Riflebirds have a distinctive shape: robust with longish neck, strongly tapered head profile, long bill, and short tail.*
Male: *All black, with a prominent iridescent blue throat shield and crown and rounded wings that rustle loudly in flight.* **Female, Juv, and Imm:** *Reddish-brown upperparts, a whitish eyebrow, and long white moustachial stripe;* buff-grey ventrally with fine dark barring. **Subspp (1):** *magnificus* (endemic).
Similar spp: See Growling R, virtually identical but for song. Twelve-wired BoP female, also in lowlands, has a black head, red iris, longer bill. Females of Black-billed and Pale-billed Sicklebills have strongly curved bills and buff or rusty tails. Superb BoP female, usually at higher elevations, is smaller, shorter-billed. **Voice:** Song is a powerful pair (sometimes 3–5) of upsweeping whistles: *wooiieet?—woiit!* First note sweeps to a higher pitch than the second note; second note sweeps higher than third note, if there is one. **Habits:** Solitary, except when at fruiting trees or in mixed-species flocks. Diet largely of arthropods, taken creeper-fashion by gleaning from bark and dead wood. Also fruit. Male seems to be territorial and sings from various canopy perches. He displays on a horizontal perch, such as a thick vine or thin log, a few meters above the ground. Nest placed at midstory level in head of pandanus or fern clump. Eggs (1–2) pale creamy pink with dark streaks. **Range:** All NG mainland except Trans-Fly and where replaced by Growling R in the E; ranges eastward in the S to Purari R basin and in the N to Ramu R, 0–1200 m (rarely higher). Also Cape York, AU. **Taxonomy:** Split from Growling R—see below.

Growling Riflebird *Ptiloris intercedens*   Pl 98
Male 34 cm; Female 28 cm. Eastern sister-form of the Magnificent R. Nearly identical and *best distinguished by growling (vs whistled) song and eastern range.* Growling R's bill is shorter and is covered by a dense mat of *feathering at the base of the upper bill covering the culmen ridge.* This feature is lacking in Magnificent R, whose culmen ridge is exposed all the way to the base of the bill. **Male:** Flank plumes are shorter than those of Magnificent, not extending beyond the tail. **Voice:** Male song is a hoarse, growling *kraay? kraoww!* **Range:** Endemic. NG: SE Pen, Huon, and Adelbert Mts, west in S to Purari R basin and in N to Ramu R, 0–1200 m (rarely higher). Both song types have been heard in the Purari R region. **Taxonomy:** Long regarded as a race of Magnificent R, *intercedens* is best treated as a species owing to consistent differences in voice, bill, and flank plumes, and by molecular distinction.

Superb Bird of Paradise *Lophorina superba*   Pl 97
Male 26 cm; Female 25 cm. A small, vocal, and common bird of paradise of mid-mountain forest, ecologically disturbed areas, and even casuarina and oak copses in highland valleys. **Male:** Black. Unique for his *blue-green breast shield shaped like a pair of wings and the long, thick, black nape-plume* (actually a folded cape). **Female, Juv, and Imm:** *Brown with fine black-and-buff barring ventrally; dark head with pale eyebrow in most races.* Like a small version of the female parotias that share its habitat. *Dark bill and iris.* **Subspp (6):** Plumage and genetics indicate 4 groups. (1) *superba* (Bird's Head and Wandammen Pen) Male lacks the black spots on breast shield present in other races, and Female has very dark head and upperparts, with a minimal pale brow trailing the eye; (2) *feminina* (W and Border Ranges) Female paler with speckled crown framed by a broad, pale wreath; (3) *latipennis* (E Ranges, Adelbert Mts, and Huon, east to Wau and Herzog Mts) variable and intermediate between the previous race and the next one; and (4) *minor* (SE Pen) Female has black head, brown back, and lacks an eyebrow or shows only an obscure one. **Similar spp:** Parotia females are larger, bulkier, bigger-headed, but otherwise very similar in plumage. There is considerable geographic variation in both Superb and the parotias, but this variation is parallel such that at most localities the female of one species looks like the other. Magnificent BoP female same size but shows a pearly blue bill and most have a pale eye-stripe (not eyebrow). All 3 of these species can be found together in fruiting trees. **Voice:** Male song over most of range is a harsh series of 5–8 grating screeches, e.g., *shree shree shree shree shree*, becoming more drawn out and emphatic toward the end. Male vocalization on Bird's Head very distinct, shriller: *yia—yia—yia-yia-yia.* Calls include a shrill, plaintive note *teer!* and a sneezy *pew!* **Habits:** Solitary. Forages for fruit and arthropods in midstory and canopy vegetation. Hops or creeps along limbs and trunks peering over the side in search of prey; gleans and excavates insects from bark, moss, epiphytes, and dead leaves. Male sings from the canopy throughout his small territory but displays to Female on a recumbent log on or near the ground. Nest is a rough cup of leaves and rootlets placed in a low palm

or pandanus crown. Eggs (1–2) brownish buff, marked with dark, longitudinal streaks and splotches. **Range:** Endemic. NG: Bird's Head and Neck (Wandammen Pen), Central Ranges, Adelbert Mts, and Huon, 1500–1900 m (occasionally 1000–2300 m).

Black-billed Sicklebill *Drepanornis albertisi* Pl 96
(**Buff-tailed Sicklebill**, *Epimachus albertisi*)
Male 35 cm; Female 33 cm. Mid-mountain forests. Patchily distributed. Uncommon where found and easily overlooked. Probes branches and takes fruit in the canopy. Smaller than the two long-tailed sicklebills, but same size as other co-occurring BoPs, from which distinguished by *distinctive sickle-shaped bill and rounded pale buff or cinnamon tail* (varies locally). Differs from Pale-billed S (lowlands only) in having a *dark bill.* **Male:** Diagnostic pattern from below—*smooth grey breast contrasts with white belly and vent*. Lavender tips to long flank plumes glint as bird moves. **Female, Juv, and Imm:** Abundant fine barring below. **Similar spp:** Females of the long-tailed Black and Brown Ss, which can share this species' habitat, differ with blackish face and throat, rufous crown, and long, narrow, ragged tail. **Voice:** Song is a long, rapidly delivered series of high piping whistled notes that usually rises and increases in volume: *dyu dyu dyu dyu dyu dyu dyu dyu dyu dyu dyu whi whi whi. . . .* Loud, with a throbbing quality at close range. Huon population gives a slower series that could be mistaken for song of a Sclater's Whistler. Display song is a high-intensity series like the preceding that ends in a string of high, hissing notes. Contact note: *whenh?* **Habits:** Solitary. Inspects branches and trunks and probes bill deep into cavities to extract insects. Sometimes climbs vertically. Also takes fruits. Males occupy large territories and are widely spaced. Male displays at a favored site—a sloping perch or thin vertical trunk a few meters up or higher in the midstory. Nests in fork of horizontal tree bough; a flattish cup of wiry stems and roots. Eggs (1–2) dull reddish cream with darker spots and dashes. **Range:** Endemic. NG: Bird's Head and Neck (Fakfak, Kumawa, and Wandammen Mts), W Ranges (Weyland Mts only), Foja Mts, E Ranges, Huon, and SE Pen, 1100–1900 m (600–2300 m). No records from most of Western and all of Border Ranges.

Pale-billed Sicklebill *Drepanornis bruijnii* Pl 96
(*Epimachus bruijnii*)
Male 35 cm; Female 34 cm. Easily located by song, but very difficult to observe in the canopy of lowland rainforest. Widespread and common within its restricted range. Joins mixed flocks of babblers, pitohuis, and other rufous birds. Unique and weird for its *long, curved, ivory-colored bill and shaved-head look,* with large, bare, lavender-colored eye-patch and brief "Mohawk haircut"— a strip of short, dark feathers atop the crown. Note *rounded, chestnut tail* in flight. Slightly larger, darker, and more compact than Black-billed S of the mountains; bald patches on face more extensive. **Male:** All dusky with black chin and very brief flank plumes showing opalescent tips.
Female, Juv, and Imm: Abundant barring on underparts. **Similar spp:** Male unique. Distinguish Female sicklebill from female Twelve-wire BoP and Magnificent Riflebird by long curved ivory bill (note all 3 are barred below and brown above and are common flock-joiners). **Voice:** Song bizarre and unique, but reminiscent of some vocalizations of Magnificent Riflebird or Lesser BoP; a loud descending series of musical, hoarse, and hollow upslurred and downslurred notes: *kooo'wi . . . koh'wi . . . koh'wi . . . kyu-kyu-kyu* often ending with weird gargling notes; contact notes similar to those of a paradisaea. **Habits:** Singing Males have been found in selectively logged forest and low karst hills as well as lowland alluvial forest. Foraging and diet similar to Black-billed S. Male holds a large territory and displays on a horizontal branch of a midstory tree. Nest and egg unknown.
Range: Endemic. NG: NW Lowlands east to W Sepik (Utai), 0–175 m.

Brown Sicklebill *Epimachus meyeri* Pl 96
Male 96 cm; Female 52 cm. The sicklebill of mossy cloud forest, above the elevations of Black S where the two co-occur. More common than Black S. **Male:** Differs from male Black S by the *dusky brown underparts, pale blue iris, and jackhammer song.* **Female, Juv, and Imm:** From female Black S by the *olive-brown wing edge (same as rest of wing) and pale blue iris*. **Subspp:** None now recognized, but becoming smaller, with Male's flank plumes paler, from E to W. **Similar spp:** Male of Black S distinguished by black breast and red iris; female exhibits bright rufous brown wing edges and dark iris. Astrapias have a short bill and are often seen with female sicklebills; they are overall darker, not as brown, and the tail tip is more blunt, less pointed. **Voice:** Male's song is far-carrying (1 km or more) and sounds like the burst of a jackhammer—percussive and unmusical: *tat-tat tat-tat-tat* or *tat-at-at-at*. Contact call a nasal *nhreh!* Juv gives a subdued honking note *ur ur.*

Habits: Often solitary; two hen-plumaged birds foraging together are presumably a Female and her offspring. Joins mixed flocks, especially with astrapias. Unwary but difficult to observe; usually seen either at a fruiting tree or searching for animal prey in moss on canopy branches. Occasionally feeds on ground. Male maintains huge territory. He displays on a horizontal branch within the canopy, presumably a habitual display site. Male sings at his display site for a long period in the morning and more sporadically while roaming his territory throughout the day. Nest situated in crotch of a small tree or pandan; a cup of mosses and small vines and rootlets. Egg (1?) pale pinkish buff with heavy blotching and streaking. **Range:** Endemic. NG Central Ranges, 1900–2900 m (1500–3200 m).

Black Sicklebill *Epimachus fastosus* Pl 96
(*E. fastuosus*)
Male 110 cm; Female 55 cm. Inhabits a narrow elevational band at the transition from mid-mountain to cloud forest. Often found in steep mountainous country. Rare and local. Seen foraging on trunks and branches in the canopy or midstory or at a fruiting tree. This species and Brown S are *large birds of paradise with a very long, pointed tail and a sickle-shaped bill*. Male's far-carrying voice best means of discovery, but the bird himself is often elusive, being shier than the Female. **Male:** Black with iridescent highlights; *black breast, red iris, and song* distinguish it from Brown S. Flight silent. **Female, Juv, and Imm:** Brown with red-brown crown, dusky face, and finely barred breast and belly. Look for the *rusty wing edges (contrasting with the rest of the upperparts) and the dark brown iris*, to separate from Brown S. **Subspp (3):** *fastosus* (Bird's Head) Male's underparts distinctly brownish (Brown S absent there); *atratus* (Bird's Neck and Central Ranges) Male's underparts blackish; *ultimus* (Foja and N Coastal Mts) like previous, but bill shorter and Female more olive. **Similar spp:** Brown S male has breast obviously brownish and paler than back, and blue iris; Brown S female very similar but for pale blue-brown iris and olive-brown edging to wing feathers. **Voice:** Male sings early in the morning. Song is a pair of powerful, quick, liquid, identical notes: variously *kwink! kwink!* or *bwik bwik!* On Bird's Head, *spoidit! spoidit!* more like Brown S. Call notes are guttural or honking, similar to those of Brown S. **Habits:** Mostly solitary; joins mixed-species flocks. Progresses through the trees in leaps and bounds. Flies slowly in flaps and glides. Probes and excavates mossy canopy branches, epiphytes, and vine clusters. Visits fruiting trees; particularly fond of the orange fruits of pandan vines and trees. Males spaced kilometers apart. Male displays from atop a vertical stub in midstage of forest interior. Nest in midstory tree and composed of orchid stems, vines, and moss.
Egg unknown. **Range:** Endemic. NG: Bird's Head and Neck (Wandammen Mts); W, Border, and E Ranges (east to Kratke Range); Foja Mts; and N Coastal Mts, mainly 1800–2200 m (1300–2600 m).

Long-tailed Paradigalla *Paradigalla carunculata* Pl 97
Male 37 cm; Female 35 cm. The paradigalla of the Arfak Mts (and possibly Tamrau and Fakfak Mts). *A slim, all-black bird with velvety plumage, long graduated tail, and yellow forehead wattles*. Gape wattles blue and red. **Male:** Iridescent scaly crown. **Female:** Like a smaller, duller version of the Male. **Juv:** No information, but see Short-tailed P. **Similar spp:** Short-tailed P, not present in range of Long-tailed P, has a very brief tail. **Voice:** Only known vocalization is a single slightly musical note on a monotone *wheeeee?* of 2 sec duration, repeated every 15 sec or so. **Habits:** This is the least-known bird of paradise. At one locality in the Arfak Mts, it was found to be common in a patch of secondary forest. It seems to be primarily a fruit eater. Mating system unknown. Nest similar to Short-tailed P. Egg, no information. **Range:** Endemic. NG: Bird's Head (Arfak Mts), presumably 1400–2100 m. Should be looked for in the Tamrau Mts and mts of the Wandammen Pen. **Taxonomy:** An unnamed form of paradigalla has been seen in the Fakfak Mts.

Short-tailed Paradigalla *Paradigalla brevicauda* Pl 97
22–23 cm. Inhabits mossy cloud forest where it is easily overlooked. *A compact, all-black bird with longish bill and short tail. Yellow forehead wattles, blue gape-wattles.* **Male:** Crown exhibits greenish, scaly iridescence. **Female:** Like Male but duller, longer-tailed, and smaller. **Juv:** Dull and lacking iridescence on crown. **Similar spp:** Long-tailed P, only in the Bird's Head and Neck to the west of Short-tailed P, has an obviously longer tail. Loria's Satinbird male much smaller and lacks yellow wattles and longish bill. Male Lawes's Parotia larger-headed, smaller-billed, and with head wires. **Voice:** Songs/calls described as a melodious *hoo-ee?* or rising *zheee*. Also a loud, powerful, high-pitched, mournful whistling of 4 clear, ascending notes, the last prolonged and rising in pitch. Calls include a throaty croak and a *churr churr churr*. **Habits:** Usually seen singly in middle and upper stages of mossy forest, often at fruiting trees; also probes moss of canopy branches for prey.

Assumed polygynous, even though the sexes look alike. Display unknown. Female alone attends the nest. Nest placed 5–10 m up in fork of sapling; a substantial cup lacking sticks but having moss, ferns, orchids, and tendrils; inner cup lined with dry leaves. Egg (1) pale pinkish buff, dark-spotted and blotched. **Range:** Endemic. NG: W, Border, and E Ranges (E extent of range unknown; no records from Kratke Mts), 1600–2400 m (1400–2650 m).

Arfak Astrapia *Astrapia nigra*   Pl 95
Male 76 cm; Female 50 cm. Bird's Head, where it is the only astrapia. Uncommon. A reclusive, blackish, long-tailed fruit eater of the cloud forest. **Male:** Entirely *blackish with green and bronze highlights; tail extremely long, broad, and round-tipped*; Male's wings rustle in flight. Differs from other male astrapias by *the lobed, black nape crest and the bronzy-bordered gorget extending up along the sides of the neck to the eye*. **Female, Juv, and Imm:** Blackish, long-tailed, with the faintest pale barring on flanks and belly. Most similar to female Huon A, but *pale barring on underparts virtually absent*, fainter than any other astrapia. **Similar spp:** Black Sicklebill has similar look, but has a sharply pointed tail and long, curved bill (may be difficult to see at first). Long-tailed Paradigalla has yellow forehead wattle and the tail isn't so long. **Voice:** Mostly silent. Call note a brief *clu-uk*. **Habits:** Little known. Forages for fruit; also takes prey by probing moss and epiphytes in the canopy of montane forest. Display similar to Huon A. Nest and egg unknown. **Range:** Endemic. NG: Bird's Head (Arfak Mts; sightings from Tamrau Mts), 1700–2450 m.

Huon Astrapia *Astrapia rothschildi*   Pl 95
(Rothschild's Astrapia)
Male 69 cm; Female 47 cm. The astrapia of Huon mountains. **Male:** All black with a *very long, broad, blunt-tipped, black tail*; note green belly and bronze breast band. Lacks bronze neck-stripe of Arfak A and has instead of nape crests a purplish nape shield. Wings make distinctive flight sound. **Female, Juv, and Imm:** *Blackish, long-tailed, with fine whitish bars on breast and belly, more pronounced than the very similar Arfak A*. **Similar spp:** Huon Parotia also long-tailed, but tail broader, shorter, and wedge-shaped. Spangled Honeyeater is large and black with a substantial but shorter tail and shows yellow facial skin and white spotting on breast. **Voice:** Mainly quiet. Call notes include a *kak kak kak kak* and also a muffled scolding *jj jj* or *jiw jiw*, a typical astrapia *wenh?* note, and a low growl. **Habits:** Similar to other astrapias. Male displays in trees on ridge crest in forest or at edge of clearing (male clustering in leks not yet observed). Male gives 2 displays. In one, Male perches erect with tail pushed forward under perch and slightly spread. In the other, Male hangs upside down from canopy branch with tail spread and raised, and in this position may sway and step side to side. Male also hops back and forth between perches. Nest a shallow cup of vines and rootlets in a foundation of leaves and moss. Egg (1) pinkish buff marked with blotches and streaks. **Range:** Endemic. NG: Huon Pen, 1500–3500 m.

Splendid Astrapia *Astrapia splendidissima*   Pl 95
Male 39 cm; Female 37 cm. The astrapia of the Western and Border Ranges. Inhabits cloud forest to timberline. Common. **Male:** Dark-bodied and compact, with unique *paddle-shaped tail, short for an astrapia. Predominantly greenish and highly iridescent, with diagnostic white basal half of tail*. Wings hiss somewhat in flight. **Female, Juv, and Imm:** Blackish, barred black-and-brown below, and with diagnostic *white in base of tail*. Tail evenly graduated. **Subspp (2):** Minor. **Similar spp:** Female Ribbon-tailed A (borders Splendid's range to E) has narrower tail usually with at least a trace of white, and the central tail feathers are by far the longest; the head has more of a greenish sheen. **Voice:** Various froglike or insect-like call notes: *to-ki*, or *jeet*, or *tch tch tch*, or *gree*. Not very vocal. **Habits:** Solitary. Diet mainly fruit. Gleans for prey in moss and epiphytes. Breeding display unknown, but lekking aggregations suspected. In one case, 4 Males seen at the edge of a natural meadow perched in high open branches ~40 m from each other; they occasionally called when Females were in the vicinity. Nest and egg, no information. **Range:** Endemic. NG: Western and Border Ranges, east to the Victor Emanuel, Hindenburg Mts, and Mt Stolle, and presumably not crossing the Strickland Gap, 1800–3450 m.

Ribbon-tailed Astrapia *Astrapia mayeri*   Pl 95
(Ribbon-tailed Bird of Paradise)
Male 125 cm; Female 53 cm. The astrapia of the western sector of the E Ranges. Common. A quiet and confiding fruit eater inhabiting the cloud forest upward to timberline. Where it co-occurs with Stephanie's A, the Ribbon-tailed inhabits the higher elevations. Hybridizes with Stephanie's. **Male:** Small dark body, *pompom atop bill*, and *stupendous pair of narrow, white, ribbonlike central*

tail feathers. Wings rustle in flight, like those of Stephanie's A. **Female, Juv, and Imm:** Blackish, with brown-and-black ventral barring and distinctive *greenish sheen on head and small tuft over the base of the bill. Central tail feathers pointed and much longer than the rest*. The central tail feathers may show some white on the shaft and as blotches near the base, and may even resemble those of the Male, but not nearly as long. **Similar spp:** Female Stephanie's has broader and more rounded tail feathers, and the tail is evenly graduated; crown has less green sheen, and there is no tuft over the base of the bill. Ribbon-tailed × Stephanie's hybrids show mixed features of both species. **Voice:** Varied, loud, froglike notes: *waugh* or whistled *whit-whit* or *wreden wheep!* Or minor contact notes: *koit?, kurrgh!, wurr?,* and *wurit?* with a questioning tone. Display vocalization of Male is also varied: *grrrow grr grr* or *hisss-sss-ssh* or a loud sharp *keaooo-ooo-ooo*. **Habits:** Solitary. Forages in trees and shrubs at any height, but mainly beneath the upper canopy. Eats fruit. Attracted to fruiting schefflera plants. Hunts for prey on canopy branches and trunks by methodically searching and picking at bark and moss. Displays at traditional sites in the treetops. Here a few Males gather and jump back and forth between branches, showing off their streaming tails. Male also jumps side to side on perch, making parted tail plumes sway back and forth. A flap-and-glide display flight undulates the tail in a sine-wave. Nests usually in an isolated sapling inside the forest at midstory level; a cup of large leaves and pieces of pandan fronds covered by wiry orchid stems. Egg (1) buffish pink or beige, marked with dark brown blotches and streaks. **Range:** Endemic. NG: E Ranges, from Mt Hagen and Mt Giluwe west to the Kaijende Highlands (Porgera) and Muller Range, 2500–3400 m (rarely down to 1800 m). Hybrids are common between the Ribbon-tailed and Stephanie's As on Mt Hagen and at the Tari Gap, below 2700 m.

Stephanie's Astrapia *Astrapia stephaniae*   Pl 95
(Princess Stephanie's Bird of Paradise)
Male 84 cm; Female 53 cm. The astrapia of NG's eastern cordillera, where common in cloud forest up to timberline. Hybridizes with Ribbon-tailed A. **Male:** Black, with iridescent green head and *spectacular all-black streamer tail*. Tail uniquely shaped: the extremely long, 2 central feathers are very narrow at the base and flare toward the tips. The tail is surprisingly stiff and held in a graceful drooping curve. Wings make a loud rustling in flight. **Female, Juv, and Imm:** Dull black with black-and-brown ventral barring. *Head all black (essentially lacking green sheen), no tuft over base of bill, and the long, black tail is evenly graduated, the feathers blunt-tipped*. **Subspp (2):** Minor. **Similar spp:** See Black Sicklebill and Brown S, with which Stephanie's often consorts. Female of Ribbon-tailed A has a distinctive greenish sheen on crown, small tuft over base of the bill, white flecking in narrower tail, pointed tips to the tail feathers, and the central pair of tail feathers is much longer than the rest of the tail. **Voice:** Mainly quiet, but calls varied. Male gives a high, shrill series: *kri kri kri kri* from high display perch. Other Male calls include *nge nge nge* and *ss ss ssw ssw*. Also a loud whistled *weert-wheet* similar to Ribbon-tailed A. Calls are catlike or froglike notes such as *whenh?* and a melidectes-like *hoo-hee-hoo-hee* . . . given by several birds at once. **Habits:** This species and Ribbon-tailed are the most studied astrapias. Solitary, but may join with sicklebills to form small foraging flocks. Eats a variety of small fruits, and often seen at schefflera plants. Forages for arthropods, and presumably small frogs and skinks, by methodically searching tree limbs and trunks, picking at or excavating the bark and epiphytes and dropping debris in the process. Males (2–5) display communally in canopy branches on ridge crest or flank, or at forest clearing. Displaying Male repeatedly leaps back and forth between 2 perches, his tail flowing behind. In low-intensity display, he stops to swing his tail forward beneath the perch in an inverted V. Nest placed 10 m up in forked branch of sapling or in climbing bamboo; nest is a cup of leaves and roots covered on the outside with orchid and fern stems. Egg (1) pale pinkish buff with dark streaks and spots. **Range:** Endemic. NG: E Ranges (from Tari eastward) and SE Pen, 1500–2800 m (1300–3500 m).

King Bird of Paradise *Cicinnurus regius*   Pl 98
Male 16 cm; Female 19 cm. The *smallest* mainland bird of paradise. Inhabits forest, edge, and second growth in lowlands and foothills. Common but inconspicuous. **Male:** *Red and white*; tail wires tipped with disk-shaped, green plumes; legs cobalt blue. **Female, Juv, and Imm:** Dull brown, darker and plain dorsally, and paler and profusely barred ventrally. *Lacks facial markings; head with a long, tapered profile and a pale yellow bill*. Bill darker in younger birds. **Subspp (2):** Minor. **Similar spp:** Female Magnificent BoP similar but slightly larger, has blue-grey bill and eye-stripe. **Voice:** Song of Male is a slow, swelling series of nasal, paradisaea-like notes: *rahn rahn rahn rahn* . . . or a rapid *ki kyer kyer kyer kyer kyer.* . . . Also twinned call notes *kyeer-kyer*. **Habits:** Solitary, but joins mixed flocks of larger black or brown insectivores. Forages at all levels. Diet a mix of fruit and insects. Male sings and displays from a subcanopy tangle of shaded vines in forest interior. Males may be stationed ~75 m apart in duos and

counter-sing. Only known nest was in a tree-hole 2 m up; the hole was filled with palm fibers. Eggs (2) creamy pink with brown streaks. **Range:** Endemic. NG (minus Trans-Fly), NW Is (Misool and Salawati; record for Batanta probably erroneous), Aru Is, and Bay Is (Yapen), 0–300 m (rarely to 850 m).

Magnificent Bird of Paradise *Diphyllodes magnificus* Pl 98
(*Cicinnurus magnificus*)
Both sexes 19 cm. In hilly country from the foot of the ranges up into lower mid-mountains. Common but seldom seen except at fruiting trees or at the display court. Best located by voice. The only small mainland bird of paradise with a *bright blue-grey bill and legs*. Also note the *blue-grey eye-stripe* in all plumages. **Male:** Weird-looking in life, colors generally muted in the forest interior; look for *reflective tail wires and contrasting dark ventral and light dorsal pattern*. **Female, Juv, and Imm:** Dull brown above, finely barred below. **Subspp (3):** Minor variation. **Similar spp:** Female King BoP quite similar but for ivory bill and no eye-stripe. **Voice:** Male calls mostly from display court. Typical song is a rapid, rolling series of ~8 loud, downslurred *churr* notes, growing louder and more insistent. Also a single note *kyerng!* Near court, Male gives a cussing *ksss—ks ks ks—ksss*. **Habits:** Solitary, but joins mixed insectivore flocks. Diet mostly fruit, plus some insects and spiders. Look for this bird at fruiting trees, which it shares with other species of BoPs. Male is territorial only at his display-site; neighboring male courts are a few hundred meters apart. Male display site is in a tree-fall thicket within forest interior, often on a steep slope. Court, ~1–2 m across, is cleared of leaves; standing in the court is a vertical sapling on which the Male displays. Nest a cup of roots, fibers, and moss placed in a low pandan or shrub. Eggs (1–2) pale greyish, boldly marked with streaks and blotches. **Range:** Endemic. All NG mts, Salawati (NW Is), and Yapen (Bay Is), 0–1450 (1600) m.

Wilson's Bird of Paradise *Diphyllodes respublica* Pl 98
(*Cicinnurus respublica*)
Both sexes 16 cm. The small bird of paradise of Waigeo and Batanta Is. Common in hill forest, less so in lowlands and mountains. Located by the Male's song. Both sexes instantly identified by the *bald, cobalt-blue crown* crossed with black lines. **Male:** Green below, red and yellow above; note curled tail wires. **Female, Juv, and Imm:** Dull brown above, finely barred below. **Similar spp:** Magnificent BoP absent from Wilson's range. **Voice:** Similar to that of Magnificent BoP. A rolling trill: *chu chu chu chu chu chu;* a sharp explosive *ketch*; a soft, high-pitched *teel*. **Habits:** Habits similar to Magnificent BoP. Nest and egg unknown. **Range:** Endemic. NW Is (Waigeo and Batanta), 0–1200 m.

Blue Bird of Paradise *Paradisornis rudolphi* Pl 99
(*Paradisaea rudolphi*)
Both sexes 30 cm. Mid-mountain oak forests and regrowth of the eastern Central Ranges. Uncommon, and best located at Male's morning calling perch or by observation at a favored fruiting tree. *An unmistakable BoP, dark with distinctive broken white eye-ring, prominent white bill, and blue wings* in all plumages. **Male:** Blue breast and blue-and-amber flank plumes; long black tail-streamers. **Female, Juv, and Imm:** Brown breast, variably barred. **Subspp (2):** Minor. **Voice:** Song is a slow, cadenced, and plaintive series of notes, more nasal and higher pitched than those of a typical paradisaea advertisement series: *wahr wahr wahr wahr wahr wahr wahr*. When moving around from song post to other sites, the Male gives low growling notes, a rolling: *brrn brrn. . . .* During inverted display a low, rhythmic, machine-like grating or humming noise is produced by the Male. **Habits:** Solitary and intolerant, driving other BoP from fruiting trees. Diet is fruit and arthropods extracted from bark and epiphytes on branches in usual BoP fashion. Male is perhaps territorial. He sings from a high, open perch and displays to the Female near the ground in a concealed site. Nest placed in the fork of a low branch in forest or scrub; a flat cup that varies in materials used (strips of pandanus leaves, or tendrils and vines, or orchid stems). Eggs (1–2) pale pink-buff with darker streaks. **Range:** Endemic. NG: E Ranges and SE Pen, from the Tari and Kompiam areas southeastward to the central Owen Stanley Range (to about the Kokoda region), 1400–1800 m (1100–2000 m). Apparently patchily distributed and absent in some areas that appear suitable for the species.

Emperor Bird of Paradise *Paradisaea guilielmi* Pl 99
(Emperor of Germany Bird of Paradise)
Male 33 cm; Female 31 cm. Mid-mountain forests of the Huon Pen, mainly above the elevational range of the similar-looking Raggiana and Lesser BoPs. Common. **Male:** *Crown and extensive breast-bib dark green; the mostly white flank plumes are short and sparse*; also a white patch on lower belly.

Female, Juv, and Imm: *Dark face includes entire crown, and yellow nape extends far onto back. Iris dark, unlike Raggiana or Lesser.* **Similar spp:** Male Lesser BoP with yellow-and-white plumes and brown breast; female Lesser with white breast and belly. Female Raggiana with more limited yellow on the back of the head and neck and with distinctive chocolate breast band. **Voice:** Some songs sound similar to those of Raggiana and Lesser but others very distinct: *kar kar kar kar kar kar* (series descending and melancholy); or *whok whok whok* . . . or an excited group call by 2 or more Males given with speed, cadence, and energy: *pio-pio-pio-pio-pio-p-p-p-p*; or *weer weer weer weer weer* (high and plaintive); also a lazy *kah—kah.* **Habits:** Similar to Raggiana and Lesser BoPs but perhaps even more gregarious. Nest is a deep cup with a foundation of leaves and made of tendrils and vines. Eggs (1–2) whitish to pale pinkish marked with dark streaks. **Range:** Endemic. NG: Huon Pen, 700–1400 m (350–1500 m). Replaced at lower elevations by the Raggiana in E Huon, and by the Lesser in W Huon, and hybridizes with both.

Red Bird of Paradise *Paradisaea rubra* Pl 99

Male 33 cm; Female 30 cm. The red-plumed bird of paradise of Waigeo and Batanta Is. Widespread and common; inhabits canopy of lowland and hill forest. Bill yellow, unusual for a paradisaea. **Male:** *Curved, red flank plumes and a pair of prominent, long, curling, black tail-streamers.* Green face with rounded, cushiony crest above each eye. **Female, Juv, and Imm:** *Chocolate-faced and with broad breast band of dull yellow* that connects to the yellow nape. **Similar spp:** None on these islands. **Voice:** Song a loud, clear *wak wak wak* . . . ; a throaty and guttural *work—wok, wok, wak, wak, wak* . . . ; or a loud and clear *work—wau wau wau wau*; a high-piched *ca-ca-ca-ca-ca-ca-ca*. Call a low throaty rolling growl: *grrrrr.* **Habits:** Similar to other paradisaeas. Nest unknown. Eggs (1–2) pale pinkish buff, presumably with streaks and spotting of darker color. **Range:** Endemic. NW Is of Batanta and Waigeo, also nearby Gam and Saonek, 0–600 m.

Goldie's Bird of Paradise *Paradisaea decora* Pl 99

Male 33 cm; Female 29 cm. The plumed BoP representative on Fergusson and Normanby Is, where uncommon and patchy. Inhabits hill forest and regrowth; rarely in the lowlands. **Male:** *Red flank plumes, yellow crown and mantle, and unique grey breast.* **Female, Juv, and Imm:** *Dull yellow crown, chocolate throat, and finely barred breast,* unusual for a paradisaea. **Similar spp:** None in these islands. **Voice:** *Wok wok wok* . . . or *wark wark wark* . . . or *whick-whick;* or a Male duetting of *waak* notes; an explosive and sharp *whit whit whit.* . . . **Habits:** Similar to Raggiana BoP, but forages from the canopy to the ground. Nest and egg unknown. **Range:** Endemic. SE Is: Fergusson and Normanby of D'Entrecasteaux group, 0–600 m.

Lesser Bird of Paradise *Paradisaea minor* Pl 100

Male 32 cm. The plumed BoP of the North and West. Locally common in lowland and hill forest, edge, and regrowth, including near human settlement. Noisy; heard persistently through forest in its range. **Male:** Abundant *yellow-and-white flank plumes* and *yellow crown, nape, and mantle* merging into the brown back. From Greater BoP by smaller size, flank plumes typically more white than yellow, yellow mantle and wing-coverts, and absence of blackish breast cushion. **Female, Juv, and Imm:** Unique *snowy white breast and belly.* **Subspp (2):** Minor. **Similar spp:** Male Greater BoP exhibits a sharply separated brown mantle and black cushion on breast. Raggiana BoP female shows maroon-buff underparts, without white. **Voice:** Songs of displaying Males variable, given separately or in a chorus: *whik wang—wau wau wau wau;* or *wak wak wak* . . . ; or *waik waik waik.* . . . Notes are more liquid, with greater range between high and low notes, than those of Raggiana BoP. **Habits:** Singly or in small parties. Often in mixed-species flocks with other BoP species and large insectivores. Forages in midstory and lower canopy. Diet includes fruit and animal prey, gleaned from canopy branches. Males display in canopy lek of 1–7 plumed individuals, plus many unplumed Males. Nest a cup of twigs, vine stems, and leaves set in fork of tree branch. Eggs (1–2) creamy pink, spotted and streaked. **Range:** Endemic. Northern NG (Bird's Head to Huon), Misool (NW Is), and Yapen (Bay Is), 0–1000 m (locally to 1600 m). Meets Greater on the SE Bird's Neck somewhere between Lakahia/Etna Bay and Mimika R; no hybrids reported. Meets Raggiana in the upper Ramu R and on N flank of Huon Pen, where the two hybridize. Also hybridizes with Emperor BoP.

Greater Bird of Paradise *Paradisaea apoda*   Pl 100

Male 43 cm; Female 35 cm. The plumed BoP of SW NG and Aru Is. **Male:** Abundant *yellow-and-white flank plumes, prominent blackish breast cushion, and brown back sharply contrasting with yellow nape.* **Female, Juv, and Imm:** *Entirely rich brown*, darker on head and breast and paler on belly. Note bright yellow eye and pale blue bill. **Similar spp:** Lesser BoP male exhibits a yellow mantle and wing-coverts, lacking in Greater. Raggiana BoP female shows yellow cast on the nape offsetting the dark face, and paler breast and belly. **Voice:** Advertisement song *wank wank wank wank wok wok wok* and variations of this, much like Raggiana. **Habits:** Habits much like those of Lesser and Raggiana. Nest unknown. Egg (1) pale pinkish buff marked with streaks and splotches. **Range:** Endemic. NG and Aru Is, 0–950 m. NG: Western S Lowlands between Mimika and Fly Rs. Meets Raggiana at the Fly R (Kiunga and Bensbach) where hybrids are known. May meet the Lesser in the West, but range overlap and hybrids are unknown.

Raggiana Bird of Paradise *Paradisaea raggiana*   Pl 100
(Count Raggi's Bird of Paradise)

Male 34 cm; Female 33 cm. E NG. A characteristic species of the lowland and hill forests, extending farther into the mountains than Greater and Lesser BoPs. Relatively unwary; can be found in forest, edge, gardens, and regrowth. This bird is featured on the national emblem of PNG. **Male:** *Abundant red or orange flank plumes and narrow yellow forecollar.* **Female, Juv, and Imm:** Exhibit diagnostic *yellowish crown and nape framing the dark face, dark brown breast, and pale maroon-brown belly.* **Subspp (4):** *salvadorii* (eastern S Lowlands, E Ranges, and southern SE Pen, east to Cloudy Bay) brown mantle and brick-red flank plumes; *raggiana* (far SE Pen, east of Cloudy and Goodenough Bays) Male with yellow mantle and orange-red flank plumes; *intermedia* (N coast of SE Pen, Mambare R to Collingwood Bay) like previous but much yellow dorsally; *augustaevictoriae* (Ramu R east to Mambare R in northern SE Pen) yellow from nape to rump, and pale orange plumes. **Similar spp:** Greater BoP female dark-headed and without the distinct dark breast band. Lesser BoP female white ventrally and dark-crowned. Emperor BoP female with dark brown underparts, dark crown, and dark iris. **Voice:** Song is a loud *wau wau wau wau wau!* and variations of this; or a dozen or more loud caw notes that rise in pitch and decelerate; display call is a high-pitched and rapid keening series that rises in pitch and intensity: *ki ki ki ki ki ki ki.* **Habits:** Similar to Lesser and Greater BoPs. Nest a bulky cup of fibers, rootlets, vine tendrils, and dry leaves set in the fork of a branch in a small tree. Eggs (1–2) pale pinkish buff with streaky markings. **Range:** Endemic. NG: eastern S Lowlands (west to Ok Tedi R and Trans-Fly), E Ranges, Huon, and SE Pen, 0–1400 m. Rare in the flat country and most common between 500 and 1000 m. Meets both Lesser and Greater BoP and hybridizes with them; see those accounts.

AUSTRALASIAN ROBINS: Petroicidae

Though superficially similar in many respects to the robins and flycatchers of Eurasia and Africa, the Australasian robins are not closely related to these birds and stand as a well-defined and separate family. The 46 species live in Australia, New Guinea, and nearby islands. The New Guinea Region is home to 25 species, most of which are confined to New Guinea itself. Robins appear to be poor colonizers of islands—only 6 species occur on the satellite islands, and many satellite islands have no robins at all. Robins occupy all habitats, from the mangroves reaching into the sea, through all types of forest to the highest peaks, far above timberline.

This is a diverse group of insect eaters that includes terrestrial ground-robins and scrub-robins and flycatchers that live high in the forest canopy, and one that exclusively frequents streamsides. However, most dwell in the forest interior close to the ground, where they search for prey while perched almost motionless, clinging to the side of a vertical stem or trunk and darting down to snatch up their food. They forage independently and do not flock, although some will occasionally join mixed feeding flocks. Each pair or family group maintains a territory from which it excludes others of its own kind and other similar robin species. Perhaps because of this territoriality, each species tends to occupy a characteristic elevational band to which it is perhaps better suited than neighboring robin species. As a result, one encounters a succession of different "thicket-robin" species when proceeding up a mountain.

Most robin species have conspicuous musical whistled songs. They build cup nests on a stem or forked branch. The flycatchers build an especially small, cup-shaped nest adhered to a branch; not only does the nest blend in perfectly with the bark, but so does the mottled chick, which adopts a head-up pose like a miniature frogmouth. The juvenile plumage of all robins is quickly lost, and because this plumage is rarely seen (indeed it is undescribed for some species), we do not present it for most species. The juvenile plumage is usually brown, often with light tips to the wing feathers.

While some robin species are easy to find and observe, those living close to the ground can be shy and retiring and are best located by listening for their song and then approaching with care or using playback recordings.

Greater Ground-Robin *Amalocichla sclateriana* Pl 65
20 cm. *Largest robin, robust and long-billed.* Inhabits subalpine forest floor. Usually rare, local, and inconspicuous, but nevertheless vocal. Note long stilt legs and short, droopy tail. **Adult:** Olive brown above with dingy white throat, *grey breast and flanks* with *vague, dark scalloping on throat and breast*. White loral spot obscure. **Juv:** Rufous with dark mottling and scaling. **Subspp (2):** *occidentalis* (W Ranges) browner and underparts more mottled; *sclateriana* (SE Pen) more olive. Huon population undetermined. **Similar spp:** Lesser GR at lower elevation is half as big, with lighter proportions and brownish breast and flanks. Island Thrush juv, of more open habitats, is darker below and with medium tail. **Voice:** Song is a brief, somber tune of mellow, humanlike whistles; 2–5 notes, typically 3 notes in SE. Readily imitated. Call a sharp scolding. **Habits:** Skulks in the drenched, mossy, branch-tangled understory where feeds on invertebrates, such as beetles. Solitary or paired, wary and missed if not for song. Little known. No nesting information. **Range:** Endemic. NG, disjunct: W Ranges, Huon, and SE Pen (Mt Kumbak near Wau east to Mt Suckling), 2500–3900 m. Seemingly absent in Border and E Ranges, but perhaps overlooked.

Lesser Ground-Robin *Amalocichla incerta* Pl 65
14–15 cm. Similar to the preceding species but smaller and more dainty, with smaller, thinner bill; widespread and common. Also lives on montane forest floor but at lower elevation. **Adult:** *Diagnostic combination of brownish breast and flanks and conspicuous white forehead spot and throat.* **Juv:** Buffy plumage with dark scalloping. **Subspp (3):** Regional color shifts in need of better documentation. **Similar spp:** Greater GR twice as big and at higher elevations; scaly grey breast and flanks. Mountain Mouse-Warbler half as small, shorter legs, and lacks white loral spot. **Voice:** Song is a beautiful tune of usually 6–7 clear, piping notes of higher pitch than Greater GR. The notes jump around in pitch and usually end on a low note. Mountain Mouse-Warbler song lower pitched and coarser (slurred or burry); also, Mountain MW tends to repeat phrases in its song and also to repeat song type before switching to a fresh type. Could also be confused with Papuan Logrunner. Call notes include a brief, buzzy *engk* and a high-pitched *eeez* followed by a harsh chatter. **Habits:** Similar to Greater GR. Runs or hops, pausing after each few steps to scan for invertebrate prey. Nest on ground or in tree hole 1 m up; nest an open cup of moss and other materials. Egg (1) whitish with spotting. **Range:** Endemic. NG, all mts within its elevational range (no records from Fakfak, Kumawa, and Cyclops Mts), 1750–2500 m (1400–2700 m extremes).

Green-backed Robin *Pachycephalopsis hattamensis* Pl 103
15 cm. A common mid-mountain robin of forest understory in W NG, but shy and easily overlooked. **Adult:** A unique *white-eyed* robin exhibiting whistler-like plumage pattern with *grey head, white throat, and green body.* **Juv:** No information. **Subspp (4):** Minor; subtle color shifts. **Similar spp:** Banded Yellow R lacks white throat and has prominent dark breast bar. White-faced R is obviously smaller, dark-eyed, and with a clear yellow breast. **Voice:** Song similar to Black-capped R: a series of clear, piercing, short whistled notes, 1/sec, either regularly or with occasional brief pauses. Calls, similar to White-eyed R, include a piercing upslurred disyllabic whistle and a rising series of ~5 buzzy slurs. **Habits:** Habits similar to better-known White-eyed R. Where the 2 species co-occur, Green-backed replaces White-eyed at higher elevations. One nest in liana 0.5 m up; no other nesting information. **Range:** Endemic. NG and Bay Is (Yapen), 700–1600 m. Bird's Head east in South to E Ranges, in North to Foja Mts.

White-eyed Robin *Pachycephalopsis poliosoma* Pl 102
15–16 cm. The common grey robin of mid-montane forest floor and understory; vocal, but furtive and difficult to observe. **Adult:** Generally *grey with white iris and throat.* **Juv:** Similar, but tinged brown. **Subspp (4):** Subtle color differences, including lack of buff in undertail (SE Pen, Huon), amount of white in face, amount of brown and darkness in wings, tail, or head. **Similar spp:** No other grey robin has white eye, but see dark-eyed Smoky and Blue-grey Rs, which abut White-eye's elevational range. **Voice:** Song similar to Green-backed R, slower than Black-capped R. Call is a single, loud, ringing, whistled note, rising in inflection: *periwee?* or *kerwee?*; also 4–7 buzzy notes, each upslurred and about a half tone higher than the preceding note. **Habits:** Generally above the elevational ranges of Black-sided R and Banded Yellow R and below those of Black-capped R and Blue-grey R. Often heard, rarely seen. Perches motionless on a horizontal stem; descends to ground to pounce on prey or launches off with rustling wings. Also, hops on the ground or log, stopping to listen with head cocked. Nest built in a sapling, ~1 m up; nest a cup of rootlets and green mosses. Egg (1) buff white and covered with dark blotches. **Range:** Endemic. NG: Central Ranges, N Coastal Mts, Adelbert Mts, and Huon, 1000–1700 m (400–2150 m).

Garnet Robin *Eugerygone rubra* Pl 103
10–11 cm. A petite, fantail-like robin, inconspicuous and uncommon in montane cloud forest. Behavior first clue to identity: perches horizontally, nervously jerks about and slowly flashes wings, showing *white blaze across flight-feathers and white in tail.* Note narrow white eye-ring. **Male:** *Maroon-red upperparts,* with grey face and underparts. **Female:** Same pattern, but green replaces red. **Juv:** Reddish brown, paler below. **Similar spp:** Red Male unique. Gerygones, Fantailed Berrypecker female, and fantails lack white in wing. **Voice:** Song a rising and falling tinkling series of ~10 very high notes, incessantly repeated, often with a segment of trilled notes. Call is a delicate, upslurred, mewing note: *uwee, uweeo* or *deeyu*. **Habits:** An inconspicuous resident of forest interior, foraging from understory to subcanopy. Usually alone or with mixed feeding flock. Gleans insects from foliage, often by sallying. No nesting information. **Range:** Endemic. NG: Bird's Head and Neck (Kumawa Mts), Central Ranges, Foja Mts, and Huon, 1700–2500 m (1400–3700 m).

Subalpine Robin *Petroica bivitatta* Pl 102
(**Mountain** or **Alpine Robin**)
11–12 cm. A small black-and-white robin of the highest mountains; rare and local in subalpine forest and shrubland. *Black with white breast and forehead spot. Narrow white wing bar.* **Male:** Black. **Female:** Sooty black. **Juv:** No information. **Subspp (2):** *caudata* (W Ranges) little white in tail; *bivittata* (E Ranges, SE Pen) outer tail feathers extensively white; Huon population undetermined. **Similar spp:** White-winged R is all black below. **Voice:** Generally quiet. Song a loud, deliberate *dad dad dad. . . .* **Habits:** Keeps to the canopy of stunted alpine forest. Solitary, paired, or small parties. Active and acrobatic—perches upright on horizontal branch and sallies out to fly-catch and hover-glean. Slowly progresses through the forest from one foraging perch to the next. No nest information. Egg (clutch?) grey-brown with a ring of spots. **Range:** Endemic. NG, where sketchily known through Central Ranges and mts of Huon Pen, 2700–3500 m.

Snow Mountain Robin *Petroica archboldi* Pl 102
(**Snow Robin**)
14 cm. Restricted to a few peaks in the W Ranges, where locally common; inhabits rocky alpine scree at the highest elevation of any NG bird. Conspicuous and noisy, seen perched on rocks. **Adult:** *Sooty black with red breast patch.* **Juv:** Lacks red breast and white forehead. **Similar spp:** None. **Voice:** Song unknown. Calls are loud chips and slurred notes suggestive of a Willie Wagtail and harsh scolding notes. **Habits:** The only avian denizen of rock faces and boulder slopes. In small parties of 3–6. Stolid, perches on rock waiting and looking about; flies down to ground or rock to capture arthropod prey and returns to same perch or flies off to new one. No nesting information. **Range:** Endemic. NG: W Ranges (Mts Carstensz/Jaya and Wilhelmina/Trikora), 3800–4200 m. Might be expected on summits of Mts Goliath/Yamin and Juliana/Mandala.

Flycatchers or Flyrobins. The 6 New Guinea species sort out spatially by habitat. Three species live in forests and segregate by elevation: Olive F in the lowlands, Yellow-legged F in the hills, and Papuan F in the mountains. Lemon-bellied F occurs patchily in open habitats and savannah, and Jacky Winter is confined to the Port Moresby savannahs. The distinctive Torrent F lives only along stony creeks and rivers.

Yellow-legged Flycatcher *Kempiella griseoceps* Pl 103
(**Yellow-legged Flyrobin**, **Yellow-footed Flycatcher**, *Microeca griseoceps*)

12–13 cm. Local, uncommon, and inconspicuous in mid-mountain forest where it frequents subcanopy and midstory; also Trans-Fly. **Adult:** A small, cryptic robin with unique combination of *white throat, pale yellow breast, and yellow legs*. Head grey, but this color difficult to see. Lower bill is yellow. **Juv:** Brown with pale spotting. **Similar spp:** No other flycatcher has white throat and grey head. White-faced Robin shares habitat but is bright yellow below and has white face. Grey Whistler somewhat similar but larger and has dark legs; gleans casually and does not sally. **Voice:** Song is a fairly loud succession of 4–8 sweet notes, some trilled, similar to song of Black-fronted White-eye. Call a high-pitched lisping: *chh—chchchchch*—the second series descending in pitch; repeatedly delivered. **Habits:** Mainly in mid-mountain forest, but also occupies lowland forest in Trans-Fly. May venture into forest openings. Similar to Papuan F, but keeps to the canopy mostly. Noticed only when making short, dashing flights to a new perch or to capture prey. An occasional cooperative breeder (AU data). Nest on midstory branch; a small, neat cup of rootlets decorated with bark and lichen (AU data). Eggs (2) pale blue, finely speckled. **Range:** NG and Aru Is. Most NG mts (no records from Kumawa, Wandammen, and Foja Mts, and Huon), 600–1600 m; also, locally in E Trans-Fly, 0–100 m. Also Cape York, AU.

Olive Flycatcher *Kempiella flavovirescens* Pl 103
(**Olive Flyrobin**, *Microeca flavovirescens*)

13–14 cm. Common in midstory of lowland and hill forest. **Adult:** A small, yellowish-olive robin, *plain but for its yellow lower bill and legs*. Has the most obvious eye-ring of any flycatcher. **Juv:** Brown above with pale spotting; note white eye-ring and yellowish legs. **Similar spp:** Yellow-legged F, also with yellow lower bill and legs, has white throat and grey head. Lemon-bellied F has black lower bill and legs, occupies open habitat. **Voice:** Song is a quiet, monotonous, rapid whistling, the disyllabic notes 6/sec: *uliuliuliuliuliuli....* Calls include a chatter and a hinge note. **Habits:** Occupies forest midstory; may venture into forest openings. Joins mixed feeding flocks. Flips or shivers tail. Forages in typical flycatcher fashion. Nest built on horizontal branch in midstory; a small neat cup decorated with moss. Eggs (2) undescribed. **Range:** Endemic. NG, NW Is (Batanta, Misool, Waigeo), Bay Is (Yapen), and Aru Is, 0–1000 m (rarely to 1500 m).

Papuan Flycatcher *Devioeca papuana* Pl 103
(**Canary Flycatcher** or **Flyrobin**, **Yellow Flycatcher**, *Microeca papuana*)

12–13 cm. Common in interior of high mountain cloud forest. **Adult:** A small robin with *rich yellow underparts and bright orange legs*. **Juv:** Brown and green above, heavily pale-spotted, yellow legs. **Similar spp:** None in its habitat, but be aware of 2 small, yellow-legged robins of mid-mountain forest: Yellow-legged F has white throat and is paler yellow below; White-faced Robin has white face. **Voice:** Song is a high-pitched, formless, brief tinkling warble within a narrow range, descending in pitch and reminiscent of that of the White-shouldered Fairywren. Call is a weak, sibilant *tseet*. Also a repeated dry scold note. **Habits:** Singly or in pairs, tame. Perches quietly mainly in midstory and above, sallying for flying insects. Periodically flutters wings and tail while perched. Nest on midstory branch; a small neat cup. Egg, no information. **Range:** Endemic. NG, all mts high enough (no records Kumawa and Cyclops Mts; absent from Adelbert Mts), 1700–3500 m.

Jacky Winter *Microeca fascinans* Pl 101
(*Microeca leucophaea*)

14 cm. Scarce and local resident of Port Moresby savannah. Apparently no recent records. **Adult:** A small, upright flycatcher, nondescript dingy brown above, whitish below, with diagnostic *white outer tail feathers*, especially evident in flight. **Juv:** Like Adult, but with pale spotting above and dark spotting below. **Subspp (1):** *zimmeri* (endemic). **Similar spp:** Other related flycatchers are yellow or olive below—only Yellow-bellied F shares its habitat. Pied Bushchat female has a dark breast and white rump. White-winged Triller female has white in wing and obscure barring on sides of the breast.

Voice: A cheerful, repeated, medium-high note or pair of notes. (Performs song-flights in AU.)
Habits: Locally distributed in small colonies. Singly, in pairs, or in family parties. Flycatches from stump, fence, or low tree, often returning to the same perch and foraging close to ground. Takes small insects midair or from the ground. Switches tail from side to side. Nest built on a horizontal branch fork; nest a small cup camouflaged with bark and lichen (AU data). Eggs (2) bluish, dark-spotted.
Range: NG: Port Moresby savannah region of SE Pen, lowlands. Also AU.

Lemon-bellied Flycatcher *Microeca flavigaster* Pl 103
(Lemon-bellied Flyrobin)
12–13 cm. A patchily distributed but common savannah species that also occupies towns and large clearings in the forest, from lowlands up into deforested mountain valleys. **Adult:** Differs from other *yellow-breasted* flycatchers by its *all-black legs and bill* (may be paler at base of lower bill). Throat whitish; note yellowish eyebrow and no eye-ring. **Juv:** Brown and pale spotted above; white with brown markings below. **Subspp (3):** Poorly defined but varying in brightness of yellow. **Similar spp:** Other yellow flycatchers have yellow/orange legs and live mainly in forest interior. Grey Whistler larger, forages by foliage gleaning. **Voice:** Song energetic medley of repeated elements (slurs or brief single notes), each element successively repeated several times; abrupt pauses and restarts within each song. Sings perched, but also performs a high, circling song-flight, larklike. **Habits:** In savannah trees, mangroves, gardens, towns, and road clearings through the forest. Singly, in pairs, or family parties; the cooperative breeding behavior observed at AU nests has not yet been reported from NG. Perches conspicuously in open canopy, singing repetitively and sallying. Nest built on branch fork in tree crown; nest a tiny, neat cup decorated with bark and nearly covered by sitting bird. Eggs (1–2) bluish with dark speckles and blotches (AU data). **Range:** NG and Aru Is, 0–1500 m. Also AU.

Torrent Flycatcher *Monachella muelleriana* Pl 101
(Torrent Flyrobin)
14–15 cm. Locally common. Forages over rocky streams and rivers from foothills up to the transition with cloud forest, both in primary forest and disturbed habitat. **Adult:** *A whitish robin with bold black cap, wings, and tail.* **Juv:** Shares this pattern but is missing the white supraloral spot, upperparts are faintly pale-spotted, and underparts are brownish mottled. **Subspecies (1):** *muelleriana* (endemic). **Similar spp:** Torrentlark may share habitat but has black back and throat and is larger, noisier, and more nervous. **Voice:** Calls frequently. Song a clear, high-pitched whistled series of up to 8 peeping notes. Call is a single such note. **Habits:** Lives in small groups that occupy territories along fast-flowing, rocky streams. Perches on rock, gravel bar, or overhanging branch, and sallies out to take insects, usually returning to the same perch. Bobs or wags tail. Nest on a thin branch; nest a wide cup of thin rootlets and lichens adhered to its substrate with mud. Eggs (2) undescribed. **Range:** NG, Bird's Head to Central Ranges, also Foja, N Coastal, and Huon Mts, foothills to 1800 m. Also New Britain I.

Papuan Scrub-Robin *Drymodes beccarii* Pl 63
(formerly included in **Northern Scrub Robin**, *D. superciliaris*)
20 cm. A lanky, thrushlike, terrestrial robin that inhabits forest interior mainly in hills and lower mountains, also locally in S Lowlands. **Adult:** *Black vertical bar through eye,* white throat, *white wing bars,* long tail, and *long pink legs.* **Juv:** Crown feathers edged black, throat and breast flecked brown. **Subspp (3, in 2 groups):** *beccarii* (Bird's Head and Neck and N NG east to Sepik-Ramu) upperparts blackish brown; *brevirostris* (S Lowlands, Aru Is, SE Pen) upperparts cinnamon. **Similar spp:** None, but compare with Russet-tailed Thrush. **Voice:** Song unique, a slow series of ~5 high-pitched, minor key, drawn-out whistles (1/sec), each note dropping slightly, then maintaining pitch in the final segment. Other songs described as having fewer, but similar drawn-out whistled notes. Can be confused with quail-thrush or jewel-babbler song. Scold is a quiet *skusssh.* **Habits:** Difficult to observe. Quietly forages alone on ground, usually under cover. Takes invertebrates. Hops a few times, then pauses to look around. Slowly raises and lowers tail. Occasionally raises tail up to a 45-degree angle, then flashes its wings, with head erect. Nest (AU data) built in depression next to a tree or log; nest a stick cup lined with dead leaves and fibers. Eggs (2) pale grey, thickly spotted and blotched. **Range:** Endemic. NG and Aru Is, 0–1400 m. NG: throughout, although no record from Fakfak and Kumawa Mts (Bird's Neck) and Huon. **Taxonomy:** Split from Northern Scrub-Robin (*D. superciliaris*) of Australia on DNA differences of subsp *brevirostris*. Subsp group *beccarii* of equally distinct appearance has not been sampled for DNA.

Ashy Robin *Heteromyias albispecularis* Pl 102
(**Grey-headed Robin**, *Poecilodryas albispecularis*)
15–16 cm. The Bird's Head/Neck representative of the Black-capped Robin. **Adult:** An overall *greyish and much duller* version of Black-capped, lacking the black cap and white eyebrow. No other robin within its range shows the *white cheek patch connecting with the eye*. **Voice:** Rapid bursts of 8–12 identical *pee* notes at a steady rate. **Habits:** Presumably similar to Black-capped R. **Range:** Endemic. NG: Bird's Head (Arfak and Tamrau Mts) and Neck (Kumawa Mts), 1600–2400 m. **Taxonomy:** The unnamed *Heteromyias* robin of the Foja Mts resembles this species, but its relationships are unknown at this time.

Black-capped Robin *Heteromyias armiti* Pl 102
(formerly **Grey-headed** or **Ashy Robin**, *Heteromyias/Poecilodryas albispecularis*)
15–18 cm. Frequents forest floor and understory at the transition from mid-mountain forest to cloud forest; locally common but usually rare. Heard, but difficult to see. A medium-large robin with *long, pale legs*. **Adult:** *Prominent white throat patch* that connects to the eye. Distinctive *black cap and white eyebrow*. *White window in wing* noticeable in flight. Most of upperparts olive, underparts tinted grey and ochre. **Male:** Distinctive white bill tip. **Juv:** Rufous with hint of Adult pattern. **Subspp (2):** *rothschildi* (W Ranges to E Ranges) olive-brown back; *armiti* (SE Pen, Huon, Adelbert Mts) greenish-olive back. **Similar spp:** None in its habitat. Black-chinned R (lowlands), also with white eyebrow, is black and white. **Voice:** Song is a rapid, monotonous series (~15 sec) of musical, medium to high-pitched whistled notes, 4–5/sec delivered at a jerky, uneven rate. Another song is an ascending series of paired notes, the second note dropping in pitch. Call is a soft, pleasant whistle, *eeya—eeyuwee*. Alarm call of several harsh notes. **Habits:** Singly or in pairs; very shy and elusive. Clings motionless to a thin, vertical stem waiting for prey or hops about on ground. Eats small invertebrates. Nest built 1–3 m up in a sapling near the stem; nest a cup composed of rootlets, tendrils, twigs, leaves, and moss. Egg (1) buff and heavily marked. **Range:** Endemic. NG Central Ranges, Adelbert Mts, and Huon, 1600–2400 m. An undescribed form resembling Ashy R inhabits the Foja Mts. **Taxonomy:** Formerly treated as conspecific with Ashy R and Grey-headed Robin (*H. cinereifrons*) of AU. DNA divergence for subspp *rothschildi* and *armiti* suggests the two may be separate species.

Black-chinned Robin *Heteromyias brachyurus* Pl 101
(*Poecilodryas brachyura*)
14–15 cm. An uncommon robin of forest interior in northern lowlands and hills. Note *pale legs*, compared with dark legs of the more common Black-sided R. **Adult:** Striking black and white with *thick white eyebrow and black chin*. **Juv:** Like Adult but heavily smudged with brown. **Subspp (2):** *brachyurus* (Bird's Head and Neck, far W Ranges) back dark grey; *dumasi* (NW Lowlands and Sepik-Ramu; Yapen I) back black. **Similar spp:** Black-sided R co-occurs (mainly in forest edge and second growth) but the white eyebrow does not extend much beyond the eye, and the chin is white, not black; also has a black mark on side of breast and dark legs. **Voice:** Song a rapidly descending series of 9 whistled notes; plaintive and resembles song of Fan-tailed Cuckoo. **Habits:** Occupies forest midstory. Singly or in pairs. Forages by perching motionless, sometimes on vertical perches, and sallying out to snatch insects. No nesting information. **Range:** Endemic. Northern NG and Yapen (Bay Is), 0–650 m. NG: Bird's Head and Neck (Wandammen), NW Lowlands, and Sepik-Ramu.

Black-throated Robin *Plesiodryas albonotata* Pl 101
(*Poecilodryas albonotata*)
18–19 cm. Uncommon in subcanopy and midstory in transition from mid-mountain to cloud forest. **Adult:** A large, distinctive robin of upright posture and *black face interrupted by a white horizontal streak*. **Juv:** Pale cinnamon; said to show white neck streak. **Subspp (3):** *albonotata* (Bird's Head) and *correcta* (SE Pen, Huon) underparts mostly white; *griseiventris* (W, Border, and E Ranges) underparts mostly grey. **Similar spp:** Not to be confused with a cuckooshrike. **Voice:** Song unique—an extremely high-pitched but quite loud, insect-like, thin upslur, repeated 2–4 times; like an electronic tone and easily overlooked. Also a ringing note typical of other robins. **Habits:** Seen singly. Perches upright and motionless, may slowly raise and lower tail; sallies out to pluck insects from foliage or midair. No nesting information. **Range:** Endemic. NG mts (Bird's Head, Central Ranges, Huon), 1600–2300 m (rarely to 2700 m).

The *Poecilodryas* and *Peneothello* robins. These 6 species are similar in size and in their perch-and-pounce foraging technique, and all live in the understory and lower midstory and segregate by habitat. A 5-species altitudinal sequence is formed by Black-sided Robin in the lowlands, then White-rumped R in hill forest, Smoky and Blue-grey Rs in mid-mountain forest, and White-winged Robin in cloud forest. Smoky R is a western species, Blue-grey R is widespread, with a geographic range of overlap within which Smoky lives at elevations above Blue-grey. Lastly, Mangrove R is named for its main habitat.

Black-sided Robin *Poecilodryas hypoleuca* Pl 101
13–15 cm. The common robin of lowland and hill forest. **Adult:** A black-and-white robin with *distinctive facial pattern showing a dark eye and lores bracketed by white*; there is also a black mark on the side of the breast. **Juv:** Brown, with white belly and undertail-coverts. **Subspp (3):** Minor. **Similar spp:** see Black-chinned R. **Voice:** Song a descending series of 4 clear whistled notes, with a pause after the third note. Call a loud, disyllabic *whi-chew*, the first note staccato and higher pitched, the second more explosive. **Habits:** Seen singly, shy. No nesting information. **Range:** Endemic. NG and NW Is (Batanta, Misool, Salawati, Waigeo), 0–300 m, but to as much as 1200 m in absence of other understory robins. Most of NG mainland, except E half of SE Pen and Trans-Fly.

White-rumped Robin *Peneothello bimaculata* Pl 101, 104
13–14 cm. Uncommon and local in hill forest understory. **Adult:** *Black with white on rump, undertail-coverts, and bend of wing*. **Juv:** Sooty brown with reduced amounts of white. **Subspp (2):** *bimaculata* (species range except next) belly white; *vicaria* (Adelbert Mts, Huon, and N slopes of SE Pen) Male's belly black, Female's with some white. **Similar spp:** White-winged R has dark rump, lives at higher elevations. Pied Bushchat larger and occupies grassland. **Voice:** Song regionally variable. A pleasant, ringing medium-pitched series of 6 whistled notes, lasting 2 sec, the first note at higher pitch; quality suggestive of mouse-warbler song. Also a series of 9 or more rapidly delivered rolling notes, becoming slightly higher pitched and speeding up toward the end. **Habits:** Elevational range above Black-sided R and below White-eyed R. Habits similar to other thicket-robins. No nesting information. **Range:** Endemic. NG and Bay Is (Yapen), 700–1100 m (300–1700 m). NG: Bird's Head, Central Ranges (no records from most of N slope), Foja Mts, Adelbert Mts, and Huon.

White-winged Robin *Peneothello sigillata* Pl 102
14–15 cm. Common in cloud forest and at timberline of high mountains. **Adult:** A compact, velvety *black robin with a white patch on the inner flight-feathers*. **Juv:** Plumage similar to Adult but heavily overlaid with brown spotting and streaking. **Subspp (3, in 2 groups):** *sigillata* (species range except next); *quadrimaculata* (W Ranges) white patch on side of breast. **Similar spp:** None in its habitat. White-rumped R (hill forest) and Pied Bushchat (grassland and scrub) have white rump and undertail-coverts. **Voice:** Song a beautiful piping trill, sometimes rising, sometimes gradually falling in pitch. Calls include a sharp note *peek!* and thin metallic ones. **Habits:** Singly or in pairs; tame and easily observed. Perches quietly on mossy branches in midstory or understory, pouncing on moving arthropods. Nest built on the fork of a forest sapling, 1–3 m up; nest a bulky cup of rootlets and moss. Egg (1) light olive with sparse, dark markings and darker cap. **Range:** Endemic. NG Central Ranges and Huon, 2400–3900 m.

Mangrove Robin *Peneothello pulverulenta* Pl 101
(*Eopsaltria* or *Poecilodryas pulverulenta*)
14–15 cm. Common but patchily distributed along NG coast and some rivers in mangroves and thickets. **Adult:** *Grey, with white throat, white in tail*, and all-dark wings. **Juv:** Brown above, whitish below, streaked or flecked throughout. **Subspp (2):** *pulverulenta* (mainland NG) head and neck darker than back; *leucura* (Aru Is and AU) head and neck greyer, not contrasting with back, breast greyer. **Similar spp:** Broad-billed and Leaden Flycatchers lack white throat. **Voice:** Song a mournful, 3-note whistle, all notes on same pitch—easily imitated to draw the bird in. Also a whistled monotone lasting ~1 sec, usually followed by a similar note at lower pitch. Call a sharp whistle. Also a churring and chattering. **Habits:** Usually in mangrove interior, but also melaleuca swamp forest near Port Moresby and in reed beds and mangrove-like shrub-thickets along Sepik R. Singly or in pairs. Perches in understory and midstory, often on a vertical stem, dropping down on the ground for prey. Nest built on midstory branch; a compact cup of bark strips and fiber. Eggs (2) green with dark spots. **Range:** NG and Aru Is, sea level. NG: more prevalent on S coast, though absent from Trans-Fly; local on N coast; inland on Sepik and Fly R and Lake Daviumbu in S Lowlands. Also AU.

Smoky Robin *Peneothello cryptoleuca* Pl 102
14–15 cm. A common robin of mid-mountain and cloud forest in the West. **Adult:** Upperparts *smoky grey*, underparts vary geographically. **Juv:** No information. **Subspp (3):** *cryptoleuca* (Bird's Head and Foja Mts) underparts entirely grey; *maxima* (Bird's Neck: Kumawa Mts) underparts entirely white; *albidior* (W Ranges) underparts grey except belly greyish white. **Similar spp:** Blue-grey R has distinct bluish wash, lacks pale belly, and has a different voice. **Voice:** Song is an ascending, decelerating whistled trill, much slower and louder than that of Blue-grey R, ending in a disyllabic flourish. Call is a dry, 1- to 2-note chip. **Habits:** Similar to the more widespread Blue-grey R, but Smoky occurs at higher elevations than Blue-grey where the ranges of the two overlap. Nest 2 m up in a forest sapling; a mossy cup lined inside with ferns and plant fiber. Egg (1) pale brown wreathed with reddish-brown splotching. **Range:** Endemic. NG: Bird's Head and Neck (Kumawa), far W Ranges, and Foja Mts, 1400–2200 m.

Blue-grey Robin *Peneothello cyanus* Pl 102
(Slaty Robin)
14–15 cm. The common understory robin at the transition between mid-mountain and cloud forests. **Adult:** *Uniform bluish grey.* **Juv:** Grey with brown shaft streaking and spotting. **Subspp (2):** Minor. **Similar spp:** Smoky R. **Voice:** Distinctive and varied series of loud, musical songs. One is a loud, cheery *teeder teeder teeder. . . .* Another is a series of softly whistled, mellow, trilled notes, running together and ending with an excited upslurred note: *dudududududududududu-WHI!* A similar series gradually rises in pitch, and lacks the final explosive note. Call is a staccato, musical *whik* or *chwink*. **Habits:** Elevational range generally above that of White-eyed R and below White-winged R, or Smoky R in W NG. Singly or in pairs. Active but shy; moves through thickets and under and middle stories of forest. Rarely observed in the open. Will approach an observer, watch from an open perch for a few seconds, then flee with whirring wings. Cocks tail when perched. Forages in a manner similar to other thicket-robins. Nest as high as 6 m in a sapling, built of rootlets and moss and covered with fern leaves. Egg (1) light olive with sparse markings. **Range:** Endemic. NG: nearly all mts (no record from Kumawa Mts), mostly 1500–2400 m.

Banded Yellow Robin *Gennaeodryas placens* Pl 103
(Olive-yellow Robin, *Poecilodryas* or *Eopsaltria placens***)**
14–15 cm. Locally common but rarely encountered because of its patchy distribution and restriction to a narrow elevational band in hill forest. Frequents understory. **Adult:** Unusual color pattern with *brilliant yellow half-collar framed by an olive breast band.* **Juv:** Similar but tinged brown. **Similar spp:** Difficult to confuse other yellow-breasted robins or whistlers. **Voice:** Song a beautiful, slow, flutelike series of 5 notes with the first 1 or 2 notes detached. **Habits:** Behavior similar to other understory robins that feed mainly from the ground. Its elevational range is sandwiched between Black-sided R in lowlands and other hill forest robins above (e.g., White-faced and White-eyed Rs); may or may not co-occur with White-rumped R. Nest built 1–2 m up on a branch of a forest sapling; nest a cup of roots, etc. with moss on outside. Egg, no information. **Range:** Endemic. NG and NW Is (Batanta), 500–1100 m (0–1450 m). Known from scattered localities on Bird's Neck, far W Ranges, S slope of E Ranges, E Sepik-Ramu and Adelbert Mts, and SE Pen.

White-faced Robin *Tregellasia leucops* Pl 103
11–13 cm. A common and characteristic bird of the understory and midstory of mid-mountain forests, also a population in S Lowlands. Inconspicuous; scolds but rarely sings. **Adult:** *A small, white-faced, green-and-yellow robin.* Note orange-yellow legs. Bill mostly orange in E Ranges and S Lowlands. **Juv:** Extensively smudged with brown; only a hint of Adult facial pattern.
Subspp (9, in 4 groups) vary mainly in facial pattern and bill color: *leucops* (Bird's Head to S slope of W Ranges) white only in front of eye, divided by black forehead line; *melanogenys* (mts of NW Lowlands to Huon) entire face white, bill mostly black; *wahgiensis* (E Ranges and western SE Pen, locally S Lowlands) white face, bill mostly orange; *albifacies* (eastern SE Pen) white face, black forehead stripe and bill mostly black. **Similar spp:** Green-backed R also has white face, but is larger and has a green breast, brown wings, and white eye. Flycatchers lack white in face and perch upright. Fairy Gerygone is the only other bird with white in the face, but is slimmer and an active foliage gleaner. **Voice:** Song, rarely heard, is a muted, musical, whistled 5-note phrase. Calls include a plaintive, descending scolding: *ch-ch-ch . . .* ; a soft, thin *see*, repeated 4–5 times; a squeaking sound like a wet cork rubbing on glass, usually given twice; and a quiet *tsip* repeated indefinitely. **Habits:** Singly or in pairs. Nervous and shy, but readily observed; often seen perched on side of tree trunk, scolding.

Nest in sapling fork, 3–4 m up; a large, neat cup of grasslike leaves and fiber. Eggs (1–2) white with dark spots and blotches. **Range:** NG (all mts) and Yapen (Bay Is), 600–1600 m; at 0–100 m in S Lowlands (mouth of Fly R). Also AU.

LARKS: Alaudidae

The larks (98 spp) occur nearly worldwide, but most species inhabit Africa and Eurasia. Only 1 Asian species reaches New Guinea and Australia. Larks are small to medium-sized, cryptic, ground-dwelling birds of open grasslands. They walk or run over the ground, searching for insects and seeds. They are best known for their lengthy, breathless songs given on the wing during special song-flights.

Horsfield's Bushlark *Mirafra javanica* Pl 104
(Singing Bushlark, Australasian Lark)
13 cm. Patchily distributed and locally common. Seen along roadsides and airfields, more generally in openings of short grass amid taller grassland habitats. Lowlands to mid-mountains. **All plumages:** A small, chunky, brownish or blackish-brown ground bird with a *unique, hesitant, mothlike flight*. Note the *conical, stubby bill*; much streaking; *rufous patches on wings*; and *white outer tail feathers*. **Subspp (1):** *timoriensis* (also Wallacea: Timor I). **Similar spp:** Pipits are slimmer, have longer legs and thin bills, and lack rufous on flight-feathers. House Sparrow female is larger, lacks dark streaking on the head, and shows a small, white shoulder patch. **Voice:** Sings in the air for long periods, hovering on fluttering wings; also sings from atop a bush or other perch. Song is a typical larklike, high-pitched, accelerating trilled song, including whistled upslurs. Call: *pk* note. **Habits:** Solitary, in small parties, or seasonally in large congregations on S coast. Picks up seeds and insects from the ground. Perches readily on small shrubs or rocks, often crouches or runs for cover rather than flushing. Nest is a small cup of grass stems in a depression on the ground in grassland. Eggs (2–4) white with dark spots and blotches (AU data). **Range:** NG: Sepik-Ramu to northern SE Pen, mid-mountain valleys of E Ranges and western SE Pen, Port Moresby area, and Trans-Fly, 0–1700 m. Also SE Asia to AU. **Taxonomy:** Singing Bushlark (*M. cantillans*) of Africa and India has been split from *M. javanica*.

BULBULS: Pycnonotidae

The 151 species of bulbuls are a family of Asian and African songbirds. Not native to the New Guinea Region, 1 species has been recently introduced. Bulbuls are slender songbirds of forest and open country. Many species are crested. Generalist feeders, they capture invertebrate prey by gleaning and flycatching, and many species also take fruit and nectar. They are monogamous breeders and build a cup nest among the foliage of a tree or shrub. Often abundant in human-dominated environments.

Sooty-headed Bulbul *Pycnonotus aurigaster* Pl 107
15–20 cm. Introduced to Biak I and Jayapura area. First reported in 1995. Seems to be established, yet true status unknown, and needs to be confirmed with photos or specimens. A conspicuous, garrulous, easily recognized bird of towns and countryside. *Grey with black head and tail; pale rump; tail with white tip*. **Adult:** Unique *black face, short crest, and white ear-coverts*. **Juv:** Face and cap brownish. **Subspp:** Presumably *aurigaster* (Java, Bali); birds sighted on Biak exhibit the diagnostic yellow undertail. **Voice:** Song is a varied, high-pitched, halting warble, *kwi-kwi-kwikwikwi-kwikwi-kwikwi* . . . etc. Single- and double-noted calls. **Habits:** Singly, in pairs, or in small groups. Feeds in trees and also descends to ground. Nests 1–3 m up in shrub or small tree; nest a compact, flimsy cup. Eggs (2–4) (nest and egg, Asian data). **Range:** Established on Biak (Bay Is); sighting from Lake Sentani near Jayapura (NW Lowlands). Originally S China, Indochina, and Java (where possibly introduced).

SWALLOWS AND MARTINS: Hirundinidae

The swallows and martins are a large and distinctive family of songbirds (88 spp) adapted to life on the wing. The family is most diverse in Asia and Africa and is placed with the insectivorous passerid bird families, such as warblers and tits, centered on those continents. In a biogeographic sense, swallows and martins are recent arrivals to New Guinea. Only one species is known to breed in our region (Pacific Swallow), whereas 6 species are overwintering migrants—three species from northern Asia and 3 from Australia. The names swallow and martin are used in our region for genera either with tail-streamers or without, respectively, but the names do not indicate relationship among the genera. Swallows and martins resemble swifts but are not related to them. They share the same trim, aerodynamic shape, long pointed wings, a broad head on a short neck, and small bill, legs, and feet. Unlike swifts, they can perch on branches and wires, and in flight the wings bend more at the wrist. Their diet is insects caught in flight. The New Guinea species avoid forests and are generally found flying over open habitats such as grasslands, gardens, and towns, mainly in the lowlands. All species are drawn to wetlands wherever aerial prey is abundant. Swallows and martins are sociable birds, congregating in flocks, sometimes of more than one species. This is useful for finding the rare species, which tend to associate with the more common ones. They are moderately vocal, giving chirping and twittering calls.

Sand Martin *Riparia riparia* Pl 58
(Bank Swallow)
12 cm. Migratory vagrant from Asia. Favors open country, especially over or near bodies of freshwater or along the seacoast. **Adult:** A small *brown* swallow with *clean white underparts divided by a dark breast band*. **Juv:** Similar, but upperparts pale-scalloped and throat buff-tinged. **Subspp (1):** presumably *ijimae* (breeds in E Asia; winters in SE Asia). **Similar spp:** The brown swiftlet species have grey-brown underparts and lack a breast band. **Voice:** Call note a brief *tschr* or *schrrp*. **Habits:** To be looked for during austral summer. **Range:** 2 records from NG: E Ranges and Huon, 0–1800 m. Also Eurasia, Africa, N and S Amer.

Barn Swallow *Hirundo rustica* Pl 58
(Common Swallow)
15–18 cm. A rare to locally common Asian migrant. Favors wetlands, mainly in lowlands; also visits highlands. **Adult:** Similar to Pacific Swallow but note *long tail wires, clean white underparts, and darker, more reddish face* with dark rufous throat separated from white breast by an *obvious black throat band* (rarely missing). The contrast between the dark throat band and white breast is sharp, unlike the Pacific S's gradual transition from brown throat to grey breast. **Male** tail wires longer. Beware that growing tail wires may be short like Pacific S. **Juv:** Tail short, but outer pair of feathers thin; overall duller, with face paler rufous. **Subspp (1):** *gutturalis* (breeds in E Asia; winters in SE Asia to NG and AU). **Similar spp:** Welcome S adult also has long tail wires but lacks black throat band and has dingy grey underparts. These field marks less well defined in juv Welcome S. **Voice:** Calls are chipping notes and twittering. **Habits:** Overwinters during austral summer. Singly or in flocks, often with other swallows. Hawks insects over marshes and other wetlands. **Range:** Widespread in the NG Region, though no records yet from SE Is, 0–2000 m. Also Eurasia, Africa, N and S Amer.

Pacific Swallow *Hirundo tahitica* Pl 58
13 cm. The only resident NG swallow, common in cities and towns. **Adult:** Upperparts glossy black, throat rufous brown blending into grey breast and belly. *Tail short and notched but lacking streamers*, although outer pair of tail feathers pointed; essentially the same in Male and Female. **Juv:** Not glossy; outer tail feathers rounded. **Subspp (2):** *frontalis* (NG range except that of next subsp) darker underparts, tail spots very small; *albescens* (Trans-Fly to SE Pen) paler underparts, tail spots larger. **Similar spp:** See Barn S and Welcome S—both have long tail wires. **Voice:** Song is twittering and chattering. Call *swee*. **Habits:** Has expanded with human settlement, nesting on wooden buildings and other structures and feeding in open areas. Natural habitats include cliffs for nesting, as along the seacoast and offshore islets. Avoids expanses of closed forest. In pairs or small flocks. May associate with other swallows when feeding. Some local movement and disperses readily to colonize new nesting sites. Nests on sea cliffs, building eaves, bridge trestles, etc., selecting a sheltered, darkened

recess; nest a mud cup lined with grass and feathers, plastered on a ledge or corner. Eggs (2–4) white or pale pink, spotted. **Range:** Throughout NG Region, but in SE Is on D'Entrecasteaux Is and Misima I only, 0–2000 m, rarely higher. Also India, SE Asia, Sunda Is, Philippine Is, eastward through Melanesia to W Polynesia.

Welcome Swallow *Hirundo neoxena* Pl 58

13–17 cm. Australian vagrant to NG south coast. **Adult:** *Long tail wires* separate most Welcome S from otherwise very similar Pacific S. **Male** tail wires longer than **Female's**; thus, Female more likely to be confused with Pacific S. White subterminal tail band broader in Welcome than Pacific. Lacks faint dark striations on flanks and belly as Pacific is said to have. **Juv:** Probably not reliably separable from Pacific. **Subspp (1):** presumably *neoxena* (E AU). **Similar spp:** Also see Barn S. **Voice:** Similar to Pacific S. **Habits:** Similar to Pacific. **Range:** A few records from SE Pen (Port Moresby area). Also AU and NZ. **Taxonomy:** Sometimes treated as a subsp of Pacific Swallow.

Red-rumped Swallow *Cecropis daurica* Pl 58
(*Hirundo daurica*)

16–17 cm. An uncommon and irregular Asian migrant. *A fork-tailed swallow with conspicuous orange rump and pale underparts with dark streaking.* **Adult:** Back glossy blue-black, rump orange, long tail wires (slightly longer in Male). **Juv:** Back brownish black without gloss, rump buff, outer tail feathers pointed and protruding but not forming long wires. **Subspp (1):** presumably *japonica* (breeds in E Asia; wintering range uncertain, but includes NG Region and N AU). **Similar spp:** Barn S lacks streaking on white underparts, no rufous rump patch. House and Fairy Martins have short, notched tails without wires, and the rump patch is white. **Voice:** Call *reep*; twittering. **Habits:** Frequents open areas, mainly in the lowlands, less often in mountain valleys. Overwinters during austral summer. Singly or in small groups, often with other swallows. Flight slower and more buoyant than other swallows; wings appear broader. **Range:** Perhaps throughout, but most records are PNG sightings, 0–1500 m. Also S Eurasia and N Africa.

Tree Martin *Petrochelidon nigricans* Pl 58
(*Hirundo nigricans*)

12 cm. A common and widespread AU migrant during austral winter. Possibly uncommon resident breeder. *A small black-and-white swallow with a conspicuous white rump patch.* Forehead with small rufous patch, visible only at close range. Tail notched. **Adult:** Forehead bright rufous, back glossy black. **Juv:** Forehead paler buff, back dull brownish black. **Subspp (2 reported):** *nigricans* (breeds in Tasmania) rufous forehead redder, rump and underparts more saturated with ochre, more heavily streaked below; *neglecta* (uncertain whether present in NG Region; breeds in E AU) rufous forehead less reddish, rump and underparts paler, narrower streaking below. **Similar spp:** Pacific Swallow lacks white rump patch. Fairy M has rufous cap. **Voice:** Calls *drrt drrt* and twitterings. **Habits:** Nonbreeding resident in austral winter. Frequents open habitats, drawn to wetlands, mainly in lowlands and foothills, singly or in groups, often with other swallows. Flocks build during migration. (Nesting should be looked for in NG savannahs and other open habitats.) Nests in small hollow of tall tree. **Range:** NG and Aru Is, also records from Gag (NW Is) and Karkar I, 0–2000 m. Also Wallacea, Melanesia, AU.

Fairy Martin *Petrochelidon ariel* Pl 58
(*Hirundo ariel*)

12 cm. A rare AU migrant to S NG during austral winter. *Small swallow with white rump and rufous cap.* Tail notched. **Adult:** Crown bright rufous, back glossy blue-black. **Juv:** Crown paler with some noticeable black streaking, back dull brownish black. **Similar spp:** Tree M lacks rufous cap, has rufous forehead patch instead. Pacific Swallow has all-dark upperparts, lacks rufous cap and whitish rump. **Voice:** Flight call, *djrrr*, distinct from Tree M. **Habits:** Similar to Tree M but much less common. **Range:** NG: sightings from Trans-Fly, S Lowlands, E Ranges, and SE Pen, 0–1500 m. Also AU.

LEAF-WARBLERS AND ALLIES: Phylloscopidae

Once part of the much larger family of "Old World Warblers," the Sylviidae, the leaf-warblers include 77 species of small, foliage-gleaning insectivores. The family is mainly Eurasian, with 3 species occurring in the New Guinea Region.

Island Leaf-Warbler *Phylloscopus poliocephalus*   Pl 105, 77
(**Mountain Leaf-Warbler**, *P. trivirgatus*)
9–10 cm. Common in mountain forests, generally below the cloud forest zone. Tiny warbler of the forest canopy with a distinctive and persistent song. **All plumages:** Note *yellowish wash on underparts, dark eye-line and crown, pale eyebrow and crown-stripe* (latter absent in Bird's Head race). **Subspp (5, in 3 groups):** *poliocephalus* (Bird's Head and Neck) grey crown lacking median stripe; *giulianettii* (NG mainland west to Bird's Neck; also Karkar I) dark grey crown with pale median stripe; *hamlini* (D'Entrecasteaux Is) blackish crown with pale mid-stripe. Racial identity of Fakfak, Kumawa, Yapen, and Foja populations to be determined. **Similar spp:** Other forest warblers lack the striped facial markings. **Voice:** Song, varied yet distinctive, is a repeated, very high-pitched, rapid warble, often ending with a very high upslur seeming to rise out of human auditory range. Often given by 2 or more birds at once, this song frequently signals the presence of a flock of montane insect eaters. **Habits:** Usually in small flocks of its own kind or with white-eyes, gerygones, and other small insectivores. Forages by gleaning deliberately and unobtrusively in outer foliage of forest canopy, occasionally sallying, taking small insects and spiders. Calls incessantly while foraging. Nest is a mossy globe set in a mossy bank or tree hole. Eggs (2) white. **Range:** NG (all ranges), Yapen (Bay Is), Karkar I, and SE Is (D'Entrecasteaux Is: Fergusson, Goodenough); 1200–1800 m (600–2400 m). Also Moluccas east to Solomon Is. **Taxonomy:** Split from the Mountain Leaf-Warbler (*P. trivirgatus*) of SE Asia. *P. poliocephalus* once included the Numfor and Biak LWs. **Extralimital spp: Arctic Warbler** (*P. borealis*), a Palearctic migrant, winters in Wallacea; also vagrant to islands just north of NG Region (Kaniet or Anchorite Is). Grey-green above, *greyish white below* with white eyebrow and dark eye-line and *thin white wing bar* on greater wing-coverts.

Numfor Leaf-Warbler *Phylloscopus maforensis*   Pl 105
(**Island Leaf Warbler**, *P. poliocephalus*)
9–10 cm. The only leaf-warbler on Numfor I. Inhabits forest. **All plumages:** *A dull grey warbler with greenish wings and tail*; grey head with obscure pattern. **Similar spp:** None within its range. Other leaf-warblers very different. **Voice:** Song high-pitched and varied, similar to preceding species. **Habits:** Presumably similar to other leaf-warblers. **Range:** Endemic. Numfor (Bay Is). **Taxonomy:** Formerly a subsp of Island LW.

Biak Leaf-Warbler *Phylloscopus misoriensis*   Pl 105
(**Island Leaf Warbler**, *P. poliocephalus*)
9–10 cm. The only leaf-warbler on Biak I. Inhabits forest. **All plumages:** Olive green above, yellow below with *orange legs and bill*. Relatively long-legged. **Similar spp:** None on Biak. **Voice:** Song a long, rapid, breezy series; the notes high-pitched and slurred; recognizable as a leaf-warbler song. **Habits:** Presumably similar to other leaf-warblers. **Range:** Endemic. Biak (Bay Is).

WHITE-EYES: Zosteropidae

Long classified in their own family, the white-eyes were recently revealed to be a branch of the much larger Asian babbler assemblage. Pending a revision of the babblers, the white-eyes are still kept as a separate family (~130 spp). White-eyes evolved in Asia and in the short span of 2 million years spread and diversified prolifically over the Old World tropics, from Africa through southern Asia to Australia and many Pacific islands. Their success as invaders is aided by strong dispersal ability, flocking behavior, a generalist diet, and high fecundity enhanced by a well-protected nest placed out on twigs beyond the reach of most predators. Although 9 species have been recorded from the New Guinea Region, only 4 inhabit the mainland, the rest being restricted to various islands. The actual number of white-eye species in the region is open to question. Affinities and species limits are problematic for several species. This could be best resolved with molecular phylogenetic analysis of the entire suite of forms within the region.

In New Guinea, white-eyes constitute a uniform group of small, warbler-like birds of green plumage with underparts green, yellow, grey, or white. Their namesake feature, the white eye-ring, is present in all but 1 species. Sexes are alike. The juvenile may have a narrower eye-ring than the adult but is otherwise similar. As the birds themselves are often hard to see high in the treetops, their high-pitched musical songs are important for discovery and recognition. White-eyes inhabit forest and second growth from sea level up into the lower cloud forest. They travel in vocal flocks, feeding on fruit, insects, and nectar. In the mountains of mainland New Guinea, flocks occasionally include 2 white-eye species. The nest is a thin-walled, tightly woven cup suspended from a horizontal fork. Both parents share nesting duties.

Of the 9 species of the New Guinea Region, 5 (Tagula, Oya Tabu, Biak, Louisiade, and Lemon-bellied White-eyes) are confined to islands, and no island has more than 1 of these species. The other 4 species occupy New Guinea mainland forests: 2 widespread in hill forests but absent in the flat lowlands (Green-fronted WE in the Northwest; Black-fronted everywhere else), and Capped and New Guinea White-eyes in mountain forests. The distributions of Capped and New Guinea WEs are mostly geographically separate, Capped in the West and New Guinea mainly in the East, but present in the Kumawa Mts and Bird's Head. Oddly, the New Guinea WE also locally inhabits the lowlands in both North and South.

Lemon-bellied White-eye *Zosterops chloris* Pl 106
(Yellow-bellied White-eye)

12 cm. On a few islets and small islands around the Aru and NW Is. It is the only white-eye where it occurs; abundant and ever present. **All plumages:** *Bright yellow underparts.* **Subspp (1):** *chloris* (also Moluccas). **Similar spp:** No other white-eye in the region has all-yellow underparts. **Voice:** Song unknown for NG Region; in Asia, a beautiful and varied twittering with rapid change of phrases. Contact call is a downslurred *chee*. **Habits:** A tramp species that specializes in seeking out and colonizing tiny islands. Inhabits strand vegetation, mangroves, coastal forest, villages. Nests in a shrub or tree crown; nest is a pendent cup (Asian data). Eggs (2–3) pale blue to white. **Range:** NW Is (small islands off Misool and Batanta) and Aru Is (Babi, Karang, Enu), sea level. Also Sunda Is and Wallacea.
Extralimital spp: Pale White-eye (*Z. citrinellus*) of Lesser Sundas and islands of the Torres Strait is sometime classified as the same species; similar, but breast and belly ashy white.

Black-fronted White-eye *Zosterops atrifrons* Pl 106
(*Z. minor*)

11 cm. Common in hill and lower montane forests. One of 3 look-alike *white-breasted* mainland species. **All plumages:** *Rich orange-yellow throat sharply delineated from white breast* distinguishes this species from the very similar New Guinea WE in eastern NG; the *black lores and forehead* are diagnostic, especially on Bird's Head where New Guinea WE also has a sharply defined yellow throat (and a greyish throat band, where Black-fronted is all white). Black-fronted usually has brighter white flanks, breast, and belly, whereas those of New Guinea WE are washed dingy greyish olive or buff.
Subspp (3, in 2 groups): *chrysolaemus* (Bird's Head; S slope of W, Border, and E Ranges; Adelbert Mts, Huon, and northern SE Pen) eye-ring of medium width, lores blackish, and forehead variably black or dark green; *delicatulus* (SE Pen S slope and E tip, plus D'Entrecasteaux Is) broad eye-ring, black mask, throat yellow (not orange-yellow of other races), upperparts olive. **Similar spp:** See New Guinea WE.

Green-fronted WE replaces it in the North, lacks a broad eye-ring. Capped WE is all green-yellow below. **Voice:** Song diagnostic, a series of ~8 notes trending downscale and always with the quality of a turning squeaky wheel; varies geographically, that of E birds sounds like *dyudyu dyodyo dee dyo dyu*. Calls include an inquiring, sweet, high upslur *tswee*; a hoarse, slightly downslurred *chee*; or *weedit* repeated by many birds. **Habits:** In vocal and active flocks in forest canopy, regrowth, and gardens. In some locations it flocks with the New Guinea WE. No nesting information. **Range:** NG (excluding range of Green-fronted White-eye in NW Lowlands and Sepik-Ramu) and SE Is (3 main D'Entrecasteaux Is), 400–1450 m. Questionable record for Aru Is. Also Wallacea. **Taxonomy:** Subspecies from NG Region sometimes treated as belonging in *Z. minor*. Tagula WE may be subsp of Black-fronted WE.

Green-fronted White-eye *Zosterops minor* Pl 106
(formerly included in **Black-fronted White-eye**, *Z. atrifrons*)

11 cm. Replaces Black-fronted WE in the North; common in hill forests. **All plumages:** Differs from Black-fronted by *eye-ring absent or much reduced, lores and forehead green, and upperparts with bright greenish yellow*. **Subspp (2):** Minor. **Similar spp:** Black-fronted WE meets Green-fronted WE at the Bird's Neck and somewhere around the Ramu Valley. New Guinea WE co-occurs in the East, Capped WE in the West; both have an obvious white eye-ring. **Voice:** Similar to Black-fronted WE. **Habits:** Similar to Black-fronted WE. **Range:** Endemic. NG and Yapen (Bay Is), 400–1450 m. NG: N slopes of W, Border, and E Ranges (E extent unknown); also Foja, Cyclops, and N Coastal Mts. **Taxonomy:** See Black-fronted White-eye. There is evidence for hybridization in the Weyland and Adelbert Mts.

Tagula White-eye *Zosterops meeki* Pl 106
(**White-throated White-eye**)

11 cm. The only white-eye on Tagula I; sister species to Black-fronted WE. Uncommon in hill forests. **All plumages:** *All-white throat and breast*. **Similar spp:** No other white-eye has snow-white throat. **Voice:** Song and calls similar to those of Black-fronted WE, subsp *delicatulus*. **Habits:** As for Black-fronted. Keeps high in the canopy—a difficult bird to see. No nesting information. **Range:** Endemic. SE Is (Tagula), 0–700 m, most numerous at mid-elevations. **Taxonomy:** A member of Black-fronted WE complex, closest to subsp *delicatulus*.

Biak White-eye *Zosterops mysorensis* Pl 106

11 cm. The only white-eye on Biak I. Uncommon; prefers undisturbed forest, occurs locally in regrowth. **All plumages:** *No white eye-ring*, and without it this species looks nondescript; however, the *combination of green upperparts, whitish underparts, and yellow vent is unique*. **Similar spp:** Gerygones and leaf-warblers on Biak lack the evenly whitish underparts. **Voice:** Song recognizable as a white-eye's; based on a 5-note pattern. Calls, typical white-eye. **Habits:** Travels in flocks together with gerygones and other small insectivores. Mainly insectivorous. No nesting information. **Range:** Endemic. Biak (NW Is), 0–650 m.

Capped White-eye *Zosterops fuscicapilla* Pl 106
(**Western Mountain White-eye**)

11 cm. The uniformly olive-green white-eye of W mountains, where locally common. Inhabits forest, regrowth, and gardens. **All plumages:** *Olive green with black cap*. Gleaming white eye-ring stands out against dark face. Separable from Oya Tabu WE by its mainland range, narrower eye-ring, and brown iris. **Similar spp:** The other mainland white-eyes—Black-fronted and New Guinea WEs—have white breast and belly. **Voice:** Song (?) begins with a bushchat-like, slurred, harmonic phrase *weecher*, then rolls into a rapid-fire, halting series of chips suggestive of a House Sparrow. Flocking call is a repeated *tsyew* note drier in quality than calls of other mainland species. **Habits:** Similar to other white-eyes. Often in large flocks. Nest is a typical, suspended cup of plant fiber and arthropod silk. Eggs, no information. **Range:** Endemic. NG: Bird's Head and Neck (Fakfak and Wandammen Mts), W Ranges to western E Ranges (Tari); also Foja, Cyclops, and N Coastal Mts; 1200–2200 m (down to 750 m in N Coastal Mts). **Taxonomy:** See Oya Tabu WE.

Oya Tabu White-eye *Zosterops crookshanki* Pl 106
(formerly included in **Capped/Western Mountain White-eye**, *Z. fuscicapilla*)

11 cm. The montane, cloud forest white-eye in the D'Entrecasteaux Is, where locally common. **All plumages:** An *olive-green* white-eye differing from Capped WE by its *wider eye-ring and silver-grey*

iris. Crown and face dark green on Goodenough I, blackish on Fergusson I. **Similar spp:** Black-fronted WE, the only other white-eye in its range, has a white breast, entirely different song, and lives below the elevational range of Oya Tabu WE. **Voice:** Song, a beautiful, loud, reedy warble of high notes, some liquid, others harmonic, suggestive of a bushchat. Call is a recognizable white-eye *ti . . . ti . . . ti . . .*, softer than Black-fronted. **Habits:** Replaces Black-fronted WE in forests above 800 m elevation. Not as social as other white-eyes; found singly, in pairs, or in small flocks. Mostly in canopy, but descends lower to forage. Feeds on insects and fruit, including figs. No nesting information. **Range:** Endemic. SE Is (D'Entrecasteaux: Goodenough and Fergusson Is only); 800–1400 m, perhaps both lower and higher. **Taxonomy:** Split from Capped WE on new information from the recently discovered population on Fergusson. Previously known only from 2 Female specimens taken on Goodenough. Its affinities are open to question, whether with Capped WE, New Guinea WE, or species in Bismarck or Solomon Is.

New Guinea White-eye *Zosterops novaeguineae*   Pl 106
(Papuan White-eye)

11 cm. A *white-bellied* mainland white-eye of ordinary appearance that lacks distinctive markings and is difficult to characterize. Noteworthy for its puzzling distribution. **All plumages:** Separated from Black-fronted WE with difficulty. Where the two meet *in the East, New Guinea's paler yellow throat blends into the white breast* (but is sharply divided on Bird's Head and Neck and in the lowlands, like that of Black-fronted). Where they meet in the West on Bird's Head and Neck, Black-fronted has a green forehead as does New Guinea WE, so look for New Guinea's greenish lores (not black) and greyish breast band (breast and belly white in Black-fronted). **Subspp (6, in 4 groups):** *novaeguineae* (mts of Bird's Head and Neck) narrow eye-ring, yellow throat sharply defined from white breast, greyish breast band; *wuroi* (lowlands of Trans-Fly and Aru Is) similar but smaller and eye-ring is wide; *magnirostris* (coast near Ramu R delta at Awar, known only from the type specimen) bill thicker and breast with prominent yellow streak down center; *crissalis* (mts of eastern E Ranges, Huon, and SE Pen) eye-ring narrow in E Ranges and wider elsewhere, yellow forehead (not green) on SE Pen and Huon, grey upper half of eye-ring on Huon, yet throughout the yellow throat grades into white breast. The lowland races are smaller (wing < 55 mm). **Similar spp:** Black-fronted WE. **Voice:** Song on Bird's Head consists of several notes on different pitches followed by sweet trill. In East, song is lengthy and sweet, composed of whistles and upslurs suggesting the song of the Pied Bushchat. Call (in East), a sweet descending *tsyew*, often given by many flock members; also, a short, dry, upslurred trilled note. **Habits:** Habitat in the highlands includes mid-mountain and cloud forest, casuarinas, and gardens; habitat in the lowlands is monsoon or coastal forest. Absent from intervening hill forest inhabited by Black-capped WE, although the two meet in the mid-mountains and flock together there. Habits identical to Black-fronted and Capped WEs. Nest, hung from a forked twig, is a thin-walled cup of fiber lined on the outside with moss. Eggs (2) pale blue. **Range:** Endemic. NG and Aru Is. NG: Bird's Head (above 1000? m) and Neck (Kumawa, 600–1200 m), Trans-Fly and Sepik-Ramu (at Awar), 0–100 m, and eastern E Ranges, Huon, and SE Pen, 1200–2400 m (750–2600 m). **Taxonomy:** Perhaps a species in concept only and actually represents a convenient grouping of obscure but phylogenetically distinct forms.

Louisiade White-eye *Zosterops griseotinctus*   Pl 106

11–13 cm. A small-island species of the SE Is—present and abundant on some, absent from the others. Does not share its islands with any other white-eye species. Found from coastal forests and villages up into the hills. *A large white-eye with a disproportionately large bill.* **All plumages:** Olive green with *yellowish-green breast*. Bill brown with pale mandible. **Subspp (3, in 2 groups):** *griseotinctus* (small islands between SE Pen and Rossel I) pinkish mandible and greyish legs, plumage variably olive to yellow-green; *pallidipes* (Rossel I) orangish mandible and legs, plumage yellow-green. **Similar spp:** Capped and Oya-tabu WEs are also greenish yellow below, but are smaller, with shorter bills; both are montane. Lemon-bellied WE, on small islands far to the west, is truly yellow below and has a shorter bill. **Voice:** An able songster. Song louder, lower pitched, and slower than other white-eyes, is composed of varied slurred, trilled, and harmonic phrases, bushchat-like. Call is a repeated, mellow trill, suggestive of a police whistle, *jeeo*. As heard from a boat offshore, the chorus made by these white-eyes is memorable. **Habits:** Lives in small flocks. Behavior similar to other white-eyes. Nest hangs from a forked twig and is loosely constructed of rootlets, fiber, and fine grasses, covered outside with plant wool, bark, and insect cocoons. Eggs (3) pale blue. **Range:** SE Is (many small islands spread across the subregion, plus larger Rossel I), 0–300 m. Also some small islands of the Admiralty and Bismarck Is. **Taxonomy:** Sometimes included with other forms from the Solomons.

REED-WARBLERS AND ALLIES: Acrocephalidae

This is another family split from the "Sylviidae." It is a fairly uniform group of mostly brownish warblers (60 spp) that inhabit reeds and other rank growth. Reed-warblers are mainly Eurasian and African, with 2 species reported from the New Guinea Region, one a Palearctic migrant, the other a resident. The classification of reed-warbler species has proved very difficult, and it is not known for certain to which species the NG resident population belongs.

Oriental Reed-Warbler *Acrocephalus orientalis* Pl 105
(**Great Reed-Warbler**, *A. arundinaceus*)
18 cm. Rare Palearctic migrant; status uncertain because of confusion with Australian RW. In cane, reeds, sedges, or mangroves bordering wetlands. **All plumages:** Distinguish with care from Australian RW by (1) *song (rarely sings on wintering grounds) is of lower pitch and quieter overall, with quality harsher, more grating, lacking the explosive, cleaner quality of Australian RW song*, which has fewer grating notes; (2) diagnostic *faint dusky streaking on sides of throat and upper breast*; (3) bill very slightly longer and more robust; (4) larger size, but not noticeably so in the field. **Similar spp:** See Australian RW and Gray's Grasshopper-Warbler. Grassbirds are streaked. **Voice:** Song, *crot-crot-crot scratchy-scratchy chet-chet* and so forth. **Habits:** Usually singly, skulking. **Range:** Winters marginally in NG Region, with only a few, widely separated records from the lowlands. Breeds E Asia, winters Indochina to Wallacea. **Taxonomy:** Split from Great Reed-Warbler (*A. arundinaceus*) of Europe and W Asia.

Australian Reed-Warbler *Acrocephalus australis* Pl 105
(**Clamorous Reed-Warbler**, *A. stentoreus*)
16–17 cm. The resident reed-warbler in the NG Region, although some birds may be migrants from AU. Patchily distributed. A retiring but exuberant songster inhabiting tall reeds and other grasses, often near water. **All plumages:** *A plain brown warbler, paler below, with faint pale eyebrow and lores and dark legs.* **Subspp (1):** *sumbae* (Moluccas to Bismarck and Solomon Is). **Similar spp:** See Oriental RW and Gray's Grasshopper-Warbler. Grassbirds are streaked. **Voice:** Song is a series of varied, rapidly repeated phrases or single notes; each element is repeated 3 or more times before switching to the next; elements vary from sweet and liquid to guttural and harsh. Calls include a *chep* note and churring. **Habits:** Solitary. Gleans insects. Nest, built in reeds, is a cup of grasses. Eggs (2–3) whitish with dark spots and blotches. **Range:** NG, 0–2300 m. Also Moluccas to Solomon Is; Australia. **Taxonomy:** Taxonomic status of NG population has been controversial. Currently, on the basis of DNA evidence, the Australian RW (*A. australis*) is considered distinct from Clamorous RW (*A. stentoreus*) of Eurasia, Africa, and Wallacea. The taxon to which the NG population is assigned is *sumbae*. Samples of *sumbae* from the Solomon Is are genetically closest to the Australian RW rather than Clamorous RW.

GRASSBIRDS AND ALLIES: Locustellidae

The grassbirds and allies (58 spp) constitute yet another family split from the former "Sylviidae." Eurasian and African in origin, this family exhibits a secondary center of diversity in Australasia, represented by several genera. Five species occur in the NG Region: 4 species of resident grassbirds and 1 species of grasshopper-warbler, a Palearctic migrant. Members of this family inhabit low, thick vegetation, often near water. They are typically brown in color and often sport a long tail.

Gray's Grasshopper-Warbler *Locustella fasciolata* Pl 105
(**Gray's Warbler**)
17–18 cm. Rare Palearctic migrant, mainly on the Indonesian side. Skulks in dense underbrush and grass thickets. Distinguish from reed-warblers with great care. Gray's GW not so confined to tall grass and reed beds as reed-warblers, so any bird resembling a reed-warbler seen at the forest edge or in shrubby second growth should arouse suspicion. Differs from reed-warblers in shape, with *stocky build and heavy, rounded tail*; note *pale, greyish-pink legs* (grey-black for reed-warblers). Note *indistinct grey breast band*. **Adult:** Thick grey eyebrow, pale greyish below. **Juv:** Yellowish below.

Subspp: No information. **Similar spp:** Reed-warblers. **Voice:** Call, a harsh, incessant *tik!* **Habits:** Solitary. Shy, but inquisitive if call is imitated; perches openly at edge of cover with raised tail switched back and forth. Feeds on small invertebrates. **Range:** NG, NW Is (Gebe, Misool, Salawati, Waigeo), and Bay Is (Biak, Yapen), 0–1800 m. NG: Bird's Head and Neck and far W Ranges; potentially throughout, but only a few records in PNG. Breeds from Siberia to Japan; winters in the Philippines, Wallacea, and W NG.

Tawny Grassbird *Megalurus timoriensis* Pl 105

18–21 cm. Trans-Fly and adjacent S Lowlands. Inhabits grassland and other rank vegetation. **Adult:** *Compared with Papuan GB, Tawny GB is tawnier overall* and shows *streaking on the orange-rufous cap.* It is smaller and has a proportionately shorter tail. **Juv:** From Adult by reduced eyebrow and more diffuse streaking. From juv Papuan by lack of yellowish wash on underparts. **Subspp (1):** *muscalis* (endemic). **Similar spp:** Papuan GB's range abuts Tawny's in the S Lowlands. Fly River GB, also relatively tawny, is smaller, has a shorter and less tapered tail, and lacks streaking on the head. Do not confuse with much smaller cisticolas. **Voice:** Not described for NG population. In AU, sings both from a perch and in a song-flight display. Song is composed of loud, varied notes in rapid succession, beginning with squeaky notes, followed by a reeling series of notes gradually descending in pitch, and finishing with clucking notes. Calls include a harsh *jk* note and scolding *tic-tic-tic . . .* , and others. **Habits:** As for Papuan GB. Nest and eggs also similar (AU data). **Range:** NG: Trans-Fly. Also AU and E Wallacea; may include Philippine grassbird populations. **Taxonomy:** Formerly included the races now split off as Papuan GB.

Papuan Grassbird *Megalurus macrurus* Pl 105
(formerly included in Tawny Grassbird, *M. timoriensis*)

20–23 cm. Widespread and at all elevations, although most common in mid-mountain valleys. This is the *streak-backed warbler with long, ragged tail* in grassland, garden edge, and alpine moors. Tail tapers to fine point, more acute than in other grassbirds. **Adult:** *Rufous cap is unstreaked; breast is white.* **Juv:** Yellow wash below; cap and breast diffusely streaked. **Subspp (5?, in 2 groups):** *macrurus* (widespread lowland and mid-mountain race) not streaked below; *alpinus* (Central Ranges above 2500 m) streaked flanks. **Similar spp:** See Tawny GB. Fly River and Little GBs have restricted ranges, are smaller, and have much shorter tails that are less pointed. **Voice:** Sings from a perch or in song-flight display. Song begins with a few clucking notes then changes to more rapid, higher-pitched staccato notes. Characteristic call is a loud, downslurred *cheyup!*, repeated at 5–10-sec intervals; far-carrying yet can be difficult to pinpoint. Scolding note, *buk-buk-buk.* . . . **Habits:** Usually singly. Secretive, but noisy. Calls from a perch atop grass clump, where its white breast stands out. When flushed, makes a short flight from one clump of grass to another. Walks and runs, mouselike. Gleans insects, less so seeds. Nest is built in grass, ~0.5 m above the ground; nest is a cup of grass. Eggs (2–4) pinkish with darker spots. **Range:** NG (plus sighting on Goodenough, SE Is), 0–4000 m. Also Bismarck Is. **Taxonomy:** Split from Tawny Grassbird (*M. timoriensis*). The relationships of the montane *alpinus* subspecies group, considered a separate species by some authorities, are confusing both within the group and with mid-mountain and lowland forms.

Fly River Grassbird *Megalurus albolimbatus* Pl 105

14–15 cm. Trans-Fly marshes; very local and habitat-specific, but common where found. Usually accessible only by boat. Secretive but curious; attracted to "pishing." **All plumages:** From Tawny GB by *clean white underparts*; *cap unstreaked, bright tawny rufous*; *prominent white eyebrow*; *shorter tail, with tip somewhat rounded* rather than sharply tapered, often held cocked; tertials black with *obvious white margins*; longer, more slender bill; smaller size; *mournful whistle.* **Similar spp:** Tawny GB also in Trans-Fly, but generally not out over water, has streaked cap, long pointed tail, buff edges to tertials. Little GB has dark crown and is more streaked overall. Golden-headed Cisticola with similar color pattern, but bird is much smaller. White-shouldered Fairywren female lacks rufous cap and shows broken white eye-ring. **Voice:** Most vocal at dusk, also the best time to see the birds. Song is a complex mixture of harsh and sweet notes, far more attractive than song of Tawny GB and more complex than simple repertoire of Little GB. (No song-flight reported.) Call is a mournful single note whistle, *zeee.* Alarm call is an easily imitated *tchit* or *tchit-tchit—churrr.* **Habits:** Lives in inundated or floating grasses, sedges, reeds, or lotus along watercourses and lakes. Habitat threatened by introduced Rusa Deer. An active, pert bird with jerky movements and fairly strong, direct flight; much less sluggish than Tawny GB. Tail cocked and wings drooped. Feeds by gleaning for insects and spiders.

Nests in dense vegetation, the nest ~0.6 m above water; nest globular with side entrance near the top, woven of grass blades. Eggs (3) pinkish with spots. **Range:** Endemic. NG Trans-Fly. Known from 2 localities: Bensbach and middle Fly River (lakes Daviumbu, Owa, and Pangua). An unpublished record from Wasur National Park needs confirmation.

Little Grassbird *Megalurus gramineus* Pl 105

14–15 cm. Status in NG Region poorly known. Reported from only 2 far-flung localities—a lake in the W Ranges and unconfirmed record from the Fly River marshes. Retiring; hides in grassy marsh vegetation and cane thickets. **Adult:** A *small* grassbird with *dark, streaked crown* and faint, pale eyebrow. **Juv:** More straw-colored overall. **Subspp (1):** *papuensis* (Paniai Lakes); unknown which subsp occurs in Fly River. **Similar spp:** Tawny and Fly River GBs both have rufous caps and distinctive calls. Cisticolas are smaller, with shorter, pale-tipped tail. **Voice:** Unknown in NG. In AU, song is a series of mournful whistles or piping, *p-peee-peee*, all on the same pitch. Alarm call is a dry scolding rattle. **Habits:** Singly or in pairs; creeps mouselike through vegetation, also runs and hops. Gleans for small invertebrates. Nest usually in thick cover over water; nest is globular or a deep cup, made of grass leaves (AU data). Eggs (2–4) pinkish, dark markings. **Range:** NG: W Ranges (Paniai or Wissel Lakes, 1600 m); detections in S Lowlands (middle Fly R) reported with no details, need confirmation. Also AU.

CISTICOLAS AND ALLIES: Cisticolidae

This is the fourth and last family of the former "sylviid" warblers in the NG Region. Nearly all the 160 species inhabit Eurasia and Africa, with only 2 reaching Australasia. These are small warblers with cocked tails. Many build complex nests.

Zitting Cisticola *Cisticola juncidis* Pl 105
(Fan-tailed Warbler)

10–11 cm. Trans-Fly only; uncommon in open, grassy floodplains, where clumps of grass form hummocks. In both cisticola species, Males have a distinct breeding plumage with short tail; the male Nonbreeding plumage resembles the longer-tailed Female. *Distinguish Zitting C from its congener with care by voice; more heavy streaking above; usually duller, less buffy plumage; and more conspicuous pale tipping on the tail* (whitish vs buffy in Golden-headed). **Breeding male:** Crown not streaked; tail short. From breeding male Golden-headed by brownish head with conspicuous pale eyebrow and dark eye-stripe. **Female and Nonbreeding male:** Crown streaked; longer tail; less buffy than most Golden-headeds. **Juv:** Resembles Female but with slight yellowish wash and streaking less distinct. **Subspp (1):** undetermined, possibly *laveryi* (also NE AU). **Similar spp:** Golden-headed C. **Voice:** Song, *zit-zit-zit* . . . , given in song-flight, the bird circling with bouncing flight high above the grass. Song of perched bird sounds like a ticking watch, a few notes in each series, *click-click*. Buzzing call in alarm. **Habits:** Solitary or in pairs. Usually keeps hidden in the grass. Flight jerky and bouncing; fans tail when alighting. Runs on the ground. Nests in a grass clump, the leaves drawn together and held in place with spiderwebs to form a cup or urn with the entrance facing upward, and the inside is formed of dried grass (AU data). Eggs (3–5) pale blue with fine dark markings. **Range:** NG: Trans-Fly. Also Eurasia, Africa, AU.

Golden-headed Cisticola *Cisticola exilis* Pl 105

10–11 cm. The common, widespread cisticola in NG. A tiny, long-legged warbler in open grassland (but not in really tall grass). Note *honey-buff on nape and flanks*. **Breeding male:** Head golden buff, no streaking; tail short. **Female and Nonbreeding male:** Crown streaked; longer tail. **Juv:** Resembles Female but streaking less distinct. **Subspp (1):** *diminutus* (NG and NE AU). **Similar spp:** See Zitting C. **Voice:** Song is a series of 4 monarch-like buzzing notes followed by a chirp, *bjj-bjj-bjj-bjj-trreep!* Sings from perch or in song-flight, in manner similar to that of Zitting C. Call is a peevish buzzing note *queh*. **Habits:** Seen alone or paired. Evident only when breeding Male sings and displays; otherwise remains hidden in the grass. Forages for insects and other small invertebrates. Nest built suspended from stems in grass; the neat domed nest is composed of plant down and a few leaves. Eggs (2–3) pale green-blue with darker markings. **Range:** NG, Manam I, and SE Is (3 main D'Entrecasteaux Is), 0–2200 m. Also S Asia to Bismarck Is and AU.

STARLINGS: Sturnidae

This family of 123 species is primarily Eurasian and African, with few species in Australasia. New Guinea supports 7 resident species. The widespread Common Starling has been recorded as a vagrant, and the Common Myna is hypothetical for the region. These are slim to robust birds with a moderately heavy bill and strong legs. New Guinea species are predominately black, although the mynas have patches of yellow and white, and immature starlings have pale underparts with blackish streaking. Sexes are alike in plumage, but males are somewhat larger. New Guinea species inhabit forest, woodland, edge, and gardens, from lowlands to middle altitudes. All resident species are primarily fruit eating, although insects are also taken. They are vocal and travel in pairs or flocks that can sometimes be quite large. The native species are all strictly arboreal, but the introduced Common Starling feeds mainly on the ground. Most nest in tree cavities, although the Metallic and Yellow-eyed Starlings weave pendent nests, with hundreds in a tall colony tree. Two groups: glossy starlings are glossy black and slim, and often fly in tight, vocal flocks. The mynas are chunky, short-tailed, black and yellow, with wings marked by a white patch obvious in flight.

Metallic Starling *Aplonis metallica* Pl 107
(Shining Starling)
22–24 cm. *The common, widespread, pointed-tail glossy starling.* Typically in mobile flocks. In lowland and hill forest and edge. Most birds are resident, although some migrate to AU for the austral summer. **Adult:** *Black with green-and-purple gloss; hackles on neck; red iris.* **Juv:** Dark brown above, white below with black streaking; iris dark brown. **Imm:** Upperparts like Adult, underparts like Juv; iris red. **Subspp (2):** *metallica* (widespread, minus Biak and Numfor); *inornata* (Bay Is of Biak and Numfor) smaller and less iridescent. **Similar spp:** See Yellow-eyed S. Singing S is short-tailed and duller black. Also see Moluccan S (NW Is) and Long-tailed S (Bay Is). **Voice:** Song is a high-pitched, fluty warble, sometimes including mimicry. Flight call is a downslurred nasal note, *skier.* Various other calls are nasal, whistling, or wheezing. Nesting colonies can be located by their continuous twittering. **Habits:** Highly gregarious. Travels in a tight, swift, low-flying flock, a stream of black streaks on a roller-coaster course over and around the contours of the forest canopy. Active and nervous. Feeds at fruiting and flowering trees and at caterpillar outbreaks or termite eruptions. Also hawks over rivers for emerging mayflies, the birds so low to the water they sometimes touch it. Roosts communally. Breeds in pairs. Nests colonially with many, sometimes hundreds, of globular nests constructed in a tall isolated tree; spherical nest is woven of vines, twigs, grass, and strips of plant material, with the entrance on the side toward the top; neighboring nests may coalesce. Eggs (1–4) greenish or bluish with darker spots. **Range:** Entire NG Region except Gebe in NW Is and no record from Manam I, 0–1200 m. Also Moluccas to Solomon Is and AU.

Yellow-eyed Starling *Aplonis mystacea* Pl 107
20 cm. Restricted mainly to the S Lowlands; resident in lowlands, visits foothills. An uncommon, close relative of Metallic S, with which it associates, even nesting in the same colony. Difficult to separate; look for the *yellowish-white iris.* Other differences harder to see: smaller size; shorter, thicker bill; sparse forehead tuft of bristly feathers; darker head and neck essentially black without much color; no hackles; and shorter tail. **Adult, Juv, and Imm:** Same plumages as Metallic S. **Voice:** Similar to Metallic S but said to include a short, harsh note and a rather bell-like note. **Habits:** Indistinguishable from Metallic S. Flocks with its own kind and with Metallic S. Nests colonially, often in association with Metallic. Nest similar to Metallic's. Egg, no information. **Range:** Endemic. NG: S Lowlands between Bird's Neck (Wanggar, Cenderawasih Bay) and southwestern SE Pen (Lakekamu R), 0–300 (700) m.

Singing Starling *Aplonis cantoroides* Pl 107
19–22 cm. Widespread in lowlands. Common in towns, deforested areas, seacoast, forest edge, occasionally in disturbed habitats in montane valleys. *The only glossy starling with a short, square tail.* **Adult:** Glossy black with only short hackles and without much iridescent color; *iris orange-red.* **Juv:** Brown above, whitish or pale buff below with dark streaks; iris dark brown. **Imm:** Similar, but darker above (somewhat glossy) and more heavily marked; iris yellowish or orange-red. **Similar spp:** Other glossy starlings show a longer, graduated tail. **Voice:** Calls include a sweet,

downslurred whistle *tsiiew* repeated after a pause, incessantly; a soft note, *jyu*; and creaking notes. **Habits:** Solitary, in pairs, or in large postbreeding flocks; generally less gregarious than Metallic S, with which it sometimes associates. Likewise somewhat nomadic. Feeds mainly on fruit. Nests as solitary pair or in small colonies. Nest built in a tree cavity, in coconut palm crown, on rock face, or on a building; nest is a ball of grass and other plant material. Eggs (2–3) pale blue, dark spotted. **Range:** NG, NW Is (Batanta, Misool, Salawati, Waigo), Aru Is, Karkar I, and the SE Is (Normanby, Woodlark, Misima, Tagula), 0–1700 m. Range extends to Admiralty, Bismarck, and Solomon Is.

Long-tailed Starling *Aplonis magna*   Pl 107
28–41 cm. Bay Is only, which it shares with Metallic S. Abundant throughout. *A large, glossy starling with a very long, lax tail. Dark iris.* **Adult:** Glossy black with full-length tail. Female smaller than Male. **Juv:** Also all black, but duller and tail shorter. **Subspp (2):** *magna* (Biak I) larger, with longer tail; *brevicauda* (Numfor I) smaller, with shorter tail. **Similar spp:** Metallic S is smaller, has a shorter tail and red iris. **Voice:** Calls are a series of loud, rather incomplete warbles and a shrill descending note, *cheew*. **Habits:** In pairs or small parties. Active in fruiting trees. Nests high in trees at forest edge or in gardens. **Range:** Endemic. Biak and Numfor (Bay Is), 0–650 m.

Moluccan Starling *Aplonis mysolensis*   Pl 107
20 cm. NW Is only, and mainly a small island species. Range shared with Metallic and Singing Ss. Uncommon and patchy in any forested and open habitat with trees. *Note wedge-shaped tail and dark eye in all plumages.* **Adult:** Glossy black with greenish iridescence. **Juv (and Imm?):** Brown above, white below with dark streaking. **Similar spp:** Metallic S more iridescent, with bright red iris and a more sharply pointed tail. Singing S has a short, squared tail. **Voice:** Call, *tseu!*, higher pitched and more strident than that of Singing S. Nasal whistling, squealing, chattering. **Habits:** Poorly known. In small parties. May associate with other glossy starlings. Feeds on fruit, probably also insects and nectar. A colonial nester; nests in tree hole (Moluccan data). Egg, no information. **Range:** NW Is (all except Gag), 0–1000 m. Also Moluccas.

Yellow-faced Myna *Mino dumontii*   Pl 108
25–26 cm. Common and easily observed in lowland and hill forest canopy or in flight overhead, particularly along edges such as clearings or river courses. A chunky, noisy, *black bird with yellow eye-patch and belly; white patches on wings and rump.* **Adult:** Bright yellow-orange facial skin. **Juv:** Facial skin paler. **Similar spp:** Golden M has black face, yellow neck, yellow rump. **Voice:** Unmistakable. Loud, humorous, guttural exclamations: *AH?!* or *OW!* or *Gwalaow!* and so on. **Habits:** Typically in pairs or small parties, occasionally large roosting flocks of more than a hundred. Direct flight with rapid wingbeats. Perches in the treetops, calling regularly. Eats fruit. Also gleans arthropods and occasionally sallies for insects. Roosts communally in a tree, or pair or family may roost in nest hole. Breeds in monogamous pairs; a single helper has been reported. Nests in a high tree hole or in base of huge bird's nest fern; nest of sticks and green leaves. Eggs (1–2) pale blue with faint dark spots. **Range:** Endemic. NG, NW Is (Batanta, Salawati, Waigeo), Bay Is (Yapen, plus single sightings from Biak), and Aru Is, 0–750 m (rarely to 1800 m). **Taxonomy:** Long-tailed Myna (*M. kreffti*) of Bismarck and Solomon Is was recently split off.

Golden Myna *Mino anais*   Pl 108
24 cm. Uncommon in lowland forest; usually less numerous than Yellow-faced M. **Adult:** A chunky *black bird with orange neck* and *white wing patch* and obvious *yellow rump*. **Juv:** Head black, underparts mottled black and yellow. **Imm:** Shows Adult pattern but yellow parts paler and mottled black. **Subspp (3, in 2 groups):** *anais* (NW Is, W Bird's Head) head black, yellow patch behind eye; *orientalis* (remainder of species' range) yellow crown, black nape (usually missing in S Lowlands, SE Pen), no eye-patch. **Similar spp:** Yellow-faced M lacks yellow neck, has yellow eye-patch; rump white, not yellow. **Voice:** Song is a short series (2–6 notes) of whistled and squeaky notes. Calls include a variety of hoarse, nasal, whistled, or squeaking notes: *chaaah!* or *peeshur-raah*; a squeal like that of Black-capped Lory or Red-cheeked Parrot, *kyur-lee!* **Habits:** Similar to Yellow-faced M, with which it sometimes associates; however, more often found in extensive areas of primary lowland forest. Nest site also similar, but egg not described. **Range:** Endemic. NG (except Trans-Fly, Huon, and N coast of SE Pen), Salawati (NW Is), and possibly Yapen (Bay Is), 0–300 (600) m. **Taxonomy:** Bird's Head form is a potential split.

Common Myna *Acridotheres tristis* Pl 107
(*Sturnus tristis*)
23–25 cm. Hypothetical. Urban and agricultural habitats, seen foraging along roadsides and on lawns, also perched on wires. First reported in 1997, without details, from Alotau Harbor (SE Pen), with subsequent sightings there. Anticipated to invade and settle the NG Region. Brown with black head, white patch in wing, yellow bill and legs. **Adult:** Broad white tail tip, obscure in **Juv**. **Subspp:** No information. **Similar spp:** Yellow-faced and Golden Ms are black rather than brown and show patches of yellow plumage. They do not feed on lawns. **Voice:** Noisy. A variety of repeated phrases of clear or harsh notes. **Habits:** Lives in pairs that gather in flocks to roost in a tree. Forages on the ground for insects and scraps of discarded human food; visits trees to take fruit. Nests in a cavity in building or tree; nest a messy cup of grass and rubbish; eggs blue. **Range:** NG: Alotau Harbor (SE Pen). Native to S and SE Asia, introduced to AU, Solomon Is, and widely around the world.

Common Starling *Sturnis vulgaris*
21 cm. Vagrant to Port Moresby. Urban and agricultural habitats. *An all-dark, short-tailed starling with long, narrow, pointed bill and wedge-shaped head*; flight profile resembles woodswallow, with pointed wings and squared tail. Dark iris. **Adult:** Glossy black, with white speckles in fresh plumage; bill yellow. **Juv:** Dull brownish grey; bill black. **Subspp:** No information. **Similar spp:** Singing S is not as compact, is strictly arboreal, has a red eye. **Voice:** Song complex, composed of bursts of loud wheezing, whistling, warbling, often interspersed with mimicry. Flight call, *prurp*. **Habits:** Singly or with others of its kind. Usually seen perched on electrical wire or building, or foraging on the ground. Walks about on lawns and fields, gleaning and probing for insects. Also feeds in fruiting trees. **Range:** NG: 2 sightings from Port Moresby (SE Pen). Native to Eurasia, N Africa. Introduced widely, including AU.

THRUSHES: Turdidae

Some 185 species worldwide; only 2 species from New Guinea Region. The thrushes are nearly cosmopolitan, with greatest diversity in Eurasia and Africa. Some species formerly treated as thrushes have been moved to the Muscicapidae. The New Guinea species inhabit hill forest (Russet-tailed Thrush) and mountaintops (Island Thrush). They forage mainly on the ground for invertebrates.

Russet-tailed Thrush *Zoothera heinei* Pl 63
(Scaly Thrush, *Z. dauma*)
21–23 cm. Uncommon and very rarely observed; inhabits the forest floor in hills and mid-mountains. Note *robust, elongate shape and horizontal posture. Large, dark eye and pink legs*. **All plumages:** Heavily *scaled all over*; brown above with buffy wing bars; whitish below. **In flight:** Underwing with black-and-white patch, white outer tail feathers. **Subspp (1):** *papuensis* (NG). **Similar spp:** Papuan Logrunner, the only other scaly ground bird, is smaller, has a spine-tipped tail, grey cheek, dark legs, and not so scaly below. Papuan Scrub-Robin lacks scaling. **Voice:** No information for the region. Song and calls known from AU. Song is a pair of whistles, the second on a lower pitch. Calls include a *tsee-ip* and chattering. **Habits:** Alone or in pairs. In AU, this and a related species seem to be encountered most often at dawn and dusk or in heavily overcast weather. Flushes from the ground and alights on a low perch to watch the intruder. Forages on the ground by running forward, then pausing motionless while waiting to detect prey; also hops about; gleans, flicks leaves, and probes ground (AU data). Takes invertebrates. Nest built on a mossy branch in the midstory; nest cup-shaped with mossy exterior (AU data). Eggs (2) green or blue with fine spots. **Range:** NG: Central Ranges and Kumawa, Foja, N Coastal, Adelbert, and Huon Mts, 500–1700 m. Also AU and Mussau I (Bismarcks). **Taxonomy:** Whether the New Guinea form belongs in *Z. heinei* of AU, or constitutes its own species, has yet to be settled.

Island Thrush *Turdus poliocephalus* Pl 63
23–25 cm. High mountains, where common in alpine shrubland and subalpine woods.
Adult: *Dark brown with distinctive orange bill, eye-ring, and legs*—unmistakable. **Juv:** Dark above, spotted and streaked with rufous; below, rufous spotted with black. **Subspp (3, in 2 groups):** *papuensis* (NG) all brownish black; *canescens* (D'Entrecasteaux Is: Goodenough) pale grey head. **Similar spp:** None. **Voice:** Song is typical of the genus: *chirrup-churrup-cherrio*. . . . Calls include

a short rasp, a squawk, and a thin, high note. Alarm call a series of a few sharp notes, typical of a blackbird. **Habits:** Solitary, shy. Often noticed perched on an exposed branch. Forages on the ground for invertebrates in a typical, thrushlike, run-and-stop mode. Plucks fruit in trees and shrubs. Nest is a simple, bulky cup of plant material placed a few meters up on a branch in dense shrub. Eggs, no information for the region, elsewhere (2) bluish green with dark speckles. **Range:** NG (Central Ranges and Huon, 2700–4100 m, rarely lower), Karkar I, and SE Is (D'Entrecasteaux Is: Goodenough, 1600–2600 m). Also Sunda Is, Formosa, Philippines, Wallacea to Oceania. **Taxonomy:** The Island Thrush could some day be split into many species, with 1 or more in NG.

CHATS, OLD WORLD FLYCATCHERS: Muscicapidae

This is a large assemblage of nearly 300 Old World species ranging from Eurasia south to Africa and barely reaching Australasia. Many species formerly classified as thrushes have been moved into this family. Some species closely resemble unrelated Australasian robins, Petroicidae. Three muscicapid species recorded from New Guinea are migrants from N Asia; one other is a tropical Asian species of open, man-made habitats that has extended its breeding range to the NG Region.

Siberian Rubythroat *Luscinia calliope* Pl 104
(*Erithacus calliope*)
14–16 cm. Vagrant, Palearctic migrant. Overwinters in open country and thickets in tropical lowlands and mountains below the cloud forest; seen along roadsides, not expected in forest. Dull grey-brown; cocks or flicks rufous brown tail; diagnostic *white eyebrow and moustachial stripe*. **Male:** *Ruby-red throat*. **Female and Juv:** White throat. **Similar spp:** Lesser Ground-Robin has similar color pattern (white spot on lores is larger) but is more pitta-like with compact body on stilt legs (rubythroat spindle-shaped); inhabits different habitat—forest interior, usually at higher elevation. **Habits:** Solitary. Feeds on or near the ground, taking small invertebrates, fruit. **Range:** NG: 1 sighting near Wau (SE Pen) at 1750 m. Breeds in N Asia, winters in SE Asia and Philippine Is.

Pied Bushchat *Saxicola caprata* Pl 104
(Pied Chat)
14–15 cm. Common in grasslands, airstrips, gardens, edges of towns; regularly near human habitation; mainly in mountains, but also lowlands in North and SE Pen. **Male:** *Black with white rump, undertail, and shoulder* (last sometimes hard to see). **Female:** *Dull brownish grey, streaky, with white rump and undertail.* May show white wing bars. **Juv:** Like Female but with buff spots above, dark scaling below, and buff wing bars (male shows some white in wing). **Subspp (1):** *aethiops* (NG and New Britain). **Similar spp:** White-shouldered Fairywren male is much smaller, with cocked tail, no white on rump or undertail. White-rumped Robin dwells in forest. White-winged Triller female resembles female chat but lacks white rump. **Voice:** Song is a musical, reedy and burry warble; it is composed of repeated series; each series of 4–8 notes is repeated several times, then the bird begins another series. **Habits:** Alone or in pairs. Perches atop a rock, shrub, tall grass, wire, or building. Flicks wings nervously. Takes insect food from the ground and by flycatching. Territorial. Nest built in a natural cavity on a rock face, bank, or building, or under grass clump; a cup of dried grass and plant fiber. Eggs (2–4) blue-green and dark-spotted. **Range:** NG (range has recently increased): Central Ranges, Huon, also patchy in lowlands from NW Lowlands to SE Pen, 0–2850. Also S Asia to Bismarck Is.

Blue Rock-thrush *Monticola solitarius* Pl 104
20–23 cm. Vagrant, Palearctic migrant. Frequents rocky places in open country, cliffs, buildings, shores. Assumes a distinctive sleek posture; note long bill. **Male:** Unique—*dark grey-blue with dusky rufous breast and belly*. **Female:** *Grey-brown with heavy scaling*. **Juv:** Body plumage and wing-coverts with pale tips. **Subspp (1):** probably *philippensis* (breeds N Asia and Japan, winters SE China, Philippines, and Wallacea). **Similar spp:** Russet-tailed Thrush resembles Female but has warm brown and rufous tones and strong, pale wing bars. **Habits:** Solitary. Takes food mostly from the ground (small animals, both invertebrate and vertebrate), sallies for insects, and plucks fruit. **Range:** NG: records from Bird's Head (Manokwari) and SE Pen (Port Moresby); could turn up anywhere in suitable habitat. Breeds in Eurasia and N Africa; winters in Africa and S Asia to Wallacea and Philippines.

Grey-streaked Flycatcher *Muscicapa griseisticta* Pl 104
13–14 cm. Palearctic migrant that overwinters in W New Guinea Region during austral summer; common from lowlands to mid-mountains. Frequents high, open perches at forest edge, such as tracks and roads, and in gardens. *A small, drab flycatcher with white spot before eye and pale underparts with dark streaking.* **Adult:** Single wing bar, on greater wing-coverts. **Juv:** Row of white spots above the wing bar on the greater coverts. **Similar spp:** No other flycatcher has the streaked white breast. **Voice:** Silent on wintering grounds. **Habits:** Solitary. Sallies out to capture flying insects. Cocks and flicks tail. **Range:** NG (mainly Bird's Head and Neck; rare in NW Lowlands and W and Border Ranges), NW Is (all except Kofiau), and Yapen (Bay Is), 0–1500 m. Breeds in Asia (Siberia), winters in Philippines, Borneo, east to NG.

SUNBIRDS AND FLOWERPECKERS: Nectariniidae

Sunbirds and flowerpeckers are a large family (187 spp) of small, colorful birds. Sunbirds have a long, narrow, downcurved bill used to probe flowers for nectar, whereas flowerpeckers have a short pointed bill for plucking and swallowing fruit. Both also take insects and spiders. Sometimes classified as separate families, the sunbirds and flowerpeckers have recently been shown by molecular and morphological studies to be closely related. Their greatest diversity and probable place of origin are in tropical Asia, although there are also many sunbirds in Africa. There are 3 sunbirds and 3 flowerpeckers in New Guinea, where the family is widely distributed. Males are brightly colored and iridescent, whereas females generally lack these adornments and are rather plain. Their small size is matched by their active behavior and high-pitched songs and calls. They construct pendent nests with side entrances and usually include spiderwebs as building material. Sunbirds are often first met in village gardens where they are attracted by ornamental plantings. Flowerpeckers are best seen at the forest edge—look for them visiting mistletoes and *Piper* spp, two favorite fruits.

Olive-crowned Flowerpecker *Dicaeum pectorale* Pl 79
9 cm. Replaces Red-capped F in the far W. *Upperparts entirely olive, no red cap or rump.* **Male:** Red breast, lacking in **Female. Juv:** Resembles Female, base of bill bright orange-red. **Subspp (2):** Minor. **Voice and Habits:** Similar to Red-capped F. **Range:** Endemic. NG (Bird's Head and Neck) and NW Is (all), 0–1700 (2400) m.

Red-capped Flowerpecker *Dicaeum geelvinkianum* Pl 79
(Once part of the more inclusive **Papuan Flowerpecker**, *D. pectorale*)
9 cm. The widespread NG flowerpecker, often heard but hard to see. Lives high in the forest canopy from lowlands to mid-mountains. Common. *A tiny, stubby bird with red cap and rump,* but otherwise variable. **Male:** Red breast. **Female:** Upperparts duller and no red on breast. **Juv:** Lacks any red patches, otherwise resembles Female, but base of bill bright orange-red. **Subspp (11, in 2 groups):** (1) *geelvinkianum* (NG, Bay Is, and D'Entrecasteaux Is): in Bay Is and NW Lowlands upperparts olive, somewhat glossy, throat white (*maforense*, others); in S Lowlands upperparts black with blue-purple gloss, throat red (*albopunctatum*, *rubrigulare*); intermediate in Sepik-Ramu to SE Pen (*rubrocoronatum*) and D'Entrecasteaux (*violaceum*). (2) *nitidum* (SE Is: Misima, Tagula, Rossel; *rosseli* illustrated) with green-blue gloss and yellow-green underparts. **Similar spp:** Olive-capped F lacks red cap and rump. Mistletoebird has red vent. Neither co-occurs with Red-capped. **Voice:** Song poorly known. Call notes of 2 types, given singly: (1) a short, insect-like buzz and (2) a high upslur, similar to that of Black Sunbird. **Habits:** Solitary or in pairs. Feeds energetically on fruit, such as mistletoe, and small arthropods taken by gleaning or hovering. Nest a suspended pouch with a side entrance near the top, attached to a thin branch a few meters up, and built of fine plant material and spider silk. Eggs (2) white. **Range:** Endemic. NG (minus Bird's Head and Neck), Bay Is (Biak, Numfor, Yapen), Manam and Karkar Is, and SE Is (3 main D'Entrecasteaux Is; Misima, Rossel, and Tagula Is); 0–1700 (2400) m. **Taxonomy:** Subsp group *nitidum* is sometimes split as a separate species, Louisiade Flowerpecker (*D. nitidum*).

Mistletoebird *Dicaeum hirundinaceum*   Pl 79
9 cm. *Aru Is only*, where this is the only flowerpecker. *A tiny, stubby bird with red vent.*
Male: Blue-black with brilliant ruby red throat; black streak down center of breast. **Female:** Olive with *dingy white streak down center of breast.* **Juv:** Resembles Female but lacks red in undertail; base of bill orange-red. **Subspp (1):** *ignicolle* (endemic). **Similar spp:** None on Aru Is. Other NG flowerpeckers lack the red vent. **Voice:** Songs of numerous types (AU data), yet easily recognized, are high pitched, clear, and penetrating, delivered from high in the treetops; often a phrase is repeated few times, e.g., *wit-wissweet, wit-wissweet*, but song may also be more complex and incorporate more than 1 type of phrase. Flight call a sharp *dzee!* or *tsew!* **Habits:** Similar to other flowerpeckers. **Range:** Aru Is. Also AU.

Black Sunbird *Leptocoma aspasia*   Pl 79
(*Nectarinia aspasia* or *sericea, Cinnyris sericeus*)
11–12 cm. A common lowland sunbird of forests and village clearings; less common in hills.
Male: *Jet black with blue and green flashes of iridescence on crown, shoulders, and tail.*
Female: Olive with grey head; *pale grey throat and breast contrasting with green-yellow belly.*
Juv: Similar to Female but with crown greenish, throat yellowish, and yellow underparts washed grey. **Subspp (10):** Minor variation. **Similar spp:** Papuan Black Myzomela male lacks iridescence, has shorter bill and white wing linings. Male Rand's S has an olive back. Olive-backed and Rand's S females have yellow throats; yellow underparts more golden, less greenish. Longbills lack the combination of grey throat and yellow belly. **Voice:** Song of high-pitched, thin, sibilant notes; e.g., a rapid series of identical upslurs followed by a slow trill at lower pitch or a rapid series of notes alternating between 2 pitches. Some calls similar to those of the Rufous-banded Honeyeater; an upslurred note like that of flowerpeckers. **Habits:** More common in primary forest than Olive-backed S. Mainly in the outer foliage. Forages for arthropods and nectar by gleaning, hovering, and probing. Generally active, solitary, and feisty. Nest pendulous and tear-shaped, composed of plant fiber and bark strips; with roofed side entrance. Eggs (2) cream or pale brown, spotted. **Range:** Throughout NG Region, except Rossel I (SE Is); 0–600 (1200) m. Also Sulawesi, Moluccas, and Bismarck Is.

Olive-backed Sunbird *Cinnyris jugularis*   Pl 79
(**Yellow-bellied Sunbird**, *Nectarinia jugularis, C. frenatus*)
11–12 cm. A common town and garden bird, also lowland and hill second growth and forest edge. *Bright yellow breast and belly* in all plumages. **Male:** *Iridescent black throat.* Nonbreeding male and molting immature Male with just a black stripe down the throat. **Female and Juv:** Olive with *prominent yellow eyebrow.* **Subspp (1):** *frenatus* (also Moluccas). **Similar spp:** See Rand's S. Black Sunbird female with grey throat. Yellow-bellied Longbill similar, especially in S Lowlands, but has longer bill, different facial pattern, and mostly occupies forest interior. **Voice:** Song a weak, fast, twittering trill. Calls include an inquiring, weak, high upslur and soft sweet, *tsip*. **Habits:** Similar to Black S. Nest also similar but usually covered with spiderweb and feathers. Eggs (2) white with dark mottling. **Range:** Throughout NG Region, except Gebe I (NW Is) and Trobriands and Woodlark, Misima, Tagula, and Rossel Is (SE Is), 0–600 m, rarely 1200 m. Status in NW Lowlands and Sepik-Ramu unclear with respect to Rand's Sunbird. Also from Indochina east to the Solomons and south to AU.

Rand's Sunbird *Cinnyris idenburgi*   Pl 79
(**Olive-backed** or **Yellow-bellied Sunbird**, *Cinnyris* or *Nectarinia jugularis*)
11 cm. Swampy second growth in the large river basins of NW Lowlands and Sepik-Ramu. Uncommon, local, and little known. Its range seems entirely surrounded by Olive-backed S, a coastal species where their ranges meet (whether the 2 species co-occur needs clarification). Slightly smaller than Olive-backed and darker above in all plumages. **Male:** *Underparts entirely glossy black; note unique orange flank tufts.* **Female and Juv:** From Olive-backed S by *darker, more brownish upperparts; whitish rather than yellow throat* (and undertail-coverts?)*; eyebrow obscure; and greyer, more obscure tips to outer tail feathers.* **Similar spp:** Olive-backed S. **Voice:** Song (Sepik: Karawari) a quick, monotonous series of ~6 high-pitched notes, delivered 2–3 notes/sec, *sit, sit, sit, sit, sit*. **Habits:** Presumably similar to Olive-backed. **Range:** Endemic. NG: River basins of the NW Lowlands (upper Mamberamo) and Sepik-Ramu (Aiome, Karawari), 0–300 m. **Taxonomy:** Previously treated as a race of Olive-backed S, *C. jugularis*. Possibly Rand's S is a race of Black-breasted S (*C. clementiae*) of Moluccas.

OLD WORLD SPARROWS: Passeridae

The sparrows need little introduction ... except in New Guinea, where they are recent arrivals. The Old World sparrows (49 spp) are native to Eurasia and Africa. Many species have adapted to man-made habitats such as farms, towns, and cities. Two of the most urbanized species have found their way to New Guinea probably as stowaways on ships, the House Sparrow in 1976 and the Eurasian Tree Sparrow in 1989. Only sparingly distributed at present, they are expected to spread rapidly. They are small, brown, seed-eating birds with conical beaks, and they feed mainly on the ground.

House Sparrow *Passer domesticus* Pl 104
(English Sparrow)
15 cm. Not native, spreading. Locally common in ports, towns, and cities. Note *dark beak* (mannikins have pale grey beak). Both sexes have unique *white bar across the shoulder*. **Male:** *Black bib, grey ear-coverts, grey-and-chestnut crown*; bib much smaller in Nonbreeding plumage. **Female and Juv:** Drab with *pale eye-stripe*. **Subspp:** No information. **Similar spp:** Eurasian Tree S has all-brown crown and black spot on ear-coverts. Horsfield's Bushlark resembles Female House S but has streaked head and breast. **Voice:** Continuous cheeping. **Habits:** Gregarious, but sometimes found alone or as a pair. Feeds mostly on the ground on grass seeds and insects, as well as scavenging for human food. Nests in buildings and trees; nest a bulky, shabby mass of plant material and rubbish, lined inside with feathers, fur, or soft material. Eggs (~3–5) white with dark speckles. **Range:** NG (Sorong on Bird's Head; E Ranges; Port Moresby; Huon and Markham Valley) and Bay Is (Biak), but probably more widely distributed. Native to Eurasia and N Africa; has spread or been introduced nearly worldwide.

Eurasian Tree Sparrow *Passer montanus* Pl 104
14 cm. Not native, spreading. Common in many ports, towns, and cities, mainly on the Indonesian side. **All plumages:** Diagnostic *white ear-coverts with black spot; chestnut crown*. **Subspp:** No information. **Similar spp:** House S lacks the black spot on its grey ear-coverts; crown is grey and chestnut. **Voice:** Chipping call. **Habits:** Similar to House S. **Range:** NG and Bay Is (Biak and Numfor). NG: Bird's Head (Sorong, Manokwari), W Ranges (Wamena), S Lowlands (Amamapare, near Timika; Kiunga), NW Lowlands (Jayapura, Sentani), SE Pen (Port Moresby). Native to Eurasia; spread or introduced to AU, Oceania, N Amer.

WAXBILLS, MANNIKINS, AND ALLIES: Estrildidae

Related to Old World Sparrows and Weaverbirds, this family of finches (142 spp) has a distribution centered in Africa, with a smaller number of species in southern Asia, Australasia, and Oceania. The 17 species in New Guinea are assigned to 3 lineages: the Australasian grassfinches (Mountain Firetail and Crimson Finch), the parrotfinches, and the mannikins (called munias in Asia, including Indonesia). These are small, gregarious seed eaters with thick pointed beaks, short wings, and often brightly colored plumage.

The lives of nearly all species revolve around grass, which is their main food and habitat. Grass seeds constitute the diet. Usually seen in small flocks, they feed from nodding grass stalks, on lawns, or on the ground. Juveniles often form flocks of their own. The nest is made of grass, except for the firetail and parrotfinches. Some mannikins use nests for both roosting and reproduction. Multiple pairs have been reported roosting in the same nest. Nests are domed in all species; the eggs are plain white. A unique feature of nestlings is the colored guiding marks on the palates and tongues.

Mannikins have speciated explosively in the region. While identification seems rather straightforward (even the plain-looking juveniles can be separated with practice), various factors cloud the picture. First, there are undescribed populations for some mannikins, so it is difficult at this time to cover the full range of species' variability. Also, many species hybridize, blurring their identity. Finally, species' ranges are expected to increasingly meet as mannikins spread to new localities following the expansion of grassy habitat.

Good places to find mannikins are along roadsides, around villages, and at soccer pitches, edges of airstrips, and grassy margins of wetlands. The forest finches are seldom seen and best detected by their calls. Unfortunately, estrildid songs are high pitched, difficult to hear, often complex, and therefore not useful for detection. Calls, however, are very helpful for locating these finches. Male and female mannikins are said to give somewhat different calls.

Mountain Firetail *Oreostruthus fuliginosus* Pl 109
13 cm. Subalpine forest, where inconspicuous but locally common. Seen on the ground or in low shrubbery in glades and edges. The only *chocolate-and-red* finch in this habitat. *Chunky.* Note *red rump* on fleeing bird. **Adult:** Red flanks and beak. Throat same dark brown as rest of head in **Male** or orange-brown in **Female**, with upper beak dark. **Juv:** Red restricted to rump only; beak black. **Similar spp:** Crimson Finch, in lowlands only, is slimmer with red face. **Voice:** Song: *swi-chi-chi-chi-chi*. Call is a peculiar, quiet, mewing note. Alarm call, an explosive and repeated *pit!* **Habits:** Solitary, pairs, or small parties. Encountered along trails and roadsides. Rather tame, but goes unnoticed until flushed, departing into the forest. Wait quietly, and disturbed birds may return to the open. Learn its odd call to locate and observe feeding birds. Forages for seeds on the ground, in grass, or in bamboo. The only nest reported was built high in a palm; a bulky, domed structure with side entrance, made of ferns, grass, ginger leaves. Eggs, no information. **Range:** Endemic. NG Central Ranges, 2800–3650 m, rarely lower.

Crimson Finch *Neochmia phaeton* Pl 109
13 cm. Lowlands, where locally common in tall grass and reeds bordering rivers, marshes, and lakes. *A slender red-and-grey finch with long, pointed, red tail* nervously switched from side to side. **Adult:** *Red face and beak*; breast red in **Male**, grey in **Female**. **Juv:** Brown head and breast; dark beak. **Subspp (1):** *evangelinae* (S Lowlands, Trans-Fly; also Cape York, AU); race of Sentani birds unknown. **Similar spp:** Mountain Firetail lives high in the mountains, is chunky with dark head. **Voice:** Song (AU data) a rapid series of 5–6 repeated nasal notes; also described as ending in a few loud, melodious, descending notes. Calls include a descending trill of 3–4 notes and a soft, musical *pit*. **Habits:** Gregarious, in small or large flocks, sometimes with mannikins, especially Grey-crowned M. Inhabits savannah, bamboo, marshy grasslands, rice fields, gardens, and tall lake-edge grasses. Feeds on grass seeds taken mostly from the plant, rarely from the ground. Nest usually built in pandanus crown among dead leaves, less often in thatched roof, tree hollow, shrub, or grasses (AU data); nest is a ball of grass blades with protruding side entrance, lined inside with downy grass heads and feathers. Eggs (4) white. **Range:** NG: S Lowlands and Trans-Fly; recent sightings from Jayapura area of NW Lowlands (Lake Sentani to Cyclops Mts), where possibly introduced. Also AU.

Blue-faced Parrotfinch *Erythrura trichroa* Pl 109
12 cm. Hills and mountains in forest and at edge. Common but nearly invisible and almost never seen; listen for its call. *This is the more common of the 2 nearly identical, green parrotfinches.* From Papuan PF by its smaller size and *smaller, narrower, straight-edged, conical beak*. Note *red tail*. **Adult:** Blue face; blue reduced in **Female**, with paler green underparts. **Juv:** Face mostly green, tail brownish red. **Subspp (1):** *sigillifera* (NG Region, Bismarck Is, AU). **Similar spp:** See Papuan PF. Small parrots have different color patterns and hooked beaks. **Voice:** Song, 2 high trills followed by a single whistle that falls then rises. Call, the only sure means of detection, is insect-like—a short, descending, high-pitched trill, *t-t-t-t*, given perched or as the bird zips by. Suggests the sound of a fingernail running across the teeth of a comb; much softer and shorter than call of Mountain Peltops. **Habits:** Singly, in pairs, or in small parties. Varies locally in abundance; probably nomadic. Usually seen at forest edge, in glades, along trails, or in gardens. Frequently heard in forest, generally at lower levels, but also ascends to canopy. Habits in the forest poorly known. Attracted to seeding bamboo, where can be abundant and easy to observe. Feeds on grass and other seeds, small fruits, and insects. Nest built high in tree, pandanus, or midstory sapling; nest pear-shaped with side entrance, made mostly of moss. Eggs (4–6?) white. **Range:** NG, Manam and Karkar Is, and SE Is (Fergusson, Goodenough, Tagula), 750–3000 m (lower on islands). NG: Bird's Head, Central Ranges, and Foja, Adelbert, and Huon Mts. Also Wallacea to Solomon Is, Oceania, AU.

Papuan Parrotfinch *Erythrura papuana* Pl 109
13 cm. One of NG's rarest and least frequently seen species. Same habitats as Blue-faced PF, with which it may co-occur, but patchily distributed. *Nearly identical to Blue-faced PF* and difficult to separate in the field. *Differs in larger size; more massive beak with a somewhat more swollen shape*; longer tail in some birds. Supposed color differences between the species are not reliable. **Voice:** Call similar to Blue-faced PF. Song (?) is an elaboration of this call—a trilling, descending series of high, musical, twittering notes. **Habits:** Behavior and ecology uncertain because of the potential for confusion with Blue-faced PF. Singly or in pairs, not in parties. In forest interior, forages in midstory or canopy at fruiting trees such as figs and at the oak *Castanopsis*. No nesting information. **Range:** Endemic. NG (Bird's Head, Central Ranges, Adelbert Mts), 1200–2600 m.

Mannikins. The 13 species in the New Guinea Region sort out such that most localities have only 1–3 species. A habitat specialist is Streak-headed Mannikin, widespread at the forest edge. Somewhat specialized in habitat are 3 species confined to the S Lowlands and Trans-Fly—White-spotted, Grey-crowned, and Black, preferring swampy grasslands to varying degrees. The remaining 9 species all live in grassland. Among them, Black-faced is confined to 2 islands of the far West. Two species are confined to alpine grassland—the Eastern and Western Alpine Mannikins. Three other species have restricted geographic ranges in mid-mountain grasslands—Grey-banded on the Bird's Head, Black-breasted in and near the Baliem Valley (W Ranges), and Grey-headed in the SE Pen, where it also occurs in the lowlands. The remaining 3 species (Grand, Hooded, and Chestnut-breasted) are widespread but patchily distributed in the lowlands and mid-mountain valleys.

Black-faced Mannikin *Lonchura molucca* Pl 110
(Moluccan Munia)
10–11 cm. A Wallacean species that barely enters the NG Region on the islands of Gag and Kofiau, where it is the only species of mannikin. Disturbed areas with grass. **Adult:** Black throat and bib; *unique white, scaly breast and rump*. **Juv:** Tan, paler below, with *cream rump (may show diagnostic faint barring) and blackish tail*. **Subspp (1):** Most likely *molucca* (Sulawesi, Moluccas). **Similar spp:** No other mannikin has a white rump. Other scaly-breasted mannikins, Chestnut-breasted and Western Alpine, have buff breast band and yellow or orange rump. **Voice:** Song a series of *peep* or *whee* notes. Call, *tissip*, similar to White-spotted M. **Habits:** In pairs or small flocks. Habits similar to other lowland mannikins, foraging low in dense scrub at edges of gardens and along roads and trails. Nests in trees or dense grass; nest a ball of grass. Eggs (4–5) white. **Range:** NW Is (recent sightings from Gag and Kofiau; should be looked for on other nearby islands), lowlands. Also Lesser Sundas and Sulawesi to Moluccas.

Streak-headed Mannikin *Lonchura tristissima* Pl 110
10 cm. The mannikin of forest edge and second growth; lowlands to mid-mountains. Uncommon. **Adult:** *Blackish brown with conspicuous yellow rump and blackish tail*. Obscure streaking on head visible at close range, more apparent in Females of some populations. Evidence of hybridization with White-spotted M locally where the S Lowlands meet the Central Ranges. **Juv:** Uniformly dull brown, no contrast between upperparts and underparts as in other juv mannikins. **Subspp (4, in 2 groups):** *tristissima* (N watershed) underparts dark brown or blackish with little or no spotting; *bigilalei* (southern SE Pen; this form needs verification) medium brown underparts with whitish spots along flanks. **Similar spp:** See White-spotted M, which co-occurs in western S Lowlands and supposedly around Port Moresby. Black M does not co-occur; is darker and lacks any pale spotting or streaking. **Voice:** Song is a complex, rambling twitter. Call is a buzzy, thin *tseed*, slightly upslurred and querulous; also a flowerpecker-like *jjb!* **Habits:** Usually in small, scattered flocks. Prefers grassy or shrubby places at forest edge, around villages, in second growth and bamboo, or in small clearings in the forest. Does not require expanses of open grasslands, as do other mannikins. Feeds on seeds of grasses, bamboo, weeds, and small berries; also insects. Nests in forest edge or interior, reportedly in rattan; nest is spherical and made of grasses. Egg, no information. **Range:** Endemic. NG and Karkar I, 0–1200 (1700) m. Widespread on NG mainland, missing from Trans-Fly, most of the S Lowlands (present in far W and around Kiunga in the upper Fly R), and tip of SE Pen. **Taxonomy:** See White-spotted M.
Extralimital spp: Scaly-breasted Mannikin (or **Nutmeg Mannikin**, *L. punctulata*), a non-native species, reported from Port Moresby in the 1960s and 1980s, but none recently. Originated either as an escaped cage bird or vagrant from AU, where introduced. Dull brown, dark face, diagnostic scaly breast.

White-spotted Mannikin *Lonchura leucosticta* Pl 109
10 cm. Uncommon. Grassy habitats in lowlands, also bamboo and forest edge. **Adult:** Warm brown with *white chin and cheek and fine white speckling on head and foreparts*. Evidence of hybridization with Streak-headed H. **Juv:** Note *spots on wing-coverts*; from juv Streak-headed M by pale chin. **In flight:** Yellow rump contrasts with black tail. **Similar spp:** Streak-headed M is darker and black-cheeked. Other mannikins in the Trans-Fly lack spotting and in flight have a yellow tail. **Voice:** Song complex string of sounds, like those of other mannikins. Call a short nasal buzz, *toot*, continually repeated in flight, quite unlike the calls of other lowland mannikins. Alarm call, *peet*. **Habits:** In small flocks. Feeding similar to other mannikins. No nesting information from the wild. Eggs in captivity (3–5) white. **Range:** Endemic. S Lowlands and Trans-Fly, lowlands. (Confirmation sought for presence in Port Moresby area of SE Pen, where known from a single specimen of doubtful provenance.) **Taxonomy:** Sometimes treated as a subsp of Streak-headed M. However, the 2 species do not intergrade, have different calls, and generally occupy different habitats.

Grand Mannikin *Lonchura grandis* Pl 110
(Great-billed Mannikin)
12 cm. Open grasslands, cane, and marshes of lowlands and lower mountains. Uncommon and local. Handsome and distinctive. Large; big-headed. Note *massive, bluish-white beak*. **Adult:** Boldly patterned with *black head and underparts*; *chestnut mantle, wings, and flanks*; fiery rump and yellow tail. **Juv:** *Dark head, big beak, smudgy streaking on breast contrasts with pale flanks and belly*. **Subspp (4):** Minor. **Similar spp:** Streak-headed M less boldly patterned; upperparts dark, dull brown. Juv Black M has streaky head and paler underparts. **Voice:** Song difficult to hear, begins with *tk tk tk tk* followed by a long *whheeeeeee. . . .* Call sharper than other mannikins', a metallic *tink . . . tink*, given sporadically. **Habits:** In small to medium-sized flocks, often with other mannikins. Feeding similar to other mannikins; diet apparently not specialized despite extreme beak size. Nests in tree, shrub, branched stump, or grass, alone or in colony of a few nests; nest a grass ball with spout-shaped entrance on the side, lined inside with fluffy grass heads. Eggs (5–6) white. **Range:** Endemic. NG: NW Lowlands, Sepik-Ramu, E Ranges, and SE Pen, 0–1300 m.

Grey-banded Mannikin *Lonchura vana* Pl 109
10 cm. Mid-mountain grasslands of Bird's Head. **Adult:** *White face*; unique *chestnut breast bordered above by a diagnostic grey band*. **Juv:** Plain tan, but note *orangish rump and belly*; indistinguishable from juv Grey-headed M. **Similar spp:** None in its range. Grey-headed M has all brownish-grey underparts, lacks breast band. **Voice:** Call, a high *ts ts ts. . . .* **Habits:** Little known. Small flocks in wet grasslands and edges of gardens. No nesting information. **Range:** Endemic. NG: Bird's Head, 1800–2100 m.

Grey-headed Mannikin *Lonchura caniceps* Pl 109
10 cm. Grasslands of lowlands and mid-mountains of SE. Common. **Adult:** *Grey head contrasts with dark beak and beady eye*. **Juv:** *Plain* tan with *dark bill*. **Subspp (3, in 2 groups):** *caniceps* (lowlands) darker, belly and flanks blackish, rump orange; *scratchleyana* (mountains) paler, belly and flanks buff, rump yellow. **Similar spp:** The other pale-headed mannikins, Grey-banded and Grey-crowned, do not co-occur; these have orange-brown breasts. Juv Chestnut-breasted M has paler bill and darker head with some dark streaking. **Voice:** Call, *seee seee seee. . . .* Also, *too*. **Habits:** In flocks of varying size, also mixes with other mannikins, especially Chestnut-breasted, in tight swarms. Occupies a variety of grasslands. Seasonal movements noted. Takes grass seeds from the plant or on the ground. Nest built in a tree, pandan or palm crown, or grass; a grassy globe with tubular side entrance, no lining. Eggs (4–6) white. **Range:** Endemic. NG: SE Pen, 0–2200 m. **Taxonomy:** Undescribed populations. The distinct montane race is possibly a separate species.

Grey-crowned Mannikin *Lonchura nevermanni* Pl 109
11 cm. Trans-Fly marshes and riverine grasses and nearby savannah. Local, but common where found. **Adult:** *White cap, black throat; rich brown underparts*; yellow tail. Some birds show a scattering of black feathers on breast and/or rump (and possibly indicate hybrid ancestry with Black M). Head often darker in **Female**. **Juv:** Tan with pale bill. **Similar spp:** No other mannikin has the black throat. Other juv Trans-Fly mannikins are darker, with more contrast between upperparts and underparts. **Voice:** Song (in captivity) begins with whispered series, then a short series of *tiks*, then ends in a complex *whheeeee*; song similar to that of Chestnut-breasted M, but last phrase bell-like. Call a typical

mannikin peeping, creating a tinkling sound in flocks. **Habits:** In flocks, highly social, often with Crimson Finch and Black M. Feeds on green grass seeds. Nests in clump of grass; no other nesting information from the wild. Eggs in captivity (3–6) white. **Range:** Endemic. NG: Trans-Fly.

Hooded Mannikin Lonchura spectabilis   Pl 110
10 cm. Grasslands from lowlands to mid-mountains. Locally common, nomadic. **Adult:** Highly variable from one population to the next, but always with the same *simple color pattern: inky black hood, brown saddle, and breast and belly of a contrasting color*. Hood sharply delineated (brown in some populations). Breast and belly vary locally from white, to buff, to rufous, rarely with barring on the flanks. **Juv:** Note *dark ear-patch*; upperparts medium brown, *underparts much paler, usually without a buff breast band*. **Subspp (??):** Current classification inadequate. **Similar spp:** Other black-headed mannikins have complex breast patterns. Other juv mannikins show less contrast between the head and upperparts vs the underparts, and breast darker than belly, e.g., Chestnut-breasted M. **Voice:** Song largely inaudible to humans, a series of whispers, clicks, and whistles. Call, a faint, sweet upslur, repeated by many birds in the flock, creating a tinkling sound. **Habits:** In flocks, usually of its own kind. Feeds on grass seed, taken from the plant or ground. Colonial nester (New Britain data). Nest built in grass or shrub; a grass ball with side spout for entrance. Eggs (3–6) white. **Range:** NG: NW Lowlands (Jayapura area), Sepik-Ramu, E Ranges to Huon and northeastern SE Pen, 0–2100 m, rarely higher. Also New Britain. **Taxonomy:** Racial taxonomy needs revision. Undescribed forms.

Chestnut-breasted Mannikin Lonchura castaneothorax   Pl 110
10 cm. Grasslands and disturbed areas, mostly in lowlands, ranging into mid-mountains. Locally common, nomadic. **Adult:** *Tan breast separated from white belly by black band*; note head pattern with *pale nape contrasting with black face*; flanks scalloped black. **Juv:** *Buff breast-band*; crown often dark-streaked; head pale buff-grey or darker grey; blue-grey bill. **Subspp (4, in 3 groups):** *sharpii* (NW Lowlands, Sepik-Ramu) grey crown; *boschmai* (Paniai Lakes, W Ranges) brown scalloping on flank, grey crown; *ramsayi* (SE Pen) head nearly all dark. **Similar spp:** Western Alpine M (does not co-occur) shares the tan breast but lacks the black breast band, has reduced black mask. Juv Hooded M lacks buff breast band. Juv Grey-headed M has a darker bill, lacks streaking on head. **Voice:** Song wheezy and mostly out of human range. Call tinkling. **Habits:** In flocks, sometimes with other species. Takes grass seed from the plant or the ground. Nest built in tall grass, pandan, or shrub; nest a ball of grass with side entrance lacking a spout. Eggs (4–6) white (AU data). **Range:** NG, Manam I, and SE Is (D'Entrecasteaux: Goodenough, Normanby), 0–1200 m and at ~1700 m at Paniai Lakes in W Ranges. NG: Patchily in North from NW Lowlands to SE Pen. Also AU; introduced Oceania.

Black Mannikin Lonchura stygia   Pl 110
11 cm. Tied to wetlands of W Trans-Fly: reeds, inundated or floating grass, nearby savannah, rice crops. Local, but common where found. **Adult:** *Black, with yellow tail and rump*. **Juv:** *Creamy white below with smudgy streaking on breast*; note *streaky head*. **Similar spp:** No other largely black mannikin inhabits Trans-Fly. Streak-headed M (to the north) has black tail and pale-streaked head. Juv Grand M has larger bill, lacks streaking on head, and has buffier underparts. Juv Grey-crowned M shows less contrast between upperparts and underparts; head lacks streaking. **Voice:** (from captives) Song typical of a mannikin but not ending in a *whheee* phrase. Calls, *tiu tiu* or *quet quet*. **Habits:** In flocks, sometimes with Grey-crowned M or Crimson Finch. Feeds on grass seed. Nests in tall, floating grass; nest is flask-shaped, made of grass leaves, lined inside with soft seed heads, minus the seeds. Eggs (4–5) white. **Range:** Endemic. NG: Trans-Fly.

Black-breasted Mannikin Lonchura teerinki   Pl 110
11 cm. Mid-mountain grasslands of W Ranges. Common. **Adult:** *Black head, breast, and flank streak*. **Juv:** Smaller than juv Western Alpine M and has streaked breast. **Similar spp:** Western Alpine M, the only other mannikin in its range (generally at higher elevation), is very different, with black mask, buff breast, and barred flanks. Eastern Alpine M (not co-occurring) has white breast. **Voice:** Song, no information. Call, variously reported as *tiu* or *tseep*. **Habits:** Flocks in man-made grasslands, old gardens, and weedy second growth in the mountains. Nest is a grass ball with side entrance. Eggs (3–5) white (nest and egg data from captivity). **Range:** Endemic. NG: central W Ranges, 1200–2200 m.

Eastern Alpine Mannikin *Lonchura monticola* Pl 110
(Alpine Mannikin)
11–12 cm. Alpine grasslands and moors of SE Pen. Locally common. **Adult:** *Black head, white breast, black lower breast band and flank-stripe.* **Juv:** *Blackish face*; faintly streaked breast band. **Similar spp:** No other mannikin co-occurs at such high elevation in the SE. Chestnut-breasted M, at lower elevations, has tan breast. Western Alpine M has black breast. **Voice:** Song, no information. Call musical, *deet deet deet. . . .* **Habits:** Usually in flocks of its own kind, although may wander down the mountain where it encounters Grey-headed and Hooded Ms. Often perches in bushes at edge of alpine scrub. A typical mannikin in its behavior. No nesting information. **Range:** Endemic. NG: SE Pen, 2700–3900 m, occasionally lower.

Western Alpine Mannikin *Lonchura montana* Pl 110
(Snow Mountain Mannikin)
11–12 cm. Alpine grassland and moors; edge of gardens at lower elevations. **Adult:** *Black crown and face, buff breast, barred flanks.* **Juv:** Lacks the obscure breast streaking of Black-breasted and Eastern Alpine Ms, but like them shows a little black around the bill; much larger than Black-breasted. **Similar spp:** Black-breasted M, at lower elevations, has black breast. Chestnut-breasted M (does not co-occur) has black breast band below the buff breast; yellow tail (rather than black). **Voice:** Call note *tyu*. **Habits:** In small flocks in boggy grasslands on the high plateau country. Feeds on seeds of grasses and herbs. Nests in grasses beside water. **Range:** Endemic. NG: W and Border Ranges, 2100–4100 m, generally above 3000 m.

WAGTAILS AND PIPITS: Motacillidae

This is another mainly Eurasian and African family, with about 70 species worldwide. Only 4 occur in New Guinea: 2 pipits as breeding birds, and 2 wagtails as northern migrants. These are fine-billed, slender songbirds that live on the ground in open habitats at various altitudes, foraging in grassy or barren areas. The 2 pipits are cryptic-colored, streaked dark brown above, paler below. The 2 wagtails generally are found in dull, greyish winter plumage, although individuals in bright yellow breeding dress have been recorded. All species show white outer tail feathers in flight and pump their tails up and down while on the ground, hence the name wagtail. They have a strong, bounding flight. Diet is generally of arthropods, with the high-altitude species also taking vegetable matter.

Eastern Yellow Wagtail *Motacilla tschutschensis* Pl 104
(Yellow Wagtail, *Motacilla flava, M. simillima*)
16–17 cm. Uncommon Palearctic migrant, mainly on Indonesian side. Overwinters in lowlands, where it is the wagtail most likely seen. Frequents short grass; not so tied to water as Grey W. *Black legs and medium-length tail* (not longer than body). *Lacks yellow rump*, which is instead the same color as back (grey or green). **Adult:** Nonbreeding—Upperparts brownish olive, underparts pale yellow; white eyebrow; **Male** sometimes brighter. Breeding—*Yellow-green back and rump, yellow throat, grey cap, white eyebrow*; **Male** sometimes brighter with more distinct head pattern. **Imm** (**Juv molts on summer grounds**): Variable; generally similar to Nonbreeding female, with pale yellow or white underparts; most show some whitish tips on retained Juv wing-coverts. **In flight:** Lacks a white stripe across flight-feathers. **Subspp (1):** *tschutschensis* (taiga and Arctic of NE Asia and Alaska). **Similar spp:** Grey W in all plumages has pale legs, yellow rump, and, in flight, white stripe across base of primaries. **Voice:** Call, an explosive, repeated *szweep*, usually given in flight. **Habits:** Singly or in groups. Forages on open ground, such as around rice fields, cattle paddocks, dried mudflats, wet meadows. Walks and runs to capture insects. Perches on low bushes. **Range:** NG, Kofiau (NW Is), and Aru Is, mainly in lowlands, rarely in mts to 4000 m. Breeds E Asia and N Amer, winters SE Asia to NG and AU. **Taxonomy and Extralimital spp:** Classification of the yellow wagtails (*M. flava*, in the broad sense) is complex and unresolved. More than 1 species is involved, but species limits have not been set. One additional form has been reported in the NG Region, the **Green-headed Yellow Wagtail** (*M. t. taivana*) of the boreal forests of NE Asia, variously considered a species or subspecies of Eastern Yellow. Because these sightings need confirmation, this form is listed as hypothetical. All plumages show

a *yellowish eyebrow* (not white or cream as in Eastern Yellow). Difficult to separate in Nonbreeding plumage, but Breeding plumage has a diagnostic green crown, same color as back.

Grey Wagtail *Motacilla cinerea* — Pl 104

18–19 cm. Palearctic migrant; the common wagtail overwintering in NG mountains. Visits lowlands on passage. Attached to water: creek beds, wet gravel or paved roads, bogs, and other open, damp habitats. *Very slim build*. Note *pale legs, long tail* (longer than body). *Yellow rump* contrasts with grey back. **Adult:** Nonbreeding—*yellow on breast and belly contrasts with white throat*. Breeding—bright yellow underparts (**Male**, *black throat*; **Female**, throat all white or partly black). **Imm (Juv molts on summer grounds):** Similar to Nonbreeding adult. **In flight:** Flashes white wing stripe and tail feathers; flight strongly undulating. **Subspp (1):** *cinerea* (mainland Eurasia). **Similar spp:** See Eastern Yellow WT. **Voice:** Call, a sharp *tzit-tzit*, given in flight. **Habits:** Solitary, occasionally in twos, and perhaps territorial on wintering grounds. Favors rocky, highland habitats throughout its range. Feeds on open ground near water, taking arthropods, snails, rarely small tadpoles and fish. Perches in trees.
Range: Probably throughout NG Region. Recorded from NG (mainly in N scarp), NW Is (Batanta, Misool, Waigeo), Bay Is (Yapen), and Karkar I, 600–2500 m. Breeds Eurasia; winters in Africa, S Asia to NG Region and Bismarck Is.

Australasian Pipit *Anthus novaeseelandiae* — Pl 104
(**Australian Pipit**, **Richard's Pipit**, *A. australis*)

15 cm. Uncommon in montane grasslands and dry lowlands of NE; frequents short grass and roads through grassland. Seen perched on rocks, posts, shrubs; note upright posture. A small, cryptic, *streaky* ground-feeding bird. *Slender profile, thin bill, long pinkish legs, white outer tail feathers. Prominently streaked white breast*. **Adult:** Upperparts dark brown. **Juv:** Pale edges to feathers of upperparts. **Subspp (1):** *exiguus* (endemic). **Similar spp:** Horsfield's Bushlark more compact, thick-billed. Alpine P darker, especially on breast, which is unstreaked in adult. **Voice:** Song a repeated, short trilling warble (AU data). Makes song-flights. Call, a brisk *chwit*. **Habits:** Solitary, in pairs, or in scattered flocks. Forages inconspicuously in short grass of pastures, airstrips, lawns, roadsides. Co-occurs locally with Alpine P along roads through alpine moors. Feeds on insects. Nests in a depression on the ground; nest a cup made of grass (AU data). Eggs (2–3) cream with spots and blotches. **Range:** NG: E Ranges and western SE Pen, 1200–2200 m (rarely to 3000 m) and down nearly to sea level in Markham and Ramu valleys. Also AU and NZ. **Taxonomy:** Split from Richard's Pipit (*A. richardi*).

Alpine Pipit *Anthus gutturalis* — Pl 104

18 cm. Uncommon in alpine grasslands. *A large, dark-breasted, robust pipit*. **Adult:** *White "spectacles" (lores and eye-ring) and throat*; black "eye shadow" and mark on side of neck; *plain buff-grey breast, without streaking*. Outer tail feathers dingy white. **Juv:** Facial pattern lacking or obscure, heavily streaked below, yet overall much darker than Australasian P. **Subspp (3?, in 2 groups):** *gutturalis* (E Ranges, Huon, SE Pen) black neck stripe; *wollastoni* (W Ranges) black on side of neck reduced or absent. **Similar spp:** Australasian P is white below, with streaked breast. Papuan Grassbird has long, pointed tail and rufous crown. Pied Bushchat female has a white rump. **Voice:** Song a succession of high, clear notes, each repeated several times and running into a trill. Call (usually on the wing) a faint, high-pitched *tsip* or *tsee*, or *tsee tsee tsee*. **Habits:** Solitary, in pairs, rarely in small flocks. Favors short grass in rolling hilly terrain, often near the cover of small shrubs. Perches atop rocks or tree ferns, surveying habitat or singing; poses with bill held high. Flight strong and low, not towering like that of Australasian P or wagtails. Walks on ground with waddling gait. Feeds on small invertebrates and seeds. Nest built into a steep, grassy bank; cup-shaped, of fine grass. Egg, no information.
Range: Endemic. NG Central Ranges and Huon, 3200–4500 m, sometimes lower.

Index

For species names, numbers in **bold** type refer to the *page number* for the plate illustration for that species, whereas plain type numbers refer to species accounts.

abbreviations list 13
Acanthiza cinerea **194**, 426
    murina **192**, 426
Acanthizidae 422
Accipiter buergersi 298
    cirrhocephalus 297
    cirrocephalus **78, 82**, 297
    doriae 298
    fasciatus **78, 82**, 296
    hiogaster **78, 82**, 296
    melanochlamys **78, 82**, 297
    meyerianus **80, 82**, 297
    novaehollandiae 296
    poliocephalus **78, 82**, 297
    soloensis **78, 82**, 295
Accipitridae 292
Aceros plicatus 394
Acridotheres tristis **254**, 506
Acrocephalidae 501
Acrocephalus arundinaceus 501
    australis **250**, 501
    orientalis **250**, 501
    stentoreus 501
Actitis hypoleucos **100**, 320
Aegotheles affinis **150**, 381
    albertisi **150**, 380
    archboldi **150**, 380
    bennettii **150**, 380
    cristatus **150**, 381
    insignis **148**, 379
    tatei **148**, 379
    wallacii **150**, 379
Aegothelidae 378
Aepypodius arfakianus **42**, 264
    bruijnii **42**, 265
Aerodramus hirundinaceus **152**, 382
    nuditarsus **152**, 383
    papuensis **152**, 383
    vanikorensis **152**, 383
Ailuroedus buccoides **256**, 396
    crassirostris 396
    melanotis **256**, 396
Alaudidae 494
Alcedinidae 392
Alcedo atthis **162**, 392
    azurea 393
    lepida 392
    pusilla 393
Alcyone azurea 393
    pusilla 393
Aleadryas rufinucha **206**, 448
Alisterus amboinensis **130**, 360
    chloropterus **130**, 360
Alopecoenas beccarii **110**, 336
    jobiensis **110**, 335
Amalocichla incerta **168**, 487
    sclateriana **168**, 487
Amaurornis cinerea 306
    ineptus 307
    moluccana **90**, 306
    olivaceus 306
Amblyornis flavifrons **170**, 398
    germana **170**, 397
    inornata **170**, 398
    macgregoriae **170**, 397
    subalaris **170**, 397
Anas acuta 272
    gibberifrons 272

    gracilis **48**, 272
    penelope **48**, 271
    querquedula **48**, 272
    superciliosa **48**, 271
    waigiuensis 270
Anatidae 269
Androphobus viridis **166**, 436
Anhinga melanogaster 291
    novaehollandiae **62**, 291
Anhingidae 291
Anous albus 326
    minutus **58**, 326
    solidus **58**, 326
Anseranas semipalmata **62**, 269
Anseranatidae 269
Anthus australis 516
    gutturalis **248**, 516
    novaeseelandiae **248**, 516
Anurophasis monorthonyx **46**, 267
Aplonis cantoroides **254**, 504
    magna **254**, 505
    metallica **254**, 504
    mysolensis **254**, 505
    mystacea **254**, 504
Apodidae 382
Aprosmictus erythropterus **130**, 360
Apus pacificus **152**, 384
Aquila audax **72, 74**, 299
    gurneyi **72, 74**, 299
    weiskei 299
Archboldia papuensis **170**, 398
    sanfordi 398
Ardea cinerea 285
    garzetta 286
    ibis **68**, 284
    intermedia **68**, 285
    modesta **68**, 285
    novaehollandiae 286
    pacifica **68**, 284
    picata 286
    purpurea 285
    sacra 286
    striata 284
    sumatrana **68**, 285
Ardeidae 282
Ardenna carneipes 276
    pacifica 275
    tenuirostris 275
Ardeola coromanda 284
    speciosa 284
    striata 284
Ardeotis australis **64**, 302
Arenaria interpres **102**, 320
Arses insularis **218**, 462
    telescophthalmus 462
    telescopthalmus **218**, 462
Artamidae 441
Artamus cinereus **200**, 441
    leucorynchus **200**, 441
    maximus **200**, 441
Aru Islands 26
Astrapia, Arfak **230**, 482
    Huon **230**, 482
    Ribbon-tailed **230**, 482
    Rothschild's 482
    Splendid **230**, 482
    Stephanie's **230**, 483
Astrapia mayeri **230**, 482

    nigra **230**, 482
    rothschildi **230**, 482
    splendidissima **230**, 482
    stephaniae **230**, 483
Aviceda subcristata **70, 82**, 292
Avocet, Red-necked 311
Aythya australis **48**, 272

Babbler, Grey-crowned **210**, 429
    Papuan **210**, 429
    Rufous 429
Barn-Owl, Australian **144**, 372
Bay Islands 25
Baza, Pacific **70, 82**, 292
Bee-eater, Blue-tailed **158**, 393
    Rainbow **158**, 394
Bellbird, Piping **208**, 449
    Rufous-naped **206**, 448
Berryhunter, Mottled **214**, 448
Berrypecker, Black **196**, 432
    Crested 435
    Eastern Crested **196**, 435
    Fan-tailed **196**, 433
    Mid-mountain **196**, 433
    Obscure **196**, 432
    Painted 435
    Spotted **196**, 433
    Streaked **196**, 433
    Tit **196**, 435
    Western Crested **196**, 435
bird, parts of a 18
bird geography for New Guinea 23
Bird of Paradise, Arfak Six-wired 476
    Blue **238**, 484
    Count Raggi's 486
    Crested 431
    Emperor **238**, 484
    Emperor of Germany 484
    Goldie's **238**, 485
    Greater **240**, 486
    King **236**, 483
    King of Saxony **234**, 476
    Lawes's Six-wired 477
    Lesser **240**, 485
    Loria's 430
    Macgregor's 412
    Magnificent **236**, 484
    Queen Carola's Six-wired 478
    Princess Stephanie's 483
    Raggiana **240**, 486
    Red **238**, 485
    Ribbon-tailed 482
    Shield-billed 431
    Six-wired 478
    Superb **234**, 479
    Twelve-wired **236**, 478
    Wahnes's Six-wired 477
    Wattle-billed 431
    Wattled 431
    Wilson's **236**, 484
    Yellow-breasted 431
bird regions for New Guinea 25
Bird's Head 25
Bird's Neck 25
Bittern, Australian Little **66**, 282
    Black **66**, 283
    Black-backed 282
    Cinnamon **66**, 283

517

Forest **66**, 282
Little 282
New Guinea Zebra 282
Von Schrenck's 283
Yellow **66**, 283
Boatbill, Black-breasted **220**, 438
Yellow-breasted **220**, 438
Boobook, Australian 374
Common 374
Jungle 374
Papuan **146**, 374
Rufous 373
Southern **146**, 374
Booby, Abbott's 289
Blue-faced 289
Brown **54**, 289
Masked **54**, 289
Red-footed **54**, 289
Border Ranges 27
Bowerbird Adelbert 399
Archbold's **170, 174**, 398
Beck's 399
Fawn-breasted **172, 174**, 400
Fire-maned **172, 174**, 399
Flame **172, 174**, 399
Golden 398
Golden-fronted **170, 174**, 398
Huon **170, 174**, 397
Lauterbach's 399
Macgregor's **170, 174**, 397
Masked **172, 174**, 398
Sanford's 398
Streaked **170, 174**, 397
Tomba 398
Vogelkop **170, 174**, 398
Yellow-breasted **172, 174**, 399
Yellow-fronted 398
Brolga **64**, 308
Bronze-Cuckoo, Golden 368
Malay 369
Bronzewing, New Guinea **110**, 335
Brush-Cuckoo, Grey-breasted 370
Brushturkey, Black-billed 265
Brown-collared 265
Bruijn's 265
Collared 265
Red-billed **42**, 265
Red-legged **42**, 265
Waigeo **42**, 265
Wattled **42**, 264
Yellow-legged **42**, 265
Bubulcus coromandus 284
Bucerotidae 394
Bulbul, Sooty-headed **254**, 494
Bulweria bulweria **52**, 276
Burhinidae 309
Burhinus grallarius **92**, 309
magnirostris 309
neglectus 309
Bushchat, Pied **248**, 507
Bush-hen 306
Bush-hen, Common 306
Rufous-tailed **90**, 306
Pale-vented 306
Bushlark, Horsfield's **248**, 494
Singing 494
Bustard, Australian **64**, 302
Butastur indicus **70, 74**, 298
Butcherbird, Black **200**, 439
Black-backed **200**, 440
Hooded **200**, 440
Louisiades 440
Tagula **200**, 440

Butorides striata **66**, 284
Buttonquail, Red-backed **46**, 308
Buzzard, Crested Honey 293
Grey-faced **70, 74**, 298
Long-tailed **80, 82**, 293
Long-tailed Honey 293
Oriental Honey 293

Cacatua galerita **122**, 349
pastinator 349
sanguinea **122**, 349
Cacatuidae 348
Cacomantis castaneiventris **140**, 370
flabelliformis **140**, 370
leucolophus 369
pallidus 369
pyrrhophanus 370
variolosus **140, 142**, 370
Calidris acuminata **104**, 323
alba **102**, 321
bairdii **104**, 322
canutus **102**, 321
ferruginea **104**, 323
fuscicollis 321
melanotos **104**, 322
minuta **102**, 322
ruficollis **102**, 321
subminuta **104**, 322
tenuirostris **102**, 320
Caliechthrus leucolophus **140**, 369
Caligavis obscura **186**, 415
subfrenata **186**, 415
Calliope calliope **248**, 507
Caloenas nicobarica **118**, 338
Calonectris leucomelas **52**, 275
Campephagidae 442
Campochaera sloetii **202**, 444
Caprimulgidae 376
Caprimulgus indicus 377
jotaka **148**, 377
macrurus **148**, 377
Carterornis chrysomela **220**, 465
Casmerodius albus 284
Cassowary, Australian 263
Double-wattled 263
Dwarf **40**, 262
Little 262
Mountain 262
Northern **40**, 263
One-wattled 263
Single-wattled 263
Southern **40**, 263
Two-wattled 263
Casuariidae 262
Casuarius bennetti **40**, 262
bennettii 262
casuarius **40**, 263
unappendiculatus **40**, 263
Catbird, Black-eared **256**, 396
Green 396
Spotted 396
White-eared **256**, 396
White-throated 396
Catharacta maccormicki 331
Cecropis daurica **154**, 496
Centropodidae 363
Centropus bernsteini **138**, 364
chalybeus **138**, 363
menbeki **138**, 363
phasianinus **138**, 364
Ceyx azureus **162**, 393
lepidus 392
pusillus **162**, 393

solitarius **162**, 392
Chaetorhynchus papuensis **254**, 467
Chaetura caudacuta 384
novaeguineae 383
Chalcites basalis **142**, 367
lucidus **142**, 368
meyerii **142**, 368
minutillus **142**, 369
osculans **142**, 367
ruficollis **142**, 367
Chalcophaps chrysochlora 338
indica **110**, 338
longirostris **110**, 338
stephani **110**, 339
Chalcopsitta atra **128, 136**, 356
duivenbodei **128, 136**, 357
scintillata **128, 136**, 357
Charadriidae 311
Charadrius alexandrinus 313
asiaticus 313
cinctus 312
dubius **94**, 312
hiaticula 313
leschenaultii **94**, 313
mongolus **94**, 313
ruficapillus 313
veredus **94**, 313
Charmosyna josefinae **126, 136**, 353
multistriata **124, 136**, 352
papou **126**, 354
placentis **124, 136**, 353
pulchella **126, 136**, 353
rubrigularis **124, 136**, 352
rubronotata **124, 136**, 353
stellae **126, 136**, 354
wilhelminae **124, 136**, 352
Chat, Pied 507
Chenonetta jubata 271
Chenorhamphus campbelli **176**, 402
grayi **176**, 401
Chlamydera cervinventris **172**, 400
lauterbachi **172**, 399
Chlidonias hybrida **60**, 330
leucopterus **60**, 330
niger 330
Choriotis australis 302
Chroicocephalus novaehollandiae **58**, 327
ridibundus **58**, 327
Chrysococcyx basalis 367
lucidus 368
malayanus 369
megarhynchus 366
meyerii 368
minutillus 369
osculans 367
ruficollis 367
Cicadabird 446
Black **204**, 447
Black-bellied **204**, 445
Common **204**, 446
Grey-headed **204**, 446
Papuan **204**, 445
Slender-billed 446
Cicinnurus magnificus 484
regius **236**, 483
respublica 484
Ciconiidae 280
Cinclosoma ajax **166**, 438
Cinclosomatidae 436
Cinnyris frenatus 509
idenburgi **198**, 509
jugularis **198**, 509
sericeus 509

518

Circus approximans **76**, 295
   melanoleucos 295
   spilothorax **76**, 295
Cisticola, Golden-headed **250**, 503
   Zitting **250**, 503
Cisticola exilis **250**, 503
   juncidis **250**, 503
Cisticolidae 503
Climacteridae 400
Climacteris placens 400
climate change 31
climate of New Guinea 21
Clytoceyx rex **158**, 388
Clytomyias insignis **176**, 401
Cnemophilidae 430
Cnemophilus loriae **170**, 430
   macgregorii **170**, 431
Cockatoo, Palm **122**, 348
   Sulphur-crested **122**, 349
Collocalia esculenta **152**, 382
   hirundinacea 382
   papuensis 383
   vanikorensis 383
   whiteheadi 383
Colluricincla ferrugineus 451
   harmonica **210**, 451
   incertus 451
   megarhyncha **180**, **210**, 450
   tenebrosa **206**, 451
Columba livia 333
   vitiensis **120**, 333
Columbidae 332
Conopophila albogularis **184**, 414
conservation 30
Coot, Eurasian **50**, 307
Coraciidae 384
Coracina boyeri **204**, 443
   caeruleogrisea **202**, 442
   incerta 445
   lineata **204**, 442
   longicauda **202**, 442
   melaena 447
   melas 447
   montana 445
   morio 445
   novaehollandiae **202**, 443
   papuensis **202**, 443
   schisticeps 446
   tenuirostris 446
Corella, Little **122**, 349
Cormobates placens **206**, 400
Cormorant, Australian Pied 290
   Great **62**, 290
   Little Black **62**, 290
   Little Pied **62**, 290
Corvidae 471
Corvus fuscicapillus **226**, 471
   orru **226**, 472
   tristis **226**, 472
Coturnix australis 268
   chinensis 268
   monorthonyx 267
   ypsilophora **46**, 268
Coucal, Biak **138**, 363
   Black-billed 364
   Greater Black **138**, 363
   Ivory-billed 363
   Lesser Black **138**, 364
   Pheasant **138**, 364
Cracticidae 439
Cracticus cassicus **200**, 440
   louisiadensis **200**, 440
   mentalis **200**, 440

   quoyi **200**, 439
   tibicen **200**, 440
Crake, Baillon's **90**, 306
   Little 306
   Marsh 306
   Red-necked **88**, 304
   Spotless **90**, 306
   White-browed **90**, 306
Crane, Australian 308
Crateroscelis murina **168**, 423
   nigrorufa **168**, 423
   robusta **168**, 423
Crow, Bare-eyed 472
   Brown-headed **226**, 471
   Grey **226**, 472
   Torresian **226**, 472
Cuckoo, Black-eared **142**, 367
   Brush **140**, **142**, 370
   Channel-billed **138**, 366
   Chestnut-breasted **140**, 370
   Fan-tailed **140**, 370
   Golden Bronze- 368
   Grey-breasted Brush-, 370
   Himalayan **140**, 371
   Horsfield's 371
   Horsfield's Bronze **142**, 367
   Little Bronze **142**, 369
   Long-billed **142**, 366
   Long-tailed **140**, 366
   Malay Bronze- 369
   Oriental **140**, 371
   Pallid **140**, 369
   Rufous-throated Bronze **142**, 367
   Shining Bronze **142**, 368
   White-crowned **140**, 369
   White-eared Bronze **142**, 368
Cuckoo-Dove, Bar-tailed 334
   Black-billed **108**, 334
   Brown **108**, 334
   Giant 334
   Great **108**, 334
   Mackinlay's **108**, 335
   Rusty 334
   Slender-billed 334
   Spot-breasted 335
Cuckooshrike, Barred **204**, 442
   Bar-tailed 334
   Black 447
   Black-faced **202**, 443
   Black-shouldered 445
   Black-tipped 446
   Boyer's **204**, 443
   Golden **202**, 444
   Hooded **202**, 442
   Large-billed 442
   Müller's 445
   New Guinea 447
   Papuan Black 447
   Sharpe's 445
   Stout-billed **202**, 442
   White-bellied **202**, 443
   Yellow-eyed 442
Cuculidae 364
Cuculus castaneiventris 370
   horsfieldi 371
   optatus **140**, 371
   pallidus 369
   pyrrhophanus 370
   saturatus 140, 371
   variolosus 370
Curlew, Beach 309
   Bristle-thighed 317
   Bush 309

   Eastern **98**, 318
   Far Eastern 318
   Little **98**, 317
Cyclopsitta diophthalma **134**, **136**, 359
   gulielmitertii **134**, 359
Cygnus atratus 270

Dacelo gaudichaud **158**, 389
   leachii **158**, 388
   tyro **158**, 389
Daphoenositta chrysoptera **206**, 447
   miranda **206**, 448
   papuensis 447
Darter 291
Darter, Australasian **62**, 291
   Oriental 291
Dendrocygna arcuata **48**, 270
   eytoni **48**, 270
   guttata **48**, 269
Devioeca papuana **246**, 489
Dicaeum geelvinianum **198**, 508
   hirundinaceum **198**, 509
   nitidum 508
   pectorale **198**, 508
Dicruridae 466
Dicrurus bracteatus **254**, 466
   hottentottus 466
diet, avian 28
Diphyllodes magnificus **236**, 484
   respublica **236**, 484
Dollarbird, Oriental **158**, 384
Domicella hypoinochroa 355
   lory 355
Dotterel, Little Ringed 312
   Oriental 313
   Red-kneed **92**, 312
Dove, Bar-shouldered **108**, 336
   Bar-tailed Cuckoo- 334
   Beautiful Fruit- **114**, 341
   Beccari's Ground 336
   Black-billed Cuckoo- **108**, 334
   Brown Cuckoo- **108**, 334
   Claret-breasted Fruit- **116**, 343
   Common Emerald **110**, 338
   Coroneted Fruit- **114**, 341
   Diadem Fruit 341
   Dwarf Fruit- **116**, 344
   Emerald 338
   Emerald Ground 338
   Giant Cuckoo- 334
   Golden-heart 335
   Great Cuckoo- **108**, 334
   Knob-billed Fruit- **116**, 344
   Mackinlay's Cuckoo- **108**, 335
   Magnificent Fruit 339
   Moluccan Fruit- **116**, 342
   Mountain Fruit- **116**, 342
   Orange-bellied Fruit- **116**, 343
   Orange-fronted Fruit- **114**, 340
   Ornate Fruit- **114**, 340
   Pacific Emerald **110**, 338
   Peaceful **108**, 336
   Pink-spotted Fruit- **114**, 340
   Rock 333
   Rose-crowned Fruit- **114**, 341
   Rusty Cuckoo- 334
   Slender-billed Cuckoo- 334
   Spot-breasted Cuckoo- 335
   Spotted **108**, 333
   Spotted Turtle-333
   Stephan's Emerald **110**, 339
   Superb Fruit- **114**, 341
   Wallace's Fruit- **114**, 341

White-bibbed Fruit- **116**, 342
White-breasted Fruit- 342
Wompoo Fruit- **114**, 339
Yellow-bibbed Fruit- **116**, 343

Dowitcher, Asian **96**, 316
   Asiatic 316
   Long-billed 316
Drepanornis albertisi **232**, 480
   bruijnii **232**, 480
Drongo, Mountain 467
   Pygmy **254**, 467
   Spangled **254**, 466
Drymodes beccarii **164**, 490
   superciliaris 490
Dryolimnas pectoralis 304
Duck, Australian Black 271
   Australian White-eyed 272
   Australian Wood-271
   Black 271
   Grass Whistle- 270
   Grey 271
   Maned 271
   Pacific Black **48**, 271
   Plumed Tree 270
   Plumed Whistling **48**, 270
   Salvadori's 270
   Spotted Tree 269
   Spotted Whistling **48**, 269
   Wandering Whistling **48**, 270
   Water Whistle- 270
   Whistling Tree 270
Ducula bicolor **118**, 347
   chalconota **120**, 346
   concinna **118**, 345
   geelvinkiana **118**, 345
   mullerii **120**, 347
   myristicivora **118**, 345
   pacifica **118**, 345
   perspicillata **118**, 344
   pinon **120**, 346
   pistrinaria **118**, 346
   rufigaster **120**, 345
   spilorrhoa **118**, 347
   zoeae **120**, 347
Dupetor flavicollis 283

Eagle, Gurney's **72, 74**, 299
   Kapul 298
   Little 299
   New Guinea 298
   New Guinea Harpy-**80, 82**, 298
   Papuan 298
   Pygmy **70, 72, 74**, 299
   Wedge-tailed **72, 74**, 299
   White-bellied Sea-**72, 74**, 294
   White-breasted Sea 294
Eastern Ranges 27
Eclectus roratus **122**, 360
Edolisoma incertum **204**, 445
   melas **204**, 447
   montanum **204**, 445
   schisticeps **204**, 446
   tenuirostre **204**, 446
Egret, Cattle **68**, 284
   Eastern Cattle 284
   Eastern Great **68**, 285
   Eastern Reef-**68**, 286
   Great 285
   Intermediate **68**, 285
   Lesser 285
   Little **68**, 286
   Pacific Reef 286

   Pied 286
   Plumed 285
Egretta alba 285
   coromanda 284
   garzetta **68**, 286
   intermedia 285
   novaehollandiae **68**, 286
   picata **68**, 286
   sacra **68**, 286
Elanus axillaris 293
   caeruleus **70, 86**, 293
Elevation, importance of 27
Entomyzon cyanotis **184**, 411
Eopsaltria placens 493
   pulverulenta 492
Eos cyanogenia **128, 136**, 357
   squamata **128, 136**, 357
Ephippiorhyncus asiaticus **64**, 280
Epimachus albertisi 480
   bruijnii 480
   fastosus **232**, 481
   fastuosus 481
   meyeri **232**, 480
equipment for field 33
Erithacus calliope 507
Erythrogonys cinctus **92**, 312
Erythropitta erythrogaster **164**, 395
Erythrotriorchis buergersi **80, 82**, 298
Erythrura papuana **258**, 512
   trichroa **258**, 511
Esacus giganteus 309
   magnirostris **92**, 309
   neglectus 309
Estrildidae 510
etiquette for visitors 35
Euaegotheles insignis 379
   tatei 379
Eudynamys orientalis **140**, 365
   scolopacea 365
   taitensis 366
Eugerygone rubra **246**, 488
Eulabeornis castaneoventris **88**, 305
   plumbeiventris 305
Eulacestoma nigropectus **206**, 449
Eulacestomatidae 449
Eulipoa wallacei **44**, 266
Eupetes caerulescens 437
   castanonotus 437
   leucostictus 436
Eupodia veredus 313
Eupodotis australis 302
Eurostopodus albogularis 376
   archboldi **148**, 377
   argus **148**, 376
   guttatus 376
   mystacalis **148**, 376
   papuensis **148**, 377
Eurystomus orientalis **158**, 384
Excalfactoria chinensis **46**, 268

Fairywren, Broad-billed **176**, 401
   Campbell's **176**, 402
   Emperor **176**, 402
   Orange-crowned **176**, 401
   Wallace's **176**, 401
   White-shouldered **176, 250**, 402
Falco berigora **84, 86**, 301
   cenchroides **84, 86**, 300
   hypoleucos , 301
   longipennis **84, 86**, 301
   moluccensis **84, 86**, 300
   peregrinus **84, 86**, 302
   severus **84, 86**, 301

Falcon, Brown **84, 86**, 301
   Grey 301
   Little 301
   Peregrine **84, 86**, 302
Falconidae 300
Fantail, Arafura **224**, 470
   Black **224**, 470
   Black Thicket- **222**, 468
   Chestnut-backed 469
   Chestnut-bellied **222**, 470
   Dimorphic **224**, 469
   Friendly **222**, 470
   Mangrove **222**, 471
   Mangrove Grey 471
   Northern **222**, 471
   Rufous **224**, 469, 470
   Rufous-backed **224**, 469
   Sooty Thicket- **222**, 469
   White-bellied Thicket- **222**, 468
field research in New Guinea 34
Figbird 458
   Australasian **256**, 458
Fig-Parrot, Double-eyed **134, 136**, 359
   Edward's **132**, 358
   Large **132, 136**, 358
   Orange-breasted **134**, 359
   Salvadori's **132**, 358
Finch, Crimson **258**, 511
Firetail, Mountain **258**, 511
Flowerpecker, Louisiade 508
   Olive-crowned **198**, 508
   Papuan 508
   Red-capped **198**, 508
Flycatcher, Biak 460
   Biak Black **216**, 460
   Broad-billed **216**, 460
   Canary 489
   Common Shining 461
   Grey-streaked **248**, 508
   Leaden **216**, 460
   Lemon-bellied **246**, 490
   Olive **246**, 489
   Paperbark 461
   Papuan **246**, 489
   Restless **216**, 461
   Satin **216**, 461
   Shining **216**, 461
   Torrent **242**, 490
   Yellow 489
   Yellow-footed 489
   Yellow-legged **246**, 489
Flyrobin, Canary 489
   Lemon-bellied 490
   Olive 489
   Torrent 490
   Yellow-legged 489
Forest-Rail, Chestnut **88**, 303
   Forbes's **88**, 304
   Mayr's **88**, 304
   White-striped **88**, 304
Fregata andrewsi 288
   ariel **56**, 288
   minor **56**, 288
Fregatidae 288
Fregetta grallaria 278
   tropica **52**, 278
Friarbird, Brass's **182**, 410
   Helmeted **182**, 411
   Little **182**, 410
   Meyer's **182**, 410
   New Guinea 411
   Noisy **182**, 411
Frigatebird, Christmas Island 288

Great **56**, 288
Lesser **56**, 288
Frogmouth, Marbled **144**, 375
  Papuan **144**, 375
Fruit-Dove, Beautiful **114**, 341
  Claret-breasted **116**, 343
  Coroneted **114**, 341
  Dwarf **116**, 344
  Knob-billed **116**, 344
  Moluccan **116**, 342
  Mountain **116**, 342
  Orange-bellied **116**, 343
  Orange-fronted **114**, 340
  Ornate **114**, 340
  Pink-spotted **114**, 340
  Rose-crowned **114**, 341
  Superb **114**, 341
  Wallace's **114**, 341
  White-bibbed **116**, 342
  White-breasted 342
  Wompoo **114**, 339
  Yellow-bibbed **116**, 343
Fulica atra **50**, 307

Gallicolumba rufigula **110**, 335
Gallinago hardwickii **96**, 315
  megala **96**, 316
  stenura **96**, 316
Gallinula olivacea 306
  tenebrosa **50**, 307
Gallirallus castaneoventris 305
  pectoralis 304
  philippensis **90**, 305
  torquatus **90**, 305
Garganey **48**, 272
Garritornis isidorei **210**, 429
Gavicalis versicolor **186**, 418
Gelochelidon nilotica **60**, 327
Gennaeodryas placens **246**, 493
Geoffroyus geoffroyi **130**, 361
  simplex **130**, 361
geographic scope 14
geology of New Guinea 20
Geopelia humeralis **108**, 336
  placida **108**, 336
  striata 336
Gerygone, Ashy 426
  Brown-breasted **194**, 428
  Fairy **194**, 428
  Green-backed **194**, **250**, 427
  Grey 426
  Large-billed **194**, 428
  Mangrove **194**, 428
  Mountain 426
  Treefern 428
  White-throated **194**, 427
  Yellow-bellied **194**, 427
Gerygone chloronota **194**, **250**, 427
  chloronotus 427
  chrysogaster **194**, 427
  cinerea 426
  levigaster **194**, 428
  magnirostris **194**, 428
  olivacea **194**, 427
  palpebrosa **194**, 428
  ruficollis **194**, 428
Glareola maldivarum **106**, 325
Glareolidae 325
Glycichaera fallax **180**, 405
Godwit, Bar-tailed **96**, 317
  Black-tailed **96**, 317
Goldenface **212**, 422
Goose, Cotton Pygmy **50**, 271

Green Pygmy **50**, 271
Magpie **62**, 269
White Pygmy-271
Goshawk, Australasian 296
  Australian 296
  Black-mantled **78**, **82**, 297
  Brown **78**, **82**, 296
  Buergers's 298
  Chestnut-shouldered **80**, **82**, 298
  Chinese 295
  Collared **78**, **82**
  Doria's 298
  Grey 296
  Grey-headed **78**, **82**, 297
  Meyer's **80**, **82**, 297
  Variable **78**, **82**, 296
  Varied 296
Goura cristata **112**, 337
  victoria **112**, 338
  scheepmakeri **112**, 337
Grallina cyanoleuca **242**, 463
  bruijnii **242**, 462
Grassbird, Fly River **250**, 502
  Little **250**, 503
  Papuan **250**, 502
  Tawny **250**, 502
Grasshopper-Warbler, Gray's **250**, 501
Grass-Owl, Eastern **144**, 373
Grebe, Australasian **50**, 279
  Little 278
  Tricoloured **50**, 278
Greenshank, Common **98**, 318
Ground-Dove, Bronze **110**, 336
  Cinnamon **110**, 335
  White-bibbed **110**, 335
  White-breasted 335
Ground-Pigeon Thick-billed **110**, 336
Ground-Robin Greater **168**, 487
  Lesser **168**, 487
Gruidae 308
Grus rubicunda **64**, 308
Gull, Black-headed **58**, 327
  Common Black-headed 327
  Silver **58**, 327
Gygis alba **58**, 326
Gymnocorvus tristis 472
Gymnocrex plumbeiventris **88**, 305
Gymnophaps albertisi **120**, 348

habitats of New Guinea 21
Habroptila inepta 307
Haematopodidae 310
Haematopus longirostris **92**, 310
Halcyon chloris 390
  macleayii 390
  megarhyncha 391
  nigrocyanea 389
  sancta 390
  saurophaga 390
  torotoro 390
Halcyonidae 385
Haliaeetus leucogaster **72**, **74**, 294
Haliastur indus **70**, **74**, 294
  sphenurus **70**, **74**, 294
Halocyptena matsudairae 277
Hardhead **48**, 272
Harpy-Eagle, New Guinea **80**, **82**, 298
Harpyopsis novaeguineae **80**, **82**, 298
Harrier, Papuan 295
  Pied 295
  Swamp **76**, 295
Hawk, Australian Sparrow 297
  Bat **70**, **86**, 293

Brown 301
Crested **292**
Doria's **80**, **82**, 298
Frog 295
Hawk-Owl, Papuan **146**, 374
hazards to avoid in New Guinea 35
health in New Guinea 34
Hemiprocne mystacea **152**, 381
Hemiprocnidae 381
Henicopernis longicauda **80**, **82**, 293
Henicophaps albifrons **110**, 335
Heron, Great-billed **68**, 285
  Grey 285
  Javan Pond- 284
  Mangrove 284
  Nankeen Night- **66**, 283
  New Guinea Tiger 282
  New Guinea Zebra 282
  Pacific 284
  Pied **68**, 286
  Purple 285
  Rufous Night- 283
  Striated **66**, 284
  White-faced **68**, 286
  White-necked **68**, 284
Heteromyias albispecularis **244**, 491
  armiti **244**, 491
  brachyurus **242**, 491
Heteroscelis brevipes 319
  incanus 319
Heteroscenes pallidus **140**, 369
Hieraaetus morphnoides 299
  weiskei **70**, **72**, **74**, 299
Himantopus himantopus **92**, 310
  leucocephalus 310
Hirundapus caudacutus **152**, 384
Hirundinidae 495
Hirundo ariel 496
  daurica 496
  neoxena **154**, 496
  nigricans 496
  rustica **154**, 495
  tahitica **154**, 495
historical changes within avifauna 30
Hobby, Australian **84**, **86**, 301
  Oriental **84**, **86**, 301
Honeyeater, Arfak 412
  Belford's 418
  Black 404
  Black-backed 407
  Black-backed Streaked 407
  Black-throated **186**, 415
  Blue-faced **184**, 411
  Brown **184**, 408
  Brown-backed **184**, 415
  Carbon 404
  Common Smoky **186**, 413
  Dusky 404
  Dwarf 434
  Forest 420
  Giant Wattled **186**, 412
  Graceful 421
  Green-backed **180**, 405
  Grey-streaked **180**, 407
  Guise's 407
  Hill Forest 421
  Huon Wattled 417
  Leaden **180**, 406
  Long-bearded **188**, 416
  Long-billed **182**, 413
  Louisiades 422
  Macgregor's 412
  Marbled **182**, 408

521

Mayr's 407
Mayr's Streaked **180**, 407
Meek's 406
Mimic 421
Mountain Red-headed 405
Mountain Yellow-eared 421
Obscure **186**, 415
Olive **184**, 408
Olive-streaked 406
Orange-cheeked **186**, 416
Ornate 417
Plain **182**, 407
Puff-backed 419
Pygmy 434
Red 404
Red-collared 403
Red-headed 405
Red-throated 404
Red-tinted 404
Rufous-backed **180**, 407
Rufous-backed Streaked 407
Rufous-banded **184**, 414
Rufous-sided **180**, 406
Scarlet-bibbed 405
Scarlet-throated 405
Scrub 422
Short-bearded **188**, 416
Silver-eared **184**, 409
Smoky 413
Sooty **188**, 416
Spangled **186**, 413
Spot-breasted 420
Spotted **186**, 409
Streak-headed **182**, 408
Tawny-breasted **184**, 409
Varied **186**, 418
Wattled Smoky **186**, 413
Western Smoky **186**, 412
White-chinned 403
White-throated **184**, 412
Yellow-browed 418
Yellow-gaped 420
Yellow-streaked **180**, 406
Yellow-tinted **184**, 419
Hornbill, Blyth's **138**, 394
    Papuan 394
Huon Peninsula 27
Hydrobates matsudairae 277
Hydrobatidae 277
Hydrochous gigas 383
Hydroprogne caspia **58**, 327

Ibis, Australian White **64**, 280
    Glossy **64**, 281
    Sacred 280
    Straw-necked **64**, 281
identifying birds in field 34
Ieracidea berigora 301
Ifrit, Blue-capped **206**, 466
Ifrita kowaldi **206**, 466
Ifritidae 466
Irediparra gallinacea **106**, 314
island biogeography 24
Ixobrychus cinnamomeus **66**, 283
    dubius **66**, 282
    eurhythmus 283
    flavicollis **66**, 283
    minutus 282
    sinensis **66**, 283

Jabiru 280
Jacana, Comb-crested **106**, 314
Jacanidae 314

522

Jacky Winter **242**, 489
Jaeger, Arctic **56**, 331
    Long-tailed **56**, 332
    Parasitic 331
    Pomarine **56**, 331
Jewel-babbler, Blue **166**, 437
    Brown-capped 437
    Chestnut-backed **166**, 437
    Dimorphic **166**, 437
    Lowland 437
    Mid-Mountain 437
    Spotted **166**, 436

Kempiella flavovirescens **246**, 489
    griseoceps **246**, 489
Kestrel, Australian 300
    Moluccan 300
    Nankeen **84**, **86**, 300
    Spotted **84**, **86**, 300
Kingfisher, the 392
Kingfisher, Australian Paradise 387
    Azure **162**, 393
    Beach **160**, 390
    Biak Paradise- **156**, 386
    Black-sided 389
    Blue-black **160**, 389
    Brown-backed Paradise 387
    Brown-headed Paradise- **156**, 387
    Buff-breasted Paradise- **156**, 387
    Collared **160**, 390
    Common **162**, 392
    Common Paradise- **156**, 385
    Forest **160**, 390
    Greater Yellow-billed 391
    Hook-billed **158**, 388
    Lesser Yellow-billed 391
    Little **162**, 393
    Kofiau Paradise- **156**, 386
    Little Paradise- **156**, 386
    Lowland 391
    Mangrove 390
    Mountain **162**, 391
    Mountain Yellow-billed 391
    Numfor Paradise- **156**, 386
    Papuan Dwarf **162**, 392
    Pink-breasted Paradise 387
    Red-breasted Paradise- **156**, 387
    River 392
    Rossel Paradise- **156**, 386
    Sacred **160**, 391
    Shovel-billed 388
    Variable Dwarf 392
    White-collared 390
    White-headed 390
    White-tailed Paradise 387
    Yellow-billed **162**, 391
King-Parrot, Moluccan **130**, 360
    Papuan **130**, 360
Kite, Black **70**, **74**, **76**, 294
    Black-shouldered 293
    Black-winged **70**, **86**, 293
    Brahminy **70**, **74**, 294
    Whistling **70**, **74**, 294
Knot, Great **102**, 320
    Red **102**, 321
Koel, Common 365
    Dwarf **142**, 365
    Eastern **140**, 365
    Indian 365
    Long-tailed 366
    Pacific 365
    White-crowned 369
Kookaburra, Blue-winged **158**, 388

Gaudichaud's 389
Hook-billed 388
Rufous-bellied **158**, 389
Shovel-billed **158**, 388
Spangled **158**, 389

Lalage atrovirens **202**, 444
    leucomela **202**, 445
    leucoptera **202**, 445
    sueurii 444
    tenuirostris 446
    tricolor **202**, 444
Landrail, Banded *305*
Laniidae 458
Lanius cristatus **248**, 458
    schach **248**, 458
Lapwing, Masked **92**, 311
Laridae 326
Lark, Australasian 494
Larus novaehollandiae 327
    ridibundus 327
Leaf-Warbler, Biak **250**, 497
    Island **194**, **250**, 497
    Mountain 497
    Numfor **250**, 497
Leatherhead 411
Leptocoma aspasia **198**, 509
Lewinia pectoralis **90**, 304
Lichenostomus chrysogenys 416
    flavescens 419
    obscurus 415
    subfrenatus 415
    versicolor 418
Lichmera alboauricularis **184**, 409
    argentauris **184**, 408
    indistincta **184**, 408
Lily-trotter 314
Limicola falcinellus **104**, 323
Limnodromus scolpaceus 316
    semipalmatus **96**, 316
Limosa lapponica **96**, 317
    limosa **96**, 317
Lobibyx miles 311
Loboparadisea sericea **172**, 431
Locustella fasciolata **250**, 501
Locustellidae 501
Logrunner 430
    Papuan **166**, 430
Lonchura caniceps **258**, 513
    castaneothorax **260**, 514
    grandis **260**, 513
    leucosticta **258**, 513
    molucca **260**, 512
    montana **260**, 515
    monticola **260**, 515
    nevermanni **258**, 513
    spectabilis **260**, 514
    stygia **260**, 514
    teerinki **260**, 514
    tristissima **260**, 512
    vana **258**, 513
Longbill, Dwarf 434
    Grey-bellied 434
    Pygmy **198**, 434
    Slaty-chinned 434
    Slaty-headed **198**, 434
    Spectacled **198**, 434
    Yellow-bellied **198**, 434
Lophorina superba **234**, 479
Loria loriae 430
Loriculus aurantiifrons **134**, 359
Lorikeet, Emerald 355
    Fairy **126**, **136**, 353

Goldie's **124, 136**, 356
Josephine's **126, 136**, 353
Little Red 353
Musschenbroek's 354
Papuan **126**, 354
Plum-faced **124, 136**, 352
Pygmy **124, 136**, 352
Orange-billed **124, 136**, 355
Rainbow **128, 136**, 357
Red-chinned **124, 136**, 352
Red-flanked **124, 136**, 353
Red-fronted **124, 136**, 353
Stella's **126, 136**, 354
Streaked 352
Striated **124, 136**, 352
Whiskered 352
Yellow-billed **124, 136**, 354
Lorius hypoinochrous **126, 136**, 355
  lory **126, 136**, 355
Lory, Biak Red 357
  Black **128, 136**, 356
  Black-capped **126, 136**, 355
  Black-winged **128, 136**, 357
  Brown **128, 136**, 357
  Dusky **128, 136**, 356
  Eastern Black-capped 355
  Fairy 354
  Greater Streaked 357
  Moluccan Red 357
  Purple-bellied **126, 136**, 355
  Rajah 356
  Violet-necked **128, 136**, 357
  Western Black-capped 355
  Yellow-streaked **128, 136**, 357
  Yellowish-streaked 357
Lotusbird 314
Luscinia calliope **248**, 507

Macgregoria pulchra **186**, 412
Machaerhamphus alcinus 293
Machaerirhynchidae 438
Machaerirhynchus flaviventer **220**, 438
  nigripectus **220**, 438
Macheiramphus alcinus **70, 86**, 293
Macronectes giganteus 273
Macropygia amboinensis **108**, 334
  mackinlayi **108**, 335
  nigrirostris **108**, 334
Magpie, Australian **200**, 440
Magpielark **242**, 463
Maluridae 401
Malurus alboscapulatus **176, 250**, 402
  cyanocephalus **176**, 402
  grayi 401
Mannikin, Alpine 515
  Black **260**, 514
  Black-breasted **260**, 514
  Black-faced **260**, 512
  Chestnut-breasted **260**, 514
  Eastern Alpine **260**, 515
  Grand **260**, 513
  Great-billed 513
  Grey-banded **258**, 513
  Grey-crowned **258**, 513
  Grey-headed **258**, 513
  Hooded **260**, 514
  Snow Mountain 515
  Streak-headed **260**, 512
  Western Alpine **260**, 515
  White-spotted **258**, 513
Manucode, Crinkle-collared **226**, 475
  Curl-crested **226**, 474
  Glossy **226**, 476

Glossy-mantled 476
Jobi **226**, 475
Trumpet **226**, 474
Manucodia ater **226**, 476
  atra 476
  chalybata 475
  chalybatus **226**, 475
  comrii **226** , 474
  jobiensis **226**, 475
  keraudrenii 474
map key 17
Martin, Fairy **154**, 496
  Sand **154**, 495
  Tree **154**, 496
Mearnsia novaeguineae **152**, 383
measurements 17
Megacrex inepta **90**, 307
Megalampitta gigantea **208**, 473
Megaloprepia magnifica 339
Megalurus albolimbatus **250**, 502
  gramineus **250**, 503
  macrurus **250**, 502
  timoriensis **250**, 502
Megapode, Moluccan 266
Megapodiidae 264
Megapodius affinis 267
  decollatus **44**, 267
  eremita **44**, 266
  freycinet **44**, 266
  geelvinkianus **44**, 266
  reinwardt **44**, 267
  wallacei 266
Megatriorchis doriae **80, 82**, 298
Melampitta, Greater **208**, 473
  Lesser **164**, 472
Melampitta gigantea **208**, 473
  lugubris **164**, 472
Melampittidae 472
Melanocharis arfakiana **196**, 432
  longicauda **196**, 433
  nigra **196**, 432
  striativentris **196**, 433
  versteri **196**, 433
Melanocharitidae 432
Melanorectes nigrescens **170, 206**, 452
Melidectes, Belford's **188**, 418
  Cinnamon-browed **188**, 417
  Huon **188**, 417
  Long-bearded 416
  Ornate **188**, 417
  Short-bearded 416
  Sooty 416
  Vogelkop **188** 417
  Yellow-browed **188**, 418
Melidectes belfordi **188**, 418
  foersteri **188**, 417
  fuscus 416
  leucostephes **188**, 417
  ochromelas **188**, 417
  nouhuysi 416
  princeps 416
  rufocrissalis **188**, 418
  torquatus **188**, 417
Melidora macrorrhina **158**, 388
Melilestes megarhynchus **182**, 413
Melionyx fuscus **188**, 416
  nouhuysi **188**, 416
  princeps **188**, 416
Meliphaga, Elegant **190**, 420
  Forest White-eared 420
  Graceful **190**, 421
  Louisiades 422
  Mimic **190**, 421

Mottled **190**, 420
Mountain **190**, 421
Puff-backed **190**, 419
Scrub **190**, 422
Scrub White-eared 422
Spot-breasted 420
Tagula **190**, 422
White-eared **190**, 420
Yellow-gaped **190**, 420
Meliphaga albonotata **190**, 422
  analoga **190**, 421
  aruensis **190**, 419
  chrysogenys 416
  chrysotis 409
  cinereifrons **190**, 420
  flavescens 419
  flavirictus **190**, 420
  flaviventer 409
  gracilis **190**, 421
  mimikae **190**, 420
  montana **190**, 420
  obscura 415
  orientalis **190**, 421
  polygramma 409
  subfrenata 415
  versicolor 418
  vicina **190**, 422
Meliphagidae 403
Melipotes, Arfak 412
  Common 413
Melipotes ater **186**, 413
  carolae **186**, 413
  fumigatus **186**, 413
  gymnops **186**, 412
Melithreptus albogularis **184**, 412
Meropidae 393
Merops ornatus **158**, 394
  philippinus **158**, 393
Microcarbo melanoleucos **62**, 290
Microdynamis parva **142**, 365
Microeca fascinans **242**, 489
  flavigaster **246**, 490
  flavovirescens 489
  griseoceps 489
  leucophaea 489
  papuana 489
Micropsitta bruijnii **134**, 363
  geelvinkiana **134**, 362
  keiensis **134**, 362
  pusio **134**, 362
migration, intercontinental 29
  within New Guinea 29
Milvus migrans **70, 74, 76**, 294
Mino anais **256**, 505
  dumontii **256**, 505
Mirafra javanica **248**, 494
Misocalius osculans 367
Mistletoebird **198**, 509
molt 29
Monachella muelleriana **242**, 490
Monarch, Biak **218**, 464
  Black 463
  Black and Gold 465
  Black-faced **220**, 465
  Black-winged **220**, 466
  Fantailed **218, 224**, 463
  Frilled **218**, 462
  Golden **220**, 465
  Hooded **218**, 463
  Islet **220**, 465
  Kofiau **218**, 464
  Ochre-collared **218**, 462
  Rufous **218**, 463

523

Spectacled **218**, 464
Spot-winged **218**, 464
Monarcha alecto 461
   axillaris 463
   brehmii 464
   chrysomela 465
   cinerascens **220**, 465
   frater **220**, 466
   guttula 464
   julianae 464
   manadensis 463
   melanopsis **220**, 465
   rubiensis 463
   trivirgatus 464
Monarchidae 460
Monticola solitarius **248**, 507
Moorhen, Dusky **50**, 307
Motacilla cinerea **248**, 516
   flava 515
   simillima 515
   taivana **248**, 515
   tschutschensis **248**, 515
Motacillidae 515
Mountain-Pigeon, Papuan **120**, 348
Mouse-Warbler, Bicoloured **168**, 423
   Mountain **168**, 423
   Rusty **168**, 423
Mudlark 463
Munia, Moluccan 512
Muscicapa griseisticta **248**, 508
Muscicapidae 507
Myiagra alecto **216**, 461
   atra **216**, 460
   cyanoleuca **216**, 461
   inquieta **216**, 461
   nana 461
   ruficollis **216**, 460
   rubecula **216**, 460
Myiolestes megarhynchus 450
Myna, Common **254**, 506
   Golden **256**, 505
   Yellow-faced **256**, 505
Myzomela, Dusky **178**, 404
   Elfin **178**, 405
   Papuan Black **178**, 404
   Red **178**, 404
   Red-collared **178**, 403
   Red-headed **178**, 405
   Red-throated **178**, 404
   Ruby-throated **178**, 404
   Sclater's **178**, 405
   White-chinned **178**, 403
Myzomela adolphinae **178**, 405
   albigula **178**, 403
   cruentata **178**, 404
   eques **178**, 404
   erythrocephala **178**, 405
   nigrita **178**, 404
   obscura **178**, 404
   rosenbergii **178**, 403
   sclateri **178**, 405

natural resource development 30
Nectarinia aspasia 509
   jugularis 509
   sericea 509
Nectariniidae 508
Needletail, White-throated **152**, 384
Neneba striativentris 433
Neochmia phaeton **258**, 511
Neopsittacus musschenbroekii **124**, **136**, 354
   pullicauda **124**, **136**, 355

Neositta chrysoptera 447
   papuensis 447
Neosittidae 447
nesting 29
Nettapus coromandelianus **50**, 271
   pulchellus **50**, 271
New Guinea bird regions 25
New Guinea region, definition 14
New Guinean culture and birds 35
Night-Heron, Nankeen **66**, 283
   Rufous 283
Nightjar, Archbold's **148**, 377
   Grey **148**, 377
   Jungle 377
   Large-tailed **148**, 377
   Mountain 377
   Papuan **148**, 377
   Spotted **148**, 376
   White-throated **148**, 376
Ninox boobook 374
   connivens **146**, 374
   novaeseelandiae **146**, 374
   rufa **146**, 373
   theomacha **146**, 374
Noddy, Black **58**, 326
   Brown **58**, 326
   Common 326
   White 326
   White-capped 326
Northwestern Islands 25
Northwestern Lowlands 26
Notophoyx novaehollandiae 286
   picata 286
Numenius madagascariensis **98**, 318
   minutus **98**, 317
   phaeopus **98**, 317
   tahitiensis 317
Nycticorax caledonicus **66**, 283

Oceanites oceanicus **52**, 277
Oceanitidae 277
Oceanodroma leucorhoa 277
   matsudairae **52**, 277
Oedistoma iliolophus **198**, 434
   pygmaeum **198**, 434
Onychoprion anaethetus **58**, 328
   fuscatus **58**, 328
   lunatus **58**, 328
Opopsitta diophthalma 359
   gulielmi III 359
Oreocharis arfaki **196**, 435
Oreoicidae 448
Oreopsittacus arfaki **124**, **136**, 352
Oreornis chrysogenys **186**, 416
   obscurus 415
   subfrenatus 415
Oreostruthus fuliginosus **258**, 511
Oriole, Brown **256**, 459
   Green **256**, 459
   Olive-backed **256**, 459
   Yellow 459
Oriolidae 458
Oriolus flavocinctus **256**, 459
   sagittatus **256**, 459
   szalayi **256**, 459
Ornorectes cristatus **208**, 449
Orthonychidae 430
Orthonyx novaeguineae **166**, 430
   temmincki 430
Osprey 291
Osprey, Eastern **72**, **74**, 219
Otididae 302
Otidiphaps nobilis **112**, 337

Otus beccarii **146**, 373
   magicus 373
Owl, Australian Barn-**144**, 372
   Australian Masked **144**, 372
   Barking **146**, 374
   Barn 372
   Biak Scops- **146**, 373
   Eastern Barn 372
   Grass 373
   Greater Sooty **144**, 372
   Jungle Hawk 374
   Masked 372
   Moluccan Scops- 373
   Papuan Hawk-**146**, 374
   Rufous **146**, 373
   Sooty **144**, 372
Owlet-nightjar, Allied **150**, 381
   Archbold's **150**, 380
   Australian **150**, 381
   Barred **150**, 380
   Feline **148**, 379
   Large 379
   Mountain **150**, 380
   Spangled 379
   Starry **148**, 379
   Vogelkop 381
   Wallace's **150**, 379
Oystercatcher, Australian Pied **92**, 310
   Pied 310

Pachycare flavogriseum **212**, 422
Pachycephala aurea **212**, 454
   balim **212**, 453
   citreogaster 453
   collaris **212**, 453
   griseiceps 455
   hyperythra **214**, 455
   leucogastra **214**, 456
   lorentzi **212**, 454
   melanura **212**, 454
   meyeri **212**, 454
   modesta **214**, 455
   monacha **214**, 456
   pectoralis 453
   phaionota **214**, 455
   rufinucha 448
   rufiventris 456
   schlegelii **212**, 452
   simplex **214**, 455
   soror **212**, 453
   tenebrosa 451
   umbrina 451
Pachycephalidae 450
Pachycephalopsis hattamensis **246**, 487
   poliosoma **244**, 488
Pachyptila belcheri 273
   turtur 273
   vittata 273
Pandion cristatus **72**, **74**, 291
   haliaetus 291
Pandionidae 291
Papasula abbotti 289
Paradigalla, Long-tailed **234**, 481
   Short-tailed **234**, 481
Paradigalla brevicauda **234**, 481
   carunculata **234**, 481
Paradisaea apoda **240**, 486
   decora **238**, 485
   guilielmi **238**, 484
   minor **240**, 485
   raggiana **240**, 486
   rubra **238**, 485
   rudolphi 484

Paradisaeidae 473
Paradise-Kingfisher, Biak **156**, 386
    Brown-headed **156**, 387
    Buff-breasted **156**, 387
    Common **156**, 385
    Kofiau **156**, 386
    Little **156**, 386
    Numfor **156**, 386
    Red-breasted **156**, 387
    Rossel **156**, 386
Paradisornis rudolphi **238**, 484
Paramythia montium **196**, 435
    olivacea **196**, 435
Paramythiidae 435
Parotia, Bronze **228**, 478
    Carola's **228**, 478
    Eastern **228**, 477
    Foja 478
    Lawes's **228**, 477
    Queen Carola's 478
    Wahnes's **228**, 477
    Western **228**, 476
Parotia berlepschi **228**, 478
    carolae **228**, 478
    helenae **228**, 477
    lawesii **228**, 477
    sefilata **228**, 476
    wahnesi **228**, 477
Parrot, Blue-collared **130**, 361
    Brehm's Tiger- **132**, 351
    Buff-faced Pygmy **134**, 362
    Desmarest's Fig 358
    Double-eyed Fig- **134, 136**, 359
    Eclectus **122**, 360
    Edward's Fig- **132**, 358
    Geelvink Pygmy **134**, 362
    Great-billed **130**, 362
    Green-winged King 360
    Large Fig- **132, 136**, 358
    Lilac-collared 361
    Madarasz's Tiger- **132**, 351
    Modest Tiger- **132**, 351
    Moluccan King- **130**, 360
    New Guinea Vulturine **122**, 349
    Orange-breasted Fig- **134**, 359
    Orange-fronted Hanging **134**, 359
    Painted Tiger- **132**, 351
    Papuan Hanging 359
    Papuan King- **130**, 360
    Pesquet's 349
    Red-breasted Pygmy **134**, 363
    Red-cheeked **130**, 361
    Red-winged **130**, 360
    Salvadori's Fig- **132**, 358
    Vulturine 349
    Yellow-capped Pygmy **134**, 362
Parrotfinch, Blue-faced **258**, 511
    Papuan **258**, 512
parts of a bird 18
Passer domesticus **248**, 510
    montanus **248**, 510
Passeridae 510
Pelagodroma marina 278
Pelecanidae 287
Pelecanus conspicillatus **62**, 287
Pelican, Australian **62**, 287
Peltops, Highland 439
    Lowland **200**, 439
    Mountain **200**, 439
Peltops blainvillii **200**, 439
    montanus **200**, 439
Peneothello bimaculata **242, 248**, 492
    cryptoleuca **244**, 493
    cyanus **244**, 493

    pulverulenta **242**, 492
    sigillata **244**, 492
Pernis ptilorhynchus 293
Petrel, Beck's **52**, 275
    Black-bellied Storm- **52**, 278
    Bulwer's **52**, 276
    Collared 274
    Cook's 274
    Gould's 274
    Herald **52**, 274
    Kermadec 274
    Leach's Storm- 277
    Matsudaira's Storm- **52**, 277
    Phoenix 275
    Providence 274
    Pycroft's 274
    Southern Giant 273
    Stejneger's 274
    Tahiti **52**, 275
    White-bellied Storm- 278
    White-faced Storm- 278
    Wilson's Storm- **52**, 277
Petrochelidon ariel **154**, 496
    nigricans **154**, 496
Petroica archboldi **244**, 488
    bivittata **244**, 488
Petroicidae 486
Phaethon lepturus **54**, 279
    rubricauda **54**, 279
Phaethontidae 279
Phalacrocoracidae 290
Phalacrocorax carbo **62**, 290
    melanoleucos 290
    sulcirostris **62**, 290
    varius 290
Phalarope, Grey 324
    Northern 324
    Red 324 (note)
    Red-necked **106**, 324
Phalaropus fulicarius 324 (note)
    lobatus **106**, 324
Phasianidae 267
Philemon brassi **182**, 410
    buceroides **182**, 411
    citreogularis **182**, 410
    corniculatus **182**, 411
    novaeguineae 411
    meyeri **182**, 410
Philomachus pugnax **106**, 324
Phonygammus keraudrenii **226**, 474
Phylloscopidae 497
Phylloscopus maforensis **250**, 497
    misoriensis **250**, 497
    poliocephalus **194, 250**, 497
    trivirgatus 497
Piezorhynchus alecto 461
Pigeon 333
    Bare-eyed 348
    Brown 334
    Collared Imperial **120**, 347
    Common 333
    D'Albertis 348
    Domestic 333
    Elegant Imperial **118**, 345
    Feral 333
    Floury Imperial **118**, 346
    Geelvink Imperial **118**, 345
    Grey Imperial 346
    Island Imperial 346
    Jungle Bronzewing 335
    Magnificent Ground 337
    Metallic 333
    Mountain 348
    Mueller's Imperial 347

    Nicobar **118**, 338
    Nutmeg 347
    Pacific Imperial **118**, 345
    Papuan Mountain- **120**, 348
    Pheasant **112**, 337
    Pied Imperial **118**, 347
    Pinon's Imperial **120**, 346
    Purple-crowned 341
    Purple-tailed Imperial **120**, 345
    Rose-crowned 341
    Rufescent Imperial **120**, 346
    Southern Crowned **112**, 337
    Spectacled Imperial **118**, 344
    Spice Imperial **118**, 345
    Stephan's Green-winged Pigeon 339
    Thick-billed Ground- **110**, 336
    Torresian Imperial **118**, 347
    Torres Strait 347
    Victoria Crowned **112**, 338
    Western Crowned **112**, 337
    White-capped Ground 335
    White-eyed Imperial 344
    White-throated **120**, 333
    Wompoo 339
    Zoe's Imperial **120**, 347
Pintail, Northern 272
Pipit, Alpine **248**, 516
    Australasian **248**, 516
    Australian 516
    Richard's 516
Pitohui, Black **170, 206**, 452
    Crested 449
    Hooded **208**, 457
    Northern Variable **208**, 456
    Raja Ampat **208**, 457
    Rusty **210**, 451
    Southern Variable **208**, 457
    White-bellied **210**, 451
Pitohui cerviniventris **208**, 457
    cristatus 449
    dichrous **208**, 457
    ferrugineus 451
    incertus 451
    kirhocephalus **208**, 456
    nigrescens 452
    uropygialis **208**, 457
Pitta, Blue-breasted 395
    Hooded **164**, 395
    Noisy **164**, 396
    Red-bellied **164**, 395
Pitta sordida **164**, 395
    versicolor **164**, 396
Pittidae 395
Platalea flavipes **64**, 281
    regia **64**, 281
plate credits 12
Plegadis falcinellus **64**, 281
Plesiodryas albonotata **244**, 491
Ploughbill, Wattled **206**, 449
Plover, Black-bellied 312
    Common Ringed 313 (note)
    Greater Sand- **94**, 313
    Grey **94**, 312
    Kentish 313
    Large Sand- 313
    Lesser Golden 312
    Lesser Sand- **94**, 313
    Little Ringed **94**, 312
    Masked 311
    Mongolian 313
    Oriental **94**, 313
    Pacific Golden **94**, 312
    Red-capped 313
    Spur-winged 311

Pluvialis dominica 312
  fulva **94**, 312
  squatarola **94**, 312
Podargidae 375
Podargus ocellatus **144**, 375
  papuensis **144**, 375
Podiceps novaehollandiae 279
  ruficollis 278
Podicipedidae 278
Poecilodryas albispecularis 491
  albonotata 491
  brachyura 491
  hypoleuca **242**, 492
  placens 493
  pulverulenta 492
Poliolimnas cinereus 306
Pomareopsis bruijnii 462
Pomatostomidae 429
Pomatostomus isidorei 429
  temporalis **210**, 429
Pond-Heron, Javan 284
population growth, human 31
Porphyrio porphyrio **50**, 307
Porzana cinerea **90**, 306
  tabuensis **90**, 306
  pusilla **90**, 306
Pratincole, Australian **106**, 325
  Oriental **106**, 325
Prion, Broad-billed 273
  Fairy 273
  Slender-billed 273
Pristorhamphus versteri 433
Probosciger aterrimus **122**, 348
Procellariidae 273
protected areas in New Guinea 32
Pseudeos fuscata **128, 136**, 356
Pseudobulweria becki **52**, 275
  rostrata **52**, 275
Pseudorectes ferrugineus **210**, 451
  incertus **210**, 451
Psittacella brehmii **132**, 351
  madaraszi **132**, 351
  modesta **132**, 351
  picta **132**, 351
Psittaculidae 350
Psittaculirostris desmarestii **132, 136**, 358
  edwardsii **132**, 358
  salvadorii **132**, 358
Psitteuteles goldiei **124, 136**, 356
Psittrichas fulgidus **122**, 349
Psittrichasidae 349
Psophodidae 436
Pteridophora alberti **234**, 476
Pterodroma alba 275
  arminjoniana 274
  brevipes 274
  cookii 274
  heraldica **52**, 274
  leucoptera 274
  longirostris 274
  neglecta 274
  pycrofti 274
  rostrata 275
  solandri 274
Ptilinopus, aurantiifrons **114**, 340
  bellus **116**, 342
  coronulatus **114**, 341
  insolitus **116**, 344
  iozonus **116**, 343
  magnificus **114**, 339
  nainus **116**, 344
  nanus 344
  ornatus **114**, 340

  perlatus **114**, 340
  prasinorrhous **116**, 342
  pulchellus **114**, 341
  regina **114**, 341
  rivoli **116**, 342
  solomonensis **116**, 343
  superbus **114**, 341
  viridis **116**, 343
  wallacii **114**, 341
Ptilonorhynchidae 396
Ptiloprora erythropleura **180**, 406
  guisei **180**, 407
  mayri **180**, 407
  meekiana **180**, 406
  perstriata **180**, 407
  plumbea **180**, 406
Ptiloris intercedens **236**, 479
  magnificus **236**, 479
Ptilorrhoa caerulescens **166**, 437
  castanonota **166**, 437
  geislerorum **166**, 437
  leucosticta **166**, 436
Ptilotula flavescens **184**, 419
Puffinus bailloni **52**, 276
  carneipes **52,** 276
  griseus 276
  heinrothi **52**, 276
  huttoni 52 (note), 276
  lherminieri 276
  nativitatis 276
  pacificus **52**, 275
  tenuirostris **52**, 275
Pycnonotidae 494
Pycnonotus aurigaster **254**, 494
Pycnopygius cinereus **182**, 408
  ixoides **182**, 407
  stictocephalus **182**, 408
Pygmy-Goose, White 271

Quail, Asian Blue 268
  Blue-breasted 268
  Brown **46**, 268
  Chinese 268
  King **46**, 268
  Snow Mountain **46**, 267
Quail-thrush, Painted **166**, 438
Rail, Bare-eyed **88**, 305
  Barred **90**, 305
  Buff-banded **90**, 305
  Chestnut **88**, 305
  Chestnut Forest- **88**, 303
  Chestnut-bellied 305
  Forbes's Chestnut Rail 304
  Forbes's Forest- **88**, 304
  Lewin's **90**, 304
  Mayr's Chestnut Rail 304
  Mayr's Forest- **88**, 304
  New Guinea Flightless **90**, 307
  Red-necked 304
  Slate-breasted 304
  Sooty 306
  White-striped Chestnut Rail, 304
  White-striped Forest- **88**, 304
Rallicula forbesi **88**, 304
  leucospila **88**, 304
  mayri **88**, 304
  rubra **88**, 303
Rallidae 303
Rallina forbesi 304
  leucospila 304
  mayri 304
  rubra 303
  tricolor **88**, 304

Rallus pectoralis 304
  philippensis 305
  torquatus 305
Ramsayornis modestus **184**, 415
Recurvirostra novaehollandiae 311
Recurvirostridae 310
Redshank, Common **98**, 318
  Spotted 318
Reed-Warbler, Australian **250**, 501
  Clamorous 501
  Great 501
  Oriental **250**, 501
Reef-Egret, Eastern **68**, 286
Reinwardtoena reinwardti **108**, 334
  reinwardtsi 334
Rhagologidae 448
Rhagologus leucostigma **214**, 448
Rhamphocharis crassirostris **196**, 433
Rhamphomantis megarhynchus **142**, 366
Rhipidura albolimbata **222**, 470
  atra **224**, 470
  brachyrhyncha **224**, 469
  dryas **224**, 470
  hyperythra **222**, 470
  leucophrys **222**, 468
  leucothorax **222**, 468
  maculipectus **222**, 468
  phasiana **222**, 471
  rufidorsa **224**, 469
  rufifrons **224**, 469, 470
  rufiventris **222**, 471
  threnothorax **222**, 469
Rhipiduridae 467
Rhyticeros plicatus **138**, 394
Riflebird, Growling **236**, 479
  Magnificent **236**, 479
Riparia riparia **154**, 495
Robin, Alpine 488
  Ashy **244**, 491
  Banded Yellow **246**, 493
  Black-capped **244**, 491
  Black-chinned **242**, 491
  Black-sided **242**, 492
  Black-throated **244**, 491
  Blue-grey **244**, 493
  Garnet **246**, 493
  Greater Ground- **168**, 487
  Green-backed **246**, 487
  Grey-headed 491
  Lesser Ground- **168**, 487
  Mangrove **242**, 492
  Mountain 488
  Northern Scrub 490
  Olive-yellow 493
  Papuan Scrub-**164**, 490
  Slaty 493
  Smoky **244**, 493
  Snow 488
  Snow Mountain **244**, 488
  Subalpine **244**, 488
  White-eyed **244**, 488
  White-faced **246**, 493
  White-rumped **242, 248**, 492
  White-winged **244**, 492
Rock-thrush, Blue **248**, 507
Roller, Broad-billed 384
Rubythroat, Siberian **248**, 507
Ruff **106**, 324

Salvadorina waigiuensis **50**, 270
Sanderling **102**, 321
Sandpiper, Baird's **104**, 322
  Broad-billed **104**, 323

Buff-breasted 324
Common **100**, 320
Curlew **104**, 323
Green 319
Marsh **98**, 318
Pectoral **104**, 322
Sharp-tailed **104**, 323
Terek **100**, 320
White-rumped 322
Wood **100**, 319
Sand-Plover, Greater **94**, 313
    Large 313
    Lesser **94**, 313
Satinbird, Crested **170**, 431
    Loria's **170**, 430
    Yellow-breasted **172**, 431
Sauromarptis gaudichaud 389
    tyro 389
Saxicola caprata **248**, 507
Scolopacidae 314
Scolopax rosenbergii **96**, 315
    saturata 315
Scops-Owl, Biak **146**, 373
    Moluccan 373
Scrubfowl, Biak **44**, 266
    Common 266
    Dusky **44**, 266
    Melanesian **44**, 266
    Moluccan **44**, 266
    New Guinea **44**, 267
    Orange-footed **44**, 267
    Wallace's 266
Scrub-Robin, Papuan **164**, 490
Scrubwren, Beccari's 424
    Buff-faced **192**, 425
    Grey-green **192**, 426
    Large **192**, 425
    Pale-billed **192**, 424
    Papuan **192**, 426
    Perplexing 424
    Tropical **192**, 424
    Vogelkop **192**, 425
Scythrops novaehollandiae **138**, 366
Sea-Eagle, White-bellied **72**, **74**, 294
selected references 36
Seleucidis melanoleucus **236**, 478
Sepik-Ramu 26
Sericornis arfakianus **192**, 426
    beccarii **192**, 424
    nouhuysi **192**, 425
    papuensis **192**, 426
    perspicillatus **192**, 425
    rufescens **192**, 425
    spilodera **192**, 424
    virgatus 424
Sericulus ardens **172**, 399
    aureus **172**, 398
    bakeri **172**, 399
Shearwater, Audubon's 276
    Christmas 276
    Flesh-footed **52**, 276
    Heinroth's **52**, 276
    Hutton's 276
    Short-tailed **52**, 275
    Sooty 276
    Streaked **52**, 275
    Tropical **52**, 276
    Wedge-tailed **52**, 275
Shelduck, Burdekin 270
    Raja **50**, 270
    White-headed 270
Shieldbill, Clicking 439
    Singing 439
Shrike, Black-capped 458

    Brown **248**, 458
    Long-tailed **248**, 458
    Schach 458
Shrikethrush, Grey **210**, 451
    Little **180**, **210**, 450
    Rufous 450
    Sooty **206**, 451
Shriketit, Wattled 449
Sicklebill, Black **232**, 481
    Black-billed **232**, 480
    Brown **232**, 480
    Buff-tailed 480
    Pale-billed **232**, 480
Sipodotus wallacii **176**, 401
Sittella, Black **206**, 448
    Papuan 447
    Varied **206**, 447
Skua, Arctic 331
    Long-tailed 332
    Pomarine 331
    South Polar **56**, 331
Snipe, Chinese 316
    Japanese 315
    Latham's **96**, 315
    Marsh 316
    Pin-tailed **96**, 316
    Swinhoe's **96**, 316
Southeastern Islands 27
Southeastern Peninsula 27
Southern Lowlands 26
Sparrow, English 510
    House **248**, 510
    Eurasian Tree **248**, 510
Sparrowhawk, Chinese **78**, **82**, 295
    Collared **78**, **82**, 297
species count, New Guinea 14
Sphecotheres vieilloti **256**, 458
    viridis 458
Spizaetus gurneyi 299
Spoonbill, Royal **64**, 281
    Yellow-billed **64**, 281
Starling, Common 506
    Long-tailed **254**, 505
    Metallic **254**, 504
    Moluccan **254**, 505
    Shining 504
    Singing **254**, 504
    Yellow-eyed **254**, 504
status treatment 17
Stercorariidae 330
Stercorarius longicaudus **56**, 332
    maccormicki **56**, 331
    parasiticus **56**, 331
    pomarinus **56**, 331
Sterna albifrons 328
    anaethetus 328
    bengalensis 328
    bergii 327
    caspia 327
    dougallii **60**, 329
    hirundo **60**, 329
    nilotica 327
    paradisaea 330
    sumatrana **60**, 329
Sternula albifrons **60**, 328
Stilt, Black-wnged **92**, 310
    White-headed 310
Stiltia isabella **106**, 325
Stint, Little **102**, 322
    Long-toed **104**, 322
    Red-necked **102**, 321
    Rufous-necked 321
Stone-curlew, Beach **92**, 309
    Bush **92**, 309

Stork, Black-necked **64**, 280
Stormbird 366
Storm-Petrel, Black-bellied **52**, 278
    Leach's 277
    Matsudaira's **52**, 277
    White-bellied 278
    White-faced 278
    Wilson's **52**, 277
Straightbill, Olive **180**, 414
    Tawny **180**, 414
Streptopelia chinensis **108**, 333
Strigidae 373
Sturnidae 504
Sturnus tristis 506
    vulgaris 506
subspecific treatment 16
Sula abbotti 289
    dactylatra **54**, 289
    leucogaster **54**, 289
    sula **54**, 289
Sulidae 289
Sunbird, Black **198**, 509
    Olive-backed **198**, 509
    Rand's **198**, 509
    Yellow-bellied 509
Swallow, Bank 495
    Barn **154**, 495
    Common 495
    Pacific **154**, 495
    Red-rumped **154**, 496
    Welcome **154**, 496
Swamphen, Purple **50**, 307
Swan, Black 270
Swift, Fork-tailed **152**, 384
    New Guinea Spine-tailed 383
    Pacific 384
    Papuan Spine-tailed **152**, 383
    White-rumped 384
Swiftlet, Bare-legged **152**, 383
    Glossy **152**, 382
    Lowland 383
    Mountain **152**, 382
    Three-toed **152**, 383
    Uniform **152**, 383
    Waterfall 383
    Whitehead's 383
Syma megarhyncha **162**, 391
    torotoro **162**, 391
Symposiachrus axillaris **218**, **224**, 463
    brehmii **218**, 464
    guttula **218**, 464
    julianae **218**, 464
    manadensis **218**, 463
    rubiensis **218**, 463
    trivirgatus **218**, 464
Synoicus australis 268
systematic treatment 15

Tachybaptus novaehollandiae **50**, 279
    ruficollis 278
    tricolor **50**, 278
Tadorna radjah **50**, 270
Talegalla cuvieri **42**, 265
    fuscirostris **42**, 265
    jobiensis **42**, 265
Tanygnathus megalorynchos **130**, 362
Tanysiptera carolinae **156**, 386
    danae **156**, 387
    ellioti **156**, 386
    galatea **156**, 385
    hydrocharis **156**, 386
    nympha **156**, 387
    riedelii **156**, 386
    rosseliana **156**, 386

sylvia **156**, 387
Tattler, Grey-tailed **100**, 319
  Polynesian 319
  Siberian 319
  Wandering **100**, 319
taxonomic treatment 15
Teal, Grey **48,** 272
  Salvadori's **50,** 270
Tern, Arctic 330
  Black 330
  Black-naped **60,** 329
  Bridled **58,** 328
  Brown-winged 328
  Caspian **58,** 327
  Common **60,** 329
  Crested **60,** 327
  Fairy 326
  Great-crested 327
  Grey-backed **58,** 328
  Gull-billed **60,** 327
  Least 328
  Lesser Crested **60,** 328
  Little **60,** 328
  Marsh 330
  Roseate **60,** 329
  Sooty **58,** 328
  Spectacled 328
  Swift 327
  Whiskered **60,** 330
  White **58,** 326
  White-winged **60,** 330
  White-winged Black 330
Thalasseus bengalensis **60,** 328
  bergii **60,** 327
Thicket-Fantail, Black **222,** 468
  Sooty **222,** 469
  White-bellied **222,** 468
Thick-knee, Beach 309
  Bush 309
Thornbill, Grey **194,** 426
  Mountain 426
  New Guinea **192,** 426
  Papuan 426
threatened bird species 31
Threskiornis aethiopicus 280
  moluccus **64,** 280
  spinicollis **64,** 281
Threskiornithidae 280
Thrush, Island **164,** 506
  Russet-tailed **164,** 506
  Scaly 506
Tiger-Parrot, Brehm's **132,** 351
  Madarasz's **132,** 351
  Modest **132,** 351
  Painted **132,** 351
Timeliopsis fulvigula **180,** 414
  griseigula **180,** 414
Todiramphus chloris **160,** 390
  macleayii **160,** 390
  nigrocyaneus **160,** 389
  sanctus **160,** 391
  saurophagus **160,** 390
Todopsis cyanocephala 402
  wallacii 401
Torrentlark **242,** 462
Toxorhamphus iliolophus 434
  novaeguineae **198,** 434
  poliopterus **198,** 434
Trans-Fly 26
Treecreeper, Papuan **206,** 400
Treeswift, Moustached **152,** 381
Tregellasia leucops **246,** 493
Trichoglossus goldiei 356

haematodus **128, 136,** 357
Triller, Australian 444
  Biak **202,** 445
  Black-browed **202,** 444
  Varied **202,** 445
  White-winged **202,** 444
Tringa brevipes **100,** 319
  erythropus 318
  glareola **100,** 319
  hypoleucos 320
  incana **100,** 319
  nebularia **98,** 318
  ochropus 319
  stagnatilis **98,** 318
  terek 320
  totanus **98,** 318
Tropicbird, Red-tailed **54,** 279
  White-tailed **54,** 279
Trugon terrestris **110,** 336
Trumpetbird 474
Tryngites subruficollis **324**
Turdidae 506
Turdus poliocephalus **164,** 506
Turkey, Plains 302
Turnicidae 308
Turnix maculosus **46,** 308
Turnstone, Ruddy **102,** 320
Turtle-Dove, Spotted 333
Tyto alba 372
  capensis 373
  delicatula **144,** 372
  javanica 372
  longimembris **144,** 373
  novaehollandiae **144,** 372
  tenebricosa **144,** 372
Tytonidae 372

Urodynamis taitensis **140,** 366
Uroglaux dimorpha **146,** 374

Vanellus miles **92,** 311
  novaehollandiae 311
vegetation, importance of 28
vegetation of New Guinea 21
visiting New Guinea 34

Wagtail, Eastern Yellow **248,** 515
  Green-headed Yellow **248,** 515
  Grey **248,** 516
  Willie **222,** 468
  Yellow 515
Warbler, Australian Reed- **250,** 501
  Biak Leaf- **250,** 497
  Bicoloured Mouse- **168,** 423
  Black-throated 428
  Clamorous Reed- 501
  Fan-tailed 503
  Gray's 501
  Great Reed- 501
  Island Leaf- **194, 250,** 497
  Mountain Leaf- 497
  Mountain Mouse- **168,** 423
  Numfor Leaf- **250,** 497
  Oriental Reed- **250,** 501
  Rusty Mouse- **168,** 423
  Treefern 428
web sources for New Guinea birds 39
Western Ranges 26
when and where to visit 33
Whimbrel **98,** 317
  Little 317
Whipbird, Papuan **166,** 436
Whistle-Duck, Grass 270

Water 270
Whistler, Baliem **212,** 453
  Bismarck 453
  Black-headed **214,** 456
  Brown 455
  Brown-backed **214,** 455
  Black-tailed 454
  Common Golden 453
  Dwarf 422
  Golden 453
  Golden-backed **212,** 454
  Grey **214,** 455
  Grey-headed 455
  Hill Golden 453
  Island **214,** 455
  Lorentz's **212,** 454
  Louisiade **212,** 453
  Mangrove Golden **212,** 454
  Mottled 448
  Regent **212,** 452
  Rufous 456
  Rufous-naped 448
  Rusty **214,** 455
  Schlegel's 452
  Sclater's **212,** 453
  Sooty 451
  Vogelkop **212,** 454
  White-bellied **214,** 456
White-eye, Biak **252,** 499
  Black-fronted **252,** 498, 499
  Capped **252,** 499
  Green-fronted **252,** 499
  Lemon-bellied **252,** 498
  Louisiade **252,** 500
  New Guinea **252,** 500
  Oya Tabu **252,** 499
  Papuan 500
  Tagula **252,** 499
  Western Mountain 499
  White-throated 499
  Yellow-bellied 498
Wigeon, Eurasian **48,** 271
Winter, Jacky **242**
Woodcock, East Indian 315
  New Guinea **96,** 315
  Rufous 315
Wood-Duck, Australian 271
Woodswallow, Black-faced **200,** 441
  Great **200,** 441
  New Guinea 441
  White-breasted **200,** 441

Xanthomelus aureus 398
  bakeri 399
Xanthotis chrysotis 409
  flaviventer **184,** 409
  polygrammus **186,** 409
Xenorhynchus asiaticus 280
Xenus cinereus **100,** 320

Zonerodius heliosylus **66,** 282
Zoothera dauma 506
  heinei **164,** 506
Zosteropidae 498
Zosterops atrifrons **252,** 498
  chloris **252,** 498
  crookshanki **252,** 499
  fuscicapilla **252,** 499
  griseotinctus **252,** 500
  meeki **252,** 499
  minor **252,** 498, 499
  mysorensis **252,** 499
  novaeguineae **252,** 500